The Study of Agricultural Geography

A Scholarly Guide and Bibliography

Thomas A. Rumney

The Scarecrow Press, Inc.
Lanham, Maryland • Toronto • Oxford
2005

SCARECROW PRESS, INC.

Published in the United States of America
by Scarecrow Press, Inc.
A wholly owned subsidiary of
The Rowman & Littlefield Publishing Group, Inc.
4501 Forbes Boulevard, Suite 200, Lanham, Maryland 20706
www.scarecrowpress.com

PO Box 317
Oxford
OX2 9RU, UK

British Library Cataloguing in Publication Information Available

Library of Congress Cataloging-in-Publication Data

Rumney, Thomas A.
The study of agricultural geography : a scholarly guide and bibliography / Thomas A. Rumney.
p. cm.
Includes bibliographical references and index.
ISBN 0-8108-5702-2 (hardcover : alk. paper)
1. Agricultural geography–Bibliography. I. Title.

Z5074.G45R86 2005
[S494.5.G46]
016.63'09–dc22 2005019598

The paper used in this publication meets the minimum requirements of American National Standard for Information Sciences—Permanence of Paper for Printed Library Materials, ANSI/NISO Z39.48-1992.
Manufactured in the United States of America.

Contents

Preface

The spatial elements and processes of the study of agriculture have long interested and stimulated geographers. In fact, the study of agricultural activities and distributions from a geographical viewpoint provides students, scholars, policy makers, and others with a unique opportunity to combine and synthesize the usually diffuse and disparate subfields of applied and academically based geography. For example, agricultural geography can be studied from an environmental standpoint with examinations of soils, climates, hydrology, and biotic systems of both plants and animals. Agricultural geography can also provide a large number of possibilities for the cultural, settlement, and historical geographer as agriculture has formed, and continues to form, a major influence upon the creation of humanized landscapes and interaction processes.

Economic geographical systems associated with agricultural commodities, processing, land use, and consumption patterns of food, fiber, and other products have been studied and analyzed throughout the world by geographers and others. Agriculture must also be recognized as a great force for change among the multitude of human spatial dimensions. The expansion of settlement frontiers, the growth and development of human systems, and resultant migrations from lands of want to lands of plenty are commonly rooted in agricultural dimensions of production, technology, land use and tenure, and short-term "human" decisions. This compendium attempts to present much of the available geographical literature dealing with agricultural topics.

The organization of this bibliography follows familiar pathways, yet occasionally travels on roads less taken by other bibliographies. To begin, there is a large section of works that are either of a general, macroscale character or that have discussions on several world regions. Part 1 concludes with a "theoretical" section not related to any specific area of the world. As shown in the lengthy table of contents, the entries are grouped by topic. These categories include atlases and graphic presentations, general works, crops and cropping patterns, animal husbandry, dairying, farm land issues, agricultural populations and social issues, environmental and conservation topics, irrigation, aquaculture, agricultural marketing and distribution, transportation topics, and agricultural labor works. The entries are listed by their dates with the earliest coming first. In the event of more than one work per year listed, the entries are then arranged alphabetically by author's last name.

After the first part we move on to the more region-specific divisions, which encompass the majority of entries for this bibliography. They begin with North America and gradually evolve into sections on the United States and its respective areas and states. The same goes for Canada, Latin America, Europe, Africa, Asia, and Australasia-Oceania. The topical organization is applied to each regional collection, as is the chronological arrangement of entries. Entries are listed in this fashion to provide the reader with a picture of the development and directions of geographical research on agricultural geography from its academic beginnings to the present. This format presents not only research on agricultural geography but also a history of the development of this research.

Included are atlases, books and texts, chapters, articles from scholarly journals, master's theses, and doctoral dissertations. There are entries written in English, French, Spanish, Italian, German, Dutch, Swedish, and other languages, with English entries that were written in another language in their respective source or journal. The goal has been to include as many geographical publications in this collection as possible, however, some might have been missed. Regardless, this bibliography will prove helpful to those interested and involved in the study of agricultural geography.

Part 1

GENERAL AND THEORETICAL WORKS

General and Theoretical Works

ATLASES AND GRAPHIC PRESENTATIONS

Hainsworth, Reginald G. *A Graphic Summary of World Agriculture.* Washington, D.C. Office of Foreign Agricultural Relations, United States Department of Agriculture, 1949.

Fenton, E.W. "Vegetation and Agricultural Activities as Shown by Aerial Photographs." *Scottish Geographical Magazine*, 67, No. 2 (1951): 105-109.

Gregor, Howard F. "A Map of Agricultural Adjustment." *Professional Geographer*, 16 (1964): 16-19.

Church, B.M., et al. "A Type of Farming Map Based on Agricultural Census Data." *Outlook on Agriculture*, 5 (1968): 191-196.

International Association of Agricultural Economists. *World Atlas of Agriculture.* Novara, Italy: Istituto Geografico de Agostini, 1969. 4 volumes.

Hart, John Fraser. "A Map of the Agricultural Implosion." *Proceedings, Association of American Geographers*, 2 (1970): 68-71.

Ryerson, Robert A. "A Model to Provide a Measure of Agricultural Productivity Using Remote Sensing Techniques." Master's Thesis, McMaster University, Hamilton, Ontario, 1970.

Bertin, Jacques, Jean-Jacques Hemardinquer, Michael Keel, and W.G.L. Randles. *Atlas of Food Crops.* Paris, France, and New York, N.Y.: Mouton and Walter de Grayter, 1971.

Boserup, Ester. "Present and Potential Food Production in Developing Countries." Pp. 100-113 in *Geography and a Crowding World*, edited by Wilbur Zelinisky, L.A. Kosinski, and R.M. Prothero. New York, N.Y.: Oxford University Press, 1971.

n.a. *World Atlas of Agriculture.* Novara, Italy: Instituto Geografico de Agostini, 1973.

Enyedi, Gyorgy, ed. *Agrarian-Industrial Complexes in Modern Agriculture. Papers Presented for the Symposium K23 (Rural Planning and Development) of the 23 International Geographical Congress, Odessa, USSR, 20-26 July 1976.* Budapest, Hungary: Research Institute of Geography, Hungarian Academy of Sciences, 1976.

Jeanneret, F., and P. Vauthier. *Kartierung der Klimaeigung fur die Landwirtschaft.* Bern, Switzerland: Jahrbuch der Geographischen Gesellschaft Bern, Beiheft 4, 1977.

Murdych, Z. "The Possibilities of Photo Interpretation in the Geography of Agriculture and Forestry" (in Czech). *Acta Universitatis Carolinae Geographica*, 12, No. 1 (1977): 47-52.

Johnson, Hugh. *A World Atlas of Wine: A Complete Guide to the Wines and Spirits of the World*, revised edition. New York, N.Y.: Simon & Schuster, 1981; 3rd edition, 1985.

Curran, Paul J. "Aerial Photography for the Assessment of Crop Condition: A Review." *Applied Geography*, 5 (1985): 347-360.
Curran, Paul J. "Remote Sensing in Agriculture: An Introductory Review." *Journal of Geography*, 86, No. 4 (1987): 147-155.
Price, Frank, Rosemary George, and Michael Edwards, eds. *Le grand atlas du vin*. Paris, France: Editions Atlas, 1992.
Lowe, Philip, Terry K. Marsden, and Sarah Whatmore. *Regulating Agriculture*. London, England: David Fulton, 1994.

GENERAL TEXTS, ARTICLES, AND THESES

n.a. "The Agricultural Potentialities of Tropical Lands." *Scottish Geographical Magazine*, 28, No. 12 (1912): 617-623.
Finch, Vernar C. "Geography of the World's Agriculture." Ph.D. Dissertation, University of Washington, Seattle, Washington, 1916.
Finch, Vernar C., and O.E. Baker. *Geography of the World's Agriculture*. Washington, D.C.: U.S. Government Printing Office, 1917.
Huntington, Ellsworth, and Sumner W. Cushing. "The Nature and Possibilities of Tropical Agriculture." *Bulletin of the Geographical Society of Philadelphia*, 17, No. 3 (1919): 83-97.
Grime, Sarah L. "Agriculture in the Time of Vergil." *Journal of Geography*, 22, No. 2 (1923): 66-74.
Marbut, C.F. "Agriculture in the United States and Russia." *Geographical Review*, 21, No. 4 (1931): 598-612.
Perpillou, A. "Un fleau agricole: Le doryphone." *Annales de Geographie*, 42 (1933): 113-126.
Buttner, Harold. *Die Landwirtschaftlichen Betriebsformen im Ried, an der Bergstrasse und im vorderen Odenwald in ihrer geographische Bedingtheit*. Frankfurt, Germany: H.L. Bronner, 1934.
Hartshorne, Richard, and Samuel N. Dicken. "A Classification of the Agricultural Regions of Europe and North America on a Uniform Statistical Basis." *Annals of the Association of American Geographers*, 25 (1935): 99-120.
Whittlesey, Derwent. "Major Agricultural Regions of the Earth." *Annals of the Association of American Geographers*, 26 (1936): 199-240.
Sievers, Angelika. *Schriften zur Landwirtschaftsgeographie: Ein Uberblick uber das letzte Jahrzeht*. Leipzig, Germany: K.F. Koehler, 1944.
Cholley, A. "Problems de structure agraire et d'economie rurale." *Annales de Geographie*, 55 (1946): 81-101.
Bacon, Lois B. *Agricultural Geography of Europe and the Near East*. Washington, D.C.: U.S. Government Printing Office, 1948.
Zimmerman, Josef. *Bochenkultur und Landschaft der Erftniederung*. Bonn, Germany: Bonner Geographische Abhandlungen, Heft 3, 1949.
Aschmann, Homer. "A Consumer Oriented Classification of the Products of Tropical Agriculture." *Economic Geography*, 28, No. 2 (1952): 143-150.
Aartsen, J.P. von. "Agricultural Geography." *Tijdschrift voor Economische en Sociale Geographie*, 44 (1953): 27-28.

Bowen, E.G. "The Geographer's Approach to Agriculture." *Association of Agriculture Review*, 2 (1953): 10-14.

Barrere, Pierre. "Les agrumes dans le monde." *Les Cahiers d'Outre Mer*, 7 (1954): 155-194; Deuxieme partie, 258-301.

Dunn, Edgard S. *The Location of Agricultural Production*. Gainesville, Fla.: University of Florida Press, 1954.

McCarty, H.H. "Agricultural Geography." Pp. 258-274 in *American Geography: Inventory and Prospect*, edited by Preston E. James and Clarence F. Jones. Syracuse, N.Y.: Syracuse University Press, 1954.

Van Royen, William. *The Agricultural Resources of the World*. New York, N.Y.: Prentice-Hall, 1954.

Vischer, Stephen S. "Comparative Agricultural Potentials of the World's Regions." *Economic Geography*, 31, No. 1 (1955): 82-86.

Champier, L. "La recherche agraire en matiere de geographie agraire depuis un quart de siecle." *Revue de Geographie de Lyon*, 31 (1956): 319-327.

Garrison, William L., and Duane F. Marble. "The Spatial Structure of Agricultural Activities." *Annals of the Association of American Geographers*, 46, No. 2 (1957): 137-144.

Helburn, Nicholas. "The Bases for a Classification of World Agriculture." *Professional Geographer*, 9 (1957): 2-7.

Kostrowicki, Jerzy. "Remarks on the Geography of Agriculture in English Speaking Countries" (in Polish). *Przeglad Geograficzny*, 29, No. 1 (1957): 47-66.

von Wissman, Hermann. "Ursprungsherde und Ausbrietungswege von Pflanzen-und Tierzuckt und ihre Abhangigkeit von der Klimageschichte." *Erdkunde*, 11, No. 2 (1957): 81-94.

Zabko-Potopowicz, A. "The Development of the Geography of Agriculture Since World War II." *Przceglad Geograficzny*, 29, No. 1 (1957): 21-46.

Weaver, John C. "A Design for Research in the Geography of Agriculture." *Professional Geographer*, 10 (1958): 2-8.

Blaunt, James M. "Microgeographic Sampling: A Quantitative Approach to Regional Agricultural Geography." *Economic Geography*, 35, No. 1 (1959): 79-88.

Buchanon, R.O. "Some Reflections on Agricultural Geography." *Geography*, 44, No. 1 (1959): 1-13.

Duckham, A.N. "The Current Agricultural Revolution." *Geography*, 44 (1959): 71-78.

Fenelon, Pierre. "Structure agraire d'un Finage perigoudin." *Norois*, 5 (1958): 21-40.

Chisholm, Michael. "Agricultural Production, Location and Rent." *Oxford Economic Papers*, 13 (1961): 342-359.

Fenelon, Pierre. "Vocabulaire de geographie agraire." *Norois*, 8 (1961): 230-242.

Panhuysen, Helene. *Die Entwicklung der Agrarlandschaft im Raume Straelenseit 1800 unter*. Bonn, Germany: F. Dummler, 1961.

Strong, T.H. "Agriculture in Relation to Economic Development." *Bulletin, Ghana Geographical Association*, 6, No. 2 (1961): 13-21.

Fenelon, Pierre. "Vocabulaire de Geographie agraire." *Norois*, 9 (1962); "de Cabal a Cedrat." 41-56; "de Celeri a Charrue." 199-212; "de Chase a Chocheron." 333-341; "de chou a Complanteur." 386-396.

Fenelon, Pierre. "Vocabulaire de Geographie agraire." *Norois*, 10 (1962). "Compoix a Czarnoziem." 40-55; "Lettre D." 275-294.

Nawloka-Bielcka, Krystyna. "Investigations of John C. Weaver on the Spatial Structure of Agriculture" (in Polish). *Przeglad Geograficzny*, 34, No. 4 (1962): 773-779.

Uhlig, Harold. *Beitrage zur Geographie tropischer und subtropischer Entwicklungslander.* Giessen, Germany: Giessener Geographische Schriften, Heft 2, 1962.

Jones, G.E. "The Diffusion of Agricultural Innovations." *Journal of Agricultural Economics*, 15 (1963): 387-405.

Spencer, Joseph E., and Ronald J. Horvath. "How Does an Agricultural Region Originate?" *Annals of the Association of American Geographers*, 53, No. 1 (1963): 74-92.

Butzer, Karl W. "The Significance of Agricultural Dispersal into Europe and Northern Africa." Pp. 438-460 in *Environment and Archeology: An Introduction to Pleistocene Geography*, edited by Kark W. Butzer. Chicago, Ill.: Aldine, 1964.

Chisholm, Michael. "Problems in the Classification and Use of Farm-Type Regions." *Transactions, Institute of British Geographers*, 35 (1964): 91-103.

Fenelon, Pierre. "Vocabulaire de geographie agraire." *Norois*, 11 (1964); "E a Emouse." 51-64; "Emaillage a Eyzines." 349-369.

Hart, John Fraser. "Geographic Co-variants of Types of Farming Areas." Pp. 7-10 in *Agricultural Geography: IGU Symposium*, edited by E.S. Simpson. Liverpool, England: University of Liverpool, 1965.

Healy, E.O., and A.C. Egbert. "Regional Programming of Efficient Agricultural Production Patterns." *Econometrica*, 32 (1964): 374-386.

Heller, Charles F. "The Use of Model Farms in Agricultural Geography." *Professional Geographer*, 16 (1964): 20-23.

Keunig, H.J. "Agrarische Geografie: Doelstelling, Ontwikkeling, Methoden." *Tijdschrift von het Koninklijk Nederlandsche Aardrijkskindig Genootschap*, 81, No. 1 (1964): 10-19.

Kostrowicki, Jerzy. "Geographical Typology of Agriculture, Principles and Methods." *Geographia Polonica*, 1 (1964): 111-146.

Barnes, F.A. "The Geographical Typology of Agriculture: A Discussion." Pp. 59-74 in *Agricultural Geography IGU Symposium*, edited by E.S. Simpson. Liverpool, England: University of Liverpool, Department of Geography, Research Paper No. 3, 1965.

Ditt, Heldgard. *Struktur und Wandelwestfalischer Agrarlandschaften*. Munster, Germany: Aschendorffsche Verlagsbuchhandlung, 1965.

Enyedi, Gyorgy. *Geographical Types of Agriculture*. Budapest, Hungary: Akademiai Kiado Foldrajzi Tanulmanyoi, No. 4, 1965.

Enyedi, Gyorgy. "The Geographical Types of World Agriculture." *Foldrajzi Kozlemenyek*, 13, No. 3 (1965): 239-264.

Fenelon, Pierre. "Vocabulaire de geographie agraire." *Norois*, 12 (1965); "Fa-Flo." 224-239; "Flo-Fuye." 368-386.

Jones, G.E. "The Diffusion of Agricultural Innovations." Pp. 475-492 in *Readings in Resource Management and Conservation*, edited by Ian Burton and Robert W. Kates. Chicago, Ill.: University of Chicago Press, 1965.

Simpson, E.S., ed. *Agricultural Geography: IGU Symposium*. Liverpool, England: University of Liverpool, Department of Geography, Research Paper No. 3, 1965.

Belshaw, D.G., and B.G. Jackson. "Type-of-Farm Areas: The Application of Sampling Methods." *Transactions, Institute of British Geographers*, 38 (1966): 89-93.

Coloquio sobre Geografia Agraria. Salamanca, Spain: Universidad de Salamanca, 1966.

Fenelon, Pierre. "Vocabulaire des geographie agraire." *Norois*, 13 (1966): 239-257.

Kostrowicki, Jerzy. "Tipologia geografica de la Agricultura mundial. Principos y metodos." *Union Geografica International, Conferencia Regional Latino-America, Mexico, D.F.*, Vol. 2 (1966): 793-807.

Oshiro, Kenji K. "The Development of Agricultural Innovations in Earth Space." *B.C. Geographical Series*, 7 (1966): 105-110.

Topp, Manfred. *Agrargeographie von Ingelheim: Einewirtschaftsgeographische Untersuchung*. Bad Godesberg, Germany: Forschungen zur deutschen Landeskunde, Vol. 155, 1966.

Dunn, Edgard S. *The Location of Agricultural Production*. Gainesville, Fla.: University of Florida Press, 1967.

Henshall, Janet D. "Models of Agricultural Activity." Pp. 425-458 in *Models in Geography*, edited by Richard J. Chorley and Peter Haggett. London, England: Methuen, 1967.

Jones, G.E. "The Adoption and Diffusion of Agricultural Practices." *World Agricultural Economies, Rural Section Abstracts, Review Articles*, 6 (1967): 1-34.

Velcea, I. "Can Agricultural Geography Be Considered as an Applied Science?" (in Romanian). *Studia Universitatis Babes-Boblyai, Serie geologia-geographia*, 12, No. 2 (1967): 359-363.

Brookfield, Harold C. "New Directions in the Study of Agricultural Systems in the Tropical Areas." Pp. 413-439 in *Evolution and Environment*, edited by E.T. Drake. New Haven, Conn.: Yale University Press, 1968.

Chang, Jen-hu. "The Agricultural Potential of the Humid Tropics." *Geographical Review*, 58, No. 3 (1968): 333-361.

Coppock, J.T. "The Geography of Agriculture." *Journal of Agricultural Economics*, 19 (1968): 153-175.

George, Pierre. *Geographie agricole du monde*. Paris, France: Presses Universitaires de France, 1968.

Kostrowicki, Jerzy. "Agricultural Typology, Agricultural Regionalization, Agricultural Development." *Geographia Polonica*, 14 (1968): 265-274.

Manshard, Walther. *Tropical Agriculture: A Geographical Introduction and Appraisal*. London, England: Longman, 1968.

Boichard, J. "Gestion agricole et geographe rurale." *Revue de Geographie de Lyon*, 44 (1969): 323-374.

Floyd, Barry. "Toward a More Specific Geography of Traditional Agriculture in the Tropics: Or Goodbye to Machete and Dibble Stick." *Professional Geographer*, 21, No. 4 (1969): 248-251.

Grigg, David B. "The Agricultural Regions of the World: Review and Reflections." *Economic Geography*, 45, No. 2 (1969): 95-132.

Kostrowicki, Jerzy. "Agricultural Typology." *Bulletin, IGU*, 20 (1969): 36-40.

Kostrowicki, Jerzy. "Agricultural Typology: Principles, Criteria, Methods, and Techniques." *Przeglad Geograficzny*, 41, No. 4 (1969): 599-621.

Kuznetsov, A.A. "Application of Mathematical Methods in Economic-Geographical Research on the Location of Agricultural Production" (in Russian). Pp. 166-177 in *Sbornik: Ekon. geografiya (Collection: Economic Geography)*. Moscow, 1969.

Laut, Peter. *Agricultural Geography, Vol. 1: Systems, Subsistence and Plantation Agriculture*. Don Mills, Ontario: Nelson, 1969.

Laut, Peter. *Agricultural Geography, Vol. 2: Mid-Latitude Commercial Agriculture*. Don Mills, Ontario: Nelson, 1969.

Wilbanks, Thomas J. "Accessibility and Agricultural Development: A Systems Approach to an Aspect of the Economic Development Process." Ph.D. Dissertation, Syracuse University, Syracuse, N.Y., 1969.

Duckham, A.N., and G.B. Masefield. *Farming Systems of the World*. London, England: Chatto and Windus, 1970.

Finch, Vernar C., and O.E. Baker. *Geography of the World's Agriculture*. Port Washington, N.Y.: Kennikat Press, 1970.

Floyd, Barry N. "No Easy Harvest: The Challenge to Agricultural Geographers." *East Lakes Geographer*, 6 (1970): 48-55.

Gregor, Howard F. *Geography of Agriculture: Themes in Research*. Englewood Cliffs, N.J.: Prentice-Hall, 1970.

Grigg, David B. *The Harshlands: A Study in Agricultural Development*. London, England: Macmillan, 1970.

Hoekveld, G.A. "Agriculture in the Developed Countries" (in Dutch). *Geografisch Tijdschrift*, 4 (1970): 102-116.

Isaac, Erich. *Geography of Domestication*. Englewood Cliffs, N.J.: Prentice-Hall, 1970.

Kostrowicki, Jerzy, and W. Tyszkiewicz, eds. "Agricultural Typology: Selected Methodological Materials." *Dokumentacja Geograficzna*, 1 (1970): 1-60.

Munton, R.J.C. "Diffusion and Agriculture." *Area*, 2, No. 3 (1970): 74-75.

Robert, B.K. "Agricultural Improvement." *Area*, 2, No. 3 (1970): 67-73.

Rutherford, J. "Agricultural Geography as a Discipline: A Suggested Conceptual Foundation for University Work." *Jerusalem Studies in Geography*, 1 (1970): 37-105.

Wheeler, K.E., A.M. Ladley, and F.G. Leong. *Studies in Agricultural Geography*. London, England: Blond Educational, 1970.

Bunce, Michael F., and G.T. McDonald. *Themes in Rural Geography: A Bibliographic Review*. Toronto, Ontario: University of Toronto Department of Geography, Discussion Paper No. 11, 1971.

Clout, Hugh D. *Agriculture*. London. England: Macmillan, 1971.

Jackson, W.A.D., ed. *Agrarian Policies and Problems in Communist and Non-Communist Countries*. Seattle, Wash.: University of Washington Press, 1971.

Klatt, Werner. "Successes and Failures of Communist Farming: Causes and Consequences." Pp. 462-478 in *Agrarian Policies and Problems in Communist and Non-Communist Countries*, edited by W.A.D. Jackson. Seattle, Wash.: University of Washington Press, 1971.

Kostrowicki, Jerzy. "The Methodological Bases for the Typology of World Agriculture." Pp. 141-173 in *Symposium of Agricultural Typology and Agricultural Settlements,* edited by G. Krajko, I. Penses, and J. Toth. Szeged, Hungary: 1971.

Kostrowicki, Jerzy. "Commission on Agricultural Typology." *IGU Bulletin*, 22, No. 1 (1971): 16-21.

Morgan, W.B., and R.J.C. Munton. *Agricultural Geography*. London, England: Methuen, 1971.

Pfeifer, Gottfried, ed. *Symposium zur Agrargeographie anlassliche des 80. Geburtstages von Leo Waibel am 22. Februar 1968*. Heidelberg, Germany: Heidelberger Geographische Arbeiten, Heft 36, 1971.

Piersma, A. "Agricultural Geography Versus Rural Geography" (in Dutch). *Geographische Tijschrift*, 5, No. 1 (1971): 25-32.

Waites, Bryon, K.S. Wheeler, and J.A. Giggs. *Patterns and Problems in World Agriculture*. Leicester, England: Blond Educational Ltd., 1971.

Wittfogel, Karl A. "Communist and Non-Communist Agrarian Systems, with Special Reference to the U.S.S.R and Communist China: A Comparative Approach." Pp. 3-60 in *Agrarian Policies and Problems in Communist and Non-Communist Countries,* edited by W.A.D. Jackson. Seattle, Wash.: University of Washington Press, 1971.

Blazoni, Mario. "Das Problem der Welternahrung eine Skizze angewandter Sozialgeographie." *Mitteilungen des Osterreichischen Geographischen Gesellschaft*, 114, No. 1-2 (1972): 197-201.

Costa Marques, Cecilia, Jorge de Sa, and Pierre Desquin. "Une measure de la perseverance dans l'agriculture." *Revue Belge de Geographie*, 96 (1972): 5-32.

Dubel, Krystyna. "Agricultural-Geographical Micro-Scale Research as a Basis for Analyses and Agricultural Progress Programming" (in Polish). *Przeglad Geograficzny*, 44, No. 2 (1972): 299-305.

Harris, David R. "The Origins of Agriculture in the Tropics." *American Scientist*, 60 (1972): 180-193.

Kostrowicki, Jerzy. "An Attempt at the Typology of World Agriculture" (in Polish). *Przeglad Geograficzny*, 44, No. 3 (1972): 395-435.

Kostrowicki, Jerzy. "A Preliminary Attempt at a Typology of World Agriculture." Pp. 1097-1100 in *International Geography 1972*, Vol. 2, edited by W.P. Adams and F.M. Helleiner. Montreal, Quebec: IGU, 1972.

Le Coz, Jean. "Integration ternaire de l'agriculture et amenagement de l'espace rural." *L'Espace Geographique*, 1, No. 4 (1972): 217-229.

Lehmann, Heinrich. *Die Agrarlandschaft in den Linken Nebent alern des oberen Mittelrhules und ihr Strukturwandel*. Bonn-Bad Godesberg, Germany: Selbsturlag der Bundesforschungsanstalt fur Landeskunde und Raumordnung, 1972.

Lidstone, Michael. "An Example of an Eastern Mediterranean Agricultural Region." *Swansea Geographer*, 10 (1972): 57-60.

Munton, R.J.C. "Farm Systems Classification: A Use of Multivariate Analysis." Pp. 89-106 in *Agricultural Typology and Land Utilisation*, edited by C. Vanzetti. Verona, Italy: Center for Agricultural Geography, 1972.

Piasecki, Z. "Proposal of a New Identification–Verification Method to Be Used in Agricultural Typology." Pp. 269-276 in *Agricultural Typology and Land Utilisation*, edited by C. Vanzetti. Verona, Italy: Center for Agricultural Geography, 1972.

Prillevitz, F.C. "On the Quality and Quantity of Our Future Agricultural Area" (in Dutch). *Geografisch Tijdschrift*, 6, No. 3 (1972): 193-210.

Schulze-Von Hanxleden, Peter. *Extensivierungserscheinungen in der Agrarlandschaft des Dillgebietes*. Marburg, Germany: Marburger Geographische Schriften, Heft 54, 1972.

Tkocz, Jan. "An Attempt to Determine the Features of the Agricultural Aggregate" (in Polish). *Przeglad Geograficzny*, 44, No. 2 (1972): 289-298.

Becker, David. "Spatial Variations of Factors Involved in Farm Recreational Income." Master's Thesis, Mankato State College, Mankato, Minn., 1973.

Bowler, Ian R. "Co-operation: A Note on Governmental Promotion of Change in Agriculture." *Area*, 4, No. 3 (1973): 169-172.

Colin, Georges. "Repenser la geographie agraire." *Travaux de l'Institut de Geographie de Reims*, 16 (1973): 3-12.

Grigg, David B. "Geographical Studies of Economic Development with Special Reference to Agriculture." Pp. 18-82 in *Studies in Human Geography*, edited by Michael Chisholm and B. Rogers. London, England: Heinemann, 1973.

Ilbery, Brian W. "Decision-Making in Farming: The Importance of Sociological Factors." *Swansea Geographer*, 11 (1973): 39-44.

Kostrowicki, Jerzy. "The Typology of World Agriculture: A Preliminary Scheme." Pp. 2-52 in *Agricultural Typology and Land Use*, edited by Lloyd G. Reeds. Hamilton, Ontario: Commission for Agricultural Typology, McMaster University, 1973.

Lacombe, P., and R. Lifron. "Structures agricoles et espaces geographiques." *Bulletin de la Societe Languedocienne de Geographie*, 7, No. 2 (1973): 169-220.

Ruppert, Karl. *Agrargeographie*. Darmstadt, Germany: Wissenschaftliche Buchgesellschaft, 1973.

Semevskiy, B.N. "Problems of Agricultural Geography at the Twenty-Second International Geographical Congress" (in Russian). *Vestnik Leningradskogo Universiteta, Geologiya-Geografiya*, 24, No. 4 (1973): 73-82.

Spencer, Joseph E., and Norman R. Stewart. "The Nature of Agricultural Systems." *Annals of the Association of American Geographers*, 63, No. 4 (1973): 529-544.

Begley, Bryan W. "Agricultural Strategy in a Hot-Wet, Hot-Dry Environment in Tropical Central America and Australia: The Aborigine Tragedy." Master's Thesis, University of Michigan, Ann Arbor, Mich., 1974.

Bhattacharya, Anima. "Farm Activity and Its Place in Planned Economy." *Journal of the North East India Geographical Society*, 6, No. 1-2 (1974).

Coyaud, Louis-Marie. "Richesse agricole et urbanisation." *Revue Geographique de l'Est*, 14, No. 3-4 (1974): 183-214.

Dubel, Krystyna. "Micro-Scale Studies in Agricultural Geography as a Basis for Territorial Analysis and Programming Agricultural Progress." Pp. 377-381 in *Regional Studies: Methods and Analyses*, edited by Imre Bencze and Gyula Bora. Budapest, Hungary: Akademiai Kiado, 1974.

Gilbank, Gerald. *Introduction a la geographie generals de l'agriculture*. Paris, France: Editions Massons, 1974.

Manshard, Walther. *Tropical Agriculture: A Geographical Introduction and Appraisal*. London, England: Longman, 1974.

Morgan, W.B., and R.J.C. Munton. *Agricultural Geography*. London, England: Methuen, 1974.

Munton, R.J.C. "Farming on the Urban Fringe." Pp. 201-223 in *Suburban Growth*, edited by James H. Johnson. New York, N.Y: Wiley, 1974.

Tarrant, J.R. *Agricultural Geography*. New York, N.Y.: Halsted-Wiley, 1974.

Windhorst, Hans-Wilhelm. "Agrarformation." *Geographische Zeitschrift*, 62, No. 4 (1974): 272-294.

Andreae, Bernd. "Problems of Improving the Productivity in Tropical Farming." *Applied Sciences and Development*, 3 (1975): 124-142.

Kryuchkov, V.G., and V.S. Tikunov. "A Method for the Automatic Classification of Farming Production-Types" (in Russian). *Vestnik Moskovskogo Universiteta, Seriya Geografiya*, 2 (1975): 36-45.

McLeay, P. *Register of Agricultural Geography Research Interests*. London, England: Institute of British Geographers, 1975.

Adams, Ian H. *Agrarian Landscape Terms: A Glossary for Historical Geography*. London, England: Institute of British Geographers, 1976.

Bowler, Ian R. "The Adoption of Grant Aid in Agriculture." *Transactions, Institute of British Geographers*, 1, No. 2 (1976): 143-158. (New Series)

Ceron, Antonio O. "Conceitas economicas basicas para a geografia da agricultura." *Geografia, Rio Claro*, 1, No. 1 (1976): 35-51.

Davidson, G., and G. Rich, eds. *Agricultural Planning Issues*. Waterloo, Ontario: University of Waterloo, 1976.

Dunlop, Stewart. *Farming and the Countryside*. London, England: Heinemann, 1976.

Enyedi, Gyorgy. *Agrarian-Industrial Complexes in Modern Agriculture*. Budapest, Hungary: Hungarian Academy of Sciences, Research Institute of Geography, 1976.

Fuller, Anthony M., and Julius A. Mage, eds. *Part-Time Farming: Problem or Resource in Rural Development. Proceedings of the First Rural Geography Symposium, University of Guelph 1975*. Norwich, England: Geo Abstracts, 1976.

Kryuchkov, V.G., and Andre N. Rakitnikov. "Complex Problems of the Development of Agriculture as Related to the Geographical Sciences" (in Russian). *Voprosy Geografii*, 100 (1976): 135-145.

Olszewski, Tadeuz. "The State of Agriculture and Food in the World." *Czasopismo Geograficzne*, 47, No. 2 (1976): 119-142.

Rakitnikov, Andre N. *The Problem of the Agricultural Typology: The Influence of Agrarian-Industrial Integration on the Spatial Organization of the Rural Areas* (in Russian). Moscow, Russia: Commission of Agricultural Typology, 1976.

Unon-Lepagnot, L. "L'espace agricole est-il un espace atone?" *Bulletin de l'Association Geographes Francais*, 53 (1976): 55-62.

Van der Vliet, E. "Farm Size as a Factor in Agricultural Decision Making." *Swansea Geographer*, 12 (1976): 50-53.

Alvarado, Ivonne, and Glenda Mendoza. "Areas potenciales para desorrollo agropecuario Enfoque metodologico." *Sintesis Geograficas*, 1, No. 2 (1977): 8-11.

Bielecka, K., M. Paprzyck, and Z. Piasecki. "Badania nad stosowalnoscia metod ilosciowych w typologii rolnictwa: Reflksje methodologiczne." *Przeglad Geograficzny*, 44 (1977): 807-818.

Carter, George F. "A Hypothesis Suggesting a Single Origin of Agriculture." Pp. 89-133 in *Origins of Agriculture*, edited by Charles A. Reed. The Hague, Netherlands: Mouton, 1977.

Chiviya, Esau M. "A Comparative Analysis of Two Agricultural Projects as Strategies for Rural Development." Master's Thesis, University of Wisconsin, Madison, Wis., 1977.

Grigg, David B. *The Agricultural Systems of the World: An Evolutionary Approach*. London, England: Cambridge University Press, 1977.

Harris, David R. "Alternative Pathways Toward Agriculture." Pp. 179-244 in *Origins of Agriculture*, edited by Charles A. Reed. The Hague, Netherlands: Mouton, 1977.

Ilbery, Brian W. "Point Score Analysis: A Methodological Framework for Analyzing the Decision-Making Process in Agriculture." *Tijdschrift voor Economische en Sociale Geographie*, 68 (1977): 66-71.

Kostrowicki, Jerzy. "Agricultural Typology: Concept and Method." *Agricultural Systems*, 2 (1977): 33-45.

Mace, Georges. "L'espace agricole mayenais vu a travers la methode Bertin." *Bulletin de l'Association Geographes Francais*, 24 (1977): 405-413.

Nicolas-Obadia, Georges. "La theorie des noyaux regionaux agricoles." *L'Espace Geographique*, 6, No. 1 (1977): 25-40.

Symons, Leslie J. "The Role of Aviation in Agriculture." *Swansea Geographer*, 15 (1977): 45-47.

Yapa, Lakshman. "The Green Revolution: A Diffusion Model." *Annals of the Association of American Geographers*, 67, No. 3 (1977): 350-359.

Ali, Mohammad. *Studies in Agricultural Geography*. New Delhi, India: Rajesh Publications, 1978.

Berry, David. "Effects of Urbanization on Agricultural Activities." *Growth and Change*, 9 (1978): 2-8.

deBolos, Maria. "Elements a predre en compte pour la classification des paysages agraires en relation avec les tendances recentes." *Geographia Polonica*, 38 (1978): 23-24.

George, Pierre. *Precis de geographie rurale*. Paris, France: Presses Universitaires de France, 1978.

Ilbery, Brian W. "Agricultural Decision-Making: A Behavioral Perspective." *Progress in Human Geography*, 2 (1978): 448-466.

Mann, S., and J. Dickinson. "Obstacles to the Development of a Capitalist Agriculture." *Journal of Peasant Studies*, 5 (1978): 466-481.

Mohammad, Ali, ed. *Dynamics of Agricultural Development*. New Delhi, India: Concept Publishing Company, 1978.

Morgan, W.B. *Agriculture in the Third World: A Spatial Analysis*. Boulder, Colo.: Westview, 1978.

Singh, B.B. "Agricultural Geography: Definition, Approach and Methods" (in Hindi). *Geographical Observer*, 14 (1978): 19-23.

Turner, Billie L., II, and William E. Doolittle. "The Concept and Measure of Agricultural Intensity." *Professional Geographer*, 30, No. 3 (1978): 297-301.

Andreae, Bernd. "Agrarprobleme der Dritten Walt: Irrtum und Wahrbeit zwischen Tradition und Fortschrift." *Geographische Rundschau*, 31, No. 10 (1979): 390-395.

Bowler, Ian R. *Government and Agriculture: A Spatial Perspective*. London, England: Longman, 1979.

Coppock, J.T., ed. *Agriculture and Food Supply in Developing Countries*. Edinburgh, Scotland: University of Edinburgh, Department of Geography, 1979.

Dorney, R.S., and D.W. Hoffman. "Development of Landscape Planning Concepts and Management Strategies for an Urbanizing Agricultural Region." *Landscape Planning*, 6 (1979): 151-177.

Escudier, J.L., and A. Soulier. "L'agriculture d'arriere-pays et ses paradoxes." *Bulletin de la Societe Languedocienne de Geographie*, 13, No. 3-4 (1979): 243-274.

Grigg, David B. "Ester Boserup's Theory of Agrarian Change." *Progress in Human Geography*, 5, No. 1 (1979): 64-84.

Guermond, Yves. "Methodes statistiques en geographie agricole: Un regard en arriere." Pp. 1019-1037 in *Recherches de geographie rurale: Hommage au Professeur Frans Dussort*. Liege, Belgium: Bulletin de la Societe de Geographie de Liege, 2, 1979.

Manshard, Walther. *Tropical Agriculture: A Geographical Introduction and Appraisal*. London, England: Longman, 1979.

Rakitnikov, Andre N. "Developpement des etudes sur la typologie de l'agriculture et la planification de la production agricole." *Geographia Polonica*, 40 (1979): 23-28.

Shynne Hammond, Charles. "Agriculture." Pp. 57-75 in *Elements of Human Geography*, edited by Charles Hammond Shynne. London, England: George Allen and Unwin, 1979.

Symons, Leslie J. *Agricultural Geography*. Boulder, Colorado: Westview, 1979.

Visser, Sent. "The Spatial Dynamics of Agricultural Intensity." Unpublished Ph.D. Dissertation, Ohio State University, Columbus, Ohio, 1979.

Andreae, Bernd. *The Economics of Tropical Agriculture*. Farnham Royal, England: Commonwealth Agricultural Bureaux, 1980.

Bowler, Ian R. "Source Materials for Teaching Agricultural Geography." *Teaching Geography*, 6 (1980): 80-82.

Clark, Gordon, and D. Gordon. "Sampling for Farm Studies in Geography." *Geography*, 65 (1980): 101-106.

Hart, Patrick W.E. "Problems and Potentialities of the Behavioural Approach to Agricultural Location." *Geografiska Annaler B*, 62B, No. 1 (1980): 43-56.

Huggett, Richard, and Iain Meyer. *Geography: Theory and Practice, Agriculture*. New York, N.Y.: Harper & Row, 1980.

Kostrowicki, Jerzy. "Uklad hierarchiczny typow rolnictwaswiata." *Przeglad Geograficzny*, 52 (1980): 271-302.

Newbury, Paul A.R. *A Geography of Agriculture*. Plymouth, England: MacDonald & Evans, 1980.

Simpson, E.S., ed. *The Rural-Agricultural Sector*. Newcastle, England: Institute of British Geographers Developing Areas Study Group, 1980.

Tarrant, J.R. *Food Policies*. Chichester, England: John Wiley & Sons, 1980.

Ahnert, Frank, and Reinhart Zschoche, eds. *Festschrift fur Felix Monheimzum 65*. Aachen, Germany: Aachener Geographische Arbeiten, Heft 14, 1981.

Andreae, Bernd. *Farming Development and Space: A World Agricultural Geography*. New York, N.Y.: Walter de Gruyter, 1981.

Bhattacharya, Anima. "Origin and Evolution of Agriculture." Pp. 141-154 in *Perspectives in Agricultural Geography*, vol. 1, edited by Noor Mohammad. New Delhi, India: Concept Publishing Company, 1981.

Bowler, Ian R. "Self-Service Down on the Farm." *Geography*, 66, No. 2 (1981): 147-150.

Bowler, Ian R. "Regional Specialization in the Agricultural Industry." *Journal of Agricultural Economics*, 32 (1981): 43-54.

Casetti, Emilio. "Correlates and Determinants of Agricultural Productivity Growth: A Test of Some Hypotheses." Pp. 387-400 in *Perspectives in Agricultural Geography*, vol. 4, edited by Noor Mohammad. New Delhi, India: Concept Publishing Company, 1981.

Dhillon, S.S. "Origins and Development of Agriculture." Pp. 101-140 in *Perspectives in Agricultural Geography*, vol. 1, edited by Noor Mohammad. New Delhi, India: Concept Publishing Company, 1981.

Dhillon, S.S. "The Origins of Agriculture." *Asian Profile*, 9, No. 4 (1981): 371-382.

Kostrowicki, Jerzy. "A Hierarchy of World Types of Agriculture." Pp. 165-208 in *Perspectives in Agricultural Geography*, vol. 1, edited by Noor Mohammad. New Delhi, India: Concept Publishing Company, 1981.

Krishnan, Gopal. "The Concept of Agricultural Development." Pp. 77-88 in *Perspectives in Agricultural Geography*, vol. 4, edited by Noor Mohammad. New Delhi, India: Concept Publishing Company, 1981.

Mohammad, Noor, ed. *Perspectives in Agricultural Geography*. New Delhi, India: Concept Publishing Company, 1981. 5 volumes.

Mohammad, Noor. "Origin, Diffusion and Development of Agriculture." Pp. 11-62 in *Perspectives in Agricultural Geography*, vol. 1, edited by Noor Mohammad. New Delhi, India: Concept Publishing Company, 1981.

Mohammad, Noor. "Trends of Diffusion of Agricultural Innovations." Pp. 359-382 in *Perspectives in Agricultural Geography*, vol. 5, edited by Noor Mohammad. New Delhi, India: Concept Publishing Company, 1981.

Obara, Dunston A. "Origin and Evolution of Agriculture: A Geographical Appraisal." Pp. 63-84 in *Perspectives in Agricultural Geography*, Vol. 1, edited by Noor Mohammad. New Delhi, India: Concept Publishing Company, 1981.

Pulyarkin, V.A. "Some Geographical Problems of Agricultural Intensification in the Developing Countries" (in Czech). *Prace Geograficzne Instytut Geografii i Przestrzennego Zagospodarowana PAN*, 14, No. 1 (1981): 101-108.

Raza, Mehdi. "Origin and Early Dispersal of Agriculture." Pp. 85-100 in *Perspectives in Agricultural Geography*, Vol. 1, edited by Noor Mohammad. New Delhi, India: Concept Publishing Company, 1981.

Spedding, C.R.W. "Agricultural Potential in the Developing Countries." *Applied Geography*, 1, No. 3 (1981): 157-166.

Troughton, Michael J. "The Industrialization of Agriculture: The Process and Its Implications for Rural Resource Management." *Geographical Inter-University Resource Management Seminars*, 11 (1981): 74-99.

Upmeier, Helga. *Der Agrarwirtschaftsraum der Poebene: Eignung, Agrarstruktur und regionale Differenzierung; ein geographischer Beitrag zur landwirtschaftlichen Strukturforschung einer industriell gepragten EG-Agrarregion*. Tubingen, Germany: Tubinger Geographische Studien, Heft 82, 1981.

Balaam, David N., and Michael J. Carey. *Food Politics: The Regional Conflict*. Totowa, N.J.: Rowman & Littlefield, 1982.

Buttel, F.H. "The Political Economy of Part-Time Farming." *GeoJournal*, 6, No. 4 (1982): 293-300.

Cavazzoni, A., and Anthony M. Fuller. "International Perspectives on Part-Time Farming: A Review." *GeoJournal*, 6, No. 4 (1982): 383-390.

Christodoulou, D. "Part-Time Farming in the Developing World: A Case of Hobson's Choice of the Privilege of Half a Loaf." *GeoJournal*, 6, No. 4 (1982): 373-380.

Clark, Gordon. *The Agricultural Census: United Kingdom and United States.* Norwich, England: Geo Abstracts, Conceptual Techniques in Modern Geography, No. 35, 1982.

Enyedi, Gyorgy, and I. Volgyes, eds. *The Effect of Modern Agriculture on Rural Development.* New York, N.Y.: Pergamon, 1982.

Grigg, David B. *The Dynamics of Agricultural Change: The Historical Experience.* New York, N.Y.: St. Martin's, 1982.

Grigg, David B. "Agricultural Geography." *Progress in Human Geography*, 6 (1982): 242-246.

Kaminiski, Z. *Przestrzenna dyfuzja innowacji rolniczych.* Uniwersytet im. Adama Mickiewicza, Seria Geografia, No. 25, 1982.

Louault, F. "Le delimitation des regions agricoles." *Norois*, 29 (1982): 345-364.

Mage, Julius A. "The Geography of Part-Time Farming: A New Vista for Agricultural Geographers." *GeoJournal*, 6, No. 4 (1982): 301-312.

Newbury, Paul A.R. *A Geography of Agriculture.* Estover, England: MacDonald & Evans, 1982.

Aitchison, J.W. "Model Types of World Agriculture: Problems of Definition and Case Identification." *Geographia Polonica,* 46 (1983): 175-186.

Bowler, Ian R. "Structural Change in Agriculture." Pp. 46-74 in *Progress in Rural Geography*, edited by Michael Pacione. London, England: Croom Helm, 1983.

Grigg, David B. "Agricultural Geography." *Progress in Human Geography*, 7, No. 2 (1983): 255-260.

Jones, Donald W. "Location, Agricultural Risk and Farm Income Diversification." *Geographical Analysis*, 15, No. 3 (1983): 231-246.

Kostrowicki, Jerzy, and Wieslawa Tyszliewicz. *Agricultural Typology and Rural Development.* Warsaw, Poland: Polish Scientific Publishers, 1983.

Munton, R.J.C. "Agriculture and Conservation: What Room for Compromise?" Pp. 353-371 in *Conservation in Perspective,* edited by F.B. Goldsmith. London, England: Wiley, 1983.

Sick, Wolf D. *Agrargeographie*. Braunschweig, Germany: Westermann, 1983.

Sisler, Daniel G. *Readings for Agricultural Geography for Use in the Economics of Agricultural Geography.* Ithaca, N.Y.: Cornell University, Department of Agricultural Economics, 1983.

Skotnicki, M. "Typologie de l'agriculture mondiale et realistes du Tiers-monde." *Bulletin de la Societe Languedocienne de Geographie*, 17, No. 1-2 (1983): 241-248.

Smil, Vaclav. *Energy Analysis in Agriculture.* Boulder, Colo.: Westview, 1983.

Bowler, Ian R. "Agricultural Geography." *Progess in Human Geography*, 8 (1984): 256-262.

Bryant, Christopher R. "Agriculture in the Urban Fringe: A Systems Perspective." *Rural Systems*, 2, No. 1 (1984): 1-15.

Clark, Gordon. "The Meaning of Agricultural Regions." *Scottish Geographical Magazine*, 100, No. 1 (1984): 34-44.

Crabb, P. "Agricultural Change and Agricultural Systems." *Geography Bulletin*, 15, No. 4 (1984): 261-274.

Dagarne, A., and J.P. Paulet, eds. *Geographies l'aujour d'hui: Melanges offerts en hommage au doyen F. Gay*. Nice, France: Universite de Nice, Laboratoire d'analyse spatiale Raoul Blanchard, 1984.

Doolittle, William E. "Agricultural Change as an Incremental Process." *Annals of the Association of American Geographers*, 74, No. 1 (1984): 124-137.

Ehlers, Eckert. *Die agrares Siedlungsgrenzen der Erde*. Stuttgart, Germany: Erkkundliches Wissen, Heft 69, 1984.

Grigg, David B. *An Introduction to Agricultural Geography*. London, England: Hutchinson, 1984.

Hill, Brian E. *The Common Agricultural Policy: Past, Present, and Future*. New York, N.Y.: Methuen, 1984.

Hill, Ronald D. "Agriculture and Structural Change in the World's Economies." *Asian Geographer*, 3, No. 2 (1984): 147-162.

Norton, William. "Agricultural Evolution on the Frontier." *Professional Geographer*, 36, No. 1 (1984): 18-27.

Rindos, David. *The Origins of Agriculture: An Evolutional Perspective*. Orlando, Fla.: Academic Press, 1984.

Ruppert, Karl. "Agrargeographie im Wandel." *Geographica Helvetica*, 39 (1984): 168-172.

Weischet, Wolfgang. "Agrarwirtschaft in den feuchten Tropen." *Geographische Rundschau*, 36, No. 7 (1984): 344-351.

Deffontaines, Jean-Pierre. "Etude de l'activite agricole en analyse du paysage." *L'espace Geographie*, 14, No. 1 (1985): 37-48.

Ehlers, Eckert. "Die agraren Siedlungsgrenzen der Erde: Ein Uberlick." *Geographische Rundschau*, 37 (1985): 330-338.

Gilg, Andrew. "The Regional Development of Agriculture." *Geographia Polonica*, 51 (1985): 301-312.

Grigg, David B. "World Patterns of Agricultural Productivity." *Tijdschrift voor Economische en Sociale Geographie*, 76, No. 4 (1985): 253-260.

Wallace, Iain. "Towards a Geography of Agribusiness." *Progress in Human Geography*, 9, No. 4 (1985): 491-514.

Arnold, A. *Agrargeographie*. Paderhorn, Germany: Ferdinand und Schonigh, 1986.

Blanc, M. "La diversite dans l'agriculture perspectives." *Revue Geographique des Pyrenees et du Sud-Ouest*, 57, No. 1 (1986): 9-35.

Grigg, David B. "World Patterns of Agricultural Output." *Geography*, 71, No. 3 (1986): 240-245.

Kain, Roger J.P., and Roland Allison. "Agriculture in the Past." *Teaching Geography*, 11, No. 2 (1986): 58-63.

Kayser, B. "De la differenciation des agriculture a la diversite des agriculteurs." *Revue Geographique des Pyrenees et du Sud-Ouest*, 57, No. 1 (1986): 37-46.

LaForge, J. "Pour une analyse dialectique des concepts de developpement agricole et de developpement rural." *Revue Geographique des Pyrenees et du Sud-Ouest*, 57, No. 1 (1986): 61-72.

LeBeau, Rene. *Les grands types de structures agraires dans le monde*. Paris, France: Masson, 1986.

Manshard, Walther. "Agrarforschung, Agrargeographische und rurale Entwicklungspraxis in den Tropen." *Geographische Zeitschrift*, 74, No. 2 (1986): 63-73.

Marsden, Terry K., et al. "The Restructuring Process and Economic Centrality in Capitalist Agriculture." *Journal of Rural Studies*, 2, No. 4 (1986): 271-280.

Mohamden, M. *The Principal and the Scope of Agricultural Geography* (in Arabic). Saudi Arabia: Graduate Publishing House, 1986.

Newbury, Paul A.R. *A Geography of Agriculture*. Harlow, England: Longman, 1986.

Perrot, E. "Les nouvelles constraintes de l'agriculture productiviste." *Revue Geographique des Pyrenees et du Sud-Ouest*, 57, No. 1 (1986): 47-59.

Retaille, D. "L'agriculture dans le monde: Le probleme de l'evolution par Etat." *L'Information Geographique*, 50, No. 2 (1986): 64-79.

Yamamoto, S., and A. Tezuka. "The Domain of Agricultural Geography" (in Japanese). *Tsukuba Studies in Human Geography*, 10 (1986): 1-15.

Bowler, Ian R. "Agricultural Geography." *Progress in Human Geography*, 11, No. 3 (1987): 425-432.

Bowler, Ian R. "The Regional Restructuring of Production in Capitalist Agriculture." *Rural Systems*, 5, No. 1 (1987): 17-26.

Bowler, Ian R., and Brian W. Ilbery. "Redefining Agricultural Geography." *Area*, 19, No. 4 (1987): 327-332.

Brush, Stephen B., and Billie L. Turner II. "The Nature of Farming Systems." Pp. 11-54 in *Comparative Farming Systems*, edited by Billie L. Turner II and Stephen B. Brush. New York, N.Y.: Guilford, 1987.

Bryant, Christopher R., and Thomas R.R. Johnston. "Agricultural Adoption: The Prospects for Sustaining Agriculture Near Cities." Pp. 9-21 in *Sustaining Agriculture Near Cities*, edited by William Lockerretz. Ankeny, Iowa: Social and Water Conservation Society, 1987.

Charlton, Clive A. "Problems and Prospects for Sustainable Agricultural Systems in the Humid Tropics." *Applied Geography*, 7, No. 2 (1987): 153-174.

Clawson, David L. "Teaching Traditional Tropical Agriculture." *Journal of Geography*, 86, No. 5 (1987): 204-210.

Dube, R.S. *Agricultural Geography: Issues and Applications*. Delhi, India: Gian Publishers, 1987.

Eizner, N. "Les Montagnes siches dans les etats generaux du developpement agricole." *Bulletin de la Societe Languedocienne de Geographie*, 18, No. 3-4 (1987): 513-520.

Bowler, Ian R. "Agricultural Geography." *Progress in Human Geography*, 12, No. 4 (1988): 538-548.

Briggs, David, and Frank Countney. *Agricultural Systems*. New York, N.Y.: Wiley, 1988.

Chalmin, Philippe. "L'agro-business: Reponse a une demande alimentaire?" *Annales de Geographie*, 89, No. 3 (1988): 359-372.

Conte, Analia S. *Geografia Agricola*. Buenos Aires, Argentina: SENOC—Asociacion para la Promocian de Sistemas Educativos no Convencionales, 1988.

Ilbery, Brian W. "Farm Diversification and the Restructuring of Agriculture." *Outlook on Agriculture*, 17, No. 1 (1988): 35-39; 108-121.
Kostrowicki, Jerzy. "Le developpement recent des etudes comparatives de l'agriculture mondiale." Pp. 412-416 in *Campagnes et littoraux d'Europe: Melanges offerts a Pierre Flatres*. Lille, France: Institut de Geographie de Lille: Hommes et Terres du Nord, 1988.
Kostrowicki, Jerzy. "Comparative Research in World Agriculture" (in Polish). *Przeglad Geograficzny*, 60, No. 4 (1988): 511-572.
LeHeron, Richard B. "Food and Fibre Production Under Capitalism: A Conceptual Agenda." *Progress in Human Geography*, 12, No. 3 (1988): 409-430.
Marsden, Terry K. "Exploring Political Economy Approaches in Agriculture." *Area*, 20, No. 4 (1988): 315-322.
Morgan, W.B. "Agrarian Structure." Pp. 77-113 in *The Geography of the Third World*, edited by Michael Pacione. London, England: Routledge, 1988.
Rother, K. "Agrargeographie." *Geographische Rundschau*, 40, No. 2 (1988): 36-41.
Saez, L.M. *Geografia agraria: Introduccion a las paisajis rurales*. Madrid, Spain: Editorial Sintesis, 1988.
Singh, V.R., and N.K. Singh, eds. *Perspectives in Agricultural Typology*. Varanasi, India: Star, 1988.
Bowler, Ian R. "Revising the Research Agenda on Agricultural Policy in Developed Market Economies." *Journal of Rural Studies*, 5, No. 4 (1989): 385-394.
Edwards, C.A. "The Importance of Integration in Sustainable Agricultural Systems." *Agriculture, Ecosystems and Environment*, 27 (1989): 25-35.
Ehlers, Eckert. "Agricultural Frontiers." *Terra*, 101, No. 1 (1989): 46-53.
Harris, J., and B. Harris. "Agrarian Transformation in the Third World." Pp. 258-278 in *Horizons in Human Geography*, edited by David Gregory and R. Walford. Basingstoke, England: Macmillan, 1989.
Jedut, R. "Geografia produktow spozy wezyeh swiata-tluszeze." *Czasopismo Geograficzne*, 60, No. 1 (1989): 43-63.
Leeming, Frank. "Rural Change and Agricultural Development." *Geography*, 74, No. 4 (1989): 348-351.
Munton, R.J., Sarah C. Whatmore, and Terry K. Marsden. "Part-Time Farming and Its Implications for the Rural Landscape: A Preliminary Analysis." *Environment and Planning A*, 21, No. 4 (1989): 523-536.
Napton, Darrell. "Contemporary Agriculture and Rural Land Use." Pp. 333-350 in *Geography in America*, edited by Gary L. Gaile and Cort J. Willmott. Columbus, Ohio: Merrill, 1989.
Spielmann, Hans O. *Agrargeographie in Stichwarten*. Unterageri: Hirt, 1989.
Windhorst, Hans-Wilhelm. *Die Industrialisierung der Agrarwirtschaft: Ein Vergleich ablaufender Prozesse in den USA und der Bundesrepublik Deutschland*. Frankfurt, Germany: Strothe, 1989.
Windhorst, Hans-Wilhem. *Industrialisierte Landwirtschaft und Agrarindustrie Entwicklungen, Strukturen und Probleme*. Vechta, Germany: Vechtaer

Druckerie und Verlag, Vechtaer Arbeiten zur Geographie und Regionalwissenschaft, Band 8, 1989.

Zuka, M. *Agricultural Geography* (in Arabic). Alexandria, Egypt: University of Alexandria Press, 1989.

Bebbington, Anthony, and Judith A. Carney. "Geography in the International Research Centers: Theoretical and Practical Concerns." *Annals of the Association of American Geographers*, 80, No. 1 (1990): 314-348.

Blanc-Pamand, Chantel, and Andre Lericollais, eds. *Dynamique des systemes agraires: A travers champs, agronomes et geographes*. Paris, France: Editions de l'ORSTOM, 1991.

Bowler, Ian R. "Agricultural Geography." *Progress in Human Geography*, 14, No. 4 (1990): 569-578.

Guellac, A. "La succession dans l'agriculture: L'heritage et les acteurs—la maniere pour demain." *L'Information Geographique*, 54, No. 1 (1990): 6-15.

Khan, S.A. "Role of Agriculture in the Economic Development of the Third World Countries." *National Geographer*, 25, No. 2 (1990): 141-148.

Mohr, Bernhard, Konrad Sonntag, and Jorg Stadelbauer. *Raumliche Strukturen im Wandel: Festschrift fur Wolf-Dieter Sick. Teil B, Beitrage zur Agrarwirtschaft der Tropen*. Freiberg, Germany: Freiberger Geographische Arbeiten, Heft 30, 1990.

Munton, R.J.C., Terry K. Marsden, and Sarah Whatmore. "Technological Change in a Period of Agricultural Adjustment." Pp. 104-126 in *Technological Change and the Rural Environment*, edited by Philip Lowe, Terry K. Marsden, and S. Whatmore. London, England: David Fulton, 1990.

Pierce, John T. *The Food Resource*. London, England: Longman, 1990.

Brklacich, Michael, Christopher R. Bryant, and Barry E. Smit. "A Review and Appraisal of the Concept of Sustainable Food Production Systems." *Environmental Management*, 15, No. 1 (1991): 1-14.

Christiansen, D. "Bio-Dynamic Agriculture, World View and Practice." Pp. 87-91 in *Proceedings, Sixteenth New Zealand Geography Conference*, edited by David Hawke. Auckland, New Zealand: New Zealand Geographical Society, 1991.

Fenelon, Pierre. *Dictionnaire d'histoire et de geographie agraires*, second edition. Paris, France: Conseil International de la Longue Francaise, 1991.

Sharma, B.L. *Applied Agricultural Geography*. Jaipur, India: Rawat Publications, 1991.

Troughton, Michael J. "Agriculture and Rural Resources." Pp. 53-84 in *Resource Management and Development*, edited by Bruce Mitchell. Toronto, Ontario: Oxford University Press, 1991.

Whatmore, Sarah. "Agricultural Geography." *Progress in Human Geography*, 15, No. 3 (1991): 303-310.

Wilken, G.C. *Sustainable Agriculture Is the Solution, But What Is the Problem?* Washington, D.C.: USAID, Board for International Food and Agricultural Development and Economic Cooperation, Occasional Paper No. 14, 1991.

Aitchinson, John. "Farm Types and Agricultural Regions." Pp. 109-133 in Ian R. Bowler, ed. *The Geography of Agriculture in Developed Market Economies*. London, England: Longman, 1992.

Bowler, Ian R., ed. *The Geography of Agriculture in Developed Market Economies*. London, England: Longman, 1992.

Bowler, Ian R., Christopher R. Bryant, and M. Duane Nellis, eds. *Contemporary Rural Systems in Transition: Volume 1, Agriculture and Environment*. Wallingford, England: C.A.B. International, 1992.

Bryant, Christopher R. "Farming at the Urban Fringe." Pp. 275-304 in *The Geography of Agriculture in Developed Market Economies*, edited by Ian R. Bowler. London, England: Longman, 1992.

Bryant, Christopher R., and Thomas R.R. Johnston. *Agriculture in the City's Countryside*. Toronto, Ontario: University of Toronto Press, 1992.

Bulman, Teresa L. "Water Management Implications of Modern Viticultural Practices." Pp. 73-86 in *Viticulture in Geographic Perspective: Proceedings of the Miami AAG Symposium*, edited by Harm de Blij. Miami, Fla.: Miami Geographical Society, 1992.

Clark, Gordon. "Data Sources for Studying Agriculture." Pp. 32-55 in *The Geography of Agriculture in Developed Market Economies*, edited by Ian R. Bowler. London, England: Longman, 1992.

Edwards, Christopher. "Changing Farm Enterprises." Pp. 134-161 in *The Geography of Agriculture in Developed Market Economies*, edited by Ian R. Bowler. London, England: Longman, 1992.

Grigg, David B. "World Agriculture: Production and Productivity in the Late 1980s." *Geography*, 77, No. 2 (1992): 97-108.

Grigg, David B. "Agriculture in the World Economy: An Historical Geography of Decline." *Geography*, 77, No. 3 (1992): 210-222.

Hoggart, Keith, ed. *Agricultural Change, Environment and Economy: Essays in Honour of W.B. Morgan*. London, England: Mansell, 1992.

Marsden, Terry K., Jonathan Murdoch, and S. Williams. "Regulating Agriculture in Deregulating Economies: Emerging Trends in the Uneven Development of Agriculture." *Geoforum*, 23, No. 3 (1992): 333-346.

Mohammad, Noor, ed. *New Dimensions in Agricultural Geography*. New Delhi, India: Concept Publishers, 1992.

Munton, R.J.C. "Factors of Production in Modern Agriculture." Pp. 56-84 in *The Geography of Agriculture in Developed Economies*, edited by Ian R. Bowler. London, England: Longman, 1992.

Munton, R.J.C. "The Uneven Development of Capitalist Agriculture." Pp. 25-48 in *Agricultural Change, Environment and Economy: Essays in Honour of W.B. Morgon*, edited by Keith Hoggart. London, England: Mansell, 1992.

Parry, Martin. "Agriculture as a Resource System." Pp. 207-238 in *The Geography of Agriculture in Developed Market Economies*, edited by Ian R. Bowler. London, England: Longman, 1992.

Pierce, John T. "The Policy Agenda for Sustainable Agriculture." Pp. 221-236 in *Contemporary Rural Systems in Transition: Volume 1, Agriculture and*

Environment, edited by Ian R. Bowler, Christopher R. Bryant, and M. Duane Nellis. Wallingford, England: C.A.B. International, 1992.

Tarrant, John R. "Agriculture and the State." Pp. 239-274 in *The Geography of Agriculture in Developed Market Economies*, edited by Ian R. Bowler. London, England: Longman, 1992.

Wallace, Iain. "International Restructuring of the Agri-food Chain." Pp. 15-28 in *Contemporary Rural Systems in Transition: Volume 1, Agriculture and Environment*, edited by Ian R. Bowler, Christopher R. Bryant, and M. Duane Nellis. Wallingford, England: C.A.B. International, 1992.

Bryan, Deborah. "Modeling Risk in Agriculture." Master's Thesis, Ohio State University, Columbus, Ohio, 1993.

Evans, N.J., and Brian W. Ilbery. "The Pluriacticity, Part-Time Farming and Farm Diversification Debate." *Environment and Planning A*, 25, No. 7 (1993): 945-960.

Hoggart, Keith. "Global Economic Structures and Agricultural Change." Pp. 1-24 in *Agricultural Change, Environment, and Economy: Essays in Honour of W.B. Morgan*, edited by Keith Hoggart. London, England: Mansell, 1993.

LeHeron, Richard B. *Globalized Agriculture: Political Choice*. New York, N.Y.: Pergamon Press, 1993.

Pierce, John T. "Agriculture, Sustainability and the Imperatives of Policy Reform." *Geoforum*, 24, No. 4 (1993): 381-396.

Smit, Barry E., and J. Smithers. "Sustainable Agriculture: Interpretations, Analyses and Prospects." *Canadian Journal of Regional Science*, 16, No. 2 (1993): 399-524.

Smith, William. "Sustainability: Strategies for Agriculture." Pp. 278-283. in *Proceedings, Seventeenth New Zealand Geography Conference*, edited by William Whitaker. Christchurch, New Zealand: New Zealand Geographical Society, 1993.

Ufkes, Frances M. "The Globalization of Agriculture." *Political Geography*, 12, No. 3 (1993): 194-197.

Bonnamour, J. "A propos des accords agricoles du GATT et de la PAC: Les geographes et l'actualite." *L'Information Geographique*, 58, No. 1 (1994): 1-10.

Cai, Y., and Barry E. Smit. "Sustainability in Agriculture: A General Review." *Agriculture, Ecosystems and Environment*, 49 (1994): 299-307.

Lowe, Philip, Terry K. Marsden, and Sarah Whatmore. "Changing Regulatory Orders: The Analysis of the Economic Governance of Agriculture." Pp. 1-30 in *Regulating Agriculture*, edited by Philip Lowe, Terry K. Marsden, and Sarah Whatmore. London, England: David Fulton, 1994.

Paarlbert, Robert L. *Countryside at Risk: The Political Geography of Sustainable Agriculture*. Washington, D.C.: Overseas Development Council, 1994.

Peltre, Jean. "Permanence, changement et necessite de la geographie agraire historique." *Herodote*, 74-75 (1994): 86-94.

Bryant, Christopher R. "The Role of Local Actors in Transferring the Urban Fringe." *Journal of Rural Studies*, 11, No. 3 (1995): 255-267.

Marsden, Terry K., and A. Arce. "Constructing Quality: Emerging Food Networks in the Rural Transition." *Environment and Planning A*, 27, No. 8 (1995): 1261–1280.

Totte, M. "Reflexions sur l'apport de la teledetection a l'etude des systems agraires." *Revue Belge de Geographie*, 119, No. 1-2 (1995): 51-72.

Whatmore, Sarah. "From Farming to Agribusiness: The Global Agro-Food System." Pp. 36-49 in *Geography of Global Change: Remapping the World in the Late Twentieth Century*, edited by R.J. Johnston, Peter J. Taylor, and M.J. Watts. Oxford, England: Blackwell, 1995.

Whatmore, Sarah, Terry K. Marsden, and Philip Lowe, eds. *Regulating Agriculture*. New York, N.Y.: John Wiley, 1995.

Christiansen, Sofus. "Concentrational Agriculture: Types, Functions, and Derivation." *Geografisk Tidsskrift*, 96 (1996): 123-128.

Grigg, David B. "Famines and Surpluses in World Food Production." Pp. 677-701 in *Companion Encyclopedia of Geography: The Environment and Humankind*, edited by Ian Douglas, Richard Huggett, and Mike Robinson. London, England: Routledge, 1996.

Ilbery, Brian W., and Ian R. Bowler. "Industrialization and World Agriculture." Pp. 228-248 in *Companion Encyclopedia of Geography: The Environment and Humankind*, edited by Ian Douglas, Richard Huggett, and Mike Robinson. London, England: Routledge, 1996.

Watts, Michael J. "Development III: The Global Agrofood System and Late Twentieth Century Development (or Kautsky Redux)." *Progress in Human Geography*, 20, No. 2 (1996): 230-245.

Bryant, Christopher R. "L'Agriculture periurbaine: L'economie politique d'une espace innovateur." *Cahiers Agricultures*, 6 (1997): 125-130.

Donkin, R.A. "A Servant of Two Masters." *Journal of Historical Geography*, 23, No. 3 (1997): 247-266.

Goodman, D, and M. Watts. *Globalizing Food*. London, England: Routledge, 1997.

Ilbery, Brian W., and L.E. Holloway. "Responses to the Challenge of Productivist Agriculture." *Built Environment*, 23 (1997): 184-191.

Marsden, Terry K. "Creating Space for Food: The Distinctiveness of Recent Agrarian Development." Pp. 169-191 in *Globalizing Food*, edited by D. Goodman and Michael J. Watts. London, England: Longman, 1997.

Mereene-Scholmaker, B. "La localisation des productions agricoles: Introduction methodologique." *Bulletin de la Societe de Liege*, 33 (1997): 129-138.

Napton, Darrell. "Restructuring for Rural Sustainability: Overcoming Scale Conflicts and Cultural Biases." Pp. 329-340 in *Agricultural Restructuring and Sustainability: A Geographical Perspective*, edited by Brian W. Ilbery, Quentin Chiotti, and Timothy Rickard. Wallingford, England: C.A.B. International, 1997.

Robinson, Guy M. "Greening and Globalizing: Agriculture in the 'New Times.'" Pp. 41-54 in *Agricultural Restructuring and Sustainability: A Geographical Perspective*, edited by Brian W. Ilbery, Quentin Chiotti, and Timothy Rickard. Wallingford, England: C.A.B. International, 1997.

Durbianco, C. "Les zones 'naturelles' et agricoles des espaces periurbains mediterranees: Une nouvelle problematique." *Mediterranee*, 89, No. 2-3 (1998): 47-52.

Ilbery, Brian W., and Ian R. Bowler. "From Agricultural Production to Post-Production." Pp. 57-84 in *The Geography of Rural Change,* edited by Brian W. Ilbery. London, England: Longman, 1998.

Marsden, Terry K. "Agriculture Beyond the Treadmill? Issues for Policy, Theory and Research Practice." *Progress in Human Geography*, 22, No. 2 (1998): 265-275.

Smith, C.S., and G.T. McDonald. "Assessing the Sustainability of Agriculture at the Planning Stage." *Journal of Environmental Management*, 52, No. 1 (1998): 15-37.

Douence, H. "Approche spatiale des transformations de l'agriculture." *L'Information Geographique*, 63, No. 3 (1999): 139-141.

Mannion, A.M. "Domestication and the Origins of Agriculture: An Appraisal." *Progress in Physical Geography*, 23, No. 1 (1999): 37-56.

Morris, C., and N.J. Evans. "Research on the Geography of Agricultural Change: Redundant or Revitalized?" *Area*, 31, No. 4 (1999): 349-358.

Pilleboue, Jean. "La qualite agro-alimenataire et ses territoires productifs avant-propos." *Sud-Ouest Europeen*, 6 (1999): 1-8.

Poinsot, Yves. "L'incidence geographique des risques agricoles (une formulation theorique a partir de cas andins et africains)." *Revue de Geographie Alpine*, 87, No. 3 (1999): 31-50.

Murdoch, Jonathan, Terry K. Marsden, and Jo Banks. "Quality, Nature, and Embeddedness: Some Theoretical Considerations in the Context of the Food Sector." *Economic Geography*, 76, No. 2 (2000): 107-1245.

CROPS AND CROPPING PATTERNS

Reed, J.H. "Cotton-Growing Within the British Empire." *Scottish Geographical Magazine*, 27, No. 5 (1911): 242-256.

Baker, O.E. "The Potential Supply of Wheat." *Economic Geography*, 1, No. 1 (1925): 15-52.

Rice, Phyllis. "Some Geographic Factors Affecting the Coffee Industry." Master's Thesis, University of Nebraska, Lincoln, Nebr., 1925.

Popenoe, Paul. "The Distribution of the Date Palm." *Geographical Review*, 16 (1926): 117-121.

Trewartha, Glenn T. "The Tea Crop." *Journal of Geography*, 28, No. 1 (1929): 1-24.

Robertson, Curtis J. "Geographical Trends in Sugar Production." *Geographical Review*, 22 (1932): 120-130.

Strong, Helen M. "Export Wheat Producing Regions." *Economic Geography*, 8, No. 2 (1932): 161-190.

Albright, W.D. "Crop Growth in High Latitudes." *Geographical Review*, 23, No. 4 (1933): 608-620.

Smith, James W. "Geographical Aspects of World Cotton." Master's Thesis, University of North Carolina, Chapel Hill, N.C., 1935.

Buchanon, Ogilvie. "Some Features of World Wheat Production." *Scottish Geographical Magazine*, 52, No. 5 (1936): 313-324.

Deasy, George F. "Location Factors in the Commercial Coconut Industry." *Economic Geography*, 17, No. 2 (1941): 130-140.

Delong, George C. "The Uses and Production of Tung Oil." Master's Thesis, Columbia University, New York, N.Y., 1941.

Fanti, Mary. "The Mattress-Growing Tree: Kapok." *Journal of Geography*, 42, No. 3 (1943): 96-103.

Simmons, Anna G.E. "Babassa Nut." *Economic Geography*, 19, No. 3 (1943): 279-282.

Mitchell, John R. "The Geographic Interpretations of Jute." Master's Thesis, University of Illinois, Urbana, Ill., 1947.

Nuttonson, Michael Y. *International Cooperation in Crop Improvement Through the Utilization of the Concept of Agroclimatic Analogues*. Prague, Czech Republic: Institute for International Collaboration in Agriculture and Foresty, 1947.

Sinadinos, Betty B. "The Geographic Aspects of Coconuts." Master's Thesis, University of Illinois, Urbana, Ill., 1947.

Barrere, Pierre. "Le marche mondial de la banane." *Les Cahiers d'Outre Mer*, 3 (1950): 343-369.

Lemaire, M.E. "Manioc." *Journal of Geography*, 49, No. 9 (1950): 376-385.

Randall, Lawrence E. "Some Geographical Aspects of Ramie Fiber Production." Master's Thesis, Clark University, Worcester, Mass., 1950.

Viers, Georges. "Le Cacao dans le Monde." *Les Cahiers d'Outre Mer*, 6 (1953): 297-351.

von Aartsen, J.P. "Geographic Aspects of Rice Growing." *Tijdschrift voor Economische en Sociale Geographie*, 45 (1954): 57-65.

Grist, D.H. "The Rice Problem." *Journal of Tropical Geography*, 5 (1955): 20-25.

Keso, Edward E. "Production and Use of Tagua Nut." *Proceedings, Oklahoma Academy of Science*, 37 (1957): 144-147.

Mountjoy, A.B. "Vegetable Oils and Oilseeds: Aspects of Production and Use." *Geography*, 42, No. 1 (1957): 37-44.

Farrell, Bryan H. "A Coefficient of Intercultivation." *Professional Geographer*, 10 (1958): 15-16.

Gay, F. "Aspects de l'agriculture burichonne au Moyen Age d'apres les textes des inventaires des establissements religieux." *Norois*, 5 (1958): 191-201.

Jonasson, O. "The Potential Areas of Coffee-Growing." *Geografiska Annaler*, 40, No. 2 (1958): 89-100.

Pan American Union. *Plantation Systems of the World*. Washington, D.C.: Pan American Union, Social Science Monographs, No. 7, 1959.

Thompson, Edger T. "The Plantation as a Social System." *Revista Geografica*, 51 (1959): 41-56.

Vasseur, L.V. "Un mode original de-mise en valeur agricole: La non-culture dans les plantations arbustives." *Les Cahiers d'Outre Mer*, 12 (1959): 56-63.

Bhatia, Shyam S. "An Index of Crop Diversification." *Professional Geographer*, 12 (1960): 3-4.

Blyn, George. "A Diagram for Demonstrating Certain Concepts in Economic Crop Geography." *Professional Geographer*, 12 (1960): 1-3.

Stewart, J.H. "Perspectives on Plantations." *Revista Geografica*, 26 (1960): 77-85.

Burrill, Robert M. "A Classification of Hybrid Corns Using Thornthwaite's Development Unit." Master's Thesis, University of Chicago, Chicago, Ill., 1961.

Borden, Charles B. "Rice and Economic Development." Master's Thesis, Columbia University, New York, N.Y., 1962.

Brunn, Stanley D. "A Cultural Plant Geography of the Quince." *Professional Geographer*, 15 (1963): 16-18.

Chateau, Robert. "La riziculture dans le monde." *Les Cahiers d'Outre Mer*, 16 (1963): 394-412.

Seig, Louis. "The Spread of Tobacco: A Study in Cultural Diffusion." *Professional Geographer*, 15, No. 1 (1963): 17-21.

Vogt, Jean. "Notes agraires: L'anciennete de l'assolement biennal al sacien." *Revue Geographique de l'Est*, 3, No. 3 (1963): 241-244.

Chateau, Robert. "La riziculture dans le monde (second article)." *Les Cahiers d'Outre Mer*, 17 (1964): 149-174.

Huetz de Lemps, Alain. "Deux vignobles de masse: Algerie et Argentine." *Les Cahiers d'Outre Mer*, 15 (1964): 85-88.

Miller, E.J.W. "Hybrid Corn and More Food for More People." *Journal of Geography*, 63, No. 9 (1964): 413-417.

Gregor, Howard F. "The Changing Plantation." *Annals of the Association of American Geographers*, 55, No. 2 (1965): 221-238.

Gregor, Howard F. "Plantation Farming: Its Wider Application in Agricultural Classification." *California Geographer*, 6 (1965): 25-28.

Nelson, Clare J. "Geography of the Sugar Cane Industry." Master's Thesis, University of Nebraska, Lincoln, Nebr., 1966.

Blume, Helmut. "Zuckerrohr und Zuckerrube im Subtropischen Trockengurtel der Altenn Welt." *Erdkunde*, 21 (1967): 111-132.

Harris, David R. "New Light on Plant Domestication and the Origins of Agriculture: A Review." *Geographical Review*, 57, No. 2 (1967): 90-107.

Sauer, Jonathan D. "The Grain Amaranths and Their Relatives: A Revised Taxonic and Geographic Survey." *Annals of the Missouri Botantical Garden*, 54 (1967): 103-137.

Veit, Manfred. "Der Weizen: Ein Produkt und seine geographische Erscheinung." *Zeitschrift fur Wirtschaftsgeographie*, 11, No. 5 (1967): 129-138.

Higman, B.W. "Plantations and Typological Problems in Geography: A Review Article." *Australian Geographer*, 11 (1969): 192-203.

Jackson, James C. "Towards an Understanding of Plantation Agriculture." *Area*, 1, No. 4 (1969): 36-40.

Miller, Grant W. "Soybean Diffusion and World Population Growth." Master's Thesis, Arizona State University, Tempe, Ariz., 1969.

Tiller, James W. "The Winter Garden Vegetable Industry: A Study in Agricultural Geography." Ph.D. Dissertation, University of Oklahoma, Norman, Okla., 1969.

Anderson, David L. "A Simple Growth Model for the Diffusion of Hybrid Corn." *Proceedings of the Association of American Geographers*, 2 (1970): 1-4.

Chang, Jen-hu. "Sugar Cane in Hawaii and Taiwan: Contrasts in Ecology, Technology and Economics." *Economic Geography*, 46, No. 1 (1970): 39-52.

Hung, F. "Application of the Systems Concept to a Comparative Study of the Tea Industry." Pp. 165-172 in *Proceedings, Canadian Association of Geographers*. Winnipeg, Manitoba: University of Manitoba Department of Geography, 1970.

Lentnek, Barry, George P. Pattern, and Richard C. Jones. "A Spatial Production Function Analysis of Corn Yield." *Proceedings of the Association of American Geographers*, 2 (1970): 85-88.

Igbozurike, Matthias U. "Against Monoculture." *Professional Geographer*, 23, No. 2 (1971): 113-117.

Mitchell, Robert D. "The Grass Pea: Distribution, Diet and Disease." *Yearbook, Association of Pacific Coast Geographers*, 33 (1971): 29-46.

Pinard, Jacques. "Le cognac—du vignoble a la vente: Une geographie des stocks." *Annales de Geographie*, 80, No. 1 (1971): 65-83.

Singh, L.R. "Agricultural Origins and Dispersal of Rice." *National Geographical Journal of India*, 17, No. 2-3 (1971): 113-119.

Holtgrieve, Donald G. "Geography of the Vine: The Origin and Diffusion of Viticulture." *Oregon Geographer*, 6, No. 1 (1972): 1-24.

Menk, Lothar. *Landwirtschaftliche Sonderkulturen im unteren Werratal*. Marburg, Germany: Marburger Geographische Schriften, Heft 55, 1972.

Schacht, Siegfried. "Zur Geographie des Reisenbaus." *Zeitschrift fur Wirtschaftsgeographie*, 16, No. 2 (1972): 33-38.

Stevens, Thomas J. "An Evaluation of the Effects of Areal Variations in Climatological Characteristics on the Greenhouse Tomato Industry." Ph.D. Dissertation, Indiana State University, Terre Haute, Ind., 1972.

Shih, Tinmoth. "Some Basic Problems of Transforming Shifting Cultivation in the Humid Tropics." Master's Thesis, Illinois State University, Bloomington, Ill., 1973.

Andreae, Bernd. "Naturfaserproduktion in der Weltlandwirtschaft." *Geographische Rundschau*, 8 (1974): 297-307.

Bohm, W. "Die Geographie der Kulturpflanzes aus der Sicht der Pflanzenbauwissenschaft." *Geographische Zeitschrift*, 62 (1974): 105-114.

Delvert, J. "Projet de classement des types de riziculture." *Bulletin de l'Association Geographes Francais*, 51 (1974): 269-270.

Huetz de Lemps, Alain. *Types d'agriculture commerciale: La canne a sucre en Afrique du Sud, le cafe en Nouvelle Caledonia*. Talence, France: Travaux et documents de geographie tropicale, No. 12, 1974.

Wagner, Philip L. "Wines: Grape Vines and Climate." *Scientific American*, 230, No. 6 (1974): 106-115.

Wishart, David. "The Opium Poppy: The Forbidden Crop." *Journal of Geography*, 73, No. 1 (1974): 14-25.

Zvorykin, K.V., and V.A. Uglov. "The Stability of the Yield of Agricultural Crops" (in Russian). *Vestnik Moskovskogo Universiteta, Geografiya*, 3 (1974): 45-51.

Kelly, James L. "The Role of Crop Yields in Shifting Cultivation." Ph.D. Dissertation, University of Kansas, Lawrence, Kans., 1975.

Stanislowski, Daniel. "Dionyseus Westward: Early Religion and the Economic Geography of Wine." *Geographical Review*, 65 (1975): 427-442.

Dragovich, Deidre. "Adaption and Spread of Kikuyu, an Improved Pasture Grass." *Professional Geography*, 28, No. 3 (1976): 277-282.

Chang, Jen-hu. "Tropical Agriculture: Crop Diversity and Crop Yields." *Economic Geography*, 53, No. 3 (1977): 241-254.

Graeve, W.D. "Welwirtschaftliche Prospekte der Getreideproduktion." *Geographische Berichte*, 85 (1977): 258-279.

Jackson, Richard H., et al. "The Historical Diffusion of Alfalfa." *Journal of Agronomic Education*, 6 (1977): 13-19.

Dayal, Edison. "A Measure of Cropping Intensity." *Professional Geographer*, 30, No. 3 (1978): 289-296.

Hart, Patrick W.E. "Geographical Aspects of Contract Farming with Special Reference to the Supply of Crops to Processing Plants." *Tijdschrift voor Economische en Sociale Geographie*, 69, No. 4 (1978): 205-215.

Huetz de Lemps, Alain, ed. *Geographie historique des vignobles*. Paris, France: Centre National de la Recherche Scientifique, 1978. 2 volumes.

Gresse, Michel. "Recherches sur une filiere agro-alimentaire: La cas des fruits et legumes frais." *Annales de Geographie*, 88, No. 1 (1979): 65-84.

Cabantac, D.E. "Multiple Cropping Is Essentially Crop Diversification." *Philippine Geographical Journal*, 24, No. 3 (1980): 97-101.

Courtenay, P.P. *Plantation Agriculture*. Boulder, Colo.: Westview, 1980.

Bishnoi, O.P. "Crop Ecology." Pp. 343-362 in *Perspectives in Agricultural Geography*, vol. 3, edited by Noor Mohammad. New Delhi, India: Concept Publishing Company, 1981.

Burt, James E., et al. "A Parametric Crop Water Use Model." *Water Resources Research*, 17, No. 4 (1981): 1095-1108.

Chardon, Roland E., ed. *Studies in Historical Geography: Diffusion of Plantation Traits in the New World*. Baton Rouge, La: Louisiana State University, Department of Geography, 1981.

deBlij, Harm J. *Geography of Viticulture*. Miami, Fla.: Miami Geographical Society, 1981.

Leaman, J. Harold. "The Global Implications of Fuel Crops for World Food Supplies." *Proceedings, Middle States Division, Association of American Geographers*, 15 (1981): 48-52.

Majeed, Abdul. "Crop Combinational Analysis: A Review of Methodology." Pp. 159-168 in *Perspectives in Agricultural Geography*, vol. 4, edited by Noor Mohammad. New Delhi, India: Concept Publishing Company, 1981.

Mohammad, Noor, and Rabinder Singh. "Measurement of Crop Productivity: A Review." Pp. 159-168 in *Perspectives in Agricultural Geography*, vol. 4, edited by Noor Mohammad. New Delhi, India: Concept Publishing Company, 1981.

Tawde, M.D. "Geography of Fruit Farming." Pp. 229-240 in *Perspectives in Agricultural Geography*, vol. 1, edited by Noor Mohammad. New Delhi, India: Concept Publishing Company, 1981.

Atanasiu, Nicolae, and Joachim Alkamper. "Possibility of Increasing the Production of Rice." *Applied Geography and Development*, 20 (1982): 88-111.

Bondad, N.D. "Origin and Distribution of Mango." *Philippine Geographical Journal*, 26, No. 1 (1982): 44-52.

Wikan, G. "Absenteeism: Crop Cultivation and Standard of Living in Developing Countries." *GeoJournal*, 6, No. 6 (1982): 561-569.

Chardon, Roland E., ed. *Plantation Traits in the New World.* Baton Rouge, La.: Lousiana State University Department of Geography and Anthropology, 1983.

deBlij, Harm J. *Wine: A Geographic Appreciation.* Totowa, N.J.: Rowman & Littlefield, 1983.

Dove, Michael A. "Theories of Swidden Agriculture and the Political Economy of Ignorance." *Agroforestry Systems*, 1 (1983): 85-99.

Meshcherskya, A.V. "The Long Term Relationship Between Grain Yields and the Heat-Moisture Index." *Soviet Geography*, 24, No. 3 (1983): 231-243.

Michaels, P.J. "Climate and High-Yielding Variety Wheat Yield." *Geoforum*, 14, No. 4 (1983): 441-446.

Schumacher, Klaus-Dieter. "The Production of Palm Oil and Palm Kernals and the International Trade in Palm Oil, Palm Kernals and Palm-Kernal Oil." *Applied Geography and Development*, 22 (1983): 58-73.

Baresh, H., and F. Schrader. "Ressourcen-und effektorientierte Ansatze zur Landwirtschaftsdiagnose und-progres in Havellandischen Obstaugebiet." *Petermann's Geographische Mitteilungen*, 128 (1984): 19-24.

Berleant-Schiller, Riva. "The Plantation in the New World: A Bibliography Essay." *Choice*, 22 (1984): 219-239.

Brookfield, H.C. "Intensification Revisited." *Pacific Viewpoint*, 25 (1984): 15-44.

Gourou, Pierre. *Riz et civilization*. Paris, France: Fayard, 1984.

Rumney, Thomas A. "A Look at World Soybean Production: Distribution, Uses, and Dilemmas." *Geographical Perspectives*, 53 (1984): 64-67.

Andrae, G., and B. Beckman. *The Wheat Trap*. London, England: Zed Books, 1985.

Blume, Helmut. *Geography of Sugar Cane: Environmental Structure and Economic Aspects of Sugar Cane Production*. Berlin, Germany: Verlag Dr. Albert Bartens, 1985.

deBlij, Harm J. *Wine Regions of the Southern Hemisphere.* Totowa, N.J.: Rowman & Littlefield, 1985.

deBlij, Harm J. "Wine Quality and Climate." *Focus*, 35, No. 2 (1985): 10-15.

deBlij, Harm J. "Nine Canons of the Geography of Viticulture." *East Lakes Geographer*, 21 (1986): 1-10.

Mohler, Robert C., et al. "An Automatic Procedure for Estimating Spring Small Grains Acreage for Landsat Data." *Professional Geographer*, 38, No. 4 (1986): 375-382.

Goldthorpe, L.C. "A Definition and Typology of Plantation Agriculture." *Singapore Journal of Tropical Geography*, 8, No. 1 (1987): 26-43.

Jones, Donald W., and John R. Krummel. "The Location Theory of the Plantation." *Journal of Regional Science*, 27, No. 2 (1987): 157-182.

Knutsen, H.M. "Plantations: Is Their Form of Production Changing?" *Norsk Geografisk Tidsskrift*, 41, No. 3 (1987): 131-142.

Bulman, Teresa L. "The Origin and Diffusion of Area Catechu L. (Arecaceae), the Betel Nut Palm." *Malaysian Journal of Tropical Geography*, 17 (1988): 11-17.

Kononova, N.K. "Circulation Factors of Fluctuations in the Yields of Grain Crops" (in Russian). *Izvestiya-Akademiya Nauk SSSR, Seriya Geografischekaya*, 1 (1988): 15-26.

Miah, M.M. "On the Use of Inputs and Yield of Potato." *Oriental Geographer*, 32 (1988): 27-32.

Peters, Gary L., and Harm J. deBlij. "Location, Scale and Climate: Geographic Dimensions of Viticulture." *Mid-South Geographer*, 4 (1988): 1-16.

Dickenson, John P. "Nine Axioms for the Geography of Viticulture: A Countervailing Perspective." *East Lakes Geographer*, 24 (1989): 1-12.

Harris, David R., and Gordon C. Hillman, eds. *Foraging and Farming: The Evolution of Plant Exploitation*. London, England: Unwin Hyman, 1989.

Le Berre, Maryvonne, and Patrice Uvietta. "Vigne et contraintes climatiques: Un modile de simulation." *L'Espace Geographique*, 18, No. 1 (1989): 55-68.

O'Kelly, M.E. "Equilibrium in a Two-Crop Model with a Low-Cost Transportation Route." *Environment and Planning A*, 21, No. 3 (1989): 385-396.

Ridgley, M.A. "Applying Geographical Queueing Theory: A Crop Distribution Case." *Geoforum*, 20, No. 4 (1989): 397-408.

Rigg, Jonathan. "The New Rice Technology and Agrarian Change: Guilt by Association?" *Progress in Human Geography*, 13, No. 3 (1989): 374-400.

Delvert, J. "La plantation dans le monde tropical. Un point de vue asiatique." *Bulletin de l'Association Geographes Francais*, 67, No. 3 (1990): 213-220.

Dickenson, John P. "Viticultural Geography: An Introduction to the Literature in English." *Journal of Wine Research*, 1, No. 1 (1990): 5-24.

Dion, Roger. *Le paysage et la vigne: Essais de geographie historique*. Paris, France: Payot, 1990.

Cozato-Giotart, J.P. "Exploitaties et proprietes viticoles: Exemples du permanences et de mutation." *L'Information Geographique*, 54, No. 1 (1990): 1-5.

Harris, David R. "Vavilov's Concept of Centers of Origin of Cultivated Plants: Its Genesis and Its Influence on the Study of Agricultural Origins." *Biological Journal of the Linnean Society*, 39 (1990): 7-16.

Huetz de Lemps, Alain, J.R. Pitte, X. de Phanhel, and P. Roudie. *Les vins de l'impossible*. Grenoble, France: Glenat, 1990.

Buell, Thea. "A Geographical Analysis of Grain Amaranth as a Food Crop in the Real World." Master's Thesis, Brigham Young University, Provo, Utah, 1991.

Maloney, B.K. "Rice Agricultural Origins: Recent Advances." *GeoJournal*, 23, No. 2 (1991): 121-124.

Shueb, S., and P. Atkins. "Crop Area Estimation: A Comparison of Remote Sensing and Census Methods." *Geography*, 76, No. 3 (1991): 235-240.

Carbone, Genevieve. "La vanille ou le privilege d'une orchidie." *Les Cahiers d'Outre Mer*, 45 (1992): 327-336.

De Koninck, Rodolphe. "La geographie de la vigne et du vin: Quoi lire?" *Cahiers de Geographie de Quebec*, 37, No. 102 (1992): 555-567.

Dickenson, John P. "Progress in Viticultural Geography." *Journal of Wine Research*, 3, No. 3 (1992): 159-162.

Dickenson, John P. "Aristocratic Versus Popular Wine: Trends in the Geography of Viticulture." Pp. 23-44 in *Viticulture in Geographic Perspective: Proceedings of the Miami AAG Symposium, Miami*, edited by Harm de Blij. Miami, Fla.: Miami Geographical Society, 1992.

Heaps, T., J.M. Munro, and C.S. Wright. "A Location Model of Grain Production and Transportation." *Annals of Regional Science*, 26, No. 2 (1992): 111-134.

Huetz de Lemps, Christian. "Le roy des fruits: L'bananas." *Les Cahiers d'Outre Mer*, 45 (1992): 337-348.

Huetz de Lemps, Alain. "Le cacao: Richese des pays pauvres." *Les Cahiers d'Outre Mer*, 45 (1992): 357-372.

Vavilov, N.I. (translated by Doris Love). *Origin and Geography of Cultivated Plants*. New York, N.Y.: Cambridge University Press, 1992.

Zimmerer, Karl. "The Loss and Maintenance of Native Crops in Mountain Agriculture." *GeoJournal*, 27, No. 1 (1992): 61-72.

Fold, Niels. *Linking Agriculture and Industry in Developing Countries: A Study of the Vegetable Oil Industry in Malaysia and Zimbabwe*. Copenhagen, Denmark: University of Copenhagen, Institute of Geography, Geographica Hafniensia AE, 1993.

Maloney, B.K. "Palaeoecology and the Origin of the Coconut." *GeoJournal*, 31, No. 4 (1993): 355-362.

Moran, Warren. "The Wine Appelation as Territory in France and California." *Annals of the Association of American Geographers*, 83, No. 4 (1993): 694-717.

Sauer, Jonathan D. *Historical Geography of Crop Plants: A Select Roster*. Boca Raton, Fla.: CRC Press, 1993.

Arraudeau, M.A. "Upland Rice: Challenges and Opportunities in a Less Favorable Ecosystem." *GeoJournal*, 35, No. 3 (1995): 325-328.

Bennett, J. "Biotechnology and the Future of Rice Production." *GeoJournal*, 35, No. 3 (1995): 333-336.

Cassman, K.G., and P.L. Pingali. "Intensification of Irrigated Rice Systems: Learning from the Past to Meet Future Challenges." *GeoJournal*, 35, No. 3 (1995): 299-306.

Denning, G.L., and F.A. Bernardo. "Partnership in Rice Research: Rationale, Experience and Implications." *GeoJournal*, 35, No 3 (1995): 374-384.

Huggan, R.D. "Co-Evolution of Rice and Humans." *GeoJournal*, 35, No. 3 (1995): 262-266.

Khush, G.S. "Modern Varieties: Their Real Contribution to Food Supplies and Equity." *GeoJournal*, 35, No. 3 (1995): 275-285.

Khush, G.S. "Breaking the Yield Frontier of Rice." *GeoJournal*, 35, No. 3 (1995): 329-332.

Lampe, K. "Rice Research: A Major World Life Force." *GeoJournal*, 35, No. 3 (1995): 253-261.

Nguyen, Dac Simone A. "Situation et perspectives mondiales du riz." *L'Information Geographique*, 59 (1995): 57-61.

Goldberg, D.M., et al. "Regional Differences in Resverattral Isomar Concentrations of Wines from Various Cultivars." *Journal of Wine Research*, 7, No. 1 (1996): 13-24.

Tabeaud, M. "L'algoculture dans le monde et ses contraints." *Annales de Geographie*, 105, No. 4 (1996): 480-494.

Johnson, Stephen R., Jr. "The Spatial Variability of Seed Populations of White Corn Based on Soil Moisture Holding Capacity." Master's Thesis, Oklahoma State University, Stillwater, Okla., 1997.

Smil, Vaclav. "Some Unorthodox Perspectives on Agricultural Biodiversity: The Case of Legume Cultivation." *Agricultural Ecosystems and Environment*, 62 (1997): 135-144.

Brush, Stephen B. "Crop Diversity in Mountain Areas and Conservation." *Revue de Geographie Alpine*, 86, No. 4 (1998): 115-130.

Zimmerer, Karl S. "The Ecogeography of Major Crops: Versatility and Conservation." *BioScience*, 48, No. 6 (1998): 445-454.

Collins, Lyndhurst. "Renewable Energy from Energy Crops: A Global Perspective." *Geography*, 84, No. 2 (1999): 169-179.

Roche, M.M., M.K. McKenna, and Richard B. LeHeron. "Making Fruitful Comparisons: Southern Hemisphere Producers and the Global Apple Industry." *Tijdschrift voor Economische en Sociale Geographie*, 90, No. 4 (1999): 410-426.

Sato, R. "Current Trends in Studies of Swidden Agriculture in the Tropics: Ecological Aspects and Historical Contexts" (in Japanese). *Jimbun Chiri/Human Geography*, 51, No. 4 (1999): 375-395.

Bailly, A.S. "Le temps des cepages: Du terroir ou systeme-monde." *Annales de Geographie*, 109, No. 4-5 (2000): 516-524.

Carney, Judith A. *Black Rice: The African Origins of Rice Cultivation in the Americas*. Cambridge, Mass.: Harvard University Press, 2000.

Guichard, F. "Le dit et le non-dit du vin: Le langage des etiquettes." *Annales de Geographie*, 109, No. 4-5 (2000): 364-381.

Lee, R. "Shelter from the Storm? Geographies of Regard in the Worlds of Horticultural Consumption and Production." *Geoforum*, 31, No. 2 (2000): 159-174.

Skaggs, Barbara L. "Comparing Factors That Affect Yield in Viticulture." Master's Thesis, Oklahoma State University, Stillwater, Okla., 2000.

ANIMAL HUSBANDRY

Davidson, F.A. "Relation of Taurine Cattle to Climate." *Economic Geography*, 3, No. 4 (1927): 466-485.

Jones, Wellington D., and Derwent S. Whittlesey. "Nomadic Herding Regions." *Economic Geography*, 8, No. 4 (1932): 378-385.

Mikesell, Marvin N. "Notes on the Dispersal of the Dromedary." *Southwestern Journal of Anthropology*, 11 (1955): 231-245.

Leland, Robert P. "Beef Potentialities in the Humid Cotton Belt." Master's Thesis, Columbia University, New York, N.Y., 1956.

Wodzicka, Maria A. "A Study of Some Environmental Factors Affecting Wool Growth and Heat Tolerance of Sheep." Ph.D. Dissertation, Johns Hopkins University, Baltimore, Md., 1959.

Hofmeister, B. "Wesen und Erscheinungsformen der Transhumance." *Erdkunde*, 15, No. 3 (1961): 121-135.

Harris, David R. "The Distribution and Ancestry of the Domestic Goat." *Proceedings of the Linnean Society of London*, 173 (1962): 79-91.

Johnson, Charles W. "Background of Major Changes in the Beef Industry." *Journal of Geography*, 64, No. 2 (1965): 53-58.

Otremba, Erich. *Die Stellung der Viehwirschaft im Agrarraum der Erde: Forschungsstand und Forschung Saufgaben*. Wiesbaden, Germany: Frans Steiner Verlag, 1965.

Moule, G.R. "The Improvement of Animals Through Introduction and Breeding." Pp. 363-385 in *Arid Lands: A Geographical Appraisal*, edited by E.S. Hills. London, England: Methuen, 1966.

de Phanhol, Xavier. "Nomades et pasteurs." *Revue Geographique de l'Est*, 8, No. 3-4 (1968): 363-374.

Hartman, Ora M. "The Origin and Dispersion of the Domestic Fowl: A Study in Cultural Diffusion." Master's Thesis, University of Oregon, Corvallis, Oreg., 1973.

Kelly, James L. "Development Through the Domestication of Animals: The Case of the Musk Ox and the Eland." *Geographical Bulletin*, 10 (1975): 31-36.

Andreae, Bernd. "Entwicklungsstufen der Ranchbetriebe in den Trockenzonen der Kontinente." *Tropenlandwirt*, 77 (1976): 7-24.

Asztalos, Istvan. "Geographical Research Work in the Interests of Developing Stock-Breeding" (in Hungarian). *Foldrajzi Ertesito*, 25, No. 2-4 (1976): 373-382.

Crossley, J.C. "The Location of Beef Processing." *Annals of the Association of American Geographers*, 66, No. 1 (1976): 60-75.

Gomez-Ibanez, Daniel A. "Energy, Economies, and the Decline of Transhumance." *Geographical Review*, 67 (1977): 284-298.

Graeve, W.D., and H. Mertens. "Globale und regionale Aspeckte der Viehwirtschaft." *Geographische Berichte*, 91 (1979): 127-140.

Von Maydell, Hans-Jurgen, and Gunter Spatz. "Effects of Goat Husbandry on the Development of the Landscape." *Applied Geography and Development*, 18 (1981): 30-44.

Bozon, Pierre. *Geographie mondiale de l'elevage*. Paris, France: Librairies Techniques, 1983.

Kent, Robert B. "Agriculture and Ranching." Pp. 136-143 in *Borderlands Sourcebook: A Guide to the Literature on Northern Mexico and the American Southwest*, edited by Ellyn R. Stoddard, Richard L. Nostrand, and Jonathan P. West. Norman, Okla.: University of Oklahoma Press, 1983.

Ryder, M.L. *Sheep and Man*. London, England: Duckworth, 1983.

Loup, J. "Pasteurs et agriculteurs." *Revue de Geographie Alpine*, 73, No. 2 (1985): 87-112.

Malassis, L. "Production et consommation de produits d'origine animale dans le Tiers Monde." *Revue de Goegraphie de Lyon*, 62 (1987): 411-418.

Rinschede, Gisbert. "Transhumance in European and American Mountains." Pp. 96-108 in *Human Impact on Mountains*, edited by Nigel J.R. Allen, Gregory W. Knapp, and Christoph Stadel. Lanham, Md.: Rowman & Littlefield, 1988.

Seifert, H.S.H. "Ranching: Eine Alternative zum Nomadism?" *Die Erde*, 119 (1988): 269-274.

Galaty, John G., and Douglas L. Johnson. *The World of Pastoralism: Herding Systems in Comparative Perspective*. New York, N.Y.: Guilford, 1990.

Virchez, J. *Tropical Beef Cattle Production*. Tolucca, Mexico: Schumager Editores, 1992.

Grigg, David B. "The Role of Livestock Products in World Food Consumption." *Scottish Geographical Magazine*, 109, No. 2 (1993): 66-74.

Anderson, Kay. "A Walk on the Wild Side: A Critical Geography of Domestication." *Progress in Human Geography*, 21, No. 4 (1997): 463-485.

Raish, C. "Agricultural Intensification and the Economic Role of Domesticated Animals." *Papers and Proceedings of the Applied Geography Conferences*, 21 (1998): 137-147.

Windhorst, Hans-Wilhelm. "Pigs and Space: Hog Farming and Pork Production in the European Union and the United States in Transition." *Erdkunde*, 52, No. 3 (1998): 232-249.

Meuner, M. "Weidewirtschafs und Viehhaltung-eine okologische perspektive." *Geographische Rundschau*, 51, No. 5 (1999): 230-235.

DAIRYING

Barrada, Mohamed S.E. "Responses of Dairy Cattle to Hot Environments with Special Emphasis on Respiratory Reactions." Ph.D. Dissertation, Johns Hopkins University, Baltimore, Md., 1957.

Fielding, Gordon J. "Dairying in Cities Designed to Keep People Out." *Professional Geographer*, 14 (1962): 12-17.

Lewthwaite, Gordon R. "Wisconsin and the Waikato: A Comparative Study of Dairy Farming in the United States and New Zealand." *Annals of the Association of American Geographers*, 54, No. 1 (1964): 59-87.

Simoons, Frederick J. "The Antiquity of Dairying in Asia and Africa." *Geographical Review*, 61, No. 3 (1971): 431-439.

FARM LAND

Smith, J. Russell. "Grassland and Farmland as Factors in the Cyclical Development of Eurasian History." *Annals of the Association of American Geographers*, 33 (1943): 135-161.

Montelius, S. "The Burning of Forestland for the Cultivation of Crops." *Geografiska Annaler*, 35 (1953): 41-54.

Pelzer, Karl J. "Land Utilization in the Humid Tropics: Agriculture." *Proceedings of the Nineth Pacific Science Congress*, 20 (1957): 124-143.

Gerasimov, I.P. "The Geographical Study of Agricultural Land Use." *Geographical Journal*, 124, No. 4 (1958): 452-463.

Farmer, B.H. "On Not Controlling Subdivision in Paddy Lands." *Transactions of the Institute of British Geographers*, 28 (1960): 225-235.

Kostrowicki, Jerzy. "Land Utilization Survey as a Basis for Geographical Typology of Agriculture." *Przeglad Geograficzny*, 32, Supplement (1960): 169-183.

Spencer, Joseph E., and G.A. Hale. "The Origin, Nature and Distribution of Agricultural Terracing." *Pacific Viewpoint*, 2 (1961): 1-40.

Steiner, Dieter. *Die Jahreszeit als Faktor bei der Landnutzungsinterpretation auf panchromatischen Luftbilder, gezeigt am Beispiel des schweizerishcen Mittellandes*. Bad Godesberg, Germany: Selbstverlag der Bundesanstalt fur Landeskunde und Raumforschung, 1961.

Zvorykin, K.V. "Scientific Principles for an Agro-Production Classification of Lands." *Soviet Geography*, 4, No. 9 (1963): 3-9.

Stamp, L. Dudley. *Man and the Land*. London, England: Colliers, 1964.

Gasson, R. "The Influence of Urbanization on Farm Ownership and Practice." *Studies in Rural Land Use*, 7. Wye, England: Wye College, 1966.

Grigg, David B. "The Geography of Farm Size: A Preliminary Survey." *Economic Geography*, 42, No. 3 (1966): 205-235.

Harvey, David. "Theoretical Concepts and the Analysis of Agricultural Land-Use Patterns in Geography." *Annals of the Association of American Geographers*, 56, No. 3 (1966): 361-374.

Mellen, Philip G. "Land Tenure and Agrarian Reform." Master's Thesis, George Washington University, Washington, D.C., 1966.

Pavlov, A.V., and G.P. Ustenko. "The Heat Balance and Radiation Regime of Corn Crops." *Soviet Geography*, 7, No. 2 (1966): 44-57.

Whyte, R.O. "The Use of Arid and Semi-Arid Land." Pp. 301-361 in *Arid Lands: A Geographical Appraisal,* edited by E.S. Hill. London, England: Methuen, 1966.

Gasson, R. "Part-Time Farming and Farm-Size." *Sociologica Ruralis*, 7 (1967): 176-189.

Singh, B.B. "Land Use, Cropping Pattern and Their Ranking." *National Geographical Journal of India*, 13, No. 1 (1967): 1-13.

Domoy, Francis M. "An Agricultural Efficiency and Land Value Relationsip Methodology." Master's Thesis, State University of New York at Buffalo, 1968.

Field, N.C. "Environmental Quality and Land Productivity: A Comparison of the Agricultural Land Base of the U.S.S.R. and North America." *Canadian Geographer*, 12, No. 1 (1968): 1-14.

Lozano, Eduardo E. *Location and Regions: Agricultural Land Use in an Integrated Economy*. Cambridge, Mass.: Harvard University, Department of City and Regional Planning, 1968.

Paquette, Romain. "Lot Cultivation: Its Role in Adjustment to Tropical Urban Life." Ph.D. Dissertation, McGill University, Montreal, Quebec, 1968.

Uhlig, Harald. "Probleme un Aufgaben der Internationalen Arbeitsgruppe fur die Terminologie der Agrarlandschaft." *Geographische Zeitschrift*. Erdkundliches Wissen Beiheft 18 (1968): 176-187.

Vedenichev, P.F., and P.P. Marakulin. "Basic Principles of Economic Evaluation of Agricultural Lands." *Soviet Geography*, 9, No. 3 (1968): 172-179.

Taubmann, Wolfgang. "Islands Landwirtschaft: Grundzuge und neuere Wandlungens." *Erdkunde*, 23 (1969): 30-47.

Chang, Jen-hu. "Potential Photosynthesis and Crop Productivity." *Annals of the Association of American Geographers*, 60, No. 1 (1970): 92-100.

Igbozurike, Matthias U. "Fragmentation in Tropical Agriculture: An Overrated Phenomenon." *Professional Geographer*, 22, No. 6 (1970): 321-325.

Chang, Jen-hu. "A Critique of the Concept of Growing Season." *Professional Geographer*, 23, No. 4 (1971): 337-340.

Found, William C. *A Theoretical Approach to Rural Land Use Patterns*. London, England: Edward Arnold, 1971.

Igbozurike, Matthias U. "Fragmentation in Tropical Agriculture: Concept, Process, Result." Ph.D. Dissertation, University of Florida, Gainesville, Fla., 1971.

Mathews, Thomas E. "Detection of Areas of Agricultural Land-Use Change." Master's Thesis, University of Michigan, Ann Arbor, Mich., 1972.

Sitwell, O.F.G. "The Relationship Between Farm Area and the Quality of Land." *Area*, 3, No. 4 (1972): 234-236.

Wilfang, Richard T., and Nancy D. Hughes. "Significance of Second Waves in Breaking Seedling Dormancy in Cereal Grain Crops and Its Application to Northern Crop Production." *Virginia Geographer*, 7, No. 2 (1972): 115-16.

Gilg, Andrew. "A Study in Agricultural Disease Diffusion." *Transactions, Institute of British Geographers*, 59 (1973): 77-98.

Huang, Y. "On Some Determinants of Farm Size Across Countries." *American Journal of Farm Economics*, 55 (1973): 89-92.

Kamp, A. "Et forslag til landbrukstypologi." *Geografisk Orientiring*, 3 (1973): 134-137.

Reeds, Lloyd G., ed. *Agricultural Typology and Land Use: Proceedings of the Agricultural Typology Meeting*. Hamilton, Ontario: International Geographical Union Commission for Agricultural Typology, McMaster University, 1973.

Seavoy, Ronald E. "The Shading Cycle in Shifting Cultivation." *Annals of the Association of American Geographers*, 63, No. 4 (1973): 522-528.

Tkocz, Jan. "General Types of Field Patterns in Agricultural Typology" (in Polish). *Przeglad Geograficzny*, 45, No. 3 (1973): 631-635.

Wheeler, P.T. "Agriculture and Agrarian Land Planning." Pp. 34-56 in *Evaluating the Human Environment: Essays in Applied Geography*, edited by John A. Dawson and John C. Doornkamp. New York, N.Y.: St. Martin's, 1973.

Csaki, Norbert. *Land Supply and International Specialization in Agriculture*. Budapest, Hungary: Akademiai Kiado, 1974.

Denevan, William M., and Billie L. Turner, II. "Forms, Functions and Associations of Raised Fields in the Old World Tropics." *Journal of Tropical Geography*, 39 (1974): 24-33.

Grigg, David B. "The Growth and Distribution of the World's Arable Land 1870–1970." *Geography*, 59, No. 2 (1974): 104-110.

Kellerman, Martin. "Some Implications of Biotic Interactions for Sustained Tropical Agriculture." *Proceedings of the Association of American Geographers*, 6 (1974): 142-145.

Lindroth, Marna J. "Distance as a Farm Management Problem." Master's Thesis, Kansas State University, Manhattan, Kans., 1974.

Nichol, Kenneth J. *Models of Soil Loss, Land and Water Use, Spatial Agricultural Structure and the Environment*. Ames, Iowa: Center for Agricultural and Rural Development, Iowa State University, 1974.

Bryant, Christopher R. "The Conversion of Arable Land to Orchards: A Simulation Experiment Using Simple Probability Surfaces." *Geografiska Annaler B*, 57B (1975): 88-99.

Hart, John Fraser. *The Look of the Land*. Englewood Cliffs, N.J.: Prentice-Hall, 1975.

McCardeff, Nora E. "Agricultural Success: An Exploratory Study of Man, Land and Interrelationships." Ph.D. Dissertation, University of Waterloo, Waterloo, Ontario, 1975.

Rakitnikov, A.N. "Application of the Method of Landscape-Geographical Analogues to Substantiating the Rational Utilization of Land Resources" (in Russian). *Voprosy Geografii*, 99 (1975): 74-95.

Lindroth, Mara L. "Spatial Aspects of Farmland Acquisitions." *Transactions, Kansas Academy of Science*, 79, No.1-2 (1976): 35-42.

Berry, David, and T. Plaut. "Retaining Agricultural Activities Under Urban Pressures: A Review of Land Use Conflicts and Policies." *Policy Studies*, 9 (1978): 153-178.

Ishamuratov, B.M. "The Geographical Basis for Differentiating the Size of Farms." *Soviet Geography*, 19, No. 7 (1978): 474-483.

Kellerman, A. "Determinants of Rent from Agricultural Land Around Metropolitan Areas." *Geographical Analysis*, 10, No. 1 (1978): 1-12.

Firoz Khan, M., and Fakhruddin. "Spatial Analysis of Agricultural Land Use: A New Approach." *The Geographer, Aligarh*, 26, No. 2 (1979): 25-34.

Vogeler, Ingolf. "Agricultural Land Use on the Rural-Urban Fringe: Sinclair's Model Applied." *Ecumene*, 11 (1979): 63-67.

Yapa, Laksman. "Ecopolitical Economy of the Green Revolution." *Professional Geographer*, 31, No. 4 (1979): 371-376.

Beaven, C.W., B.K. Barton, and Benjamin Moulton. "Soil Productivity Ratings as the Basis for Determining Assessed Value of Farmland." *Geographical Perspectives*, 45 (1980): 25-34.

Blair, A.M. "Compulsory Purchases: A Neglected Factor in the Agricultural Land Loss Debate." *Area*, 12, No. 3 (1980): 183-190.

Murray, Robert L. "A Geographical Analysis of Three Approaches for Classifying Prime Agricultural Land." Master's Thesis, University of Akron, Akron, Ohio, 1980.

Schaffer, Gerhard. "Ensuring Man's Food Supplies by Developing New Land and Preserving Cultivated Land." *Applied Geography and Development*, 16 (1980): 7-27.

Simmons, I.G. "Ecological-Functional Approaches to Agriculture in Geographical Contexts." *Geography*, 65, No. 4 (1980): 305-316.

Coughlin, Robert, and John Keene. *The Protection of Farmland.* Amherst, Mass.: National Agricultural Land Study, 1981.

Majeed, Abdul. "Approaches to the Land Use Survey: A Global Perspective." Pp. 209-230 in *Perspectives in Agricultural Geography*, Vol. 3, edited by Noor Mohammad. New Delhi, India: Concept Publications Company, 1981.

Parker, Douglas S. "A Monitoring System for Identifying and Recording Land Conversion in Agricultural Lands with SCS Soil Capability Classes I-IV." Master's Thesis, Oregon State University, Corvallis, Oreg., 1981.

Pofadi, R.M. "Land Use Survey and Planning." Pp. 185-208 in *Perspectives in Agricultural Geography*, Vol. 3, edited by Noor Mohammad. New Delhi, India: Concept Publishing Company, 1981.

Singh, R.P. "Concept of Land Use." Pp. 167-184 in *Perspectives in Agricultural Geography*, Vol. 3, edited by Noor Mohammad. New Delhi, India: Concept Publishing Company, 1981.

Smit, Barry E. "Prime Land, Land Evaluation and Land Use Policy." *Journal of Soil and Water Conservation*, 36, No. 4 (1981): 209-212.

Clawson, David L., and Raymond Crist. "Evolution of Land-Use Patterns and Agricultural Systems." *Mountain Research and Development*, 2, No. 3 (1982): 265-272.

Furuseth, Owen J. "Agricultural Land Conversion: Background and Issues." *Journal of Geography*, 81, No. 3 (1982): 84-93.

Furuseth, Owen J., and John T. Pierce. *Agricultural Land in an Urban Society.* Washington, D.C.: Association of American Geographers, Resource Publications, 1982.

Lapping, Mark B., and J.F. Simons. "Beyond the Land Issue: Farm Viability Strategies." *GeoJournal*, 6, No. 6 (1982): 519-524.

Mundie, R.M. "Evaluating the Effectiveness of Local Government Farmland Protection Programs." *GeoJournal*, 6, No. 6 (1982): 513-518.

Ross, Charles R. "Agricultural Land Conversion: A National Perspective and a Local Level Multiple Objective Planning Application." Master's Thesis, University of Washington, Seattle, Wash., 1982.

Bowers, J.K., and Paul Cheshire. *Agriculture, the Countryside and Land Use*. New York, N.Y.: Methuen, 1983.

Furuseth, Owen J. "A Model for Rural Land Conversion." *Rural Systems*, 1, No. 3 (1983): 177-188.

Kostrowicki, Jerzy. "Land Use Survey, Agricultural Typology, and Land Use Systems: Introducing Remarks." *Rural Systems*, 1, No. 1 (1983): 1-24.

Muller-Wille, Wilhelm. *Probleme und Ergebnisse geographischer Landesxforschung und Landerkunde: Gesammelte Beitrage 1936–1979*. Munster, Germany: Westfaliche Geographische Studien, Heft 39-40, 1983. 2 volumes.

Phipps, M. "Application of Information Theory to the Study of Agricultural Land Use Adjustment to Natural Environments." *Papers and Proceedings of the Applied Geography Conferences*, 6 (1983): 61-72.

Shafi, M. "Agricultural Land and Labour Productivity in Developing Countries in Relation to Food Supplies." *The Geographer, Aligarh*, 30, No. 1 (1983): 38-52.

Bryant, Christopher R. "The Recent Evolution of Farming Landscapes in Urban-Centred Regions." *Landscape Planning*, 11 (1984): 307-326.

Gerasimov, I.P. "Land Resources of the World: Their Use and Reserves Definition of the Lands' Quality and Their Agricultural Potential by Soil Survey Maps." *Rural Systems*, 2, No. 1 (1984): 1-14.

Hamilton, P., and J. Woolcock. *Agricultural Landscapes: An Approach to Their Improvement*. Cheltenham, England: Countryside Commission Paper 169, 1984.

Hill, Ronald D. "Some Questions of Land Use Change." *Rural Systems*, 2, No. 4 (1984): 219-224.

Jaranov, T. "Geography of the World's Arable Land and Its Use" (in Bulgarian). *Problemi na Geografijata*, 1 (1984): 10-21.

Maas, J.H.M. *Landbauw en ruimte: Theorie en praktijk von de agrarische lokatie. (Farming and Space: The Theory and Practice of Agricultural Location.)* Assen, Netherlands: Van Garcum, 1984.

Marsden, T.K., and David G. Symes. "Land Ownership and Farm Organization: Evolution and Change in Capitalist Agriculture." *International Journal of Urban and Regional Research*, 8 (1984): 388-401.

Romanova, E.P., and A.M. Ryabchikov. "Agrolandscape Regionalization of Dryland of the Globe" (in Russian). *Vestnik-Moskovskogo Universiteta, Seriya Geografiya*, 6 (1985): 18-24.

Flaherty, M.S., Barry E. Smit, M.J. Webber, and Lloyd G. Reeds. "Agricultural Land-Use Flexibility: The Concept and Some Measures." *Resource Management and Optimization*, 6, No. 2 (1987): 29-51.

Furuseth, Owen J. "Public Attitudes Toward Local Farmland Protection Programs." *Growth and Change*, 18, No. 3 (1987): 49-61.

Grigg, David B. "The Industrial Revolution and Land Transformation." Pp. 79-110 in *Land Transformation in Agriculture*, edited by M.G. Wolman and F.G.A. Fournier. Chichester, England: Wiley, 1987.

Lewan, Nils. "Marginal Land in Agricultural and Environmental Policy." *Rural Systems*, 5, No. 2 (1987): 95-102.

Achtnich, W. "Die Notwendigkeit ackerbaulicher und gartnerische Bodennutzung im Bereich nomadischer Viehhaltung." *Die Erde*, 119 (1988): 275-280.

Conception, Rogelio N. "Land Use Evaluation Framework on Resource-Based Crop Zonification and Diversification Strategies." *Philippine Geographical Journal*, 32, No. 4 (1988): 108-134.

Hortge, Karl H. "The Problem of Compaction on Agricultural Land." *Applied Geography and Development*, 32 (1988): 44-50.

Osunade, M.A.A. "Soil Suitability Classification by Small Farmers." *Professional Geographer*, 40, No. 2 (1988): 194-201.

Jensen, Frank S. "The Farmer and the Marginal Land." *Geografisk Tidsskrift*, 89 (1989): 44-48.

Napton, Darrell. "Contemporay Agriculture and Rural Land Use." Pp. 333-350 in *Geography in America*, edited by Gary L. Gaile and Cort J. Willmott. Columbus, Ohio: Merrill, 1989.

Smit, Barry E., and Johanne Kristjonson. "Parametric Approaches to Rating the Importance of Land for Agriculture." *Journal of Environmental Management*, 29, No. 4 (1989): 345-362.

Cabrido, Candido A. "An Overview of the Methods and Technical Considerations in the Valuation of Agricultural Lands." *Philippine Geographical Journal*, 34, No. 1 (1990): 27-40.

Maitland, D. "A Conceptual Model of Migration: The Example of Farmland Conversion." *Tijdschrift voor Economische en Sociale Geographie*, 82, No. 5 (1990): 381-387.

Marsden, Terry K., and R.J.C. Munton. "The Farmed Landscape and the Occupancy Change Process." *Environment and Planning A*, 23, No. 5 (1991): 663-676.

Munton, R.J.C., and Terry K. Marsden. "Occupancy Change and the Farmed Landscape: An Analysis of Farmlevel Trends, 1970–85." *Environment and Planning A*, 23, No. 4 (1991): 499-510.

Roccon, Marie-Helene. "De l'espace cultive a l'espace culte: Logiques de evolutions." *Revue de Geographie de Lyon*, 66, No. 1 (1991): 23-26.

Bowler, Ian R. "The Agricultural Significance of Farm Size and Land Tenure." Pp. 85-108 in *The Geography of Agriculture in Developed Market Economies*, edited by Ian R. Bowler. London, England: Longman, 1992.

Hiley, J.C., W.W. Pettapiece, and R.D. McConnall. "A Method to Evaluate Land Productivity Rating Systems." *Operational Geographer*, 10, No. 4 (1992): 15-18.

MacMillan, W.D. "Risk and Agricultural Land Use: A Reformation of the Portfolio-Theoretic Approach to the Analysis of a Von Thunen Economy." *Geographical Analysis*, 24, No. 2 (1992): 142-158.

Fearnside, Philip M. "Forests or Fields? A Response to the Theory That Tropical Forest Conservation Poses a Threat to the Poor." *Land Use Policy*, 10, No. 2 (1993): 108-121.

Van Vuuren, W., and S. Sappong Kuman-Kumale. "The Cost of Agricultural Land Preservation and the Siting of Urban Development." *Geography Research Forum*, 13 (1993): 84-95.

Turner, Billie L., II, and Patricia Benjamin. "Fragile Lands and Their Management." Pp. 101-145 in *Agriculture, Environment and Health: Toward Sustainable Development in the Twenty-First Century*, edited by Vernon W. Ruttan. Minneapolis, Minn.: University of Minnesota Press, 1994.

Wang, F. "The Use of Artificial Neural Networks in a Geographical Information System for Agricultural Land-Suitability Assessment." *Environment and Planning A*, 26, No. 2 (1994): 265-284.

Clarke, William C. "The Sustainability of Sustenance: Land and Agricultural Production in the Third World." Pp. 651-676 in *Companion Encyclopedia of Geography: The Environment and Humankind*, edited by Ian Douglas, Richard Huggett, and Mike Robinson. London, England: Routledge, 1996.

Schroeder-Lanz, Hellmut. "Do 'Infertile' Coastal Dunes Represent One of the Last Reserves of Agricultural Land That Can Be Used by Overpopulated and Underdeveloped States." *Applied Geography and Development*, 48 (1996): 52-68.

Cocklin, Christopher, Greg Blunden, and Warren Moran. "Sustainability, Spatial Hierarchies and Land-Based Production." Pp. 25-40 in *Agricultural Restructuring and Sustainability: A Geographical Perspective*, edited by Brian W. Ilbery, Quentin Chiotti, and Timothy Rickard. Wallingford, England: C.A.B. International, 1997.

Hilts, Steward G. "Achieving Sustainbility in Rural Land Management Through Landowner Involvement in Stewardship Programmes." Pp. 267-178 in *Agricultural Restructuring and Sustainability: A Geographical Perspective*, edited by Brian W. Ilbery, Quentin Chiotti, and Timothy Rickard. Wallingford, England: C.A.B. International, 1997.

AGRICULTURAL POPULATIONS AND SOCIAL ISSUES

Oran, P.A., L.W. Bowden, P. Pelissier, and Homer Aschmann. "Historical, Geographical and Sociocultural Aspects." Pp. 1-111 in *Proceedings of an International Symposium on Rainfed Agriculture in Semi-Arid Regions*, edited by G.H. Connell. Riverside, Calif.: University of California, undated.

Demongeon, A. "Economie agricole et peuplement rural." *Annales de Geographie*, 43 (1934): 1-21.

Shaw, Earl B. "Geography of Mass Feeding." *Economic Geography*, 16, No. 3 (1940): 234-249.

Bennett, Merrill K. "International Contrasts in Food Consumption." *Geographical Review*, 31 (1941): 365-376.

Hagood, M.J. "Statistical Methods for Delineation of Regions Applied to Data on Agriculture and Population." *Social Forces*, 21 (1943): 287-297.

Conklin, Harold C. "An Ethnoecological Approach to Shifting Agriculture." Pp. 457-464 in Philip L. Wagner and Marvin W. Mikesell, eds. *Readings in Cultural Geography*. Chicago, Ill.: University of Chicago Press, 1954.

Thompson, Edgar T. "The Plantation as a Social System." *Revista Geografica*, 51 (1959): 41-56.

Kiefer, Wayne E. "Modeling and Remodeling the Agricultural Landscape-Settlement: Feature Analysis by Form and Function." Master's Thesis, Indiana University, Bloomington, Ind., 1963.

Stewart, Norman R. "Foreign Agricultural Colonization as a Study in Cultural Geography." *Professional Geographer*, 15 (1963): 1-5.

Williams, W.M. "The Social Study of Family Farming." *Geographical Journal*, 129 (1963): 63-75.

Hecklau, Hans. *Die Gliederung der Kukturlandschaft im Gebiet von Schriescheim/Bergstrasse: Ein Beitrag zur Methodick der Kulturlandschaftsforschung*. Berlin, Germany: Freie Universitat, Geographisches Institut Abhandlungen, Band 8, 1964.

Hunter, John M. "Ascertaining Population Carrying Capacity Under Traditional Systems of Agriculture in Developing Countries." *Professional Geographer*, 18 (1966): 151-154.

Irving, Robert M. *Amenity Agriculture*, B.C. Geographical Series, 11. Vancouver, British Columbia: Tantalus, 1966.

Kariel, Herbert G. "A Proposed Classification of Diet." *Annals of the Association of American Geographers*, 56, No. 1 (1966): 68-79.

Kohs, John. "Locational Aspects of Cultural Ecology: The Case of the Goal in Non-Western Agriculture." *Geographical Review*, 56, No. 4 (1966): 577-584.

Blaunt, James M. "Geography and the Development of Peasant Agriculture." Pp. 200-220 in *Problems and Trends in American Geography*, edited by Saul B. Cohen. New York, N.Y.: Basic Books, 1967.

Darnell, Bernard W. "Socio-Economic Parameters in Farm Performance, 1969." Master's Thesis, McMaster University, Hamilton, Ontario, 1969.

Harris, David R. "Agricultural Systems, Ecosystems and the Origins of Agriculture." Pp. 3-15 in *The Domestication and Exploitation of Plants and Animals*, edited by Peter J. Ucko and T.W. Dimbleby. Chicago, Ill.: Aldine, 1969.

Carriere, P., D. Archambault-Leconte, and E. Raymond. "Les societes d'interet collectif agricole dans le Gardi." *Bulletin de la Societe Languedocienne de Geographie*, 4, No. 2 (1970): 183-212.

Grigg, David B. "Trends in the World's Agricultural Population." *Geography*, 56, No. 4 (1971): 320-324.

Josuweit, Werner. *Die Betriebsgrosse als agrarraumlicher Steuerungsfakter im heutigen kulturlandschaftgefuge: Analyse dreier Gemarkungen im Mittleren Leinetal*. Gottingen, Germany: Gottinger Geographische Abhandlungen, Heft 57, 1971.

McLoughlin, Peter F.M. "Increasing Agricultural Output to Help with the Population Problem." Pp. 83-94 in *Geography and a Crowding World*, edited by Wilbur Zelinsky, L.A. Kosinski, and R.M. Prothero. New York, N.Y.: Oxford University Press, 1971.

Krajko, G., I. Penzes, and J. Toth, eds. *Agricultural Typology and Agricultural Settlements*. Szeged, Hungary: Papers of a Symposium Held at Szeged and Pecs, 15-19 August 1971, 1972.

Pulyarkin, V.A. "Nomadism in the Modern World." *Soviet Geography*, 13, No. 3 (1972): 161-173.

Szani, Ferenc. "Problems of Food-Supply in Developing Countries" (in Hungarian). *Acta Geographica Debrecina*, 11 (1972).

Beguin, Hubert. "Densite de population, productivite et developpement agricole." *L'Espace Geographique*, 3, No. 4 (1974): 267-272.

Dabrowski, P.H. "Les changements survences dans les annees 1960–1970 dans le niveau et la structure de la production agricole commercialisee des exploitations paysannes." *Geographia Polonica*, 29 (1974): 219-249.

Grigg, David B. "Agricultural Populations and Economic Development." *Tijdschrift voor Economische en Sociale Geographie*, 65 (1974): 414-420.

Grzeszczak, J. "Le role de l'industrie dans les processes de la modernisation campagnes et de l'agriculture (a la lumiere de l'acrroissement de la population a double activite)." *Geographia Polonica*, 29 (1974): 71-80.

Pollock, Adrian. "Spatial Theory and Agricultural Location in Tribunal Society." Master's Thesis, University of Michigan, Ann Arbor, Mich., 1974.

Birch, Brian P. "Pattern and Process in Frontier Farm Settlement." *National Geographical Journal of India*, 21, No. 3-4 (1975): 166-171.

Symanski, Richard, Ian R. Manners, and R.J. Bromley. "The Mobile-Sedentary Continuum." *Annals of the Association of American Geographers*, 65 (1975): 461-471.

Grigg, David B. "Population Pressure and Agricultural Change." Pp. 133-176 in *Progress in Geography*, 8, edited by C. Board, et al. London, England: Arnold, 1976.

Knight, C. Gregory, and R. Paul Wilcox. *Triumph or Triage: The World Food Problem in Geographical Perspective*. Washington, D.C.: Association of American Geographers, Resource Paper No. 75-3, 1976.

Townsend, J.G. "Farm Failures: The Application of Personal Constructs in the Tropical Rainforest." *Area*, 8, No. 3 (1976): 219-224.

Vermeer, Donald E. "Food, Farming and the Future: The Role of Traditional Agriculture in the Developing Areas of the World." *Social Science Quarterly*, 57 (1976): 383-396.

Casetti, Emilio, and Howard L. Gauthier. "A Formalization and Test of the Hallow Frontier Hypothesis." *Economic Geography*, 53, No. 1 (1977): 70-78.

Wagner, Philip L. "The Concept of Environmental Determinism in Cultural Evolution." Pp. 49-74 in *Origins of Agriculture*, edited by Charles A. Reed. The Hague, Netherlands: Mouton, 1977.

Layton, Robert L. "The Operational Structure of the Hobby Farm." *Area*, 10, No. 4 (1978): 242-246.

Blanes, M., and G. Allaine. "Types d'exploitations et couches sociales dans l'agriculture." *Revue Geographique des Pyrenees et du Sud-Ouest*, 50, No. 2 (1979): 343-369.

Carevet, C. "De la polyculture paysanne a l'integration: Les couches sociales dans l'agriculture." *Norois*, 26 (1979): 507-522.

Chisholm, Michael. *Rural Settlement and Land Use*. London, England: Hutchinson, 1979.

Dabrowski, P.H. "Regionalizacja rolnicza swiata na podstawic struktury zawodowej Ludnosci." *Przeglad Geograficzny*, 49 (1979): 731-740.

Layton, Robert L. "Hobby Farming." *Geography*, 65 (1979): 220-223.

Grigg, David B. *Population Growth and Agrarian Change: An Historical Perspective*. Cambridge, England: Cambridge Geographical Studies, 1980.

Tarrant, John R. *Food Policies*. Chichester, England: Wiley, 1980.

Abuzur, Mohammad. "Models of Growing Population for the Analysis of Traditional Agriculture." Pp. 35-44 in *Perspectives in Agricultural Geography*, vol. 5, edited by Noor Mohammad. New Delhi, India: Concept Publishing Company, 1981.

Bhattacharya, Anima. "Population Pressure and Agricultural Resources: A Global Perspective." Pp. 21-34 in *Perspectives in Agricultural Geography*, vol. 5, edited by Noor Mohammad. New Delhi, India: Concept Publishing Company, 1981.

Maos, J.O., and P. Levinger. "Adopting Collective Patterns to Cooperative Family-Farm Settlements." *Journal of Rural Cooperation*, 9, No. 2 (1981): 163-183.

Obara, Dunstan A. "Environmental Perception and Behavioral Approach in Agricultural Geography: Application of Repertory-Grid Technique." Pp. 205-222 in *Perspectives in Agricultural Geography*, vol. 5, edited by Noor Mohammad. New Delhi, India: Concept Publishing Company, 1981.

Shafi, M. "Population Growth and Food Supply in the World." Pp. 13-20 in *Perspectives in Agricultural Geography*, vol. 5, edited by Noor Mohammad. New Delhi, India: Concept Publishing Company, 1981.

Wollner, Shaun A. "The Suburban Cultiver Community: An Alternative Development for Agricultural Land." Master's Thesis, California State University at Chico, Chico, Calif., 1981.

Marsden, Terry K. "Capitalist Farming and the Farm Family: A Case Study." *Sociology*, 18 (1984): 205-224.

Shafi, M. "Food Production and Consumption in Developed and Developing Countries." *Rural Systems*, 2, No. 4 (1984): 191-198.

Behfarooz, Fatemah. "The Role of Rural Development and Farmers' Preferences in Changing the Viability and Spatial Distribution of Central Villages." Master's Thesis, Michigan State University, East Lansing, Mich., 1985.

Cant, Garth. "Administrators, Planners, Scientists and Pastoral Nomads: Conflicting Perception of a Productive System." *New Zealand Journal of Geography*, 78 (1985): 2-6.

Shaw, Anthony B. "Constraints on Agricultural Innovation Adaption." *Economic Geography*, 61, No. 1 (1985): 25-45.

Ehlers, Eckert. "The Limits of Agricultural Settlements on the Earth." *Applied Geography and Development*, 27 (1986): 106-112.

Ilbery, Brian W. *Agricultural Geography: A Social and Economic Analysis*. New York, N.Y: Oxford University Press, 1986.

Jungarius, P.D. *Perception and Use of the Physical Environment in Peasant Societies*. Reading, England: University of Reading Department of Geography, 1986.

Norton, William. "Comparative Cultural Analysis of Nineteenth Century Agricultural Frontiers: Rationales and Prospects." *National Geographical Journal of India*, 32, No. 2 (1986): 85-92.

Whatmore, Sarah, R.J.C. Munton, and Terry K. Marsden. "Internal and External Relations in the Transformation of the Family Farm." *Sociologia Ruralis*, 26, No. 3 (1986): 396-398.

Muldavin, Joshua. "The Individual Responsibility System: Present and Future Implications with Special Reference to Agricultural Sustainability." Master's Thesis, University of California at Berkeley, 1987.

Walker, Gerald E. *The Farmers' Class*. North York, Ontario: York University Department of Geography, Discussion Paper No. 34, 1987.

Iberall, A., and D. White. "Evidence for a Long Term Process Scale for Social Change in Modern Man Settled in Place Via Agriculture and Engaged in Trade and War." *GeoJournal*, 17, No. 3 (1988): 311-340.

Manshard, Walther, and W.B. Morgan, eds. *Agricultural Expansion and Pioneer Settlements in the Humid Tropics: Selected Papers Presented at a Workshop Held in Kuala Lumpur, 17–21 September 1985*. Tokyo, Japan: United Nations University Press, 1988.

Morgan, W.B. "Agricultural Production Frontiers in the Humid Tropics: Aspects of Spontaneous and Planned Settlement." Pp. 9-28 in *Naturliche Resoursen und landliche Entwicklungsprobleme der Tropen: Festschrift fur Walther Manshard*, edited by R. Mackel and Wolf D. Sick. Stuttgart, Germany: Franz Steiner, Verlag, 1988.

Santos, Paterno R. "Geographic and Socioeconomic Considerations in Agribusiness." *Philippine Geographical Journal*, 32, No. 4 (1988): 151-154.

Kuhn, D. "Bearbeitungstypen Landwirtschaftlicher Nutzflachen." *Petermann's Geographische Mitteilungen*, 133, No. 1 (1989): 23-36.

Wiese, B. "Plantagen und Bauernwirtschaften in den Tropen: Vom Konflict zur Cooperation?" *Geographische Rundschau*, 41, No. 7-8 (1989): 406-412.

Anifowoshe, T.O. "Food-Production: Problems and Prospects." *GeoJournal*, 20, No. 3 (1990): 243-248.

Little, J. "Women's Employment in the Food System." Pp. 141-156 in *Political, Social and Economic Perspectives in the International Food System*, edited by Terry K. Marsden and J. Little. Aldershot, England: Avebury, 1990.

Marsden, Terry K., and J. Little, eds. *Political, Social and Economic Perspectives in the International Food System*. Aldershot, England: Avebury, 1990.

Parafina, Sofronio S. "Inside the Black Box: Household Relations and Agricultural Decision Making." Master's Thesis, University of Florida, Gainesville, Fla., 1991.

Shucksmith, M., and L. Watkins. "Housebuilding on Farmland: The Distributional Effects in Rural Areas." *Journal of Rural Studies*, 7, No. 3 (1991): 153-168.

Symes, David G. "Changing Gender Roles in Productionist and Post-Productionist Capitalist Agriculture." *Journal of Rural Studies*, 7, No. 1-2 (1991): 85-90.

Tarrant, John R. *Farming and Food*. New York, N.Y.: Oxford University Press, 1991.

Townsend, J.G. "Geografia y genero en la colonizacion agricola." *Documents d'Analisi Geografia*, 18 (1991): 89-99.

Whatmore, Sarah. *Farming Women: Gender, Work and Family Experience*. London, England: Macmillan, 1991.

Whatmore, Sarah. "Life Cycle or Patriarchy? Gender Divisions in Family Farming." *Journal of Rural Studies*, 7, No. 1-2 (1991): 71-76.

Zhikharevich, B.S. "The Role Played by Agro-Recreational Activities in the Formation of an Integral Rural-Urban Environment." *Geoforum*, 22, No. 1 (1991): 99-104.

Bowler, Ian R. "Sustainable Agriculture as an Alternative Path of Farm Business Development." Pp. 237-253 in *Contemporary Rural Systems in Transition: Vol. 1, Agriculture and Environment*, edited by Ian R. Bowler, Christopher R. Bryant, and M. Duane Nellis. Wallingford, England: C.A.B. International, 1992.

Bowler, Ian R. "Sustainable Agriculture in the Food Supply System." Pp. 4-13 in *Geographical Perspectives on the Social and Economic Restructuring of Rural Areas*, edited by M. Duane Nellis. Manhattan, Kans.: Kansas State University, 1992.

Delange, Yves. "Contribution de la flare americain aux cultures d'ornament." *Les Cahiers d'Outre Mer*, 45 (1992): 479-488.

Munton, R.J.C., Terry K. Marsden, and N. Ward. "Uneven Agrarian Development and the Social Relations of Farm Households." Pp. 61-73 in *Contemporary Rural Systems in Transition: Vol. 1, Agriculture and Environment*, edited by Ian R. Bowler, Christopher R. Bryant, and M. Duane Nellis. Wallingford, England: C.A.B. International, 1992.

Arce, A., and Terry K. Marsden. "The Social Construction of International Food: A New Research Agenda." *Economic Geography*, 69, No. 3 (1993): 293-311.

Johnston, Thomas R.R., Richard B. LeHeron, and M.M. Roche. "Farm Women and Work." Pp. 292-304 in *Geographic Perspectives on the Social and Economic Restructuring of Rural Areas*, edited by M. Duane Nellis. Manhattan, Kans.: International Geographical Union Commission on Changing Rural Systems, 1993.

McMichael, Philip. "World Food System Restructuring Under a GATT Regime." *Political Geography*, 12, No. 3 (1993): 198-214.

Grigg, David B. "Food Expenditure and Economic Development." *GeoJournal*, 33, No. 4 (1994): 377-382.

Lund, R., and S. Mishra Panda. "The Role of Gender in the Quest for Sustainable Agricultural Development." *Norsk Geografisk Tidsskrift*, 48, No. 3 (1994): 113-122.

Cocklin, Christopher. "Agriculture, Society, and the Environment: Discourse on Sustainability." *International Journal of Sustainable Development and World Ecology*, 2 (1995): 240-256.

Evans, N.J., and Brian W. Ilbery. "Exploring the Influence of Pluriactivity on Gender Relations in Capitalist Agriculture." *Sociologica Ruralis*, 36 (1996): 74-92.

Holloway, L.E., and Brian W. Ilbery. "Farmers' Attitudes Towards Environmental Change, Particularly Global Warming, and the Adjustment of Crop Mix and Farm Management." *Applied Geography*, 16, No. 2 (1996): 159-172.

Curry-Roper, Janel M. "Community-Level Worldviews and the Sustainability of Agriculture." Pp. 101-116 in *Agricultural Restructuring and Sustainability: A Geographial Perspective*, by Brian W. Ilbery, Quentin Chiotti, and Timothy Rickard, eds. Wallingford, England: C.A.B. International, 1997.

Ehlers, Eckert. "Traditional Environmental Knowledge and Consciousness and the Problem of Sustained Agricultural Development." *Applied Geography and Development*, 49 (1997): 79-95.

Roberts, Rebecca, and Gail M. Hollander. "Sustainable Technologies, Sustainable Farms: Farms, Households and Structural Change." Pp. 55-72 in *Agricultural Restructuring and Sustainability: A Geographical Perspective*, by Brian W. Ilberry, Quentin Chiotti, and Timothy Rickard, eds. Wallingford, England: C.A.B. International, 1997.

Drangert, J.O. "Urine Blindness and the Use of Nutrients from Human Excreta in Urban Agriculture." *GeoJournal*, 45, No. 3 (1998): 201-208.

Henderson, George. "Nature and Fictitious Capital: The Historical Geography of an Agrarian Question." *Antipode*, 30, No. 2 (1998): 73-118.

ENVIRONMENTAL AND CONSERVATION ISSUES

Reed, William G., and H.R. Trolley. "Weather as a Business Risk in Farming." *Geographical Review*, 2 (1916): 48-53.

Smith, J. Warren. "Agricultural Meteorology." *Journal of Geography*, 20, No. 5 (1921): 161-175.

Azzi, G. "The Problem of Agricultural Ecology." *Geografiska Annaler*, 4 (1922): 137-146.

Klages, Karl H.W. *Ecological Crop Geography*. New York, N.Y.: Macmillan, 1942.

Bennett, Hugh H. "Adjustment of Agriculture to Its Environment." *Annals of the Association of American Geographers*, 33 (1943): 163-189.

Papadakis, J. *Agricultural Geography of the World: Climate, Growth Rate and Rhythm, Vegetation, Soils, Crops, Agricultural Regions*. Buenos Aires, Argentina: Author, 1952.

Molga, M. "Agroclimatological Problems in Studies Related to the Regional Distribution of Agricultural Production" (in Polish). *Przeglad Geograficzny*, 27, No. 1 (1955): 81-102.

Bartlett, H.H. "Fire, Primitive Agriculture and Grazing in the Tropics." Pp. 692-720 in *Man's Role in Changing the Face of the Earth*, edited by W.L. Thomas. Chicago, Ill.: University of Chicago Press, 1956.

Blaunt, James M. "The Ecology of Tropical Farming Systems." *Revista Geografica*, 54 (1961): 47-67.

Smith, S.I. "Climatic Control of Distribution and Cultivation of Sugar Cane." Ph.D. Dissertation, McGill University, Montreal, Quebec, 1963.

Kraikow, Julius. "Industrial and Municipal Waste as a Source of Agricultural Humus." Pp. 25-28 in *Agricultural Geography I.G.U. Symposium*, edited by E.S. Simpson. Liverpool, England: University of Liverpool, Department of Geography, Research Paper No. 3, 1965.

Waggoner, Paul E., et al. *Agricultural Meteorology*. Boston, Mass.: American Meteorological Society, Meteorological Monograph Vol. 6, No. 28, 1965.

Chang, Jen-hu. *Climate and Agriculture: An Ecological Survey*. Chicago, Ill.: Aldine, 1968.

Chang, Jen-hu. "Progress in Agricultural Climatology." *Professional Geographer*, 20, No. 5 (1968): 317-320.

Granovskaya, A. "Specialization of Agriculture in Relatively Similar Natural Environments of the U.S.S.R. and the United States." *Soviet Geography*, 9, No. 10 (1968): 830-837.

Harris, David R. "Agricultural Systems, Ecosystems, and the Origins of Agriculture." Pp. 3-15 in *The Domestication and Exploitation of Plants and Animals*, edited by Peter J. Ucko and G.W. Dimbleby. London, England: Duckworth, 1969.

Van Eimern, Josef. "Die Bedentung von Energiestramen fur das Klima und die Agrarischen Standartbedingungen." *Erdkunde*, 23 (1969): 10-19.

Berky, L. "Hygenic Problems of Agricultural Regional Planning." *Geographica Medica*, 2 (1971): 15-22.

Birch, J.W. *Farming Systems as Resource Systems*. Leeds, England: University of Leeds, Department of Geography, Working Paper No. 2, 1971.

Igbozurike, Matthias U. "Ecological Balance in Tropical Agriculture." *Geographical Review*, 61, No. 4 (1971): 519-529.

Willan, G.C. "Microclimate Management by Traditional Farmers." *Geographical Review*, 62, No. 4 (1972): 544-560.

Mahleer, P.J. "Agricultural Development and the Environment." *Geoforum*, 10 (1972): 53-70.

Harris, David R. "The Prehistory of Tropical Agriculture: An Ethnoecological Model." Pp. 391-418 in *The Explanation of Culture Change: Models in Prehistory*, edited by Colin Renfrew. London, England: Duckworth, 1973.

Bryson, R.A., J.E. Ross, R.W. Hougas, and L.E. Engelbert. *Climatic Change and Agricultural Responses*. Madison, Wis.: University of Wisconsin, Institute for Environmental Studies, IES Report No. 20, 1974.

Huke, Robert E. "San Bartolome and the Green Revolution." *Economic Geography*, 50, No. 1 (1974): 47-58.

Kellman, Martin. "Some Implications of Biotic Interactions for Sustained Tropical Agriculture." *Proceedings, Association of American Geographers*, 6 (1974): 142-145.

Stephens, J.S. "A Field Study of Infiltration on Various Agricultural Soils." Master's Thesis, Wilfrid Laurier University, Waterloo, Ontario, 1974.

Merchant, James W., and Glen A. Marotz. "Agricultural Water Pollution: Some Problems Associated with Its Control." *Journal of Geography*, 74, No. 8 (1975): 419-429.

Parry, Martin L. "Secular Climatic Change and Marginal Agriculture." *Transactions, Institute of British Geographers*, 64 (1975): 1-13.

Harvey, Nigel. *Fields, Hedges and Ditches.* Aylesbury, England: Shire Publications, 1976.

Manners, Ian R. "The Environmental Impacts of Modern Agricultural Technologies." *Ekistics*, 39 (1976): 56-64.

Bailey, William G. "Atmospheric and Surface Control on Evapotranspiration During Soybean Maturation." Ph.D. Dissertation, McMaster University, Hamilton, Ontario, 1977.

Dregne, H.E. "Desertification of Arid Lands." *Economic Geography*, 53, No. 4 (1977): 322-331.

Scott, Geoffrey. "The Importance of Old Field Sucession Biomass Increments to Shifting Cultivation." *Great Plains–Rocky Mountain Geographical Journal*, 6, No. 2 (1977): 318-327.

Parry, Martin L. *Climatic Change, Agriculture and Settlement.* Folkestone, England: Dawson-Archon Books, Studies in Historical Geography, 1978.

Van Voorthuizen, E.G. "Global Desertification and Range Management." *Journal of Range Management*, 31, No. 5 (1978): 378-380.

Cox, George W., and Michael D. Atkins. *Agricultural Ecology: An Analysis of World Food Production Systems.* San Francisco, Calif.: W.H. Freeman, 1979.

Hare, F. Kenneth. "Food, Climate, and Man." Pp. 1-11 in *Food, Climate and Man*, edited by M.R. Biswas and A.K. Biswas. New York, N.Y.: Wiley Interscience, 1979.

O'Rourke, Patricia. "A Simple Crop Energy Budget Model and Aspects of Worldwide Potential Productivity." Ph.D. Dissertation, University of California at Los Angeles, 1979.

Pawlik, Laszlo, and Mihaly Morton. "Researches for the Prevention of Dangers of Pesticides Used in Agriculture." *Geographica Medica*, 9 (1979): 86-108.

Thorne, D. Wynne, and Marlowe D. Thorne. *Soil, Water and Crop Production.* Westport, Conn.: AVC Publishing Company, 1979.

Burt, James E. "Simulations Assessing the Utility of Leaf Photosynthesis Models in Estimating Crop Productivity." Ph.D. Dissertation, University of California at Los Angeles, 1980.

Hare, F. Kenneth. "Climate and Agriculture: The Uncertain Future." *Journal of Soil and Water Conservation*, 35, No. 3 (1980): 112-115.

Uhlig, Harold. "Agricultural Landscapes in Tropical Karst Regions." *Applied Geography and Development*, 15 (1980): 7-27.

Ahmad, Enagat. "Geomorphology and Agriculture." Pp. 59-68 in *Perspectives in Agricultural Geography*, vol. 2, edited by Noor Mohammad. New Delhi, India: Concept Publishing Company, 1981.

Armstrong, Patrick. "The Ecosystem Concept as a Framework for Studies in Agricultural Geography." Pp. 269-292 in *Perspectives in Agricultural Geography*, vol. 1, edited by Noor Mohammad. New Delhi, India: Concept Publishing Company, 1981.

Bryson, R.A. "Practical Climatic Forecasting for Agriculture." Pp. 11-34 in *The Environment: Chinese and American Views*, edited by Laurence J.C. Ma and Allen G. Noble. New York, N.Y.: Methuen, 1981.

Burbridge, Peter, John Dixon, and Bedjo Soewordi. "Forestry and Agriculture: Options for Resource Allocation in Choosing Lands for Transmigration Development." *Applied Geography*, 1, No. 4 (1981): 237-258.

De, N.K. "Fertility and Productivity of Soil." Pp. 205-224 in *Perspectives in Agricultural Geography*, vol. 2, edited by Noor Mohammad. New Delhi, India: Concept Publishing Company, 1981.

Jager, Heinrich. "New Dams: The Key to Modern Agriculture?" *Applied Geography and Development*, 17 (1981): 7-16.

Kar, Anita, and Amal Kar. "Relevance of Geomorphology in Agricultural Planning: Some Global Examples." Pp. 69-76 in *Perspectives in Agricultural Geography*, vol. 2, edited by Noor Mohammad. New Delhi, India: Concept Publishing Company, 1981.

Kent, Martin, and Nicholas Smart. "A Method for Habitat Assessment in Agricultural Landscapes." *Applied Geography*, 1, No. 1 (1981): 9-30.

Lamb, H.H. "Climatic Changes and Food Production: Observation and Outlook in the Modern World." *GeoJournal*, 5, No. 2 (1981): 101-112.

Lauer, Wilhelm, and Peter Frankenberg. *Untersuchungen zur Humiditat: Das Koncept einer potentiellen Landschaftsverdunstung*. Bonn, Germany: Bonner Geographische Abhandlungen, Heft 66, 1981.

Monteith, J.L. "Climatic Variation and the Growth of Crops." *Quarterly Journal of the Royal Meteorological Society*, 107, No. 454 (1981): 760-774.

Parry, Martin L. "Climatic Change and the Agricultural Frontier: A Research Strategy." Pp. 319-336 in *Climate and History: Studies in Past Climates and Their Impact on Man*, edited by T.M.L. Wigley, M.J. Ingram, and G. Farmer. Cambridge, England: Cambridge University Press, 1981.

Schmiedechen, Wolfgang. "Humidity and Cultivated Plants: An Attempt at Parallelizing Zones of Humidity and Optimal Locations of Selected Cultivated Plants in the Tropics." *Applied Geography and Development*, 17 (1981): 45-57.

Yao, Augustin Y.M. "Agricultural Climatology." Pp. 189-298 in *World Survey of Climatology*, edited by Helmut R. Landsberg. Amsterdam, Netherlands: Elsevier, 1981.

Yapa, Lakshman. "Ecopolitical Economy of Agricultural Innovations." Pp. 118-137 in *Agriculture, Rural Energy and Development*, edited by R.S. Ganapathy. Ann Arbor, Mich.: University of Michigan, Graduate School of Business Administration, 1981.

Bayliss-Smith, Timothy P. *The Ecology of Agricultural Systems*. Cambridge, England: Cambridge University Press, 1982.

Bunting, A.H., et al. "Climate and Crop Distribution." Pp. 42-75 in *Food, Nutrition and Climate*, edited by Keith Blaxter and Leslie Fawden. London, England: Applied Science Publishers, 1982.

Eldred, Cecilia A. "A Global Survey of Organic Agriculture." Master's Thesis, University of Texas, Austin, Tex., 1982.

Floyd, Barry N. "The Rain Forest and the Farmer." *GeoJournal*, 6, No. 5 (1982): 433-442.

Kryuchkov, V.G. "A Systems Approach to the Study of Interactions Between Agriculture and the Environment" (in Russian). *Vestnik Moskovskogo Universiteta, Seriya Geografiya*, 2 (1982): 10-14.

Cusack, David F. *Agroclimate Information for Development: Reviving the Green Revolution*. Boulder, Colo.: Westview, 1983.

Dregne, H.E. *Desertification of Arid Lands*. New York, N.Y.: Harwood, 1983.

Weller, F. "Agrarokologische und Naturraumliche Gliederung: Ein Vergleich." *Verhandlungen-Gesellschaft fur Okologie*, 11 (1983): 445-454.

Bayliss-Smith, Timothy P., and S. Wanmali, eds. *Understanding Green Revolutions: Agrarian Change and Development Planning*. Cambridge, England: Cambridge University Press, 1984.

Bruneau, Michel, and Jean Kilion. "Inventaires agroecologiques, paysages et teledetection en milieu tropical." *L'Espace Geographique*, 13, No. 3 (1984): 215-224.

Dabrowski, P.H. "La specialisation regionale et sectorielle de l'agriculture et la conservation des resources naturelles." *Geographia Polonica*, 49 (1984): 161-170.

Gennadiyev, A.N., et al. "Agroecological Investigations by FAO" (in Russian). *Vestnik Moskovskogo Universiteta, Seriya Geografiya*, 2 (1984): 32-37.

Scholz, U. "Ist die Agrarproduktion der Tropen okologisch benachteiligt?" *Geographische Rundschau*, 36, No. 7 (1984): 360-366.

Symons, Leslie J. "Bio-Aeronautics: The Use of Aircraft for Crop Protection and Land Improvement." *Cambria*, 11, No. 2 (1984): 87-96.

Von Maydell, Hans-Jurgen. "Possibilities of Increasing the Human-Ecological Carrying Capacity of Semiarid Tropical and Subtropical Mountain Regions by Agroforestry Land-Use Practices." *Applied Geography and Development*, 24 (1984): 102-111.

Wu, Shan-mei. "Evaluations of Methods to Study Nitrogen Cycling in Agroecosystems." Master's Thesis, University of Hawaii, Manoa, Hawaii, 1984.

Briggs, David, and Frank Courtney. *Agriculture and Environment: The Physical Geography of Temperate Agricultural Systems*. London, England: Longman, 1985.

Gjedrem, Trygve. "Improvement of Productivity Through Breeding Systems." *GeoJournal*, 10, No. 3 (1985): 233-244.

Huke, Robert E. "The Green Revolution." *Journal of Geography*, 84, No. 6 (1985): 248-254.

Mishra, P.L., ed. *Environmental Influence on Agriculture*. Hapur, India: Ritu Prakshan, 1985.

Nellis, M. Duane. "Remote Sensing Techniques for Identifying Regional Agrohydrologic Parameters." *Papers and Proceedings of the Applied Geography Conferences*, 8 (1985): 117-124.

Parry, Martin L. "The Impact of Climatic Variations on Agricultural Margins." Pp. 351-367 in *Climatic Impact Assessment: Studies of the Interaction of Climate and Society*, edited by Robert W. Kates, et al. Chichester, England: John Wiley & Son, 1985.

Simon, K.H. "Zur Abschatzung der Spatfrostgefahrdang von Landwirtschaftlichen Nutzflachen." *Petermann's Geographische Mitteilungen*, 129, No. 3 (1985): 177-179.

Sivanappan, R.K., and R.M Panchanathan. "Watershed-Based Dryland Farming." *Land Use Policy*, 2, No. 4 (1985): 345-349.

Gale, S.J., and C.O. Hunt. "The Hydrological Characteristics of a Floodwater Farming System." *Applied Geography*, 6, No. 1 (1986): 33-42.

Bayliss-Smith, Timothy. *The Ecology of Agricultural Systems*. Cambridge, England: Cambridge University Press, 1987.

Bruins, H.J., M. Evernari, and A. Rogel. "Run-Off Farming Management and Climate." Pp. 3-14 in *Progress in Desert Research*, edited by L. Berkofsky and M.G. Wurtele. Totowa, N.J.: Rowman & Littlefield, 1987.

Hallberg, G.R. "The Impacts of Agricultural Chemicals on Ground Water Quality." *GeoJournal*, 15, No. 3 (1987): 283-296.

Hofmeyer, A. "Bewertung der naturlichen Agrarressmurcen: Ein Vergleich ausgewahlter Untersuchungen aus methodischer Sicht." *Wirtschaftsgeographische Studien*, 15-16 (1987): 17-38.

Moss, Michael R., and L. Strath Davis. *Ecological Crop Productivity and Climatic Variability*. Guelph, Ontario: University of Guelph, Department of Geography, Occasional Papers in Geography No. 12, 1987.

O'Riordan, T. "Agriculture and Environmental Protection." *Geography Review*, 1 (1987): 35-40.

Reddy, S.J. "An Agro-Climatic Classification of the Semi-Arid Tropics." Ph.D. Dissertation, Australian National University, Canberra, Australia, 1987.

Smith, Nigel J.H. "Genebanks: A Global Payoff." *Professional Geographer*, 39, No. 1 (1987): 1-8.

Tivy, Joy. "Nutrient Cycling in Agroecosystems." *Applied Geography*, 7, No. 2 (1987): 93-113.

Allen, P., and D. Van Dusen, eds. *Global Perspectives on Agroecology and Sustainable Agricultural Systems*. Santa Cruz, Calif.: University of California, 1988.

Newton, Gary A. "Farm Planning for Wildlife Resource Enhancement." Master's Thesis, Southwest Missouri State University, Springfield, Mo., 1988.

Parry, Martin L., T.R. Carter, and N.T. Konjin, eds. *The Impact of Climatic Variations on Agriculture, Volume 2: Assessments in Semi-Arid Regions*. Dordrecht, Netherlands: Kluwer, 1988.

Parry, Martin L., and T.R. Carter. "The Assessment of Effects of Climatic Variations on Agriculture: Aims, Methods and Summary of Results." Pp. 11-95 in *The Impact of Climatic Variations on Agriculture, Volume 1: Assessments in Cool Temperate and Cold Regions*, edited by Martin L. Parry, T.R. Carter, and N.T. Konijn. Dordrecht, Netherlands: Kluwer, 1988.

Simmons, I.G. "Ecological Functional Approaches to Agriculture in Geographical Contexts." *Geography*, 65 (1988): 305-316.

Smith, Barry E., Laurie L. Ludlow, and Michael Brklacich. "Implications of a Global Climatic Warming for Agriculture: A Review and Appraisal." *Journal of Environmental Quality*, 17, No. 4 (1988): 519-527.

Warrick, R.A. "Carbon Dioxide, Climatic Change and Agriculture." *Geographical Journal*, 154, No. 2 (1988): 221-233.

Jackson, I.J. *Climate, Water and Agriculture in the Tropics*, Second Edition. London, England: Longman, 1989.

Mohan Rao, E.V.M. "Problems and Perspectives in Agroclimatological Studies in the Semi-Arid Tropics." *Transactions, Institute of Indian Geographers*, 11, No. 2 (1989): 1-6. (New Series)

Hariss, Robert C. "Agricultural Production Versus Climate Protection: An Emerging Conflict in the 1990s?" *Land Use Policy*, 7, No. 2 (1990): 173-175.

Jauneau, Jean-Claude, and Sylvie Chedin. "Agriculture: Le vercars en friche? Et alors." *Revue de Geographie Alpine*, 78, No. 4 (1990): 31-42.

Parry, Martin L., and T.R. Carter. "An Assessment of the Effects of Climatic Change on Agriculture." Pp. 61-84 in *Climatic Change and Plant Genetic Resources*, edited by M. Jackson, et al. London, England: Belhaven, 1990.

Parry, Martin L. *Climate Change and World Agriculture*. London, England: Earthscan, 1990.

Redchift, Michael. "Developing Sustainability: Designating Agroecological Zones." *Land Use Policy*, 7, No. 3 (1990): 202-216.

Spitthover, M. "Flachen stillegungen und agrarstrukturelle Wandel." *Geographie und Schule*, 12, No. 63 (1990): 37-41.

Williams, Michael. "Agricultural Impacts in Temperate Wetlands." Pp. 181-216 in *Wetlands: A Threatened Landscape*, edited by Michael Williams. Oxford, England: Institute of British Geographers, Special Publication No. 25, 1990.

Burt, Timothy, and Nick Haycock. "Farming and Nitrate Pollution." *Geography*, 76, No. 1 (1991): 60-63.

Carr, S., and J. Tait. "Farmers' Attitudes to Conservation." *Built Environment*, 16 (1991): 218-231.

Gaskell, P.T., and M.F. Tanner. "Agricultural Change and Environmentally Sensitive Areas." *Geoforum*, 22, No. 1 (1991): 81-90.

Gilg, Andrew. "Planning for Agriculture: The Growing Case for a Conservation Component." *Geoforum*, 22, No. 1 (1991): 75-80.

Myers, Michael. "A Cultural Ecology of Crop Disease: Agricultural Practices and the Solamum-Phytophthara Infestations Pathosystem." Master's Thesis, University of Texas, Austin, Tex., 1991.

Porter, J.H., Martin L. Parry, and T.R. Carter. "The Potential Effects of Climatic Change on Agricultural Insect Pests." *Agricultural and Forest Meteorology*, 57 (1991): 221-240.

Sharma, B.L. *Dryland Farming: Perspectives and Prospects.* Delhi, India: Daya Publications, 1991.

Troughton, Michael J. "Ecological Assessment of Modern Agriculture." Pp. 141-202 in *Limits to Rural Land*, edited by G. Van Dort, et al. Wageningen, Netherlands: Pudoc, 1991.

Zepp, Harold, and Sylvia Stein. "Zur Problematik geokologischer Kartierung in intensiv genutzten Agrarlandschaften. Ein Diskussionsbeitrag zur Hemerobiestufenerfassung im Rahamen der geookologischen Landschaftsaufnahme." *Geographische Zeitschrift*, 79, No. 2 (1991): 94-112.

Brklacich, Michael. "Changes in Food Production Under Conditions of Climatic Change." Pp. 25-29 in *Global Climate Change: Health Issues and Priorities.* Ottawa, Ontario: Health and Welfare Canada, 1992.

Croll, Elisabeth, and David Parkin, eds. *Bush Base: Forest Farm Culture, Environment and Development.* London, England: Routledge, 1992.

Hoggart, Keith, ed. *Agricultural Change, Environment and Economy.* London, England: Mansell, 1992.

Hurni, Hans, and Kebede Tato, eds. *Erosion, Conservation, and Small-Scale Farming.* Marceline, Mo.: Walsworth, 1992.

Mathier, L., and A.G. Roy. "Comparison des hydrogrammes de crues et du transport des sediments en milieu agricole lors detrois precipitations naturelles." *Revue des Sciences de l'Eau*, 5 (1992): 1-21.

Mohammad, Noor, ed. *New Dimensions in Agriculture: The Ecology of Agricultural Systems.* New Delhi, India: Concept Publishing Company, 1992.

Roberts, Rebecca. "Nature, Uneven Development, and the Agricultural Landscape." Pp. 119-130 in *Contemporary Rural Systems in Transition: Volume 1, Agriculture and Environment*, edited by Ian R. Bowler, Christopher R. Bryant, and M. Duane Nellis. Wallingford, England: C.A.B. International, 1992.

Smith, Nigel J.H., et al. *Tropical Forests and Their Crops.* Ithaca, N.Y.: Cornell University Press, 1992.

Tricart, Jean. *Ecogeography and Rural Management: A Contribution to the International Geosphere–Biosphere Programme.* Harlow, England: Longman, 1992.

Hall, J.A., and R. Burkholder. "The Impact of Climate Change on Agriculture." Pp. 69-81 in *The Impact of Climate Change on Water in the Grand River Basin, Ontario*, edited by Marie Sanderson. Waterloo, Ontario: University of Waterloo, Department of Geography, 1993.

Healy, Robert G. "Forest or Fields: A Land Allocation Perspective." *Land Use Policy*, 10, No. 2 (1993): 122-126.

Lightfoot, C., and R. North. "A Participatory Experiment in Sustainable Agriculture." *Journal for Farming Systems Research-Extension*, 4 (1993): 11-34.

Weischet, Wolfgang, and Cesar N. Caviedes. *The Persisting Ecological Constraints of Tropical Agriculture.* London, England: Edward Arnold, 1993.

Wood, David. "Forest to Fields: Restoring Tropical Lands to Agriculture." *Land Use Policy*, 10, No. 2 (1993): 91-107.

Yapa, Lakshman. "What Are Improved Seeds? An Epistemology of the Green Revolution." *Economic Geography*, 69, No. 3 (1993): 254-273.

Glantz, Michael H., ed. *Drought Follows the Plow: Cultivating Marginal Areas*. New York, N.Y.: Cambridge University Press, 1994.

Kalluri, Satayanaragana. "Modelling Evapotranspiration and Water Stress in Crops from Satellite and Ground Measurements." Ph.D. Dissertation, University of Maryland, College Park, Md., 1994.

Lightfoot, Dale R. "Morphology and Ecology of Lithic-Mulch Agriculture." *Geographical Review*, 84, No. 2 (1994): 172-185.

Lightfoot, Dale R., and F.M. Eddy. "The Agricultural Utility of Lithic-Mulch Gardens: Past and Present." *GeoJournal*, 34, No. 4 (1994): 425-438.

Smit, Barry E., and J. Smithers. "Sustainable Agriculture and Agroecosystem Health." Pp. 31-38 in *Agroecosystem Health: Proceedings of an International Workshop*, edited by N.O. Nielson. Guelph, Ontario: University of Guelph, 1994.

Couturier, A., and Barry E. Smit. *Natural Areas as Components of Agroecosystems: Identification and Characteristics*. Guelph, Ontario: University of Guelph, Agroecosystem Health Project, Discussion Paper No. 11, 1995.

Couturier, A., and Barry E. Smit. *Spatial Dynamics of Natural Areas as Indicators of Agroecosystem Health*. Guelph, Ontario: University of Guelph, Agroecosystem Health Project, Discussion Paper No. 12, 1995.

Hantschel, R.E., T. Kamp, and F. Buse. "Increasing the Soil Temperature to Study Global Warming Effects on the Soil Nitrogen Cycle in Agroecosystems." *Journal of Biogeography*, 22, No. 2-3 (1995): 375-380.

Heong, K.L., P.S. Teng, and K. Moody. "Managing Rice Pests with Less Chemicals." *GeoJournal*, 35, No. 3 (1995): 337-350.

Jackson, M.T. "Protecting the Heritage of Rice Biodiversity." *GeoJournal*, 35, No. 3 (1995): 267-274.

Jauneau, Jean-Claude. "Objectifs environnementaux et objectifs de soutien a l'agriculture: Une convergence possible dans les zones de moyenne montagne menacees de deprise?" *Revue de Geographie Alpine*, 83, No. 3 (1995): 83-100.

Keiffer, Cyril. "A Model Agricultural Hazardous Materials Response Plan." Master's Thesis, University of Toledo, Toledo, Ohio, 1995.

Ladha, J.K., and P.M. Reddy. "Extension of Nitrogen Fixation to Rice-Necessity and Possibilities." *GeoJournal*, 35, No. 3 (1995): 363-373.

Neue, H.V., L.H. Ziska, R.B. Matthews, and Q. Dai. "Reducing Global Warming: The Role of Rice." *GeoJournal*, 35, No. 3 (1995): 351-362.

Ryszkowski, L., and A. Kedziora. "Modification of the Effects of Global Climate Change by Plant Cover Structure in an Agricultural Landscape." *Geographia Polonica*, 65 (1995): 5-34.

Spaling, H., and Barry E. Smit. "A Conceptual Model of Cumulative Environmental Effects of Agricultural Land Drainage." *Agriculture, Ecosystems and Environment*, 53 (1995): 99-108.

Ward, Neil. "Technological Change and the Regulation of Pollution from Agricultural Pesticides." *Geoforum*, 26, No. 1 (1995): 19-34.

De Leval, J. "Y-a-t-il une place pour la haie dans l'agro-ecosystem de l'an 2000?" *Bulletin de la Societe Geographique de Liege*, 32 (1996): 281-286.

Farres, P.J., and J. Muchena. "Spatial Patterns of Soil Crusting and Their Relationship to Crop Cover." *Catena*, 26, No. 3-4 (1996): 247-260.

Lightfoot, Dale R. "The Nature, History and Distribution of Lithic- Mulch Agriculture: An Ancient Technique of Dryland Agriculture." *Agricultural History Review*, 44, No. 2 (1996): 206-222.

Mannion, A.M. *Agriculture and Environment*. New York, N.Y.: John Wiley, 1996.

Smit, Barry E., David McNabb, and J. Smithers. "Agricultural Adaptation to Climatic Variation." *Climatic Change*, 33, No. 1 (1996): 7-29.

Brklacich, Michael. "Assessing the Impacts of Climatic Change on Agriculture." Pp. 197-209 in *Ecosystem Effects of Atmospheric Change*, edited by J. Carey, H. Vaughan, and F. Crisp. Burlington, Ontario: Environment Canada, National Water Research Institute, 1997.

Mannion, A.M. *Agriculture and Environmental Change: Temporal and Spatial Dimensions*. New York, N.Y.: John Wiley, 1997.

Marsden, Terry K. "Reshaping Environments: Agriculture and Water Interaction and the Creation of Vulnerability." *Transactions, Institute of British Geographers*, 22, No. 3 (1997): 321-337. (New Series)

Nkemdrian, L.C. "The Tropics: Agriculture in a Changing Climate." Pp. 137-142 in *Land Resource Management*, edited by A.L. Singh. Delhi, India: R.R. Publications, 1997.

Quine, Timothy, et al. "Erosion Processes and Landform Evolution in Agricultural Land: New Perspectives from Caesium 137 Measurements and Topographic-Based Erosion Modelling." *Earth Surface Processes and Landforms*, 22, No. 9 (1997): 799-816.

Smithers, J., and Barry E. Smit. "Agricultural System Response to Environmental Stress." Pp. 167-184 in *Agricultural Restructuring and Sustainability: A Geographical Perspective*, edited by Brian W. Ilbery, Quentin Chiotti, and Timothy Rickard. Wallingford, England: C.A.B. International, 1997.

Tao, Y., et al. "Analysis of the Relationship Between NDVI and Rainfall in Subhumid Agricultural Areas and the Influence of Environmental Factors." *Papers and Proceedings of the Applied Geography Conferences*, 20 (1997): 310-317.

Mandal, C., and D.K. Mandal. "Agro-Climatic Database for Geographical Research in Agriculture." *Geographical Review of India*, 60, No. 3 (1998): 310-342.

Potter, Clive. "Conserving Nature: Agri-Environmental Policy Development and Change." Pp. 85-105 in *The Geography of Rural Change*, edited by Brian W. Ilbery. London, England: Longman, 1998.

Rasmussen, M.S. "Developing Simple, Operational, Consistent NDV1-Vegetation Models by Applying Environmental and Climatic Information.

Part II: Crop Yield Assessment." *International Journal of Remote Sensing*, 19, No. 1 (1998): 119-140.

Sadhukhan, Shyamali, and Sunil Sadhukhan. "Ecofarming: An Antipode of Green Revolution." *Geographical Review of India*, 60, No. 2 (1998): 155-161.

Soegaard, Henrik, and Halldor Thorgirson. "Carbon Dioxide Exchange at Leaf and Canopy Scale for Agricultural Crops in the Boreal Environment." *Journal of Hydrology*, 212-213 (1998): 51-61.

Zimmerer, Karl S. "The Ecogeography of Major Crops: Versatility and Conservation." *BioScience*, 48, No. 6 (1998): 445-454.

Holloway, L. "Understanding Climate Change and Farming: Scientific and Farmers' Constructions of 'Global Warming' in Relation to Agriculture." *Environment and Planning A*, 31, No. 11 (1999): 2017-2032.

Jakubiak, Bogumil. "Main Features of Global Climate Models and the Impact of Predicted Climate Changes for Agriculture and Forestry." *Geographia Polonica*, 72, No. 2 (1999): 15-26.

Lipa, Jerzy J. "Do Climate Changes Increase the Threat to Crops by Pathogens, Weeds, and Pests?" *Geographia Polonica*, 72, No. 2 (1999): 99-110.

Pilleboue, Jean. "La qualite agro-alimentaire et ses territoires productifs: Avant-propos." *Sud-Ouest Europeen*, 6 (1999): 1-8.

Smil, Vaclav. "Crop Residues: Agriculture's Largest Harvest." *Bioscience*, 49 (1999): 299-308.

Morgan, K., and Jonathan Murdoch. "Organic Versus Conventional Agriculture: Knowledge, Power and Innovation in the Food Chain." *Geoforum*, 31, No. 2 (2000): 175-190.

Takken, Ingrid, et al. "The Effect of Tillage-Induced Roughness on Runoffs and Erosion Patterns." *Geomorphology*, 37, No. 1-2 (2001): 1-14.

Tate, A.B. "Global Warming's Impact on Wine." *Journal of Wine Research*, 12, No. 2 (2001): 95-109.

IRRIGATION

Bryan, Kirk. "Flood-Water Farming." *Geographical Review*, 19 (1929): 444-456.

Thorne, D. Wynne, and D.A. Anderson. "Irrigation and Permanent Agriculture." *Yearbook, Association of Pacific Coast Geographers*, 8 (1945): 6-14.

Labasse, J. "L'economie des oasis, ses difficultes et ses chances." *Revue de Geographie de Lyon*, 32 (1957): 307-320.

Bowden, L.W. *The Diffusion of the Decision to Irrigate*. Chicago, Ill.: University of Chicago, Department of Geography, Research Paper No. 97, 1965.

Highsmith, Richard M. "Irrigated Lands of the World." *Geographical Review*, 55, No. 3 (1965): 382-389.

Faudrey, D. "Les agriculteurs et les consequences economiques globales de l'extention des perimetred d'irrigation." *Revue de Geographie de Lyon*, 41 (1966): 257-286.

Green, Herbert. "Irrigation in Arid Lands." Pp. 255-270 in *Arid Lands: A Geographical Appraisal*, edited by E.S. Hills. London, England: Methuen, 1966.

Parodi, M. "La decision d'acceptation ou de refus de l'irrigation par aspersion." *Revue de Geographie de Lyon*, 41 (1966): 197-223.

Cantor, Leonard M. *A World Geography of Irrigation*. London, England: Oliver and Boyd, 1967.

Mather, John R. "Irrigation Agriculture in Humid Areas." *Yearbook, Association of Pacific Coast Geographers*, 30 (1968): 107-122.

Cantor, Leonard M. *A World Geography of Irrigation*. New York, N.Y.: Praeger, 1970.

Atkinson, K. "Saline Water Irrigation." *Geography*, 57, No. 3 (1972): 244-245.

Andreae, Bernd. "Types of Irrigation Farming." *Applied Sciences and Development*, 6 (1975-76): 77-93.

White, Gilbert F. "The Main Effects and Problems of Irrigation." Pp. 1-71 in *Arid Land Irrigation in Developing Countries*, edited by E.B. Worthington. Oxford, England: Pergamon Press, 1977.

Munson, Brian E. "The Effect of Irrigation on Available Void Capacity of Disturbed Soils in Semiarid Regions." Master's Thesis, University of Denver, Denver, Colo., 1978.

White, Gilbert F. *Environment Effects of Arid Land Irrigation in Developing Countries*. Paris, France: UNESCO, 1978.

Todhunter, Paul. "A Computer Model Validation of the Simulation of Crop Water Requirements and Irrigation Needs." Master's Thesis, University of California at Los Angeles, 1981.

Wolff, P. "Technical Aspects of Mechanization in Irrigation Farming: A Land Improvement Specialist's View." *Applied Geography and Development*, 19 (1982): 7-21.

Bajwa, R.S. *Humid Region Irrigation: Geographic and Economic Perspectives*. Ann Arbor, Mich.: Michigan Geographical Publication No. 28, 1983.

Kutsch, Horst. "Currently Used Techniques in Rain-Fed Water-Concentrating Culture." *Applied Geography and Development*, 21 (1983): 108-117.

Pigram, J.J. "Irrigation Agriculture and the Rural System." *Rural Systems*, 1, No. 2 (1983): 121-134.

Kraus, Pavel. "Irrigation Ecology." Master's Thesis, University of Chicago, Chicago, Ill., 1984.

Mosonyi, Emil, Alexandre Lindner, and Fritz Fracke. "Modelling Techniques in the Planning of Irrigation Systems." *Applied Geography and Development*, 23 (1984): 94-108.

Walker, Hans H. "Problems of the Organization of Irrigation Projects: A Systems Oriented Study of Irrigation Projects in Developing Countries." *Applied Geography and Development*, 23 (1984): 20-45.

Gale, S.J., and C.O. Hunt. "The Hydrological Characteristics of a Floodwater Farming System." *Applied Geography*, 6, No. 1 (1986): 33-42.

Gilbertson, D.D. "Runoff Floodwater Farming and Rural Water Supply in Arid Lands." *Applied Geography*, 6, No. 1 (1986): 5-11.

Towne, Douglas T. "The Relationship Between Irrigated Farmland Decline and Physical Landscape Factors: A Spatial Analysis." Master's Thesis, University of Arizona, Tucson, Ariz., 1986.

Bethemont, J. "Sur la dynamique de l'irrigation dans les pays en voie de developpement." *Revue de Geographie de Lyon*, 65 (1990): 5-8.

Chen, Zhang-quan. "A Simple Model of an Irrigation System for Oasis Fields." *GeoJournal*, 20, No. 4 (1990): 381-384.

Fazzi, P. "Les developpement de l'irrigation dans la diagonale aride entre logique productive et logique strategique." *Revue de Geographie de Lyon*, 65 (1990): 21-26.

Middleton, Nick. "The Environmental Cost of Irrigation." *Geography Review*, 9, No. 2 (1995): 15-17.

Vorshney, R.S. "Modern Methods of Irrigation." *GeoJournal*, 35, No. 1 (1995): 61-66.

AQUACULTURE

Lucas, A. "L'aquaculture dans les our tempereis." *Norois*, 27 (1980): 237-250.

Bordach, John E. "The Role of Aquaculture in Human Nutrition." *GeoJournal*, 10, No. 3 (1985): 221-232.

Hepher, B. "Aquaculture Intensification Under Land and Water Limitations." *GeoJournal*, 10, No. 3 (1985): 253-260.

Moore, Lori B. "The Role of Feeds and Feeding in Aquatic Animal Production." *GeoJournal*, 10, No. 3 (1985): 245-252.

Muir, J.F. "The Role of Aquaculture in the Industrial Regions of the Temperate Zone." *GeoJournal*, 10, No. 3 (1985): 277-298.

Rosenthal, Harold. "Constraints and Perspectives in Aquaculture Development." *GeoJournal*, 10, No. 3 (1985): 305-324.

Doumenge, F. "La revolution aquacole." *Annales de Geographie*, 95 (1986): 445-482, 527, and 586.

Galante, John. "Design and Modeling of a Site-Specific Aquaculture Pond in a Systems Analytic Framework." Master's Thesis, Pennsylvania State University, University Park, Pa., 1986.

Sornin, J.M. "Overfarming and Deterioration of the Shellfish Culture Grounds: Consequences on the Farming Systems—Examples in France and in Japan." *Norois*, 34 (1987): 51-61.

Wood, L.J., K. Anutha, and A. Peschken. "Aquaculture: Marine Farming of Atlantic Salmon." *Geography*, 75, No. 3 (1990): 211-221.

Coull, James R. "Will a Blue Revolution Follow the Green Revolution? The Modern Upsurge of Aquaculture." *Area*, 25, No. 4 (1993): 350-357.

Miller, Craig. "Salmon Aquaculture in Shetland and New Brunswick: A Comparative Study of Resource Regions within a Moral Perspective." Master's Thesis, University of Western Ontario, London, Ontario, 1995.

Barton, Jonathan R., and Dan Staniford. "Net Deficits and the Case for Aquacultural Geography." *Area*, 30, No. 2 (1998): 145-156.

AGRICULTURAL MARKETING AND DISTRIBUTION

Marbut, C.F. "Russia and the United States in the World's Wheat Market." *Geographical Review*, 21, No. 1 (1931): 1-21.

Marble, Duane F. "The Spatial Structure of the Farm Business." Master's Thesis, University of Washington, Seattle, Wash., 1956.

Volf, M.B. "Tendencies of World Agricultural Production and the Market for the Supply 1975-1985" (in Russian). *Izvestiya Vsesoyuznogo Geograficheskogo Obshchestra*, 100, No. 5 (1968): 420-427.

Delbos, C., and M. Furestier. "Cares cooperatives et commercialisation du vin en Sommierois et Vaunage." *Bulletin de la Societe Languedocienne de Geographie*, 4, No. 2 (1970): 167-182.

Lundgren, Jan. "Function and Geographical Relationships in Agricultural Marketing Mechanisms." *Geographical Survey*, 1, No. 2 (1972): 34-45.

Smith, W. "The Farm-Market Linkage: A Note." *Area*, 7, No. 3 (1975): 180-182.

Huetz de Lemps, Alain. "Canne a sucre et sucre de canne: Un bilon." *Les Cahies d'Outre Mer*, 33 (1980): 199-226. Part II, 319-334.

Maillard, Jean-Claude. "Le commerce international de la banane, caracteres originaux et evolution, recente d'un marche." *Les Cahiers d'Outre Mer*, 33 (1980): 259-284.

Grigg, David B. "Agricultural Perspectives Under Free Trade." Pp. 24-34 in *The Changing Countryside*, edited by J. Bluden and N. Curry. London, England: Methuen, 1985.

Bowler, Ian R. "Direct Supply Control in Agriculture: Experience in Western Europe and North America." *Journal of Rural Studies*, 2, No. 1 (1986): 19-30.

Rozlucki, W. "Spatial Structure of the Grain Trade in the Mediterranean: The Ancient Pattern Reversed" (in Polish). *Przeglad Geograficzny*, 60, No. 4 (1988): 585-598.

Bowler, Ian R. "Revising the Research Agenda on Agricultural Policy in Developed Market Economies." *Journal of Rural Studies*, 5 (1989): 385-394.

Geneau, Isabelle de Camarliere. "Le marche commun agricole: Essai de bilon geographique." *Revue Geographique de l'Est*, 29, No. 2 (1989): 97-116.

Susman, Paul. "Exporting the Crisis: U.S. Agriculture and the Third World." *Economic Geography*, 65, No. 4 (1989): 293-313.

Kopp, Daniela, and Iain Wallace. "The Wheat Imports of Non-Traditionally Wheat-Producing Countries." *Geography*, 75, No. 2 (1990): 148-151.

LeHeron, Richard B. "Good Food Worldwide? Internationalisation and Performance of Goodman Fielder Wattie Ltd." Pp. 100-109 in *The Corporate Firm in a Changing World Economy*, edited by M. DeSmidt and E. Wever. London, England: Routledge, 1990.

Richardson, Katherine. "Consumers, the Decision-Making Proces, and Organic Farming." Master's Thesis, University of Chicago, Chicago, Ill., 1990.

Schmidt-Wulffen, Wulf. "Ernahrungssicherung und technischer Fortschrift durch Marktliberalisierung? Zur Akzeptanz des freien Getreide markets durch Bauern in Mal." *Geographische Zeitschrift*, 79, No. 3 (1991): 168-180.

Hart, Patrick W.E. "Marketing Agricultural Produce." Pp. 162-206 in *The Geography of Agriculture in Developed Market Economies*, edited by Ian R. Bowler. London, England: Longman, 1992.

Ufkes, Frances M. "Trade Liberalization, Agro-Food Politics and the Globalization of Agriculture." *Political Geography*, 12, No. 3 (1993): 215-231.

Windhorst, Hans-Wilhelm. "Raumliche Verbundsysteme in der agrarischen Produktion." *Erdkunde*, 47, No. 2 (1993): 118-130.

Flynn, Andrew, Terry K. Marsden, and Neil Ward. "Retailing, the Food System and the Regulatory State." Pp. 90-103 in *Regulating Agriculture*, edited by Philip Lowe, Terry K. Marsden, and Sarah Whatmore. London, England: David Fulton, 1994.

Lowe, Philip, Terry K. Marsden, and Sarah Whatmore, eds. *Regulating Agriculture*. London, England: David Fulton, 1994.

Whatmore, Sarah. "From Farming to Agri-Business: The Global Agro-Food System." Pp. 36-49 in *Geographies of Global Change: Remapping the World in the Late Twentieth Century*, edited by R.J. Johnston, Peter J. Taylor, and Michael J. Watts. Cambridge, England: Blackwell, 1995.

Le Gars, C., and J.C. Hinnewinlal. "Le commerce mondial des vins a la fin du XXe siecle." *Annales de Geographie*, 109, No. 4-5 (2000): 381-394.

TRANSPORTATION AND MECHANIZATION

Anaya Garduno, Manuel. "Technology and Desertification." *Economic Geography*, 53, No. 4 (1977): 407-412.

Visser, Sent. "Technological Change and the Spatial Structure of Agriculture." *Economic Geography*, 56, No. 4 (1980): 311-319.

Joffe, G.V., and A.I. Shoyter. "The Transport Accessibility of a Town and the Differentiation of the Factors of Farm Production." *Soviet Geography*, 22, No. 2 (1981): 94-98.

Mehrotra, Naresh. "The Constraints in the Transfer of Rainfed Technology and the Priority in Agricultural Development." Pp. 427-444 in *Perspectives in Agricultural Geography*, vol. 2, edited by Noor Mohammad. New Delhi, India: Concept Publishing Company, 1981.

Symons, Leslie J. "Technological Innovation in Twentieth Century Agriculture: The Case of Bio-Aeronautics." Pp. 279-304 in *Perspectives in Agricultural Geography*, vol. 5, edited by Noor Mohammad. New Delhi, India: Concept Publishing Company, 1981.

Wood, L.J. "Energy and Agriculture: Some Geographical Implications." *Tijdschrift voor Economische and Sociale Geographie*, 72, No. 4 (1981): 224-234.

Troughton, Michael J. "Process and Response in the Industrialization of Agriculture." Pp. 213-228 in *The Effect of Modern Agriculture on Rural Development*, edited by Gyorgy Enyedi and I. Volgyes. New York, N.Y.: Pergamon, 1982.

Rothbert, Debbi L. "Aircraft Monitoring of Agricultural Crops in the Thermal Infrared: A Comparison of Corn and Soybean Canopy Response Patterns." Master's Thesis, Columbia University, New York, N.Y., 1983.

Troughton, Michael J. "The Impacts of Industrializing Agriculture on the Rural Physical and Socio-Economic Environments." Pp. 199-219 in *Papers:*

Colloque international d'amenagement rural, vol. 2. Liege, Belgium: 1983. pp. 199-219.

Stoms, David. "Preliminary Design of a Farm Monitoring Geographic Information System." Master's Thesis, University of California at Santa Barbara, 1986.

Troughton, Michael J. "The Nature and Environmental Impacts of the Industrialization of Agriculture." Pp. 161-173 in *Environmental Challenges: Learning for Tomorrow's World*, edited by P.F Wilkinson and M. Wymon. London, England: Althouse Press, 1986.

Shaw, Anthony B. "Approaches to Agricultural Technology Adaption and Consequences of Adaption in the Third World: A Critical Review." *Geoforum*, 18, No. 1 (1987): 1-20.

Windhorst, Hans-Wilhelm. "Die Industrialisierung der Agrarwirtschaft als Heraus forderung an die Agrargeographie." *Geographische Zeitschrift*, 77, No. 3 (1989): 136-153.

Munton, R.J.C., Terry K. Marsden, and Sarah Whatmore. "Technological Change in a Period of Agricultural Adjustment." Pp. 104-126 in *Technological Change and the Rural Environment*, edited by Philip Lowe, Terry K. Marsden, and Sarah Whatmore. London, England: David Fulton, 1990.

Deslauriers, Pierre, Christopher R. Bryant, and Claude Marois. "Une revue de la litterature scientifique sur l'etalement urbaine et sur relations urbaines-agricoles dans la frange urbaine." *Espaces, Populations, et Societes*, 7, No. 2 (1991): 325-334.

Smil, Vlacov. "Agricultural Energy Costs: National Analysis." Pp. 85-99 in *Energy in World Agriculture: Volume 6, Energy in Farm Production*, edited by R.C. Fluck. Amsterdam, Netherlands: Elsevier, 1991.

Bowler, Ian R. "The Industrialization of Agriculture." Pp. 7-31 in *The Geography of Agriculture in Developed Market Economies*, edited by Ian R. Bowler. London, England: Longman, 1992.

Smil, Vlacov. "Agricultural Energy Costs: National Analyses." Pp. 85-100 in *Energy in Farm Production*, edited by R.C. Fluck. Amsterdam, Netherlands: Elsevier, 1992.

AGRICULTURAL LABOR STUDIES

Earle, Carville, and Ronald Hoffman. "The Foundation of the Modern Economy: Agriculture and the Costs of Labor in the United States and England, 1800–60." *American Historical Review*, 85 (1980): 1055-1094.

Laurent, C. "Multiple Jobholding Farmers in Agricultural Policy." *GeoJournal*, 6, No. 4 (1982): 287-292.

Clark, Gordon. "People Working in Farming: The Changing Nature of Farmwork." Pp. 67-83 in *People in the Countryside*, edited by T. Champion and C. Watkins. London, England: Paul Chapman, 1991.

Walford, Nigel. "Agriculture in the Context of the Restructuring of Rural Employment." Pp. 187-200 in *Contemporary Rural Systems in Transition: Volume 2, Economy and Society*, edited by Ian R. Bowler, Christopher R.

Bryant, and M. Duane Nellis. Wallingford, England: C.A.B. International, 1992.
Das, Raju J. "The Green Revolution, Agrarian Productivity and Labor." *International Journal of Urban and Regional Research*, 22, No. 1 (1998): 133-135.
Manz, Beatriz, and Paul Spoonley. "Gendered Work: Female Labour in Pipfruit Production in New Zealand and Chile." *New Zealand Geographer*, 57, No. 1 (2001): 8-15.

THEORETICAL WORKS

In addition to the regional and general works lists in other sections of this bibliography, the following titles are offered as potentially and actually related to the study of agricultural geography. They are very much theoretical in content and seem to be without specific regional focus; however, they do not comfortably fit in with the general topics titles.

Krymoneski, R. "Graphical Presentations of Thunen's Theory of Intensity." *Journal of Farm Economics*, 10 (1928): 461-482.
Johnson, Hildegard B. "A Note on Thunen's Circles." *Annals of the Association of American Geographers*, 52, No. 2 (1962): 213-220.
Hall, Peter, ed. *Von Thunen's Isolated State: An English Edition of "Der Isoliete Staat"* by Johann Heinrich von Thunen. London, England: Pergamon, 1966.
Peucker, Thomas K. "Eine Erweiterung des Thunen-Modells um klimafaktoren und technischen Fortschriff: Raumliche Beziehung als wirtschaftsgeographischen Problem." Ph.D. Dissertation, Universitat Heidelberg, Heidelberg, Germany, 1966.
Clark, C. "Von Thunen's 'Isolated State.'" *Oxford Economic Papers* (New Series), 19 (1967): 370-377.
Howes, R. "A Test of a Linnear Programming Model for Agriculture." *Papers of the Regional Science Association*, 19 (1967): 123-140.
Sinclair, R. "Von Thunen and Urban Sprawl." *Annals of the Association of American Geographers*, 57, No. 1 (1967): 72-88.
Stevens, B.H. "Location Theory and Programming Models: The Von Thunen Case." *Papers and Proceedings of the Regional Science Association*, 21 (1968): 29-34.
Day, R.H., and E.H. Tinney. "A Dynamic Von Thunen Model." *Geographical Analysis*, 1, No. 2 (1969): 137-151.
Peet, Richard J. "Von Thunen Theory and the Dynamics of Agricultural Expansion." *Explorations in Economic History*, 8, No. 2 (1970): 181-201.
Morgan, W.B. "The Doctrine of the Rings." *Geography*, 58 (1973): 301-312.
Webber, M.J. "Equilibrium of Location in an Isolated State." *Environment and Planning*, 5, No. 6 (1973): 751-759.
Katzman, M. "The Von Thunen Paradigm, the Industrial-Urban Hypothesis, and the Spatial Structure of Agriculture." *American Journal of Agricultural Economics*, 56 (1974): 683-696.

McDonald, G.T. "The Fellowship of the Ring." Pp. 33-42 in *Proceedings of the International Geographical Union Regional Conference and the Eighth New Zealand Geography Conference*, edited by William Brockie, Richard LeHeron, and Evelyn Stokes. Palmerston North, New Zealand: New Zealand Geographical Society, 1975.

Jones, Richard C. "Cautious Optimization Versus Spatial Satisfaction: Alternative Agricultural Programming Models." *Proceedings of the Association of American Geographers*, 8 (1976): 56-62.

Jones, Richard C. "Testing Macro-Thunen Models by Linear Programming." *Professional Geographer*, 28, No. 4 (1976): 353-361.

Kellerman, A. "The Pertinence of the Macro-Thunian Analysis." *Economic Geography*, 53, No. 3 (1977): 255-264.

Datoo, B.A. "Toward a Reformation of Boserup's Theory of Agricultural Change." *Economic Geography*, 54, No. 2 (1978): 135-144.

Jones, A.P., W.J. McGuire, and A.D. Witte. "A Reexamination of Some Aspects of Von Thunen's Model of Spatial Location." *Journal of Regional Science*, 18 (1978): 1-16.

Mesquita, Olindina V. "O modelo de Von Thunen: Uma discussao." *Revista Brasileira de Geografia*, 40, No. 2 (1978): 60-130.

Rakitnikov, A.N. "Von Thunen and the Significance of His Work for the Geography of Agriculture." *Soviet Geography*, 19, No. 8 (1978): 567-571.

Chisholm, Michael. "Von Thunen Anticipated." *Area*, 11, No. 1 (1979): 37-40.

Norton, William. "The Relevance of Von Thunen Theory to Historical and Evolutionary Analysis of Agricultural Land Use." *Journal of Agricultural Economics*, 30, No. 1 (1979): 39-47.

Cromley, Robert G. "The Isolated Estate: An Agricultural Location Game." *Journal of Geography*, 79, No. 8 (1980): 230-234.

O'Sullivan, Patrick, and Bruce Ralston. "On the Equivalence of Consumer Surplus and Von Thunen Rent." *Economic Geography*, 56, No. 1 (1980): 73-88.

Dayal, Edison. "Agricultural Location Theory: A Review." Pp. 13-32 in *Perspectives in Agricultural Geography*, vol. 4, edited by Noor Mohammad. New Delhi, India: Concept Publishing Company, 1981.

Huff, James O. "Rich Man–Poor Man in Von Thunen's Isolated State." *Economic Geography*, 57, No. 2 (1981): 127-133.

Cromley, Robert G. "The Von Thunen Model and Environmental Uncertainty." *Annals of the Association of American Geographers*, 72, No. 3 (1982): 404-410.

Visser, Sent. "On Agricultural Location Theory." *Geographical Analysis*, 14, No. 2 (1982): 167-176.

Kellerman, A. "Economic and Spatial Aspects of Von Thunen's Factor Intensity Theory." *Environment and Planning A*, 15, No. 11 (1983): 1521-1530.

Barnes, Trevor. "Theories of Agricultural Rent Within the Surplus Approach." *International Journal of Regional Science*, 9, No. 2 (1984): 125-140.

Portugali, J. "On Relevance in Geography: Thunen's Isolated State in Relation to Agriculture and Political Economy." *Geoforum*, 15, No. 2 (1984): 201-208.

Jones, Donald W., and J.R. Frummell. "The Location Theory of the Plantation." *Journal of Regional Science*, 27 (1987): 157-182.

Wilson, A.G., and M. Birkin. "Dynamic Models of Agricultural Location in a Spatial Interaction Framework." *Geographical Analysis*, 19, No. 1 (1987): 31-58.

O'Kelly, Morton E. "Aggregate Rent and Surplus Measurement in a Von Thunen Model." *Geographical Analysis*, 20, No. 3 (1988): 187-197.

Cromley, Robert G., and D.M. Hanink. "A Financial-Economic Von Thunen Model." *Environment and Planning A*, 21, No. 6 (1989): 951-960.

Kellerman, A. "Agricultural Location Theory. 1: Basic Models." *Environment and Planning A*, 21, No. 10 (1989): 1381-1396.

Kellerman, A. "Agricultural Location Theory. 2: Relaxation of Assumptions and Applications." *Environment and Planning A*, 21, No. 11 (1989): 1427-1446.

Jones, Donald W. "An Introduction to the Thunen Location and Land Use Model." *Research in Marketing*, 5 (1991): 35-70.

Macmillan, W.D. "Risk and Agricultural Land Use: A Reformation of the Portfolio Theoretic Approach to the Analysis of a Von Thunen Economy." *Geographical Analysis*, 24, No. 2 (1992): 142-158.

Chendrayudu, N., and Y.V. Ramanaiah. "Agricultural Typology: A Theoretical Perspective." *Transactions, Institute of Indian Geographers*, 16, No. 2 (1994): 115-122.

O'Kelly, Morton E., and Deborah Bryan. "Agricultural Location Theory: Von Thunen's Contribution to Economic Geography." *Progress in Human Geography*, 20, No. 4 (1996): 457-475.

Sang, Y.L. "Agricultural Location Theory and Behaviors in Marked Based Von Thunen's Model." Ph.D. Dissertation, University of Arizona, Tucson, Ariz., 1996.

Part 2

NORTH AMERICA

North America

GENERAL TEXTS, ARTICLES, AND THESES

Baker, O.E. "Agricultural Regions of North America." *Economic Geography*: Part I, 2 (1926): 459-493; Part II, 3 (1927): 50-86; Part III, 3 (1927): 309-339; Part IV, 3 (1927): 447-465; Part V, 5 (1928): 44-73; Part VI, 4 (1928): 399-433; Part VII, 5 (1929): 36-69; Part VIII, 6 (1930): 166-190; Part IX, 6 (1930): 278-308; Part X, 7 (1931): 109-153; Part XI, 7 (1931): 325-364; Part XII, 8 (1932): 325-377; Part XIII, 9 (1933): 167-197.

Sauer, Carl O. "American Agricultural Origins: A Consideration of Nature and Culture." Pp. 278-297 in *Essays in Anthropology Presented to A.L. Kroeber in Celebration of His Sixtieth Birthday, June 11, 1936*. Berkeley, Calif.: University of California Press, 1936.

Moodie, D. Wayne, and Barry Kaye. "The Northern Limit of Indian Agriculture in North America." *Geographical Review*, 59, No. 4 (1969): 513-529.

Alford, John J. "Extinction as a Possible Factor in the Invention of New World Agriculture." *Professional Geographer*, 22, No. 2 (1970): 120-123.

Reitsma, Hendrik J. "Crop and Livestock Production in the Vicinity of the United States–Canada Border." *Professional Geographer*, 23, No. 3 (1971): 216-223.

DeLisle, David. "Effects of Distance Internal to the Farm: A Challenging Subject for North American Geographers." *Professional Geographer*, 30, No. 3 (1978): 278-288.

Bryant, Christopher R. "Agriculture in the Urban Fringe: A Systems Perspective." *Rural Systems*, 2, No. 1 (1984): 15-26.

Troughton, Michael J. "Industrialization of U.S. and Canadian Agriculture." *Journal of Geography*, 84, No. 6 (1985): 255-263.

Wallace, Iain, and W. Smith. "Agribusiness in North America." Pp. 57-74 in *The Industrialisation of the Countryside*, edited by Michael J. Healey and Brian W. Ilbery. Norwich, England: GeoBooks, 1985.

Pierce, John T., and Owen J. Furuseth. "Constraints to Expanded Food Production: A North American Perspective." *Natural Resources Journal*, 26, No. 1 (1986): 15-39.

Doolittle, William E. "Agriculture in North America on the Eve of Contact: A Reassessment." *Annals of the Association of American Geographers*, 82, No. 3 (1992): 386-401.

Vest, Jay H.C. "The Wild and the Tame: Understanding Wilderness and Agriculture in Native America." *National Geographical Journal of India*, 39 (1993): 215-229.

CROPS AND CROPPING PATTERNS

Gumprecht, Thaddaus E. "Der Weinbau in den Vereinigten Staaten von Nord-Amerika." *Petermann's Geographische Mitteilungen*, 2, (1856): 222-229.

Brigham, Albert P. "The Development of Wheat Culture in North America." *Geographical Journal*, 35 (1910): 42-55.

Hammett, John E. "Wheat Production, Transportation and Marketing for Anglo-America." Master's Thesis, University of Chicago, Chicago, Ill., 1958.

Thorp, Thomas E. "Crop Protection in Native North America." Master's Thesis, University of California, Berkeley, Calif., 1959.

Seig, Louis. *Tobacco, Peacepipes and Indians.* Palmer Lake, Colo.: Filler Press, 1971.

Jackson, Richard H., et al. "The Historical Diffusion of Alfalfa." *Journal of Agronomic Education*, 6 (1977): 13-19.

ANIMAL HUSBANDRY

McKnight, Tom L. *Feral Livestock in Anglo-America.* Berkeley, Calif.: University of California Publications in Geography, No. 16, 1964.

Jordan, Terry G. *North American Cattle-Ranching Frontiers: Origins, Diffusion and Differentiation.* Albuquerque, N.Mex.: University of New Mexico Press, 1993.

Walton, J.R. "Pedigree and Productivity in the British and North American Cattle Kingdoms Before 1930." *Journal of Historical Geography*, 25, No. 4 (1999): 441-462.

FARM LAND

Donkin, R.A. *Agricultural Terracing in the Aboriginal New World.* Tucson, Ariz.: University of Arizona Press, Viking Fund Publications in Anthropology, No. 56, 1979.

Bryant, Christopher R., and Lorne H. Russwurm. "North American Farmland Protection Strategies in Retrospect." *GeoJournal*, 6, No. 6 (1982): 501-512.

Furuseth, Owen J., and John T. Pierce. "A Comparative Analysis of Farmland Preservation Programs in North America." *Canadian Geographer*, 26, No. 3 (1982): 191-206.

Mage, Julius A., Heather A. Clemson, and Grant Lee. *Foreign Ownership of Farmland in Canada and the United States: Legislative Controls, Literature Review and Annotated Bibliography.* Guelph, Ontario: University of Guelph School of Rural Planning and Development, 1983.

Pierce, John T., and Owen J. Furuseth. "Assessing the Adequacy of North American Agricultural Land Resources." *Geoforum*, 14, No. 4 (1983): 413-426.

Bunce, Michael F. "Agricultural Land as a Real Estate Commodity: Implications for Farmland Preservation in the North American Urban Fringe." *Landscape Planning*, 12 (1985): 177-192.

Beazley, Susan L., and Kenneth B. Beesley. "Preserving Agricultural Land in New England and the Maritimes: A Comparative Analysis." *Proceedings, New England–St. Lawrence Valley Division, Association of American Geographers*, 25 (1995): 15-27.

Bunce, M. "Thirty Years of Farmland Preservation in North America: Disclosures and Ideologies of a Movement." *Journal of Rural Studies*, 14, No. 2 (1998): 233-247.

Doolittle, William E. *Cultivated Landscapes of Native North America*. New York, N.Y.: Oxford University Press, 2000.

AGRICULTURAL POPULATIONS AND SOCIAL ISSUES

Alanen, Arnold, and William Tishler. "Finnish Farmstead Organization in Old and New World Settings." *Journal of Cultural Geography*, 1, No. 1 (1980): 68-81.

ENVIRONMENTAL AND CONSERVATION ISSUES

Unstead, J.F. "The Climatic Limit of Wheat Cultivation with Special Reference to North America." *Geographical Journal*, 39 (1912): 347-366; 421-446.

Spath, Hans-Joachim. *The Agro-Ecological Dry Boundary of Non-Irrigated Agriculture in the Central Great Plains of North America*. Wiesbaden, Germany: Franz Steiner Verlag, Erdwissenschaftliche Forschung, Band 15, 1980.

Michaels, P.J. "Atmospheric Pressure Patterns, Climatic Change and Winter Wheat Yields in North America." *Geoforum*, 13, No. 3 (1982): 263-274.

AGRICULTURAL MARKETING AND DISTRIBUTION

Stephen, L. LeMar. "Agricultural Surpluses in Relation to Trade Between the United States and Canada." *Journal of Geography*, 41, No. 3 (1942): 95-100.

The United States as a Whole

ATLASES AND GRAPHIC PRESENTATIONS

Stire, O.C., and O.E. Baker. *Atlas of American Agriculture, Part V: The Crops, Cotton*. Washington, D.C.: United States Department of Agriculture, 1918.

Goldenweiser, E.A. *Atlas of American Agriculture, Part IX: Rural Population and Organization*. Washington, D.C.: United States Department of Agriculture, 1919.

Marbut, C.F. *Atlas of American Agriculture, Part III: Soils of the United States*. Washington, D.C.: Government Printing Office, 1935.

Gregor, Howard F. *Industrialization of U.S. Agriculture: An Interpretive Atlas*. Boulder, Colo.: Westview, 1982.

Pillsbury, Richard, and John Florin. *Atlas of American Agriculture: The American Cornucopia*. New York, N.Y.: Prentice-Hall, 1996.

GENERAL TEXTS, ARTICLES, AND THESES

Stewart, Robert. "The Farm Problem." *Economic Geography*, 5 (1929): 358-358.

Myers, W.I. "Current Aspects of American Agriculture." *Economic Geography*, 19 (1943): 372-379.

Carter, George F. "Origins of American Indian Agriculture." *American Anthropologist*, 48, No. 1 (1946): 1-21.

Gibson, Lyle E. "American Corn and Dairy Belts: A Study in Regional Margins." Ph.D. Dissertation, University of Iowa, Iowa City, Iowa, 1947.

Gibson, Lyle E. "Characteristics of a Regional Margin of the Corn and Dairy Belts." *Annals of the Association of American Geographers*, 38 (1948): 244-270.

Trewartha, Glenn T. "Some Regional Characteristics of American Farmsteads." *Annals of the Association of American Geographers*, 38, No. 2 (1948): 169-225.

Bartz, Fritz. "Strukturwandlungen in der US–Amerikanischen Landwirtschaft." *Erdkunde*, 6 (1952): 132-145.

Doerr, Arthur H. "A Conservative Approximation of Potential Agricultural Production of Some Sample Commodities Along Railroad Rights-of-Way in the United States." *Professional Geographer*, 6 (1954): 17-18.

Mather, Eugene C., and John Fraser Hart. "Fences and Farms." *Geographical Review*, 44, No. 2 (1954): 301-323.

Durand, Loyal. "The American Centralizer Belt." *Economic Geography*, 31 (1955): 301-320.

Haystead, Ladd, and Gilbert C. Fite. *The Agricultural Regions of the United States*. Norman, Okla.: University of Oklahoma Press, 1955.

Mather, Eugene C., and John Fraser Hart. "The Geography of Manure." *Land Economics*, 32, No. 1 (1956): 25-38.

Weaver, John C. "The County as a Spatial Average in Agricultural Geography." *Geographical Review*, 46, No. 4 (1956): 536-565.

Harris, Chauncy D. "Agricultural Production in the United States: The Past Fifty and the Next." *Geographical Review*, 47, No. 2 (1957): 174-193.

Hart, John Fraser. *America's Farms and Ranches*. Garden City, N.Y.: Nelson Doubleday, 1957.

Higbee, Edward C. *American Agriculture: Geography, Resources, Conservation*. New York, N.Y.: John Wiley, 1958.

Gregor, Howard F. "Agricultural Intensity and Its Regionalization." *Yearbook, Association of Pacific Coast Geographers*, 22 (1960): 7-27.

Hart, John Fraser, and Eugene C. Mather. "The Character of Tobacco Barns and Their Role in the Tobacco Economy of the United States." *Annals of the Association of American Geographers*, 51 (1961): 274-293.

Burton, Ian. "Types of Agricultural Occupance of Flood Plains in the United States." Ph.D Dissertation, University of Chicago, Chicago, Ill., 1962.

Higbee, Edward C. *Farms and Farmers in an Urban Age*. New York, N.Y.: The Twentieth Century Fund, 1963.

Kovalevskiy, V.P. "Some Trends in Changes of the Structure and Location of United States Agriculture." *Soviet Geography*, 4, No. 3 (1963): 30-38.

Hart, John Fraser. "Indices of Agricultural Underdevelopment." *Proceedings of the Indiana Academy of the Social Sciences* (1964): 17-21.

Olmstead, Clarence W., and Vaughan P. Manley. "The Geography of Input, Output, and Scale of Operation in American Agriculture." Pp. 37-47 in *Agricultural Geography I.G.U. Symposium*, edited by E.S. Simpson. Liverpool, England: International Geographical Union and University of Liverpool, Department of Geography, Research Paper No. 3, 1965.

Buchholz, H.E., and G.G. Judge. "Ein Standortmodell der tierischen Produktion der Verstaigten Staaten von Amerika." *Berischte uber Landwirtschaft*, 2 (1966): 392-431.

Paterson, J.H. "American Agriculture and the Decline of the Agrarian Ideal." Pp. 344-351 in *Melanges de geographie physique, humaine, economique, applique, offerts a M. Omer Tulippe,* vol. 1, edited by Jose A. Sporck. Gembloux, Belgium: J. Duculot, 1967.

Gregor, Howard F. "Farm Structure in Regional Comparison: California and New Jersey Vegetable Farms." *Economic Geography*, 45, No. 3 (1969): 209-225.

Peet, J. Richard. "The Spatial Expansion of Commercial Agriculture in the 19th Century: A Von Thunen Interpretation." *Economic Geography*, 45, No. 3 (1969): 283-301.

Burghardt, Andrew F. "The Relative Productivity of Differing Types of American Agriculture." *Journal of Geography*, 69, No. 4 (1970): 225-231.

Enyedi, Gyorgi. *Farms and Farmers: The American Agriculture* (in Hungarian). Budapest, Hungary: Mexogazdasayi Kiado, 1970.

Gregor, Howard F. "The Large Industrialized American Crop Farm: A Mid-Latitude Plantation Variant." *Geographical Review*, 60, No. 2 (1970): 151-175.

Muller, Peter O. "Thunian Analysis at the Macro-Scale: A Trend-Surface Analysis of the Spatial Organization of American Agriculture." Ph.D. Dissertation, Rutgers University, New Brunswick, N.J., 1970.

Clark, Andrew H. "Suggestions for the Geographical Study of Agricultural Change in the United States, 1790–1840." *Agricultural History*, 46, No, 1 (1972): 155-172.

Muller, Peter O. "Trend Surfaces of American Agricultural Patterns: A Macro-Thunian Approach." *Economic Geography*, 49 (1973): 228-242.

Frederic, Paul B. "Geography and Living Historical Farm Sites." *Agricultural History*, 48, No. 1 (1974): 5-10.

Aiken, Charles S. "Expressions of Agrarianism in American Geography: The Cases of Isaiah Bowman, J. Russell Smith, and O.E. Baker." *Professional Geographer*, 27, No. 1 (1975): 19-20.

Dorel, Gerard. "La penetration du grand capitalisme dans l'agriculture des Etats-Unis. Le corporate farming." *Travaux de l'Institut de Geographie de Reims*, 21-22 (1975): 47-71.

Raitz, Karl. "The Impact of Federal Agricultural Policy on Rural America." *Antipode*, 7, No. 3 (1975): 3-10.

Windhorst, Hans-Wilhelm. *Die landwirtschaft der Vereinigten Staaten: Strukturelle und regionale Dynamik.* Wiesbaden, Germany: Steiner Verlag, 1975.

Chominot, Albert. "Quelques tendances nouvelles de l'agriculture des U.S.A." *Annales de Geographie*, 85, No. 1 (1976): 28-47.

Vogel, Philip E. "American Agriculture: Some Factors Influencing Decisions." *Geographical Perspectives*, 41 (1978): 25-35.

Windhorst, Hans-Wilhelm. "Struktur und raumliche Ordnung der US–amerikanischen Agrarwirtschaftgegen Ende der siebzier Jahr." *Berichte uber Landwirtschaft*, 57, No. 2 (1979): 302-332.

Dorel, Gerard. "La grande agriculture aux Etats Unis." These de doctorat d'etat, Universite de Paris, Paris, France, 1980.

Smith, Everett G. "America's Richest Farms and Ranches." *Annals of the Association of American Geographers*, 70, No. 4 (1980): 528-541.

Winsberg, Morton D. "Concentration and Specialization in United States Agriculture." *Economic Geography*, 56, No. 2 (1980): 183-189.

Gregor, Howard F. "Agricultural Capitalization as a Spatial Complex." Pp. 407-426 in *Perspectives in Agricultural Geography*, vol. 2, edited by Noor Mohammad. New Delhi, India: Concept Publishing Company, 1981.

Vogeler, Ingolf. *The Myth of the Family Farm: Agribusiness Dominance of U.S. Agriculture.* Boulder, Colo.: Westview, 1981.

Winsberg, Morton D. "Agricultural Specialization in the United States Since World War II." *Agricultural History*, 56, No. 4 (1982): 692-701.

Napton, Darrell. "Incentives to the Landowner: Unexpected Results of Farm Policy." *Geographical Perspectives*, 55 (1985): 65-68.

Daniels, T.L. "Hobby Farming in America: Rural Development or Threat to Commercial Agriculture?" *Journal of Rural Studies*, 2, No. 1 (1986): 31-40.

Windhorst, Hans-Wilhelm. "Die fianzielle Situation der amerikanischen Farmer in der Mitte der achtzinger Jahre." *Zeitschrift fur Agrargeographie*, 4, No. 2 (1986): 97-120.

Lemon, James T. "Agriculture and Society in Early America." *Agricultural History Review*, 35, No. 1 (1987): 76-94.

Windhorst, Hans-Wilhelm. "Die Rolle der Grosformen in der US–amerikanischen Agrarwirtschaft." *Geographische Zeitschrift*, 75, No. 4 (1987): 228-244.

Windhorst, Hans-Wilhelm. "Die US–amerikanischen Agrarwirtschaft auf dem Wege zu einer dualen Struktur." *Zeitschrift fur Agrargeographie*, 5, No. 4 (1987): 283-335.

Windhorst, Hans-Wilhelm. "Ein deutscher Agrarunternchmer investiert in den USA." *Praxis Geographie*, 17, No. 7-8 (1987): 6-12.

Windhorst, Hans-Wilhelm. "Kapitalinvestitionen Sudoldenburger Agrarunternchment in den USA und deren Ruckwirkungen auf die soziolkonomische Struktur ihrer Standraume." *Erdkunde*, 41, No. 1 (1987): 14-29.

Windhorst, Hans-Wilhelm. "Konzentrationprozesse in der US–amerikanischen Agrarwirtschaft und ihre raumlichen Konsequenzen." *Geographische Rundschau*, 39, No. 9 (1987): 474-480.

Lawrence, Henry W. "Changes in Agricultural Production in Metropolitan Areas." *Professional Geographer*, 40, No. 2 (1988): 159-174.

Windhorst, Hans-Wilhelm. "Megafarms: Grossformen und agrarindustrielle Unternehmen in der USA." *Berichte uber Landwirtschaft*, 62, No. 3 (1988): 490-507.

Windhorst, Hans-Wilhelm. "The Role of Large-Scale Farms in U.S. Agriculture." *Mitteilungsblatt des Arbeitskreises USA im Zentraverband der Deutschen Geographen*. Vechta, Germany: Vechtaer Arbeiten zur Geographie und Regionalwissenschaft, Heft 8, 1988: 41-62.

Gregor, Howard F. "The Regionalization of Rationalization in U.S. Agricultural Production." *Geographia Polonica*, 57 (1989): 69-80.

Lawrence, Henry W. "Intensification of Small-Area Farms in U.S. Agriculture." *East Lakes Geographer*, 24 (1989): 63-82.

Windhorst, Hans-Wilhelm. "Industrialization of Agricultural Production and the Role of Large-Scale Farms in United States Agriculture." *Tijdschrift voor Economische en Sociale Geographie*, 80, No. 5 (1989): 270-283.

Fitzsimmons, Margaret I. "The Social and Environmental Relations of U.S. Agricultural Regions." Pp. 8-32 in *Technological Change and the Rural Environment*, edited by Philip D. Lowe, Terry K. Marsden, and Sarah Whatmore. London, England: Fulton, 1990.

Windhorst, Hans-Wilhelm. "Agrarindustrielle Unternehmungen: Entstehungsbedingungen, Strukture, raumliche Organisation." *Vechtaer Arbeiten zur Geographie und Regionalwissenschaft*, 8 (1989): 89-104.

Klohn, Werner. "Landerbericht: Die Farmer-Genossenschaften in den USA." *Zeitschrift fur das gesamte Genossenschaftswesen*, 41, No. 1 (1991): 48-61.

McQuillan, D. Aidan. "Migration and Adaptation: Developing New Farming Strategies on the American Grasslands." Pp. 111-131 in *Migrants in Agricultural Development*, edited by J.A. Mollet. London, England: Macmillan, 1991.

Curry-Roper, Janel M. "Alternative Agriculture and Conventional Paradigms in U.S. Agriculture." Pp. 254-264 in *Contemporary Rural Systems in Transition: Vol. 1, Agriculture and Environment*, edited by Ian R. Bowler, Christopher R. Bryant, and M. Duane Nellis. Wallingsford, England: C.A.B. International, 1992.

Hart, John Fraser. "Nonfarm Farms." *Geographical Review*, 82, No. 2 (1992): 166-179.

Napton, Darrell. "Farm Diversification in the United States." Pp. 87-99 in *Contemporary Rural Systems in Transition: Vol. 1, Agriculture and Environment*, edited by Ian R. Bowler, Christopher R. Bryant, and M. Duane Nellis. Wallingford, England: C.A.B. International, 1992.

Rickard, Timothy. "Agricultural Policies for Urban Fringes in the United States." Pp. 166-181 in *Contemporary Rural Systems in Transition: Vol. 1,*

Agriculture and Environment, edited by Ian R. Bowler, Christopher R. Bryant, and M. Duane Nellis. Wallingford, England: C.A.B. International, 1992.

Mims, Patricia A. "Agricultural Change in the Urban-Rural Fringe: A Test of the Permetropolitan Bow Wave Model." Master's Thesis, Virginia Polytechnic and State University, Blacksburg, Va., 1994.

Windhorst, Hans-Wilhelm. "Die Agrarwirtschaft der USA zu Beginn der 70er Jahre." *Der Erdkundeunterricht*, 46, No. 7-8 (1994): 271-282.

Klohn, Werner, and Hans-Wilhelm Windhorst. *Die Landwirtschaft der USA*. Vechta, Germany: Vechtaer Materialien zum Geographieunterricht, Heft 1, 1995.

Winsberg, Morton D. "Geographical Change in the Distribution of the Nation's Agricultural Production Between 1929 and 1992." *Agricultural History*, 70, No. 3 (1996): 525-536.

Duram, Leslie A., and Kelli L. Larson. "Agricultural Research and Alternative Farmers' Information Needs." *Professional Geographer*, 53, No. 1 (2001): 84-96.

CROPS AND CROPPING PATTERNS

Anderson, Esther S. "Geography of the Beet Sugar Industry." Master's Thesis, University of Nebraska, Lincoln, Nebr., 1917.

Clark, Ivan S. "Geographic Aspects of the Tobacco Industry: An Economic Study." Master's Thesis, University of North Carolina, Chapel Hill, N.C., 1926.

Klages, Karl H.W. "Geographical Distribution of Variability in the Yield of Field Crops in the States of the Mississippi Valley." *Ecology*, 11 (1930): 293-306.

Landon, Charles E. "The Westward Movement of Cotton Growing in the United States." *Journal of Geography*, 29, No. 5 (1930): 215-219.

Reed, William G. "Competing Cottons and United States Production." *Economic Geography*, 8, No. 3 (1932): 282-298.

Primmer, George H. "United States Soybean Industry." *Economic Geography*, 15 (1939): 205-221.

Deasy, George F. "Geography of the United States Flaxseed Industry." *Journal of Geography*, 39, No. 9 (1940): 226-231.

Deasy, George F. "Geography of the United States Cottonseed Oil Industry." *Economic Geography,* 17, No. 4 (1941): 345-352.

Deasy, George F. "Geography of the United States Soybean-Oil Industry." *Journal of Geography*, 40, No. 1 (1941): 1-7.

Primmer, George H. "The United States Flax Industry." *Economic Geography*, 17, No. 1 (1941): 24-30.

Weaver, John C. "American Barley Production: A Study in Agricultural Geography." Ph.D. Dissertation, University of Wisconsin, Madison, Wis., 1942.

Meigs, Peveril. "Exploring American Orchards." *Yearbook, Association of Pacific Coast Geographers*, 6 (1943): 3-5.

Weaver, John C. "Barley in the United States." *Geographical Review*, 33, No. 1 (1943): 56-73.

Parsons, Ruben L. "Geographic Factors in American Tung Culture." Ph.D. Dissertation, Clark University, Worcester, Mass., 1944.

Weaver, John C. "United States Malting Barley Production." *Annals of the Association of American Geographers*, 34 (1944): 97-131.

Garland, John H. "Hemp: A Minor American Fiber Crop." *Economic Geography*, 22, No. 2 (1946): 126-132.

Allen, Marcia J. "Some Geographic Aspects of the Mint Industry." Master's Thesis, Syracuse University, Syracuse, N.Y., 1947.

Carter, George F. "Sweet Corn Among the Indians." *Geographical Review*, 38, No. 2 (1948): 206-221.

Fellman, Jerome D. "A Comparative Study of the Major Commercial Apple Producing Areas of the United States." Master's Thesis, University of Chicago, Chicago, Ill., 1948.

Munn, Alvin A. "Production and Utilization of the Soybean in the United States." *Economic Geography*, 26, No. 3 (1950): 223-234.

Weaver, John C. *American Barley Production: A Study in Agricultural Geography*. Minneapolis, Minn.: Burgess, 1950.

Klalaf, Lamon F.S. "The Palm Date Industry of the United States." Master's Thesis, University of Chicago, Chicago, Ill., 1952.

Naqavi, S.H.H. "The Peanut Crop in the United States." Master's Thesis, University of North Carolina, Chapel Hill, N.C., 1954.

Grotewold, Andreas. *Regional Changes in Crop Production in the United States from 1890 to 1949*. Chicago, Ill.: University of Chicago, Department of Geography, Research Paper No. 40, 1955.

Olmstead, Clarence W. "American Orchard and Vineyard Regions." *Economic Geography*, 32, No. 3 (1956): 189-236.

Warntz, William. "An Historical Consideration of the Terms 'Corn' and 'Corn Belt' in the United States." *Agricultural History*, 31 (1957): 40-45.

Assiff, Lee. "Regional Changes in Oat Distribution in the United States, 1850–1954." Master's Thesis, Illinois State University, Bloomington, Ill., 1960.

Large, David C. "Hybrid Corn Planting in the United States." *Geography*, 45 (1960): 113-114.

McLure, John A. "A Regional Study of United States Peanut Production." Master's Thesis, George Washington University, Washington, D.C., 1960.

Schnell, George A. "A Geographical Analysis of Factors Affecting Distribution of Unirrigated Cotton Cultivation in the United States." Master's Thesis, Pennsylvania State University, University Park, Pa., 1960.

Landing, James E. "The Cultivation of Peppermint and Spearmint." *Southeastern Geographer*, 3 (1963): 28-33.

Suurnakki, Jorma. "Resources for and Development of Potato Production in the United States of America" (in Finnish). *Terra*, 77, No. 4 (1965): 150-160.

Bucksar, Richard G. "Some Recent Changes in the American Mushroom Industry." *Journal of Geography*, 65, No. 3 (1966): 114-124.

Garanson, Leonard D. "An Analysis of the Location and the Spatial Arrangements of the United States Beet Sugar Industry." Ph.D. Dissertation, University of Iowa, Iowa City, Iowa, 1966.

Durrenberger, Robert. "Evolution of the American Lemon-Growing Industry." *California Geographer*, 8 (1967): 103-113.

Cross, Ralph D. "The Measurement and Spatial Variation of Modal Wheat Concentrations in the American Hard Winter Wheat Belt in 1964." Ph.D. Dissertation, Michigan State University, East Lansing, Mich., 1968.

Earney, Fillmore C. "Mushrooms and Mines: A Study in Horticulture." *Journal of Geography*, 67, No. 1 (1968): 42-48.

Landing, James E. "Peppermint and Spearmint in the United States." *Journal of Geography*, 67, No. 9 (1968): 548-553.

Scott, Steven L. "The Spatial Distribution of Dry Edible Bean Production in the United States, 1879–1959." Master's Thesis, Western Illinois University, Macomb, Ill., 1968.

Jumper, Sidney R. "The Fresh Vegetable Industry in the U.S.A.: An Example of Dynamic Interregional Dependency." *Tijdschrift voor Economische en Sociale Geographie*, 60 (1969): 308-318.

Kolbrak, Lawrence J. "Geographical Survey of the Cranberry Industry in the United States." Master's Thesis, Northern Illinois University, DeKalb, Ill., 1969.

Landing, James E. *American Essence: A History of the Peppermint and Spearmint Industry in the United States*. Kalamazoo, Mich.: Michigan Public Museum, 1969.

Lee, Jonathan J. "The Demand of United States Rice: An Economic-Geographic Analysis." Ph.D. Dissertation, University of Washington, Seattle, Wash., 1970.

Vatter, Arnold. "Neuerungen bei der Raumwollproduktion in U.S.A." *Zeitschrift fur Wirtschaftsgeographie*, 14, No. 6 (1970): 180-183.

Aiken, Charles S. "An Examination of the Role of the Eli Whitney Cotton Gin in the Origin of the United States Cotton Regions." *Proceedings of the Association of American Geographers*, 3 (1971): 5-9.

Hadley, Edgar M. "Hops in the United States." *Geographical Bulletin*, 2 (1971): 37-51.

Desire, E.P. "La culture du maisaux Etats-Unis." *L'Information Geographique*, 36, No. 5 (1972): 220-232.

Sawvell, Robert D. "The Influence of Government Policy on the Amount of Land Devoted to Sugar Production in the Continental United States, 1890 to the Present." Ph.D. Dissertation, University of Oklahoma, Norman, Okla., 1973.

Siniard, L. Arnold. "Certain Spatial Aspects of American Soybean Culture." *Southeastern Geographer*, 13 (1973): 1-11.

Rowe, James E. "A Delimitation of the Optimum Peanut Producing Areas in the United States." *Virginia Geographer*, 10, No. 1 (1975): 3-12.

Bailey, William H. "Cotton Allotment Transfer: An Indicator of Location Changes in Free Market Cotton Production in the United States." Ph.D. Dissertation, University of Tennessee, Knoxville, Tenn., 1976.

Hoff, Lawrence G. "The Domestication of Wild Rice as a Commercial Crop." Master's Thesis, Mankato State University, Mankato, Minn., 1977.

Bailey, William H. "Government Policy and the Location of Cotton Production." *Southeastern Geographer*, 18, 2 (1978): 93-101.

Sander, Hans-Jorg. "Reisanbau und Reisforschung in den USA." *Zeitschrift fur Wirtschaftsgeographie*, 23, No. 6 (1979): 176-183.

Bassett, Thomas J. "Reaping on the Margins: A Century of Community Gardening in American." *Landscape*, 25, No. 2 (1981): 1-8.

DeLisle, David. "Effects of Distance on Cropping Patterns Internal to the Farm." *Annals of the Association of American Geographers*, 72, No. 1 (1982): 88-98.

Goodenough, Richard. "The Great American Crop Surplus: 1983 Solution." *Geography*, 69, No. 4 (1984): 351-353.

Fitzsimmons, Margaret. "The New Industrial Agriculture: The Regional Integration of Specialty Crop Production." *Economic Geography*, 62, No. 4 (1986): 334-353.

Konnecke, Michael. "Baum Kulturen in den winterfeuchlen Subtropen der Neuen Welt." *Die Erde*, 118 (1987): 5-20.

Klohn, Werner. *Die Farmer-Genossenschaften in der USA: Eine agrargeographische Untersuchung.*: Vechta, Germany: Vechtaer Arbeiten zur Geographie und Regionalwissenschaft, Heft 9, 1990.

Bailey, William G., K.B. Von Dalfsen, and Y. Guo. "The Role of Ginseng Drying in the Harvest and Post-Harvest Production System for American Ginseng." Pp. 155-163 in *Proceedings of the Sixth International Ginseng Symposium.* Seoul, South Korea: Korea Ginseng and Tobacco Research Institute, 1993.

Baxevanis, John. *The Wine Regions of America: Geographical Reflections and Appraisals.* Stroudsburg, Pa.: Vinifera Wine Growers Journal and Association, 1993.

Holly, Brian P. "Organizational Structure of Wine Production in the U.S." *Journal of Wine Research*, 5, No. 2 (1994): 91-101.

MacDonald, Gerald M., and Denyse Lemaire. "American Viticultural Areas: A Problem in Regional Geography." *Journal of Geography*, 94, No. 2 (1995): 376-385.

Rumney, Thomas A. "The Birth and Spread of a New Crop: Canola in the United States Since the 1970s." *North American Geographer*, 2, No. 2 (2000): 80-88.

ANIMAL HUSBANDRY

Brayer, Herbert O. "British Cattle Ranches in America." *Geographical Magazine*, 20, No. 4 (1947): 125-131.

Fox, K.A. "A Spatial Equilibrium Model of Livestock-Feed Economy of the United States." *Econometrica*, 21 (1953): 547-566.

Stomper, Clifford M. "The Thoroughbred Breeding Industry." Master's Thesis, University of Tennessee, Knoxville, Tenn., 1953.

McKnight, Tom L. "The Ferral Burro in the United States: Distributions and Problems." *Journal of Wildlife Management*, 22, No. 2 (1958): 163-179.

Post, Lauren C. "Cattle Breeds: An Aspect of Regional Geography." *Yearbook, Association of Pacific Coast Geographers*, 21 (1959): 51-58.

Brinkman, Leonard. "The Historical Geography of Improved Cattle in the United States to 1870." Ph.D. Dissertation, University of Wisconsin, Madison, Wis., 1964.

Mulmat, Kenneth M. "Trends in the Egg Industry." Master's Thesis, Northern Illinois University, DeKalb, Ill., 1965.

Gomez-Ibanez, Daniel A. "The Rise and Decline of Transhumance in the United States." Master's Thesis, University of Wisconsin, Madison, Wis., 1967.

Bucksar, Richard G. "Significant Changes in the American Egg Industry." *Journal of Geography*, 67, No. 1 (1968): 36-41.

Vining, James W. "Domestic Rabbit Production in the United States." *Geographical Perspectives*, 40 (1970): 17-25.

Lord, J. Dennis. "The Growth and Localization of the United States Broiler Chicken Industry." *Southeastern Geographer*, 11 (1971): 29-42.

Jordan, Terry G. "The Origin and Distribution of Open Range Cattle Ranching." *Social Science Quarterly*, 53 (1972): 105-121.

Lord, J. Dennis. "Interregional Competition and Locational Advantage in the United States Broiler Industry: A Linear Programming Approach." *Professional Geographer*, 24, No. 4 (1972): 330-334.

Doughty, Robin. "Ostrich Farming American Style." *Agricultural History*, 47, No. 2 (1973): 133-145.

Leeman, Albert. "Rindfleisch in den U.S.A." *Zeitschrift fur Wirtschaftsgeographie*, 18, No. 1 (1974): 21-23.

Galiegue, A. "L'elevage aux Etats-Unis." *L'Information Geographique*, 47, No. 1 (1983): 4-11.

Windhorst, Hans-Wilhelm. "Structurwandlungen und raumliche Prozesse in der US–amerikanische Geflugelwirtschaft." *Geographische Zeitschrift*, 72, No. 2 (1984): 225-244.

Howard, Bruce F. "The Changing Role of Draft Animals in American Agriculture." Master's Thesis, University of Texas, Austin, Tex., 1986.

Charvet, J.P., and Gerard Dorel. "La filiere cereales-viande bovine aux Etats-Unis." *Revue de Geographie de Lyon*, 62 (1987): 363-386.

Windhorst, Hans-Wilhelm. "Die Struktur der amerikanischen Broilerproduktion und des Broilermarktes." *Deutsche Geflugelwirtschaft und Schweineproduktion*, 39, No. 8 (1987): 211-216.

Windhorst, Hans-Wilhelm. "Rockingham Poultry: Auflosung einer Farmergenossenschaft in den USA." *Deutsche Geflugelwirtschaft und Schweineproduktion*, 40, No. 28 (1988): 803-806.

Windhorst, Hans-Wilhelm. "Con Agra: Portrail eines Agrarkonzens." *Deutsche Geflugelwirtschaft und Schweineproduktion*, 41, No. 2 (1989): 41-45.

Furuseth, Owen J. "Sustainability Issues in the Industrialization of Hog Production in the United States." Pp. 293-312 in *Agricultural Restructuring and Sustainability: A Geographical Perspective*, edited by Brian Ilbery, Quentin Chiotti, and Timothy Rickard. Wallingford, England: C.A.B. International, 1997.

DAIRYING

Coulter, John W. "Contrasts in Dairying in Two Dissimilar Areas of the United States." *Geographical Review*, 17 (1927): 605-610.

Hartshorne, Richard. "A New Map of the Dairy Areas of the United States." *Economic Geography*, 11, No. 4 (1935): 347-355.

Durand, Loyal. "Italian Cheese Production in the American Dairy Region." *Economic Geography*, 24, No. 3 (1948): 217-230.

Durand, Loyal. "The American Dairy Region." *Journal of Geography*, 48, No. 1 (1949): 1-19.

Durand, Loyal. "The Migration of Cheese Manufacture in the United States." *Annals of the Association of American Geographers*, 42, No. 3 (1952): 263-282.

Durand, Loyal. "Historische Geographie und Westwanderung der Kase: Erzeugung im amerikanischen Milchwirtschaftgebiet." *Die Erde*, 85, No. 2 (1953): 123-135.

Gregor, Howard F. "Industrialized Drylot Dairying: An Overview." *Economic Geography*, 39, No. 4 (1963): 299-318.

Foster, Stephen. "Size and Shape of Functional Areas in the Dairy Industry." Master's Thesis, State University of New York at Buffalo, 1971.

Windhorst, Hans-Wilhelm. "Formen und regionale Verteilung der Putenproduktion in den USA." *Deutsche Geflugelwirtschaft und Schweineproduktion*, 27, No. 5 (1975): 101-102.

Dorel, Gerard. "La laitue aux Etats-Unis: Complementarite des espaces productifs et strategie des firmes de l'agri-business." *Annales de Geographie*, 87, No. 2 (1978): 129-151.

Mattingly, Paul F. "Disadoption of Milk Cows by Farmers: 1930–1978." *Professional Geographer*, 36, No. 1 (1984): 40-50.

Windhorst, Hans-Wilhelm. "Sektorale und regionale Wandlungsprozesse in der US-amerikanischen Milchwirtschaft." *Berichte uber Landwirtschaft*, 72 (1994): 641-656.

Pritchard, W. "Shifts in Food Regions, Regulation and Producer Cooperatives: Insights from the Australian and U.S. Dairy Industries." *Environment and Planning A*, 28, No. 5 (1996): 857-876.

FARM LAND

Hartshorne, Richard. "Agricultural Land in Proportion to Agricultural Populations in the United States." *Geographical Review*, 29, No. 4 (1939): 488-492.

Stewart, Charles L. "Farm Realty Changes in the Eastern and Western United States." *Economic Geography*, 19, No. 3 (1943): 196-205.

Hewes, Leslie. "Drained Land in the United States in the Light of the Drainage Census." *Professional Geographer*, 5 (1953): 6-12.

Linder, Peter J. "Intensity of Agricultural Land-Use in the United States." Master's Thesis, Syracuse University, Syracuse, N.Y., 1954.

Wooten, Hugh H., and James R. Anderson. *Agricultural Land Resources in the United States.* Washington, D.C.: United States Department of Agriculture, 1955.

Higbee, Edward C. *The American Oasis: The Land and Its Uses.* New York, N.Y.: Alfred A. Knopf, 1957.

Wooten, Hugh H., and James R. Anderson. *Major Uses of Land in the United States.* Washington, D.C.: United States Department of Agriculture, 1957.

Burton, Ian. *Types of Agricultural Occupance of Flood Plains in the United States.* Chicago, Ill.: University of Chicago, Department of Geography, Research Paper No. 75, 1962.

Higbee, Edward. "Agricultural Land on the Urban Fringe." Pp. 57-66 in *Metropolis on the Move: Geographers Look at Urban Sprawl*, edited by Jean Gottman and Robert Harper. New York, N.Y.: John Wiley, 1967.

Fisher, James S. "Federal Crop Allotment Programs and Responses by Individual Farm Operators." *Southeastern Geographer*, 10 (1970): 47-58.

Fravega, Elizabeth M. "Abandonment of Farmland: Problem or Polemic?" *Southeastern Geographer*, 10 (1970): 55-62.

Mattingly, Paul F. "Intensity of Agricultural Land Use Near Cities: A Case Study." *Professional Geographer*, 24, No. 1 (1972): 7-10.

Platt, Rutherford H. "The Loss of Farmland: Evolution of Public Response." *Geographical Review*, 67, No. 1 (1977): 93-101.

Fishel, William A. "The Urbanization of Agricultural Land: A Review of the National Agricultural Lands Study." *Land Economics*, 58, No. 2 (1982): 236-259.

Fletcher, W.W., and Charles Little. *The American Cropland Crisis: Why U.S. Farmland Is Being Lost and How Citizens and Governments Are Trying to Save What Is Left.* Bethesda, Md.: American Land Forum, 1982.

Furuseth, Owen J. "The Impact of Local Infrastructural and Development Policies on Agricultural Land Alienation." *Landscape Planning*, 9, No. 3-4 (1983): 209-226.

Furuseth, Owen J. "A Model for Rural Land Conversion." *Rural Systems*, 1, No. 3 (1983): 177-188.

Hart, John Fraser. "Cropland Change in the United States, 1944–78." Pp. 224-249 in *The Resourceful Earth: A Response to Global 2000*, edited by Julian L. Simon and Herman Kahn. New York, N.Y.: Basil Blackwell, 1984.

Redfield, Sarah E. *Vanishing Farmland: A Legal Solution for the States.* Lexington, Mass.: Lexington Books, 1984.

Stopp, G. Harry. "The Destruction of American Agricultural Land." *Geography*, 69, No. 1 (1984): 64-65.

Lewis, Michael E., and Rebecca Roberts. "Would an Effective Cropland Retirement Program Threaten Rural Communities? Some Preliminary Results." *Papers and Proceedings of the Applied Geography Conferences*, 8 (1985): 30-41.

Platt, Rutherford H. "The Farmland Conversion Debate: NALS and Beyond." *Professional Geographer*, 37, No. 4 (1985): 433-442.

Volkman, Nancy J. "Vanishing Lands in the USA: The Use of Agricultural Districts as a Method to Preserve Farm Land." *Land Use Policy*, 4, No. 1 (1987): 14-30.

Steiner, F. "The Evolution of Federal Agricultural Land Use Policy in the United States." *Journal of Rural Studies*, 4, No. 4 (1988): 349-364.

Kawalek, Charles E. "Data Quality and Farmland Assessment." Master's Thesis, Western Illinois University, Macomb, Ill., 1994.

Greene, Richard P., and John M. Harlin. "Threat to High Market Value Agricultural Lands from Urban Encroachment: A National and Regional Perspective." *Social Science Journal*, 32, No. 2 (1995): 137-156.

AGRICULTURAL POPULATIONS AND SOCIAL ISSUES

Baker, O.E. "Population, Food Supply and American Agriculture." *Geographical Review*, 18 (1928): 353-373.

Amedeo, Douglas M. "Analysis of the Distribution of Farm Operator Levels of Living: 1950 and 1959." Ph.D. Dissertation, University of Iowa, Iowa City, Iowa, 1967.

Thompson, Kenneth. "The Perception of the Agricultural Environment." *Agricultural History*, 49, No. 1 (1975): 230-237.

Vogeler, Ingolf. "Agrarian Capitalists and Agritourism." *Antipode*, 7, No. 3 (1975): 37-42.

Foust, J. Brady, and Ingolf K. Vogeler. "Prolegomenon to Future Research on Black Agriculture." *West Georgia College Studies in the Social Sciences*, 18 (1979): 85-92.

Gregor, Howard F. "Large-Scale Farming as a Cultural Dilemma in U.S. Rural Development: The Role of Capital." *Geoforum*, 13, No. 1 (1982): 1-10.

Vail, David. "Family Farms in the Web of Community: Exploring the Rural Political Economy of the United States." *Antipode*, 14, No. 3 (1982): 26-38.

Williams, Michael. "Pioneer Farmlife and Forest Use." *Encylopedia of American Forest and Conservation History*, 2 (1983): 529-534.

Klohn, Werner. "Die Farmer-Genossenschaften in den USA." *Berichte uter Landwirtschaft*, 62, No. 3 (1988): 477-489.

Swanson, Louis E. *Agriculture and Community Change in the U.S.* Boulder, Colo.: Westview, 1988.

Fitzsimmons, Margaret I. "The Social and Environmental Relations of U.S. Agricultural Regions." Pp. 8-32 in *Technological Change and the Rural Environment*, edited by Philip D. Lowe, Terry K. Marsden, and Sarah Whatmore. London, England: Fulton, 1990.

ENVIRONMENTAL AND CONSERVATION ISSUES

Cragoc, E.J. "The Value of Rainfall in Selected Crop Areas of the United States." *Journal of Geography*, 14 (1915–1916): 1-6.

Brigham, Albert P. "Environment in the History of American Agriculture." *Journal of Geography*, 21, No. 2 (1922): 41-49.

Lemons, Hoyt. "Hail in American Agriculture." *Economic Geography*, 18, No. 4 (1942): 363-378.

Kaikow, J. "Industrial and Municipal Wastes as a Source of Agricultural Humus." Pp. 25-28 in *Agricultural Geography: I.G.U. Symposium*, edited by E.S. Simpson. Liverpool, England: University of Liverpool and International Geographical Union, 1964.

Fleming, Lois F. "Some Effects of Agricultural Technology on Soils and Vegetation: A Comparison of Mixed Farming and Cash-Grain Farming." Ph.D. Dissertation, Indiana State University, Terre Haute, Ind., 1972.

Climate and Food: Climatic Fluctuation and U.S. Agricultural Production. Washington, D.C.: National Academy of Sciences, 1976.

Rydant, Albert L. "Factors Affecting the Adoption of Crop-Hail Insurance." Master's Thesis, Southern Illinois University at Carbondale, 1976.

Gersmehl, Philip J. "No-Till Farming: The Regional Applicability of a Revolutionary Agricultural Technology." *Geographical Review*, 68, No. 1 (1978): 66-79.

Rydant, Albert L. "Adjustments to Natural Hazards: Factors Affecting the Adoption of Crop-Hail Insurance." *Professional Geographer*, 31, No. 3 (1979): 312-320.

Napton, Darrell. "Soil Quality and Agricultural Zoning." *Journal of the Minnesota Academy of Science*, 47, No. 3 (1981): 8-10.

Trimble, Stanley W. "Perspectives on the History of Soil Erosion Control in the Eastern United States." *Agricultural History*, 59, No. 2 (1985): 162-180.

Gersmehl, P.J., and D.A. Brown. "Geographic Differences in the Validity of a Linear Scale of Innate Soil Productivity." *Journal of Soil and Water Conservation*, 45, No. 3 (1990): 379-382.

Kraus, James. "An Assessment of Tillage Practices on the Reduction of Particulate-Phosphate from Agricultural Lands." Master's Thesis, University of Akron, Akron, Ohio, 1990.

Nellis, M. Duane. "Agricultural Externatilities and the Environment in the United States." Pp. 131-141 in *Contemporary Rural Systems in Transition: Volume 1, Agriculture and Environment*, edited by Ian R. Bowler, Christopher R. Bryant, and M. Duane Nellis. Wallingford, England: C.A.B. International, 1992.

Napton, Darrell. "Nontraditional Farm and Ranch Diversification: Strategies to Cope with Stress." Pp. 287-290 in *Biostress Mechanisms, Responses, Management: Proceedings of the Biostress Symposia*, edited by Nels H. Granholm. Brookings, S.Dak.: South Dakota State University, 1993.

Lane, Natalie. "Nutrient Export Analysis of an Agricultural Watershed Utilizing Geographic Information Systems." Master's Thesis, Murray State University, Murray, Ky., 1994.

Gomez, Basil. "Assessing the Impact of the 1985 Farm Bill on Sediment-Related Nonpoint Source Pollution." *Journal of Soil and Water Conservation*, 50, No. 4 (1995): 374-377.

Nellis, M. Duane, Lisa M.B. Harrington, and Jason Sheehy. "Policy, Sustainability and Scale: The US Conservation Reserve Program." Pp. 219-232 in *Agricultural Restructuring and Sustainability: A Geographical Perspective*, edited by Brian W. Ilbery, Quentin Chiotti, and Timothy Rickard. Wallingford, England: C.A.B. International, 1997.

IRRIGATION

Smart, George R. "Supplemental Irrigation in the Humid Eastern United States: A Geographical Appraisal." Master's Thesis, Ohio State University, Columbus, Ohio, 1959.

Mather, John R. "Irrigation in Humid Areas." *Yearbook, Association of Pacific Coast Geographers*, 30 (1969): 107-123.

Krzemien, Gabrielle. "Strukturen und Probleme des Bewasserungsfeldbaus in der USA und seine Entwicklung in den Industrialisierungsprocess der Landwirtschaft." *Vechtaer Arbeiten zur Geographie und Regionalwissenschaft*, 8 (1989): 63-88.

Stadler, S.T. "Assessing Short-Term Statewide Irrigation Needs Using Radar and Mesoscale Surface Data." *Papers and Proceedings of Applied Geography Conferences*, 18 (1995): 115-124.

AQUACULTURE

Wells, Alice B. "The Concentration of Catfish Culture." Master's Thesis, Western Illinois University, Macomb, Ill., 1973.

AGRICULTURAL MARKETING AND DISTRIBUTION

Kniffen, Fred. "The American Agricultural Fair: The Pattern." *Annals of the Association of American Geographers*, 39, No. 2 (1949): 264-282.

Kniffen, Fred. "The American Agricultural Fair: Time and Place." *Annals of the Association of American Geographers*, 41, No. 1 (1951): 42-57.

Jumper, Sydney R. "The Disposition of Oranges Produced in the United States." *Southeastern Geographer*, 2 (1962): 19-31.

Collier, James E. "Sources of Fresh Vegetables in United States Markets." *Journal of Geography*, 69, No. 6 (1970): 342-343.

Lord, J. Dennis. "Regional Marketing Patterns and Locational Advantages in the United States Broiler Industry." Ph.D. Dissertation, University of Georgia, Athens, Ga., 1970.

Folkerts, Kenneth L. "Grain Elevator Differences Along the Case Grain Boundary: A Restructuring of the Boundary." Master's Thesis, Eastern Illinois University, Charlestown, Ill., 1973.

Habrich, Wulf. "Die Agrarproduktion der USA und ihre Auswirkungen auf den Binnenmarkt und die Aussenwirschaft-ein Unterrichts-beispiel fur Klass 9/10 der Sekundarstafe I." *Geographische Rundschau*, 4 (1974): 132-141.

Shakow, Don. "The Municipal Farmer's Market as an Urban Service." *Economic Geography*, 57, No. 1 (1981): 68-77.

Berardi, G.M. "Agricultural Export Demand and Farm Policies: Implications for Soil Loss in the U.S." *Papers and Proceedings of the Applied Geography Conferences* 8 (1985): 42-56.

Grant, Richard. "Against the Grain: Agricultural Trade Policies of the U.S., the European Community and Japan at the GATT." *Political Geography*, 12, No. 3 (1993): 247-262.

TRANSPORTATION AND MECHANIZATION

Pfeifer, Gottfried. "Zur Geographie der Technisierung der USA-Landwirtschaft." *Technik und Wissenschaft*, 1, No. 3 (1939): 49-50.

Patterson, Ethel S. "A Geographical Study of Power Farming in the United States, 1925–1945." Ph.D. Dissertation, University College of the Southwest (England), 1953.

Hilliard, Samuel B. "The Dynamics of Power: Recent Trends in Mechanization on the American Farm." *Technology and Change*, 13, No. 1 (1972): 1-24.

AGRICULTURAL LABOR STUDIES

Gregor, Howard F. "Wage Labor as a Primary Plantation Indicator: The Case of American Cotton Farms." *California Geographer*, 8 (1967): 89-102.

The Northeastern United States

GENERAL TEXTS, ARTICLES, AND THESES

Lefferts, Walter. "New England Farming: Its Present and Future." *Journal of Geography*, 11 (1912–1913): 190-194.

Davis, I.G. "Agricultural Production in New England." Pp. 118-167 in *New England's Prospect, 1933*, edited by John K. Wright. New York, N.Y.: American Geographical Society, 1933.

Gardula, Robert A. "Trends in the Nature and Intensity of Agricultural Activity in New England for the Period 1930–1960." Master's Thesis, Clark University, Worcester, Mass., 1962.

Birch, J.W. "On the Stability of Farming Systems with Particular Reference to the Connecticut Valley." Pp. 223-246 in *Essays in Geography for Austin Miller*, edited by J.B. Whittaw and P.D. Wood. Reading, England: University of Reading Department of Geography, 1965.

Berry, David. *The Farmer's Response to Urbanization: A Study of the Middle Atlantic States.* Philadelphia, Pa.: Regional Science Research Institute, Paper No. 92, 1976.

Schnell, George A., and Mark Monmonier. "Agricultural Change in the Northeast, 1954–1974." *Proceedings, Middle State Division, Association of American Geographers*, 12 (1978): 85-89.

Dalton, R.T. "Agricultural Trends in New England." *Geography*, 71, No. 4 (1986): 362-365.

Bell, Michael. "Did New England Go Downhill?" *Geographical Review*, 79, No. 4 (1989): 450-466.

CROPS AND CROPPING PATTERNS

Dahlberg, Richard E. "A Geographic Analysis of the Chautauqua–Erie Grape Industry." Ph.D. Dissertation, University of Wisconsin, Madison, Wis., 1959.

Seig, Louis. "Tobacco Culture in the Connecticut Valley." Master's Thesis, Louisiana State University, Baton Rouge, La., 1961.

Lewis, Thomas R. "Recent Changes in the Connecticut Valley Tobacco Industry." *Journal of Geography*, 68, No. 1 (1969): 46-49.

Lewis, Thomas R. "Declining Cigar Tobacco Production in Southern New England." *Journal of Geography*, 79, No. 3 (1980): 108-111.

ANIMAL HUSBANDRY

Keinath, William F., Jr. "Egg Specialization Among Poultry Farmers in the Northeastern United States." Master's Thesis, University of Iowa, Iowa City, Iowa, 1956.

DAIRYING

Hassan, Mohamed I. "Geographic Elements in the Dairy Trends in New England." Ph.D. Dissertation, Clark University, Worcester, Mass., 1949.

Durand, Loyal. "The Major Milksheds of the Northeastern Quarter of the United States." *Economic Geography*, 40, No. 1 (1964): 9-32.

Monmonier, Mark S., and George A. Schnell. "Dairying, Urbanization and Recent Agricultural Change in the Northeast." *Proceedings, Middle States Division of the Association of American Geographers*, 14 (1980): 28-33.

FARM LAND

Greenley, Roland B. "Part-Time Farming and Recreational Land Use in New England." *Economic Geography*, 19, No. 2 (1942): 146-152.

Logan, Richard F. "Abandonment of Agricultural Lands in the Uplands of Western New England." Ph.D. Dissertation, Harvard University, Boston, Mass., 1948.

Dahlberg, Richard E. "Agricultural Land Utilization in the Lake Erie Grape Belt." Master's Thesis, University of North Carolina, Chapel Hill, N.C., 1952.

Hook, John C. "The Relationship Between Roughness of Terrain and Phenomena Related to Agriculture in the Northeastern United States." Ph.D. Dissertation, University of Iowa, Iowa City, Iowa, 1955.

Hart, John Fraser. "Loss and Abandonment of Cleared Farmland in the Eastern United States." *Annals of the Association of American Geographers*, 58, No. 3 (1968): 417-440.

Berry, David. "The Effect of Property Taxes on the Loss of Farmland: The Potential Effectiveness of Preferential Assessment." *Proceedings, Middle States Division of the Association of American Geographers*, 9 (1975): 29-34.

Monmonier, Mark S., and George A. Schnell. "Land Quality and Agricultural Change in the Northeast." *Proceedings, New England–St. Lawrence Division of the Association of American Geographers*, 9 (1979): 24-28.

AGRICULTURAL POPULATIONS AND SOCIAL ISSUES

Schnell, George A., and Mark S. Monmonier. "Changing Age Structure of Farm Operators in the Northeast, 1959–1974." *Proceedings, Middle States Division of the Association of American Geographers*, 12 (1979): 59-64.

ENVIRONMENTAL AND CONSERVATION ISSUES

Smith, David C., et al. "Salt Marshes as a Factor in the Agriculture of Northeastern North America." *Agricultural History*, 63, No. 2 (1989): 270-294.

IRRIGATION

Bajwa, Rajinder S. *Humid Region Irrigation: Geographic and Economic Perspectives*. Ann Arbor, Mich.: University of Michigan Department of Geography, Michigan Geographical Publication No. 28, 1983.

TRANSPORTATION AND MECHANIZATION

Monmonier, Mark S., and George A. Schnell. "Spatial-Temporal Trends in Farm Mechanization: The Tractor in New York and Pennsylvania, 1930–1964." *Proceedings, New England–St. Lawrence Valley Division of the Association of American Geographers*, 1 (1971): 2-9.

Monmonier, Mark S., and George A. Schnell. "The Tractor in New York and Pennsylvania, 1930–1964: A Study of Geographic Trends in Agricultural Mechanization." *Proceedings of the Association of American Geographers*, 4 (1972): 70-75.

Maine

CROPS AND CROPPING PATTERNS

Wilson, Ella M. "The Aroostock Valley: A Study in Potatos." *Geographical Review*, 16 (1926): 196-205.

Carlson, A.S., and J. Weston. "The Sweet Corn Industry of Maine." *Economic Geography*, 10, No. 4 (1934): 382-394.

Wallach, Bret. "The Potato Landscape: Aroostock County, Maine." *Landscape*, 23, No. 1 (1979): 15-22.

FARM LAND

Frederic, Paul B. "Protecting Farmland Protection: The Case of Auburn, Maine." *Proceedings, New England–St. Lawrence Valley Division of the Association of American Geographers*, 17 (1987): 27-32.

ENVIRONMENTAL AND CONSERVATION ISSUES

Smith, David C., and Anne Bridges. "Salt Marsh Dykes as a Factor in Eastern Maine Agriculture." *Maine Historical Society Quarterly*, 21, No. 4 (1982): 219-226.

Vermont

GENERAL TEXTS, ARTICLES, AND THESES

Anderson, S. Axel, and Florence M. Woodard. "Agricultural Vermont." *Economic Geography*, 8, No. 1 (1932): 12-42.

Illick, J. Rowland. "Agricultural Trends in the Southern Champlain Valley of Vermont, 1940–1980." *Proceedings, New England–St. Lawrence Valley Division of the Association of American Geographers*, 13 (1983): 12-21.

Sunderman, Frederick W. "High-Altitude Agriculture Prior to 1890 in Lincoln, Vermont and the Central Green Mountains." Master's Thesis, University of Vermont, Burlington, Vt., 1991.

DAIRYING

Skahill, John. "Dairy Farms with Noncontiguous Land Parcels in Southern Chittendon County, Vermont, 1964 to 1977." *Vermont Geographer*, 3 (1984): 31-37.

FARM LAND

Lapping, Mark B., and Frederick O. Sargent. "Determining the Economic Viability of Agricultural Lands: A Vermont Case Study." *Papers and Proceedings of the Applied Geography Conferences*, 2 (1979): 269-279.

Morrison, Victor I. "Farmland Conservation: Westford, Vermont." Master's Thesis, University of Vermont, Burlington, Vt., 1981.

Skahill, John. "Farms with Non-Contiguous Land Parcels in Southern Chittenden County, Vermont from 1964 to 1977." Master's Thesis, University of Vermont, Burlington, Vt., 1981.

Lindahl, David. "Development and Farmland Loss in a Vermont Township: Separating Myth from Reality." *Proceedings, New England–St. Lawrence Valley Division of the Association of American Geographers*, 17 (1987): 33-42.

ENVIRONMENTAL AND CONSERVATION ISSUES

Demeritt, David. "Climate, Cropping and Society in Vermont, 1820–1850." *Vermont History*, 59, No. 3 (1991): 133-165.

TRANSPORTATION AND MECHANIZATION

Czizmesin, Frederich. "Influences of the Garden Tractor Upon the Agriculture of Vermont." Master's Thesis, Clark University, Worcester, Mass., 1960.

Massachusetts

CROPS AND CROPPING PATTERNS

Mason, Carol Y. "The Cranberry Industry in Massachusetts." *Economic Geography*, 2, No. 1 (1926): 59-69.

LaFleur, Albert. "A Geographic Analysis of the Tobacco Industry of the Connecticut Valley Lowland in Massachusetts." Ph.D. Dissertation, Clark University, Worcester, Mass., 1929.

Casarjian, Conrad K. "A Geographical Analysis of Cranberry Bog Distribution in Massachusetts." Master's Thesis, University of Colorado, Boulder, Colo., 1967.

Simonds, Sue C. "The Microgeography of Cranberry Farming in Plymouth County, Massachusetts." Master's Thesis, Clark University, Worcester, Mass., 1971.

DAIRYING

Helmberg, Kenneth E. "Dairy Farms in Transition: Land Use Change in Three Rural Massachusetts Towns, 1973–1992." Master's Thesis, University of Massachusetts, Amherst, Mass., 1994.

FARM LAND

Kyi, Ma Thin. "Changes in Agricultural Land Utilization in Hampshire County, Massachusetts." Ph.D. Dissertation, Clark University, Worcester, Mass., 1952.

Al-Mayad, Ali M. "Agricultural Land Use, Town of Holden, Worcester County, Massachusetts." Master's Thesis, Clark University, Worcester, Mass., 1956.

ENVIRONMENTAL AND CONSERVATION ISSUES

Millington, B.R. "Glacial Topography and Agriculture in Central Massachusetts." *Economic Geography*, 6, No. 4 (1930): 408-415.

Connecticut

GENERAL TEXTS, ARTICLES, AND THESES

Rickard, Timothy J. "Prospects for the Survival of Agriculture in Connecticut." *The Connecticut Review*, 14, No. 2 (1992): 25-37.

FARM LAND

Monroe, Charles B. "Farmland Versus Development: Alterations for Rural Land Use Planning in Connecticut." *Proceedings, New England–St. Lawrence Valley Division of the Association of American Geographers*, 7 (1977): 62-67.

Lewis, Thomas R., and John E. Harmon. "Bounty from the Soil." Pp. 147-164 in *Connecticut: A Geography*, edited by Thomas R. Lewis and John E. Harmon. Boulder, Colo.: Westview, 1986.

Rickard, Timothy J. "Problems in Implementing Farmland Preservation Policies in Connecticut." *Journal of Rural Studies*, 2, No. 3 (1986): 197-208.

Rickard, Timothy J. "Town Efforts to Preserve Agricultural Land: A Case Study in Hartford's Urban Fringe." *Proceedings, New England–St. Lawrence Valley Division of the Association of American Geographers*, 17 (1987): 17-26.

AGRICULTURAL MARKETING AND DISTRIBUTION

Rickard, Timothy J. "Direct Marketing as Agricultural Adaptation in Megalopolitan Connecticut." Pp. 77-88 in *Limits to Rural Land Use*, edited by G.M.R.A. van Oort and A.H.H.M. Kempers. Wageningen, Netherlands: Centre for Agricultural Publishing and Documentation, 1991.

New York

ATLASES AND GRAPHIC PRESENTATIONS

Beach, J. Richard, and Malcolm Fairweather. *Agricultural Atlas of New York State*. Plattsburgh, N.Y.: State University of New York–Plattsburgh, Department of Geography, 1979.

GENERAL TEXTS, ARTICLES, AND THESES

Epstein, Bart. "The Agricultural Geography of Elbridge Township, Onondoga County, New York." Master's Thesis, George Washington University, Washington, D.C., 1953.

Williams, Henry G., Jr. "Oneida County: Agricultural Geography." Master's Thesis, Syracuse University, Syracuse, N.Y., 1958.

Bresson, Marcelle M. "L'evolution agricole de Dutchess County, New York." *Annales de Geographie*, 71, No. 387 (1962): 500-511.

Leaman, J. Harold. "The Growth of Agricultural Specialization in Western New York State: 1840–1860." Master's Thesis, State University of New York, Buffalo, N.Y., 1970.

Burke, Jack D. "Agricultural Health in New York State: A First Approximation." *Proceedings, Middle States Division of the Association of American Geographers*, 15 (1981): 8-14.

CROPS AND CROPPING PATTERNS

Walling, Fitzroy. "A Geographic Approach to the Problems Confronting Fruit Growers in the Middletown Apple Area." Master's Thesis, Syracuse University, Syracuse, N.Y., 1949.

Hymes, Henry H. "Geographical Aspects of Market Gardening in the Syracuse Area." Master's Thesis, Syracuse University, Syracuse, N.Y., 1950.

Dunbar, Gary S. "Economic Geography of the New York Hop Industry." Master's Thesis, University of Virginia, Charlottesville, Va., 1953.

Dahlberg, Richard E. "The Concord Grape Industry of the Chatauqua–Erie Area." *Economic Geography*, 37, No. 2 (1961): 150-169.

Hirya, Yoswa W. "The Spatial Distribution and Association of the Sugar Beet Industry in Central New York State." Master's Thesis, Syracuse University, Syracuse, N.Y., 1967.

Schultz, Paul E. "The Grape and Wine Industry of the Naples Valley, New York: A Geographic Interpretation." Master's Thesis, University of Oklahoma, Norman, Okla., 1967.

Toy, Terrance J. "The Effect of the Physical Environment and Management on Concord Grape Productivity in Chautauqua County, New York." Master's Thesis, State University of New York, Buffalo, N.Y., 1970.

Newman, James L. "Vines, Wines and Regional Identity in the Finger Lakes Region." *Geographical Review*, 76, No. 3 (1986): 301-316.

Newman, James L. "Decline and Development in the Finger Lakes Wine Region of New York State." *Journal of Wine Research*, 3, No. 2 (1992): 79-96.

Newman, James L. "The Survival of the Fittest: Small Farm Wineries in the Finger Lakes Region of New York State." Pp. 87-96 in *Viticulture in Geographic Perspective: Proceedings of the Miami AAG Symposium*, edited by Harm deBlij. Miami, Fla.: Miami Geographical Society, 1992.

Rumney, Thomas A. "A Search for Economic Alternatives: Hops in Franklin County, New York During the Nineteenth Century." *Middle States Geographer*, 31 (1998): 23-34.

DAIRYING

Spicola, Sebastian T. "The Study of the Niagara Frontier Milkshed and Milk Market." Master's Thesis, State University of New York, Buffalo, N.Y., 1966.

Durand, Loyal. "The Historical and Economic Geography of Dairying in the North Country of New York State." *Geographical Review*, 57, No. 1 (1967): 24-47.

FARM LAND

Vaughan, Lawrence M. "Abandoned Farm Areas in Southern New York." Ph.D. Dissertation, Cornell University, Ithaca, N.Y., 1928.

Thaung, P. "Agricultural Land Use in Cayuga County, New York." Ph.D. Dissertation, Syracuse University, Syracuse, N.Y., 1955.

Silvernail, George R. "The Agricultural Land Use of the Western Finger Lakes Region of New York, 1600–1959." Ph.D. Dissertation, University of North Carolina, Chapel Hill, N.C., 1960.

Ohl, Clyde E. "The Usurpation of Agricultural Land by Non-Productive Uses: A Case Study of the Ninemile Creek Watershed, Onondoga County, New York." Master's Thesis, Syracuse University, Syracuse, N.Y., 1961.

Ruth, Gerald. "The Soils and Agricultural Land Use of the Eden Valley Region of New York State." Master's Thesis, State University of New York, Buffalo, N.Y., 1965.

Campney, Thomas. "Agricultural Land Use Theory, An Empirical Test: Upper New York State." Master's Thesis, Queen's University, Kingston, Ontario, 1969.

Pompi, Louis W. "A Quantitative Analysis of Agricultural Land Use in Western New York State: 1850." Master's Thesis, State University of New York, Buffalo, N.Y., 1970.

Amato, Jon P. "The Physical Environment Factor in the Analysis of Agricultural Land Use Patterns: A Case Study in Western New York." Ph.D. Dissertation, University of Toronto, Toronto, Ontario, 1976.

Brownell, Joseph W. "Continuing Farmland Abandonment in the Adirondack Fringe." *Proceedings, New England–St. Lawrence Valley Division of the Association of American Geographers*, 20 (1990): 16-24.

Monahan, Kelly M., Kyle Miller, Kim N. Irvine, and John Whitney. "Analysis of the 1993 Farmland Protection Survey for Erie County, New York." *Middle States Geographer*, 28 (1995): 28-36.

Rumney, Thomas A. "Changing Patterns in New York State's Agricultural Land Use: 1969–1992." *Pennsylvania Geographer*, 33, No. 2 (1995): 93-108.

Coffey, Brian, and Edward Dempsey. "Tax Shift Issues in Agricultural Land Preservation: The Example of New York State." *Small Town*, 29, No. 1 (1998): 4-13.

ENVIRONMENTAL AND CONSERVATION ISSUES

Mordoff, Richard A. "The Climate of New York State and Its Relation to Agriculture." Ph.D. Dissertation, Cornell University, Ithaca, N.Y., 1921.

Bain, Keith. "Land Use, Farm Size, and Farm Tenure in Relation to Soil Damage in Yorktown, New York, 1954." Master's Thesis, Syracuse University, Syracuse, N.Y., 1957.

Hall, Jerry. "A Correlation of the Physical Environment and Orchard Land Use in the Western Fruit Belt." Master's Thesis, State University of New York, Buffalo, N.Y., 1965.

IRRIGATION

Lorelli, James T. "The Effect of Supplemental Irrigation on the Land Use: Hay Crops in Onondoga County, New York." Master's Thesis, Syracuse University, Syracuse, N.Y., 1962.

AGRICULTURAL MARKETING AND DISTRIBUTION

Langdale, John. "Impact of the Telegraph on the Buffalo Agricultural Commodity Market: 1846–1848." *Professional Geographer*, 31, No. 2 (1979): 165-169.

TRANSPORTATION AND MECHANIZATION

Tennant, John L. "The Economic Relationship Between Roads and Agriculture in New York." Ph.D. Dissertation, Cornell University, Ithaca, N.Y., 1928.

Leaman, J. Harold, and E.C. Conkling. "Transport Change and Agricultural Specialization." *Annals of the Association of American Geographers*, 65, No. 3 (1975): 425-432.

New Jersey

GENERAL TEXTS, ARTICLES, AND THESES

Schinkel, William E. "Agricultural Geography of Hunterdon County, New Jersey." Master's Thesis, Columbia University, New York, N.Y., 1968.

Kelland, Frank S., and Marylin C. Kelland. "Agriculture." Pp. 131-146 in *New Jersey: Garden or Suburb? A Geography of New Jersey*, edited by Frank S. Kelland and Marylin C. Kelland. Dubuque, Iowa: Kendall-Hunt, 1978.

Vining, James W. "Early American Geographers and New Jersey Agriculture." *Directions*, 10, No. 1 (1981): 10-20.

Stansfield, Charles A., Jr. "Agriculture." Pp. 121-138 in *New Jersey: A Geography*, edited by Charles A. Stansfield. Boulder, Colo.: Westview, 1983.

CROPS AND CROPPING PATTERNS

Farace, Eugene D. "The Tomato Industry in the New Jersey–Delmarva Area: A Study in Localization." Master's Thesis, University of Maryland, College Park, Md., 1956.

FARM LAND

Sickler, Harry A. "Changes in Agricultural Land in Gloucester County, New Jersey." Master's Thesis, Clark University, Worcester, Mass., 1964.

Hearn, Peter B. "A Quantitative Analysis of Farmland Loss in New Jersey with Implications for Public and Private Farmland Preservation Efforts in the Metropolitan New York Region." Master's Thesis, Columbia University, New York, N.Y., 1982.

AGRICULTURAL POPULATIONS AND SOCIAL ISSUES

Wacker, Peter O. "A Preliminary View of the Possible Associations Between Cultural Background and Agriculture in New Jersey During the Latter Part of the Eighteenth Century." *Proceedings, New York–New Jersey Division, Association of American Geographers*, 4 (1974): 41-57.

Pennsylvania

GENERAL TEXTS, ARTICLES, AND THESES

James, Henry F. "Agricultural Industry in Southeastern Pennsylvania: A Study in Economic Geography." Ph.D. Dissertation, University of Pennsylvania, Philadelphia, Pa., 1927.

James, Henry F. "The Agricultural Industry of Southeastern Pennsylvania: IV—Colonial Settlers; and VI—Historical Introduction." *Bulletin of the Philadelphia Geographical Society*, 26, No. 1 (1928): 29-59; 26, No. 2 (1928): 123-147; and 26, No. 3 (1928): 208-229.

James, Henry F. "The Agricultural Industry of Southeastern Pennsylvania: VI—Historical Introduction." *Bulletin of the Philadelphia Geographical Society*, 26, No. 3 (1928): 87-126; 167-192.

Kuntz, Leonard I. "Nippenose Valley: A Geographic Study of an Agricultural Valley in North Central Pennsylvania." Master's Thesis, University of Pittsburgh, Pittsburgh, Pa., 1954.

Fuller, Theodore E. "A Geographic Analysis of the Agricultural Regions of Adams County, Pennsylvania." Master's Thesis, Pennsylvania State University, University Park, Pa., 1956.

Glass, Joseph W. "Agricultural Regions of Lancaster County, Pennsylvania." Master's Thesis, Pennsylvania State University, University Park, Pa., 1959.

Leavy, Thomas A. "Agricultural Region of Clarion County, Pennsylvania." Master's Thesis, Pennsylvania State University, University Park, Pa., 1962.

Misra, R. "Agricultural Development in a Traditional Society: A Theoretical and Empirical Analysis of Processes of Agricultural Development and Change in Lancaster County, Pennsylvania." Ph.D. Dissertation, University of Maryland, College Park, Md., 1964.

Justham, Stephen A. "Agricultural Relationships in the McIntyre, Pennsylvania Area." Master's Thesis, Indiana University of Pennsylvania, 1970.

Vining, James W. "Early Views of Pennsylvania Agricultural Geography." *Pennsylvania Geographer*, 21, No. 1-2 (1983): 20-27.

Miller, E. Willard. "Agriculture." Pp. 183-202 in *A Geography of Pennsylvania*, edited by E. Willard Miller. University Park, Pa.: Pennsylvania State University Press, 1995.

Griffith, Daniel A. "Small Geographic Area Estimation Contributions to Federal Agricultural Data: A Pennsylvania Case Study." *Pennsylvania Geographer*, 38, No. 1 (2000): 3-29.

CROPS AND CROPPING PATTERNS

Brockmyer, Irene. "The Market-Gardening Industry in the Troy Hill Area of Pittsburgh, Pennsylvania." Master's Thesis, University of Pittsburgh, Pittsburgh, Pa., 1949.

Emery, Earl B. "The Cultivated Mushroom Industry of Chester County, Pennsylvania." Master's Thesis, University of Virginia, Charlottesville, Va., 1953.

Heppell, Roger C. "Agricultural Geography of the Cigar Tobacco Industry of the Lancaster, Pennsylvania Area." Ph.D. Dissertation, Pennsylvania State University, University Park, Pa., 1953.

Merriam, W.B. "The Mushroom Industry at Kennett Square, Pennsylvania." *Journal of Geography*, 61, No. 2 (1962): 68-70.

Good, Daniel B. "The Localization of Tobacco in Lancaster County, Pennsylvania." Master's Thesis, University of Tennessee, Knoxville, Tenn., 1967.

Monmonier, Mark S. "The Ecology of Commercial Sour Cherry Production in Adams County, Pennsylvania." *Proceedings of the Pennsylvania Academy of Science*, 41 (1967): 123-128.

Good, David. "The Localization of Tobacco Production in Lancaster County." *Pennsylvania History*, 49, No. 3 (1982): 190-200.

Dickenson, John P. "Dimensions of a Wine Geography: The Pennsylvania Experience." *Ontario Geography*, 39 (1992): 23-34.

DAIRYING

Connors, Robert. "A Geographical Analysis of the Dairy Industry of Crawford County, Pennsylvania." Master's Thesis, Pennsylvania State University, University Park, Pa., 1952.

Gault, Thomas G., and Robert W. Thomas. "Philadelphia's and Pittsburgh's Milksheds, 1962." *Proceedings of the Pennsylvania Academy of Science*, 39, No. 2 (1966): 234-239.

FARM LAND

Dreikorn, Edward L. "An Agricultural Land Use Study of Cambria County, Pennsylvania." Master's Thesis, Indiana University, Indiana, Pa., 1965.

Fitzgerald, Edward P. "A Study of Agricultural Land Uses: Warren County, Pennsylvania." Master's Thesis, Indiana University, Indiana, Pa., 1966.

Slocum, Warren D. "An Areal Account of Agricultural Land Abandonment in Pennsylvania, 1945–1959." Master's Thesis, Pennsylvania State University, University Park, Pa., 1969.

Berry, David. "Idling of Farmland in the Philadelphia Region, 1930–1970." *Regional Science Research Institute*, Discussion Paper No. 88 (1976).

Conn, William J. "Solving the Agricultural Land Use Dilemma: Zoning Techniques in York County." *Pennsylvania Geographer*, 19, No. 2 (1981): 9-21.

Fox, Richard G. "The Effectiveness of Zoning to Retain Agricultural Land: A Case Study of the Capital Region." Master's Thesis, Shippensburg State University, Shippensburg, Pa., 1982.

Rumney, Thomas A. "Patterns in Pennsylvania's Agricultural Land on the Eve of the Twenty-First Century: Changes and Processes." *Pennsylvania Geographer*, 30, No. 2 (1992): 55-68.

Timoney, Brian. "The Market for Farmland in Lancaster County, Pennsylvania." Master's Thesis, Pennsylvania State University, University Park, Pa., 1999.

AGRICULTURAL POPULATIONS AND SOCIAL ISSUES

Kollmorgen, Walter M. "The Pennsylvania German Farmer." Pp. 27-55 in *The Pennsylvania Germans,* edited by Rudoph Wood. Princeton, N.J.: Princeton University Press, 1942.

Kollmorgen, Walter M. "The Agricultural Stability of the Old Order Amish and the Old Order Mennonites of Lancaster County, Pennsylvania." *American Journal of Sociology*, 44 (1943): 233-241.

Lemon, James T. "The Agricultural Practices of National Groups in Eighteenth-Century Southeastern Pennsylvania." *Geographical Review*, 56 (1966): 467-496.

Davis, Charles H. "A Quantitative Index of Population and Agricultural Fluctuations in Bradford County Farms over the Last Century." Master's Thesis, State University of New York, Binghamton, N.Y., 1973.

Siegel, Charles I. "Agriculture, Culture, Technology: Amish Agriculture in Pennsylvania's Kishacoquillas Valley." Master's Thesis, Pennsylvania State University, University Park, Pa., 1987.

ENVIRONMENTAL AND CONSERVATION ISSUES

Kehs, Stephen L. "Agricultural Preservation: The Chester County Approach." *Pennsylvania Geographer*, 19, No. 2 (1981): 1-4.

Thompson, Robert. "The Chester County Agricultural Development Council: A Model for Agricultural Preservation." *Pennsylvania Geographer*, 19, No. 2 (1981): 5-9.

Dellinger, Denise, and John E. Benhart. "The Mason-Dixon Farm: Energy and Agricultural Efficiency Using Anaerobic Digestion." *Pennsylvania Geographer*, 23, No. 1-2 (1985): 22-29.

Eheman, Jane H. "Managing Agricultural Resources in Gettysburg National Military Park." *Pennsylvania Geographer*, 25, No. 1-2 (1987): 1-7.

Dilley, F.B. "The Statistical Relationship Between Weather-Type Frequencies and Corn (Maize) Yields in Southwestern, Pennsylvania, USA." *Agricultural and Forest Meteorology*, 59 (1992): 149-164.

AGRICULTURAL MARKETING AND DISTRIBUTION

Lewandowski, James P. "Pennsylvania's Agricultural Exports to Mexico and the NAFTA: Structural Changes and Estimates of Volume Increases Resulting from Mexico's Import Policy Change." *Pennsylvania Geographer*, 33, No. 1 (1995): 23-43.

AGRICULTURAL LABOR STUDIES

Holt, J.S., and R.H. Warland. "Characteristics and Labor Force Behavior of Migrant Agricultural Workers in Pennsylvania." *Proceedings of the Pennsylvania Academy of Science*, 47 (1973): 31-34.

Maryland, Delaware, and the District of Columbia

GENERAL TEXTS, ARTICLES, AND THESES

Gemmill, Paul F. "The Agriculture of the Eastern Shore Country." *Economic Geography*, 2, No. 2 (1926): 197-212.

Love, Ann G. "A Century of Agricultural Progress in a Rural Community: Sandy Spring, Maryland, 1844–1949." Master's Thesis, University of Maryland, College Park, Md., 1950.

Al-Shalash, Ali H. "A Geographical Analysis of Agriculture in Carroll County, Maryland." Ph.D. Dissertation, University of Maryland, College Park, Md., 1962.

Whisler, O.C. "Changes in Agriculture: Calvert County, Maryland, 1940–1960." Master's Thesis, University of Maryland, College Park, Md., 1965.

DiLisio, James E. "Agricultural Activities." Pp. 64-81 in *Maryland: A Geography*, edited by James E. DiLisio. Boulder, Colo.: Westview, 1983.

CROPS AND CROPPING PATTERNS

Weinrich, F.R. "Distribution of Tobacco Production in Southern Maryland." Master's Thesis, Catholic University, Washington, D.C., 1949.

Altai, M.H. "Geographical Analysis of Washington County, Maryland, and Its Fruit Industry." Ph.D. Dissertation, University of Maryland, College Park, Md., 1953.

ANIMAL HUSBANDRY

Martin, Kenneth R. "Analysis of the Broiler Chicken Industry on the Delmarva Peninsula." Ph.D. Dissertation, University of Wisconsin, Madison, Wis., 1956.

DAIRYING

Nuth, Viswambhar. "A Geographic Study of the Washington Milk Shed." Ph.D. Dissertation, University of Maryland, College Park, Md., 1949.

FARM LAND

Kelly, I.B. "An Agricultural Land Use Study of Western Montgomery County, Maryland." Master's Thesis, University of Maryland, College Park, Md., 1953.

Von Mayer, G.W. "The Effect of Urbanization upon the Agriculture of Howard County, Maryland." Master's Thesis, University of Maryland, College Park, Md., 1967.

Shollenberger, Craig. "Protecting Farmland in Maryland: A Review of the Agricultural Land Preservation Program." *Geographical Bulletin*, 23, No. 1 (1983): 41-49.

Kucera, Abigail. "An Analysis of Factors Influencing the Value of Farms on the Eastern Shore of Maryland, 1983–1987." Master's Thesis, University of Akron, Akron, Ohio, 1990.

ENVIRONMENTAL AND CONSERVATION ISSUES

DeNardo, Anthony. "Cloud Seeding Frequencies and Corn and Soybean Water Stress Periods in Delaware." Master's Thesis, University of Delaware, Newark, Del., 1981.

Kalkstein, Lawrence S., and William C. Liebhardt. "Comparative Climate–Corn Yield Relationships in Three Delmarva Counties." *Southeastern Geographer*, 21, No. 1 (1981): 26-39.

Kaltenbacher, Teresa. "Agricultural Drought Mitigation in Carroll County, Maryland." *Geographical Bulletin*, 36, No. 1 (1994): 23-30.

IRRIGATION

Maleady, Thomas G. "The Geographic Distribution of Supplemental Irrigation in Maryland." Master's Thesis, Catholic University, Washington, D.C., 1961.

AGRICULTURAL MARKETING AND DISTRIBUTION

Brune, Basel. "The Changing Spatial Organization of Early Tobacco Marketing in the Patuxent River Basin." Pp. 78-89 in *Geographical Perspectives on*

Maryland's Past, edited by Robert D. Mitchell and Edward K. Muller. College Park, Md.: University of Maryland Department of Geography, Occasional Paper No. 4, 1979.

West Virginia

GENERAL TEXTS, ARTICLES, AND THESES

Fesharki, P. "A Geographical Reconaissance of Jefferson County, West Virginia, with Emphasis upon Agricultural Aspects of the County." Master's Thesis, University of Maryland, College Park, Md., 1964.

The Southeastern United States

ATLASES AND GRAPHIC PRESENTATIONS

Hilliard, Samuel B. *Atlas of Antebellum Southern Agriculture*. Baton Rouge, La.: Louisiana State University Press, 1984.

GENERAL TEXTS, ARTICLES, AND THESES

Harper, Roland M. "Development of Agriculture in the Pine Barrens of the Southeastern United States." *Journal of Geography*, 15, No. 2 (1916): 42-48.

Parkins, Alman E. "Southern Agriculture." Pp. 52-79 in *Culture in the South*, edited by W.T. Counch. Chapel Hill, N.C.: University of North Carolina Press, 1933.

Kollmorgen, Walter M. "A Reconaissance of Some Culture–Agricultural Islands in the South." *Economic Geography*, 17, No. 4 (1941): 409-420.

Strack, Charles M. "Agricultural Changes in the TVA Area, 1930–1945." Ph.D. Dissertation, University of Iowa, Iowa City, Iowa, 1950.

Mather, Eugene C., and John Fraser Hart. "Agriculture in the Deep South and the Border States." *Tidjschrift voor Economische en Sociale Geographie*, 45, No. 9-10 (1954): 161-166.

Prunty, Merle C., Jr. "The Renaissance of the Southern Plantation." *Geographical Review*, 45, No. 4 (1955): 459-491.

Prunty, Merle C., Jr. "The Woodland Plantation as a Contemporary Occupance Type in the South." *Geographical Review*, 53, No. 1 (1963): 1-21.

Anderson, James R. "Specialized Agriculture in the South." *Southeastern Geographer*, 10 (1970): 13-27.

Coffman, John E. "Agricultural Diversification and Specialties in the Traditional Cotton Belt Since Mid-Century." Ph.D. Dissertation, University of California, Los Angeles, Calif., 1972.

Mitchell, Robert D. "Agricultural Regionalization: Origins and Diffusions in the Upper South Before 1860." Pp. 740-742 in *International Geography 1972, Vol. 1, Section 5, Historical Geography*, edited by W.P. Adams and F.M. Helleiner. Toronto, Ontario: University of Toronto Press, for the International Geographical Union, 1972.

Anderson, James R. *A Geography of Agriculture in the United States' Southeast.* Budapest, Hungary: Akademiai Kiado, 1973.

Lenz, K. "Bevolkerungsstruktur und Bevolkerungsmobilitat bie unterschiedlichen agrarwirtschaftlichen Varaussetzungen, dargestellt an Gadsden County (Fla.) und Grady County (Ga.) in Sudoisten der USA." *Die Erde*, 104 (1973): 176-184.

Hilliard, Samuel B. "The Tidewater Rice Plantation: An Ingenious Adaptation to Nature." Pp. 57-66 in *Coastal Resources*, edited by H. Jesse Walker. Baton Rouge, La.: Louisiana State University, School of Geoscience, *Geoscience and Man*, Vol. 12, 1975.

Maury, R. "Les exploitations agricoles du Nord de la Touraine assujetties a la TVA." *Bulletin de l'Association Geographes Francais*, 53 (1976): 139-145.

Schmitz, Mark. "Farm Interdependence in the Antebellum Sugar Section." *Agricultural History*, 52, No. 1 (1978): 93-103.

Hart, John Fraser. "The Role of the Plantation in Southern Agriculture." *Proceedings of the Tall Timbers Ecology and Management Conference*, 16 (1979): 1-19.

Hilliard, Samuel B. "The Plantation in Antebellum Southern Agriculture." *Proceedings of the Tall Timbers Ecology and Management Conference*, 16 (1979): 127-139.

Prunty, Merle C., Jr. "Persistence of the Plantation and National Agricultural Problems." *Proceedings of the Tall Timbers Ecology and Management Conference*, 16 (1979): 89-110.

Winsberg, Morton D. "Intensity of Agricultural Production and Distance from the City: The Case of the Southeastern United States." *Southeastern Geographer*, 21, No. 1 (1981): 54-63.

Hilliard, Samuel B. "Plantations and the Moulding of the Southern Landscape." Pp. 104-126 in *The Making of the American Landscape*, edited by Michael P. Conzen. Boston, Mass.: Unwin Hyman, 1990.

Pudup, Mary B. "The Limits of Subsistence Agriculture and Industrialization in Central Appalachia." *Agricultural History*, 64, No. 1 (1990): 61-89.

CROPS AND CROPPING PATTERNS

Browne, W.A. "The Dark-Fired Tobacco Region of the North Highland Rim." *Economic Geography*, 14, No. 1 (1938): 55-67.

Stephen, L. LeMar. "Peanut Production in the Southeastern United States." *Economic Geography*, 21, No. 2 (1945): 183-191.

Erickson, Franklin C. "The Broken Cotton Belt." *Economic Geography*, 24, No. 3 (1948): 263-268.

Seitz, Kerlin M. "Wheat in the Southeastern States." *Economic Geography*, 24, No. 2 (1948): 170-173.

Ogilvie, Bruce C. "Development of the Tobacco Belt of Georgia and Florida: A Study in Economy Geography." Master's Thesis, Clark University, Worcester, Mass., 1949.

Prunty, Merle C., Jr. "Soybeans in the Lower Mississippi Valley." *Economic Geography*, 26, No. 4 (1950): 301-314.

Durand, Loyal, and Elsie T. Bird. "The Burley Tobacco Region of the Mountain South." *Economic Geography*, 26, No. 3 (1950): 274-300.

Prunty, Merle C., Jr. "Recent Quantitative Changes in the Cotton Regions of the Southeastern United States." *Economic Geography*, 27, No. 2 (1951): 189-208.

Asselanis, S.K. "Areal Trends and Regional Densities in the Southeastern Cotton Distribution, 1943–1952." Master's Thesis, University of Georgia, Athens, Ga., 1955.

Jumper, Sidney R. "A Geographical Analysis of the Production and Marketing of Fresh Fruits and Vegetables in Florida, Georgia and South Carolina." Ph.D. Dissertation, University of Tennessee, Knoxville, Tenn., 1960.

Vanderhill, Burke G. "Hairy Indigo in the Southeast." *Journal of Geography*, 63, No. 4 (1964): 149-154.

Mingret, P. "L'evolution de la culture du coton dans la vallee du Mississippi." *Revue de Geographie de Lyon*, 43 (1968): 179-224.

Siniard, L. Arnold. "Changing Cropping Patterns in the Midsouth." *Southeastern Geographer*, 10 (1970): 59-71.

Prunty, Merle C., Jr., and Charles S. Aiken. "The Demise of the Piedmont Cotton Region." *Annals of the Association of American Geographers*, 62, No. 2 (1972): 283-306.

Hilliard, Samuel B. "The Tidewater Rice Plantation: An Ingenious Adaptation to Nature." *Geoscience and Man*, 12 (1975): 57-66.

Siniard, L. Arnold. "Dominance of Soybean Cropping in the Lower Mississippi River Valley." *Southeastern Geographer*, 15, No. 1 (1975): 17-32.

Stapp, George H. "An Analysis of the Relationship Between Soybean Agriculture and Land-Use Transformation in the South." Ph.D. Dissertation, Louisiana State University, Baton Rouge, La., 1975.

Earle, Carville, and Ronald Hoffman. "Staple Crops and Urban Development in the 18th Century South." *Perspectives in American History*, 10 (1976): 7-78.

Hart, John Fraser. "The Demise of King Cotton." *Annals of the Association of American Geographers*, 67, No. 3 (1977): 307-322.

Hart, John Fraser. "Cropland Concentrations in the South." *Annals of the Association of American Geographers*, 68, No. 4 (1978): 505-517.

Hilliard, Samuel B. "Antebellum Tidewater Rice Culture in South Carolina and Georgia." Pp. 91-115 in *European Settlement and Development in North America: Essays on Geographical Change in Honor and Memory of An-*

drew Hill Clark, edited by J.R. Gibson. Toronto, Ontario: University of Toronto Press, 1978.

de Blij, Harm. "Viticulture and Viniculture in the Southeastern United States." *Southeastern Geographer*, 27, No. 2 (1987): 115-130.

Charvet, J.P., and Gerard Dorel. "Du cotton belt au poultry belt." *Bulletin de l'Association de Geographes Francais*, 65, No. 2 (1988): 165-174.

Hugill, Peter. "The Macro-Landscape of the Wallerstein World Economy: 'King Cotton' and the American South." *Geoscience and Man*, 25 (1988): 77-84.

Chestung, Ennis L. "Tobacco Industry." Pp. 773-754 in *Encyclopedia of Southern Culture*, edited by Charles R. Wilson and William Ferris. Chapel Hill, N.C.: University of North Carolina Press, 1989.

Rehder, John B. "Sugar Plantations." Pp. 576-577 in *Encyclopedia of Southern Culture*, edited by Charles R. Wilson and William Ferris. Chapel Hill, N.C.: University of North Carolina Press, 1989.

Rumney, Thomas A. "Regional Agricultural Change in Virginia, Maryland and Delaware." *Virginia Geographer*, 22, No. 2 (1990): 33-52.

Lee, Jeon. *The Historical Geography of Rice Culture in the American South*. Seoul, South Korea: American Studies Institute, Seoul National University, 1991.

Hart, John Fraser. "The Metempsychosis of King Cotton." *Southeastern Geographer*, 30, No. 1 (2000): 93-105.

ANIMAL HUSBANDRY

Shaw, Earl B. "Geography of Mast Feeding." *Economic Geography*, 16, No. 3 (1940): 233-249.

Lamb, Robert B. "The Role of the Mule in Southern Agriculture." Ph.D. Dissertation, University of California, Los Angeles, Calif., 1959.

Stansfield, Charles A. "A Geographical Analysis of the Cattle Industry of the Southeastern United States." Master's Thesis, Pennsylvania State University, University Park, Pa., 1961.

Lamb, Robert B. *The Mule in Southern Agriculture*. Berkeley, Calif.: University of California Publications in Geography, No. 15, 1963.

Hilliard, Samuel B. "Pork in the Ante-Bellum South: The Geography of Self-Sufficiency." *Annals of the Association of American Geographers*, 59, No. 3 (1969): 461-480.

Gersmehl, Philip. "Factors Leading to Mountaintop Grazing in the Southern Appalachians." *Southeastern Geographer*, 10 (1970): 67-71.

Lord, J. Dennis. "Poultry." Pp. 28-29 in *Encyclopedia of Southern Culture*, edited by Charles R. Wilson and William Ferris. Chapel Hill, N.C.: University of North Carolina Press, 1989.

Bays, Brad A. "Evolution of Cherokee Open Range Cattle Herding, 1761–1861: A Spatial Expression of Cultural Change." Master's Thesis, University of Tennessee, Knoxville, Tenn., 1991.

DAIRYING

Hatcher, Helene. "Dairying in the South." *Economic Geography*, 20, No. 1 (1944): 54-64.

Gross, Leo V. "Dairy Cattle and Climate in the Southern United States." Master's Thesis, University of Maryland, College Park, Md., 1963.

Durand, Loyal. "Dairying in the South." *Southeastern Geographer*, 10 (1970): 28-38.

FARM LAND

Prunty, Merle C., Jr. "Land Management at Antebellum Hopeton Plantation in the Light of the Soil Exhaustion Thesis." *Memorandum Folio, Southeastern Divsion of the Association of American Geographers*, 11 (1959): 97-101.

Aiken, Charles S. "The Fragmented Neoplantation: A New Type of Farm Operation in the Southeast." *Southeastern Geographer*, 11 (1971): 43-51.

Hart, John Fraser. "Land Rotation in Appalachia." *Geographical Review*, 67, No. 2 (1977): 148-166.

Christopher, J.F., and N.S. Sternitzke. "Land Clearing in the Lower Mississippi Valley." *Southeastern Geographer*, 10 (1970): 63-66.

Ainsley, W. Frank. "Pulsating Patterns of Land Occupancy: The Impacts of Farm Colonization Experiments on the Rural South." *Pioneer America Society Transactions*, 10 (1987): 43-52.

Pyle, Lizbeth A. "Part Owners and the Changing Spatial Structures of Farms in the South: A Case Study." *Southeastern Geographer*, 31, No. 1 (1991): 44-51.

AGRICULTURAL POPULATIONS AND SOCIAL ISSUES

Kollmorgen, Walter M. "Immigrant Settlements in Southern Agriculture: A Commentary on the Significance of Cultural Islands in Agricultural History." *Agricultural History*, 19, No. 2 (1945): 69-77.

Fisher, James S. "Negro Farm Ownership in the South." *Annals of the Association of American Geographers*, 63, No. 3 (1973): 478-489.

Higgs, Robert. "The Boll Weevil, the Cotton Economy and Black Migration, 1910–1930." *Agricultural History*, 50, No. 2 (1976): 335-350.

Aiken, Charles S. "The Decline of Sharecropping in the Lower Mississippi Valley." *Geoscience and Man*, 7 (1978): 151-165.

Foust, J. Brady, and Ingolf Vogeler. "Prolegomenon to Future Research on Black Agriculture in the South." *West Georgia College Studies in the Social Sciences*, 18 (1979): 85-92.

ENVIRONMENTAL AND CONSERVATION ISSUES

Hall, Arthur. "Terracing in the Southern Piedmont." *Agricultural History*, 23, No. 2 (1949): 96-106.

Earle, Carville. "The Myth of the Southern Soil Miner: Macrohistory, Agricultural Innovation and Environmental Change." Pp. 175-210 in *The Ends of the Earth: Perspectives on Modern Environmental History*, edited by Donald Worster. Cambridge, England: Cambridge University Press, 1988.

Cooter, Ellen J. "The Impact of Climatic Change on Continuous Corn Production in the Southern USA." *Climatic Change*, 16, No. 1 (1990): 53-82.

Rohli, Robert V., and Jeffrey C. Rogers. "Atmospheric Teleconnections and Citrus Freeze in the Southern United States." *Physical Geography*, 14, No. 1 (1993): 1-15.

IRRIGATION

Bunn, David W., and Arthur J. Hawley. "Irrigation in the South." *Southeastern Geographer*, 10 (1970): 6-12.

Peart, Robert M., et al. "Energy and Irrigation in South Eastern U.S. Agriculture Under Climate Change." *Journal of Biogeography*, 22, No. 4-5 (1995): 635-642.

AQUACULTURE

Cox, Gary C. "The Catfish Farming Industry of Southeastern United States: A Geographic Inquiry Into the Patterns and Processes of Regional Aquaculture." D.A. Dissertation, University of Northern Colorado, Greeley, Colo., 1972.

Kilburn, John W. "Catfish Farming in the Mid-South: Its Distribution and Potential Growth." Master's Thesis, University of Southern Mississippi, Hattiesburg, Miss., 1972.

Grace, Gordon R. "The Freshwater Mussel Industry of the Lower Tennessee River: Ecology and Future." Master's Thesis, Western Kentucky University, Bowling Green, Ky., 1974.

AGRICULTURAL MARKETING AND DISTRIBUTION

Hilliard, Samuel B. "Hogmeat and Hoecake: A Geographical Viewe of Food Supply in the Heart of the Old South, 1840–1860." Ph.D. Dissertation, University of Wisconsin, Madison, Wis., 1966.

TRANSPORTATION AND MECHANIZATION

Prunty, Merle C., Jr. "Some Quantitative Aspects of Farm Mechanization in the Southeastern Cotton Habitat, 1924–1949." *Memorandum Folio, Southeastern Division of the Association of American Geographers*, 5 (1953): 53-59.

Aiken, Charles S. "The Evolution of Cotton Ginning in the Southeastern United States." *Geographical Review*, 63, No. 2 (1973): 196-224.

AGRICULTURAL LABOR STUDIES

Aiken, Charles S. "The Decline of Sharecropping in the Lower Mississippi Valley." Pp. 151-165 in *Man and Environment in the Lower Mississippi Valley*, edited by Samuel B. Hilliard. Baton Rouge, La.: *Geoscience and Man*, Vol. 19, 1978.

Virginia

GENERAL TEXTS, ARTICLES, AND THESES

Davis, George O., Jr. "Geographic Foundations of Agriculture in Accomac County." Master's Thesis, University of Virginia, Charlottesville, Va., 1953.

Mitchell, Robert D. "Agricultural Change and the American Revolution: A Virginia Case Study." *Agricultural History*, 47, No. 2 (1973): 119-132.

Moon, Henry E., Jr., and Karl J. Raitz. "A Virginia Agricultural System in Transition." *Virginia Geographer*, 24, No. 2 (1992): 19-28.

Hardin, David S. "Alterations They Have Made at This Day: Environment, Agriculture and Landscape Change in Colonial Tidewater Virginia." Ph.D. Dissertation, University of Maryland, College Park, Md., 1995.

CROPS AND CROPPING PATTERNS

Wolf, William S. "A Geographical Study of the Orchards of the Crozet District, Albemarle County, Virginia." Master's Thesis, University of Virginia, Charlottesville, Va., 1950.

Gouger, James B. "The Northern Neck of Virginia: A Tidewater Grain-Farming Region in the Antebellum South." *West Georgia College Studies in Social Science*, 16 (1977): 73-90.

Gruber, John N. "The Northern Virginia Wine Industry." *Virginia Geographer*, 14, No. 1 (1981): 1-10.

Armstrong, Thomas F. "Tobacco Growth and Urban Strategy: Antebellum Lynchburg, Virginia." *Virginia Geographer*, 14, No. 1 (1982): 1-12.

ANIMAL HUSBANDRY

Gildea, Ray Y., Jr. "Geographical Aspects of Cattle Production in Albermarle County, Virginia." Master's Thesis, University of Virginia, Charlotttesville, Va., 1950.

FARM LAND

Wise, Stephen L. "Farmland Preservation Through Transferable Development Rights in Loudon County, Virginia." *Virginia Geographer*, 16, No. 2 (1984): 1-14.

TRANSPORTATION AND MECHANIZATION

Richetto, Jeffrey, and Henry E. Moon, Jr. "Railroad Development in Central Virginia and Its Effects on the Redistribution of Tobacco Farming, 1840–1880." *Virginia Geographer*, 17, No. 1 (1985): 1-23.

North Carolina

GENERAL TEXTS, ARTICLES, AND THESES

Jones, Reece A. "The Inner Coastal Plain of North Carolina: An Agricultural Region." Ph.D. Dissertation, University of North Carolina, Chapel Hill, N.C., 1954.

Jenning, James M. "The Agricultural Geography of Wilson County, North Carolina." Master's Thesis, University of North Carolina, Chapel Hill, N.C., 1957.

Larson, Albert J. "Agricultural Geography of New Hanover and Perry Counties, North Carolina." Master's Thesis, University of North Carolina, Chapel Hill, N.C., 1958.

Wildinson, Kenneth. "Agricultural Changes in Hyde County, North Carolina." Master's Thesis, East Carolina University, Greenville, N.C., 1968.

Hart, John Fraser, and Ennis L. Chestung. "Rural Revolution in East Carolina." *Geographical Review*, 68 (1978): 435-458.

CROPS AND CROPPING PATTERNS

Landon, Charles E. "The Tobacco Growing Industry of North Carolina." *Economic Geography*, 10, No. 3 (1934): 239-253.

Picklesimer, Parnell W. "The New Bright Tobacco Belt of North Carolina: A Study in Human Geography." Ph.D. Dissertation, George Peabody University, Nashville, Tenn., 1937.

Browne, W.A. "Dark-Fired Tobacco Region of the North Highland Rim." *Economic Geography*, 14, No. 1 (1938): 55-67.

Phillips, Coy T. "The Culture of Flue-Cured Tobacco in North Carolina." Master's Thesis, University of Colorado, Boulder, Colo., 1938.

Emory, Samuel T. "Bright Tobacco in the Agriculture, Industry and Foreign Trade of North Carolina." Ph.D. Dissertation, University of Chicago, Chicago, Ill., 1939.

Picklesimer, Parnell W. "Forest in the New Bright Tobacco Belt of North Carolina." *Economic Geography*, 19, No. 3 (1943): 276-278.

Erickson, Franklin C. "The Cotton Belt of North Carolina." *Economic Geography*, 20, No. 1 (1944): 4-8.

Picklesimer, Parnell W. "The New Bright Tobacco Belt of North Carolina." *Economic Geography*, 20, No. 1 (1944): 14-19.

Erickson, Franklin C. "The Tobacco Belt of North Carolina." *Economic Geography*, 21, No. 1 (1945): 58-61.

Dudley, Dennis. "The Role of Flue-Cured Tobacco in the Rural Geography of Pitt County, North Carolina." Master's Thesis, East Carolina University, Greenville, N.C., 1967.

Ellis, Robert F. "Spatial Pattern of the Flue-Cured Tobacco in North Carolina Related to Potential Out-Migration of Allotments." Master's Thesis, University of North Carolina, Chapel Hill, N.C., 1970.

Josit, Donald W., and Gerald E. Tyner. "Mountain Agriculture: Commercial Cabbage Production in North Carolina." *Journal of Geography*, 78, No. 1 (1979): 12-16.

Chestung, Ennis. "The Great Cotton Explosion." Pp. 175-177 in *Snapshots of the Carolinas: Landscapes and Cultures*, edited by D. Gordon Bennett. Washington, D.C.: Association of American Geographers Field Guide, 1996.

Hart, John Fraser, and Ennis L. Chestung. "Tobacco in North Carolina." Pp. 179-182 in *Snapshots of the Carolinas: Landscapes and Cultures*, edited by D. Gordon Bennett. Washington, D.C.: Association of American Geographers Field Guide, 1996.

ANIMAL HUSBANDRY

Arndt, Carroll B. "Locational Considerations in the North Carolina Broiler Industry." Master's Thesis, University of North Carolina, Chapel Hill, N.C., 1970.

Hart, John Fraser, and Ennis L. Chestung. "Vertical Integration and Confinement Production of Broilers, Turkeys and Hogs." Pp. 183-187 in *Snapshots of the Carolinas: Landscapes and Cultures*, edited by D. Gordon Bennett. Washington, D.C.: Association of American Geographers Field Guide, 1996.

Furuseth, Owen J. "Hog Farming in Eastern North Carolina." *Southeastern Geographer*, 41, No. 1 (2001): 53-64.

FARM LAND

Fuson, Robert H. "A Land-Use Study, with Emphasis on Forest Grazing of Craven County, North Carolina." Master's Thesis, Florida State University, Tallahassee, Fla., 1951.

Emory, Samuel T., Jr. "The Durham Triassic Basin: Agricultural Land Use in a Predominantly Industrial Region." Master's Thesis, University of North Carolina, Chapel Hill, N.C., 1959.

Lovingood, Paul E. "Agricultural Land Use in the Asheville Basin of North Carolina." Ph.D. Dissertation, University of North Carolina, Chapel Hill, N.C., 1963.

Holder, Gerald L. "The Tobacco Farm Complex in Lee County, North Carolina, 1938–1965." Master's Thesis, University of Georgia, Athens, Ga., 1966.

Mason, Dorothy S. "Agricultural Land Use Trends in Guilford County, North Carolina." Ph.D. Dissertation, University of North Carolina, Chapel Hill, N.C., 1967.

Gesler, Wilbert M. "The Willingness of Farmowners to Sell Land in Ashe County, North Carolina." Master's Thesis, University of North Carolina, Chapel Hill, N.C., 1974.

Stammerjohn, Bettie. "Conversion of Farmland to Urban Uses in Mechlenburg County." Master's Thesis, University of North Carolina at Charlotte, Charlotte, N.C., 1980.

Rumney, Thomas A. "A Geographic Examination of North Carolina Farms and Farmlands: 1969 to the Present." *North Carolina Geographer*, 8 (2000): 1-13.

AGRICULTURAL POPULATIONS AND SOCIAL ISSUES

Cherry, Angela E. "Patterns of Movement of Migrant Farm Workers in North Carolina." Pp. 131-134 in *Snapshots of the Carolinas: Landscapes and Cultures*, edited by D. Gordon Bennett. Washington, D.C.: Association of American Geographers Field Guide, 1996.

Birdsall, Stephen S. "Tobacco Farmers and Landscape Change in North Carolina's Old Belt Region." *Southeastern Geographer*, 41, No. 1 (2001): 65-73.

ENVIRONMENTAL AND CONSERVATION ISSUES

Mausel, Paul W. "North Carolina Coastal Plain Terraces as Soils and Agricultural Regions." Ph.D. Dissertation, University of North Carolina, Chapel Hill, N.C., 1966.

Barnett, James. "The Influence of Soils and Agricultural Practices on Nutrient Runoff in a Coastal Plain Watershed." Master's Thesis, East Carolina University, Greenville, N.C., 1992.

IRRIGATION

Ross, Thomas E. "Irrigation in South-Central North Carolina." *North Carolina Geographer*, 4 (1995): 41-51.

AQUACULTURE

Purgason, Frank. "Catfish Farming in North Carolina." Master's Thesis, East Carolina University, Greenville, N.C., 1985.

AGRICULTURAL MARKETING AND DISTRIBUTION

Morgan, John T., Jr. "The Role and Status of the Tobacco Ordering Pit in Eastern North Carolina." Master's Thesis, East Carolina University, Greenville, N.C., 1976.

Austin, T.E. "Tobacco Marketing Warehouses and Their Location in the Urban Landscape of the Eastern Tobacco Belt of North Carolina." Master's Thesis, East Carolina University, Greenville, N.C., 1977.

TRANSPORTATION AND MECHANIZATION

Johnson, James H. "Innovations in Flue-Cured Tobacco Farming in the North Carolina Coastal Plain and Their Implications for Energy Use." Master's Thesis, University of Wisconsin, Madison, Wis., 1977.

South Carolina

ATLASES AND GRAPHIC PRESENTATIONS

Cowen, D.J., and Paul E. Lovingood. *Atlas of Agriculture: South Carolina, 1960–1970.* Clemson, S.C.: Clemson University, 1972.

GENERAL TEXTS, ARTICLES, AND THESES

Salter, Paul S. "Changing Agricultural Patterns on the South Carolina Sea Islands." *Journal of Geography*, 67, No. 4 (1968): 223-228.

Kovacik, Charles E., and John J. Winberry. "Agriculture and Fisheries." Pp. 159-180 in *South Carolina: The Making of a Landscape*, edited by Charles E. Kovacik and John J. Winberry. Columbia, S.C.: University of South Carolina Press, 1987.

CROPS AND CROPPING PATTERNS

Pearson, James E. "The Carolina Rice Industry." Master's Thesis, University of Oklahoma, Norman, Okla., 1949.

Holmes, William M. "The Rice Industry of South Carolina." Master's Thesis, University of Tennessee, Knoxville, Tenn., 1959.

Stough, Roger R. "The Spatial Expansion of Soybean Cultivation in South Carolina for the Period 1949–1967: An Analysis." Master's Thesis, University of South Carolina, Columbia, South Carolina, Columbia, S.C., 1971.

Carpenter, James G. "The Rice Plantation Lands of Georgetown County, South Carolina: A Historical Geographic Study." Master's Thesis, University of South Carolina, Columbia, S.C., 1973.

Mason, Robert E. "A Historical Geography of South Carolina's Sea Island Cotton Industry." Master's Thesis, University of South Carolina, Columbia, S.C., 1976.

Pett, Linda M. "Changing Spatial Patterns of Tobacco Production in South Carolina." Master's Thesis, University of South Carolina, Columbia, S.C., 1976.

Kovacik, Charles F. "South Carolina Rice Coast Landscape Changes." *Proceedings of the Tall Timbers Ecology and Management Conference*, 16 (1979): 47-65.

Winberry, John J. "Reputation of Carolina Indigo." *South Carolina Historical Magazine*, 80, No. 3 (1979): 242-250.

Winberry, John J. "Indigo in South Carolina: A Historical Geography." *Southeastern Geographer*, 19, No. 2 (1979): 91-102.

Kovacik, Charles F., and Robert E. Mason. "Changes in the South Carolina Sea Island Cotton Industry." *Southeastern Geographer*, 25, No. 2 (1985): 77-104.

Savitsky, Basil C. "Agricultural Remote Sensing in South Carolina: A Study of Crop Identification Capabilities Uitlizing Landsat Multispectral Scanner Data." Master's Thesis, University of South Carolina, Columbia, S.C., 1986.

Carney, Judith A., and Richard Percher. "Geographies of the Past: Rice, Slaves and Technological Transfer in South Carolina." *Southeastern Geographer*, 33, No. 2 (1993): 127-147.
Locke, Christopher. "Estimating Biophysical Properties of Soybeans Using Field Data, Crop Modeling and Remote Sensing." Master's Thesis, University of South Carolina, Columbia, S.C., 1999.
Walcott, Susan M. "Tea Production in South Carolina." *Southeastern Geographer*, 39, No. 1 (1999): 46-60.
Carney, Judith. "The African Origins of Carolina Rice Culture." *Ecumene*, 7, No. 2 (2000): 125-149.

ANIMAL HUSBANDRY

Otto, John S. "The Origins of Cattle-Ranching in Colonial South Carolina, 1670–1715." *South Carolina Historical Magazine*, 87, No. 2 (1986): 117-124.
Yoder, Michael. "The Feeder Cattle Industry in Saluda County, South Carolina." Pp. 189-192 in *Snapshots of the Carolinas: Landscapes and Cultures*, edited by D. Gordon Bennett. Washington, D.C.: Association of American Geographers Field Guide, 1996.

FARM LAND

Fairey, Daniel A. "Decline in Cropland in Relation to Land Quality and Change in Farm Production in South Carolina." Master's Thesis, University of South Carolina, Columbia, S.C., 1973.
Kovacik, Charles. "Land Use Change on the South Carolina Rice Plantation Lands." Pp. 64-67 in *International Geography '76, Section 9, Historical Geography*. Moscow, Russia: International Geographical Congress, 1976.
Pyle, Lizbeth A. "A Geographical Analysis of Part Ownership and Farm Fragmentation: Calhoun County, South Carolina, 1976." Master's Thesis, University of South Carolina, Columbia, S.C., 1977.

Georgia

ATLASES AND GRAPHIC PRESENTATIONS

Myers, Robert R., David C. Weaver, and John C. Upchurch. *Atlas of Georgia Agriculture*. Carrolton, Ga.: West Georgia College, 1977.

GENERAL TEXTS, ARTICLES, AND THESES

Black, Blanton E. "The Rise and Decline of Plantation Agriculture in Coastal Georgia." Master's Thesis, University of Chicago, Chicago, Ill., 1937.
Johnson, Irene. "Agricultural Diversification in Georgia." Ph.D. Dissertation, Ohio State University, Columbus, Ohio, 1965.

Putnam, Elizabeth W. "An Analysis of Spatial Variations in Selected Agricultural Practices in the Georgia Piedmont." Master's Thesis, University of Washington, Seattle, Wash., 1966.

Johnson, Irene. "Agricultural Diversification in Georgia, 1939–1959." *Southeastern Geographer*, 7 (1967): 34-49.

Lin, Gong-yuh. "The Delineation of Agricultural Regions in Georgia, 1945–1954, and 1964." Master's Thesis, University of Georgia, Athens, Ga., 1968.

Pfeifer, Gottfried. "Die Wirtschaftsformation des Baumwollgurtels in Georgia und deren Entwicklung in Raum und Zeit." Pp. 150-169 in *Der Wirtschaftsraum: Beitrage zu Methoden und Anwendung eines geographischen Forschung sonsatzes*. Wiesbaden, Germany: Franz Steiner Verlag, Erdkundliches Wissen, Beihefte zur Geographischen Zeitschrift, Heft 41, 1975.

Wilms, Douglas C. "Agrarian Progress in the Cherokee Nation Prior to Removal." *West Georgia College Studies in the Social Sciences*, 16 (1977): 1-16.

CROPS AND CROPPING PATTERNS

Imperatore, William A. "Effects of Federal Controls on the Basic Geographic Characteristics of Cotton Production in Georgia." Master's Thesis, University of Georgia, Athens, Ga., 1963.

Farr, James T. "Shifts in the Spatial Distribution of Soybean Production in Georgia: 1929–1967." Master's Thesis, University of Georgia, Athens, Ga., 1969.

Wilms, Douglas C. "The Development of Rice Culture in 18th Century Georgia." *Southeastern Geographer*, 12 (1972): 45-57.

Clout, Hugh D. "L'heritage de la production cotonniere en Georgia, USA." *Norois*, 21 (1974): 585-605.

Fung, Devlin. "Modeling Cannabis Cultivation in North Georgia." Ph.D. Dissertation, University of Georgia, Athens, Ga., 1993.

Moseley, William G., and Carl F. Jordan. "Measuring Agricultural Sustainability: Energy Analysis of Conventional Till and No-Till Maize in the Georgia Piedmont." *Southeastern Geographer*, 41, No. 1 (2001): 105-116.

DAIRYING

Fisher, James C. "Georgia's Arbitory Dairy Region." Master's Thesis, University of Georgia, Athens, Ga., 1964.

FARM LAND

Nunnally, Nelson R. "Size-Group and Functional Variations in the Land Occupance Patterns of Part-Time Farms in Walton County, Georgia." Master's Thesis, University of Georgia, Athens, Ga., 1960.

Bederman, Sanford H. "Recent Changes in Agrarian Farm Use in Georgia." *Southeastern Geographer*, 10 (1970): 72-82.

IRRIGATION

Myers, Robert R. "Irrigation in Southwestern Georgia: Agricultural Land Use Change in the Coastal Plain." *Papers and Proceedings of the Applied Geography Conferences*, 3 (1980): 98-105.

Yates, Donald F. "The Impact of Center Pivot Irrigation Systems on Landscape Morphology and Land Use Practices in Southwest Georgia." Master's Thesis, University of Georgia, Athens, Ga., 1983.

Florida

GENERAL TEXTS, ARTICLES, AND THESES

Harper, Roland M. "Agricultural Conditions in Florida in 1925." *Economic Geography*, 3, No. 4 (1927): 340-353.

Dicken, Samuel N. "Central Florida Farm Landscape." *Economic Geography*, 11, No. 2 (1935): 173-182.

Chardon, Roland E. "An Examination of Some Problems of Agricultural Diversification in Gadsden County, Florida." Master's Thesis, Florida State University, Tallahassee, Fla., 1954.

Ford, Robert N. "A Resource Use Analysis and Evaluation of the Everglades Agricultural Area." Ph.D. Dissertation, University of Chicago, Chicago, Ill., 1956.

Greene, Robert E.L., G.N. Rose, and D.L. Brooke. *Location of Agricultural Production in Florida*. Gainesville, Fla.: University of Florida Agricultural Experiment Station, Institute of Food and Agricultural Sciences, 1969.

Anderson, James R. "Plantation Agriculture in the Middle Suwanee Basin of Florida, 1825–50." *Proceedings of the Tall Timbers Ecology and Management Conference*, 16 (1979): 34-45.

Dorel, Gerard. "Nouvelles strategies du grand capitalisme agro-industriel: Les Everglades de Florida." *L'Espace Geographique*, 11, No. 1 (1982): 17-29.

Birnbaum, Shira. "Florida Agriculture and the Collapse of the State Farmer's Alliance, 1880–1891." *Florida Geographer*, 27 (1996): 81-98.

CROPS AND CROPPING PATTERNS

Hicks, W.T. "The Development of the Tung Oil Industry in Florida." *Journal of Geography*, 31, No. 1 (1932): 27-35.

Wann, John L. "Where Florida Truck Crops Are Grown." *Economic Geography*, 9, No. 2 (1933): 85-103.

Delameter, Howard. "The Citrus Industry in Pinelles County, Florida." Master's Thesis, University of Colorado, Boulder, Colo., 1939.

Strain, Warren. "The Florida Citrus Crop." *Economic Geography*, 18, No. 1 (1942): 17-28.

Stephen, L. LeMar. "Vegetable Production in the Northern Everglades." *Economic Geography*, 10, No. 1 (1944): 79-101.

Stephen, L. LeMar. "A Little-Known Sugar Bowl in Florida." *Journal of Geography*, 43, No. 2 (1944): 49-61.

Ficarrotta, Frank R., Jr. "The Citrus Industry of West Central Florida." Master's Thesis, University of Florida, Gainesville, Fla., 1955.

Crist, Raymond. "The Citrus Industry in Florida." *American Journal of Economics and Sociology*, 15, No. 1 (1956): 1-11.

Moore, Fleming S. "The Role of Floriculture in the Agriculture of Florida." Ph.D. Dissertation, University of Washington, Seattle, Wash., 1959.

Roth, Richard. "Prospects for Expansion in Florida's Citrus Industry." Master's Thesis, Florida State University, Tallahasse, Fla., 1959.

Hanson, Louise F. "Some Aspects of the Tung Oil Industry, with Special Emphasis on Florida." Master's Thesis, Florida State University, Tallahassee, Fla., 1963.

Neel, Earl M. "Central Highlands Citrus District of Florida." Ph.D. Dissertation, University of Tennessee, Knoxville, Tenn., 1963.

Vest, Banks B., Jr. "South Florida Sugar Production: A Geographic Analysis." Master's Thesis, University of Florida, Gainesville, Fla., 1963.

Found, William C. "The Distribution of Citrus Land in Orange County, Florida." Master's Thesis, University of Florida, Gainesville, Fla., 1964.

Chestung, Ennis. "The Shade-Grown Cigar Wrapper Tobacco District of Gadsden County, Florida." Ph.D. Dissertation, Indiana University, Bloomington, Ind., 1965.

Found, William C. "The Relation of the Distribution of Citrus to Soil Type and Winter Temperature in Orange County, Florida." *Canadian Geographer*, 9 (1965): 63-73.

Jumper, Sidney R. "Changes in the Areal Distribution of Commercial Vegetable Production in Florida." *Southeastern Geographer*, 5 (1965): 24-32.

Jumper, Sidney R. "The Florida Truck Crop Industry: Some Characteristics, Problems and Prospects." *Virginia Geographer*, 4, No. 2 (1969): 1-9.

Potke, P.M. "Citrusbau und Citrus-verarbeitung in Florida." *Zeitischrift fur Wirtschaftsgeographie*, 18, No. 5 (1974): 160-162.

Marcus, Robert B., and Edward A. Fernald. "Citrus, Vegetables and Specialty Crops." Pp. 132-148 in *Florida: A Geographical Approach*, edited by Robert B. Marcus and Edward A. Fernald. Dubuque, Iowa: Kendall-Hunt, 1975.

Marcus, Robert B., and Edward A. Fernald. "Field Crops and General Farms." Pp. 149-157 in *Florida: A Geographical Approach,* edited by Robert B. Marcus and Edward A. Fernald. Dubuque, Iowa: Kendall-Hunt, 1975.

Sheldon, Samuel. "Spatial and Temporal Perspectives on Florida's Sugarcane Industry." *Florida Geographer*, 15 (1981): 1-6.

Sheldon, Samuel. "Sugarcane Cultivation in East Florida: Perceptions and Realities." *Florida Geographer*, 19 (1985): 28-35.

Weischet, Wolfgang, and Cesar N. Caviedes. "Citrus in Florida: Ecological Management and Nature's Latest Intervention Through Freeze." *Erdkunde*, 41, No. 3 (1988): 210-225.

Edwards, Doris M., and Cesar N. Caviedes. "Citrus Growing in North Central Florida: Perceptions and Change." *Florida Geographer*, 30 (1999): 70-78.

ANIMAL HUSBANDRY

Tym, Alice L. "Changes in Land Utilization Accompanying the Advent of Thoroughbred Horse Farming in Marion County, Florida." Master's Thesis, University of Florida, Gainesville, Fla., 1966.

Anderson, James R., and Alice L. Tym. "Thoroughbred Horse Farming in Florida." *Southeastern Geographer*, 7 (1967): 50-61.

Mealor, William T. "Open-Range Ranching and Its Contemporary Successors in South Florida." Ph.D. Dissertation, University of Georgia, Athens, Ga., 1972.

Frost, Melvin J. "Florida Marshlands to Cattle Ranches: The Changing Landscape." *Ecumene*, 6 (1974): 19-24.

Harris, James. "The Florida Cattle Industry: A Transition Economy, 1840–1890." Master's Thesis, University of Chicago, Chicago, Ill., 1974.

Marcus, Robert B., and Edward A. Fernald. "Animal Industries." Pp. 158-170 in *Florida: A Geographical Approach*, edited by Robert B. Marcus and Edward A. Fernald. Dubuque, Iowa: Kendall-Hunt, 1975.

Mealor, William T., and Merle C. Prunty, Jr. "Open-Range Ranching in Southern Florida." *Annals of the Association of American Geographers*, 66, No. 3 (1976): 360-376.

Otto, John S. "Open-Range Cattle Herding in Antebellum South Florida, 1842–1860." *Southeastern Geographer*, 26, No. 1 (1986): 55-67.

Otto, John S. "Open-Range Cattle Herding in Southern Florida." *Florida Historical Quarterly*, 65, No. 3 (1987): 317-337.

Clare, Darry K. "Spatial Patterns of the Cattle and Calf Industry in Florida: Perceptions and Realities." Ph.D. Dissertation, University of Florida, Gainesville, Fla., 1996.

DAIRYING

Fry, Virginia K. "An Analysis of the Dairying Industry of Leon County, Florida." Master's Thesis, Florida State University, Tallahassee, Fla., 1966.

AGRICULTURAL POPULATIONS AND SOCIAL ISSUES

Edwards, Doris M. "Changing Perceptions Among Citrus Growers on Florida's Northern Fringe of the Citrus Belt." Master's Thesis, University of Florida, Gainesville, Fla., 1995.

Kirkconnell, Cindy. "Efficiency, Equity, and the Landscape in the Everglades Agricultural Area." Master's Thesis, University of South Carolina, Columbia, S.C., 1999.

ENVIRONMENTAL AND CONSERVATION ISSUES

Hifny, Mohammed B.E. "Climate and Agriculture in Florida." Ph.D. Dissertation, Clark University, Worcester, Mass., 1950.

Ward, Robert M. *Cold and Wind Hazard Perception by Orange and Tomato Growers in Central and South Florida*. Ann Arbor, Mich.: University of Michigan, Department of Geography, Michigan Geographical Publication No. 9, 1973.

Renner, M.E. "Probability of Drought-Induced Corn Crop Failure in North Florida." Master's Thesis, University of Florida, Gainesville, Fla., 1982.

Stowers, Dewey M., and M.L. LeVasseur. "The Effects of Fluoride Pollutants on Agriculture in West Central Florida." *Florida Geographer*, 18, No. 1 (1984): 1-7.

Hollander, Gail M. "Agroenvironmental Conflict and World Food System Theory: Sugarcane in the Everglades Agricultural Area." *Journal of Rural Studies*, 11, No. 3 (1995): 309-318.

Wood, James. "Agriculture and Water Quality in Madison County, Florida: An Assessment of Conservation Practices." Master's Thesis, Florida State University, Tallahassee, Fla., 1996.

Hollander, Gail M. "Raising Cane in the Glades: Regional Development and Agro-Environmental Conflict in South Florida." Ph.D. Dissertation, University of Iowa, Iowa City, Iowa, 1999.

Letson, David, et al. "Florida's Agriculture and Climate Variability: Reducing Vulnerability." *Florida Geographer*, 32 (2001): 38-57.

Alabama

GENERAL TEXTS, ARTICLES, AND THESES

Totten, Bertha J. "Agriculture in Alabama: Changing Aspects of Diversification and Specialization." Master's Thesis, Eastern Illinois University, Charlestown, Ill., 1973.

Kimbrel, Steven. "Agricultural Occupance Types in South Baldwin County, Alabama." Master's Thesis, University of Southern Mississippi, Hattiesburg, Miss., 1976.

CROPS AND CROPPING PATTERNS

Tower, J. Allen. "Alabama's Shifting Cotton Belt." *Alabama Review*, 1 (1948): 27-38.

Tower, J. Allen. "Cotton Change in Alabama, 1879–1946." *Economic Geography*, 26, No. 1 (1950): 6-28.

Reynolds, William R. "Recent Cotton Production in the Tennessee Valley of North Alabama, 1950–1960." Master's Thesis, University of Tennessee, Knoxville, Tenn., 1961.

Perry, John L. "Crop Combination Regions in Alabama, 1934–1964." Master's Thesis, University of Alabama, Tuscaloosa, Ala., 1968.

Bevis, Douglas L. "Commercial Peach Production in Alabama: A Geographic Analysis." Master's Thesis, University of Alabama, Tuscaloosa, Ala., 1982.

ANIMAL HUSBANDRY

Stickney, Hazel L. "The Conversion from Cotton to Cattle Economy in the Alabama Black Belt, 1930–1960." Ph.D. Dissertation, Clark University, Worcester, Mass., 1961.

Lancaster, Joel R. "A Geographic Analysis of the Broiler Poultry Industry of Cullman County, Alabama." Master's Thesis, University of Tennessee, Knoxville, Tenn., 1968.

FARM LAND

Morris, William R. "Alabama: Agricultural Regions and Land Use Past and Present." Master's Thesis, University of Michigan, Ann Arbor, Mich., 1974.

Rumney, Thomas A. "The Vine and Olive Land Grants of Alabama: Geographical Ideal Versus Reality." *Southeastern Geographer*, 20, No. 2 (1980): 120-133.

Moon, Henry E., Jr. "The Loss of Farmland: The Process of Urbanization in Alabama's Black Belt." Master's Thesis, University of Alabama, Tuscaloosa, Ala., 1984.

AGRICULTURAL POPULATIONS AND SOCIAL ISSUES

Kollmorgen, Walter M. *The German Settlement in Cullman County, Alabama: An Agricultural Island in the Cotton Belt.* Washington, D.C.: Department of Agriculture, 1941.

Mississippi

CROPS AND CROPPING PATTERNS

Thomas, J.A. "An Analysis of Selected Crop Yields in Mississippi During El Nino, La Nina, and Neutral Phases." Master's Thesis, University of Southern Mississippi, Hattiesburg, Miss., 1999.

ANIMAL HUSBANDRY

Israel, Kenneth D. "A Geographical Analysis of the Cattle Industry in Southeastern Mississippi from Its Beginning to 1860." Master's Thesis, University of Southern Misssissippi, Hattiesburg, Miss., 1970.

DAIRYING

Kohn, Clyde F. "Development of Dairy Farming in Mississippi." *Economic Geography*, 19, No. 2 (1943): 188-195.

FARM LAND

Gilbert, Dennis W. "Agricultural Land Use Changes in De Soto County, Mississippi, 1949–1969." Master's Thesis, Memphis State University, Memphis, Tenn., 1970.

Kelley, Arthell. "Small Farm Land Tenure in the Yazoo Basin." *Mississippi Geographer*, 3, No. 1 (1975): 37-42.

McKnight, James P. "A Geographical Study of Selected Factors in the Conversion of Prime Agricultural Land to Non-Agricultural Uses in Northeast Lamar County, Mississippi." Master's Thesis, University of Southern Mississippi, Hattiesburg, Miss., 1981.

ENVIRONMENTAL AND CONSERVATION ISSUES

Sumi, Lisa S. "Seasonal Availability and Mineralization of Nitrogen in Cotton and Soybean Cropping Systems in Mississippi." Master's Thesis, University of Toronto, Toronto, Ontario, 1999.

Tennessee

GENERAL TEXTS, ARTICLES, AND THESES

Glendinning, Robert M., and E.N. Torbert. "Agricultural Problems in Granger County, Tennessee." *Economic Geography*, 14, No. 2 (1938): 159-166.

Prunty, Merle C., Jr. "Evolution of the Agricultural Geography of Dyer County, Tennessee." Ph.D. Dissertation, Clark University, Worcester, Mass., 1944.

CROPS AND CROPPING PATTERNS

Plumacher, D. "Zur Geschichte des Weinbaus in Tennessee." *Das Ausland*, 59, No. 12 (1886): 226-227.

Bird, Elsie T. "Tobacco in East Tennessee." Master's Thesis, University of Tennessee, Knoxville, Tenn., 1948.

Hill, James E., Jr. "The Strawberry Industry in the Hilly and Mountainous Regions of Tennessee." Master's Thesis, University of Tennessee, Knoxville, Tenn., 1959.

Foust, J.B. "The Eastern Dark-Fired Tobacco Region of Kentucky and Tennesse." Master's Thesis, University of Tennessee, Knoxville, Tenn., 1966.

Jumper, Sidney R. "Tennessee's Place in the National Fresh Vegetable Industry." *Virginia Geographer*, 6, No. 1 (1971): 8-10.

DAIRYING

Kohland, William F. "The Geography of the Knoxville Milk Shed." Master's Thesis, University of Tennessee, Knoxville, Tenn., 1953.

McMahan, W.O. "The Geography of the Nashville Milkshed." Master's Thesis, University of Tennessee, Knoxville, Tenn., 1953.

Marshall, Henry F. "The Memphis Milkshed: A Geographical Analysis." Master's Thesis, Memphis State University, Memphis, Tenn., 1968.

FARM LAND

Fravega, Elizabeth M. "Changes in the Patterns of Agricultural Land Use in West Tennessee." Master's Thesis, Memphis State University, Memphis, Tenn., 1967.

Shankman, David. "Forest Regeneration on Abandoned Agricultural Fields in Western Tennessee." *Southeastern Geographer*, 30, No. 1 (1990): 36-47.

AGRICULTURAL POPULATIONS AND SOCIAL ISSUES

Kollmorgen, Walter M. "Observations on Cultural Islands in Terms of Tennessee Agriculture." *East Tennessee Historical Society Publications*, 16 (1944): 65-78.

ENVIRONMENTAL AND CONSERVATION ISSUES

Prunty, Merle C., Jr. "Physical Bases of Agriculture in Dyer County, Tennessee." *Journal of the Tennessee Academy of Science*, 23 (1948): 215-235.

Kung, Hsiang-te. "Relationships of Agricultural Land use to Slope and Soil Characteristics in Portions of the Valley of East Tennessee." Master's Thesis, University of Tennessee, Knoxville, Tenn., 1972.

Emison, John A. "Participation of Farm Operators in Public Land Conservation Programs in the Cypress Basin of Crockett County, Tennessee." Ph.D. Dissertation, University of Tenn., Knoxville, Tennessee, 1979.

Bright, Edward A. "An Evaluation of Levee Protection of Agricultural Land in West Tennessee." Master's Thesis, University of Tennessee, Knoxville, Tenn., 1982.

Smith, Larry S. "Targeting Areas Prone to Agricultural Non-Point Source Pollution in Bent Creek, Hamblen County, Tennessee: A Comparison of AGNPS and Conventional Methods." Master's Thesis, University of Tennessee, Knoxville, Tenn., 1991.

AQUACULTURE

Minton, James O. "Catfish Farming: A New Tennessee Farm Crop." Master's Thesis, University of Tennessee, Knoxville, Tenn., 1971.

TRANSPORTATION AND MECHANIZATION

Buckwalter, Donald W. "Effects of Early Nineteenth Century Transportation Disadvantage on the Agriculture of Eastern Tennessee." *Southeastern Geographer*, 27, No. 1 (1987): 18-37.

Kentucky

GENERAL TEXTS, ARTICLES, AND THESES

Greenberg, Arthur. "The Bluegrass Tobacco Barn and the Physiological Subdivisions of the Kentucky Bluegrass: A Case Study in Areal Association." Master's Thesis, University of Cincinnati, Cincinnati, Ohio, 1965.

Raitz, Karl B. "Gentleman Farms in Kentucky's Inner Bluegrass: A Problem in Mapping." *Southeastern Geographer*, 15, No. 1 (1975): 33-46.

CROPS AND CROPPING PATTERNS

Whitaker, J. Russell. "The Development of the Tobacco Industry in Kentucky: A Geographical Interpretation." *Bulletin of the Philadelphia Geographical Society*, 27, No. 1 (1929): 15-42.

Reeves, Hershel C. "A Geographical Study of the Production and Marketing of Burley Tobacco in Kentucky." Master's Thesis, University of Kentucky, Lexington, Ky., 1956.

FARM LAND

Hart, John Fraser. "Abandonment of Farm Land in Kentucky." *Southeastern Geographer*, 4 (1964): 1-10.

Hart, John Fraser. "Abandonment of Farm Land on the Appalachian Fringe of Kentucky." Pp. 352-360 in *Melanges de Geographie: Physique, Humaine, Economique, Appliquee, offerts a M. Omer Tulippe*, edited by L. Dethier. Gembloux, Belgium: J. Duculot, 1968.

ENVIRONMENTAL AND CONSERVATION ISSUES

Collier, James E. "Topographic and Soil Factors in Kentucky's Agriculture." Master's Thesis, University of Cincinnati, Cincinnati, Ohio, 1938.

Tucker, Nancy. "Nonpoint Agriculture Pollution in a Karst Aquifer: Lost River Groundwater Drainage Basin, Warren County, Kentucky." Master's Thesis, Western Kentucky University, Bowling Green, Ky., 1982.

Raitz, Karl B., and Nancy O'Malley. "An Index of Soil Production Potential as Applied to Historic Agricultural Adaptation: A Kentucky Example." *Historical Methods*, 18, No. 4 (1985): 137-146.

Louisiana

CROPS AND CROPPING PATTERNS

Foscue, Edwin J., and E. Foscue. "Sugar Plantations of the Irish Bend District, Louisiana." *Economic Geography*, 12, No. 4 (1936): 373-380.

Chatterton, H.J. "The Development of the Pepper Industry of Louisiana." *Journal of Geography*, 38, No. 9 (1939): 364-369.

Post, Lauren C. "The Rice Country of Southwestern Louisiana." *Geographical Review*, 30, No. 4 (1940): 574-590.

Blume, Helmut. *Zuckerrohranbau am unteren Mississippi (Louisiana Sugar Bowl)*. Regensberg, Germany: M.L. Kallmunz, Muchner Geographische Hefte, No. 5, 1954.

Blume, Helmut. "El cultivo de la cana de azucar en Andalucia, comparado con el cultivo de la cana en Luisiana." *Estudios Geograficos*, 19 (1958): 87-120.

Bederman, Sanford H. "The Citrus Fruit Industry in Louisiana." *Southeastern Geographer*, 1 (1961): 1-8.

Babia, Edward C. "The Riverine District: The Economic Geography of Sugar Cane Production in Southern Louisiana." Master's Thesis, University of Arkansas, Fayetteville, Ark., 1967.

Hernandez, T.D. "The Economic Development of Soybean Production in Natchitoches Parish, Louisiana." Master's Thesis, Northwest Louisiana State University, Natchitoches, La., 1971.

Rehder, John B. "Sugar Plantations in Louisiana: Origin, Dispersal and Responsible Location Factors." *West Georgia College Studies in the Social Sciences*, 12 (1973): 78-93.

Shea, Philip. "The Louisiana Sugar Cane Industry, 1751–1970: A Study in Geographical Relationships." Ph.D. Dissertation, Michigan State University, East Lansing, Mich., 1974.

Babin, Edward C. "A Functional Typology of Louisiana Sugar Cane Production Units in 1969." Ph.D. Dissertation, University of Georgia, Athens, Ga., 1975.

Rehder, John. "Diagnostic Landscape Traits of Sugar Plantations in Southern Louisiana." *Geoscience and Man*, 19 (1978): 135-150.

Schmitz, Mark. "Farm Interdependence in the Antebellum Sugar Sector." *Agricultural History*, 52, No. 1 (1978): 93-103.

Hilliard, Samuel B. "Site Characteristics and Spatial Stability of the Louisiana Sugarcane Industry." *Agricultural History*, 53, No. 1 (1979): 254-269.

Rehder, John B. "Sugar Plantations in Louisiana." *Proceedings of the Tall Timbers Ecology and Management Conference*, 16 (1979): 111-123.

Mitchum, Samuel. "The Origins and Evolution of the Southwestern Louisiana Rice Region, 1880–1920." Ph.D. Dissertation, University of Tennessee, Knoxville, Tenn., 1986.

Rehder, John B. *Delta Sugar: Louisiana's Vanishing Plantation Landscape*. Baltimore, Md.: Johns Hopkins University Press, 1999.

ANIMAL HUSBANDRY

Post, Lauren C. "The Old Cattle Industry in Southwest Louisiana." *McNeese Review*, 9 (1957): 43-55.

Post, Lauren C. "Cattle Branding in Southwest Louisiana." *McNeese Review*, 10 (1958): 101-117.

DAIRYING

Lindsey, Allen D. "The Geography of Dairying in North Louisiana Parishes." Master's Thesis, University of Arkansas, Fayetteville, Ark., 1971.

Walker, Ronald J. "Factors in the Regionalization of the Dairy Industry in Louisiana." Master's Thesis, Northwest Louisiana State University, Natchitoches, La., 1971.

FARM LAND

Kollmorgen, Walter M., and Robert W. Harrison. "The Place of French Speaking Farmers of Southern Louisiana in Future Land Development and Reclamation." *Land Economics*, 22 (1946): 223-231.

Arkansas

GENERAL TEXTS, ARTICLES, AND THESES

Corbet, John H. "The Arkansas Grand Prairie: Delineation and Resource Use of an Agricultural Region." Ph.D. Dissertation, University of Florida, Gainesville, Fla., 1966.

Pecotte, Michael J. "Buffalo Island: A Cultural/Agricultural Island in the St. Francis Basin of Arkansas." Master's Thesis, University of Tennessee, Knoxville, Tenn., 1976.

Stroud, Hubert B., and Gerald T. Hansen. "Agriculture in Arkansas." Pp. 63-68 in *Arkansas Geography: The Physical Landscape and the Historical-Cultural Setting*, edited by Hubert B. Stroud and Gerald T. Hansen. Little Rock, Ark.: Rose, 1981.

CROPS AND CROPPING PATTERNS

Jenks, George F. "New Land Development in the Rice Economy of the Weiner Area, Arkansas." Ph.D. Dissertation, Syracuse University, Syracuse, N.Y., 1950.

Hewes, Leslie. "Tontitown: Ozark Vineyard Center." *Economic Geography*, 29, No. 2 (1953): 125-143.

Lyons, Arland W. "Geography of the Nashville–Highland Peach District of Southwest Arkansas." Master's Thesis, University of Arkansas, Fayetteville, Ark., 1967.

LaFevers, James. "A Geographic Analysis of Precision Leveling of Rice Fields in Lonoke County, Arkansas." Master's Thesis, University of Southern Mississippi, Hattiesburg, Miss., 1970.

AGRICULTURAL POPULATIONS AND SOCIAL ISSUES

Bratton, Sam T. "Agricultural Communities in the St. Francis Basin." *Bulletin of the Geographical Society of Philadelphia*, 26 (1928): 157-167.

ENVIRONMENTAL AND CONSERVATION ISSUES

Reid, James W., Jr. "The Problem of Droughts in Arkansas Agriculture." Ph.D. Dissertation, George Peabody University, Nashville, Tenn., 1938.

IRRIGATION

Green, D. Brooks. "Irrigation Experiment in Arkansas: A Preliminary Investigation." *Arkansas Historical Quarterly*, 45, No. 3 (1986): 261-268.

The Midwestern United States

GENERAL TEXTS, ARTICLES, AND THESES

Shirley, A.E. "A Typical Farm in the Corn Belt: A Project." *Journal of Geography*, 24, No. 6 (1925): 220-226.

King, Voris V. "The Geography of a Corn-Belt Farm." Master's Thesis, Illinois State University, Bloomington, Illinois, 1949.

Kollmorgen, Walter M. "Significance of Agriculture." Pp. 40-51 in *The North American Midwest: A Regional Geography*, edited by John H. Garland. New York, N.Y.: John Wiley, 1955.

Kollmorgen, Walter M. "Farms and Farming in the American Midwest." Pp. 75-91 in *Problems and Trends in American Geography*, edited by Saul B. Cohen. New York, N.Y.: Basic Books, 1967.

Akin, Wallace E. "Agricultural Regions of the North Central States." Pp. 72-95 in *The North Central United States*, edited by Wallace E. Akin. Princeton, N.J.: Van Nostrand, 1968.

Wishart, David. "Agriculture at the Trading Posts on the Upper Missouri Prior to 1843." *Agricultural History*, 47, No. 1 (1973): 57-62.

Hart, John Fraser. "Two Corn Belt Farms." *Journal of Cultural Geography*, 5, No. 1 (1984): 7-18.

Hart, John Fraser. "Change in the Corn Belt." *Geographical Review*, 76, No. 1 (1986): 51-72.

Roberts, Rebecca, and David R. Lighthall. "The Political Economy of Agriculture, Ground Water Quality Management and Agricultural Research." *Water Resources Bulletin*, 27, No. 3 (1991): 437-446.

Santer, Richard A. "Food and Agriculture in the Economy." Pp. 297-314 in *Geography of Michigan and the Great Lakes Basin*, edited by Richard A. Santer. Dubuque, Iowa: Kendall-Hunt, 1993.

Lighthall, David R., and Rebecca Roberts. "Towards an Alternative Logic of Technological Change: Insights from Corn Belt Agriculture." *Journal of Rural Studies*, 11, No. 3 (1995): 319-334.

CROPS AND CROPPING PATTERNS

Rose, John K. "Corn Yield and Climate in the Corn Belt." *Geographical Review*, 26, No. 1 (1936): 88-102.

Karnes, Lowry B. "A Geographical Analysis of the Corn and Winter Wheat Belt of the United States." Ph.D. Dissertation, Ohio State University, Columbus, Ohio, 1940.

Weaver, John C. "Crop-Combination Regions in the Middle West." *Geographical Review*, 44, No. 2 (1954): 175-200.

Weaver, John C. "Crop-Combination Regions for 1919 and 1929 in the Middle West." *Geographical Review*, 44, No. 4 (1954): 560-572.

Wallace, Henry A. "Corn and the Midwestern Farmer." *Landscape*, 6, No. 3 (1957): 9-12.

Hidore, John J. "The Relationship Between Cash-Grain Farming and Flat Land in the Western Midwest." Master's Thesis, University of Iowa, Iowa City, Iowa, 1958.

Marple, Robert P. "The Transition Zone Between the Corn Belt and Hard Winter Wheat Belt." Ph.D. Dissertation, University of Nebraska, Lincoln, Nebr., 1958.

Roepke, Howard G. "Changes in Corn Production on the Northern Fringes of the Corn Belt." *Agricultural History*, 33, No. 3 (1959): 126-132.

Hoag, Leverett P. "Locational Determinants for Cash-Grain Farming in the Corn Belt." *Professional Geographer*, 14 (1962): 1-7.

Heathcote, R.L. "Bread or Cake: A Geographer and a Historian on the Nineteenth Century Wheat Frontier: A Review." *Economic Geography*, 39, No. 2 (1963): 176-182.

Fentem, Arlin D. "Cash Grain Farming in the North American Corn Belt." Ph.D. Dissertation, University of Wisconsin, Madison, Wis., 1974.

Persi, P. "Il Corn Belt, una regione agraria in continzione?" *Bolletin della Societa Geografica Italiana*, Serie 10, No. 9 (1980): 461-478.

Hartnett, Sean. "Harvesting the New Land: A Geographical Appraisal of the Wheat Frontier in the Upper Midwest, 1835–1885." Master's Thesis, University of Wisconsin, Madison, Wis., 1981.

Nellis, M. Duane. "The Emergence of an American Oilseed Sunflower Region." *Geographical Perspectives*, 54 (1984): 47-52.

ANIMAL HUSBANDRY

Shaw, Earl B. "Swine Production in the Corn Belt of the United States." *Economic Geography*, 12, No. 4 (1936): 359-372.

Weaver, John C., Leverett P. Hoag, and Barbara L. Fenton. "Livestock Units and Combination Regions in the Middle West." *Economic Geography*, 32, No. 3 (1956): 239-259.

Beach, J. Richard. "The Sheep Industry in the Upper Ohio Area, 1770–1973: A Geographic Analysis." Ph.D. Dissertation, University of Pittsburgh, Pittsburgh, Pa., 1975.

Beach, J. Richard. *Two Hundred Years of Sheep Raising in the Upper Ohio Valley*. Monongahela, Pennsylvania: Bicentennial Commission of Washington County, Pa., 1976.

DAIRYING

Lewthwaite, Gordon R. "Cows in the Corn: The Emergence of the Tri-State Butter Region." *Proceedings of the Association of American Geographers*, 7 (1975): 113-117.

FARM LAND

Hart, John Fraser. "Cleared Farm Land in the Upper Lakes States, 1910–1964." *Journal of the Minnesota Academy of Science*, 36 (1969): 29-31.

Smith, Everett G. "Fragmented Farms in the Central States." *Annals of the Association of American Geographers*, 65, No. 1 (1975): 58-70.

Hart, John Fraser. "Part-Ownership and Farm Enlargement in the Midwest." *Annals of the Association of American Geographers*, 81, No. 1 (1991): 66-79.

AGRICULTURAL POPULATIONS AND SOCIAL ISSUES

Lewthwaite, Gordon R. "Midwestern Swiss Migrants and Foreign Cheese." *Yearbook of the Association of Pacific Coast Geographers*, 34 (1972): 41-60.

Hart, John Fraser. "People and Farms." Pp. 46-79 in *Upper Coulee Country*, edited by Eugene C. Mather, John Fraser Hart, and Hildegard Binder Johnson. Prescott, Wis.: Trimbelle Press, 1975.

Hart, John Fraser. "The Persistence of Family Farming Areas." *Journal of Geography*, 86, No. 5 (1987): 198-203.

ENVIRONMENTAL AND CONSERVATION ISSUES

Rose, John K. "Corn Yield and Climate in the Corn Belt." Ph.D. Dissertation, University of Chicago, Chicago, Ill., 1935.

Albrecht, James. "A Climatic Model for Evaluating Agricultural Potential in the Missouri River Basin." Ph.D. Dissertation, Southern Illinois University, Carbondale, Ill., 1968.

Limbird, Arthur. "The Influence of Contrasting Adjacent Soil Regions on the Agricultural Systems of Selected Areas in Illinois, Indiana and Ohio." Ph.D. Dissertation, Michigan State University, East Lansing, Mich., 1971.

Moran, Joseph M., W.A. Dugas, and R. Olson. "Agricultural Implications of Climatic Change." *Journal of Soil and Water Conservation*, 32 (1977): 80-83.

Burkholder, Martin J. "The Life Cycles of Farmland Terraces in Selected Iowa and Illinois Counties." Master's Thesis, Western Illinois Univesity, Macomb, Ill., 1982.

Poncet, Robert A. "A Comparison of Agro-Climatic Advantages of Two Regions: The Mississippi Delta and Central Midwest." Master's Thesis, Mississippi State University, Miss., 1990.

Lighthall, David R. "Chemicals, Class and Sustainable Agriculture: A Corn Belt Case Study." Ph.D. Dissertation, University of Iowa, Iowa City, Iowa, 1993.

McCorvie, Mary R., and Christopher L. Lant. "Drainage District Formation and the Loss of Midwestern Wetlands, 1850–1930." *Agricultural History*, 67, No. 4 (1993): 13-39.

Prince, Hugh. "Floods in the Upper Mississippi River Basin, 1993: Newspapers, Official Views and Forgotten Farmlands." *Area*, 27, No, 2 (1995): 118-126.

Garrity, Colleen. "Climatological Constraints of a Wheat-Soybean Double-Cropping System in the Great Lakes Region." Master's Thesis, Michigan State University, East Lansing, Mich., 1999.

Knox, James C. "Agricultural Influence on Landscape Sensitivity in the Upper Mississippi River Valley." *Catena*, 42, No. 2-4 (2001): 193-224.

IRRIGATION

Hidore, John J. "Supplemental Irrigation in the Upper Mississippi Valley." *Professional Geographer*, 19, No. 4 (1967): 184-188.

AGRICULTURAL MARKETING AND DISTRIBUTION

Johnson, James F. "The Influence of Cost Distance Factors on the Overseas Export of Corn from the United States Midwest." *Economic Geography*, 45, No. 2 (1969): 170-179.

Erickson, Frank A. "Location of a System of Storage-Distribution Terminals of Agricultural Ammonia in the Corn Belt." Ph.D. Dissertation, University of Illinois, Urbana, Ill., 1973.

TRANSPORATATION AND MECHANIZATION

Johnson, Paul E. "The Diffusion of Commercial Fertilizer in the Corn Belt." Master's Thesis, University of Tennessee, Knoxville, Tenn., 1978.

Ohio

GENERAL TEXTS, ARTICLES, AND THESES

Russ, Frances. "Agricultural Geography of Darke County, Ohio." Master's Thesis, Miami University, Miami, Ohio, 1933.

Doughton, Helen R. "The Agricultural Geography of the Maumee Plain." Master's Thesis, Ohio State University, Columbus, Ohio, 1941.

Stephen, L. LeMar. "An Agricultural Survey of Portage County, Ohio." *Economic Geography*, 19, No. 2 (1943): 156-170.

Hamm, Lewis R. "The Agriculture of Meigs County." Master's Thesis, Ohio State University, Columbus, Ohio, 1951.

Kakela, Peter J. "A Study of Conservation Practices on Two Indian Creek Demonstration Area Farms." Master's Thesis, Miami University, Miami, Ohio, 1964.

Noble, Allen G. "The Agriculture of Ohio." Pp. 54-72 in *Ohio: An American Heartland*, edited by Allen G. Noble and Albert J. Korsok. Columbus, Ohio: State of Ohio Department of Natural Resources, 1975.

Southwick, Stacy. "A Case Study of Wood County Agricultural Decision Making." Master's Thesis, University of Toledo, Toledo, Ohio, 1992.

CROPS AND CROPPING PATTERNS

Reisinger, Ann. "The Sugar Beet Industry of Ohio." Master's Thesis, Miami University, Miami, Ohio, 1929.

Morrison, Paul C. "The Grape Industry of Ohio." Master's Thesis, Ohio State University, Columbus, Ohio, 1933.

Owens, Olive R. "A Geographical Interpretation of the Miami Valley Tobacco Industry." Master's Thesis, Miami University, Miami, Ohio, 1934.

Morrison, Paul C. "Viticulture in Ohio." *Economic Geography*, 12, No. 1 (1936): 71-85.

Amann, Richard J. "Changes in Crop Agriculture in Ohio, 1939–1949." Master's Thesis, Miami University, Miami, Ohio, 1954.

Mulhauser, John E. "Commercial Greenhouse Horticulture in Cuyahoga County, Ohio." Master's Thesis, Kent State University, Kent, Ohio, 1961.

Jenks, W.R. "Burley Tobacco Agriculture in Ohio." Master's Thesis, Miami University, Miami, Ohio, 1968.

Williams, R.J. "Analysis of Some Factors Influencing the Introduction, Regionalization and Expansion of Soybean Production in Northeast Ohio." Master's Thesis, Bowling Green State University, Bowling Green, Ohio, 1969.

Jenks, W.R. "Ohio's Burley Tobacco Agriculture: A Primary Regional Cash Crop." *Ohio Journal of Science*, 72, No. 5 (1972): 285-291.

Sifritt, Susan K. "The Ohio Wine and Wine-Grape Industries." Ph.D. Dissertation, Kent State University, Kent, Ohio, 1976.

Rumney, Thomas A. "Ohio Soybean Production: A Geographic Inquiry." *Ohio Geographer*, 16 (1988): 57-67.

ANIMAL HUSBANDRY

White, C. Langdon, and Clyde E. Cooper. "The Sheep Industry of Southeastern Ohio." *Economic Geography*, 7 (1931): 263-272.

Heinlein, Paul C. "Cattle Driving from the Ohio Country, 1800–1850." *Agricultural History*, 28, No. 2 (1954): 83-94.

Heinlein, Paul C. "Shifting Range-Feeder Patterns in the Ohio Valley Before 1860." *Agricultural History*, 31 (1957): 1-12.

Stover, Stephen L. "The Changing Regionalization of Sheep Husbandry in Ohio." Ph.D. Dissertation, University of Wisconsin, Madison, Wis., 1960.

Stover, Stephen L. "Early Sheep Husbandry in Ohio." *Agricultural History*, 36, No. 2 (1962): 101-107.

Stover, Stephen L. "Ohio's Sheep Year: 1868." *Agricultural History*, 38, No. 2 (1964): 102-107.

Knutsen, Glade W. "Development and Decline of the Sheep Industry in a Marginal Area: A Case Study of Athens County, Ohio: 1840–1900." Master's Thesis, Ohio University, Athens, Ohio, 1992.

DAIRYING

Frazier, Nelson V. "The Development of the Dairy Industry in Belmont County, Ohio." Master's Thesis, Ohio State University, Columbus, Ohio, 1959.

Parish, Richard J. "A Study of Selected Dairy Farms and Changing Land Use Patterns in Homer and Spencer Townships, Medina County, Ohio." Master's Thesis, Kent State University, Kent, Ohio, 1960.

FARM LAND

Sharp, Daniel W. "Agricultural Land Use: Potential Soil Productivity Relationships in a Southwestern Ohio Township." Master's Thesis, Miami University, Miami, Ohio, 1974.

Williams, Michael. "Ohio: Microcosm of Agricultural Clearing in the Midwest." Pp. 3-13 in *Global Deforestation and the Nineteenth Century World Economy*, edited by R.P. Tucher and J.F. Richard. Durham, N.C.: Duke University Press, 1983.

Okey, Brian W. "Agricultural Intensification and Landscape Pattern in the Upper Four Mile Creek Watershed, Preble County, Ohio." Master's Thesis, Miami University, Miami, Ohio, 1992.

ENVIRONMENTAL AND CONSERVATION ISSUES

Garrigus, Woolford M. "Average Slope and Agricultural Productivity in Ohio." Ph.D. Dissertation, Clark University, Worcester, Mass., 1958.

Sauder, Robert A. "A Geographic Analysis of the Impact of Blackbird Feeding upon Changes in Corn Average in Ottawa and Sandusky Counties, Ohio, 1957–1966." Master's Thesis, Bowling Green State University, Bowling Green, Ohio, 1968.

Limbird, A. "Wisconsin Versus Illinoian Age Soils: Their Effect on Agricultural Types in Ohio." *Ohio Journal of Science*, 80, No. 6 (1980): 273-279.

Fritsch, Thomas. "The Effect of Climate on Soybean Yields in Northwestern Ohio." Master's Thesis, University of Akron, Akron, Ohio, 1981.

Orndorff, Maurice R. "The Use of No-Till Agriculture in Geauga County, Ohio." Master's Thesis, Kent State University, Kent, Ohio, 1988.

Zimmerman, Roger D. "An Assessment of the Usage of an Agricultural Non-Point Pollution Model in Summit County, Ohio." Master's Thesis, University of Akron, Akron, Ohio, 1991.

Rollins, Yvonne P. "Sediment Budget Analysis for an Agricultural Watershed in Southwestern Ohio." Master's Thesis, Miami University, Miami, Ohio, 1995.

AGRICULTURAL MARKETING AND DISTRIBUTION

Sunderman, John W. "Factors Relating to the Location and Source Areas of Alfafa Mills in Henry and Wood Counties, Ohio." Master's Thesis, Bowling Green State University, Bowling Green, Ohio, 1974.

Indiana

GENERAL TEXTS, ARTICLES, AND THESES

Carter, George F. "Agricultural Geography of Southwest Indiana." Ph.D. Dissertation, University of California, Berkeley, Calif., 1942.

Bracken, Lawson E. "Some Geographical Factors in a Quarter Century of Indiana Harvests, 1926–50." *Proceedings of the Indiana Academy of Science*, 62 (1952): 256-265.

Frederic, Paul B. "Agricultural Landscape Transformation: Parke County, Indiana." *Proceedings, New England–St. Lawrence Valley Division of the Association of American Geographers*, 3 (1973): 7-14.

Showalter, Gerald R. "An Analysis of the Spatial Structure of Differences of Agriculture in Southern Indiana." Ph.D. Dissertation, University of Illinois, Urbana, Ill., 1973.

CROPS AND CROPPING PATTERNS

Zierer, Clifford M. "Utilization of Indiana Muck Lands for Specialty Crops." *Journal of Geography*, 31, No. 9 (1932): 313-323.

Thornbury, William D. "The Tomato Industry of Indiana." *Economic Geography*, 18, No. 1 (1942): 68-78.

Alcorn, Nina. "The Tomato Industry of Indiana." Master's Thesis, Miami University, Miami, Ohio, 1949.

Landing, James E. "Analysis of the Decline of the Commerical Mint Industry of Indiana Since World War II." Master's Thesis, Pennsylvania State University, University Park, Pa., 1963.

Cott, Neal E. "A Study of the Processes and Problems of Watermelon Production in Knox County, Indiana." Master's Thesis, Indiana State University, Terre Haute, Ind., 1971.

Hartig, Robert J. "Trends in Apple and Peach Orcharding in the Lower Wabash Valley of Indiana." Master's Thesis, Eastern Illinois University, Charlestown, Ill., 1973.

Cott, Neal E., and George Webb. "Watermelon Production in Southwestern Indiana." *Indiana State University Department of Geography Papers*, 10 (1978): 11-27.

Weber, Neil. "Modeling Predictive Indices for Indiana Corn Production: 1960–1969." *Indiana State University Department of Geography Papers*, 10 (1978): 3-10.

FARM LAND

Kasambira, T.S. "Diffusion of the Decision to Drain Agricultural Land in Delaware County, Indiana." *Geographical Bulletin*, 18 (1979): 12-20.

Venugopal, G., and Paul W. Mausel. "Application of Remotely Sensed Data to Measure Agricultural Land Conversion in Metropolitan Fringe of Indianapolis, USA." *Indian Geographical Journal*, 67, No. 1 (1992): 40-68.

ENVIRONMENTAL AND CONSERVATION ISSUES

Rose, John K. "Climate and Corn Yield in Indiana, 1887–1930." *Proceedings of the Indiana Academy of Science*, 41 (1932): 317-321.

Mausel, Paul W. "Agricultural Productivity of Prairie, Forest and Prairie-Forest Transition Soils in Indiana." *Indiana State University Department of Geography Papers*, 4 (1973): 44-67.

Michigan

GENERAL TEXTS, ARTICLES, AND THESES

Whitaker, J. Russell. "The Relation of Agriculture to Mining in the Upper Peninsula of Michigan." *Journal of Geography*, 25, No. 1 (1926): 21-29.

Vanderhill, Burke G. "The Principle Farming Types of Antrim County, Michigan." *Papers of the Michigan Academy of Science, Arts and Letters*, 34 (1948): 195-206.

Deasy, George F. "Agriculture in Luce County, Michigan, 1880 to 1930: A Study of Agricultural Development in the Upper Great Lakes Region." *Agricultural History*, 34, No. 1 (1950): 29-41.

Daoust, Charles F. "Transition in Central Michigan: Agriculture to Industry." Master's Thesis, University of Chicago, Chicago, Ill., 1954.

Heimonen, Henry S. "Agricultural Trends in the Upper Peninsula." *Michigan History*, 41, No. 1 (1957): 45-52.

Meeks, Harold A. "Marginal Agriculture in Michigan's Copper Country." Ph.D. Dissertation, University of Minnesota, Minneapolis, Minn., 1964.

Sommers, Lawrence A. "Agriculture." Pp. 128-134 in *Michigan: A Geography*, edited by Lawrence A. Sommers. Boulder, Colo.: Westview, 1984.

CROPS AND CROPPING PATTERNS

Ketchum, Hazel. "Fruit Growing in Michigan." *Journal of Geography*, 16, No. 3 (1917-18): 90-96.

Stilgenbauer, F.A. "The Michigan Sugar Beet Industry." *Economic Geography*, 3, No. 4 (1927): 486-506.

Stilgenbauer, F.A. "Chicory: Michigan's Infant Monopoly Crop." *Economic Geography*, 7, No. 1 (1931): 84-100.

Hudgins, Bert. "Bean Production in Michigan." *Economic Geography*, 9, No. 3 (1933): 265-272.

Brown, E. Lucille. "A Study of the Cherry Industry of the Michigan Fruit Belt." Master's Thesis, Wayne State University, Detroit, Mich., 1941.

Hull, Gordon C. "A Geographical Analysis of the Possibility for Expansion of the Cherry Industry of Benona Township, Oceana County, Michigan." Master's Thesis, Syracuse University, Syracuse, N.Y., 1948.

Deasy, George F. "The Blueberry Industry in an Upper Michigan County." *Journal of Geography*, 48, No. 6 (1949): 239-247.

Olmstead, Clarence W. "Patterns of Orchards in Michigan." Ph.D. Dissertation, University of Michigan, Ann Arbor, Mich., 1951.

Ball, John M. "Changes in Sugar Beet Production in Michigan, 1899–1958." *Papers of the Michigan Academy of Science, Arts, and Letters*, 45 (1960): 145-153.

Senninger, Earl J. "The Chicory Industry of Michigan." *Papers of the Michigan Academy of Science, Arts, and Letters*, 45 (1960): 145-153.

Westinghouse, Barbara. "The Grape Industry in Southwest Michigan: A Geographic Study." Master's Thesis, Michigan State University, East Lansing, Mich., 1961.

Landing, James E. "The Decline of the Commercial Mint Industry in Michigan." *Professional Geographer*, 15 (1963): 21-25.

Grossa, John. "Viticulture in Michigan." Master's Thesis, University of Michigan, Ann Arbor, Mich., 1967.

Yli-Jokipii, Pentii. "Obstabaugebeit am Michigansee." *Zeitschrift fur Wirtschaftsgeographie*, 15, No. 4 (1971): 112-116.

Peters, Bernard C. "Comments on the Distribution of Garden Beds in Kalamazoo County." *Ecumene*, 12 (1980): 17-25.

Bogue, Margaret. "The Lake and the Fruit: The Making of Three Farm-Type Areas." *Agricultural History*, 59, No. 4 (1985): 493-522.

Heller, Charles F., and John T. Houdek. "When Wheat Was King: An Analysis of Michigan's Wheat Belt in 1883." *East Lakes Geographer*, 24 (1989): 34-47.

DAIRYING

Durand, Loyal. "The Lower Peninsula of Michigan and the Western Michigan Dairy Region: A Segment of the American Dairy Region." *Economic Geography*, 27, No. 2 (1951): 163-183.

Langschwager, Lyle M. "The Detroit Milkshed of Northeastern Michigan." *Pennsylvania Geographer*, 4 (1966): 14-18.

Calzonetti, Frank J. "Dairying in the Michigan Thumb: An Examination of the State's Dominant Dairy Sub-Region." Master's Thesis, Wayne State University, Detroit, Mich., 1974.

Henley, Ronald L., and Carl F. Ojala. "The Beet Sugar Industry of Michigan: A Geographical Analysis." *Michigan Academician*, 6, No. 3 (1974): 321-332.

FARM LAND

Ward, Robert M., and Rhonda Katz-Barnet. "Assessment of Farmland Preservation Legislation in Michigan: P.A. 116 of 1974." *Michigan Academician*, 10, No. 3 (1978): 307-320.

AGRICULTURAL POPULATIONS AND SOCIAL ISSUES

Wheeler, Jesse O., and Stanley D. Brunn. "An Agricultural Ghetto: Negroes in Cass County, Michigan, 1845–1968." *Geographical Review*, 59, No. 3 (1969): 317-329.

Houdek, John T., and Charles F. Heller. "Searching for Nineteenth-Century Farm Tenants: An Evaluation of Methods." *Historical Methods*, 19, No. 2 (1986): 55-61.

Heller, Charles F., and John T. Houdek. "Farm Tenants and Landlords in Nineteenth-Century Southern Michigan: A Study of Two Townships." *Agricultural History*, 70, No. 4 (1996): 598-625.

ENVIRONMENTAL AND CONSERVATION ISSUES

Winkler, J.A., et al. "Agroclimatic Resource Assessment: An Example for Peach Cultivation in the Lower Peninsula of Michigan." *Physical Geography*, 11, No. 1 (1990): 49-65.

TRANSPORTATION AND MECHANIZATION

Kubicki, Dale L. "An Early Intracounty Railroad Network and the Development of Fruit Culture: Van Buren County, Michigan." Master's Thesis, Western Michigan University, Kalamazoo, Mich., 1969.

Illinois

GENERAL TEXTS, ARTICLES, AND THESES

Hatch, Laura. "Geographical Factors in the Agriculture of Du Page County, Illinois." *Journal of Geography*, 13 (1914-15): 216-223.

Blanchard, O.W. "Some Geographic Factors Affecting Agriculture in Illinois, and the Agricultural Provinces of Illinois." *Journal of Geography*, 23, No. 1 (1924): 1-15.

Branom, Mendel E. "The Agricultural Economy of the American Bottoms." Ph.D. Dissertation, University of Chicago, Chicago, Ill., 1938.

Smith, Mildred. "Illinois Agriculture in the National Setting." Master's Thesis, University of Illinois, Urbana, Ill., 1940.

Dornbach, John. "An Experiment in the Delimination of Types of Farming Areas in the Big Muddy River Basin in Southern Illinois." Master's Thesis, Washington University, St. Louis, Mo., 1950.

Roepke, Howard G. "Patterns of Agricultural Change in Illinois." *Transactions of the Illinois State Academy of Science*, 49 (1956): 98-103.

Nakagawa, August T. "Statistical Mapping of Illinois Agriculture, 1925–45." Master's Thesis, University of Chicago, Chicago, Ill., 1957.

Bevenour, D.E. "Selected Aspects of Agriculture Surrounding American Metropolitan Areas: The Peoria Example." Master's Thesis, Illinois State University, Bloomington, Ill., 1970.

Werner, Carol A. "Agriculture and Urban Change: The Case of Metro East." *Bulletin, Illinois Geographical Society*, 15, No. 1 (1973): 29-40.

Price, Dalias A. "Farms and Farming." Pp. 188-210 in *Illinois: Land and Life in the Prairie State*, edited by Ronald E. Nelson. Dubuque, Iowa: Kendall-Hunt, 1978.

Meyer, Douglas K. "Types of Farming on the Illinois Frontier." *Bulletin, Illinois Geographical Society*, 21, No. 1 (1979): 9-17.

Mattingly, Paul F. "Part-Time Farming in Illinois." *Bulletin, Illinois Geographical Society*, 27, No. 1 (1985): 15-27.

Horsley, A. Doyne. "A Food Factory." Pp. 89-106 in *Illinois: A Geography*, edited by Doyne A. Horsley. Boulder, Colo.: Westview, 1986.

Yoder, Franklin L. "Beyond the Towpath: Farming Along the Illinois and Michigan Canal." Pp. 33-46 in *Settling the Upper Illinois Valley: Pattern of Change in the I&M Canal Corridor, 1830–1980*, edited by Michael P. Conzen and Melissa J. Morales. Chicago, Ill.: University of Chicago, Committee on Geographical Studies, Studies on the Illinois and Michigan Canal Corridor, No. 3, 1989.

Knox, Douglas W. "Farms Undermined: Agricultural Changes in Utica and Deer Park, 1860–1870." Pp. 33-45 in *Illinois Canal Country: The Early Years in Comparative Perspective*, edited by Michael P. Conzen and Linda S. Lim. Chicago, Ill.: University of Chicago Committee on Geographical Studies, Studies on the Illinois and Michigan Canal, No. 5, 1991.

Krueger, Joseph D. "Farming in Rutland and Fall River Townships, LaSalle County, 1860." Pp. 53-63 in *Illinois Canal Country: The Early Years in Comparative Perspective*, edited by Michael P. Conzen and Linda S. Lim. Chicago, Ill.: University of Chicago Committee on Geographical Studies, Studies on the Illinois and Michigan Canal, No. 5, 1991.

Stock, Meegan E. "Farming in Ottawa Township in 1860." Pp. 46-52 in *Illinois Canal Country: The Early Years in Comparative Perspective*, edited by Michael P. Conzen and Linda S. Lim. Chicago, Ill.: University of Chicago Committee on Geographical Studies, Studies on the Illinois and Michigan Canal, No. 5, 1991.

Keyes, Jonathan J. "Agricultural Hinterland: A Profile of the Lemont Township Community in 1860." Pp. 25-37 in *Looking for Lemont: Place and People in an Illinois Canal Town*, editing by Michael P. Conzen and Carl Z. Zimring. Chicago, Ill.: University of Chicago Committee on Geographical Studies, Studies on the Illinois and Michigan Canal Corridor, No. 7, 1994.

McManus, James M. "Estimation of Agricultural Net Primary Production in East Central Illinois, Using Satellite Remotely Sensed Data for the Years 1987 and 1988." Master's Thesis, University of Maryland, College Park, Md., 1994.

CROPS AND CROPPING PATTERNS

Colyer, Frank H. "The Cotton Industry of Southern Illinois." *Transactions of the Illinois State Academy of Science*, 17 (1924): 226-232.

Robertson, J.C. "The Ozark Orchard Center of Southern Illinois." *Economic Geography*, 4, No. 3 (1928): 253-266.

Lane, Ray. "Geographic Factors in the Commercial Non-Greenhouse Vegetable Industry of Illinois." Master's Thesis, University of Illinois, Urbana, Ill., 1935.

Jack, Elmer. "Geography of Some Important Illinois Crops." Master's Thesis, University of Illinois, Urbana, Ill., 1936.

Crompton, Mabel P. "The Soybean Crop of Illinois." *Journal of Geography*, 39, No. 4 (1940): 142-150.

Braithwaite, Gussie. "The Soybean Industry in East Central Illinois." Master's Thesis, University of Colorado, Boulder, Colo., 1941.

Trager, Louise M. "The Broomcorn Industry of Coles County, Illinois." *Economic Geography*, 17 (1941): 250-261.

Buford, J.C. "Vegetable Production and Marketing in the Chicago Area: A Geographical Study." Master's Thesis, Illinois State University, Bloomington, Ill., 1947.

Harvey, Z.R. "The Geography of Peach and Apple Production in the Unglaciated Section of Southern Illinois." Master's Thesis, Illinois State University, Bloomington, Ill., 1948.

Witzig, Fred. "The Sweet Corn Industry of Iroquois and Vermillion Counties." Master's Thesis, University of Illinois, Urbana, Ill., 1951.

Hahn, Arvin. "The Horticultural Supply of the Chicago Metropolitan Area." *Journal of Geography*, 53, No. 6 (1954): 254-256.

Williams, Thomas. "The Distribution of the Soybean in Southern Illinois." Master's Thesis, Southern Illinois University, Carbondale, Ill., 1954.

Price, Dalias A. "The Commercial Orchard Economy in Southern Illinois." Ph.D. Dissertation, University of Wisconsin, Madison, Wis., 1955.

Grotewold, Andreas. "Changing Patterns of Corn Yields per Acre: The Position of Illinois in the United States." *Transactions of the Illinois Academy of Science*, 48 (1956): 157-165.

Price, Dalias A. "Factors in the Establishment of Commercial Orcharding in Southern Illinois." *Transactions of the Illinois State Academy of Science*, 48 (1956): 149-156.

Truitt, R.A. "Changes in the Distribution of Illinois Soybean Production." Master's Thesis, Illinois State University, Bloomington, Ill., 1956.

Mobley, Leroy. "Popcorn in the Cropping System of Gallatin County." Master's Thesis, Southern Illinois University, Carbondale, Ill., 1957.

Gallion, Gene W. "The Changing Patterns of Wheat Distribution in Illinois." Master's Thesis, Illinois State University, Bloomington, Ill., 1958.

Lyon, Edward E. "Peach Production in Marion County, Illinois." Master's Thesis, Illinois State University, Bloomington, Ill., 1959.

Shireman, Myrl D. "The Commercial Orchard Economy of Calhoun County, Illinois." Master's Thesis, Eastern Illinois University, Charlestown, Ill., 1964.

Topp, Gilbert H. "Illinois Sorghum Distribution and Production." Master's Thesis, Southern Illinois University, Carbondale, Ill., 1964.

Norman, W.R. "The Role of Soybeans in the Farm Economy of McLean County, 1925–64." Master's Thesis, Illinois State University, Bloomington, Ill., 1965.

Brunn, Stanley D. "The Origin and Movement of Fresh Fruits to Chicago." *Journal of Geography*, 66, No. 3 (1967): 110-115.

Bradach, Franklin J. "Contract Tomato Farming in Northeastern Illinois." Master's Thesis, Northern Illinois University, DeKalb, Ill., 1969.

Powell, Lanny C. "An Examination of Factors Influencing the Temporal-Spatial Diffusion of Soybean Production in Illinois, 1930–1968." Ph.D. Dissertation, University of Illinois, Urbana, Ill., 1971.

Frederic, Paul B. "Feed Grain Program Influences on Cropland Changes: A Micro Study." *Geographical Perspectives*, 31 (1973): 17-20.

Harper, Charles C. "Locational Factors in the Soybean Industry." Master's Thesis, Eastern Illinois University, Charlestown, Ill., 1973.

Seeley, Mark W. "Corn Yield Prediction in Illinois Using Weather Imput: From the Test Field to the County-Wide Model." Master's Thesis, Northern Illinois University, DeKalb, Ill., 1975.

ANIMAL HUSBANDRY

Huck, Emilie. "Geographic Aspects of Meat Production in Illinois." Master's Thesis, University of Illinois, Urbana, Ill., 1937.

Whamond, R. "The Commercial Mink Fur Farming of Illinois." Master's Thesis, Illinois State University, Bloomington, Ill., 1956.

Stroebel, William. "Large Scale Beef Feeders of Northern Illinois." Master's Thesis, University of Chicago, Chicago, Ill., 1957.

Elam, William W. "A Geographic Study of an Area of Concentrated Swine Production in Northern Illinois." Master's Thesis, Northern Illinois University, DeKalb, Ill., 1959.

Walrich, Glenn A. "The Sheep Industry in Illinois." Master's Thesis, Illinois State University, Bloomington, Ill., 1962.

Stevens, George P. "Geography of Beef Feeding in Northern Illinois." Ph.D. Dissertation, University of Wisconsin, Madison, Wis., 1963.

Alderman, Ralph H. "The Beef Cattle Industry of McLean County, Illinois, 1815–1860." Master's Thesis, Illinois State University, Bloomington, Ill., 1966.

DAIRYING

Birchard, Ralph. "Some Geographic Aspects of the Dairy Industry of Illinois." Master's Thesis, University of Illinois, Urbana, Ill., 1938.

Durand, Loyal. "The Cheese Region of Southeastern Illinois." *Economic Geography*, 15, No. 3 (1939): 283-292.

Durand, Loyal. "Dairy Region of Southeastern Wisconsin and Northeastern Illinois." *Economic Geography*, 16, No. 4 (1940): 416-428.

Day, Elizabeth. "The Illinois Portion of the St. Louis Fluid Milk Region: Its Geographic Limits and Unity." Master's Thesis, University of Illinois, Urbana, Ill., 1947.

Delong, George C. "The Northeastern Illinois Dairy Region." Ph.D. Dissertation, University of Michigan, Ann Arbor, Mich., 1953.

Ferren, Robert G. "Dairying in Spring Point Township, Cumberland County, Illinois." Master's Thesis, Eastern Illinois, Charlestown, Ill., 1965.

Kolzow, David R. "Dairying in Kane County, Illinois: A Study of Agricultural Change." Master's Thesis, Northern Illinois University, DeKalb, Ill., 1966.

Frederic, Paul B. "An Analysis of the Spatial Disappearance Process: The Case of Dairying in a Case Grain Region." Ph.D. Dissertation, University of Illinois, Urbana, Ill., 1973.

Frederic, Paul B. "Spatial Variations in Dairy Cow Pricing." *Bulletin, Illinois Geographical Society*, 15, No. 1 (1973): 54-60.

Frederic, Paul B. "The Changing Role of Dairying in a Cash Grain Environment." *Proceedings of the Association of American Geographers*, 5 (1973): 66-70.

Kolzow, David R. "The Development of Commercial Dairying in Kane County, Illinois, 1850–1920." *Bulletin, Illinois Geographical Society*, 15, No. 1 (1973): 41-53.

Berry, David. "The Sensitivity of Dairying to Urbanization: A Study of Northeastern Illinois." *Professional Geographer*, 31, No. 2 (1979): 170-176.

Mattingly, Paul F., and George Aspbury. "Some Devolution Similarities to Diffusion." *Geografiska Annaler B*, 67B, No. 1 (1985): 1-6.

FARM LAND

Vogeler, Ingolf. "Effectiveness of Differential Assessment of Farmland in Metropolitan Chicago." *Geographical Survey*, 7, No. 3 (1978): 23-32.

Charvet, J.P. "The Recent Revolution of Farmland Prices in Illinois and France." *Bulletin, Illinois Geographical Society*, 24, No. 2 (1982): 3-14.

Walters, William D., and Floyd Mansberger. "Initial Field Location in Illinois." *Agricultural History*, 57, No. 3 (1983): 289-296.

Rumney, Thomas A. "Patterns of Agricultural Land Change in Illinois: 1969 to the Present." *Bulletin, Illinois Geographical Society*, 34, No. 2 (1992): 23-35.

Hannaford, Katherine W. "Three New 'R's' on the Expanding Metropolitan Fringe: Religious Retreats, Recreation, and Research as Objects of Farmland Conversion, 1920–1985." Pp. 149-159 in *Looking for Lemont: Place and People in an Illinois Canal Town*, edited by Michael P. Conzen and Carl Zimring. Chicago, Ill.: University of Chicago Committee on Geographical Studies, Studies on the Illinois and Michigan Canal Corridor, 1994.

AGRICULTURAL POPULATIONS AND SOCIAL ISSUES

Sublett, Michael D. *Farmers on the Road: Interfarm Migration and the Farming of Noncontiguous Lands in Three Midwestern Townships, 1939–1969.* Chicago, Ill.: University of Chicago, Department of Geography, Research Paper No. 168, 1975.

ENVIRONMENTAL AND CONSERVATION ISSUES

Middleton, R.K. "Conservation Farms in McKean County, Illinois." Master's Thesis, Illinois State University, Bloomington, Ill., 1966.

Changnon, Stanley A., Jr. "A Climatological-Technological Method for Estimating Irrigation Water Requirements for Maximum Crop Yields." *Journal of Soil and Water Conservation*, 24, No. 1 (1969): 12-15.

Schiel, Joseph B., Jr. "The Ecology and Distribution of Plant-Parasitic Nematodes on Three Farms in Southern Illinois." Ph.D. Dissertation, University of Illinois, Urbana, Ill., 1971.

Soper, Jan. "Impacts of Surfaced Mined Coal Production on Prime Agricultural Lands in Illinois." Master's Thesis, University of Illinois at Chicago, Chicago, Ill., 1979.

Strauss, Robert F. "Modeling Corn-Weather Relations in Illinois, 1957–1977." Master's Thesis, Southern Illinois University, Carbondale, Ill., 1981.

Isard, Scott A., Mark R. Welford, and Steven E. Hollinger. "A Simple Soil Moisture Index to Forecast Crop Yields." *Physical Geography*, 16, No. 6 (1995): 524-538.

AGRICULTURAL MARKETING AND DISTRIBUTION

Karim, Muhammad B. "Characteristics and Composition of a Farmer's Market." Master's Thesis, Eastern Illinois University, Charlestown, Ill., 1978.

TRANSPORTATION AND MECHANIZATION

Hartshorne, Richard. "The Significance of Lake Transportation to the Grain Traffic of Chicago." *Economic Geography*, 2, No. 2 (1926): 274-291.

Mattingly, Paul F. "Patterns of Horse Devolution and Tractor Diffusion in Illinois, 1920–1982." *Professional Geographer*, 39, No. 3 (1987): 298-308.

Wisconsin

GENERAL TEXTS, ARTICLES, AND THESES

Welles, W.S. "The Agricultural Industries of Wisconsin." *Journal of Geography*, 12 (1913–14): 248-253.

Seitz, Kerlin M. "Part-Time Farming in Northern Wisconsin." Ph.D. Dissertation, University of Wisconsin, Madison, Wis., 1954.

Durand, Loyal. "The Retreat of Agriculture in Milwaukee County, Wisconsin." *Transactions of the Wisconsin Academy of Sciences, Arts and Letters*, 51 (1962): 197-218.

Conzen, Michael P. *Frontier Farming in an Urban Shadow: The Influence of Madison's Proximity on the Agricultural Development of Blooming Grove, Wisconsin*. Madison, Wis.: State Historical Society of Wisconsin, 1971.

Fonstad, Todd. "The Central Wisconsin Farms Project." *Bulletin of the Wisconsin Council for Geographical Education* (Spring 1984): 44-49.

Vogeler, Ingolf. "The Southern Agricultural Region." Pp. 124-146 in *Wisconsin: A Geography*, edited by Ingolf Vogeler. Boulder, Colo.: Westview, 1986.

McKay, Thomas R., and Deborah E. Kmetz, eds. *Agricultural Diversity in Wisconsin*. Madison, Wis.: State Historical Society of Wisconsin, 1987.

Rohe, Randall E. "Lumber Company Farms and the Development of Agriculture in Wisconsin." *Wisconsin and Its Region: Proceedings of the Annual Institute of Wisconsin Studies Conference* (1989): 144-166.

CROPS AND CROPPING PATTERNS

Durand, Loyal. "The Wisconsin Cranberry Industry." *Economic Geography*, 18, No. 2 (1942): 159-172.

Haugen, Richard K. "Cranberry Growing Regions of Wisconsin." Master's Thesis, Michigan State University, East Lansing, Mich., 1959.

Belachew, Mekete. "Geography of Potato Production in Wisconsin." Master's Thesis, University of Wisconsin, Madison, Wis., 1966.

Lewison, Dale M. "Economic Geography of the Tobacco Industry of Southern Wisconsin." Master's Thesis, University of Oklahoma, Norman, Okla., 1967.

Raitz, Karl B. "The Location of Tobacco Production in Wisconsin." Ph.D. Dissertation, University of Minnesota, Minneapolis, Minn., 1970.

Raitz, Karl B. "The Government Institutionalization of Tobacco Acreage in Wisconsin." *Professional Geographer*, 23, No. 2 (1971): 123-126.

Halbach, Thomas. "The Importance of Family Food Gardens on Farms in South Central Wisconsin." Master's Thesis, University of Wisconsin, Madison, Wis., 1974.

Thoma, Marcel E. "Locational Analysis of the Green Pea and Sweet Corn Processes Industry in Wisconsin." Master's Thesis, Arizona State University, Tempe, Ariz., 1974.

Bogue, Margaret B. "The Lure of the Cherry: The Making of a Farm Type Area." Pp. 62-71 in *Agricultural Diversity in Wisconsin*, edited by Thomas R. McKay and Deborah E. Kmetz. Madison, Wis.: State Historical Society of Wisconsin, 1987.

Found, R.F. "Strawberry Fortunes and Misfortunes in Jackson County, Wisconsin, 1930–1986." *Wisconsin Geographer*, 3 (1987): 1-12.

Dean, Charles W. "An Historical Geography of Viticulture and the Wine Trade in Southcentral Wisconsin, 1850–1900: The Kehl Winery of Roxbury Township, Dane County." Master's Thesis, University of Wisconsin, Madison, Wis., 1993.

Rumney, Thomas A. "Hops Cultivation in Wisconsin: A Nineteenth Century Interlude." *Wisconsin Geographer*, Vol. 13-14 (1997–1998): 49-54.

ANIMAL HUSBANDRY

Stover, Stephen L. "Changing Regionalization of Sheep Husbandry in Wisconsin." *Transactions of the Wisconsin Academy of Sciences, Arts and Letters*, 52 (1963): 111-132.

DAIRYING

Lange, Edward G. "Dairying in Wisconsin." *Journal of Geography*, 12 (1913): 241-247.

Trewartha, Glenn T. "The Dairy Industry of Wisconsin as an Adjustment to Geographic Environment." Ph.D. Dissertation, University of Wisconsin, Madison, Wis., 1924.

Trewartha, Glenn T. "The Dairy Industry of Wisconsin as a Geographic Adjustment." *Bulletin of the Geographical Society of Philadelphia*, 23, No. 3 (1925): 93-119.

Durand, Loyal. "Geographic Regions of Wisconsin." *Wisconsin Dairying*, edited by Walter H. Ebling. Madison, Wis.: Wisconsin Crop and Livestock Reporting Service, Bulletin of the Wisconsin Department of Agriculture and Markets, No. 120, 1931.

Randall, John R. "Geography of the Central Wisconsin Dairy Region." Ph.D. Dissertation, University of Wisconsin, Madison, Wis., 1934.

Durand, Loyal. "A Profile in a Dairy Region of Southeastern Wisconsin." *Journal of Geography*, 35, No. 8 (1936): 271-278.

Durand, Loyal. *A Profile in a Dairy Section of Southeastern Wisconsin*. Madison, Wis.: Historical Society of Wisconsin Library, Pamphlet No. 57-2080, 1936.

Booth, Alfred W. "The Geography of the Southeastern Dairy Region of Wisconsin." Ph.D. Dissertation, University of Wisconsin, Madison, Wis., 1937.

Strain, Warren T. "Geography of the Northwest Dairy Region of Wisconsin." *Transactions of the Wisconsin Academy of Sciences, Arts and Letters*, 30 (1937): 179-202.

Bertrand, Kenneth J. "A Geographic Analysis of the Eastern Lake Shore Red Clay Dairy Region of Wisconsin." Ph.D. Dissertation, University of Wisconsin, Madison, Wis., 1940.

Durand, Loyal. "Dairy Barns of Southeastern Wisconsin." *Economic Geography*, 19, No. 1 (1943): 37-46.

Lewthwaite, Gordon R. "The Regionalization of the Butter and Cheese Production in Wisconsin." Ph.D. Dissertation, University of Wisconsin, Madison, Wis., 1956.

Durand, Loyal. "The Landscapes of Retreat in Milwaukee County." *Transactions of the Wisconsin Academy of Sciences, Arts, and Letters*, 52A (1963): 79-87.

Lewthwaite, Gordon R. "Wisconsin Cheese and Farm Type: A Locational Hypothesis." *Economic Geography*, 40, No. 2 (1964): 95-112.

Korb, Kenneth W. "Dairy Farming: An Analysis of an Agricultural System in the East Central Lakeshore of Wisconsin." Ph.D. Dissertation, University of Wisconsin, Madison, Wis., 1968.

Meier, Bruce W. "Urban Expansion, Property Taxes, and Dairy Farming Near Madison, Wisconsin." Ph.D. Dissertation, University of Wisconsin, Madison, Wis., 1980.

Richter, Dennis M. "Change in the Southern Wisconsin Dairy Area." *Wisconsin Geographer*, 1 (1985): 41-55.

Cross, John A. "Wisconsin's Changing Dairy Industry and the Dairy Termination Program." *Transactions of the Wisconsin Academy of Sciences, Arts, and Letters*, 77 (1989): 11-26.

Cross, John A. "1988 Drought Impacts Among Wisconsin Dairy Farmers." *Transactions of the Wisconsin Academy of Sciences, Arts and Letters*, 80 (1992): 21-34.

Cross, John A. "Agroclimatic Hazards and Dairy Farming in Wisconsin." *Geographical Review*, 84, No. 3 (1994): 277-289.

Cross, John A. "Problems Facing Wisconsin Dairy Farmers: Perceptions of the Farmers." *Wisconsin Geographer*, 10 (1994): 1-10.

Cross, John A. *Wisconsin Dairy Farmers' Evaluation of Current Stresses: Implications for Entry and Exit Behavior*. Madison, Wis.: Agricultural Technology and Family Farm Institute, University of Wisconsin, 1994.

Cross, John A. *Entry-Exit Behavior of Wisconsin Dairy Farmers*. Madison, Wis.: Agricultural Technology and Family Farm Institute, University of Wisconsin, 1994.

Cross, John A. "Dairying in an Urban Environment: The Milwaukee Metropolitan Area." *Transactions of the Wisconsin Academy of Science, Arts, and Letters*, 83 (1995): 75-86.

FARM LAND

Eighny, Thomas H. "Farm Fragmentation: A Neglected Aspect of American Farm Enlargement." *Proceedings of the Pennsylvania Academy of Science*, 39, No. 2 (1966): 289-292.

Brey, James. "Relationships Between Agricultural Land Use and Glacial Land Type in Rock and Green Counties, Wisconsin." Master's Thesis, University of Wisconsin, Madison, Wis., 1974.

Lo, Thomas H.C., Frank L. Scarpone, and Thomas M. Lillesand. "Use of Multitemporal Spectral Profiles in Agricultural Land-Cover Classification." *Photogrammetric Engineering and Remote Sensing*, 52, No. 4 (1986): 535-544

Bhatia, Shyam S. "Changing Patterns of Cropland Use in Wisconsin, 1971–1981." *Wisconsin Geographer*, 3 (1987): 25-48.

Kent, Christopher A. "The Influence of Changes in Land Cover and Agricultural Land Management Practices on Baseflow in Southwest Wisconsin, 1969–1998." Ph.D. Dissertation, University of Wisconsin, Madison, Wisc., 1999.

AGRICULTURAL POPULATIONS AND SOCIAL ISSUES

Kolb, J.H. "Rural Primary Groups: A Study of Agricultural Neighborhoods." *University of Wisconsin Agricultural Experiment Station Research Bulletin*, 51 (December 1921).

Helgeson, A.C. *Farms in the Cutover: Agricultural Settlement in Northern Wisconsin*. Madison, Wis.: State Historical Society, 1962.

Raitz, Karl B., and E. Cotton Mather. "Norwegians and Tobacco in Western Wisconsin." *Annals of the Association of American Geographers*, 61, No. 4 (1971): 684-696.

Vogeler, Ingolf. "Ethnicity, Religion and Farm Land Transfers in Western Wisconsin." *Ecumene*, 7 (1975): 6-13.

ENVIRONMENTAL AND CONSERVATION ISSUES

Johansen, Harley E. "Spatial Diffusion of Contour Strip Cropping in Wisconsin." Master's Thesis, University of Wisconsin, Madison, Wis., 1969.

Collins, Charles W. "The Influence of Drumlin Topography on Field Patterns in Dodge County, Wisconsin." *Transactions of the Wisconsin Academy of Sciences, Arts, and Letters*, 59 (1971): 55-66.

Johansen, Harley E. "Diffusion of Strip Cropping in Southwestern Wisconsin." *Annals of the Association of American Geographers*, 61, No. 4 (1971): 671-683.

Knox, James C. "Human Impacts on Wisconsin Stream Channels." *Annals of the Association of American Geographers*, 67, No. 3 (1977): 323-342.

Cross, John A. "Drought Perceptions of Wisconsin Dairy Farmers." *Wisconsin Geographer*, 6 (1990): 1-18.

Lyons, David W. "Idealism, Organic Farming Values and the Issue of Biotechnology: Narratives of Ten Wisconsin Organic Vegetable Growers." Master's Thesis, University of Wisconsin, Madison, Wis., 1993.

Fitzpatrick, Faith A., and James C. Knox. "Spatial and Temporal Sensitivity of Hydrogeomorphic Response and Recovery to Deforestation, Agriculture, and Floods." *Physical Geography*, 21, No. 2 (2000): 89-108.

Minnesota

GENERAL TEXTS, ARTICLES, AND THESES

Miller, George J. "Agriculture in Minnesota." *Journal of Geography*, 15 (1915–16): 196-201.

Irving, Robert M. "Amenity Agriculture in the Minneapolis–St. Paul Metropolitan Area (TCMA)." Ph.D. Dissertation, University of Minnesota, Minneapolis, Minn., 1966.

Rikkinen, Kalevi. "Kalevala, Minnesota: Agricultural Geography in Transition." *Acta Geographica* (Helsinki), 19, No. 5 (1969): 1-55.

CROPS AND CROPPING PATTERNS

Cady, LeRoy. "Fruit Growing in Minnesota." *Journal of Geography*, 15 (1915–16): 211-214.

Fause, A., and G.S. Corfield. "The Potato Industry in Minnesota." *Economic Geography*, 13, No. 4 (1937): 393-401.

Johnson, Hildegard B. "King Wheat in Southeastern Minnesota: A Case Study of Pioneer Agriculture." *Annals of the Association of American Geographers*, 47, No. 4 (1957): 350-362.

Tideman, Philip L. "An Example of Crop Dynamics in Minnesota: The Soybean." *Proceedings of the Minnesota Academy of Science*, 31, No. 2 (1964): 147-152.

Lindquist, Gary M. "The Maple Syrup Industry in Minnesota: A Geographical Study." Master's Thesis, Mankato State University, Mankato, Minn., 1966.

Schoenrock, Ruth E. "Hemp in Minnesota During the Wartime Emergency." Master's Thesis, Mankato State University, Mankato, Minn., 1966.

Tideman, Philip L. "Wheat on the Northern Agricultural Frontier, 1840–1920." Ph.D. Dissertation, University of Nebraska, Lincoln, Nebr., 1967.

Efteland, Jon N. "The Development of Mechanized Wild Rice Production in Minnesota." Master's Thesis, University of North Dakota, Grand Forks, N.Dak., 1973.

ANIMAL HUSBANDRY

Lanahan, Frederick A. "A Geography of the Mule in Minnesota." Master's Thesis, Mankato State University, Mankato, Minn., 1971.

DAIRYING

Washburn, R.M. "Dairying in Minnesota." *Journal of Geography*, 15 (1915–16): 206-210.

Jenkins, Robert L. "The Application of the Fox Diamond Consumption Model to a Milk Production Landscape in West Central Minnesota." Master's Thesis, University of North Dakota, Grand Forks, N.Dak., 1971.

FARM LAND

Napton, Darrell. "Regional Farmland Protection: The Twin Cities Experience." *Journal of Soil and Water Conservation*, 45, No. 4 (1990): 446-449.

AGRICULTURAL POPULATIONS AND SOCIAL ISSUES

Meeks, Harold A. "Railroad Expansion and Agricultural Settlement in Minnesota, 1860–1910." *Proceedings of the Minnesota Academy of Science*, 27 (1959): 22-37.

ENVIRONMENTAL AND CONSERVATION ISSUES

Colby, Charles C. "Agricultural Adjustments to the Natural Environment in Southeastern Minnesota during the Period of Bonanza Wheat Farming." *Transactions of the Illinois State Academy of Science*, 17 (1924): 213-225.

Burns, Bert E. "Artificial Drainage in Blue Earth County, Minnesota." Ph.D. Dissertation, University of Nebraska, Lincoln, Nebr., 1954.

Beach, Timothy. "The Fate of Eroded Soils: Sediment Sinks and Sediment Budgets of Agrarian Landscapes in Southern Minnesota, 1851–1988." *Annals of the Association of American Geographers*, 84, No. 1 (1994): 5-28.

Busket, Heidi J. "An Assessment of Soil Erodibility and Spatial Distribution of Farm Program Lands." Master's Thesis, Minnesota State University, Mankato, Minn., 2000.

AGRICULTURAL MARKETING AND DISTRIBUTION

Lively, C.E. *Growth and Decline of Farm Trade Centers in Minnesota, 1905–1930*. Minneapolis, Minn.: University of Minnesota Agricultural Experiment Station, Bulletin 287 (1932): 1-48.

TRANSPORTATION AND MECHANIZATION

Baker, Sharon F. "From Horse to Horsepower: Farm Tractorization and the Rural Landscape in Minnesota and Surrounding States." Master's Thesis, South Dakota State University, Brookings, S.Dak., 1983.

Iowa

GENERAL TEXTS, ARTICLES, AND THESES

Gosmire, R.E. "Farms, Farmsteads and Farmers in the Cash Grain Belt of Iowa." *Geographical Perspectives*, 22 (1967): 8-15.

CROPS AND CROPPING PATTERNS

Horn, Werner G. "Distribution of Soybeans in Iowa." Master's Thesis, University of Iowa, Iowa City, Iowa, 1956.

Lee, Reyong. "Modeling Corn Yields in Iowa Using Time-Series Analysis of AVHRR Data and Vegatation Phenological Metrices." Ph.D. Dissertation, University of Iowa, Iowa City, Iowa, 2000.

ANIMAL HUSBANDRY

Lu, Jonathan J. "On Confinement Beef Cattle Feeding in Iowa." *Geographical Perspectives*, 39 (1977): 17-26.

FARM LAND

Schlesselman, George W. "The Geography of the Lake Calvin Region of Iowa, with Emphasis on Land Utilization." Ph.D. Dissertation, University of Nebraska, Lincoln, Nebr., 1935.

Platt, Rutherford H. "Farmland Conversion: National Lessons for Iowa." *Professional Geographer*, 33, No. 1 (1981): 113-121.

AGRICULTURAL POPULATIONS AND SOCIAL ISSUES

Mason, Lois E. "Association of Rural-Farm Migration and Agricultural-Geographic Patterns in North-Central and South-Central Iowa." Master's Thesis, Ohio State University, Columbus, Ohio, 1965.

ENVIRONMENTAL AND CONSERVATION ISSUES

Nealson, Edward N., and George P. Malanson. "Farm Chemicals as Indicators of Sediment Sources in Iowa Rivers." *Geographical Bulletin*, 36, No. 1 (1994): 44-49.

AGRICULTURAL MARKETING AND DISTRIBUTION

Golledge, Reginald G. "A Probabilistic Model of Market Behavior: With Reference to the Spatial Aspects of Hog Marketing in Eastern Iowa." Ph.D. Dissertation, University of Iowa, Iowa City, Iowa, 1965.

Golledge, Reginald G. "A Model of Market Behavior: The Case of Hog Marketing in East-Central Iowa." *B.C. Geographical Series*, 7 (1966): 91-104.

Hellyer, David L. "Geographic Aspects of Farm Service Centers in Clinton County, Iowa: 1941–1973." Master's Thesis, Western Illinois University, Macomb, Ill., 1976.

Missouri

GENERAL TEXTS, ARTICLES, AND THESES

Werner, Carol A. "Agriculture in a Portion of the Rural–Urban Fringe of the St. Louis SMSA." Master's Thesis, Southern Illinois University, Edwardsville, Ill., 1971.

Rafferty, Milton D. "Agricultural Change in the Western Ozarks." *Missouri Historical Review*, 69, No. 3 (1975): 299-322.

Rafferty, Milton D. "Agriculture: Development and Geography." Pp. 91-104 in *Missouri: A* Geography, edited by Milton D. Rafferty. Boulder, Colo.: Westview, 1983.

CROPS AND CROPPING PATTERNS

Emerson, Gladys C. "Tobacco Culture in a 'Little Dixie' Outlier of Northwest Missouri." Ph.D. Dissertation, University of Kansas, Lawrence, Kans., 1978.

Marshal, Richard D. "An Afternoon in the Country: A Landscape Analysis of the Pick-Your-Own Produce Industry in the St. Louis Metro East." Master's Thesis, Southern Illinois University at Edwardsville, Edwardsville, Ill., 1999.

ANIMAL HUSBANDRY

Smith, Jonathan V. "Premium Standard Farms and the Transformation of Livestock Geography in Northern Missouri." *Southeastern Geographer*, 39, No. 2 (1999): 161-171.

Matisziw, Timothy C., and James D. Hipple. "Spatial Clustering and State/County Legislation: The Case of Hog Production in Missouri." *Regional Studies,* 35, No. 8 (2001): 719-730.

FARM LAND

Bratton, Sam T., and Nason N. Duncan. "Land Utilization in a Typical Mixed Farming Community: Centralia, Missouri." *Bulletin of the Geographial Society of Philadelphia*, 28 (1930): 36-42.

Johnson, Harold V. "Agricultural Land Use Patterns in the Missouri Bootheel." Master's Thesis, Southern Illinois University, Carbondale, Ill., 1966.

AGRICULTURAL MARKETING AND DISTRIBUTION

Stephenson, David E. "The Changing Spatial Organization of Soybean Marketing in the Missouri Bootheel." Master's Thesis, Southern Illinos University, Carbondale, Ill., 1965.

The Great Plains of the United States

GENERAL TEXTS, ARTICLES, AND THESES

Baker, O.E. "The Agriculture of the Great Plains Region." *Annals of the Association of American Geographers*, 13 (1923): 110-119.
Gibson, J.S. "Agriculture of the Southern High Plains." *Economic Geography*, 8, No. 3 (1932): 245-261.
Hewes, Leslie. "Transformation of an Agrarian Landscape: The Suitcase-Farming Frontier of the Central Great Plains." Pp. 432-434 in *International Geography '72: Section 5, Historical Geography*, edited by W.P. Adams and F. Helleiner. Toronto, Ontario: International Geographical Union, 1972.
Hewes, Leslie. *The Suitcase Farming Frontier: A Study in the Historical Geography of the Central Great Plains*. Lincoln, Nebr.: University of Nebraska Press, 1973.
Wishart, David J. "Agriculture at the Trading Posts on the Upper Missouri Prior to 1843." *Agricultural History*, 47, No. 1 (1973): 57-62.
Hewes, Leslie. "Early Suitcase Farming in the Central Great Plains." *Agricultural History*, 51, No. 1 (1977): 23-37.
Hewes, Leslie. "Agricultural Risk in the Great Plains." Pp. 157-186 in *The Great Plains: Environment and Culture*, edited by Brian W. Blouet and Frederick C. Luebke. Lincoln, Nebr.: University of Nebraska Press, 1979.
Roet, Jeffrey B. "Agriculture on the Dry Farming Frontier, 1900–1920." *Bulletin of the Association of North Dakota Geographers*, 29 (1979): 19-35.
Klohn, Werner. "Entwicklung und Strukture der Landwirtschaft in den Great Plains." *Vechtaer Studien zur Angewandten Geographie und Regionalwissenschaft*, 2 (1991): 39-64.

CROPS AND CROPPING PATTERNS

Anderson, Esther S. "The Sugar Beet Industry in the Intermontane and Great Plains Region of the United States." Ph.D. Dissertation, Clark University, Worcester, Mass., 1932.
Ward, Ralph E. "The Northern Great Plains as a Producer of Wheat." *Economic Geography*, 22 (1946): 231-244.

Hewes, Leslie, and A.C. Schmiedling. "Risk in the Central Great Plains: Geographical Patterns of Wheat Failure in Nebraska, 1931–1952." *Geographical Review*, 46 (1956): 375-387.

Hopple, Lee. "Ecological Crop Geography of the Sugar Beet in the Northern Great Plains." Master's Thesis, Pennsylvania State University, University Park, Pa., 1960.

Apken, James A. "The Louisiana and Texas Coastal Rice Industry: A Geographical Survey." Master's Thesis, Southern Methodist University, Dallas, Tex., 1962.

Reinbold, Leo M. "Safflower, with an Emphasis on Safflower Production in North Dakota and Montana." Master's Thesis, University of North Dakota, Grand Forks, N.Dak., 1962.

Hewes, Leslie. "Causes of Wheat Failure in the Dry Farming Region, Central Great Plains, 1939–1957." *Economic Geography*, 41, No. 3 (1965): 313-330.

Coleman, John M. "The Distribution of Sunflower in the Prehistoric Great Plains." *Professional Geographer*, 24, No. 4 (1972): 327-329.

Bowden, Martyn. "Desert Wheat Belt, Plains Corn Belt: Environmental Cognition and Behavior on the Plains Margin 1850–1910." Pp. 189-201 in *Images of the Plains*, edited by Brian W. Blouet and Merlin Lawson. Lincoln, Nebr.: University of Nebraska Press, 1975.

Adams, Russell B. "Changing Crop Combinations and Agricultural Exports: Upper Midwest Response Variability." *Geographical Perspectives*, 39 (1977): 8-16.

Spath, Hans-Joachim. "The Energy Equation for Dryland Wheat, Central Great Plains, North America." *Stuttgarter Geographische Studien*, 105 (1985): 141-154.

ANIMAL HUSBANDRY

Amedeo, Douglas. "Corn-Livestock Farming in the Midwest Prairies." *Geographical Perspectives*, 18 (1965): 18-27.

FARM LAND

Prator, M.M. "The Development of the Cattle Industry of the Great Plains." Master's Thesis, University of Chicago, Chicago, Ill., 1918.

Hewes, Leslie, and Christian L. Jung. "Early Fencing on the Middle Western Prairie." *Annals of the Association of American Geographers*, 71, No. 2 (1981): 177-201.

Hewes, Leslie. "Early Fencing on the Western Margin of the Prairie." *Annals of the Association of American Geographers*, 71, No. 4 (1981): 499-526.

Rickard, Timothy J. "The Impact of the United States–Canada Boundary on Agricultural Land Use in the Great Plains." *Proceedings, New England–St. Lawrence Valley Division of the Association of American Geographers*, 11 (1981): 73-87.

Roet, Jeffrey B. "Land Quality and Land Alienation on the Dry Farming Frontier." *Professional Geographer*, 37, No. 2 (1985): 173-182.

Baltensperger, Bradley H. "Farm Consolidation in the Northern and Central States of the Great Plains." *Great Plains Quarterly*, 7, No. 4 (1987): 256-265.

Baltensperger, Bradley H. "Larger and Fewer Farms: Patterns and Causes of Farm Enlargement on the Central Great Plains, 1930–1978." *Journal of Historical Geography*, 19, No. 3 (1993): 299-313.

AGRICULTURAL POPULATIONS AND SOCIAL ISSUES

Wright, Joseph T. "An Analysis of the Spatial Association of Rural Farm Population Density and Selected Physical Factors on the Great Plains." Master's Thesis, University of Iowa, Iowa City, Iowa, 1959.

Robinson, Arthur, James B. Lindbert, and Leonard W. Brinkman. "A Correlation and Regression Analysis Applied to Rural Farm Population Densities in the Great Plains." *Annals of the Association of American Geographers*, 51, No. 2 (1961): 211-221.

McQuillen, D. Aiden. "Farm Size and Work Ethic: Measuring the Success of Immigrant Farmers on the American Grasslands, 1875–1925." *Journal of Historical Geography*, 4, No 1 (1978): 23-44.

McQuillan, D. Aidan. "The Interface of Physical and Historical Geography: The Analysis of Farming Decisions in Response to Drought Hazards on the Margins of the Great Plains." Pp. 136-144 in *Period and Place: Research Methods in Historical Geography*, edited by Alan R.H. Baker and M. Billinge. London, England: Cambridge University Press, 1982.

Roet, Jeffrey B. "Agricultural Settlement on the Dry Farming Frontier, 1900–1920." Ph.D. Dissertation, Northwestern University, Evanston, Ill., 1982.

Baltensperger, Bradley H. "Agricultural Change Among Great Plains Russian Germans." *Annals of the Association of American Geographers*, 73, No. 1 (1983): 75-88.

Green, D. Brooks. "Agricultural Settlement: The Drylands." *AAG Field Trip Guide*. Washington, D.C.: Association of American Geographers, 1983. 34-39.

Roberts, Rebecca. "Rural Population Loss and Cropland Change in the Southern Plains: Implications for Cropland Retirement Policy." *Professional Geographer*, 39, No. 3 (1987): 275-286.

ENVIRONMENTAL AND CONSERVATION ISSUES

Smith, J. Warren. "Rainfall of the Great Plains in Relation to Cultivation." *Annals of the Association of American Geographers*, 10 (1920): 69-74.

Hewes, Leslie. "The Conservation Reserve of the American Soil Bank as an Indicator of Regions of Maladjustment in Agriculture, with Particular Reference to the Great Plains." *Festschrift Leopold G. Sheidle zum 60*. Vienna, Austria: Geburtstag II Teil (1967): 331-346.

Bridge, Daniel W. "A Simulation Model Relating Effective Climate to Spring and Winter Wheat Yields on the Northern Great Plains." Ph.D. Dissertation, Southern Illinois University, Carbondale, Ill., 1975.

Bridge, Daniel W. "A Simulation Model Approach for Relating Effective Climate to Winter Wheat Yields on the Great Plains." *Agricultural Meteorology*, 17 (1976): 185-194.

Baltensperger, Bradley H. "Agricultural Adjustments to Great Plains Drought, 1870–1970." Pp. 43-60 in *The Great Plains: Environment and Culture*, edited by Brian W. Blouet and Frederick C. Luebke. Lincoln, Nebr.: University of Nebraska Press, 1979.

Spaeth, Joachim. *The Agro-Ecological Dry Boundary of Non-Irrigated Agriculture in the Central Great Plains of North America*. Wiesbaden, Germany: Frank Steiner Verlag, 1980.

Spaeth, Joachim. "Die agro-okologische Trockengrenze: Neu-Definition und Dynamik der Trockengrenze des Regenfeldbaus in den zentralen Great Plains von Nord-Amerika." *Erdkunde*, 34, No. 3 (1980): 224-230.

Spaeth, Joachim. "Energiwirksamkeit der Weizen produktion unter semi-ariden Klimatedingungen: Zum Wechselspiel zwischen Energieeinsatz und-umstaz und geookologischer und kulturtechnische variabilitat in den zentralen Great Plains von Northamerika." *Erdkunde*, 38, No. 1 (1985): 36-44.

Carbonne, Gregory J. "Considerations of Meteorological Time Series in Estimating Regional-Scale Crop Yield." *Journal of Climate*, 6 (1993): 1607–1615.

Carbonne, Gregory J. "Issues of Spatial and Temporal Variability in Climate Impact Studies." *Professional Geographer*, 47, No. 1 (1995): 30-40.

IRRIGATION

Rickard, Timothy J. "Perceptions and Results of the Irrigation Movement in the Western U.S., 1891–1914." Ph.D. Dissertation, University of Kansas, Lawrence, Kans., 1974.

McKnight, Tom L. "Great Circles on the Great Plains: The Changing Geometry of American Agriculture." *Erdkunde*, 33, No. 1 (1979): 70-79.

Rickard, Timothy J. "The Great Plains as Part of an Irrigated Western Empire." Pp. 81-98 in *The Great Plains: Environment and Culture*, edited by Brian W. Blouet and Frederick C. Luebke. Lincoln, Nebr.: University of Nebraska Press, 1979.

Thompson, Stephen A. "Patterns and Trends in Irrigation Efficiency." *Water Resources Bulletin*, 24, No. 1 (1988): 57-63.

Kromm, David E., and Stephen E. White. "Adoption of Water-Saving Practices by Irrigators in the High Plains." *Water Resources Bulletin*, 26, No. 6 (1990): 999-1012.

Kromm, David E., and Stephen E. White. *Groundwater Exploitation in the High Plains*. Lawrence, Kans.: University Press of Kansas, 1992.

Cahoon, J. "Furrow Irrigators' Response to In-Season Precipitation and Geographic Characteristics." *Agricultural Water Management*, 23, No. 1 (1993): 41-49.

Windhorst, Han-Wilhelm, and Werner Klohn. *Die Bewasserungslandwirtschaft in den Great Plains*. Vechta, Germany: Vechtaer Druckerie und Verlag, Vechtaer Studien zur Angewandten Geographie und Regionalwissenschaft, Band, 14, 1994.

AGRICULTURAL MARKETING AND DISTRIBUTION

Samuelson, Maynard L. "Marketing Areas of the Red River Valley Potato Industry, Based on Consumer Demands and Transportation Costs." Master's Thesis, University of North Dakota, Grand Forks, N.Dak., 1965.

Texas

GENERAL TEXTS, ARTICLES, AND THESES

Foscue, Edwin J. "Agricultural Geography of the Lower Rio Grande Valley of Texas." Ph.D. Dissertation, Clark University, Worcester, Mass., 1931.

Foscue, Edwin J. "Agricultural History of the Lower Rio Grande Valley Region." *Agricultural History*, 8, No. 3 (1934): 124-137.

Browne, W.A. "Agriculture in the Llano Estacado." *Economic Geography*, 13, No. 2 (1937): 155-174.

Stevens, Richard E. "Agricultural Regions of the Northern High Plains of Texas." Master's Thesis, University of Colorado, Boulder, Colo., 1958.

Schroder, Klaus. *Agrarlandschaftsstudien im sudlichen Texas (Lower Rio Grande Valley)*. Frankfurt am Main, Germany: Frankfurter Geographische Hefte, Band 38, 1962.

Barron, Brent C. "A Geographic Study of the Agricultural Characteristics of the Eastern Cross Timbers of Texas." Master's Thesis, University of Oklahoma, Norman, Okla., 1964.

Harrison, Jenny S. "Regional Contrasts in the Agriculture of Denton County, Texas." Master's Thesis, University of Oklahoma, Norman, Okla., 1969.

Jordon, Terry G., John L. Bean, Jr., and William M. Holmes. "Agricultural Geography." Pp. 153-184 in *Texas: A Geography*, edited by Terry G. Jordon, John L. Bean, Jr., and William M. Holmes. Boulder, Colo.: Westview, 1984.

CROPS AND CROPPING PATTERNS

Browne, W.A. "Grand Prairie: A Progressive Rice Region." *Journal of Geography*, 31, No. 2 (1932): 137-147.

London, Ora L. "Geographic Aspects of the Tomato Industry of Texas." Master's Thesis, University of Missouri, Columbia, Mo., 1961.

Hill, James E., Jr. "The Lower Rio Grande Citrus Region of Texas." Ph.D. Dissertation, University of Tennessee, Knoxville, Tenn., 1963.

Cuff, David J. "The Physical Factors in the Location of Citrus Cultivation in Texas." Master's Thesis, Pennsylvania State University, University Park, Pa., 1967.

Rodriguez, Fernando A. "A Geographic Study of Grain Sorghum in the Coastal Bend Area of South Texas." Master's Thesis, University of Arkansas, Fayetteville, Ark., 1970.

Tiller, James W. "Truck Farming as a Factor in the Settlement of the Texas Winter Garden." *Ecumene*, 5 (1973): 6-15.

Wilhelm, Gene. "Dooryard Gardens and Gardening in the Black Community of Brushy, Texas." *Geographical Review*, 65, No. 1 (1975): 73-92.

Templer, Otis W. "Wine Growing in Arid and Semi-Arid West Texas." *Forum of the Association for Arid Lands Studies*, 2 (1986): 99-105.

Templer, Otis W. "East Versus West: A Survey of the Texas Wine Growing Industry." *Mid-South Geographer*, 4 (1988): 17-33.

Overfelt, Robert C. "Viticulture in West Texas: A Modern-Day Usage of Semi-Arid Lands." *Journal of the West*, 29, No. 4 (1990): 67-74.

Templer, Otis W. "The Southern High Plains: Focal Point of the Texas Wine Growing Industry." Pp. 97-110 in *Viticulture in Geographic Perspective: Proceedings of the Miami AAG Symposium*, edited by Harm deBlij. Miami, Fla.: Miami Geographical Society, 1992.

Napton, Mary L., Darrell E. Napton, and James D. Harrison. "An Economic Location Model for Texas Wineries." *Geographical Bulletin*, 35, No. 1 (1993): 33-47.

Elbaum, Nora R. "Medicinal Plants and Their Role in the Dooryard Gardens of Bryan, Texas." Master's Thesis, Texas A&M University, College Station, Tex., 1995.

ANIMAL HUSBANDRY

Chambers, W.T. "Edwards Plateau: A Combination Ranching Region." *Economic Geography*, 8, No. 1 (1932): 67-80.

Richardson, T.C. "Cattle Trade of Texas." *Texas Geographic Magazine*, 1, No. 2 (1937): 16-29.

Killam, Wilton H. "Apiculture in Texas." Master's Thesis, University of Texas, Austin, Tex., 1941.

Brown, Ralph H. "Texas Cattle Trails." *Texas Geographic Magazine*, 10 (1946).

Randell, Charles. "Espiritu Santo: An Early Texas Cattle Ranch." *Texas Geographic Magazine*, 13, No. 2 (1949): 21-25.

Stevens, Richard E. "A Geographic Study of the Cattle Feeding Industry in the Llano Estacado." Ph.D. Dissertation, University of Kansas, Lawrence, Kans., 1961.

Jordan, Terry G. "The Origin of Anglo-American Cattle Ranching in Texas: A Documentation of Diffusion from the Lower South." *Economic Geography*, 45, No. 2 (1969): 170-179.

Jordan, Terry G. "The Origins and Distribution of Open-Range Ranching." *Social Science Quarterly*, 53 (1972): 105-121.

Icenogle, David W. "Registered Beef Cattle in Texas: A Regional Approach." *Ecumene*, 5 (1973): 44-52.

Syring, George. "A Geographic Analysis of Diversified Ranching Operations on the Edwards Plateau: A Case Study of Edwards County, Texas." Ph.D. Dissertation, University of Kansas, Lawrence, Kans., 1975.

Jordan, Terry G. "Early Northeast Texas and the Evolution of Western Ranching." *Annals of the Association of American Geographers*, 67, No. 1 (1977): 66-87.

Jordan, Terry G. *Trails to Texas: Southern Roots of Western Cattle Ranching.* Lincoln, Nebr.: University of Nebraska Press, 1981.

Barnett, Douglas E. "Mohair in Texas: Livestock Experimentation on the Edwards Plateau." Master's Thesis, University of Texas, Austin, Tex., 1983.

Powell, Jeff W., and Frederick Day. "The Diffusion of Exotic Animal Ranching in Texas." *Southwestern Geographer*, 4 (2000): 30-53.

DAIRYING

Painter, Ed. "The Dairy Industry in Wise County, Texas." Master's Thesis, University of Oklahoma, Norman, Okla., 1966.

FARM LAND

Lamb, Robert B. "Agricultural Land Use of the North Grand Prairie–Eastern Cross Timbers Regions of Texas." Master's Thesis, University of California at Los Angeles, 1951.

Rowan, James E. "Agricultural Land Utilization in the Llano Estacado of Eastern New Mexico and Western Texas." Ph.D. Dissertation, University of Nebraska, Lincoln, Nebr., 1960.

AGRICULTURAL POPULATIONS AND SOCIAL ISSUES

Jordon, Terry G. *German Seed in Texas Soil: Immigrant Farmers in Nineteenth-Century Texas*. Austin, Tex.: University of Texas Press, 1966.

Thigpen, Laura L. "Brazos River Bottom Cotton Producers' Knowledge, Attitudes and Practices of Sustainable Agriculture." Master's Thesis, Texas A&M University, College Station, Tex., 1994.

ENVIRONMENTAL AND CONSERVATION ISSUES

Holmes, William R. "An Historical Geography of Dry Farming on the Northern High Plains of Texas." Ph.D. Dissertation, University of Texas, Austin, Tex., 1975.

Norwin, J., and J. Fucik. "Applied Climatology and Food Production: The Case of the Grapefruit Quality in the Rio Grande Valley of Texas." *Proceedings of the Association of American Geographers*, 7 (1975): 169-174.

Thompson, Stephen A. "Urbanization and the Middle Rio Grande Conservancy District." *Geographical Review*, 76, No. 1 (1986): 35-50.

IRRIGATION

Booth, Alfred W. "The Portales Region: A Pump Irrigation District in the Llano Estacado." *Economic Geography*, 18, No. 1 (1942): 97-105.
Collier, G. Loyd. "The Lower Rio Grande Valley of Texas: Some Problems of an Irrigated Area." Master's Thesis, University of Nebraska, Lincoln, Nebr., 1954.
Rizer, Everett. "Irrigation Potential in the Coastal Bend Region, Texas, 1957–2000 A.D." Master's Thesis, University of Texas, Austin, Texas, 1969.
Glick, Thomas F. *The Old World Background of the Irrigation System of San Antonio, Texas*. El Paso, Tex.: Texas Western Press, 1972.
Baland, Ann. "The Geographical Analysis of Sprinkler Irrigation in the Texas Panhandle." Master's Thesis, Western Illinois University, Macomb, Ill., 1977.

AGRICULTURAL MARKETING AND DISTRIBUTION

Falls, Robert H. "Agricultural Production and Marketing of the Lower Rio Grande Valley." Master's Thesis, University of Chicago, Chicago, Ill., 1956.

Oklahoma

GENERAL TEXTS, ARTICLES, AND THESES

Moore, Robert F. "Agricultural Geography of Bryan County, Oklahoma." Master's Thesis, University of Nebraska, Lincoln, Nebr., 1950.
Stephens, David T. "The Agricultural Geography of Osage County, Oklahoma." Master's Thesis, University of Oklahoma, Norman, Okla., 1966.

CROPS AND CROPPING PATTERNS

Britton, Robert L. "Regional Changes in Cotton Acreage in Oklahoma Between 1910 and 1920." *Proceedings of the Oklahoma Academy of Science*, 9 (1929): 104-105.
Self, Hubert. "The Peanut Industry in Oklahoma." Master's Thesis, Oklahoma State University, Stillwater, Okla., 1947.
Atwood, Benjamin F. "The Pecan Industry in Oklahoma." Master's Thesis, Oklahoma State University, Stillwater, Okla., 1949.
Webb, Charles E. "Distributions of Cotton Production in Oklahoma, 1907–1962." Master's Thesis, University of Oklahoma, Norman, Okla., 1963.
Bechtol, Bruce E. "The Peach Industry of the McGee–Stratford Area of East Central Oklahoma: A Geographical Interpretation." Master's Thesis, University of Oklahoma, Norman, Okla., 1965.

Garner, Robert V. "The Changing Cotton Belt of Southeast Oklahoma." *Proceedings of the Oklahoma Academy of Science*, 46 (1966): 191-193.

Williams, Cecil A. "The Production and Distribution of Grain Sorghums in Oklahoma." Master's Thesis, University of Oklahoma, Norman, Okla., 1966.

Rogers, James L. "Three Farm Types in the Hard Red Winter Wheat Belt of Woods County, Oklahoma." Master's Thesis, University of Oklahoma, Norman, Okla., 1969.

Rogers, Charles B. "A Geographical Analysis of Strawberry Production in the Stilwell District of Eastern Oklahoma." Master's Thesis, University of Arkansas, Fayetteville, Ark., 1971.

ANIMAL HUSBANDRY

Sims, Joseph V. "The Beef Cattle Industry in the Coastal Plains Region of Oklahoma." Master's Thesis, University of Oklahoma, Norman, Okla., 1962.

Calkins, Charles F. "Beef Cattle Ranching in the Arbuckle Mountains Physiographic Region of Oklahoma." Master's Thesis, University of Oklahoma, Norman, Okla., 1964.

Cavin, William B. "The Development and Distribution of Feedlots in the Oklahoma Panhandle." Master's Thesis, University of Oklahoma, Norman, Okla., 1967.

Burrill, Robert M. "The Establishment of Ranching on the Osage Indian Reservation." *Geographical Review*, 62, No. 4 (1972): 524-543.

Doran, Michael E. "Antebellum Cattle Herding in the Indian Territory." *Geographical Review*, 66, No. 1 (1976): 48-58.

DAIRYING

Cross, Ralph D. "The Oklahoma City Milk Shed." Master's Thesis, University of Oklahoma, Norman, Okla., 1961.

FARM LAND

Clark, Robert E. "Agricultural Land Utilization of the Central Washita Valley Flood Plain." Master's Thesis, University of Oklahoma, Norman, Okla., 1963.

Burrill, Robert. "The Osage Pasture Map." *Chronicles of Oklahoma*, 53, No. 2 (1975): 204-211.

Windhorst, Hans-Wilhelm, and Werner Klohn. "Ein landliches Mittelzengrum in der Krise-Perry/Oklahoma (USA)." *Praxis Geographie*, 18, No. 7-8 (1988): 43-47.

AGRICULTURAL POPULATIONS AND SOCIAL ISSUES

Lynch, Russell W. "Czech Farmers in Oklahoma: A Comparative Study of the Stability of a Czech Farm Group in Lincoln County, Oklahoma, and the

Factors Relating to Stability." Ph.D. Dissertation, Columbia University, New York, N.Y., 1942.

Lynch, Russell W. "Czech Farmers in Oklahoma: A Comparative Study of the Stability of a Czech Farm Group in Lincoln County, Oklahoma, and the Factors Relating to Its Stability." *Bulletin, Oklahoma A&M College*, 39, No. 13, 1942.

ENVIRONMENTAL AND CONSERVATION ISSUES

Bieberdorf, G.A. "History of the Distribution of the Mexican Boll Weavil in Oklahoma." *Proceedings of the Oklahoma Academy of Science*, 7 (1927): 29-34.

Gros, S.L., and Rebecca Roberts. "Groundwater Management Reform in Oklahoma: The Farmers' Verdict." *Papers and Proceedings of Applied Geography Conferences*, 8 (1985): 1-11.

IRRIGATION

Bieneman, Paul F. "Agricultural Geography of the Lugert–Altus Irrigation District of Southwestern Oklahoma." Master's Thesis, University of Oklahoma, Norman, Okla., 1967.

Kansas

GENERAL TEXTS, ARTICLES, AND THESES

Weatherbie, Wendell. "A Geographic Study of Agricultural Adjustments in the Elm Creek Watershed, Kansas." Master's Thesis, University of Colorado, Boulder, Colo., 1950.

Elliot, Harry L., Jr. "An Abrupt Agricultural Transition in Woodson County, Kansas." Master's Thesis, University of Kansas, Lawrence, Kans., 1951.

Roth, Jorge. "Mapping and Regionalizing Variable Agricultural Features in Kansas." Master's Thesis, University of Kansas, Lawrence, Kans., 1960.

Rubright, Lynnell. "Development of Farming Systems in Western Kansas, 1885–1915." Ph.D. Dissertation, University of Wisconsin, Madison, Wis., 1967.

Self, Hubert. "Agriculture." Pp. 94-126 in *Environment and Man in Kansas: A Geographical Analysis*. Lawrence, Kans.: The Regents Press of Kansas, 1978.

Hines, M. Elizabeth. "Farming and Ranching in Morton County Through 1915: The Historical Geography of a High Plains Frontier." Master's Thesis, University of Kansas, Lawrence, Kans., 1985.

CROPS AND CROPPING PATTERNS

Malin, James C. "Beginnings of Winter Wheat Production in the Upper Kansas and Lower Smokey Hill River Valleys: A Study in Adaptation to

Geographical Environment." *Kansas Historical Quarterly*, 10, No. 3 (1941): 227-259.

Malin, James C. *Winter Wheat in the Golden Belt of Kansas: A Study in Adaptation to Subhumid Geographical Environment.* Lawrence, Kans.: University of Kansas Press, 1944.

Reasons, D.R. "Soybean Expansion in Kansas: A Study in Agricultural Geography." Master's Thesis, University of Kansas, Lawrence, Kans., 1963.

Brooner, William. "Agricultural Crop Discrimination with Color Infrared Photography: A Study in Douglas County, Kansas." Master's Thesis, University of Kansas, Lawrence, Kans., 1969.

Marple, Robert P. "The Corn-Wheat Ratio in Kansas, 1879 and 1959: A Study in Historical Geography." *Great Plains Journal*, 8, No. 2 (1969): 79-86.

Shortridge, James R. "Prairie Hay in Woodson County, Kansas: A Crop Anomaly." *Geographical Review*, 63, No. 4 (1973): 533-552.

Nilchang, Sunandpattira. "The Diffusion of Grain Sorghum in Kansas." Master's Thesis, Kansas State University, Manhattan, Kans., 1981.

McGregor, Kent M. "Application of Decision Theory to Cropping Patterns in Kansas." *Papers and Proceedings of the Applied Geography Conferences*, 8 (1985): 12-18.

ANIMAL HUSBANDRY

Gustofson, Robert E. "The Supply Area of the Kansas City Cattle Market." Master's Thesis, University of Kansas, Lawrence, Kans., 1956.

Gibbs, Carol F. "Seasonal Grazing Operations in the Central Flint Hills Region of Kansas." Master's Thesis, University of Kansas, Lawrence, Kans., 1962.

Wibking, R. Kenton. "Geography of the Cattle Industry in the Flint Hills of Kansas." Ph.D. Dissertation, University of Nebraska, Lincoln, Nebr., 1963.

Kollmorgen, Walter M., and D.S. Simonett. "Grazing Operations in the Flint Hills–Bluestem Pastures of Chase County, Kansas." *Annals of the Association of American Geographers*, 55, No. 2 (1965): 260-290.

Bussing, Charles E., and Hubert Self. "Changing Structure of the Beef Industry in Kansas." *Transactions of the Kansas Academy of Science*, 84, No. 3-4 (1981): 173-186.

Hickey, Joseph V., and Charles E. Webb. "The Transition from Farming to Ranching in the Kansas Flint Hills: Two Case Studies." *Great Plains Quarterly*, 7, No. 4 (1987): 244-255.

FARM LAND

Shaklee, Ronald, Curtis Sorenson, and Charles E. Bussing. "Conversion of Agricultural Land in Wyandotte County, Kansas." *Transactions of the Kansas Academy of Science*, 87, No. 1-2 (1984): 1-11.

Rundquist, Bradley C. "A Reflection of Its Water: Agricultural Land Use Change in Finney County, Kansas, 1987–1992." Master's Thesis, Kansas State University, Manhattan, Kans., 1995.

AGRICULTURAL POPULATIONS AND SOCIAL ISSUES

Hecht, Melvin E. "Location Factors for Sidewalk Farmers in Elkhart, Morton County, Kansas." Ph.D. Dissertation, University of Kansas, Lawrence, Kans., 1962.

Merchant, James W. "Farmers' Perceptions of Soil Erosion–Water Quality Relationships in a Northeast Kansas Watershed." Master's Thesis, University of Kansas, Lawrence, Kans., 1973.

McQuillan, D. Aidan. "Adaptation of Three Immigrant Groups to Farming in Central Kansas, 1875–1925." Ph.D. Dissertation, University of Wisconsin, Madison, Wis., 1975.

McQuillan, D. Aidan. "The Mobility of Immigrants and Americans: A Comparison of Farmers on the Kansas Frontier." *Agricultural History*, 53, No. 2 (1979): 576-596.

Harrier, Donald. "An Analysis of Spatial Distribution of Rural Farm Population in Decataur County, Kansas: 1900–1988." Master's Thesis, Kansas State University, Manhattan, Kans., 1989.

Kuntz, Kara B. "Factors That Influence Kansas Farmers' Decisions to Adopt or Reject Conservation Tillage." Master's Thesis, Kansas State University, Manhattan, Kans., 1990.

Holcomb, Jason. "Aging Farmers and Land Ownership Changes in Kansas." Master's Thesis, Kansas State University, Manhattan, Kans., 1994.

ENVIRONMENTAL AND CONSERVATION ISSUES

Malin, James C. "The Adaptation of the Agricultural System to Sub-Humid Environment: Illustrating the Activities of the Wayne Township Farmers' Club of Edwards County, Kansas, 1886–1893." *Agricultural History*, 10, No. 3 (1936): 118-141.

Pechin, Edgar H. "Conservation Practices in Clark County, Kansas." Master's Thesis, Oklahoma State University, Stillwater, Okla., 1954.

Albertson, H. Carlis. "The Relationship Between Wheat Yields and Water Availability in Kansas." Master's Thesis, University of Kansas, Lawrence, Kans., 1959.

Webb, Robert M. "Relationships Between Agricultural Lowlands and Uplands in a Portion of the Big Blue Valley Region of Kansas." Ph.D. Dissertation, University of Kansas, Lawrence, Kans., 1962.

True, Marylou K. "Environmental Pollution and the Cattle Feedlot Industry in the Upper Neosho Watershed of Kansas." Master's Thesis, Kansas State University, Manhattan, Kans., 1969.

McGregor, Kent M. "Application of the Climatic Water Budget to Cropping Patterns in Kansas." *Geographical Perspectives*, 54 (1984): 53-61.

McGregor, Kent M. "The Effect of Climatic Change on Cropping Structure of Shawnee County, Kansas: A Bayesian Model." *Agricultural History*, 63, No. 2 (1989): 202-216.

Noll, Michael G. "Microbial Properties of Kenoma Silt Loam Soil Under Agriculture and Native and Reseeded Prairie." Master's Thesis, University of Kansas, Lawrence, Kans., 1993.

Duram, Leslie A. "Water Regulation Decisions in Central Kansas Affecting Cheyenne Bottoms Wetland and Neighboring Farmers." *Great Plains Research*, 5, No. 1 (1995): 5-19.

Klimstra, K.A. "Influences of Agricultural Change and the Conservation Reserve Program on Grassland Bird Populations in Southwestern Kansas." *Papers and Proceedings of the Applied Geography Conferences*, 22 (1999): 351-359.

IRRIGATION

Self, Hubert. "Irrigation Farming in Kansas." *Transactions of the Kansas Academy of Science*, 74 (1972): 310-317.

Rahtz, Daniel E. "The Impact of Irrigation on the Agricultural Geography of Finney County Since 1985." Master's Thesis, University of Kansas, Lawrence, Kans., 1985.

Wilson, Blake B. "An Assessment of the Impact of Irrigation on the Quiriva National Wildlife Refuge." Master's Thesis, Kansas State University, Manhattan, Kans., 1993.

AGRICULTURAL MARKETING AND DISTRIBUTION

Waits, David A. "A Spatial Analysis of Grain Marketing Cooperatives in Southwest Kansas." Master's Thesis, Oklahoma State University, Stillwater, Okla., 1988.

Nebraska

ATLASES AND GRAPHIC PRESENTATIONS

Anderson, Esther S. "Crop and Livestock Annual Variability Maps of Nebraska." *Yearbook, Association of Pacific Coast Geographers*, 3 (1940): 29-30.

Williams, James H., and Douglas Murfield. *Agricultural Atlas of Nebraska*. Lincoln, Nebr.: University of Nebraska Press, 1977.

GENERAL TEXTS, ARTICLES, AND THESES

Warren, Joseph A. "Agricultural Geography of Nebraska." Ph.D. Dissertation, University of Nebraska, Lincoln, Nebr., 1907.

Hayes, Frank A. "Agricultural Geography of Dawes County, Nebraska." Master's Thesis, University of Nebraska, Lincoln, Nebr., 1916.

Collett, Clinton E. "The Agricultural Geography of Fillmore County, Nebraska." Master's Thesis, University of Nebraska, Lincoln, Nebr., 1917.

Baltensperger, Bradley H. "Agricultural Patterns." Pp. 137-176 in *Nebraska: A Geography*, edited by Bradley H. Baltensperger. Boulder, Colo.: Westview, 1985.

Baltensperger, Bradley H. "Industrial Agricultural." Pp. 177-201 in *Nebraska: A Geography*, edited by Bradley H. Baltensperger. Boulder, Colo.: Westview, 1985.

CROPS AND CROPPING PATTERNS

Anderson, Esther S. "The Beet Sugar Industry of Nebraska as a Response to Geographic Environment." *Economic Geography*, 1, No. 4 (1925): 373-386.

Anderson, Esther S. "The Potato Industry of Nebraska." *Economic Geography*, 6, No. 1 (1930): 37-53.

Mumford, Mattie E. "Geographic Phases of the Wheat Industry in Nebraska." Master's Thesis, University of Nebraska, Lincoln, Nebr., 1934.

Guest, Buddy R. "Alfalfa in Nebraska." Master's Thesis, University of Nebraska, Lincoln, Nebr., 1947.

Schmiedling, Arthur C. "Geographic Patterns of Failure of Wheat and Corn in Nebraska, 1931–1952." Master's Thesis, University of Nebraska, Lincoln, Nebr., 1954.

Hewes, Leslie. "Wheat Failure in Western Nebraska, 1931–1954." *Annals of the Association of American Geographers*, 48, No. 3 (1958): 375-397.

Beisel, Donald E. "The Expansion of Grain Sorghum in Nebraska." Master's Thesis, University of Nebraska, Lincoln, Nebr., 1967.

Deal, Donald B. "Sidewalk Farming in the Specialized Wheat Region of Nebraska." Master's Thesis, University of Nebraska, Lincoln, Nebr., 1967.

Wilfang, Richard T., and Nancy D. Hughes. "Significance of Sound Waves in Breaking Seedling Dormancy in Cereal Grain Crops and Its Application to Northern Crop Production." *Virginia Geographer*, 7, No. 2 (1972): 15-16.

ANIMAL HUSBANDRY

Spear, Julia M. "Geographic Phases of the Beef Cattle Industry of Nebraska." Master's Thesis, University of Nebraska, Lincoln, Nebr., 1930.

Mather, Eugene C. "Cattle Ranching in the Sand Hills of Nebraska." Ph.D. Dissertation, University of Wisconsin, Madison, Wis., 1951.

Dewey, Donald R. "Cattle Feeding in Dawson County, Nebraska." Master's Thesis, University of Nebraska, Lincoln, Nebr., 1967.

FARM LAND

Koch, George D. "The Loess Hill Region of Northeastern Nebraska: A Geographic Study of the Land Utilization and Attending Conservation Problems." Ph.D. Dissertation, University of Nebraska, Lincoln, Nebr., 1938.

Fitzgerald, Denis P. "The Transitional Character of the Farm Landscape of Central Nebraska." Master's Thesis, University of Nebraska, Lincoln, Nebr., 1959.

Augustin, Byron D. "Changing Pattens on Cropland Use in a County on the Margin of the Hard Winter Wheat Belt and the Corn Belt: Adams County, Nebraska." Master's Thesis, University of Kansas, Lawrence, Kans., 1968.

Albrecht, Horst D. "Agricultural Land Use Changes in Cass County, Nebraska, 1860–1970." Master's Thesis, University of Nebraska, Lincoln, Nebr., 1972.

Flohr, Larry G. "The Changing Scale of Commercial Agricultural Land Base in Deuel County, Nebraska." Master's Thesis, Oregon State University, Corvallis, Oreg., 1973.

AGRICULTURAL POPULATIONS AND SOCIAL ISSUES

Bremer, Richard. "Patterns of Spatial Mobility: A Case Study of Nebraska Farmers, 1890–1970." *Agricultural History*, 48, No. 4 (1974): 529-542.

Barndt, Paul E., Jr. "Some Characteristics of the Farm Population in Nebraska Territory, 1854–1860." Master's Thesis, University of Nebraska, Lincoln, Nebr., 1977.

Baltensperger, Bradley H. "Agricultural Change Among Nebraska Immigrants, 1880–1900." Pp. 170-189 in *Ethnicity on the Great Plains*, edited by Frederick C. Luebke. Lincoln, Nebr.: University of Nebraska Press, 1980.

Baltensperger, Bradley H. "Agricultural Change Among Nebraska Immigrants, 1880–1900." *Journal of the American Historical Society of Germans from Russia*, 5, No. 4 (1982): 40-48.

Baltensperger, Bradley H. "Farms, Fields and Communities: The Rural Landscape." Pp. 201-220 in *Nebraska: A Geography*, edited by Bradley H. Baltensperger. Boulder, Colo.: Westview, 1985.

ENVIRONMENTAL AND CONSERVATION ISSUES

Wolfanger, Louisa A. "Conservation Survey of Perkins County, Nebraska." Master's Thesis, University of Nebraska, Lincoln, Nebr., 1922.

Black, Ferdinand F. "The Relationship of Corn Yields to Rainfall and Temperature Distribution in Southern Nebraska." Master's Thesis, University of Nebraska, Lincoln, Nebr., 1931.

Baker, Andrew. "The Adjustment of Agriculture to Its Environment: The Case of Antelope County, Nebraska." Master's Thesis, University of Nebraska, Omaha, Nebr., 1987.

IRRIGATION

Clute, Robert P. "Pump Irrigation in Box Butte County, Nebraska." Master's Thesis, University of Colorado, Boulder, Colo., 1950.

Kolbo, Allan D. "The Farewell Unit: Preview of an Irrigation Development in Central Nebraska." Master's Thesis, University of Nebraska, Lincoln, Nebr., 1961.

Nelson, Charles E. "Deep Well Irrigation in Hamilton County, Nebraska." Master's Thesis, Oregon State University, Corvallis, Oreg., 1961.

Brandhorst, L.C. "An Analysis of Risks in Surface-Irrigated Agriculture: The Scotts Bluff and Morrill County Area of Nebraska." Master's Thesis, University of Nebraska, Lincoln, Nebr., 1965.

Brandhorst, L.C. "The North Platte Oasis: Notes on the Geography and History of an Irrigated District." *Agricultural History*, 51, No. 1 (1977): 166-172.

Queen, Lloyd P. "Classification of Irrigated Crops in Box Butte County, Nebraska, Using Landsat." Master's Thesis, University of Nebraska, Lincoln, Nebr., 1982.

AGRICULTURAL MARKETING AND DISTRIBUTION

Wilhelm, John T. "Delimitation of the Omaha Wheat Source Supply Region." Master's Thesis, University of Nebraska, Omaha, Nebr., 1968.

TRANSPORTATION AND MECHANIZATION

Rowan, James E. "Mechanization of the Sugar Beet Industry of Scottsbluff County, Nebraska." *Economic Geography*, 24, No. 2 (1948): 174-180.

South Dakota

GENERAL TEXTS, ARTICLES, AND THESES

Dillingham, Harry L. "Suitcase Farming as a New Mode of Farm Operation in Sully County, South Dakota." Master's Thesis, University of Kansas, Lawrence, Kans., 1936.

Rawson, R.R. "The Agricultural Geography of the Dakotas." *Geography*, 25, No. 1 (1940): 6-17.

Kollmorgen, Walter M., and George F. Jenks. "Suitcase Farming in Sully County, South Dakota." *Annals of the Association of American Geographers*, 48, No. 1 (1958): 27-40.

CROPS AND CROPPING PATTERNS

Brandner, Thomas. "Factors Affecting Oat Yields in Eastern South Dakota." Master's Thesis, South Dakota State University, Brookings, S.Dak., 1976.

ANIMAL HUSBANDRY

Ballas, Donald J. "Early Agriculture and Livestock Raising Among the Teton Dakota Indians." *Bulletin, Illinois Geographical Society*, 15, No. 2 (1973): 53-62.

AGRICULTURAL POPULATIONS AND SOCIAL ISSUES

Ostergren, Robert. "European Settlement and Ethnicity Patterns on the Agricultural Frontiers of South Dakota." *South Dakota History*, 13, No. 1-2 (1983): 49-82.

ENVIRONMENTAL AND CONSERVATION ISSUES

Klages, Karl H.W. "Geographical Distribution of Variability in the Yield of Cereal Crops in South Dakota." *Ecology*, 12 (1934): 334-345.

North Dakota

GENERAL TEXTS, ARTICLES, AND THESES

Kollmorgen, Walter M., and George F. Jenks. "Sidewalk Farming in Toole County, Montana, and Trail County, North Dakota." *Annals of the Association of American Geographers*, 48, No. 2 (1958): 209-231.

Wangberg, Louis M. "The Historical Geography of Selected Farms in the Larimore, North Dakota Area." Master's Thesis, University of North Dakota, Grand Forks, N.Dak., 1964.

Huo, Song-Young. "Farm Size in North Dakota." Master's Thesis, University of North Dakota, Grand Forks, N.Dak., 1970.

CROPS AND CROPPING PATTERNS

Aho, Eino R. "Influence of Geographic Factors on the Red River Valley Potato Industry." Master's Thesis, University of North Dakota, Grand Forks, N.Dak., 1948.

Kopp, Herman J. "An Analysis of the Relationships Existing Among Selected Factors Affecting Small Grain Production in North Dakota." Master's Thesis, University of North Dakota, Grand Forks, N.Dak., 1952.

Cougill, John C. "The Sugar Beet Industry in the Red River Valley." Master's Thesis, University of North Dakota, Grand Forks, N.Dak., 1957.

Kleingartner, Elmer F. "The Soybean Industry in the Red River Valley." Master's Thesis, University of North Dakota, Grand Forks, N.Dak., 1959.

Duffy, Dennis W. "Locational Factors Influencing the Decline of Safflower Production in Western North Dakota: 1962–1972." Master's Thesis, University of North Dakota, Grand Forks, N.Dak., 1974.

Norloch, Randal L. "The Sunflower in North Dakota: A Spatial Diffusion Analysis." Master's Thesis, University of North Dakota, Grand Forks, N.Dak., 1981.

FARM LAND

Johnson, Gary E. "The Bonanza Farms and Cattle Empires: Exploitation of North Dakota's Land Resource Base." *Bulletin of the Association of North Dakota Geographers*, 27 (1977): 19-26.

The Mountain States

GENERAL TEXTS, ARTICLES, AND THESES

Meinig, Donald W. "The Growth of Agricultural Regions in the Far West, 1850–1910." *Journal of Geography*, 54, No. 5 (1955): 211-232.

Travis, Joan D. "Agrarian Problems of the Colorado River Indian Reservation." Master's Thesis, University of California, Los Angeles, Calif., 1968.

Davis, Gene P. "Effects of European Agricultural Introductions on Papago Subsistence: 1690–1965." Master's Thesis, Southern Illinois University, Carbondale, Ill., 1971.

Rinschede, Gisbert. "Formen der Wandervichwirtschaft in den westlichen USA." *Zeitschrift fur Agrargeographie*, 4 (1986): 1-41.

Matson, R.G. *The Origins of Southwestern Agriculture*. Tucson, Ariz.: University of Arizona Press, 1991.

ANIMAL HUSBANDRY

Giffiths, Thomas M. "Environmental Adaptations in the Western Sheep Industry." Master's Thesis, University of Denver, Denver, Colo., 1948.

Murphy, Donald L. "An Analysis of the Utah–Idaho Mink Industry." Master's Thesis, Brigham Young University, Provo, Utah, 1971.

Pearson, Erwin W. "Sheep Raising in the 17 Western States: Populations, Distribution and Trends." *Journal of Range Management*, 28, No. 1 (1975): 27-31.

Windhorst, Hans-Wilhelm. "Wandlungen in der Strukter und regionalen Verteilkung der Rindvielmast in den Vereinigten Staaten." *Geographische Rundschau*, 28, No. 2 (1976): 65-69.

Sullivan, Lawrence D. "The Goat and Land Use: An Historical/Environmental Analysis." Master's Thesis, Arizona State University, Tempe, Ariz., 1977.

Rinschede, Gisbert. "Saisanole nutzung des Hochgebirgraumes der westlichen USA durch Viehwirtschaft und Tourisme." *Frankfurter Wirtschafts- und Sozialgeographische Schriften*, 36 (1981): 173-212.

Rinschede, Gisbert. "The Migratory Livestock Industry in the Mountain West of the United States." *Mitteilungsblatt des Arbeitskreises USA in Zentralverband der Deutschen Geographen*, 6 (1987): 9-41.

FARM LAND

Rathburn, Arthur C. "A Framework for the Determination of the Suitability of Undeveloped Land in the Intermountain West for Use as Cropland." Ph.D. Dissertation, Oregon State University, Corvallis, Oreg., 1980.

Doolittle, William E. "The Use of Check Dams for Protecting Downstream Agricultural Land in the Prehistoric Southwest: A Contextual Analysis." *Journal of Anthropological Research*, 41 (1985): 279-305.

ENVIRONMENTAL AND CONSERVATION ISSUES

Bryan, Kirk. "Pre-Columbian Agriculture in the Southwest, as Conditioned by Periods of Alluvium." *Annals of the Association of American Geographers*, 31 (1941): 219-242.

Doolittle, William E. "The Use of Check Dams for Protecting Downstream Agricultural Lands in the Prehistoric Southwest: A Contextual Analysis." *Journal of Anthropological Research*, 41 (1985): 279-305.

Nablan, G.P. "Papago Indian Desert Agriculture and Water Control in the Sonoran Desert, 1697–1934." *Applied Geography*, 6, No. 1 (1986): 43-59.

Nablan, G.P. "Ak-cin 'Arroyo Mount' and the Environmental Setting of Papago Indian Fields in the Sonoran Desert." *Applied Geography*, 6, No. 1 (1986): 61-75.

Lightfoot, Dale R. "The Cultural Ecology of Puebloan Pebble-Mulch Gardens." *Human Ecology*, 21, No. 2 (1993): 115-124.

Lightfoot, Dale R. "The Construction and Configuration of Anasazi Pebble-Mulch Gardens in the Northern Rio Grande." *American Antiquity*, 60 (1995): 459-470.

IRRIGATION

Harper, Kelly C. "The Mormon Role in Irrigation Beginnings and Diffusions in the Western States: An Historical Geography." Master's Thesis, Brigham Young University, Provo, Utah, 1974.

Birle, Siegfried. *Irrigation Agriculture in the Southwest United States: Regional Variation of Crop Patterns*. Marburg, Germany: Marburger Geographische Schriften, Heft 67, 1976.

Colorado

GENERAL TEXTS, ARTICLES, AND THESES

Learned, Gordan W. "Effects of the Colorado-Big Thompson Project on the Agricultural Economy of Northeastern Colorado." Master's Thesis, University of Colorado, Boulder, Colo., 1940.

Gillette, N. "Geographic Analysis of Agriculture in Morgan County, Colorado." Master's Thesis, University of Colorado, Boulder, Colo., 1953.

Griffiths, Mel, and Lynnell Rubright. "Agriculture." Pp. 211-238 in *Colorado: A Geography*, edited by Mel Griffiths and Lynnell Rubright. Boulder, Colo.: Westview, 1983.

Krzemien, Gabrielle. *Formenspezialisierter Agrarwirtschaft in der Weld County, Colorado*. Vechta, Germany: Vechtaer Drukerie und Verlag, 1992.

Groves, Sally. "Agricultural-to-Urban Water Transfers and Economic and Population Changes Since 1971 in the Lower Arkansas River Valley, Colorado." Master's Thesis, University of Denver, Denver, Colo., 1994.

Pritchard, Sandra F. "The Agricultural Landscape of Summit County, Colorado." *Pennsylvania Geographer*, 33, No. 1 (1995): 3-22.

CROPS AND CROPPING PATTERNS

Gritzbaugh, R.D. "Some Geographic Aspects of the Sugar Beet Industry in the Longmont Area." Master's Thesis, University of Colorado, Boulder, Colo., 1933.

Campbell, Robert. "The Cherry Industry of the Loveland Area: An Analysis of a Geographic Adjustment in North Central Colorado." Master's Thesis, University of Colorado, Boulder, Colo., 1940.

Davis, Charles M. "Dry-Land Wheat in Northwestern Colorado." *Papers of the Michigan Academy of Science, Arts, and Letters*, 27 (1941): 345-350.

Nehrig, Edgar W. "The Flax Industry in Fresno County, Colorado." Master's Thesis, University of Colorado, Boulder, Colo., 1948.

Ramalay, Francis. "The Denver Truck Farming Area: Its Adjustment to the Physical, Cultural and Economic Landscape." Master's Thesis, University of Colorado, Boulder, Colo., 1949.

Dean, Thomas P. "The Sugarbeet Industry of the Arkansas Valley of Colorado." Master's Thesis, Oklahoma State University, Stillwater, Okla., 1954.

Thomson, Clarke W. "Crop Association in Northeastern Colorado, 1939–1949." Master's Thesis, University of Colorado, Boulder, Colo., 1958.

Bigelow, Eva L. "The Colorado Flower Industry." Master's Thesis, University of Colorado, Boulder, Colo., 1962.

Loeffler, M.J. "Beet-Sugar Production on the Colorado Piedmont." *Annals of the Association of American Geographers*, 53, No. 3 (1963): 364-390.

Spath, Hans-Joachim. "Dryland Wheat Farming on the Central Great Plains: Sedgwick County, Northeast Colorado." Pp. 313-344 in *Comparative Farming Systems*, edited by Billie L. Turner II and Stephen B. Brush. New York, N.Y.: Guilford, 1987.

ANIMAL HUSBANDRY

Oberholzer, Corabelle. "The Grazing Industry of Middle Park, Colorado." Master's Thesis, University of Colorado, Boulder, Colo., 1942.

Alberts, Fred G. "The Colorado Sheep Industry: A Geographic Analysis." Master's Thesis, University of Colorado, Boulder, Colo., 1949.

Thomas, Jerald W. "The Fur Farming Industry of Colorado." Master's Thesis, University of Colorado, Boulder, Colo., 1950.

Henderson, David A. "The Beef Cattle Industry of Colorado." Master's Thesis, University of Colorado, Boulder, Colo., 1951.

Henderson, David A. "Corn Belt Cattle Feeding in Eastern Colorado Irrigated Valleys." *Economic Geography*, 30, No. 4 (1954): 364-372.

Stock, Edward C. "A Microgeographic Study of a Colorado Foothills Ranch." Master's Thesis, University of Colorado, Boulder, Colo., 1961.

Crowley, John M. "Ranches in the Sky: A Geography of Livestock Ranching in the Mountain Parks of Colorado." Ph.D. Dissertation, University of Minnesota, Minneapolis, Minn., 1964.

Bussing, Charles E. "The Cattle Feeding Industry of the Northern Colorado Piedmont: A Geographic Appraisal." Ph.D. Dissertation, University of Nebraska, Lincoln, Nebr., 1968.

Crowley, John M. "Ranching in the Mountain Parks of Colorado." *Geographical Review*, 65, No. 4 (1975): 445-460.

Abruzzi, William S. "The Social and Ecological Consequences of Early Cattle Ranching in the Little Colorado River Basin." *Human Ecology*, 23, No. 1 (1995): 75-98.

AGRICULTURAL POPULATIONS AND SOCIAL ISSUES

Green, D. Brooks. "Irrigation Potential and Agricultural Settlement in Submarginal Land: An Example from Weld County, Colorado." D.A. Dissertation, University of Northern Colorado, Greeley, Colo., 1980.

ENVIRONMENTAL AND CONSERVATION ISSUES

Duram, Leslie A. "Great Plains Agroecologies: The Continuum from Conventional to Alternative Agriculture in Colorado." Pp. 153-166 in *Agricultural Restructuring and Sustainability: A Geographical Perspective*, edited by Brian W. Ilbery, Quentin Chiotti, and Timothy Rickard. Wallingford, England: C.A.B. International, 1997.

Duram, Leslie A. "Agriecological Behavior: The Continuum from Conventional to Alternative Agriculture in Colorado, USA." Pp. 69-80 in *Dimensions of Sustainable Rural Systems*, edited by Ian R. Bowler, Christopher R. Bryant, and P.P.P. Huigan. Utrecht: Netherlands Geographical Studies, No. 244, 1998.

IRRIGATION

Walker, Paul D. "Irrigated Agriculture: The Grand Valley Project, Mesa County, Colorado." Master's Thesis, University of Denver, Denver, Colo., 1967.

Giannetti, Claudia. "Association of the Ogallala Aquifer Depletion on Center-Pivot Irrigation Farmland in Kit Carson County, Colorado." Master's Thesis, University of Denver, Denver, Colo., 1992.

New Mexico

GENERAL TEXTS, ARTICLES, AND THESES

Gremaud, Gregory C. "Agricultural Practices at Po-shu-ouinge, a Pueblo IV Tewa Settlement in the Rio Chama Valley, Rio Ariba County, New Mexico." Master's Thesis, Southern Illinois University, Carbondale, Ill., 1977.

CROPS AND CROPPING PATTERNS

Culbert, James I. "Pinto Beans in the Estancia Valley of New Mexico." *Economic Geography*, 17, No. 1 (1941): 50-60.

ANIMAL HUSBANDRY

Culbert, James I. "Cattle Industry of New Mexico." *Economic Geography*, 17, No. 2 (1941): 155-168.

Denevan, William. "Livestock Numbers in Nineteenth Century New Mexico and the Problem of Gullying in the Southwest." *Annals of the Association of American Geographers*, 57, No. 4 (1967): 691-703.

Carlson, Alvar W. "New Mexico's Sheep Industry 1850–1900: Its Role in the History of the Territory." *New Mexico Historical Review*, 44, No. 1 (1969): 25-49.

FARM LAND

Carlson, Alvar W. "Long-Lots in Rio Ariba." *Annals of the Association of American Geographers*, 65, No. 1 (1975): 48-57.

Carlson, Alvar W. "Spanish-American Acquisition of Cropland Within the Northern Pueblo Indian Grants, New Mexico." *Ethnohistory*, 22, No. 2 (1975): 95-110.

Lightfoot, Dale R. "The Landscape Context of Anasazi Pebble-Mulched Fields in the Galisteo Basin, Northern New Mexico." *Geoarcheology*, 8 (1993): 349-370.

AGRICULTURAL POPULATIONS AND SOCIAL ISSUES

McIntire, Elliot. "Changes in an American Peasant Society: The Hopi Farmer." *The Oregon Geographer*, 1, No. 2 (1967): 9-13.

IRRIGATION

Foscue, Edwin J. "The Mesilla Valley of New Mexico: A Study of Aridity and Irrigation." *Economic Geography*, 7, No. 1 (1931): 1-27.

Utah

GENERAL TEXTS, ARTICLES, AND THESES

White, C. Langdon. "The Agricultural Geography of the Salt Lake Oasis." Ph.D. Dissertation, Clark University, Worcester, Mass., 1925.

White, C. Langdon. "The Agricultural Geography of the Salt Lake Oasis." *Journal of the Scientific Laboratories of Denison University*, 24, No. 4 (1925): 117-283.

Miller, Elbert E. "Agricultural Geography of Cache Valley, Utah." Ph.D. Dissertation, University of Washington, Seattle, Wash., 1952.

Wride, Charles H. "The Agricultural Geography of Utah County, 1849–1960." Master's Thesis, Brigham Young University, Provo, Utah, 1961.

Rathjen, M. Randall. "Evolution and Development of the Mormon Welfare Farms." Ph.D. Dissertation, Michigan State University, East Lansing, Mich., 1969.

Jackson, Richard H. "Mormon Indian Farms." Pp. 41-54 in *Geographical Perspectives on Native Americans: Topics and Resources,* edited by Jerry N. McDonald and Tony Lazewski. Washington, D.C.: Association of American Geographers, Associated Committee on Native Americans, Publication No. 1, 1976.

CROPS AND CROPPING PATTERNS

Zink, Norah E. "Dry-Farming Regions in Utah." *Economic Geography*, 15, No. 4 (1939): 421-431.

Zink, Norah E. "A Dry-Farm Region in Utah." *Economic Geography*, 16, No. 3 (1940): 281-290.

Lloyd, Roy D. "Sugar Beet Seed Production in Utah: An Economic Geography." Master's Thesis, University of Utah, Salt Lake City, Utah, 1956.

Faber, Stanley. "The Alfalfa Seed Industry in Utah." Master's Thesis, University of Utah, Salt Lake City, Utah, 1967.

Snipe, Ronald H. "A Geographical Analysis of the Fruit Tree Industry of Utah County." Master's Thesis, University of Utah, Salt Lake City, Utah, 1970.

Styler, Brent A. "Utah's Fresh Vegetable Supply: A Geographic Appraisal." Master's Thesis, Brigham Young University, Provo, Utah, 1972.

Schinzel, Robert H. "The Geography of the Pinto Bean Industry of the Great Sage Plain." Master's Thesis, University of Utah, Salt Lake City, Utah, 1973.

Brough, R. Clayton. "The Diffusion and Adaption of Alfalfa in Utah." Master's Thesis, Brigham Young University, Provo, Utah, 1975.

Budge, Thomas K. "A Multispectral Analysis of Crops in Utah Using ERTS-1 Satellite Imagery." Master's Thesis, University of Utah, Salt Lake City, Utah, 1975.

ANIMAL HUSBANDRY

White, C. Langdon. "Transhumance in the Sheep Industry of the Salt Lake Region." *Economic Geography*, 2, No. 4 (1926): 414-425.

Parker, Mayland L. "Economic Geography of Utah's Sheep Industry." Master's Thesis, University of Utah, Salt Lake City, Utah, 1951.

Hayes, Emmet R. "Geographical Analysis of the Utah Turkey Industry with Case Studies." Master's Thesis, Brigham Young University, Provo, Utah, 1971.

Reed, Carl W., Jr. "Geography of Turkey Raising in the Sanpete Valley: A Study in Time and Space." Master's Thesis, Brigham Young University, Provo, Utah, 1971.

DAIRYING

Layton, Robert L. "Historical Geography of the Salt Lake Milk Shed." *Yearbook, Association of Pacific Coast Geographers*, 14 (1951): 34-39.

Layton, Robert L. "A Geographic-Economic Study of the Salt Lake Milkshed." Master's Thesis, University of Utah, Salt Lake City, Utah, 1952.

FARM LAND

Williamson, Dee J. "An Analysis of the Impact of Interstate 15 on Agricultural and Urban Land Between Payson and Provo, Utah." Master's Thesis, Brigham Young University, Provo, Utah, 1973.

Barker, Craig C. "Consumption of Agricultural Land in Davis County, Utah, 1970–1975." Master's Thesis, University of Utah, Salt Lake City, Utah, 1978.

Buchanon, Robert M. "Agricultural Land Conversion in the Salt Lake Valley: Policies and Practices, 1950–1980." Master's Thesis, University of Utah, Salt Lake City, Utah, 1982.

AGRICULTURAL POPULATIONS AND SOCIAL ISSUES

Spencer, Joseph E. "The Development of Agricultural Villages in Southern Utah." *Agricultural History*, 14, No. 4 (1940): 181-189.

Shelley, Wayne R. "The Development and Failure of Historic Agricultural Communities of Utah: A Case Study of Johns Valley, Utah." Master's Thesis, Brigham Young University, Provo, Utah, 1989.

ENVIRONMENTAL AND CONSERVATION ISSUES

Baker, Virgil R. "Agricultural Conservation in Castle Valley, Utah." Ph.D. Dissertation, University of Utah, Salt Lake City, Utah, 1954.

Ridd, Merrill K. "The Influences of Soil and Water Conditions on Agricultural Development in the Delta Area, Utah." Ph.D. Dissertation, Northwestern University, Evanston, Ill., 1963.

IRRIGATION

Strasser, Manfred. *Die Bewasserungslandschaft der Wasatch Oase in Utah*. Freiberg, Germany: Hans Ferdinand Schulz Verlag, Freiberger Geographische Arbeiten, Vol. 4, 1972.

Jones, Clark E. "Effects of the Strawberry Valley Irrigation on Soil Salinity and Sodicity of Southern Utah Valley." Master's Thesis, Brigham Young University, Provo, Utah, 1974.

Strasser, Manfred. "Die Feldbewasserung des Mormonlands Utah in Wandel der Zeit." *Zeitschrift fur Wirtschaftsgeographie*, 19, No. 3 (1975): 87-90.

Strasser, Manfred. "Die Bewasserungswirtschaft der Mormonen." *Geographie im Unterricht*, 1, No. 7 (1976): 181-191.

Cook, Marian. "Feasibility of Future Irrigation Potentials in Snake-Valley Utah Geographical and Agricultural Provosations." Master's Thesis, Brigham Young University, Provo, Utah, 1992.

Arizona

GENERAL TEXTS, ARTICLES, AND THESES

Pecora, Aldo. *Richerche di geografia agraria nell'Arizona centro-meridionale.* Napoli, Italy: Istituto di Geografia dell'Universita, 1964.

Comeaux, Malcolm L. "Agriculture." Pp. 259-273 in *Arizona: A Geography*, edited by Malcolm L. Comeaux. Boulder, Colo.: Westview, 1981.

Perramond, Eric P. "An Historical Geography of Zuni Agriculture." Master's Thesis, Louisiana State University, Baton Rouge, La., 1994.

CROPS AND CROPPING PATTERNS

Hutsick, Paul J. "A Geographic Study of the Development in the Production of Sugar Beets and Sugar Beet Seed in Arizona." Master's Thesis, Arizona State University, Tempe, Ariz., 1965.

Capppadona, Thomas R. "Arizona's Safflower Industry." Master's Thesis, Arizona State University, Tempe, Ariz., 1966.

Shugars, Jerry G. "The Geonomics of Barley in Arizona." Master's Thesis, Arizona State University, Tempe, Ariz., 1966.

Berlin, Graydon L. "Geographical Analysis of the Leading Vegetable Crops of Maricopa County, Arizona." Master's Thesis, Arizona State University, Tempe, Ariz., 1967.

Hullihen, James D. "The Arizona Citrus Industry." Master's Thesis, Arizona State University, Tempe, Ariz., 1967.

Smith, Dwight. "The Date Industry of Arizona." Master's Thesis, Arizona State University, Tempe, Ariz., 1967.

Durfee, Lynn E. "Large Scale Vegetable Production in Central Arizona." Master's Thesis, Arizona State University, Tempe, Ariz., 1969.

Jett, Stephen C. "History of Fruit Tree Raising Among the Navajo." *Agricultural History*, 51, No. 4 (1977): 681-701.

Jett, Stephen C. "Peach Cultivation and Use Among the Canyon de Chelly Navajo." *Economic Botany*, 33, No. 3 (1979): 298-310.

Shapiro, Eric A. "Cotton in Arizona: A Historical Geography." Master's Thesis, University of Arizona, Tucson, Ariz., 1989.

Bozarth, Steven R. "Pollen and Opal Phytolith Evidence of Prehistoric Agriculture and Wild Plant Utilization in the Lower Verde River Valley, Arizona." Ph.D. Dissertation, University of Kansas, Lawrence, Kans., 1996.

ANIMAL HUSBANDRY

Fonaroff, Leonard S. "The Navajo Sheep Industry: A Study in Cross-Cultural Administration." Ph.D. Dissertation, Johns Hopkins University, Baltimore, Md., 1961.

Dunckle, John P. "Cattle Raising and Adaptations to Drought by Cattlement in the Altar Valley, Pima County, Arizona." Master's Thesis, University of Arizona, Tucson, Ariz., 1967.

Snow, Ronald W. "The Beef Cattle Industry of Arizona: A Geographical Analysis." Master's Thesis, Arizona State University, Tempe, Ariz., 1969.

Horsh, Becky J. "The Characteristics and Location of the Arizona Penfed Cattle Industry." Master's Thesis, Arizona State University, Tempe, Ariz., 1971.

Cropsey, Sharon M. "The Impact of Urbanization on the Beef and Dairy Cattle Industries in Maricopa County, Arizona." Master's Thesis, Arizona State University, Tempe, Ariz., 1975.

Jordan, Terry G. "Texas Influence in Nineteenth-Century Arizona Cattle Ranching." *Journal of the West*, 14, No. 3 (1975): 15-18.

Lin, Ning-Hui. "The Use of Large Scale Color Aerial Photography to Monitor Cattle Grazing in Mesquite Grasslands, Southern Arizona." Master's Thesis, University of Arizona, Tucson, Ariz., 1986.

ENVIRONMENTAL AND CONSERVATION ISSUES

Pifer, Joseph. "A Study of the Impact of the Watertable Decline Upon Agriculture in the Queen Creek–Chandler Heights Area, Maricopa County, Arizona." Master's Thesis, Arizona State University, Tempe, Ariz., 1968.

Pifer, Joseph. "Water Table Decline and Resultant Agricultural Patterns in the Salt River Valley of Arizona." *Journal of Geography*, 68, No. 9 (1969): 545-549.

Manera, Paul A. "Physical Factors Limiting the Agricultural Development of Butler Valley, Yuma County, Arizona." Master's Thesis, Arizona State University, Tempe, Ariz., 1970.

Koshear, Jeannine. "Hopi Agriculture: Environmental and Cultural Change." Master's Thesis, University of California, Berkeley, Calif., 1987.

Lightfoot, Dale R. "The Prehistoric Pebble-Mulched Fields of the Galisteo Anasazi: Agricultural Innovation and Adaptation to Environment." Ph.D. Dissertation, University of Colorado, Boulder, Colo., 1990.

IRRIGATION

Bowdes, Leonard W. "Pump Irrigation in Western Yuma County: An Environmental Adjustment." Master's Thesis, University of Colorado, Boulder, Colo., 1961.

Wood, Roland G. "Agricultural Utilization of Water in the Salt River Valley, Arizona." Master's Thesis, University of California, Los Angeles, Calif., 1963.

AGRICULTURAL LABOR STUDIES

Cline, Donald L. "Farm Labor in Arizona Since Termination of Public Law 78." Master's Thesis, Arizona State University, Tempe, Ariz., 1967.

Idaho

ANIMAL HUSBANDRY

Brower, Barbara. "Sheep Grazing in National Forest Wilderness: A New Look at an Old Fight." *Mountain Research and Development*, 20, No. 2 (2000): 126-129.

IRRIGATION

Luke, Timothy J. "Comparisons of Irrigated Acreage and Water Rights Records in the Big Wood River Basin." Master's Thesis, University of Idaho, Moscow, Idaho, 1992.

Wyoming

GENERAL TEXTS, ARTICLES, AND THESES

Wilson, Paul B. "Farming and Ranching on the Wind River Indian Reservation." Ph.D. Dissertation, University of Nebraska, Lincoln, Nebr., 1972.
Brown, Robert H. "Agriculture." Pp. 151-182 in *Wyoming: A Geography*, edited by Robert H. Brown. Boulder, Colo.: Westview, 1980.

ANIMAL HUSBANDRY

Calef, Wesley. "Problems of Grazing Administration in the Basins of Southern Wyoming." *Economic Geography*, 28, No. 2 (1952): 122-127.
Wallach, Bret. "Sheep Ranching in the Dry Corner of Wyoming." *Geographical Review*, 71, No. 1 (1981): 51-63.

Nevada

GENERAL TEXTS, ARTICLES, AND THESES

Bowen, Marshall E. "Elko County's Dry Farming Experimental Station." *Northeastern Nevada Historical Society Quarterly*, 2, No. 1 (1979): 35-51.

Bowen, Marshall E. "Jackrabbit Invasion of a Nevada Agricultural Community." *Ecumene*, 12 (1981): 6-16.

CROPS AND CROPPING PATTERNS

Bowen, Marshall E. "A Backward Step: From Irrigation to Dry Farming within Nevada Desert." *Agricultural History*, 63, No. 2 (1989): 231-249.

AGRICULTURAL POPULATIONS AND SOCIAL ISSUES

Bowen, Marshall E. "Turnovers of Pioneers and Property in a Marginal Nevada Farming Community." *Yearbook, Association of Pacific Coast Geographers*, 42 (1980): 45-58.

IRRIGATION

Rowley, William. "Opposition to Arid Land Irrigation in Nevada, 1890–1900." *Yearbook, Association of Pacific Coast Geographers*, 43 (1981): 113-124.

Montana

GENERAL TEXTS, ARTICLES, AND THESES

Schwartzman, Kristina H. "Case Study in Planning a Small Farm Operation in Ravilli County, Montana." Master's Thesis, University of Montana, Missoula, Mont., 1980.

FARM LAND

Struck, David G. "Conflictive Land Use: Coal Strip Mining and Agriculture, Fort Union Coal Area, Montana." Master's Thesis, Oregon State University, Corvallis, Oreg., 1974.

ENVIRONMENTAL AND CONSERVATION ISSUES

Wyckoff, William, and Katherine Hansen. "Settlement, Livestock Grazing and Environmental Change in Southwest Montana, 1860–1990." *Environmental History Review*, 15, No. 1 (1991): 45-73.

Wilson, J.P., et al. "Climate, Soil and Crop Yield Relationships in Cascade County, Montana." *Applied Geography*, 12, No. 3 (1992): 261-280.

Hansen, Katherine, William Wyckoff, and Jeff Banfield. "Shifting Forests: Historical Grazing and Forest Invasion in Southwestern Montana." *Forest and Conservation History*, 39, No. 2 (1995): 66-76.

The Pacific Region of the United States

GENERAL TEXTS, ARTICLES, AND THESES

Meinig, Donald W. "The Growth of Agricultural Regions in the Far West, 1850–1910." *Journal of Geography*, 54, No. 5 (1955): 221-232.

Gibson, James R. "Food for the Fur Traders: The First Farmers in the Pacific Northwest, 1805–1846." *Journal of the West*, 7, No. 1 (1968): 18-30.

Gregor, Howard F. "The Industrial Farm as a Western Institution." *Journal of the West*, 9, No. 1 (1970): 78-92.

Scott, James W. "Agriculture." Pp. 303-340 in *The Pacific Northwest: Geographical Perspectives*, edited by James G. Ashbaugh. Dubuque, Iowa: Kendall-Hunt, 1994.

CROPS AND CROPPING PATTERNS

Garland, John H. "The Columbia Plateau Region of Commercial Grain Farming." *Geographical Review*, 24 (1934): 371-379.

Roterus, Victor. "Spring and Winter Wheat on the Columbia Plateau." *Economic Geography*, 10, No. 4 (1934): 368-373.

Freeman, Otis W. "The Hop Industry of the Pacific Coast States." *Economic Geography*, 12, No. 2 (1936): 155-163.

Landis, Paul H. "The Hop Industry: A Social and Economic Problem." *Economic Geography*, 15, No. 1 (1939): 85-94.

Seeman, Albert L. "Cranberry Industry of the Pacific Coast." *Economic Geography*, 17, No. 2 (1941): 180-186.

Freeman, Otis W. "The Pacific Northwest Pea Industry." *Economic Geography*, 19, No. 2 (1943): 118-128.

Miller, Elbert E., and Richard Highsmith. "The Hop Industry of the Pacific Coast." *Journal of Geography*, 49, No. 2 (1950): 63-77.

Griffin, Paul F. "Some Geographic Aspects of the California and Hawaiian Sugar Industry." *Journal of Geography*, 53, No. 8 (1954): 325-335.

Brown, Robert C. "The Agricultural Geography of Peppermint in the Pacific Northwest." Master's Thesis, Oregon State University, Corvallis, Oreg., 1962.

ANIMAL HUSBANDRY

McGregor, A.C. *Country Sheep: From Open Range to Agribusiness on the Columbia Plateau*. Seattle, Wash.: University of Washington Press, 1983.

DAIRYING

Strain, Warren T. "The Geography of the Northwest Dairy Region." Ph.D. Dissertation, University of Wisconsin, Madison, Wis., 1933.

ENVIRONMENTAL AND CONSERVATION ISSUES

Meinig, Donald W. "Evolution of Understanding an Environment: Climates and Wheat Culture in the Columbia Plateau." *Yearbook, Association of Pacific Coast Geographers*, 16 (1953): 25-34.

Schroeder, Carlton R. "The Physical Geography of the Palouse, Washington and Idaho, in Relation to the Agricultural Economy." Ph.D. Dissertation, University of California, Los Angeles, Calif., 1958.

IRRIGATION

Carter, Douglas B. "The Relation of Irrigation Efficiency to the Potential Development of Irrigated Agriculture in the Pacific Northwest." Ph.D. Dissertation, University of Washington, Seattle, Wash., 1957.

Cantor, Leonard M. "Irrigated Agriculture in the Columbia Basin Project." *Geography*, 52, No. 1 (1967): 67-72.

Nellis, M. Duane. "Institutional Implications Affecting Agricultural Water Use Efficiency in the Western United States." *Geographical Perspectives*, 51 (1983): 46-48.

California

GENERAL TEXTS, ARTICLES, AND THESES

Pfiefer, Gottfried. *Raumliche Gliederung der Landwirtschaft im Nordlichen Kalifornien.* Leipzig, Germany: Wissenschaftliche Veroffentlichangen der Gesellschaft fur Erdkunde zu Leipzig, Vol. 10, 1936.

Torbert, Edward N. "The Specialized Commercial Agriculture of the Northern Santa Clara Valley." *Geographical Review*, 26, No. 2 (1936): 247-263.

Glendinning, Robert M. "The Coachella Valley, California: Some Aspects of Agriculture in a Desert." *Papers of the Michigan Academy of Science, Arts and Letters*, 35 (1949): 173-195.

Gregor, Howard F. "Changing Agricultural Patterns in the Oxnard Area, Southern California." Ph.D. Dissertation, University of California, Los Angeles, Calif., 1950.

Whittington, Roberta E. "Hemet-San Jacinto: A Geographic Survey of an Agricultural Valley." Master's Thesis, University of California, Los Angeles, Calif., 1951.

Gregor, Howard F. "Agricultural Shifts in the Ventura Lowlands of California." *Economic Geography*, 29, No. 4 (1953): 340-361.

Scott, W.E. "The Agricultural Geography of Indian Valley, Plumas County, California." Master's Thesis, California State University, Chico, Calif., 1953.

Willis, Harry W. "Evoluation of Corporate Agriculture in the San Joaquin Valley, California." Master's Thesis, University of California, Los Angeles, Calif., 1953.

Holbrook, Vinson G. "The Agricultural Geography of Shasta Valley, California." Master's Thesis, University of Oregon, Eugene, Oreg., 1955.

Cole, Chester F. "California's Agriculture: Some Realities." *Journal of Geography*, 57, No. 7 (1958): 341-345.

Griffin, Paul F., and Ronald L. Chatham. "Urban Impact on Agriculture in Santa Clara County, California." *Annals of the Associaton of American Geographers*, 48, No. 2 (1958): 195-208.

Blossom, Herbert H. "The Agricultural Development of the Coachella Valley, California." Master's Thesis, University of California, Berkeley, Calif., 1959.

Gregor, Howard F. "An Evaulation of Oasis Agriculture." *Yearbook, Association of Pacific Coast Geographers*, 21 (1959): 39-50.

Eder, H.M. "Some Aspects of the Persistence of Agriculture in the San Fernando Valley, California." Master's Thesis, University of California, Los Angeles, Calif., 1960.

Gentilcore, R. Louis. "Ontario, California and the Agricultural Boom of the 1880s." *Agricultural History*, 34, No. 2 (1960): 77-87.

Gregor, Howard F. "The Regional Primacy of San Joaquin Valley Agricultural Production." *Journal of Geography*, 61, No. 9 (1962): 394-399.

Gregor, Howard F. "Regional Hierarchies in California Agricultural Production: 1939–1954." *Annals of the Association of American Geographers*, 53, No. 1 (1963): 27-33.

Gregor, Howard F. "Urbanization of Southern California Agriculture." *Tijdschrift voor Economische en Sociale Geographie*, 54 (1963): 273-278.

Pease, Robert W. *Modoc County: A Geographic Time Continuum on the California Volcanic Tableland.* Berkeley, Calif.: University of California at Berkeley, California Publications in Geography, No. 17, 1965.

Cole, Chester, F., and Alan H. Johnson. "Die Landwirtschaft Kaliforniens." *Geographische Rundschau*, 19, No. 2 (1967): 41-47.

Getre, J.S. "The Geography of Selected Patterns of Agriculture in San Mateo County, California." Master's Thesis, California State University, San Francisco, Calif., 1968.

Kerr, William S. "Impact of Urbanization on Agriculture in Orange County, California." *Yearbook, Association of Pacific Coast Geographers*, 34 (1972): 161-170.

Gregor, Howard F. *An Agricultural Typology of California*. Budapest, Hungary: Akademiai Kiado, 1974.

Fitzsimmons, Margaret I. "Agriculture in the Pueblo of Los Angeles, 1781–1847." Master's Thesis, California State University, Northridge, Calif., 1975.

Hornbeck, David, and Mary L. Tucey. "Agriculture in Hispanic California, 1850." *California Geographer*, 15 (1975): 52-59.

Zimrick, Steven J. "The Changing Organization of Agriculture in the Southern San Joaquin Valley, California." Ph.D. Dissertation, Louisiana State University, Baton Rouge, La., 1976.

Dorel, Gerard. "Une grande region agricole californienne—les Westlands." *Revue de Geographie de Lyon*, 53 (1978): 241-264.

Thompson, Kenneth, and Richard A. Eigenheer. "The Agricultural Promise of the Sacramento Valley: Some Early Views." *Journal of the West*, 18, No. 4 (1979): 33-41.

Pryde, Philip R. "Is There Any Hope for Agriculture in California's Rapid Growth Areas?" *GeoJournal*, 6, No. 6 (1982): 533-538.

Grampp, Christopher. "Gardens for California Living." *Landscape*, 28, No. 3 (1985): 40-47.

Dilsaver, Larry. "The Development of Agriculture in a Gold Rush Region." *Yearbook, Association of Pacific Coast Geographers*, 48 (1986): 67-88.

Konnecke, Michael. "Phasen Landwirtschaftlicher Inwertsetzung des Sacramentotals." *Zeitschrift fur Wirtschaftsgeographische*, 30, No. 2 (1986): 35-48.

Klohn, Werner. *Strukturen und Probleme der Intensivlandwirtschaft in Kalifornien*. Vechta, Germany: Vechtaer Arbeiten zur Geographie und Regionalwissenschaft, Band 11, 1990.

Windhorst, Hans-Wilhelm. "Agrarstrukturwandel im Randbereich von Agglomeration: Das Beispiel der Orange County (California)." *Vechtaer Arbeiten zur Geographie und Regionalwissenschaft*, 11 (1990): 125-136.

Hart, John. *Farming on the Edge: Saving Family Farms in Marin County, California*. Berkeley, Calif.: University of California Press, 1991.

Windhorst, Hans-Wilhelm. "Industrietomatenanbau in Kalifornien." *Praxis Geographie*, 21, No. 10 (1991): 41-46.

CROPS AND CROPPING PATTERNS

Rowell, Edna E. "The Raisin Industry in California." *Journal of Geography*, 10 (1911–12): 272-275.

Colby, Charles C. "The California Raisin Industry." *Annals of the Association of American Geographers*, 14 (1924): 49-108.

Coulter, John W. "A Critical Situation in the Two One-Crop Wheat Farming Districts in California." *Economic Geography*, 5, No. 1 (1929): 87-97.

Zierer, Clifford M. "The Lima Bean Industry of the Southern California Coastal Region." *Bulletin of the Geographical Society of Philadelphia*, 27 (1929): 65-86.

Dicken, Samuel N. "Dry Farming in the San Joaquin, California." *Economic Geography*, 8, No. 2 (1932): 94-99.

Hodgson, Robert W. "The California Fruit Industry." *Economic Geography*, 9, No. 4 (1933): 337-355.

Zierer, Clifford M. "The Citrus Fruit Industry of the Los Angeles Basin." *Economic Geography*, 10, No. 1 (1934): 53-73.

Aumack, G.D. "A Geographical Study of the Tulare County Citrus Belt." Master's Thesis, University of California, Los Angeles, Calif., 1939.

Parsons, James J. "The California Hop Industry: Its Eighty Years of Development and Expansion." Master's Thesis, University of California, Berkeley, Calif., 1939.

Parsons, James J. "Hops in Early California Agriculture." *Agricultural History*, 14, No. 3 (1940): 110-116.

Meigs, Peveril. "Current Trends in California Orchards and Vineyards." *Economic Geography*, 17, No. 3 (1941): 275-286.

Eidemiller, Donald I. "Economic Geography of Avocado Growing in San Diego County." Master's Thesis, University of California, Berkeley, Calif., 1949.

Cushman, G. Raymond. "The Geography of Sugarbeet Growing in California." Master's Thesis, University of California, Berkeley, Calif., 1950.

Tow, Leonard. "Cotton Culture in the San Joaquin Valley, California." Master's Thesis, Columbia University, New York, N.Y., 1952.

McClain, Marcia. "The Distribution of Asparagus Production on the Sacramento–San Joaquin Delta." Master's Thesis, University of California, Berkeley, Calif., 1954.

Griffin, Paul F. "The Olive Industry of California." *Journal of Geography*, 54, No. 9 (1955): 429-440.

Blick, James D. "Analysis of Cotton Production in the San Joaquin Valley, California." Ph.D. Dissertation, University of California, Los Angeles, Calif., 1956.

Gregor, Howard F. "The Geographic Dynamism of California Market Gardening." *Yearbook, Association of Pacific Coast Geographers*, 18 (1956): 28-35.

Large, David C. "Cotton in the San Joaquin Valley: A Study of Government in Agriculture." *Geographical Review*, 47, No. 3 (1957): 365-380.

Williams, Charles E. "The Ecological Crop Geography of the Lemon in California." Master's Thesis, Pennsylvania State University, University Park, Pa., 1957.

Thompson, Kenneth. "Location and Relocation of a Tree Crop: English Walnuts in California." *Economic Geography*, 37, No. 2 (1961): 133-149.

Viletto, John, Jr. "The Ecological Crop Geography of the Orange in California." Master's Thesis, Pennsylvania State University, University Park, Pa., 1961.

Gammal, Farouk M. "Changes in the Distributional Patterns of Fruit Agriculture in California, 1939–59." Master's Thesis, Clark University, Worcester, Mass., 1963.

Harman, Warren W. "The Grape Industry in San Diego County, California." Master's Thesis, San Diego State University, San Diego, Calif., 1964.

Lutz, Carl L. "The San Diego County Citrus Industry." Master's Thesis, San Diego State University, San Diego, Calif., 1964.

Meleen, Nathan H. "Changing Geography of the California Prune Industry." Master's Thesis, Clark University, Worcester, Mass., 1964.

Durrenberger, Robert W. "Evolution of the American Lemon-Growing Industry." *California Geographer*, 8 (1967): 103-114.

Hawkins, Harlan. "Citrus Farming in Southern Tulare County, 1950–1966." Master's Thesis, California State University, Chico, Calif., 1967.
McInroy, Thomas E. "The Ecological Crop Geography of the Avocado in California." Master's Thesis, Pennsylvania State University, University Park, Pa., 1967.
Napier, Dennis A. "Description and Analysis of Trends in the Grape Industry of San Bernardino County, 1953 Through 1965." Master's Thesis, California State University, Los Angeles, Calif., 1967.
Jackson, Harold. "A Geography of the Sacramento Valley Peach Belt." Master's Thesis, California State University, Chico, Calif., 1968.
Mills, Madolin M. "Future Areas of Commercial Lemon Production in California." Master's Thesis, Wayne State University, Detroit, Mich., 1968.
Gibson, David J. "The Development of the Livermore Valley Wine District." Master's Thesis, University of California, Davis, Calif., 1969.
Olson, Walter. "Changing Patterns of Tomato Production in California." *California Geographer*, 10 (1969): 15-26.
Richards, Stephen. "Geographical Aspects of Rice Cultivation in California." Master's Thesis, University of California, Berkeley, Calif., 1969.
Sloper, Neal W. "California's San Joaquin Valley Fig District: A Geographic Analysis." Master's Thesis, California State University, Fresno, Calif., 1969.
Aron, Robert H. "The Changing Location of the California Almond Industry." Master's Thesis, California State University, Long Beach, Calif., 1970.
Kerr, William S., III. "The Avocado Industry in Southern California: A Study of Location, Perception and Prospect." Ph.D. Dissertation, University of Oklahoma, Norman, Okla., 1970.
Kerr, William S., III. "A Qualitative-Quantitative Synthetic Projection of Spatial Patterns in Southern California's Avocado Industry." *Professional Geographer*, 22, No. 1 (1970), 8-11.
Modry, Southard M. "The California Flower Seed Industry: A Study in Interrelationships of Commercial, Ecological and Human Factors in the Distribution of a Horticultural Activity." Ph.D. Dissertation, University of California, Los Angeles, Calif., 1970.
Claus, R.J., and F. Dalichow. "California: Centre of the United States Fruit and Vegetable Canning Industry." *Zeitschrift fur Wirtschaftsgeographie*, 15, No. 6 (1971): 165-171.
Power, James R. "A Geographical Analysis of the Pajaro Valley Apple Industry." Master's Thesis, California State University, San Jose, Calif., 1973.
Coffin, Charles T. "A Spatial Analysis of the Floriculture Industry in Northern San Diego County." Master's Thesis, San Diego State University, San Diego, Calif., 1974.
Rocca, Alvieri M. "Viticulture in Napa and Sonoma Counties, California." Master's Thesis, San Jose State University, San Jose, Calif., 1974.
White, Jack L. "Diffusion of Olive Cultivation into California with Special Reference to the Olive Industry in the San Fernando Valley." Master's Thesis, California State University, Northridge, Calif., 1974.

Hoiland, Scot. "Citrus Growing on the Northern Fringe: Butte County, A Case Study." Master's Thesis, California State University, Chico, Calif., 1975.

Tyner, Gerald B. "The Historical Development of California's Herb Industry." *California Geographer*, 16 (1976): 43-52.

Crowley, William K. "Grapes Conquer Prunes: Vineyard and Winery Expansion in Sonoma County, California." *California Geographer*, 17 (1977): 11-24.

Dorel, Gerard. "Le vignoble Californien." *Les Cahiers d'Outre Mer*, 30 (1977): 5-30.

Jensen, John R., John E. Estes, and Larry R. Tinney. "High-Altitude Versus Landsat Imagery for Digital Crop Identification." *Photogrametric Engineering and Remote Sensing*, 44, No. 6 (1978): 723-733.

Brown, Catherine E. "Dry-Farmed Grain Cultivation in the Montezuma and Dunnigan Hills, California: A LANDSAT Based Investigation, 1972–1976." Master's Thesis, California State University, Hayward, Calif., 1979.

Kunkel, Steven D. "Crop Identification in the Imperial Valley Utilizing Landsat Digital Data." Master's Thesis, San Diego State University, San Diego, Calif., 1979.

Bell, Franklin J. "Geographic Aspects of Honeybee Pollination of California Vegetable Crops." Master's Thesis, San Francisco State University, San Francisco, Calif., 1980.

Robinson, Curtis J. "Community Gardening in Davis, California: A Geographical Study." Master's Thesis, University of California, Davis, Calif., 1980.

Yagasaki, Noritaka. "Evolution of Japanese Floriculture in Southern California" (in Japanese). *Jimbun Chiri (Human Geography, Kyoto)*, 35, No. 1 (1983): 1-22.

Peters, Gary L. "Trends in California Viticulture." *Geographical Review*, 74, No. 4 (1984): 455-467.

Yassinski, Lee. "The Chappellet Vineyards: A Case Study." Master's Thesis, California State University, Northridge, Calif., 1984.

Pineo, Susan J. "The Geography of Viticulture in Sonoma County." Master's Thesis, University of California, Davis, Calif., 1985.

Olk, Daniel C. "Agriculture in a Technological Landscape: Cropping and Settlement Patterns on the West Side of Kern County, California." Master's Thesis, University of California, Davis, Calif., 1986.

Eysberg, Cees D. "Viticulture in California: Cool, Airconditioned Valleys as the Equivalent of Warm Sheltered 'Cotes.'" *GeoJournal*, 15, No. 4 (1987): 367-374.

Glasner, Alice. "Community Gardening in San Francisco." Master's Thesis, San Francisco State University, San Francisco, Calif., 1987.

Peters, Gary L. "The Emergence of Regional Cultivar Specializations in California Viticulture." *Professional Geographer*, 39, No. 3 (1987): 287-297.

Pudup, Mary Beth, and Michael J. Watts. "Growing Against the Grain: Mechanized Rice Farming in the Sacramento Valley, California." Pp. 345-384 in *Comparative Farming Systems*, edited by Billie L. Turner II and Stephen B. Brush. New York, N.Y.: Guilford, 1987.

Peters, Gary L. *Wines and Vines of California.* Belmont, Calif.: Star Publishers, 1989.

Eysberg, Cees D. *The California Wine Economy: Natural Opportunities and Socio-Cultural Constraints: A Regional Geographic Analysis of Its Origins and Perspectives.* Amsterdam, Netherlands: Koninklijk Nederlands Aardrijkskindig Genootschap, 1990.

Eysberg, Cees D. "The California Wine Economy." *Hommes et Terres du Nord*, No. 3 (1990): 187-207.

Klohn, Werner. "Sunkist Growers, Inc.: Come Grow With Us. Eine weltweit exportierende Farmer-Genossenschaft in Kalifornien." *Praxis Geographie*, 20, No. 11 (1990): 20-23.

Klohn, Werner. "Grundlagen und Strukturen des Mandelanbaues in Kalifornien." *Vechtaer Arbeiten zur Geographie und Regionalwissenschaft*, 11 (1990): 85-102.

Windhorst, Hans-Wilhelm. "Industrisierte Tomatenproduktion in Kalifornien." *Vechtaer Arbeiten zur Geographie und Regionalwissenschaft*, 11 (1990): 103-124.

Bulman, Teresa L. "Development of the Grape Monoculture of Napa County." *Yearbook, Association of Pacific Coast Geographers*, 53 (1991): 61-86.

Watkins, Russell L. "Land Suitability for Viticulture: A Geographic Information System-Derived Method for a Mediterranean Type Climate in California." Ph.D. Dissertation, University of Florida, Gainesville, Fla., 1995.

Helzer, Jennifer. "Old Traditions, New Lifestyles: The Emergence of a Cal-Ital Landscape." *Yearbook, Association of Pacific Coast Geographers*, 63 (2001): 49-62.

ANIMAL HUSBANDRY

Zierer, Clifford M. "Migratory Beekeepers of Southern California." *Geographical Review*, 22, No. 2 (1932): 260-269.

Menzel, Marion L.J. "The Historical Geography of the Sheep Industry in California in the 19th Century." Master's Thesis, University of California, Berkeley, Calif., 1944.

Burcham, Levi T. *Range Livestock Population of California, 1850–1950.* Sacramento, Calif.: California Division of Forestry, 1955.

Burcham, Levi T. "Historical Geography of the Range Livestock Industry of California." Ph.D. Dissertation, University of California, Berkeley, Calif., 1956.

Burcham, Levi T. "The Advent of Sheep in California." *California Livestock News*, 33, No. 24 (1957): 13-15.

Kennelly, Robert A. "Cattle Feeding in the Imperial Valley." *Yearbook, Association of Pacific Coast Geographers*, 22 (1960): 50-56.

McKnight, Tom L. "Survey of Feral Livestock in California." *Yearbook, Association of Pacific Coast Geographers*, 23 (1960): 28-42.

Burcham, Levi T. "Cattle and Range Forage in California, 1770–1880." *Agricultural History*, 35, No. 3 (1961): 140-149.

Burcham, Levi T. "Cattle Ranching in the Early Days of California." *American Hereford Journal*, 55, No. 5 (1964): 440-462.

Harries, Keith. "Beef Cattle and Environment in Southern California." Master's Thesis, University of California, Los Angeles, Calif., 1966.

Boswell, Thomas D. "Beef Cattle Feeding in Imperial Valley, California: A Study in Economic Geography." Master's Thesis, San Diego State University, San Diego, Calif., 1967.

Archibald, John D. "The Chicken Industry in San Diego County, California." Master's Thesis, San Diego State University, San Diego, Calif., 1968.

Emerson, Gladys C. "Geographical Aspects of the Development of Limoneira Ranch, Santa Paula, California." Master's Thesis, University of California, Los Angeles, Calif., 1968.

Mitchener, Richard B. "Seasonal Grazing of Beef in Eldorado National Forest." Master's Thesis, University of California, Davis, Calif., 1968.

White, Paul G. "Commercial Livestock Grazing on the Sierra and Los Padres National Forest, California." Master's Thesis, San Francisco State University, San Francisco, Calif., 1969.

Blanchard, Amy E. "A Geographic View of the California Equine Industry." Master's Thesis, University of California, Davis, Calif., 1970.

McDonald, Norma. "Migratory Beekeeping in California." Master's Thesis, California State University, Fresno, Calif., 1970.

DAIRYING

Copley, Richard. "An Historical Geography of the Dairy Industry of Stanislaus County, California." Master's Thesis, University of California, Berkeley, Calif., 1961.

Fielding, Gordon J. "Dairying in the Los Angeles Milkshed: Factors Affecting Character and Location." Ph.D. Dissertation, University of California, Los Angeles, Calif., 1962.

Fielding, Gordon J. "Changing Location Factors in the Los Angeles Milkshed." *California Geographer*, 3 (1962): 47-54.

Fielding, Gordon J. "The Los Angeles Milkshed: A Study of the Political Factor in Agriculture." *Geographical Review*, 54, No. 1 (1964): 1-21.

Vankampen, Carol. "From Dairy Valley to Chino: An Example of Urbanization in Southern California's Dairyland." *California Geographer*, 17 (1977): 39-48.

Shockley, Jo Lynn. "A Land Suitability Study for Dairy Operations in the Chino on Hinkley-Barstow, California Areas: A Cartographic Modeling Approach." Master's Thesis, University of California, Riverside, Calif., 1982.

FARM LAND

Lade, Charles G. "Lucerne Valley, California: Agricultural Land Use as Related to the Physical Geography." Master's Thesis, University of California, Los Angeles, Calif., 1953.

Miller, Crane. "Agriculture in an Urbanizing Environment: A Study in the Impact of Metropolitan Growth on Specialized Agriculture in Ventura County, California." Ph.D. Dissertation, University of California, Los Angeles, Calif., 1971.

Ali-Saad, Ab-Bakr. "The Land Use–Farm Size Relationship, Imperial Valley, California." Master's Thesis, University of Wisconsin, Madison, Wis., 1973.

Downs, Keith D. "The Preservation of Agricultural Lands within Riverside County." Master's Thesis, San Diego State University, San Diego, Calif., 1977.

Sandlin, Stephen H. "The Preservation of Agricultural Land: An Examination of Related Land Use Controls and a Geographical Analysis of the California Land Conservation Act of 1965." Ph.D. Dissertation, University of California, Riverside, Calif., 1977.

Hornbeck, David. "Land Tenure and Rancho Expansion in Alta California, 1784–1846." *Journal of Historical Geography*, 4, No. 4 (1978): 371-390.

Liebman, Ellen. "The Evolution of Large Agricultural Holdings in California." Ph.D. Dissertation, University of California, Berkeley, Calif., 1981.

Exline, C.H. "Sinclair's Von Thunen and Urban Sprawl: The Conversion of Agricultural Land to Suburban Uses in Sonoma County, California, 1950–1970." *California Geographer*, 23 (1983): 14-35.

Furuseth, Owen J. "Local Farmland Conservation Programmes in the U.S.: A Study of California Counties." *Applied Geography*, 5, No. 3 (1985): 211-227.

Furuseth, Owen J. "Influences on County Farmland Protection Efforts in California: A Discriminant Analysis." *Professional Geographer*, 37, No. 4 (1985): 443-451.

Konnecke, Michael. *Veranderungen Landwirtschaftlicher Nutzungsmuster im nordlichen Central Valley, Kalifornien unter besonderer Berucksichtigung des naturgeographischen Potentials: Dargestellt am Beispiel Yolo Country.* Dusseldorf, Germany: Dusseldorfer Geographische Schriften, Heft 25, 1986.

Klohn, Werner. "Auslandischer Landbesitz in Kalifornien." *Vechtaer Arbeiten zur Geographie und Regionalwissenschaft*, 11 (1990): 137-149.

Courter, William F. "Of Agriculture and Houses: Urban Growth in Southwestern Riverside County." Master's Thesis, California State University, Northridge, Calif., 1991.

Mitchell, Martin D. "Changes in Landscape Forms and Functions in the Sacramento–San Joaquin Delta, California, 1920–1993." Ph.D. Dissertation, University of Illinois, Urbana, Ill., 1993.

Baubhitz, Michael. "The Past, Present and Future of South Livermore Valley: An Historical, Political and Legal Geography of a California Wine-Producing Region and Its Preservation in a Land Trust." Master's Thesis, California State University, Hayward, Calif., 1994.

AGRICULTURAL POPULATIONS AND SOCIAL ISSUES

Tabuteau, M. "Le peuplement rural et l'exploitation agricole en Californie." *Annales de Geographie*, 62 (1953): 452-457.

Griffin, Paul F., and Ronald L. Chatham. "Population: A Challenge to California's Changing Citrus Industry." *Economic Geography*, 34, No. 3 (1958): 272-276.

Barlow, Robert S. "Historical and Regional Analysis of the Italian Role in California Viticulture and Enology." Master's Thesis, University of California, Los Angeles, Calif., 1964.

Graves, Alvin R. "Azorean Portugese: A Study of the Portugese Dairymen in the San Joaquin Valley of California." Master's Thesis, California State University, Fresno, Calif., 1969.

Stevens, D.J., and Y.I. Kim. "Japanese Agriculture in Ventura County." *Yearbook, Association of Pacific Coast Geographers*, 31 (1969): 74-79.

Seivertson, Bruce. "Pajaro Valley, California: The Sequent Occupance of a Coastal Agricultural Basin." Master's Thesis, California State University, Chico, Calif., 1970.

Quastler, I.E. "The Core Area of Images Concept and American Perception of California Agriculture, 1800–1890." *Proceedings of the Association of American Geographers*, 4 (1972): 91-95.

Eigenheer, Richard A. "Early Perceptions of Agricultural Resources in the Central Valley of California." Ph.D. Dissertation, University of California, Davis, Calif., 1976.

Graves, Alvin R. "Immigrants in Agriculture: The Portugese Californians, 1850–1970s." Ph.D. Dissertation, University of California, Los Angeles, Calif., 1977.

Yagasaki, Noritaka. "Formation of the Japanese Floriculture in Northern California: Role of the Ethnic Organization in Immigrant Agriculture" (in Japanese). *Chigaku Zasshi (Journal of Geography, Tokyo)*, 89, No. 3 (1980): 149-166.

Yagasaki, Noritaka. "Northern California Japanese Floriculture in Transition: An Ethnic Industry in the Process of Socio-Cultural, Technological and Geographical Change" (in Japanese). *Jimbun Chiri/Human Geography, Kyoto*, 32, No. 1 (1980): 23-46.

Hachimura, Keiko. "Japanese Farming in Fresno County, California: Prewar and Postwar Patterns." Master's Thesis, California State University, Fresno, Calif., 1981.

Sauder, Robert A. "The Agricultural Colonization of a Great Basin Frontier: Economic Organization and Environmental Alteration in Owens Valley, California, 1860–1925." *Agricultural History*, 64, No. 1 (1990): 78-101.

Wells, Miriam J. "Politics, Locality and Economic Restructuring: California's Central Coast Strawberry Industry in the Post-World War II Period." *Economic Geography*, 76, No. 1 (2000): 28-49.

ENVIRONMENTAL AND CONSERVATION ISSUES

Bennett, Merrill K. "Climate and Agriculture in California." *Economic Geography*, 15, No. 2 (1939): 153-164.

Carter, George F. "Lima Bean Farming and Soil Erosion in the Encinitas Area." *Yearbook, Association of Pacific Coast Geographers*, 4 (1941): 15-19.

Durrenberger, Robert W. "Climates as a Factor in the Production of Lemons in California." Ph.D. Dissertation, University of California, Los Angeles, Calif., 1955.

Aron, Robert H. "Climatic Chilling and Future Almond Growing in Southern California." *Professional Geographer*, 23, No. 4 (1971): 341-343.

Aron, Robert H. "Chilling as a Factor in Crop Location with Particular Reference to Deciduous Orchards in California." Ph.D. Dissertation, Oregon State University, Corvallis, Oreg., 1974.

Granger, Orme E. "Climatic Variations and the California Raisin Industry." *Geographical Review*, 70, No. 2 (1980): 300-313.

Rowntree, Lester, and Robert Raburn. "Rainfall Variability and California Mission Agriculture: An Analysis from Harvest and Tree Ring Data." *Yearbook, Association of Pacific Coast Geographers*, 42 (1980): 31-44.

Schmidt, Robert H., Jr. "Freeway Impact on Agricultural Areas." *Natural Resource Journal*, 20, No. 3 (1980): 587-602.

Keller, Chris. "Measuring the Compatibility of Agriculture and Transmission Lines in Central California." Master's Thesis, San Diego State University, San Diego, Calif., 1987.

Pulido, Laura. "Perceptions of Agricultural Chemicals in Kerne County, California." Master's Thesis, University of Wisconsin, Madison, Wis., 1987.

Windhorst, Hans-Wilhelm. "Agrarstruktur und Bodennutzung in der Fresno County (Kalifornien) gegen Ende der achtziger Jahre." *Vechtaer Arbeiten zur Geographie und Regionalwissenschaft*, 11 (1990): 53-84.

Elliot-Fisk, Deborah L., and Ann C. Noble. "Environments in Napa Valley, California and Their Influence on Cabernet Sauvignon Wine Flavors." Pp. 45-72 in *Viticulture in Geographic Perspective: Proceedings of the Miami AAG Symposium*, edited by Harm deBlij. Miami, Fla.: Miami Geographical Society, 1992.

Dunayevich, Bernard. "Organic Farming in Southern California: A Viable Alternative to Environmentally Damaging Agro-Industry." Master's Thesis, University of California, Los Angeles, Calif., 1992.

Elliot-Fisk, Deborah L. "Viticultural Soils of California, with Special Reference to the Napa Valley." *Journal of Wine Research*, 4, No. 2 (1993): 67-78.

Golden, Nadine. "Organic Farms as Nodes of Wildlife Corridors, Fresno County, California." Master's Thesis, San Francisco State University, San Francisco, Calif., 1994.

Johnson, L.S. "The Diffusion of Organic Agriculture in Yolo County, California: A Question of Culture." Master's Thesis, San Francisco State University, San Francisco, Calif., 1999.

Guthman, Julie. "Agrarian Dreams? The Paradox of Organic Farming in California." Ph.D. Dissertation, University of California, Berkeley, Calif., 2000.

IRRIGATION

Meier, Esther. "Irrigation in West Fresno County, California." Master's Thesis, University of North Dakota, Grand Forks, N.Dak., 1955.

Rowan, John L. "The Hermosilio Irrigation District: Problems and Prospects." Master's Thesis, University of California, Los Angeles, Calif., 1962.

Aldabbagh, Abdul W. "The Development of Agriculture in Coachella Valley, California, Before and After the Importation of the Colorado River Water." Master's Thesis, Clark University, Worcester, Mass., 1967.

McCormick, James E. "The Fresno Irrigation District: A Study of Urban and Agricultural Water Resources Management in San Joaquin Valley." Master's Thesis, Clark University, Worchester, Mass., 1968.

Thompson, Kenneth. "Irrigation as a Menance to Health in California: A Nineteenth Century View." *Geographical Review*, 59, No. 2 (1969): 195-214.

Sanger, Leslie W. "Irrigated Agricultural Developments on the Southwest Side of the San Joaquin Valley with Special Reference to Major Oil Company Holdings." Ph.D. Dissertation, University of California, Los Angeles, Calif., 1972.

McKnight, Tom L. "California's Reluctant Acceptance of Center Pivot Irrigation." *Yearbook, Association of Pacific Coast Geographers*, 41 (1979): 199-138.

Gruber, G. "Kolonisation und Bewasserung im San Joaquin Valley, Kaliforniens." *Frankfurter Wirtschafts- und Sozialgeographische Schriften*, 42 (1982): 149-192.

McKnight, Tom L. "Center Pivot Irrigation in California." *Geographical Review*, 73, No. 1 (1983): 1-14.

Franklin, R.L. "The Diffusion of Micro Irrigation Systems in California's Commercial Avocado Production Areas." Master's Thesis, California State University, Long Beach, Calif., 1987.

Helbrecht, Karen A. "Comparison of Irrigation Practices and Crop Yields in the Imperial and Mexicali Valleys." Master's Thesis, University of California, Los Angeles, Calif., 1987.

Windhorst, Hans-Wilhelm. "Mit dem Wasserkamen die Probleme: Sozio-okonomische und geokologische Probleme in der Agrarwirtschaft des Westlands Water District in Kalifornien." *Vechtaer Arbeiten zur Geographie und Regionalwissenschaft*, 11 (1990): 11-30.

Konnecke, Michael. "Regenfeldbau-Bewasserungsfeldbau: Moglichkeiten und geookologische Grenzen. Fallbeispiel aus dem Grossen longstal Kaliforniens." *Geographische Rundschau*, 43, No. 7-8 (1991): 445-452.

Sauder, Robert A. *The Lost Frontier: Water Diversion and the Destruction of Owens Valley Agriculture*. Tucson, Ariz.: University of Arizona Press, 1994.

Windhorst, Hans-Wilhelm, and Werner Klohn. *Bewasserungslandwirtschaft in Kalifornien unter Durrestress*. Vechta, Germany: Vechtaer Druckerei und Verlag, 1994.

Windhorst, Hans-Wilhelm, and Werner Klohn. "Bewasserungslandwirtschaft in Kalifornien und den Great Plains. Anpassungsstrategien an Durrejahre und die Erschopfung von Grundwasservorraten." *Erdkunde*, 50, No. 3 (1996): 252-265.

Koczot, Kathy. "Estimating Rates and Patterns of Historic Ground Water Pumpage for Irrigation Using a Geobased Numerical Model." Master's Thesis, San Diego State University, San Diego, Calif., 1997.

AQUACULTURE

Bass, Glennon. "Aquaculture at Newberry Springs." Master's Thesis, San Jose State University, San Jose, Calif., 1981.

AGRICULTURAL MARKETING AND DISTRIBUTION

Gregor, Howard F. "The Local-Supply Agriculture of California." *Annals of the Association of American Geographers*, 47, No. 2 (1957): 267-276.

Saunders, Margery H. "California Wheat: 1867–1910: Influence of Transportation on the Export Trade and the Location of Producing Areas." Master's Thesis, University of California, Berkeley, Calif., 1960.

Chapin, Edward L. "Greater Los Angeles Foodstuff Demand and Supply: A Quarter Century Consumer Region Study Based on Sources of Its Five Principal Fresh Vegetables." Ph.D. Dissertation, University of California, Los Angeles, Calif., 1971.

Bloom, Khaled. "Market-Oriented Agriculture in 19th Century San Francisco." *Yearbook, Association of Pacific Coast Geographers*, 44 (1982): 75-92.

TRANSPORTATION AND MECHANIZATION

Windhorst, Hans-Wilhelm. "Die Ausbreitung des mechanischen Vollernters fur Industrietomaten in Kalifornien: Eine agrar technologische Innovation und ihe wirkungen." *Erdkunde*, 44, No. 1 (1990): 23-37.

AGRICULTURAL LABOR STUDIES

Carlson, Alvar W. "Seasonal Farm Labor in the San Luis Valley." *Annals of the Association of American Geographers*, 63, No. 1 (1973): 97-108.

Zabin, Carol. "U.S.-Mexican Economic Integration: Labor Relations and the Organization of Work in California and Baja California Agriculture." *Economic Geography*, 73, No. 3 (1997): 337-355.

Oregon

GENERAL TEXTS, ARTICLES, AND THESES

Highsmith, Richard M., Jr. "Agricultural Geography of the Eugene Area." Ph.D. Dissertation, University of Washington, Seattle, Wash., 1951.

Jack, James T. "The Agricultural Geography of the Klamath Basin in Oregon." Master's Thesis, Oregon State University, Corvallis, Oreg., 1960.

Sutherland, John. "Umatilla Agricultural Landscapes, 1700–1973: An Historical Geography of a Region of the Oregon Country." Master's Thesis, University of Oregon, Eugene, Oreg., 1973.

Van Otten, George A. "Changing Spatial Characteristics of Willamette Valley Farmers." *Professional Geographer*, 32, No. 1 (1980): 63-71.

Wison, Jeffery. "Farming the Urban Fring: A Case Study of the River Road Farming Area." Master's Thesis, University of Oregon, Eugene, Oreg., 1981.

Gibson, James R. *Farming the Frontier: The Agricultural Opening of the Oregon Territory, 1786–1846*. Vancouver, British Columbia: University of British Columbia Press, 1985.

Lewis, Michael E. "Agricultural Evolution on Secondary Frontiers: An Oregon Model." *Geographical Perspectives*, 55 (1985): 46-52.

CROPS AND CROPPING PATTERNS

Hoffman, Charles S. "Oregon Low-Lands Suitable for Flax." *Economic Geography*, 12, No. 2 (1936): 164-166.

Goodman, Duane. "The Agricultural Geography of the Pear Industry in Jackson County, Oregon." Master's Thesis, Oregon State University, Corvallis, Oreg., 1963.

Tonack, Wayne E. "The Localization of the United States Sugar Beet Seed Industry with Emphasis on the Willamette Valley." Master's Thesis, Oregon State University, Corvallis, Oreg., 1964.

Vent, H.J., and R.E. Vent. "The Pear Industry of the Medford, Oregon Area." *Journal of Geography*, 67, No. 3 (1968): 166-171.

Dyck, Brian, "Instant Lawn: Sod Farming in Oregon." *Geographical Bulletin*, 9 (1974): 37-50.

Reynolds, W.L. "Evolution of the Grass Seed Landscape: A Historical Geography of Agriculture in the Southern Willamette Valley." Master's Thesis, University of Oregon, Eugene, Oreg., 1977.

Duerr, Richard. "Orchards of Lane County." Master's Thesis, University of Oregon, Eugene, Oreg., 1980.

Bastasch, Richard. "A Comparison of the Structure of Bottomland Commercial Agriculture in Clackamas and Jackson Counties, Oregon." Master's Thesis, Oregon State University, Corvallis, Oreg., 1984.

Keeler, Elizabeth L. "The Landscape of Horticultural Crops in the Northern Willamette Valley from 1850–1920." Ph.D. Dissertation, University of Oregon, Eugene, Oreg., 1994.

ANIMAL HUSBANDRY

Hunt, Jack H. "The Warner Valley Stock Company: A Geographic Study." Master's Thesis, Oregon State University, Corvallis, Oreg., 1964.

O'Brien, Michael. "The Distribution and Effect of Cattle in the Applegate Valley." *Oregon Geographer*, 4, No. 1 (1969): 17-19.

Henderly, Herschel H. "A Transhumance Beef Industry of Southern Oregon." Master's Thesis, Oregon State University, Corvallis, Oreg., 1975.

DAIRYING

Heintzelman, Oliver H. "The Dairy Economy of Tillamook County, Oregon." Ph.D. Dissertation, University of Washington, Seattle, Wash., 1952.

FARM LAND

Odland, John M. "An Analysis of Diversity in Agricultural Land Use in Oregon." Master's Thesis, Oregon State University, Corvallis, Oreg., 1968.

Lake, Brent L. "The Reaction of Agricultural Landowners to the Willamette River Park and Recreation System." Master's Thesis, Oregon State University, Corvallis, Oreg., 1973.

Avery, William. "The Role of the Land Developer in Converting Agricultural Land to Urban Use: Subdivisions in Tualatin, Oregon." Master's Thesis, Oregon State University, Corvallis, Oreg., 1974.

Roberts, Rebecca. "An Evaluation of Agricultural Lands for Acquisition and Consolidation: The Umatilla Indian Reservation." Master's Thesis, Oregon State University, Corvallis, Oreg., 1976.

Tillson, Gregory D. "Conversion or Retention of Agricultural Land? A Marion County, Oregon Example." Master's Thesis, Oregon State University, Corvallis, Oreg., 1977.

Van Otten, George A. "Spatial Expansion of Farm Size Changes in Polk and Linn Counties of Oregon." Ph.D. Dissertation, Oregon State University, Corvallis, Oreg., 1977.

Furuseth, Owen J., Jr. "Selected Factors Affecting the Pattern of Agricultural Land Conversion in Washington County, Oregon, 1963–1973." Ph.D. Dissertation, Oregon State University, Corvallis, Oreg., 1978.

Furuseth, Owen J., Jr. "The Structure of Agricultural Land Conversion in Washington County, Oregon." *Journal of Soil and Water Conservation*, 34 (1979): 138-141.

Furuseth, Owen J., Jr. "The Oregon Agricultural Protection Program: A Review and Assessment." *Natural Resources Journal*, 20, No. 3 (1980): 603-614.

Furuseth, Owen J., Jr. "Update on Oregon's Agricultural Protection Program: A Land Use Perspective." *Natural Resources Journal*, 21, No. 1 (1981): 57-70.

Hunter, Steven R. "The Effectiveness of the Farm Use Assessment Program in Jefferson County, Oregon: A Comparative Study." Master's Thesis, Oregon State University, Corvallis, Oreg., 1981.

Pease, J.R. "Commercial Farmland Preservation in Oregon." *GeoJournal*, 6, No. 6 (1982): 547-554.

Clinton, Patrick J. "A Potential Gross Sales Test for Farmland: The Synthesis and Application of a Rural Resource Planning Tool." Master's Thesis, Oregon State University, Corvallis, Oreg., 1993.

AGRICULTURAL POPULATIONS AND SOCIAL ISSUES

Benson, Barry R. "Agricultural Resources and Socio-Economic Affluence of Oregon Counties." Master's Thesis, Oregon State University, Corvallis, Oreg., 1975.

ENVIRONMENTAL AND CONSERVATION ISSUES

Blok, Jack H. "The Evolution of Agricultural Resource Use Strategies in the Willamette Valley." Ph.D. Dissertation, Oregon State University, Corvallis, Oreg., 1973.

Thomason, Leonora M. "Environmentally Imposed Problems of the Range Beef Cattle Industry in Eastern Oregon." Master's Thesis, Oregon State University, Corvallis, Oreg., 1974.

Aron, Robert. "Oregon's Climatic Suitability for Premium Wine Grapes." *California Geographer*, 16 (1976): 53-62.

Lewis, Michael. "Agroclimatic Hazards of the Fort Rock Basin: Perceptions and Mitigation Strategies Among Cow-Calf Operators and Cash-Crop Agriculturalists." Master's Thesis, Oregon State University, Corvallis, Oreg., 1984.

IRRIGATION

Kunnecke, Bernd H. "Distribution of Irrigation in Eastern Oregon." Master's Thesis, University of Oregon, Eugene, Oreg., 1974.

Renwick, William R., II. "Changes in Deschutes County Irrigation Agriculture Since 1950." Master's Thesis, Oregon State University, Corvallis, Oreg., 1975.

Washington

GENERAL TEXTS, ARTICLES, AND THESES

Lechner, H.J. "Some Geographic Influences of Eastern Washington in Relation to Agriculture." *Journal of Geography*, 14, No. 9 (1916): 362-364.

Earle, Frances M. "Agricultural Trends in Western Washington." *Yearbook, Association of Pacific Coast Geographers*, 2 (1939): 25-26.

Whitley, Edward C. "Agricultural Geography of the Kittitas Valley, Washington." *Pacific Northwest Quarterly*, 41, No. 1 (1950): 3-18.

Whitley, Edward C. "Agricultural Geography of the Kittitas Valley." Master's Thesis, University of Washington, Seattle, Wash., 1951.

Brandt, Theodore R. "Components of Change in the Agriculture of the Puget Sound Region, Washington, 1945–1980." *B.C. Geographical Series*, 34 (1982): 27-42.

CROPS AND CROPPING PATTERNS

Freeman, Otis W. "The Apple Industry of the Wenatchee Area." *Economic Geography*, 10, No. 2 (1934): 160-171.

Tousley, Rayburn D., and Hoyt Lemons. "The Washington Apple Industry." *Economic Geography*, 21, No. 3 (1945): 252-268.

Miller, Elbert E., and Richard M. Highsmith. "The Geography of the Fruit Industry of Yakima Valley, Washington." *Economic Geography*, 25, No. 3 (1949): 285-295.

Highsmith, Richard M., and Elbert E. Miller. "Open Field Farming in the Yakima Valley, Washington." *Economic Geography*, 28, No. 1 (1952): 74-87.

Heller, Charles F., Jr. "Rural Stability of Wheat Farmers in Adams County, Washington." Master's Thesis, University of Kansas, Lawrence, Kans., 1957.

Detering, Floyd, Jr. "The Geography of the Apple Industry of the Chelen Valley, Washington." Master's Thesis, Oregon State University, Corvallis, Oreg., 1965.

Maedler, Rick. "Preservation of Small-Scale Agriculture Near a Metropolitan Area: The Case of Market Gardening and the Pike Place Farmers Market, Seattle, Washington." Master's Thesis, Western Washington University, Bellingham, Wash., 1983.

Macinko, George. "The Ebb and Flow of Wheat Farming in the Big Bend, Washington." *Agricultural History*, 59, No. 2 (1985): 215-228.

DAIRYING

Carter, Douglas B. "The Sequin-Dungeness Lowland: A Natural Dairy Community." Master's Thesis, University of Washington, Seattle, Wash., 1948.

FARM LAND

Brandt, Theodore R. "Agricultural Land Retention in the Puget Sound Region, Washington." Master's Thesis, Western Washington University, Bellingham, Wash., 1982.

ENVIRONMENTAL AND CONSERVATION ISSUES

Meinig, Donald W. "The Evolution of Understanding an Environment: Climates and Wheat Culture in the Columbia Plateau." *Yearbook, Association of Pacific Coast Geographers*, 16 (1954): 25-34.

IRRIGATION

Highsmith, Richard M., Jr. "Irrigation Agriculture in the Yakima Valley." Master's Thesis, University of Washington, Seattle, Wash., 1947.

Smiley, William E. "The Influence of Irrigation on the Agricultural Geography of Grant County, Washington, 1950–1964." Master's Thesis, University of Kansas, Lawrence, Kans., 1973.

AGRICULTURAL MARKETING AND DISTRIBUTION

Meinig, Donald W. "Wheat Sacks Out to Sea: The Early Export Trade from the Walla Walla Country." *Pacific Northwest Quarterly*, 45 (1954): 13-18.

Jarosz, L., and J. Qazi. "The Geography of Washington's World Apple: Global Expression in a Local Landscape." *Journal of Rural Studies*, 16, No. 1 (1999): 1-11.

AQUACULTURE

Gillespie, Dennis. "An Analysis of Aquaculture in the Puget Sound." Master's Thesis, Western Washington University, Bellingham, Wash., 1992.

Hawaii

GENERAL TEXTS, ARTICLES, AND THESES

Coulter, John W. "Small Farming in Kauai: Hawaiian Islands." *Economic Geography*, 11, No. 4 (1935): 401-409.

Morgan, John T. "Agriculture: Plantations and Diversified Farming." Pp. 141-160 in *Hawaii: A Geography*, edited by John T. Morgan. Boulder, Colo.: Westview, 1983.

Huetz de Lemps, Christian. "De la primaute des plantations a economique de services: L'exemple des Hawaii." Pp. 361-398 in *Iles tropicales: Insularite, insularisme: Colloque*. Bordeaux-Talence, France: Universite Bordeaux III, Centre de recherche des espaces tropicaux, iles et archipels, No. 8, 1987.

Morgan, Joseph R. "Agriculture." Pp. 104-114 in *Hawaii: A Unique Geography*, edited by Joseph R. Morgan. Honolulu, Hawaii: Bess Press, 1996.

CROPS AND CROPPING PATTERNS

Coulter, John W. "The Oahu Sugar Cane Plantation, Waipahu." *Economic Geography*, 9, No. 1 (1933): 60-71.

Coulter, John W. "The Pineapple Industry in Hawaii." *Economic Geography*, 10, No. 3 (1934): 288-301.

Wilson, James N. "The Pineapple Industry of Hawaii." *Economic Geography*, 24, No. 3 (1948): 251-262.

Tennant, Helen F. "Sugar Cane Production in North Kohala, Hawaii." Master's Thesis, University of Colo., Boulder, Colorado, 1949.

Chen, Cheng-Siang. *Pineapple Industry of the Hawaiian Islands*. Taipei, Taiwan: Institute of Agricultural Geography, Research Report No. 40, 1953.

Suryanata, Krisna. "Products from Paradise: The Social Construction of Hawaii Crops." *Agriculture and Human Values*, 17, No. 2 (2000): 180-189.

ANIMAL HUSBANDRY

Rumens, John M. "The Changing Location of Oahu's Hog Industry: 1945–1966." Master's Thesis, University of Hawaii, Manoa, Hawaii, 1969.

DAIRYING

Durand, Loyal. "The Dairy Industry of the Hawaiian Islands." *Economic Geography*, 35, No. 3 (1959): 228-246.

FARM LAND

Ferguson, Carol A., and M.A. Khan. "Protecting Farmland Near Cities: Trade-Offs with Affordable Housing in Hawaii." *Land Use Policy*, 9, No. 4 (1992): 259-271.

ENVIRONMENTAL AND CONSERVATION ISSUES

von Maydell, Jurgen, and Gunter Spatz. "Effects of Goat Husbandry on the Development of the Landscape." *Applied Geography and Development*, 18 (1981): 30-44.

Wilson, Winnifred L. "Basal Water Quality Changes with Conversion of Sugar Cane Fields to Urban Use in South-Central Oahu." Master's Thesis, University of Hawaii, Manoa, Hawaii, 1982.

Penn, David C. "Water and Energy Flows in Hawaii Taro Pondfields." Ph.D. Dissertation, University of Hawaii, Manoa, Hawaii, 1997.

Wenger, Michael. "Development and Evolution of an Agroecosystem Ration Matrix and Its Application to Systems Analysis of a Windward Oahu Watershed and Farming System." Master's Thesis, University of Hawaii, Manoa, Hawaii, 1997.

IRRIGATION

Lateef, Abdul. "A Study of Irrigation Water Management at Honokaa Sugar Plantation, Hawaii." Master's Thesis, University of Hawaii, Manoa, Hawaii, 1966.

King, Diana L. "Impact on Ground Water Resources of the Conversion from Furrow to Drip Irrigation in the Hawaiian Sugar Industry." Master's Thesis, University of Hawaii, Manoa, Hawaii, 1988.

Freeman, William. "Agricultural Water Use Requirements for the Ho'olehua Plain Island of Moloki, Hawaii." Master's Thesis, University of Hawaii, Manoa, Hawaii, 1992.

AGRICULTURAL MARKETING AND DISTRIBUTION

Colburn, Susan A. "Farm to Market Linkages: Hawaiian Home Lands and Resort Developments, Waimea, Hawaii." Master's Thesis, University of Hawaii, Manoa, Hawaii, 1989.

TRANSPORTATION AND MECHANIZATION

Immisch, George. "Land Reclamation and the Role of the Hydro-Separator at Ewa Plantation: A Case Study of Some of the Effects of Mechanized Harvesting in the Hawaiian Sugar Industry." Master's Thesis, University of Hawaii, Manoa, Hawaii, 1964.

Alaska

GENERAL TEXTS, ARTICLES, AND THESES

Rumage, Marigalde. "A Geographic Interpretation of Alaskan Agriculture Including Type Study of the Matanuska Valley." Master's Thesis, Pennsylvania State University, University Park, Pa., 1951.

Francis, Karl E. "Outpost Agriculture: The Case of Alaska." *Geographical Review*, 57, No. 4 (1967): 496-505.

Shortridge, James. "American Perceptions of the Agricultural Potential of Alaska, 1867–1958." Ph.D. Dissertation, University of Kansas, Lawrence, Kans., 1972.

Vanderhill, Burke G. "Perspectives on Alaskan Agriculture." *Journal of Geography*, 72, 2 (1973): 38-52.

Ehlers, Eckert. "Agrarkolonisation und Agrarlandschaft in Alaska: Probleme und Entwicklungstendenzen der Landwirtschaft in hohen Breiten." *Geographische Zeitschrift*, 62 (1974): 195-219.

Moulton, Benjamin. "Agriculture in the Kenai." *Indiana State University Department of Geography Papers*, 7 (1975): 24-32.

Shortridge, James. "The Collapse of Frontier Agriculture in Alaska." *Annals of the Association of American Geographers*, 66, No. 4 (1976): 583-604.

Shortridge, James. "The Evolution of the Agricultural Potential of Alaska: 1867–1897." *Pacific Northwest Quarterly*, 68 (1977): 88-98.

Shortridge, James. "The Alaskan Agricultural Empire: An American Agrarian Vision, 1898–1929." *Pacific Northwest Quarterly*, 69 (1978): 145-158.

Lewis, Carol E., Roger W. Pearson, and W.C. Thomas. "Agricultural Development in Alaska." *Polar Record*, 23, No. 147 (1987): 673-682.

Lewis, Carol E., and Roger W. Pearson. "Three Development Models for Alaska's Agricultural Industry." *Yearbook, Association of Pacific Coast Geographers*, 52 (1991): 109-124.

CROPS AND CROPPING PATTERNS

Hill, Rodney B. "Physical Elements and Selected Cultural Controls of Crop Agriculture in the Tanana and Matanuska Valleys, Alaska." Master's Thesis, University of Alaska, Anchorage, Alaska, 1975.

FARM LAND

Rockie, W.A. "Pitting on Alaska Farm Lands: A New Erosion Problem." *Geographical Review*, 32, No. 2 (1942): 128-134.

Gjerde, John. "The Dynamics of Land Use and Values: A Potential Threat to Agriculture in Alaska's Matanuska Valley." *Indiana State University Department of Geography Papers*, 7 (1975): 33-41.

Gjerde, John. "The Dynamics of Land Use and Values: A Potential Threat to Agriculture in Alaska's Matanuska Valley." Master's Thesis, Indiana State University, Terre Haute, Ind., 1975.

Camp, C.E. "A Note on the Assessment of the Viability of Preserving Farmland in South Central Alaska Through the Purchase of Development Rights." *B.C. Studies*, 42 (1985): 25-42.

Canada as a Whole

ATLASES AND GRAPHIC PRESENTATIONS

Troughton, Michael J. *An Atlas of Canadian Agriculture*. London, Ontario: University of Western Ontario, Department of Geography, 1979.

GENERAL TEXTS, ARTICLES, AND THESES

Ruddick, T.A. "The Dairying and Fruit Industries in Canada." *Journal of Geography*, 11, No. 8 (1913): 243-245.

Gorham, Raymond P. "The Birth of Agriculture in Canada." *Canadian Geographical Journal*, 4, No. 1 (1932): 3-17.

Schott, Carl. "Die Industrialisierung der Landwirtschaft am Beispiel Kanadas." Pp. 23-54 in *Beitrage zur Geographie der Neuen Welt*, edited by Wilhelm Lover. Kiel, Germany: Schriften des Geographischen Instituts der Universitat Kiel, Heft 20, 1961.

Nicholson, N.L., and Z.W. Sametz. "As Regioes do Canada e a Nocao de Regiao." *Revista Geografica*, 59 (1963): 31-61.

Fuller, Anthony M. "The Part-Time Farm Problem: A Scheme for Geographers." *Proceedings, Canadian Association of Geographers*. Winnipeg, Manitoba: University of Manitoba, Department of Geography (1970): 121-128.

Troughton, Michael J. "Farm-Level Investigation in Agricultural Geography." Ph.D. Dissertation, University of Western Ontario, London, Ontario, 1970.

Furniss, I.F. "Post-War Productivity Gains in Canadian Agriculture." Pp. 200-215 in *Readings in Canadian Geography*, edited by Robert M. Irving. Toronto, Ontario: Holt, Rinehart and Winston, 1972.

Robinson, J. Lewis. "Agriculture in the Territories of Northwestern Canada." Pp. 1-7 in *Proceedings of the Conference on Boreal Agriculture*. Vancouver, British Columbia: University of British Columbia, Faculty of Agriculture, 1980.

Bollman, R.D. "Part-Time Farming in Canada: Issues and Non-Issues." *GeoJournal*, 6, No. 4 (1982): 313-322.

Norton, William. "Some Comments on Late Nineteenth Century Agriculture in Areas of European Overseas Expansion." *Ontario History*, 74, No. 2 (1982): 113-117.

Troughton, Michael J. *Canadian Agriculture*. Budapest, Hungary: Akademiai Kiado, 1982.

Bryant, Christopher R., Lorne H. Russwurm, and S.Y. Wang. "Agriculture in the Canadian Urban Field: An Appreciation." Pp. 12-23 in *The Pressure of Change in Rural Canada*, edited by Michael F. Bunce and Michael J. Troughton. North York, Ontario: York University Department of Geography, York Geographical Monograph No. 14, 1984.

Bunce, Michael F., and Michael J. Troughton, eds. *The Pressure of Change in Rural Canada*. North York, Ontario: York University Department of Geography, York Geographical Monograph No. 14, 1984.

Ehlers, Eckert. *Die agraren Siedlungsgrenzen der Erde: Gedonken zu ihrer Genese und Typologie am Beispel der kanadischen Waldlandes*. Wiesbaden, Germany: Steiner Verlag, 1984.

Schott, Carl. "Strukturwandel der kanadisches Landwirtschaft seit dem Zweiten Weltkrieg." *Zeitschrift der Gesellschaft fur Kanada-Studien*, 4 (1984): 5-18.

Fox, M.F. "Some Agents of Change in Canadian Agriculture." Pp. 59-69 in *Our Geographic Mosaic*, edited by David B. Knight. Ottawa, Ontario: Carleton University Press, 1985.

Nagle, Gunter. "Grenzraumer agrarische Nutzung in Kanada." *Zeitschrift der Gesellschaft fur Kanada-Studien*, 5 (1985): 85-95.

Bryant, Christopher R., and S.Y. Wang. "Changing Agricultural Structures Around Canadian Metropolitan Centres, 1941 to 1976: A Typology Approach." Pp. 174-196 in *Perspectives in Agricultural Typology*, edited by V.R. Singh and N.K. Singh. Varanasi, India: Star, 1986.

Fox, M.F. "Regional Change in Canadian Agriculture." *Geography*, 71, No. 2 (1986): 67-70.

Johnston, Thomas R.R., and Christopher R. Bryant. "Problems and Prospects of Farming in a Near-Urban Region: A Canadian Example." Pp. 56-63 in *Proceedings of the Fifteenth New Zealand Geography Conference*, edited by R. Welch. Dunedin, New Zealand: New Zealand Geographical Society, 1989.

Parson, Helen E. "The Changing Regional Pattern of Agriculture in Canada." *Operational Geographer*, 8, No. 1 (1990): 39-41.

Reeds, Lloyd G. "Canada's Agricultural Industry: Problems and Prospects." *Geografia Polonica*, 57 (1990): 111-122.

Smith, William. "Sustainable Agriculture and Canada's New Guidelines for Healthy Eating." *Proceedings, New England–St. Lawrence Division of the Association of American Geographers*, 21 (1991): 77-80.

Troughton, Michael J. *An Uncertain Future: Canadian Agriculture within the Canada–U.S. Free Trade Agreement*. Ottawa, Ontario: Carleton University, Department of Geography, 1991.

Hilts, Stewart G. "Natural Heritage and Agricultural Production in Canada." Pp. 142-152 in *Contemporary Rural Systems in Transition: Volume 1, Agriculture and Environment*, edited by Ian R. Bowler, Christopher R. Bryant, and M. Duane Nellis. Wallingford, England: C.A.B. International, 1992.

Jarochowska, M.A. *Le modile canadien de l'agriculture durable Zagadnicnia Ekoyaolymiki Rolnejle*. Warsaw, Poland: Academie Polonaise de Science, 1992.

Troughton, Michael J. "The Restructuring of Agriculture: The Canadian Example." Pp. 29-42 in *Contemporary Rural Systems in Transition: Volume 1, Agriculture and Environment*, edited by Ian R. Bowler, Christopher R. Bryant, and M. Duane Nellis. Wallingford, England: C.A.B. International, 1992.

Smit, Barry E., and J. Smithers. "Sustainable Agriculture: Interpretations, Analyses, and Prospects." *Canadian Journal of Regional Science*, 16, No. 2 (1993): 499-524.

Van Bers, C., and J.B. Robinson. "Farming in 2031: A Scenario of Sustainable Agriculture in Canada." *Journal of Sustainable Agriculture*, 4, No. 1 (1993): 41-65.

Troughton, Michael J. "Rural Canada and Canadian Rural Geography: An Appraisal." *Canadian Geographer*, 39, No. 4 (1995): 290-305.

Troughton, Michael J. "Agricultural and Rural Resources." Pp. 151-192 in *Resource and Environmental Management in Canada*, edited by Bruce Mitchell. Toronto, Ontario: Oxford University Press, 1995.

Mackinnon, Robert A. "Agricultural Change in Canada: Crisis or Renewal." *Canada–Japan Society Newsletter*, 15, No. 30 (1996): 2-6.

Found, William C. "Agriculture in a World of Subsidies." Pp. 155-174 in *Canada and the Global Economy: The Geography of Structural and Technological Change*, edited by John N.J. Britton. Montreal, Quebec: McGill-Queen's University Press, 1996.

Troughton, Michael J. "Scale Change, Discontinuity and Polarization in Canadian Farm-Based Rural Systems." Pp. 271-291 in *Agricultural Restructuring and Sustainability: A Geographic Perspective*, edited by Brian W. Ilbery, Quentin Chiotti, and Timothy Rickard. Wallingford, England: C.A.B. International, 1997.

Parson, Helen E. "Regional Trends in Agricultural Restructuring in Canada." *Canadian Journal of Regional Science*, 22, No. 3 (1999): 343-356.

CROPS AND CROPPING PATTERNS

Bishop, Avard L. "Development of Wheat Production in Canada." *Bulletin of the American Geographical Society*, 44, No. 1 (1912): 10-17.

McKay, Ian. "Trends in Canadian Soybean Production." Master's Thesis, University of Western Ontario, London, Ontario, 1966.

Leshkevich, George. "Tobacco in Ontario: A Visualization of the Growth and Marketing of a Cash Crop in Canada." Master's Thesis, Wayne State University, Detroit, Mich., 1970.

Troughton, Michael J. "Crop and Socioeconomic Types for Canadian Agriculture." *Proceedings, Association of American Geographers*, 7 (1975): 252-257.

Byrne, Joseph. "Variables That Have Influenced Sugar Beet Acreage in Canada." Master's Thesis, University of Windsor, Windsor, Ontario, 1977.

Luebbert, Scott J. "Determining Radar Systems Parameters to Optimize Crop Discrimination in Central Canada." Master's Thesis, University of Kansas, Lawrence, Kans., 1985.

Krueger, Ralph. "Trials and Tribulations of the Canadian Fruit-Growing Industry." *Canadian Geographer*, 44, No. 4 (2000): 342-354.

ANIMAL HUSBANDRY

Frederic, Paul B. "Trees to Cattle Feed: A Joint Canadian-U.S. Effort to Improve Maine's Dairy and Beef Industry." *Proceedings, New England–St. Lawrence Valley Division, Association of American Geographers*, 11 (1981): 59-61.

Stager, J.K. "Reindeer Herding as Private Enterprise in Canada." *Polar Record*, 22, No. 137 (1984): 127-136.

Diry, J.P. "Les politiques de l'elevage au Canada et leurs effets geoeconomiques." *Norois*, 35, No. 142 (1989): 151-170.

Chiotti, Quentin P. "Sectoral Adjustments in Agriculture: Dairy and Beef Livestock Industries in Canada." Pp. 43-60 in *Contemporary Rural Systems in Transition: Volume 1, Agriculture and Environment*, edited by Ian R. Bowler, Christopher R. Bryant, and M. Duane Nellis. Wallingford, England: C.A.B. International, 1992.

DAIRYING

Ramm, G. "Der Milchmarkt in Kanada." *Agrarwirtschaft*, 28 (1979): 198-206.

Sundstrom, Marvin T. "The Canadian Dairy Industry in Transition: 1950–1980." Pp. 34-53 in *The Pressure of Change in Rural Canada*, edited by Michael F. Bunce and Michael J. Troughton. North York, Ontario: York University Geographical Monograph No. 14, 1980.

Chiotti, Quentin P. "Farms in a State of Purgatory: The Canadian Dairy Farm Industry in the Aftermath of the Canada–U.S. Free Trade Agreement." Pp. 239-268 in *Rural and Urban Fringe Studies in Canada*, edited by Kenneth B. Beesley. North York, Ontario: York University Geographical Monograph No. 21, 1991.

FARM LAND

Crerar, A.D. "The Loss of Farmland and the Growth of Metropolitan Regions of Canada." Pp. 181-195 in *Resources for Tomorrow: Supplementary Volume.* Ottawa, Ontario: Queen's Printer, 1961.

Gertler, L.O., and J. Hind-Smith. *The Impact of Urban Growth on Agricultural Land.* Toronto, Ontario: Conservation Council of Ontario, 1961.

Ironside, Robert G. "Agricultural and Recreational Land Use in Canada: The Potential for Conflict or Benefit." *Canadian Journal of Agricultural Economics*, 19 (1971): 1-11.

Okafor, F.C. "Intensity of Agricultural Land Use in Canada: Von Thunen's Model Extended." Master's Thesis, University of Western Ontario, London, Ontario, 1975.

Bryant, Christopher R. *Farm-Generated Determinants of Land Use Change in the Rural–Urban Fringe, Canada, 1961–1975.* Ottawa, Ontario: Environment Canada, Lands Directorate, 1976.

Found, William C., A.R. Hill, and E.S. Spence. "Economic and Environmental Impacts of Agricultural Land Drainage in Ontario." *Journal of Soil and Water Conservation*, 31, No. 1 (1976): 20-24.

Hoffman, D.W. "Saving Farmland: A Canadian Program." *GeoJournal*, 6, No. 6 (1982): 539-546.

McCuaig, J.D., and E.W. Manning. *The Evolution of Agricultural Land Use in Canada: Process and Consequences.* Ottawa, Ontario: Environment Canada, Lands Directorate, 1982.

Cocklin, Christopher, and Barry E. Smit. "Future Urban Growth and Agricultural Land in Canada." *Applied Geography*, 3, No. 2 (1983): 91-104.

Hopson, Linda. "Absentee Ownership of Farmland in Canada." Master's Thesis, University of Toronto, Toronto, Ontario, 1984.

Bryant, Christopher R. "L'agriculture peri-urbaine au Canada: Degenerescence ou nouvelle dynamique?" *Annales de Geographie*, 98, No. 4 (1989): 403-420.

Brklacich, Michael, and K.B. MacDonald. "Prototype Agricultural Land Evaluation Systems for Canada: II. Selected Applications and Prospects." *Soil Use and Management*, 8, No. 1 (1992): 9-15.

MacDonald, K.B., and Michael Brklacich. "Prototype Agricultural Land Evaluation System for Canada: I. Overview of Systems Development." *Soil Use and Management*, 8, No. 1 (1992): 1-8.

Pierce, John T., and J. Seguin. "Exclusive Agricultural Zoning in British Columbia and Quebec: Problems and Prospects." *Progress in Rural Policy and Planning*, 3 (1993): 287-310.

Bryant, Christopher R. "Preserving Canada's Agricultural Land." *Plan Canada*, Special Edition (July 1994): 49-51.

AGRICULTURAL POPULATIONS AND SOCIAL ISSUES

Walker, Gerald E. "Farmers in the Urban Shadow: A Behavioural Profile." Pp. 185-198 in *The Rural–Urban Fringe: Canadian Perspectives*, edited by

Kenneth B. Beesley and Lorne H. Russwurm. North York, Ontario: York University, Geographical Monograph No. 10, 1981.

Troughton, Michael J. "The Dilemma of Modern Agriculture and Sustainable Rural Communities." Pp. 230-238 in *Proceedings IEEISSIT Conference: Preparing for a Sustainable Society*. Toronto, Ontario: Ryerson, 1991.

Leckie, Gloria J. "Female Farmers in Canada and the Gender Relations of a Restructuring Agricultural System." *Canadian Geographer*, 37, No. 3 (1993): 212-229.

Leckie, Gloria J. "Female Farmers in Canada, 1971–1986." *Professional Geographer*, 45, No. 2 (1993): 180-193.

Leckie, Gloria J. "They Never Trusted Us to Drive: Farm Girls and the Gender Relations of Agricultural Information Transfer." *Gender, Place and Culture*, 3, No. 3 (1996): 309-326.

Paterson, John L. "Geography and Religion: Agriculture and Stewardship in the Christian Farmers Federation of Canada." Ph.D. Dissertation, University of British Columbia, Vancouver, B.C., 1998.

ENVIRONMENTAL AND CONSERVATION ISSUES

Stewart, D.W. "Atmospheric Carbon Dioxide and Canadian Agriculture." *Climatologial Bulletin*, 18, No. 1 (1984): 3-14.

Ball, T. *The Impact of Climate Change on Sustainable Agriculture in Canada*. Ottawa, Ontario: Science Council of Canada, 1991.

Smithers, J. "Regional Agricultural Adjustment to Climatic Variability and Change: Concepts and Methods." Ph.D. Dissertation, University of Guelph, Guelph, Ontario, 1994.

Brklacich, Michael, et al. "Crop Production Risk as a Factor in Sustainable Land Management." Pp. 51-62 in *Data Reliability and Risk Assessment in Soil Interpretations*, edited by W.D. Nettleton, et al. Madison, Wis.: Soil Science Society of American, Special Publication No. 47, 1996.

Greenwood, Julie. "Ecological Farm Practices and the Family Farm." Ph.D. Dissertation, University of Western Ontario, London, Ontario, 1996.

Gillespie, T.J. "Agricultural Surfaces." Pp. 277-302 in *The Surface Climates of Canada*, edited by William G. Bailey, Timothy R. Oke, and Wayne R. Rouse. Montreal, Quebec: McGill-Queen's University Press, 1997.

Bryant, Christopher R., et al. "Adaptation in Canadian Agriculture to Climatic Variability and Change." *Climatic Change*, 45, No. 1 (2000): 181-201.

AGRICULTURAL MARKETING AND DISTRIBUTION

Maniak, S.J. "Optional Deliveries of Wheat in Canada: The Impact of CWRS Wheat Grade Multiplicity on Transportation Cost Minimization." Master's Thesis, University Alberta, Edmonton, Alberta, 1984.

Chiotti, Quentin, and Michael J. Troughton. *An Uncertain Future: Canadian Agriculture within the Canada–U.S. Free Trade Agreement*. Ottawa, Ontario: Carleton University Department of Geography, Discussion Paper No. 9, 1991.

Troughton, Michael J. "Canadian Agriculture and the Canada–U.S. Free Trade Agreement: A Critical Appraisal." *Progress in Rural Policy and Planning*, 1 (1991): 176-196.

Troughton, Michael J. "An Ill-Considered Pact: The Canada–U.S. Free Trade Agreement and the Agricultural Geography of North America." *Trade Liberalization and Rural Restructuring in Canada*. Brandon, Manitoba: The Rural Development Institute, Working Paper No. 1, 1991.

TRANSPORTATION AND MECHANIZATION

Scorgie, Edward K. "Adapting Harvestore Systems: A Study of Change, Technology and Diffusion in Agriculture." Ph.D. Dissertation, University of Waterloo, Waterloo, Ontario, 1994.

AGRICULTURAL LABOR STUDIES

Smit, Barry E. "A Regional Analysis of Employment Changes in Canadian Agriculture." Ph.D. Dissertation, McMaster University, Hamilton, Ontario, 1977.

Smit, Barry E. "Employment of Family and Non-Family Labour in Agriculture." *Canadian Journal of Agricultural Economics*, 26, No. 1 (1978): 35-42.

Eastern Canada

GENERAL TEXTS, ARTICLES, AND THESES

Randall, John R. "Agriculture in the Great Clay Belt of Canada." *Scottish Geographical Magazine*, 56, No. 1 (1940): 12-27.

Biays, Pierre. "Problemes de l'agriculture marginale dans la zone pionniere de l'Est du Canada." *Cahiers de Geographie de Quebec*, 9, No. 16 (1964): 219-230.

Sinclair, Peter. "Agricultural Colonization in Ontario and Quebec: Some Evidence from the Great Clay Belt, 1900–1945." Pp. 104-120 in *Canadian Papers in Rural History*, Vol. 5, edited by Donald Akenson. Gananoque, Ontario: Langdale Press, 1986.

Vanderhill, Burke G. "Agriculture's Struggle for Survival in the Great Clay Belt of Ontario and Quebec." *American Review of Canadian Studies*, 18, No. 4 (1988): 455-464.

Deslauriers, Pierre, Christopher R. Bryant, and Claude Marois. "Farm Business Restructuring in the Urban Fringe: The Toronto and Montreal Regions." Pp. 74-86 in *Contemporary Rural Systems in Transition: Vol. 1, Agriculture and Environment*, edited by Ian R. Bowler, Christopher R. Bryant, and M. Duane Nellis. Wallingsford, England: C.A.B. International, 1992.

CROPS AND CROPPING PATTERNS

Shaw, Anthony B. "The Emerging Cool Climate Wine Regions of Eastern Canada." *Journal of Wine Research*, 10, No. 2 (1999): 79-94.

AGRICULTURAL POPULATIONS AND SOCIAL ISSUES

Kerr, Donald, and W.J. Smith. "Agriculture, Balanced Growth and Social Change in Central Canada Since 1850." *Economic Development and Culture Change*, 28, No. 3 (1980): 615-622.

The Atlantic Provinces

GENERAL TEXTS, ARTICLES, AND THESES

Putnam, Donald F. "Distribution of Agriculture in New Brunswick." *Public Affairs*, 3, No. 1 (1939): 8-11.

Lloyd, Donald. "The Patterns and Problems of Agriculture Within Kent and Northumberland Counties, New Brunswick, Canada, 1961–1968." Ph.D. Dissertation, University of Maryland, College Park, Md., 1971.

Wynn, Graeme. "Late Eighteenth-Century Agriculture on the Bay of Fundy Marshlands." *Acadiensis*, 8, No. 2 (1979): 80-89.

Wieger, Axel. "Das mittlere St. John River Bebiet (N.B.) als hochspezialisierter agrarishcer Sonderraum in den atlantischen Provinzen Kanadas." Pp. 321-344 in *Festschrift fur Felix Monheim*, edited by Frank Ahnert and Reinhart Zschocke. Aachen, Germany: Aachener Geographische Arbeiten, Vol. 14, 1981.

Kinsman, Brian F. "Planning for Agriculture in Nova Scotia." Master's Thesis, University of Waterloo, Waterloo, Ontario, 1983.

Mackinnon, Robert A., and Graeme Wynn. "Nova Scotian Agriculture in the 'Golden Age': A New Look." Pp. 47-56 in *Geographical Perspectives on the Maritime Provinces*, edited by Douglas Day. Halifax, Nova Scotia: St. Mary's University, 1988.

Troughton, Michael J. "From Nodes to Nodes: The Rise and Fall of Agricultural Activity in the Maritime Provinces." Pp. 25-46 in *Geographical Perspectives on the Maritime Provinces*, edited by Douglas Day. Halifax, Nova Scotia: St. Mary's University, 1988.

Squires, R.J. "Farmland Conservation and Preservation in Newfoundland." *Environments*, 20, No. 1 (1989): 40-49.

Wynn, Graeme. "Exciting a Spirit of Emulation Among the Plod-Holes: Agricultural Reform in Pre-Confederation Nova Scotia." *Acadiensis*, 20, No. 1 (1990): 5-51.

Mackinnon, Robert A. "Farming the Rock: The Evolution of Commercial Agriculture Around St. John's, Newfoundland, to 1945." *Acadiensis*, 20, No. 2 (1991): 32-61.

Mackinnon, Robert A. "Agriculture and Rural Change in Nova Scotia, 1851–1951." Pp. 231-274 in *Canadian Papers in Rural History*, vol. 10, edited by Donald Akenson. Gananoque, Ontario: Langdale Press, 1996.

FARM LAND

Gentilcore, R. Louis. "Land Use and Agricultural Production in Antigonish County, Nova Scotia." Ph.D. Dissertation, University of Maryland, College Park, Md., 1950.

AGRICULTURAL POPULATIONS AND SOCIAL ISSUES

Wieger, Axel. *Agrarkolonisation, Landnutzung und kulturlandschaftsverfall in der Provinz New Brunswick (Kanada).* Aachen, Germany: Aachener Geographische Arbeiten, Heft 22, 1990.

Millward, Hugh. "Process and Pattern in the Agricultural Settlement of the Canadian Maritimes." *Proceedings, New England–St. Lawrence Valley Division, Association of American Geographers*, 23 (1993): 21-31.

Wieger, Axel. "Mythe agrarie et realite economique: La colonisation des acadiens dans la foret du nouveau Brunswick (Canada) de 1870 a 1960." *Bulletin de la Societe Geographique de Liege*, 32 (1996): 253-259.

ENVIRONMENTAL AND CONSERVATION ISSUES

Crickmer, Robin E. "An Investigation of the Role of Physical Factors in the Process of Farm Land Abandonment in Nova Scotia, 1953–1974." Master's Thesis, University of Waterloo, Waterloo, Ontario, 1976.

Bittermann, Rusty. "The Hierarchy of the Soil: Land and Labour in a Nineteenth-Century Cape Breton Community." *Acadiensis*, 18, No. 1 (1988): 33-55.

AGRICULTURAL MARKETING AND DISTRIBUTION

Mackinnon, Robert A. "The Growth of Commerical Agriculture Around St. John's, 1800–1935: A Study of Local Trade in Response to Urban Demand." Master's Thesis, Memorial University of Newfoundland, St. John's, Newfoundland, 1981.

Ontario

ATLASES AND GRAPHIC PRESENTATIONS

Keddie, Philip D., and Julius A. Mage. *Southern Ontario Atlas of Agriculture: Contemporary Patterns and Recent Changes*. Guelph, Ontario: University of Guelph, Department of Geography, Occasional Paper No. 7, 1985.

McInnis, Marvin. "Ontario Agriculture, 1851–1901: A Cartographic Overview." Pp. 290-301 in *Canadian Papers in Rural History*, vol. 5, edited by Donald Akenson. Gananoque, Ontario: Langdale Press, 1986.

GENERAL TEXTS, ARTICLES, AND THESES

Reeds, Lloyd G. "Agricultural Geography of the Lindsay-Peterborough Area." Master's Thesis, University of Toronto, Toronto, Ontario, 1942.

Forrester, James. "The Agricultural Geography of Wentworth County, Ontario." Master's Thesis, McMaster University, Hamilton, Ontario, 1954.

Reeds, Lloyd G. "The Agricultural Geography of Southern Ontario." Ph.D. Dissertation, University of Toronto, Toronto, Ontario, 1956.

May, J.A. "An Analysis of Some Aspects of the Agricultural Rehabilitation and Development Act Program,." Master's Thesis, University of Western Ontario, London, Ontario, 1963.

Glandon, Brian R. "Agriculture of the Holland Marsh, Ontario, Canada." Master's Thesis, Miami University, Miami, Ohio, 1966.

Russwurm, Lorne H. "Expanding Urbanization and Selected Agricultural Elements: Case Study, Southwestern Ontario Area, 1941–1961." *Land Economics*, 43 (1967): 101-107.

Ennals, Peter M. "The Development of Farm Barn Types in Southern Ontario During the Nineteenth Century." Master's Thesis, University of Toronto, Toronto, Ontario, 1968.

Kelly, Kenneth. "The Agricultural Geography of Simcoe County, Ontario, 1820–1880." Ph.D. Dissertation, University of Toronto, Toronto, Ontario, 1968.

Sample, K.A. "Changes in Agriculture on the Six Nations Indian Reserve." Master's Thesis, McMaster University, Hamilton, Ontario, 1968.

Reeds, Lloyd G. *Niagara Region Agricultural Research Report: Fruit Belt*. Toronto, Ontario: Ministry of Treasury, Economics and Intergovernmental Affairs, 1969.

Sterrett, David H. "A Factor of Typology of Ontario Agriculture." Master's Thesis, University of Waterloo, Waterloo, Ontario, 1969.

Kelly, Kenneth. "The Transfer of British Ideas on Improved Farming to Ontario during the First Half of the Nineteenth Century." *Ontario History*, 63, No. 2 (1971): 103-111.

Moran, Warren. "Agriculture and Urban Shadow." Pp. 75-82 in *Proceedings, Seventh New Zealand Geography Conference*, edited by Evelyn Stokes. Hamilton, New Zealand: New Zealand Geographical Society, 1972.

Stock, George E. "The Geography of Small Scale Farming in Ontario." Master's Thesis, University of Guelph, Guelph, Ontario, 1975.

Jackson, John N. "The Niagara Fruit Belt: A Resource in Jeopardy." *Contact: Journal of Urban and Environmental Affairs*, 9 (1977): 24-41; 172-176.

Dilley, Robert S. *Planning and the Future of Agriculture in Haldimand-Norfolk: A Critical View*. Hamilton, Ontario: McMaster University, Department of Geography Discussion Paper, 1979.

Hoggart, Keith. *The Determinants of Farm Linkage Patterns: An Investigation of Three Townships in Huron County, Ontario*. London, England: University of London, Kings College, Department of Geography, Paper No. 3, 1979.

Laskin, Susan L. "The Myth of a Northern Agricultural Frontier in Nineteenth Century Ontario." Master's Thesis, University of Toronto, Toronto, Ontario, 1979.

Paul, Bimal K. "Crop, Livestock and Enterprise Combination Patterns of Southern Ontario, 1951–1971." Master's Thesis, University of Waterloo, Waterloo, Ontario, 1980.

Tabayaski, A. "Agricultural Regions of Southern Ontario from 1951 to 1971." *Science Reports of the Institute of Geoscience*, Tsukuba University, 3 (1982):1-18.

Norton, William. "Agricultural Evolution on the Frontier." *Professional Geographer*, 36, No. 1 (1984): 18-27.

Casteran, Nicole. "Ecologie et agriculture preindustrielle dans l'Est Ontario." Master's Thesis, University of Ottawa, Ottawa, Ontario, 1985.

Tabayashi, A. "Changing Agricultural Regions of Southern Ontario." *Annual Report of the Institute of Geoscience*, Tsukuba University, 12 (1986): 8-14.

Tabayashi, A. "Agricultural Regions of Southern Ontario: A Review Article" (in Japanese). *Tsukuba Studies in Human Geography*, 10 (1986): 151-187.

Troughton, Michael J., and Stephen J. Harvey. "The Changing Role of Agriculture in the Algoma District, Northern Ontario." *Nordia*, 20, No. 1 (1986): 95-104.

Johnston, Thomas R.R., and Christopher R. Bryant. "Agricultural Adaptation: The Prospects of Sustaining Agriculture Near Cities." Pp. 9-22 in *Sustaining Agriculture Near Cities*, edited by William Lockeretz. Ankeny, Iowa: Soil and Water Conservation Society, 1987.

Skof, Karl J. "Agriculture in a Forest Setting: Lanark County, 1851–1871." Master's Thesis, Carleton University, Ottawa, Ontario, 1987.

Glover, E.M.A. "An Analysis of Industrial Agriculture in Middlesex County." Master's Thesis, University of Western Ontario, London, Ontario, 1988.

Hogan, Edmund N. "The Agroforestry Option in Canada with an Emphasis on Eastern Ontario." Master's Thesis, Carleton University, Ottawa, Ontario, 1988.

Bonneville, Marc J. "An Integrated Approach to the Study of Agricultural Change on Manitoulin." Master's Thesis, Carleton University, Ottawa, Ontario, 1989.

Nicholson, J.P. "Agricultural Potential on Indian Reserves in Ontario." *Operational Geographer*, 8, No. 1 (1990): 32-39.

Gayler, Hugh J. "The Demise of the Niagara Fruit Belt: Policy Planning and Development Options in the 1990s." Pp. 283-313 in *Rural and Urban Fringe Studies in Canada*, edited by Kenneth B. Beesley. North York, Ontario: York University Geographical Monograph No. 21, 1991.

Smit, Barry E., Michael Brklacich, and Truman Phillips. "Evaulating Ontario's Potential for Food Production." *Canadian Journal of Regional Science*, 14, No. 1 (1991): 1-22.

Smit, Barry E., J. Bray, and Philip D. Keddie. "Identification of Marginal Agricultural Areas in Ontario, Canada." *Geoforum*, 22, No. 3 (1991): 333-346.

Gayler, Hugh J. "Urban Sprawl and the Decline of the Niagara Fruit Belt." Pp. 128-132 in *Geographical Snapshots of North America*, edited by Donald Janelle. New York, N.Y.: Guilford, 1992.

Chapman, Paul. "Agriculture in Ontario: An Overview." Pp. 279-300 in *Niagara's Changing Landscapes*, edited by Hugh J. Gayler. Ottawa, Ontario: Carleton University Press, 1994.

Dilley, Robert S. "Farming on the Margin: Agriculture in Northern Ontario." Pp. 180-189 in *Geographic Perspectives on the Provincial Norths*, edited by Margaret Johnston. Thunder Bay, Ontario: Lakehead University, Center for Northern Studies, No. 3, 1994.

Tabayashi, A. "Sustainable Development of Agriculture in Southern Ontario" (in Japanese). *Tsukuba Studies in Human Geography*, 19 (1995): 97-134.

Walker, Gerald E. "Adjustment in Ontario Agriculture." Pp. 117-129 in *Sustainability of Rural Systems*, edited by Christopher R. Bryant and Claude Marois. Montreal, Quebec: Proceedings of the International Geographical Union Study Group on Sustainability of Rural Systems, University of Montreal, Department of Geography, 1995.

Wood, J. David. "Testing the Limits of Agriculture: The Idea of Progress Meets Northern Reality." Pp. 139-144 in *Espace et culture*, edited by Serge Courville and Normand Seguin. Sainte-Foy, Quebec: Les Presses de l'Universite Laval, 1995.

Crawford, G., et al. "Floodplains and Agricultural Origins: A Case Study in South-Central Ontario, Canada." *Journal of Field Archaeology*, 25 (1998): 123-137.

Bradshaw, Benjamin. "Resource Use Response to Subsidy Removal in Commercial Agriculture." Ph.D. Dissertation, University of Guelph, Guelph, Ontario, 1999.

Clarke, John. "A Regional Taxonomy of Agriculture: Southern Ontario in the Middle of the Nineteenth Century." Pp. 43-66 in *Themes and Issues in Canadian Geography III*, edited by Christoph Stadel. Salzburg, Austria: University of Salzburg, Salzburger Geographishe Arbeiten, Band 34, 1999.

Wood, J. David. *Making Ontario: Agricultural Colonization and Landscape Recreation Before the Railway.* Montreal, Quebec: McGill-Queen's University Press, 2000.

CROPS AND CROPPING PATTERNS

Duxbury, John A. "A Geographical Study of the Flue-Cured Tobacco Producing Areas of Allison and New Lowell, Ontario." Master's Thesis, University of Kansas, Lawrence, Kans., 1951.

Vance, R.C. "The Tobacco Economy of the New Belt." Master's Thesis, University of Western Ontario, London, Ontario, 1952.

Donahue, John J. "Survey of Tobacco Production, Durham County, Ontario." Master's Thesis, State University of New York, Buffalo, N.Y., 1964.

Helleiner, F.M. "Changes of the Distribution of Orchards in the London Region." Master's Thesis, University of Western Ontario, London, Ontario, 1966.

Retallack, Joan. "The Changing Distribution of Wheat in Southern Ontario, 1950–1890." Master's Thesis, University of Toronto, Toronto, Ontario, 1966.

Keddie, Philip. "The Expansion of Corn for Grain in Southern Ontario and Its Effect on the Farm Enterprise with Specific Reference to the Midwestern Ontario Region." Ph.D. Dissertation, University of Waterloo, Waterloo, Ontario, 1976.

McKay, Ian A. "A Note on Ontario Agriculture: The Development of Soybeans, 1893–1952." *Ontario History*, 75, No. 2 (1983): 175-186.

Gruszka, T. "Discontinuance of the White Bean Crop in Huron County." Master's Thesis, University of Guelph, Guelph, Ontario, 1984.

Tyszko, D.G. "Contract Growing of Green Peas and Sweet Corn with Green Giant and Libby's in Southwestern Ontario." Master's Thesis, University of Windsor, Windsor, Ontario, 1984.

Sheldon, Samuel R. "Ontario's Flue-Cured Tobacco Industry: The Southern United States Legacy." *American Review of Canadian Studies*, 18, No. 2 (1988): 195-212.

Huffman, Edward C. "Procedures for the Identification of Cropping Systems Using Landsat TM Imagery." Ph.D. Dissertation, University of Waterloo, Waterloo, Ontario, 1993.

Stewart, Carol. "Agricultural Restructuring of Ontario Tobacco Production." Master's Thesis, University of Western Ontario, London, Ontario, 1997.

Treitz, P.M., et al. "Agricultural Crop Classification Using SAR Tone and Texture Statistics." *Canadian Journal of Remote Sensing*, 26, No. 1 (2000): 18-29.

Ramsey, Douglas, and Barry E. Smit. "Impacts of Changes in the Flue-Cured Tobacco Sector on Farmers in Ontario, Canada." *Applied Geography*, 21, No. 4 (2001): 347-368.

Shaw, Anthony B. "Pelee Island and Lake Erie North Shore, Ontario: A Climatic Analysis of Canada's Warmest Wine Region." *Journal of Wine Research*, 12, No. 1 (2001): 19-37.

ANIMAL HUSBANDRY

Forester, J.E. "Fur Farming in Ontario." Master's Thesis, University of Western Ontario, London, Ontario, 1968.

Wind, F. "Location and Flow Analysis of the Ontario Broiler Chicken Industry." Master's Thesis, Waterloo Lutheran University, Waterloo, Ontario, 1969.

Bowler, Ian R. "The Institutional Regulation of Uneven Development: The Case of Poultry Production in the Province of Ontario." *Transactions, Institute of British Geographers*, 19, No. 3 (1994): 346-458. (New Series)

DAIRYING

Surtees, William. "The Dairy Industry of Oxford County." Master's Thesis, McMaster University, Hamilton, Ontario, 1963.

Horner, John H. "Changing Spatial Patterns in the Production and Utilization of Milk in Southern Ontario, 1910–1961." Master's Thesis, University of Toronto, Toronto, Ontario, 1967.

Sundstrom, Marvin T. "Regional and Farm Level Adjustments in Southern Ontario's Dairy Industry." Ph.D. Dissertation, McMaster University, Hamilton, Ontario, 1975.

FARM LAND

Irving, Robert M. "Agricultural Land Use in the Beatty-South Saugeen Watershed." Master's Thesis, University of Toronto, Toronto, Ontario, 1956.

Krueger, Ralph R. "Changing Land-Use Patterns in the Niagara Fruit Belt." *Transactions of the Royal Canadian Institute*, 32, Part 2 (1959): 38-140.

Krueger, Ralph R. "Changing Land-Use Patterns in the Niagara Fruit Belt." Ph.D. Dissertation, Indiana University, Bloomington, Indiana, 1959.

Siller, Nicholas J. "Land-Use Patterns in an Intensive Agricultural Area of Essex County, Ontario: The Garfield-Mersea Region: An Example." Master's Thesis, University of Western Ontario, London, Ontario, 1964.

MacDougall, E. Bruce. "Farm Number Changes in Relation to Land Capability and Alternative Employment Opportunity in the North Part of Central Ontario." Ph.D. Dissertation, University of Toronto, Toronto, Ontario, 1967.

Campbell, Neil M. "An Evaluation and Policy Proposal: Ontario Farm Enlargement and Consolidation Program." Ph.D. Dissertation, University of Waterloo, Waterloo, Ontario, 1972.

Maas, David R. "Factors Influencing Farm Sales Decisions on Toronto's Urban Fringe." Master's Thesis, McMaster University, Hamilton, Ontario, 1972.

Bunce, Michael. "Farm Consolidation and Enlargement in Ontario and Its Relevance to Rural Development." *Area*, 5, No. 1 (1974): 13-15.

Parsons, Diana Lee. "The Influence of Type of Owner on Farm Management in Chinguacousy Township." Master's Thesis, York University, North York, Ontario, 1975.

Harker, Paul T. "The Impact of Non-Resident Ownership on Farm Patterns in Glenelg and Egremont Townships: A Case Study for an Area Outside the Commuting Range of Major Ontario Cities." Master's Thesis, University of Waterloo, Waterloo, Ontario, 1976.

Crewson, Daryll M. "Study of Farmland Loss in Central Ontario, 1951–1971." Master's Thesis, McMaster University, Hamilton, Ontario, 1977.

Galloway, John. "An Evaluation of the A.R.D.A. Farm Enlargement Client Section Process." Master's Thesis, University of Guelph, Guelph, Ontario, 1977.

McLean, Jackie. "Land Capability as a Determinant of Agricultural Productivity." Master's Thesis, University of Guelph, Guelph, Ontario, 1977.

Van Der Wal, Hylke T. "Agricultural Settlement and Land Use Changes in Bosanpent Township, Ontario, 1829–1977." Master's Thesis, Eastern Michigan University, Ypsilanti, Mich., 1977.

Shewfelt, Dennis. "An Empirical Analysis of Farmland Price in Ontario, 1972–1974." Master's Thesis, University of Guelph, Guelph, Ontario, 1978.

Ali, A.M.S. "Some Factors Affecting Agricultural Land Use Change in Southern Ontario, 1971–1976." Master's Thesis, University of Windsor, Windsor, Ontario, 1979.

Fielding Jeffrey A. "Farmland Rental in an Urbanizing Environment: The Fringes of Kitchener, Waterloo and Cambridge, 1971–1978." Master's Thesis, University of Waterloo, Waterloo, Ontario, 1979.

Reid, Deborah R. "Land-Use Change in a Selected Area of the Niagara Fruit Belt." Master's Thesis, McMaster University, Hamilton, Ontario, 1980.

Cocklin, Christopher. "Future Urban Expansion and Implications for Agricultural Land in Ontario." Master's Thesis, University of Guelph, Guelph, Ontario, 1981.

DeBoer, P.J. "Agricultural Land Search Behavior: A Study of Agricultural Relocation and Expansion in the Niagara Fruit Belt." Master's Thesis, Wilfrid Laurier University, Waterloo, Ontario, 1981.

Marta, T.J. "Full-Time Farmer Decision Strategies for Farmland Manipulation in a Rural–Urban Continuum Adjacent to London, Ontario." Master's Thesis, University of Western Ontario, London, Ontario, 1981.

Gayler, Hugh J. "Conservation and Development in Urban Growth: The Preservation of Agricultural Land in the Rural–Urban Fringe in Ontario." *Town Planning Review*, 53 (1982): 321-341.

Krueger, Ralph R. "The Struggle to Preserve Specialty Cropland in the Rural–Urban Fringe of the Niagara Peninsula." *Environments*, 14, No. 3 (1982): 1-10.

Lavigne, N. "Population and Farmland Area in Ontario: An Examination of the Changes at the Township Level, 1961–1976." Master's Thesis, University of Guelph, Guelph, Ontario, 1982.

Mage, Julius A. "Absentee Foreign Ownership of Farmland: Trends, Problems, and Issues with Specific Reference to Huron-South Bruce Counties, Ontario." *Geographical Inter-University Resource Management Seminars*, 12 (1982): 1-17.

Mage, Julius A., and Thomas R.R. Johnston. "Agricultural Land Use Policy in Ontario: A Response to Perceived Crises?" *Papers and Proceedings of the Applied Geography Conferences*, 5 (1982): 101-115.

Miklos, Daniel. "Noncontiguous Farming in an Urban Pressured Environment." Master's Thesis, University of Waterloo, Waterloo, Ontario, 1982.

Osborne, Brian S. "The Farmer and the Land." Pp. 81-94 in *County of a Thousand Lakes: The History of the County of Frontenac,* edited by Bryan Rollason. Kingston, Ontario: Frontenac County Council, 1982.

Johnston, Thomas R.R. "An Evaluation of the Rationale of Farmland Preservation Policy in Ontario." Master's Thesis, University of Guelph, Guelph, Ontario, 1983.

Mage, Julius A. "Absentee Ownership of Ontario Farmland: Some Implications for Land Use Policy." *Papers and Proceedings, Applied Geography Conferences*, 6 (1983): 145-157.

Smit, Barry E., Dennis Shewfelt, and M.S. Flaherty. "Purchaser and Spatial Separability in the Market for Farmland in Ontario." *Rural Systems*, 1, No. 1 (1983): 45-60.

Krueger, Ralph R. "The Struggle to Preserve Specialty Crop Land in the Rural–Urban Fringe of the Niagara Peninsula of Canada." Pp. 292-313 in *The Pressures of Change in Rural Canada,* edited by Michael F. Bunce and Michael J. Troughton. North York, Ontario: York University Geographical Monograph No. 14, 1984.

Smit, Barry E. "Assessing Ontario's Land Resources and Food Production Prospects." *Highlights of Agricultural Research in Ontario*, 7, No. 2 (1984): 22-24.

Tsarnas, Ileanna. "Changing Agricultural Land-Use: A Case Study of South-East Ontario." Master's Thesis, Columbia University, New York, N.Y., 1984.

Bunce, Michael F. "Agricultural Land as a Real Estate Commodity: Implications for Farmland Preservation in the North American Urban Fringe." *Landscape Planning*, 12 (1985): 177-192.

Jackson, John N. "Protecting Agricultural Land from Urban Sprawl: The Niagara Fruit Belt of Ontario." *Environmental Education and Information*, 3, No. 3 (1985): 253-268.

Mage, Julius A. "Absentee Ownership of Ontario's Farmland." Pp. 197-217 in *Farming and the Rural Community in Ontario: An Introduction*, edited by Anthony M. Fuller. Toronto, Ontario: Foundation for Rural Living, 1985.

Kresovic, Walter A. "Factors Affecting Agricultural Land Use Change in a Retreating Agricultural Margin: The Case of Eastern Ontario." Master's Thesis, University of Waterloo, Waterloo, Ontario, 1986.

Ludlow, Laurie L., and Barry E. Smit. "Assessing the Implications of Environmental Change for Agricultural Production: The Case of Acid Rain in Ontario, Canada." *Journal of Environmental Management*, 25, No. 1 (1987): 27-44.

Kristjanson, Johanne A. "Methods for Rating Agricultural Land: Some New Approaches to Assist in Land Use Decision-Making." Master's Thesis, University of Guelph, Guelph, Ontario, 1988.

Bray, Justine. "A Land Use Viability Approach for the Identification of Agriculturally Marginal Areas in Ontario." Master's Thesis, University of Guelph, Guelph, Ontario, 1989.

Brklacich, Michael, Barry E. Smit, Christopher Bryant, and J. Dumanski. "Impacts of Environmental Change on Sustainable Food Production in Southwestern Ontario, Canada." *Land Degradation and Rehabilitation*, 1 (1989): 291-303.

Krushelnicki, Bruce W., and Stephen J. Bell. "Monitoring the Loss of Agricultural Land: Identifying the Urban Price Shadow in the Niagara Region, Canada." *Land Use Policy*, 6, No. 2 (1989): 141-150.

Wilson, J.P. "Soil Erosion from Agricultural Land in the Lake Simcoe-Couchiching Basin, 1800–1981." *Canadian Journal of Soil Science*, 69 (1989): 137-151.

Mackenzie, Fionna. "The Politics of Partnerships: Farm Women and Farmland, Ontario." Pp. 95-95 in *Contemporary Rural Systems in Transition: Volume 2, Economy and Society*, edited by Ian R. Bowler, Christopher R. Bryant, and M. Duane Nellis. Wallingford, England: C.A.B. International, 1992.

Hall, J. "The Impact of Climate Change on Agriculture." Pp. 69-90 in *The Impact of Climate Change on the Water in the Grand River Basin*, edited by Marie Sanderson. Waterloo, Ontario: University of Waterloo Department of Geography, Publication No. 40, 1993.

Sheldon, Samuel. "Landscape Transformation in Ontario's Norfolk Sand Plain." Pp. 325-344 in *Culture, Form and Place: Essays in Cultural and Historical Geography*, edited by Kent Mathewson. Baton Rouge, La.: Louisiana State University, Geoscience and Man Series, Vol. 32, 1993.

Xu, Wei. "Farmland Changes in Southern Ontario: 1961–1991." Master's Thesis, University of Guelph, Guelph, Ontario, 1993.

Parson, Helen E. "Reforestation of Agricultural Land in Southern Ontario Before 1931." *Ontario History*, 86, No. 3 (1994): 237-248.

Caldwell, Wayne J. "Rural Planning and Agricultural Land Preservation: The Experience of Huron County, Ontario." *Great Lakes Geographer*, 2, No. 2 (1995): 21-34.

McKay, Ian A., and Helen E. Parson. "The Impact of Farm Lot Shape on Patterns in the Rural Landscape." *Journal of Geography*, 96, No. 6 (1997): 284-292.

Xu, Wei. "Changes in Agricultural Land Use and Agroecosystem Health: A Landsat TM Approach." *Great Lakes Geographer*, 6, No. 1-2 (1999): 65-84.

AGRICULTURAL POPULATIONS AND SOCIAL ISSUES

Crothall, William R. "French-Canadian Agriculture in Ontario, 1861–1871." Master's Thesis, University of Toronto, Toronto, Ontario, 1968

Parsons, Diana Lee. "The Influence of Type of Owner on Farm Management in Chinguacousy Township." Master's Thesis, York University, North York, Ontario, 1975.

Layton, Robert L. "Hobby Farming: A Characterisation: A Case Study of the Rural–Urban Fringe of London, Ontario." Master's Thesis, University of Western Ontario, London, Ontario, 1976.

Hodgson, W.J. "The Attitudes of Farmers Towards Farm Enlargement: A Case Study in Bruce County, Ontario." Master's Thesis, University of Guelph, Guelph, Ontario, 1977.

Down, Jean. "An Examination of the Characteristics of Farm Entrants and Their Enterprises in Southern Ontario, 1966, 1971 and 1976." Master's Thesis, University of Guelph, Guelph, Ontario, 1980.

Layton, Robert L. "Attitudes of Hobby and Commercial Farms in the Rural-Urban Fringe of London, Ontario." *Cambria*, 8 (1981): 33-44.

Harvey, Stephen J. "The Role of Agriculture in the Settlement of Rural Northern Ontario: The Algoma Case Study." Master's Thesis, University of Western Ontario, London, Ontario, 1985.

Marr, William L. "The Distribution of Tenant Agriculture in Ontario, Canada, 1871." *Social Science History*, 11, No. 2 (1987): 169-186.

Leckie, Gail J. "Industrialization of Agriculture and Social Change in the Farm Community: Brooke Township, Ontario, 1965–1986." Master's Thesis, University of Western Ontario, London, Ontario, 1988.

Mage, Julius A., and Sharon A. Fletcher. *Mennonite Farming in the Mount Forest Area of Ontario: Plase II: Comparison of Mennonite and Non-Mennonite Agricultural Systems and Land Use Characteristics.* Guelph, Ontario: University of Guelph, Department of Geography, Occasional Paper No. 14, 1990.

Mansfield, Lois T. "A Macro-Scale Analysis of Hobby Farming in Southern Ontario." Master's Thesis, University of Guelph, Guelph, Ontario, 1990.

Beesley, Kenneth B., and Pamela J. Macintosh. "The Quality of Life in Urban Fringe Environments: Farm–Nonfarm Comparisons in Southwestern Ontario." *Proceedings, New England–St. Lawrence Valley Division, Association of American Geographers*, 23 (1993): 75-87.

Mage, Julius A. "The Mennonite Farmers of Waterloo County: An Enclave of Traditional Agricultural Practices." Pp. 85-96 in *Themes and Issues of Canadian Geography*, edited by Christoph Stadel and Herman Suida. Salzburg, Austria: University of Salzburg, Salzburger Geographische Arbeiten, Band 28, 1995.

Richer, Nicole. "Etude des perceptions des agriculteurs du bassin versant de la Riviere aux Raisins concernant l'impact de l'agriculture sur la qualite de l'environnement de la region." Master's Thesis, University of Ottawa, Ottawa, Ontario, 1995.

Wandel, Johannes. "An Analysis of Stability and Change in an Old Order Mennonite Farming System in the Waterloo Region, Ontario." Master's Thesis, University of Guelph, Guelph, Ontario, 1995.

Walker, Gerald E. "Sustainable Agriculture and Its Social Geographic Context in Ontario." Pp. 313-328 in *Agricultural Restructuring and Sustainability: A Geographical Perspective*, edited by Brian W. Ilbery, Quentin Chiotti, and Timothy Rickard. Wallington, England: C.A.B. International, 1997.

ENVIRONMENTAL AND CONSERVATION ISSUES

Brookstone, M. "Correlation of Oat Fields with Temperature and Precipitation in Southern Ontario." Master's Thesis, University of Toronto, Toronto, Ontario, 1940.

Tan, Kok-Chiang. "The River Valley Conservation Authority of Southwestern Ontario." Master's Thesis, University of Western Ontario, London, Ontario, 1962.

Brady, Richard F. "An Evaluation of Agricultural Outlet Drainage in Ontario in the Drain 36 Example." Master's Thesis, University of Waterloo, Waterloo, Ontario, 1982.

Caldwell, Wayne J. "Low Agricultural Soil Capability and Farming Practice: A Critical Evaluation of Class 6 Soils Located in Huron and Bruce Counties,

Ontario." Master's Thesis, University of Western Ontario, London, Ontario, 1982.

Gayler, Hugh J. "Conservation and Development in Urban Growth: The Preservation of Agricultural Land in the Urban–Rural Fringe of Ontario." *Town Planning Review*, 53 (1982): 321-341.

Moss, Michael R., and L.S. Davis. "The Potential and Actual Primary Productivity of Southern Ontario's Agro-Ecosytem." *Applied Geography*, 2, No. 1 (1982): 17-37.

Schaus, B.L. "Climatic Influences on Soybean Yields in Ontario: An Examination of Spatial Relationships." Master's Thesis, University of Western Ontario, London, Ontario, 1982.

Bardecki, M.J., and D.R. Banting. "Hydrological Implications of Agricultural Land Clearance." *Papers and Proceedings of the Applied Geography Conferences*, 7 (1984): 246-257.

Dusault, A.J. "Assessment of Soil Degradation on Row Cropped Land in Southwestern Ontario." Master's Thesis, University of Guelph, Gueph, Ontario, 1984.

Yin, Yong-Yuan. "Implications of Soil Erosion for Future Food Production." Master's Thesis, University of Guelph, Guelph, Ontario, 1985.

Ludlow, Laurie L. "Assessing the Implications of Acid Rain for Regional Agricultural Production in Ontario." Master's Thesis, University of Guelph, Guelph, Ontario, 1986.

Ludlow, Laurie L., and Barry E. Smit. "Assessing the Implications of Environmental Change for Agricultural Production: The Case of Acid Rain in Ontario, Canada." *Journal of Environmental Management*, 25, No. 1 (1987): 27-44.

Brklacich, Michael, et al. "Impacts of Environmental Change on Sustainable Food Production in Southwest Ontario." *Land Degradation and Rehabilitation*, 1 (1989): 291-303.

Smit, Barry E., et al. "Sensitivity of Crop Yields and Land Resource Potential to Climatic Change in Ontario, Canada." *Climatic Change*, 14, No. 2 (1989): 153-174.

Charbonneau, S.G. "Preconditional Factors Affecting the Adoption of Conservation Tillage Practices for Elgin, Middlesex, and Oxford Counties." Master's Thesis, University of Windsor, Windsor, Ontario, 1989.

Smithers, J. "Soil Conservation in Agriculture: Farm Level Barriers to Adoption." Master's Thesis, University of Guelph, Guelph, Ontario, 1989.

Brklacich, Michael, and Barry E. Smit. "Implications of Changes in Climatic Averages and Variability on Food Production Opportunities in Ontario, Canada." *Climatic Change*, 20, No. 1 (1992): 1-22.

McNabb, David. "Agricultural Sensitivity to Climate Change and Variability in Southern Ontario." Master's Thesis, University of Guelph, Guelph, Ontario, 1992.

Wilcox, Ian. "Environmental Regulation: An Assessment of the Practice of Agricultural Pollution Control in Ontario." Master's Thesis, University of Western Ontario, London, Ontario, 1992.

McFadden, Nanette. "Agricultural NGO's and Their Application in the Soil Erosion and Water Quality Management Problem in Ontario." Master's Thesis, University of Ottawa, Ottawa, Ontario, 1993.

Spaling, H., and Barry E. Smit. *Agroecosystem Health and Cumulative Effects Assessment: An Application to Land Drainage and Wetlands*. Guelph, Ontario: University of Guelph, Agroecosystem Project, Discussion Paper No. 7, 1994.

Blain, Robert. "Corn Hydrid Selection in Light of Variable Climatic Conditions: A Case Study in Southern Ontario." Master's Thesis, University of Guelph, Guelph, Ontario, 1995.

Blain, Robert, Philip D. Keddie, and Barry E. Smit. *Corn Hybrid Selection Under Variable Climate Conditions: A Case Study in Southern Ontario*. Guelph, Ontario: University of Guelph Department of Geography, Occasional Paper No. 22, 1995.

Couturier, A., and Barry E. Smit. *Spatial Dynamics of Natural Areas as Indicators of Agroecosystem Health*. Guelph, Ontario: University of Guelph, Agroecosystem Health Project, Discussion Paper No. 11, 1995.

Tabayashi, A. "Sustainable Development of Agriculture in Southern Ontario" (in Japanese). *Tsukuba Studies in Human Geography*, 19 (1995): 97-134.

Van Osch, Kenneth. "Considering Our Approach to Planning and Management: Insights from an Examination of Environmental Farm Planning in Ontario." Master's Thesis, University of Waterloo, Waterloo, Ontario, 1995.

Kuisma, C., and Barry E. Smit. *Forest Habitat and Public Policy in Agroecosystems in Puslinde, Ontario*. Guelph, Ontario: University of Guelph, Agroecosystem Health Project, Discussion Paper No. 39, 1996.

Shrubsole, Daniel, and Ian Wilcox. "Assessing a Regulatory Approach to Controlling Agricultural Pollution: The Experience of Ellice Township, Ontario, Canada." *Applied Geography*, 16, No. 2 (1996): 123-136.

Bradshaw, Benjamin, and Barry Smit. "Subsidy Removal and Agroecosystem Health." *Agriculture, Ecosystems and Environment*, 64 (1997): 245-260.

Brklacich, Michael, et al. "Adaptibility of Agricultural Systems to Global Climate Change: A Renfrew County, Ontario, Canada Pilot Study." Pp. 185-200 in *Agricultural Restructuring and Sustainability: A Geographical Perspective*, edited by Brian W. Ilbery, Quentin Chiotti, and Timothy Rickard. Wallingford, England: C.A.B. International, 1997.

Blay-Palmer, Alison. "Technological Innovation as a Climate Adaptation Strategy for Soybean Production in Ontario." Master's Thesis, University of Guelph, Guelph, Ontario, 1999.

Vandergaast, Gary. "Opportunities for Reducing Greenhouse Gas Emissions by Using Ethanol-Based Fuels: An Eastern Ontario (Canada) Case Study of the Potential for Grain Corn Production." Master's Thesis, Carleton University, Ottawa, Ontario, 1999.

Smithers, J., and Alison Blay-Palmer. "Technology Innovation as a Strategy for Climate Adaptation in Agriculture." *Applied Geography*, 21, No. 2 (2001): 175-197.

IRRIGATION

MacDonald, K.I. "Irrigation Water Use in Southern Ontario." Master's Thesis, Wilfrid Laurier University, Waterloo, Ontario, 1987.

AGRICULTURAL MARKETING AND DISTRIBUTION

Dilamarter, Dale F. "The Spatial Impact of a Marketing Board on Winter Wheat Production in Ontario." Master's Thesis, York University, North York, Ontario, 1971.

McIlwraith, Thomas F., Jr. "Changing Patterns of Upper Canada's Wheat Trade, 1815–1860." Ph.D. Dissertation, University of Wisconsin–Madison, 1973.

Gayler, Hugh J. "The Canada–U.S. Free Trade Agreement and the Niagara Fruit Farmer: Issues in the Current Debate." Pp. 178-195 in *L'evolution de l'utilization agricole des terres: Bilans et perspectives*, edited by J. Rouffignat. Ste. Foy, Quebec: Universite Laval, Department de Geographie, 1991.

TRANSPORTATION AND MECHANIZATION

Moyo, S.V. "A Characterization of Farmer Responses Towards Agricultural Machinery Innovation: The Case of Selected Cash Crop and Livestock Enterprises in Middlesex County, 1977." Master's Thesis, University of Western Ontario, London, Ontario, 1979.

AGRICULTURAL LABOR STUDIES

Fitzsimmons, John G. "Agricultural Labour Imput Allocation and Efficiency in North Grimsby Township." Master's Thesis, McMaster University, Hamilton, Ontario, 1969.

Benson, Richard. "Multiple Job Holding Among Farm Operations in the Rainy River District." Master's Thesis, University of Guelph, Guelph, Ontario, 1978.

Morse, R. "Seasonal Agricultural Employment in the Flue-Cured Tobacco Farming Industry of Southern Ontario." Master's Thesis, University of Guelph, Guelph, Ontario, 1985.

Smit, Barry E., Thomas R.R. Johnston, and R. Morse. "Labour Turnover on Flue-Cured Tobacco Farms in Southern Ontario." *Agricultural Administration*, 20, No. 3 (1985): 153-168.

Quebec

ATLASES AND GRAPHIC PRESENTATIONS

Atlas du Quebec: Agriculture. Quebec, Quebec: Bureau de recheches economiques, Ministere de l'Industrie et du Commerce de Quebec, 1966.

Atlas l'agro-alimentaire au Quebec. Quebec, Quebec: Ministere de l'agriculture, Service des etudes economiques, 1978.

GENERAL TEXTS, ARTICLES, AND THESES

Peattie, Roderick. "Farms and Farming in the Lower St. Lawrence Valley." *Journal of Geography*, 21, No. 5 (1922): 174-179.

Jones, Robert L. "French-Canadian Agriculture in the St. Lawrence Valley, 1815–1850." *Agricultural History*, 16, No. 3 (1942): 137-148.

Jones, Robert L. "The Agricultural Development of Lower Canada, 1850–1867." *Agricultural History*, 19, No. 4 (1945): 212-224.

Parker, W.H. "The Revolution in the Agricultural Geography of Lower Canada, 1833–1838." *Revue Canadienne de Geographie*, 11, No. 4 (1957): 189-194.

Pelletier, Raymond. "Evolution agricole de l'ile Jesus." *Revue Canadienne de Geographie*, 16, No. 1-4 (1962): 31-48.

Gagnon, Jacques. "Une economie agricole rentable dans une region marginal: L'exemple de Saint-Ubald, comte de Portneuf." *Cahiers de Geographie de Quebec*, 13, No. 28 (1969): 109-113.

Beauregard, Ludger. "Les etapes de la mise en valeur agricole de la vallee du Richelieu." *Cahiers de Geographie de Quebec*, 14, No. 32 (1970): 171-214.

Gauthier, Majella J. *L'agriculture au Lac-Saint-Jean (PQ): Etude geographique*. Chicoutimi, Quebec: Universite du Quebec a Chicoutimi, Travaux geographique du Saguenay, No. 1, 1975.

Parson, Helen E. "The Rise and Fall of Farming in a Marginal Area: The Gatineau Valley, Quebec." *Cahiers de Geographie de Quebec*, 19, No. 48 (1975): 573-582.

Dumas-Rousseau, Michele. *Les regions agricoles du Quebec*. Quebec, Quebec: Gouvernment du Quebec, Ministere de l'agriculture, 1977.

Courville, Serge. "La crise agricole du Bas-Canada, elements d'une reflexion geographique." *Cahiers de Geographie de Quebec*, 24, No. 62 (1980): 193-224.

Courville, Serge. "Le crise agricole du Bas-Canada, elements d'une reflexion geographique (deuxieme partie)." *Cahiers de Geographie de Quebec*, 24, No. 63 (1980): 385-428.

Smith, William. "A Dispersed or Concentrated Agriculture? The Role of Public Policy in Quebec." *Proceedings, New England–St. Lawrence Valley Division of the Association of American Geographers*, 11 (1981): 88-94.

Brunet, Yves, and Yolande Lepene. "Les variations spatiales de la presence exurbaine en milieu rural Quebecois." Pp. 138-159 in *The Pressures of Change in Rural Canada*, edited by Michael F. Bunce and Michael J. Troughton. North York, Ontario: York University Geographical Monograph No. 14, 1984.

Laverture, Robert. "L'histoire de l'agriculture quebecois au XIXe siecle: Une schematisation des faits et des interpretations." *Cahiers de Geographie de Quebec*, 28, No. 73-74 (1984): 275-287.

Smith, William. "The 'Vortex Model' and the Changing Agricultural Landscape of Quebec." *Canadian Geographer*, 28, No. 4 (1984): 358-372.

Smith, William. "Agricultural Change in Quebec." Pp. 54-72 in *The Pressures of Change in Rural Canada*, edited by Michael F. Bunce and Michael J.

Troughton. North York, Ontario: York University Geographical Monograph No. 14, 1984.

Desy, J., and C. Gagnon. *La developpement agro-frontier au Quebec et en Finlande*. Chicoutimi and Montreal, Quebec, Quebec: Cahiers de l'ACFAS, No. 39, 1985.

Hamelin, Louis-Edmond. "Structures agraires du Quebec." Pp. 301-318 in *Estudios em Homenagema Mariano Feio*, edited by Raquel Soeiro de Briton. Lisbon, Portugal: Instituto Nacional de Investgeao Cientifica, 1986.

Courville, Serge. "Le marche des subsistence: L'agriculture de la plaine de Montreal au debut des annees 1830: Une perspective geographique." *Revue d'Histoire de l'Amerique Francaise*, 42, No. 2 (1988): 193-239.

Martin, Jean. "De l'agriculture a l'industrie: Les communautes de scieurs au Saguenay, 1840–1880." Master's Thesis, Universite Laval, Ste. Foy, Quebec, 1990.

Bryant, Christopher R., Claude Marois, and Pierre Deslauriers. "L'agriculture dans la region metropolitaine de Montreal." Pp. 215-236 in *Le Quebec rural dans tous ses etates*, edited by B. Vachon. Montreal, Quebec: Editions Boreal, 1991.

Deslauriers, Pierre, Christopher R. Bryant, and Claude Marois. "Les processes agissant sur l'agriculture peri-urbaine montrealaise depuis 1980." Pp. 211-238 in *Rural and Urban Fringe Studies in Canada*, edited by Kenneth B. Beesley. North York, Ontario: York University Geographical Monograph No. 21, 1991.

Deslauriers, Pierre, Christopher R. Bryant, and Claude Marois. "Agriculture in Montreal's Urban Field: Recent Evolution and Future Prospects." *Proceedings, New England–St. Lawrence Valley Division of the Association of American of Geographers*, 21 (1991): 17-26.

Marois, Claude, Pierre Deslauriers, and Christopher R. Bryant. "Changements structurels dans l'agriculture de la rive-sud de Montreal, 1981–1986." Pp. 177-126 in *Montreal: Tablaux d'un espace en transformation*, edited by Frank Remiggi. Montreal, Quebec: ACFAC, Cahiers Scientifique No. 76, 1992.

Andrem, Pierre, et al. "Evolution de l'agriculture maraicheve au Quebec de 1940 a 1990: Superficies, productions, et diversite." *Canadian Geographer*, 42, No. 3 (1998): 293-305.

Beauchesne, Audrie, and Christopher Bryant. "Agriculture and Innovation in the Urban Fringe: The Case of Organic Farming in Quebec, Canada." *Tijdschrift voor Economische en Sociale Geographier*, 90, No. 3 (1999): 320-328.

CROPS AND CROPPING PATTERNS

Robert, Benoit. "Les jardiniers-maraichers de la Cote de Beauport." Master's Thesis, Universite Laval, Ste. Foy, Quebec, 1967.

Robert, Benoit. "Les cultures legumieres de la Cote de Beauport: Leurs liens avec le structure agaire et l'urbanisation." *Cahiers de Geographie de Quebec*, 16, No. 37 (1972): 31-56.

Charette, Roxanne S. "The Quebec Orchard Industry Response to the Low-Temperature Hazard." Master's Thesis, University of Waterloo, Waterloo, Ontario, 1988.

De Koninck, Rodolphe. "La vigne et la vin du Quebec: Bon gout et tenacite vigneronne." *Cahiers de Geographie de Quebec*, 37, No. 100 (1993): 79-114.

Deshaies, Laurent, and Jean-Marie Dubois. "La viti-viniculture quebecois: entre 'froidure' en vente au vignoble." *Annales de Geographie*, 105, No. 6 (1996): 603-619.

Dubois, Jean-Marie, and Laurent Deshaies. *Guide des vignobles du Quebec.* Quebec, Quebec: Les Presses de l'Universite Laval, 1997.

ANIMAL HUSBANDRY

Findlay, Marjorie C. "The Possibilities of Sheep Raising in the Ungava Bay Area." *Canadian Geographer*, 2, No. 10 (1957): 20-28.

Ouellet, Luc. "Aspects geographiques de l'elevage ovin au Quebec." Master's Thesis, Universite Laval, Ste. Foy, Quebec, 1968.

De Koninck, Rodolphe, A.M. Turcot, and Andree G. Zubrzycki. "Les pasturages communaux du Lac Saint-Pierre: De leur histoire et de leur actualite." *Cahiers de Geographie de Quebec*, 17, No. 41 (1973): 317-329.

FARM LAND

Pelletier, Raymond. "L'utilization agricole du sol de l'Ile Jesus." Ph.D. Dissertation, Universite de Montreal, Montreal, Quebec, 1962.

Clibbon, Peter B. "Changing Land Use in Terrebonne County, Quebec." *Cahiers de Geographie de Quebec*, 8, No. 15 (1964): 5-40.

Clibbon, Peter B. "Utilisation du sol et colonisation de la region des Laurentides centrales." *Geographical Bulletin*, 21 (1964): 5-20.

Maillous, A., A. Dube, and L. Tardiff. "Classement des sols selan leurs possibilities d'utilisation agricole." *Cahiers de Geographie de Quebec*, 8, No. 16 (1964): 231-250.

Clibbon, Peter B., and Jacques Gagnon. "L'evolution recente de l'utilisation du sol sur la rive nord de Saint-Laurent entre Quebec et Montreal." *Cahiers de Geographie de Quebec*, 10, No. 19 (1965–66): 55-72.

Clibbon, Peter B. *Structure and Dynamics of Land Use.* Montreal, Quebec: Les Presses de l'Universite de Montreal, 1975.

Drummond, R.N., and D. Graham. "Changes in Land Use on Either Side of the Quebec-Vermont Border, 1945–1961; 1961–1971." *Proceedings, New England–St. Lawrence Valley Division of the Association of American Geographers*, 11 (1981): 62-72.

Desy, J., and M. Mory. "Le projet d'autoraute 78 Alma-La Baie Effets, structurants et impacts previsibles sur les terres agricole, le milieu urbain et periurbain." *Le Sagamien*, 3, No. 7 (1982).

Lamoureux, D. *The Abandonment of Agricultural Land in Gaspe, Quebec: The Causes and the Impacts on Land Use.* Ottawa, Ontario: Environment Canada, Lands Directorate, Working Paper No. 29, 1984.

Tremblay, Claude. "L'utilisation agricole des sols: Relations entre des verites-terrain et des valeurs densitometriques d'images aeriannes infrarouge-coulese." Master's Thesis, Universite de Ottawa, Ottawa, Ontario, 1987.

Dupuis, Michel. "L'apport de l'humidite du sol dans la modelisation des valeurs relatives d'images radar en bande C: Le cas de sols couverts en milieu agricole." Master's Thesis, Universite de Sherbrooke, Sherbrooke, Quebec, 1988.

Rouffignat, J., ed. *L'evolution de l'utilisation des terres agricoles: Bilon et perspectives.* Quebec, Quebec: Departement de geographie, Universite Laval. Notes et documents de recherche, No. 29, 1991.

Dugas, Alain, and Denis-R. Morin. "Evaluation de quelques sources d'erreur dans un inventaire semi-automatise de l'utilisation agricole du sol: Etude de cas a Sainte-Foy." *Cahiers de Geographie de Quebec*, 37, No. 100 (1993): 35-48.

Jean, Bruno. "Terr, territoire, territorialite: Les agriculteurs et leur attachement au territoire." *Cahiers de Geographie de Quebec*, 37, No. 101 (1993): 291-308.

Lemire, Daniel. "L'impact des decisions des exploitants agricoles sur l'amenagement du territoire et le redeveloppement de l'agriculture: Le cas de la frange periurbaine de Montreal." Master's Thesis, Universite de Montreal, Montreal, Quebec, 1994.

McCallum, Charlotte A. "The Effect of Quebec's Agricultural Preservation Law on Agriculture and Rural Land Use Near Sherbrooke." Master's Thesis, Carleton University, Ottawa, Ontario, 1994.

Senecal, Gilles, Marcel Gaudreu, and Serge Des Roches. "Les mecanismes de production de la farm urbaine et la conservation des espaces agricoles et naturels dans la region de Montreal: Le case de Laval." *Cahiers de Geographie de Quebec*, 38, No. 105 (1994): 301-326.

AGRICULTURAL POPULATIONS AND SOCIAL ISSUES

Blanchard, Raoul. "Les excedents de population et l'agriculture dans la province de Quebec." *L'Actualite Economique*, 24 (1949): 635-641.

Raveneau, Jean. "Densite agricole et change humaine des terroirs: Essai methodologique avec application a quatre comtes au sud de l'estuaire du Saint-Laurent." *Cahiers de Geographie de Quebec*, 9, No. 17 (1964–1965): 11-24.

Lepine, Yolande. "Forces sociales et forces de production dans les terres noires de Napierville-Chateauguay." *Cahiers de Geographie de Quebec*, 17, No. 42 (1973): 389-414.

Courville, Serge. "Villages and Agriculture in the Seigneuries of Lower Canada: Conditions of a Comprehensive Study of Rural Quebec in the First Half of the Nineteenth Century." Pp. 121-149 in *Canadian Papers in Rural History*, vol. 5, edited by Donald Akenson. Gananoque, Ontario: Langdale, 1986.

Rialland, C. "La releve agricole du Quebec. Approche geographique et sociale." *Norois*, 41, No. 161 (1994): 169-182.

ENVIRONMENTAL AND CONSERVATION ISSUES

Lotz, J.R. "Soils and Agricultural Possibilities of the Knob Lake Area, Quebec." Master's Thesis, McGill University, Montreal, Quebec, 1957.

Bureau, Luc. "Un exemple d'adaptation de l'agriculture a des conditions ecologiques en apparence hostiles: L'Aloctiere de Lemieux." *Cahiers de Geographie de Quebec*, 14, 33 (1970): 383-394.

Dugan, Clermont. "Climat et agriculture au sud de la Baie de James et dans une partie de l'enclave argileuse de l'Ojibway (le climat et san influence sur l'agriculture abitibienne)." Master's Thesis, Universite Laval, Ste. Foy, Quebec, 1973.

Gervais, Renald, and Alfred Jaquicle. "L'Utilisation agricole de terres en friche en milieu periurbain quebecois: Les cas de Laval." *Cahiers de Geographie de Quebec*, 28, No. 75 (1984): 365-393.

Singh, Bhawan, et al. "Influence d'un changement climatique du a une hausse de gaz a effet de serveseur l'agriculture au Quebec." *Atmosphere Ocean*, 34, No. 2 (1996): 379-399.

Singh, Bhawan, et al. "Impacts of a GHG-Induced Climate Change on Crop Yields: Effects of Acceleration in Maturation, Moisture Stress, and Optimal Temperature." *Climatic Change*, 38, No. 1 (1998): 51-86.

AGRICULTURAL MARKETING AND DISTRIBUTION

Smith, William. "Agricultural Market Functions and Land Use Change: A Quebec Case Study." Master's Thesis, McGill University, Montreal, Quebec, 1972.

Smith, William. "Market-Farm Linkages and Land Use Change: A Quebec Case Study." *Cahiers de Geographie de Quebec*, 18, No. 44 (1974): 297-316.

Vachon, B. "Quelques aspects geographiques financiers et politiques du zonage agricole au Quebec." *Recherches Sociographiques*, 29, No. 2-3 (1988): 417-430.

TRANSPORTATION AND MECHANIZATION

Beaudoin, Dollard. "Motorisation et geographie agraire dans Dorchester." Master's Thesis, Universite Laval, Ste. Foy, Quebec, 1973.

Bouchard, Gerard. "L'agriculture saguenayenne entre 1840 et 1950: L'evolution de la technologie." *Revue d'histoire de l'Amerique francaise*, 43, No. 3 (1990): 353-380.

AGRICULTURAL LABOR STUDIES

Nadeau, Jean. "Soumission (de la) du travailleur de la terre a l'alienation du territoire agricole: Le cas de l'agriculture quebecois." Master's Thesis, Universite Laval, Ste. Foy, Quebec, 1984.

Western Canada

GENERAL TEXTS, ARTICLES, AND THESES

Schott, Carl. "Wandlungen der Landwirtschaft in der Kanadischen Prarieprovinzen." Pp. 83-95 in *Tagungsberichte und Wissenschaftlichen Abhandlungen des Deutschen Geographentages Essen 1953*. Wiesbaden, Germany: Franz Steiner Verlag, 1955.

Proudfoot, Bruce. "Agriculture." Pp. 51-64 in *The Prairie Provinces*, edited by P.J. Smith. Toronto, Ontario: University of Toronto Press, 1972.

Todd, Daniel, and John S. Brierley. "Farm Size and Regional Structure: A Preliminary Insight." *Environment and Planning A*, 9, No. 1 (1977): 75-84.

Moodie, D. Wayne. "Agriculture and the Fur Trade." Pp. 272-292 in *Old Trails and New Directions*, edited by Arthur J. Ray and Carol Judd. Toronto, Ontario: University of Toronto Press, 1980.

CROPS AND CROPPING PATTERNS

Lustig, Marie E. "The Northern Limit of Crop Production in Western Canada." Master's Thesis, University of Maryland, College Park, Md., 1946.

Proudfoot, B. "Toward a Rational Wheat Policy." Pp. 265-272 in *Proceedings, Canadian Association of Geographers*. Winnipeg, Manitoba: University of Manitoba, Department of Geography, 1970.

Warkentin, John H. "Dry Farmers on the Canadian Plains." *Geographical Magazine*, 45, No. 6 (1973): 443-450.

Keddie, Philip D. "The Changing Varietal Composition of Prairie Wheat Production, 1941–1985." *Prairie Forum*, 15, No. 1 (1990): 103-122.

Rumney, Thomas A. "Canola in Prairie Canada: Its Development as a Regional Crop and Its Challenge to Wheat." *Geographical Bulletin*, 43, No. 1 (2001): 28-35.

ANIMAL HUSBANDRY

Brayer, Herbert O. "British Cattle Ranches in America." *Geographical Magazine*, 20, No. 4 (1947): 125-131.

Evans, Simon M. "Stocking the Canadian Range." *Alberta History*, 26, No. 3 (1978): 1-8.

Evans, Simon M. "Canadian Beef for Victorian Britain." *Agricultural History*, 53, No. 4 (1979): 748-762.

Evans, Simon M. "The Origin of Ranching in Western Canada: American Diffusion or Victorian Transplant?" *Great Plains Quarterly*, 3, No. 2 (1983): 79-91.

Evans, Simon M. "The End of the Open Range Era in Western Canada." *Prairie Forum*, 8, No. 1 (1983): 71-87.

Evans, Simon M. "Grazing the Grasslands: Exploring Conflicts, Relationships, and Futures." *Prairie Forum*, 26, No. 1 (2001): 67-84.

AGRICULTURAL POPULATIONS AND SOCIAL ISSUES

Cobb, Douglas. "The Jamesville Bruderhof: A Hutterite Agricultural Colony." *Journal of the West*, 9, No. 1 (1970): 60-77.

Evans, Simon M. "American Cattlemen on the Canadian Range, 1874 to 1914." *Prairie Forum*, 4, No. 1 (1974): 121-135.

Katz, Yossi and John C. Lehr. "Jewish Pioneer Agricultural Settlements in Western Canada." *Journal of Cultural Geography*, 14, No. 1 (1993): 49-68.

ENVIRONMENTAL AND CONSERVATION ISSUES

Williams, G.D.V. "Applying Estimated Temperature Normals to the Zonation of the Canadian Great Plains for Wheat." *Canadian Journal of Soil Science*, 49 (1969): 263-276.

Pierce, John T., and R.J. Stathers. "Environmental Impacts on Cereal Grain Production in the Canadian Prairie Provinces: A Growth Constraint Model." *Agriculture, Ecosystems and Environment*, 21, No. 3-4 (1988): 225-243.

Ash, G.H.B, C.F. Shaykewich, and R. Raddatz. "Agroclimate Update for Manitoba and Saskatchewan." *Manitoba Agri-Forum* (1991): 20-25.

Ash, G.H.B., C.F. Shaykewich, and R. Raddatz. *Agricultural Climate for the Eastern Prairies*. Winnipeg, Manitoba: Province of Manitoba Department of Agriculture, 1992.

Ash, G.H.B., C.F. Shaykewich, and R. Raddatz. "Moisture Risk Assessment for Spring Wheat in the Eastern Prairies." Pp. 35-44 in *Soil Moisture: Modelling and Monitoring for Regional Planning*, edited by F.J. Hey, R. Gregor and L. Martin. Saskatoon, Saskatchewan: National Hydrology Research Center, Environment Canada, 1992.

Ash, G.H.B., C.F. Shaykewich, and R. Raddatz. "Moisture-Risk Assessment for Spring Wheat on the Eastern Prairies: A Water-Use Simulation Model." *Climatological Bulletin*, 26, No. 2 (1992): 65-78.

Pronovost, C., and M. Bouchard. "Influences agro-pastorales sur les caracteristiques physio-chimiques des eaux de la riviere l'Assomption (Canada) et de ses principaux affluents." Pp. 535-549 in *Farm and Erosion in Temperate Plains Environments*, edited by S. Wicherek. Amsterdam, Netherlands: Elsevier, 1992.

Sauchyn, D.J. "GIS Modelling of Agricultural Soil Loss in a Plains Glacial Landscape." *Quaternary International*, 20 (1993): 163-170.

Brklacich, Michael, and R. Stewart. "Impacts of Climate Change on Wheat Yields in the Canadian Prairies." Pp. 147-162 in *Climate Change and Agriculture: Analysis of Potential International Impacts*, edited by C. Rosenzweig and A. Iglesias. Madison, Wis.: American Society of Agronomy, ASA Special Publication No. 59, 1995.

Herman, Laura. "The Potential of Organic Farming to Enhance Soil Conservation in Saskatchewan." Master's Thesis, University of Regina, Regina, Saskatchewan, 1996.

Carlyle, William J. "The Decline of Summer Fallows on the Canadian Prairies." *Canadian Geographer*, 41, No. 3 (1997): 267-280.

IRRIGATION

Seaborne, A.A., and R.L. Blackwell. "Patterns of Irrigation Development in the South Saskatchewan River Project." *Prairie Forum*, 18, No. 1 (1993): 97-116.

TRANSPORTATION AND MECHANIZATION

Schott, Carl. "Die Auswirkungen der technischen Revolution in der Landwirtschaft nach 1945 auf die landlichen Siedlungen Oskanadas." Pp. 89-110 in *Beitrage zur Kulturgeographie von Kanada*, edited by Carl Schott. Marburg, Germany: Marburger Geographische Schriften, Heft 66, 1978.

Manitoba

GENERAL TEXTS, ARTICLES, AND THESES

Fordham, R.C. "The Structure of Manitoba's Agricultural Geography 1951–1964: Its Significance in Relation to Projected Future Trends." Master's Thesis, University of Manitoba, Winnipeg, Manitoba, 1966.

Stadel, Christoph, John C. Everitt, and G. Bugg. "A Typology of Agriculture in the Vicinity of Brandon, Manitoba." *Regina Geographical Studies*, 5 (1988): 24-41.

Carlyle, William J. "Agriculture in Manitoba." Pp. 219-236 in *The Geography of Manitoba: Its Lands and Its Peoples*, edited by John Welsted, John C. Everitt, and Christoph Stadel. Winnipeg, Manitoba: University of Manitoba Press, 1996.

CROPS AND CROPPING PATTERNS

Pattle, Calvin S. "A Geographic Analysis of the Sunflower Industry in Manitoba." Master's Thesis, University of Manitoba, Winnipeg, Manitoba, 1972.

Kaye, Barry. "The Settlers' Grand Difficulty: Haying in the Economy of the Red River Settlement." *Prairie Forum*, 9, No. 1 (1984): 1-12.

Armitage, David G. "Forage Production in the Interlake Region of Manitoba." Master's Thesis, University of Manitoba, Winnipeg, Manitoba, 1990.

ANIMAL HUSBANDRY

Payne, C.H. "Northern Animal Husbandry: A Land-Use for Northern Manitoba." Master's Thesis, University of Manitoba, Winnipeg, Manitoba, 1977.

FARM LAND

Todd, Daniel. "Regional and Structural Factors in Farm-Size Variation: A Manitoba Elucidation." *Environment and Planning A*, 11, No. 3 (1979): 257-269.

Todd, Daniel, and John S. Brierley. "Agricultural Land-Use Policies in Prairie Manitoba, and the Farm-Size Phenomena." *GeoJournal*, 6, No. 6 (1982): 525-532.

Lessier, T. "An Analysis of Farmland Ownership Versus Rental/Leasing and Management Strategies of Part-Owners in Selected Areas of Southern Manitoba." Master's Thesis, University of Manitoba, Winnipeg, Manitoba, 1992.

AGRICULTURAL POPULATIONS AND SOCIAL ISSUES

Ryan, John. "The Agricultural Operations of Manitoba Hutterite Colonies." Ph.D. Dissertation, McGill University, Montreal, Quebec, 1973.

DeLisle, David. "The Spatial Organization and Intensity of Agriculture in the Mennonite Villages of Southern Manitoba." Ph.D. Dissertation, McGill University, Montreal, Quebec, 1975.

Warouchak, Leonard B. "Agricultural and Rural Non-Farm Residential Development in the Winnipeg Region (Case Study, Rockwood): 1965 to 1985." Master's Thesis, University of Waterloo, Waterloo, Ontario, 1979.

Wiesinger, Judith P. "Agricultural Settlement in Southern Manitoba, 1872 to 1891: A Process-Form Approach." Master's Thesis, University of Manitoba, Winnipeg, Manitoba, 1981.

Wong, Fung K.K. "Determinants of Rural Out-Migration: Agro-Manitoba, 1961–1981." Master's Thesis, University of Manitoba, Winnipeg, Manitoba, 1988.

Lehr, John C. "Agricultural Settlement of German-Speaking Groups in Southern Manitoba." Pp. 185-197 in *Themes and Issues of Canadian Geography*, Volume 1, edited by Christoph Stadel and Herman Suida. Salzburg, Austria: University of Salzburg, Salzburger Geographische Arbeiten, Band 28, 1995.

ENVIRONMENTAL AND CONSERVATION ISSUES

Dunlop, Susan L. "An Agroclimatology of Southern Manitoba." Master's Thesis, University of Manitoba, Winnipeg, Manitoba, 1981.

Halpin, Janel I. "The Climatic Environment for Agriculture in the North: A Case Study of the Pasquia Land Settlement Project in Northern Manitoba." Master's Thesis, University of Manitoba, Winnipeg, Manitoba, 1985.

Gauthier, D.A., R. Widdis, and M. Lewry. *An Historical Ecological Review of Agricultural Activities in Grasslands National Park and Region 1870–1930*. Winnipeg, Manitoba: Report to Parks Canada, Prairie and Northern Region, 1993.

Saskatchewan

ATLASES AND GRAPHIC PRESENTATIONS

Fung, Ka-ju, and Stuart H. Gage. *Atlas of Saskatchewan Agriculture*. Saskatoon, Saskatchewan: University of Saskatchewan Press, 1978.

GENERAL TEXTS, ARTICLES, AND THESES

Indome, Alberta. "Changes in Farmers' Operations in Southern Saskatchewan Since 1970." Master's Thesis, University of Regina, Regina, Saskatchewan, 1999.

DAIRYING

Hardy, John A. "The Dairy Industry in Saskatchewan." Master's Thesis, University of Saskatchewan, Saskatoon, Saskatchewan, 1971.

FARM LAND

Davidson, Alex T. "Forest Versus Agricultural Land Use Along the Pioneer Fringe in Saskatchewan." Master's Thesis, University of Toronto, Toronto, Ontario, 1952.

AGRICULTURAL POPULATIONS AND SOCIAL ISSUES

Quershi, Anwar S. "Local Variation in the Rural Farm Depopulation in Southwestern Saskatchewan." Master's Thesis, University of Regina, Regina, Saskatchewan, 1977.

Clyde, Ramond E. "A Study of Saskatchewan Farmers' Perception of Hail, Drought, and Flood as Hazards to Agriculture." Master's Thesis, University of Regina, Regina, Saskatchewan, 1981.

Patterson, David. "A Study of Farmers' Perception of Land Quality Farm Practices and Government Programs as They Relate to Soil Erosion in Southwestern Saskatchewan." Master's Thesis, University of Regina, Regina, Saskatchewan, 1992.

Irwin, Robert. "The Better Sense of the Farm Population: The Partridge Plan and Grain Marketing in Saskatchewan." *Prairie Forum*, 18, No. 1 (1993): 35-52.

Bowen, Dawn. "Forward to a Farm: Land Settlement as Unemployment Relief in the 1930s." *Prairie Forum*, 20, No. 2 (1995): 207-230.

ENVIRONMENTAL AND CONSERVATION ISSUES

Sutherland, R.A., and E. Dejong. "Estimation of Sediment Redistribution Within Agricultural Fields Using Caesium-137, Crystal Springs, Saskatchewan, Canada." *Applied Geography*, 10, No. 3 (1990): 205-222.

Ash, G.H.B. "An Agroclimatic Risk Assessment of Southern Manitoba and Southeastern Saskatchewan." Master's Thesis, University of Manitoba, Winnipeg, Manitoba, 1991.

Alberta

GENERAL TEXTS, ARTICLES, AND THESES

Hozaack, W.J. "The Spatial Pattern of Farming Near Lac la Biche, Alberta." Master's Thesis, University of Alberta, Edmonton, Alberta, 1969.

Pich, G. "Agricultural Innovations, County of St. Paul, Alberta." Unpublished Master's Thesis, University of Alberta, Edmonton, Alberta, 1970.

Wonder, William C. "The Toast at the Punch Bowl." *Alberta Historical Review*, 22, No. 4 (1974): 26-30.

Irby, Charles C. "Northeast Alberta: A Marginal Agricultural Situation." Ph.D. Dissertation, Simon Fraser University, Burnaby, British Columbia, 1978.

Hamley, Will. "The Farming Frontier in Northern Alberta." *Geographical Journal*, 158, No. 3 (1992): 286-294.

Enright, P. "Agri-Processing and Rural Development: Locational Trends, Public Policy and Local Influences in Alberta, 1975–1991." Master's Thesis, University of Alberta, Edmonton, Alberta, 1994.

Chiotti, Quentin, et al. "Agricultural Response to Change: A Preliminary Investigation of Farm-Level Adaptation in Southern Alberta." Pp. 201-218 in *Agricultural Restructuring and Sustainability: A Geographical Perspective*, edited by Brian W. Ilbery, Quentin Chiotti, and Timothy Rickard. Wallingford, England: C.A.B. International, 1997.

Flower, David J. "Survival and Adaptation: An Analysis of Dryland Farming in the 1940s and 1950s in South East Alberta." Ph.D. Dissertation, University of Alberta, Edmonton, Alberta, 1997.

CROPS AND CROPPING PATTERNS

Skretkowicz, A.L., William G. Bailey, and P.F. Mills. "Legume Production and Efficiency in the Southern Peace River Region." *Western Geography*, 1, No. 1 (1990): 24-33.

ANIMAL HUSBANDRY

Lupton, Austin A. "Cattle Ranching in Alberta, 1874–1910: Its Evolution and Migration." *Albertan Geographer*, 3 (1966–1967): 48-58.

Kienholz, E. "Seasonal Cattle Movements in South-West Alberta." Master's Thesis, University of Alberta, Edmonton, Alberta, 1974.

Evans, Simon. "Spatial Aspects of the Cattle Kingdom: The First Decade." Pp. 41-56 in *Frontier Calgary: Town, City and Region 1875–1914*, edited by Anthony W. Rasporich and Henry C. Klassen. Calgary, Alberta: University of Calgary Press, 1975.

DAIRYING

Lupton, Austin A. "Some Geographic Aspects of Dairying in Alberta." Master's Thesis, University of Alberta, Edmonton, Alberta, 1965.

Sundstrom, Marvin. "Adjustments in the Primary Sector of Alberta's Dairy Industry." *B.C. Geographical Series*, 28 (1979): 49-63.

FARM LAND

Singh, J. "Land Use Change in the Eastern Irrigation District of Alberta." Master's Thesis, University of Alberta, Edmonton, Alberta, 1968.

Lupton, Austin A. "An Analysis of Some Factors Affecting the Areal Variability of Farm Size in Alberta." Ph.D. Dissertation, University of Alberta, Edmonton, Alberta, 1969.

Boylen, D. "Farm Land Tenure in Alberta: Attitudinal and Socio-Economic Variables." Master's Thesis, University of Alberta, Edmonton, Alberta, 1976.

AGRICULTURAL POPULATIONS AND SOCIAL ISSUES

Tracie, Carl J. "Agricultural Settlement in the South Peace River Area." Master's Thesis, University of Alberta, Edmonton, Alberta, 1967.

Tracie, Carl J. "An Analysis of the Three Variables Affecting Farm Location in the Process of Agricultural Settlement: The South Peace River Area." Ph.D. Dissertation, University of Alberta, Edmonton, Alberta, 1970.

Vogelsang, Roland R. "The Initial Agricultural Settlement of the Morinville-Westlock Area, Alberta." Master's Thesis, University of Alberta, Edmonton, Alberta, 1972.

Paterson, John L. "Religious Beliefs, Environmental Stewardship and Industrializing Agriculture: The Christian Farmers' Federation of Alberta." Pp. 64-68 in *Proceedings of the Fifteenth New Zealand Geography Conference*, edited by R. Welch. Dunedin, New Zealand: New Zealand Geographical Society, 1989.

ENVIRONMENTAL AND CONSERVATION ISSUES

Hayter, Roger. "The Frost Hazard for Agriculture in Northwest Alberta." Master's Thesis, University of Alberta, Edmonton, Alberta, 1970.

Jansson, M.C. "Farmer Response to Depredation by Wildlife on Agriculture in the Athabasca Area." Master's Thesis, University of Alberta, Edmonton, Alberta, 1970.

Nkemdirim, Lawrence C. "Crop Development and Water Loss: A Case Study Over a Potato Crop." *Agricultural Meteorology*, 16 (1976): 371-388.

Davis, Roderick L. "Pasture Productivity and Climate in the Peace River Region." Master's Thesis, McMaster University, Hamilton, Ontario, 1979.

Hayter, Roger. "Frost Hazard in Northeast Central Alberta." Pp. 139-162 in *Perspectives in Agricultural Geography*, vol. 2, edited by Noor Mohammad. New Delhi, India: Concept Publishing Company, 1981.

Clark, Jan K. "The Effect of Cultivation on Selected Soils in the Special Area of Alberta." Master's Thesis, University of Calgary, Calgary, Alberta, 1986.

Macquarrie, Pamela. "Evapotranspiration in an Irrigated Wheat Crop." Master's Thesis, University of Calgary, Calgary, Alberta, 1988.

Hooey, C.A. "The Perception of Climatic Variability and Its Role in Agricultural Decision-Making in West Central Alberta." *Papers and Proceedings of the Applied Geography Conferences*, 21 (1998): 148-154.

AGRICULTURAL MARKETING AND DISTRIBUTION

Backhouse, Norman A. "Optimization of Grain Elevator Service Areas." Master's Thesis, University of Calgary, Calgary, Alberta, 1973.

TRANSPORTATION AND MECHANIZATION

Brown, S.A. "The Impact of the Great Slave Lake Railway on Agricultural Land Use in the North Peace, Alberta." Master's Thesis, University of Alberta, Edmonton, Alberta, 1971.

British Columbia

GENERAL TEXTS, ARTICLES, AND THESES

Krenzlin, Anneliese. *Die Agrarlandschaft an der Nordgrenze der Besiedlung im intermontanen Britisch Columbia*. Frankfurt am Main, Germany: W. Kramer, 1965.

Winter, George R. "Agricultural Development in the Lower Fraser Valley." *B.C. Geographical Series*, 9 (1968): 101-116.

Swinnerton, G.S. "Small Farm Functions: A Study of Small Farms in the Matsqui Municipality in the Lower Mainland of British Columbia." Master's Thesis, University of British Columbia, Vancouver, British Columbia, 1969.

Reitsma, Hendrik A. "Agricultural Transboundary Differences in the Okanagan Region." *Journal of Rural Studies*, 2 (1986): 53-62.

Demeritt, David. "Visions of Agriculture in British Columbia." *B.C. Studies*, 108 (1995–1996): 29-59.

CROPS AND CROPPING PATTERNS

Reeves, Colin M. "The Establishment of the Kelowna Orcharding Area: A Study of Accommodation to Site and Situation." Master's Thesis, University of British Columbia, Vancouver, British Columbia, 1973.

Krueger, Ralph R., and N.G. Maguire. "Changing Urban and Fruit Growing Patterns in the Okanagan Valley, British Columbia." *Environments*, 16, No. 1 (1984): 1-9.

ANIMAL HUSBANDRY

Weir, Thomas R. "A Geographic Survey of the Ranching Industry Within the Interior Plateau of British Columbia." Ph.D. Dissertation, University of Wisconsin, Madison, Wis., 1951.

Weir, Thomas R. *Ranching in the Southern Interior Plateau of British Columbia.* Ottawa, Ontario: Queen's Printer, 1955.

DAIRYING

Young, Gerald L. "Dairying in the Fraser-Whatcom Lowland: A Comparative Study of Public Policy and Rural Economy." Ph.D. Dissertation, Indiana University, Bloomington, Ind., 1968.

Mogo, E. "A Study of the Spatio-Temporal Competition in the Dairy Industry: With Particular Reference to Vancouver Island." Master's Thesis, Univesity of Victoria, Victoria, British Columbia, 1978.

FARM LAND

Hardwick, Walter G. "The Moran Cam and Availability of Cultivated Land in Interior British Columbia." *Occasional Papers in Geography*, Canadian Association of Geographers, British Columbia Division, 3 (1962): 65-68.

Fumale, M. "Public Policy and the Preservation of Agricultural Land in the Southern Okanagan Valley." Master's Thesis, University of Victoria, Victoria, British Columbia, 1975.

Maguire, N.G. "The Threat to Canada's Fruit Lands: Perspectives on Land Use Control in the North Okanagan." Master's Thesis, University of Western Ontario, London, Ontario, 1979.

Dearden, P., and P. Bloodoff. "The British Columbia Agricultural Land Reserves: The Case of the Langley Appeal." *B.C. Geographical Series*, 31 (1981): 39-52.

Higginbottom, G. "Agricultural Land Preservation and Its Implications for Housing Construction Costs in Greater Victoria, British Columbia." Master's Thesis, University of Victoria, Victoria, British Columbia, 1981.

McSkimming, R.J., and D. Jones. "Realities of the Agricultural Land Reserve in the Big Bend Area, South Burnaby." *B.C. Geographical Series*, 31 (1981): 53-64.

Pierce, John T. "The B.C. Agricultural Land Commission: A Review and Evaluation." *Plan Canada*, 21, No. 2 (1981): 48-56.

Pierce, John T. "Farmland Protection Planning in British Columbia." *GeoJournal*, 6, No. 6 (1982): 555-560.

Pierce, John T. "British Columbia's Approach to Farmland Preservation." *Nordia*, 16 (1982): 149-155.

Kaspryk, Jaroslav. "Farmland Preservation in the Greater Vancouver Regional District, British Columbia: An Analysis of Agricultural Land Use Dynamics." Master's Thesis, Western Washington University, Bellingham, Wash., 1983.

Pierce, John T. "Resource and Economic Considerations in the Allocation of Agricultural Land in Periurban Areas: A Canadian Perspective." *Landscape Planning*, 10 (1983): 363-386.

Kerr, M.A. "Land Use Change Data Applied to Agricultural Zoning: A Case Study of the Okanagan Valley, British Columbia." *Papers and Proceedings of the Applied Geography Conferences*, 9 (1986): 24-35.

Patterson, Lynda. "Recreational Use of Farmland in the Duncan Agricultural District, British Columbia." Master's Thesis, University of Victoria, Victoria, British Columbia, 1988.

Hanna, Kevin S. "Regulation and Land Use Conservation: A Case Study of the British Columbia Agricultural Land Reserve." *Journal of Soil and Water Conservation*, 52, No. 3 (1997): 166-170.

AGRICULTURAL POPULATIONS AND SOCIAL ISSUES

Gibson, James R. "A Comparison of Anglo-Saxon, Mennonite and Dutch Farms on the Lower Fraser River Valley: A Methodological Study in Areal Differentiation and the Relative Influences of the Physical and Cultural Environments." Master's Thesis, University of Oregon, Eurgene, Oreg., 1959.

Irving, Robert M. "Amenity Agriculture." *B.C. Geographical Studies*, 11 (1966).

Hayward, Patricia M. "Hobby Farming in the Lower Fraser Valley." Master's Thesis, University of British Columbia, Vancouver, British Columbia, 1983.

ENVIRONMENTAL AND CONSERVATION ISSUES

Hardwick, Walter G. "The Effects of the Proposed Moran Dam on Agriculture within the Middle Fraser Region, British Columbia." Master's Thesis, University of British Columbia, Vancouver, British Columbia, 1958.

O'Riordan, Jon. "The Relationship Between Climate and Crop Water Requirements of the South Central Interior of British Columbia." Master's Thesis, University of British Columbia, Vancouver, British Columbia, 1966.

IRRIGATION

Wilson, Kenneth W. "Irrigating the Okanagan: 1860–1920." Master's Thesis, University of British Columbia, Vancouver, British Columbia, 1989.

AQUACULTURE

Porter, Richard P.R. "Pulp Mill Waste and Oyster Farming at Osborn Bay, British Columbia: A Case Study in Marine Resource Use Conflicts and Decisions." Master's Thesis, University of Calgary, Calgary, Alberta, 1969.

Nowell, D. "Fish Farming in British Columbia, Canada." *Geography*, 75, No. 1 (1990): 67-69.

The Canadian North

GENERAL TEXTS, ARTICLES, AND THESES

Gerloff, J.U. "Strategien und territoriale Organisationsformen der Erschliesung des Kanadischen Nordens." *Petermann's Geographische Mitteilungen*, 128, No. 4 (1984): 255-266.

Moodie, D. Wayne, and Barry Kaye. "Indian Agriculture in the Fur Trade Northwest." *Prairie Forum*, 11, No. 2 (1986): 171-183.

Ehlers, Eckert, and Alfred Hecht. "A Model of the Canadian Northern Agricultural Frontier." *Ontario Geography*, 39 (1992): 35-45.

CROPS AND CROPPING PATTERNS

Moodie, D. Wayne. "Gardening on Hudson Bay: The First Century." *Beaver*, 309, No. 1 (1978): 54-57.

ENVIRONMENTAL AND CONSERVATION ISSUES

Ehlers, Eckert. "Comparative Observations of the Physiognomy of the Agricultural Fringe in Northern Finland and Northwestern Canada" (in Finnish). *Terra*, 78, No. 1 (1966): 11-20.

Brklacich, Michael, and C. Tarnocai. "Agricultural Potential and Climate Change in the Mackenzie Valley." Pp. 83-90 in *Proceedings of the Fourth Meeting on Northern Climate*, edited by Geofrey Wall and L. Weber. Toronto, Ontario: Canadian Climate Program, Environment Canada, 1991.

Brklacich, Michael, P. Curran, and D. Brunt. "The Application of Agricultural Land Rating and Crop Models of CO_2 and Climatic Change Issues in Northern Regions: The Mackenzie Basin Case Study." *Agricultural and Food Science in Finland*, 5 (1996): 351-365.

Greenland

GENERAL TEXTS, ARTICLES, AND THESES

Kampp, A.H. "Agriculture in Greenland." Pp. 29-36 in *Agricultural Geography I.G.U. Symposium*, edited by E.S. Simpson. Liverpool, England: University of Liverpool Department of Geography, Research Paper No. 3, 1965.

ENVIRONMENTAL AND CONSERVATION ISSUES

Fredskild, Bent. "Erosion and Vegetational Changes in South Greenland Caused by Agriculture." *Geografisk Tidsskrift*, Vol. 92 (1992): 14-21.

Part 3

LATIN AMERICA

Latin America as a Whole

GENERAL TEXTS, ARTICLES, AND THESES

Foreign Agricultural Service. *Agricultural Geography of Latin America.* Washington, D.C.: United States Department of Agriculture, 1958.

Lentnek, Barry, and Emilio Casetti. "Aportaciones Superfluas de Capital y Mano de Obre en la Agricultura Traditional: Un Estudio Analitico del Caso." *Revista Geografica*, 71 (1969): 23-34.

Hambleton, Hugues-Georges. "La modernization de l'agriculture en Amerique latine." *Cahiers de Geographie de Quebec*, 15, No. 35 (1971): 155-170.

Valenti, Juan V. "Mano de obra y espacio Agrario en los Paises Iberoamericanos." *Estudios Geograficos*, 36 (1975): 1113-1128.

Denevan, William M. "Investigaciones recents sobre agricultura precolombina de campos elevados en America Latina." *Biotica*, 5 (1980): 63-67.

Boulbes, Diane-Colette. "L'agriculture comme base du developpement economique? Deux cas: Equador et Guadeloupe." *Acta Geographica*, 88 (1991): 28-37.

Rama, R. "Latin America and the Geographical Priorities of Multinational Agro-Industries." *Geoforum*, 27, No. 1 (1996): 39-52.

Tulet, Jean-Christian. "L'agriculture sud-americaine: Une evolution silenciense." *Bulletin de la Societe Geographique de Liege*, 33 (1997): 109-118.

CROPS AND CROPPING PATTERNS

Landini, Piero. "La Geografia del Mate." *Bolletino della Societa Geografia Italiana*, Serie 7, 2, No. 10 (1937): 733-759.

Waibel, Leo. "The Plantation System in Latin America." *Annals of the Association of American Geographers*, 31 (1941): 72-73.

Sauer, Carl O. "Cultivated Plants of South and Central America." Pp. 487-543 in *Handbook of South American Indians*, edited by J.H. Steward. Washington, D.C.: Smithsonian Institute, Bureau of American Ethnology, Bulletin No. 143, 1950.

Sauer, Carl O. "Age and Area of American Cultivated Plants." *Actas del 33 Congreso Internacional de Americanistas*, 1 (1959): 215-229.

Steward, J.H. "Perspectives on Plantations." *Revista Geographica*, 26 (1960): 77-85.

Denevan, William M. "Aboriginal Drained-Field Cultivation in the Americas." *Science*, 169 (1970): 647-654.

Dickinson, J. "Alternatives to Monoculture in the Humid Tropics of Latin America." *Professional Geographer*, 24, No. 3 (1972): 217-222.

Parsons, James J. "The Spread of African Pasture Grasses to the American Tropics." *Journal of Range Management*, 25 (1972): 13-17.

Von Hentig, Wolf-Uwe, Willi Manle, and Thomas Steib. *Gartenbau International. Beitrage zur Wirtschaftsgeographie de Gartenbaus, Teil 3: Nord-, Mittel-, und Sudamerika*. Hamburg, Germany: Verlag Paul Parey, 1973.

Kohlhupp, Gerd. "Problems of Agriculture in Latin America: Production of Food-Crops Versus Production of Energy-Plants and Exports." *Applied Geography and Development*, 27 (1986): 60-92.

Angles, Stephane. "L'olivier a la conquete du Nouveau Monde." *Les Cahiers d'Outre Mer*, 53 (2000): 95-110.

Huetz de Lemps, Alain. "Principales plantes cultivees introduites en Amerique latine depuis 1492." *Les Cahiers d'Outre Mer*, 53 (2000): 129-186.

Maillard, Jean-Claude. "Une greffe reussie en terre americaine, ou comment le bananier a conquis le Nouveau Monde." *Les Cahiers d'Outre Mer*, 53 (2000): 7-30.

Pehaut, Yves. "Le cocotier et le palmier a huile en Amerique." *Les Cahiers d'Outre Mer*, 53 (2000): 55-94.

Tulet, Jean-Christian. "La jeunesse de la cafeiculture latino-americaine." *Les Cahiers d'Outre Mer, 53* (2000): 31-54.

ANIMAL HUSBANDRY

Deffontaines, Pierre. "L'introduction du betail en Amerique latine." *Les Cahiers d'Outre Mer*, 8 (1957): 5-22.

Deffontaines, Pierre. "Transhumance et mouvements de betail en Amerique latine." *Les Cahiers d'Outre Mer*, 18 (1965): 258-294; 321-341.

Jordan, Terry G. "An Iberian Lowland/Highland Model for Latin American Cattle Ranching." *Journal of Historical Geography*, 15, No. 2 (1989): 111-125.

FARM LAND

Mathewson, Kent. "Taxonomy of Raised and Drained Fields: A Morphogenetic Approach." Pp. 835-845 in *Prehistoric Intensive Agriculture in the Tropics,* edited by I.S. Farrington. Oxford, England: British Archaeological Reports, International Series, No. 232, 1985.

Turner, Billie L., II, and William M. Denevan. "Prehistoric Manipulation of Wetlands in the Americas: A Raised Field Perspective." Pp. 16-30 in *Prehistoric Intensive Agriculture in the Tropics*, edited by I.S. Farrington. Oxford, England: British Archaeological Reports, International Series, No. 232, 1985.

AGRICULTURAL POPULATIONS AND SOCIAL ISSUES

Gonzalez, Alfonso. "Latin America: Population, Food Supply and Agricultural Dependency." *Revista Geografica*, 101 (1985): 91-96.

ENVIRONMENTAL AND CONSERVATION ISSUES

Troll, C. "The Cordilleras of the Tropical Americas: Aspects of Climate, Phytogeographical and Agrarian Ecology." *Colloquium Geographicum*, 9 (1968): 15-56.

Donkin, R.A. *Agricultural Terracing in the Aboriginal New World*. Tucson, Ariz.: University of Arizona Press, 1979.

Brush, Stephen B. "Traditional Agricultural Strategies in the Hill Lands of Tropical America." Pp. 116-128 in *Human Impact on Mountains*, edited by J.R. Nigel, Gregory W. Knapp, and Christoph Stadel. Lanham, Md.: Rowman & Littlefield, 1988.

Doolittle, William E. "Terrace Origins: Hypotheses and Research Strategies." *Yearbook, Conference of Latin Americanist Geographers*, 16 (1990): 94-97.

Williams, Lynden S. "Agricultural Terrace Evolution in Latin America." *Yearbook, Conference of Latin Americanist Geographers*, 16 (1990): 82-93.

Marsden, Terry K. "Reshaping Environments: Agricultural and Water Interactions and the Creation of Vulnerability." *Transactions, Institute of British Geographers*, 22, No. 3 (1997): 321-337. (New Series)

AGRICULTURAL MARKETING AND DISTRIBUTION

Hoy, Don R. "Trends in the Banana Export Industry of Tropical America." *Journal of Geography*, 63, No. 3 (1964): 108-115.

Clapp, Alex. "Representing Reciprocity, Reproducing Domination: Ideology and the Labor Process in Latin American Contract Farming." Master's Thesis, University of California, Berkeley, Calif., 1987.

Wiley, James. "The European Union's Single Market and Latin America's Banana Exporting Countries." *Yearbook, Conference of Latin American Geographers*, 22 (1996): 31-40.

Middle America as a Whole

GENERAL TEXTS, ARTICLES, AND THESES

Higbee, Edward C. "Agriculture in the Maya Homeland." *Geographical Review*, 38, No. 4 (1948): 457-464.

Stephens, Michael D. "The Problem of Pre-Columbian Mayan Agriculture." Master's Thesis, Johns Hopkins University, Baltimore, Md., 1962.

Wilken, G.C. "Food Producing Systems Available to the Ancient Maya." *American Antiquity*, 36, No. 4 (1971): 432-448.

Turner, Billie L., II. "Prehistoric Intensive Agriculture in the Mayan Lowlands." *Science*, 185 (1974): 118-124.

Harrison, Peter D., and Billie L. Turner, II, eds. *Pre-Hispanic Maya Agriculture*. Albuquerque, N.Mex.: University of New Mexico Press, 1978.

Turner, Billie L., II. "Agricultura prehistorica intensiva en las tierras bajas de los Mayas." *America Indigena*, 38 (1978): 105-124.

Denevan, William M. "Tipologia de configuraciones agricolas prehispanicas." *America Indigena*, 40 (1980): 619-652.

Turner, Billie L., II. "Agricultura y desarrollo del estado en las tierras bajas Mayas." *Estudios Cultura de Mayo*, 8 (1981): 285-306.

Siemens, Alfred H. "Wetland Agriculture in Pre-Historic Mesoamerica." *Geographical Review*, 73, No. 2 (1983): 166-181.

Turner, Billie L., II. "Mystery of the Maya Revealed: The Agricultural Base of a Tropical Civilization." *Focus*, 36, No. 2 (1986): 2-7.

Wilken, G.C. *Good Farmers: Traditional Agricultural Resource Management in Mexico and Central America*. Berkeley, Calif.: University of California Press, 1987.

Sluyter, Andrew. "Intensive Wetland Agriculture in Mesoamerica: Space, Time and Form." *Annals of the Association of American Geographers*, 84, No. 4 (1994): 557-584.

Dunning, Nicholas P. "A Reexamination of Regional Variability in the Prehistoric Agricultural Landscape." Pp. 53-68 in *The Managed Mosaic: Ancient Maya Agriculture and Resource Use*, edited by Scott Fedick. Salt Lake City, Utah: University of Utah Press, 1996.

CROPS AND CROPPING PATTERNS

Shaw, Earl B. "Recent Changes in the Banana Production of Middle America." *Annals of the Association of American Geographers*, 32 (1942): 371-383.

Lucero, Ernest. "A Suggested Examination of Accultural Aspects of Milpa Agriculture as Related to Resistance to Change." Master's Thesis, University of Washington, Seattle, Wash., 1959.

Johannesson, Carl L., Michael R. Wilson, and William A. Davenport. "The Domestication of Maize: Process or Event." *Geographical Review*, 60, No. 3 (1970): 393-413.

Johannesson, Carl L. "Squash Domestication." *Proceedings, Association of American Geographers*, 6 (1974): 131-133.

Turner, Billie L., II, and C.H. Miksicele. "Economic Plant Species Associated with Prehistoric Agriculture in the Maya Lowlands." *Economic Botany*, 38 (1984): 179-193.

Wiley, James. "The Banana Industries of Costa Rica and Dominica in a Time of Change." *Tijdschrift voor Economische en Sociale Geographie*, 89, No. 1 (1998): 66-81.

Stamm, A. "Kaffeewirtschaft in Zentralamerika." *Geographische Rundschau*, 51, No. 7-8 (1999): 399-407.

Whitmore, Thomas M., and Billie L. Turner, II. *Cultivated Landscapes of Middle America on the Eve of Conquest*. Oxford, England: Oxford University Press, 2001.

ANIMAL HUSBANDRY

Kent, Robert B. "Mesoamerican Stingless Beekeeping." *Journal of Cultural Geography*, 4, No. 2 (1984): 14-28.

Johannesson, Carl L. "Distribution and Use of the Black-Boned and Black-Meated Chicken in Mexico and Guatemala." *Yearbook, Conference of Latin Americanist Geographers*, 12 (1986): 43-50.

Sluyter, Andrew. "The Ecological Origins and Consequences of Cattle Ranching in Sixteenth-Century New Spain." *Geographical Review*, 86, No. 2 (1996): 161-177.

FARM LAND

Puleston, D.E. "Terracing, Raised Fields and Tree Crops in the Maya Lowlands: A New Perspective on the Geography of Power." Pp. 225-245 in *Pre-Hispanic Maya Agriculture*, edited by Peter D. Harrison and Billie L. Turner, II. Albuquerque, N.Mex.: University of New Mexico Press, 1978.

Turner, Billie L., II, and Peter D. Harrison. "Prehistoric Raised-Field Agriculture in the Maya Lowlands." *Science*, 213 (1981): 399-405.

Mathewson, Kent. "Taxonomy of Raised and Drained Fields: A Morphogenetic Approach." Pp. 835-864 in *Prehistoric Intensive Agriculture in the Tropics*, edited by I.S. Farrington. Oxford, England: British Archeological Reports, 1985.

AGRICULTURAL POPULATIONS AND SOCIAL ISSUES

Turner, Billie L., II. "The Rise and Fall of Population and Agriculture in the Central Maya Lowlands: 300 B.C. to Present." Pp. 178-211 in *Hunger in History: Food Shortage, Poverty and Deprivation*, edited by L.F. Newman. Cambridge, Mass.: Basil Blackwell, 1990.

ENVIRONMENTAL AND CONSERVATION ISSUES

Denevan, William M. "Hydraulic Agriculture in the Americas: Forms, Measures, and Recent Research." Pp. 181-203 in *Maya Subsistence*, edited by Kent V. Flannery. New York, N.Y.: Academic Press, 1982.

Turner, Billie L., II. "Rethinking the 'New Orthodoxy': Interpreting Ancient Maya Agriculture and Environment." Pp. 57-88 in *Culture, Form and Place: Essays in Cultural and Historical Geography*, edited by Kent Mathewson. Baton Rouge, La.: Louisiana State University, *Geoscience and Man*, Vol. 32, 1993.

Siemons, Alfred H. "Benign Flooding on Tropical Lowland Floodplains." Pp. 132-144 in *The Managed Mosaic: Ancient Maya Agriculture and Resource Use*, edited by Scott L. Fedick. Salt Lake City, Utah: University of Utah Press, 1996.

IRRIGATION

Doolittle, William E. "Indigenous Development of Mesoamerican Irrigation." *Geographical Review*, 85, No. 3 (1995): 301-323.

AGRICULTURAL MARKETING AND DISTRIBUTION

Brohman, John. "The Agroexport Model and Nontraditional Exports in Central America: Déjà vu or Something New?" *Yearbook, Conference of Latin Americanist Geographers*, 22 (1996): 1-16.

Mexico

GENERAL TEXTS, ARTICLES, AND THESES

Basile, David G. "New Agricultural Mexico." Master's Thesis, Columbia University, New York, N.Y., 1939.

Gorloch, Lorene A. "Agricultural Economy of the Laguna Region." *Economic Geography*, 20, No. 3 (1944): 296-304.

Britton, Gerald R. "Agriculture in Mexico." Master's Thesis, Marshall University, Huntington, W.Va., 1947.

Escarpit, Robert. "La revolution agrarie au Mexique." *Les Cahiers d'Outre Mer*, 5 (1952): 108-134.

Chardon, Roland E. "Some Geographic Aspects of Plantation Agriculture in Yucatan." Ph.D. Dissertation, University of Minnesota, Minneapolis, Minn., 1960.

Dennis, Howard W. "The Effect of Trincheras Upon Agriculture in the Pompa Basin, Sierra Madre Occidental, Mexico." Master's Thesis, University of Denver, Denver, Colo., 1967.

Nietschmann, Bernard Q. "The Hacienda and the Revolution: Changing Agriculture Occupance in the Llanos de Apon, Mexico." Master's Thesis, University of Wisconsin, Madison, Wis., 1968.

Thomas, C.H. "The Agricultural Landscape of the Tehuacan Valley, Mexico." Master's Thesis, University of Calgary, Calgary, Alberta, 1968.

Lentnek, Barry. "Economic Transition from Traditional to Commerical Agriculture: The Case of El Llano, Mexico." *Annals of the Association of American Geographers*, 59, No. 1 (1969): 65-84.

Zinty, Gerald W. "Mexican Agriculture." Master's Thesis, Eastern Michigan University, Ypsilanti, Mich., 1969.

Snaden, James. "The Annual Agricultural Cycle in a Mexican Village." *Proceedings, New England–St. Lawrence Valley Division of the Association of American Geographers*, 3 (1973): 15-20.

Snaden, James. "Petaquillas: The Agricultural Geography of a Peasant Village in Southern Mexico." Ph.D. Dissertation, University of Michigan, Ann Arbor, Mich., 1974.

Spaulding, Alan. "Agriculture in the Valle del Yaqui, Sonora, Mexico." Master's Thesis, California State University, Northridge, Calif., 1974.

Turner, Billie L., II. "Prehistoric Intensive Agriculture in the Mayan Lowlands: New Evidence from the Rio Bec Region." Ph.D. Dissertation, University of Wisconsin, Madison, Wis., 1974.

Wenzens, Gerd. "Junge Wandlungen in der Agrarlandschaft der Comarca Lagunera (Nordmexiko)." *Mitteilungen der Osterreich Geographischen Gesellschaft*, 116, No. 1-2 (1974): 22-38.

Ewald, Ursula. "The von Thunen Principle and Agricultural Zonation in Colonial Mexico." *Journal of Historical Geography*, 3, No. 2 (1977): 123-134.

Doolittle, William E. "Aboriginal Agricultural Development in the Valley of Sonora, Mexico." *Geographical Review*, 70, No. 3 (1980): 328-342.

Fish, Merle G., Jr. "El Cultivo de Roza: Shifting Cultivation in a Dry Forest of Northwest Mexico." Ph.D. Dissertation, University of California, Berkeley, Calif., 1980.

Sanches-Burgos, Guadalupe. *La region fundamental de economica compesina en Mexico*. Mexico, D.F., Mexico: Centro de Investigaciones del Desarrollo Rural, 1980.

Coll-Hurtado, Atlantida. *Es Mexico un pais agricola? Un analisis geografico*. Mexico, D.F., Mexico: Siglo Veintiuno Ediores, 1982.

Doolittle, William E. "Agricultural Expansion in a Marginal Area of Mexico." *Geographical Review*, 73, No. 2 (1983): 301-313.

Turner, Billie L., II. "Comparison of Agrotechnologies in the Basin of Mexico and Central Maya Lowlands: Formative to the Classic Maya Collapse." Pp. 13-47 in *Highland–Lowland Interaction in Mesoamerica: Interdisciplinary Approaches*, edited by A.G. Miller. Washington, D.C.: Dumbarton Oaks Research Library Collection, 1983.

Mora, E.S. "Changes in the Structure of Agricultural Production in Mexico." *Rural Systems*, 2, No. 1 (1984): 49-54.

Dahlin, B.H. "La geografia historica de la antigua agricultura Maya." Pp. 125-196 in *Historia de la Agricultura, Epoca Prehispanica-Siglo XVI*, edited by T. Rojas-Rabiela and W.T. Sanders. Mexico City, Mexico: Instituto Nacional de Antropologia e Historia, 1985.

Siemens, A.H. "Results of Air Reconnaissance over the Mexican State of Veracruz." Pp. 127-148 in *Prehistoric Intensive Agriculture in the Tropics*, edited by I.S. Farrington. Oxford, England: British Archeological Reports, 1985.

Seale, Enno. "Strukturwandlungen in der Agrarwirtschaft im Tal von Tehuacan/Puebla-Mexiko." *Erdkunde*, 40, No. 4 (1986): 283-292.

Soto Mora, C., and A. Soto Mora. "Distribucion espacial de la agricultura de temporal en Mexico." *Geografia y Desarollo*, 1, No. 2 (1988): 45-54.

Doolittle, William E. "Arroyos and the Development of Agriculture in Northern Mexico." Pp. 251-272 in *Fragile Lands of Latin America: Strategies for Sustainable Development*, edited by John O. Browder. Boulder, Colo.: Westview, 1989.

Siemens, A.H. *Tierra configurada: Investigaciones de los vestigios de agricultura Precolumbiana en tierra inundatales costeras des de el norte de Veracruz hasta Belice*. Mexico City, Mexico: Consejo Nacional para la Cultura y las Artes, 1989.

Sota Mora, C. "Impacto de las Empresas Transnaciones en la Agricultura Mexicana." *Investigaciones Geograficas: Boletin del Instituto de Geografia*, Universidad Nacional Autonoma de Mexico, 21 (1990): 57-73.

Wohlgemuth, Neusa H.M. "Intensification of Agriculture Within the Basin of Mexico During the Aztec Period." Master's Thesis, University of California, Berkeley, Calif., 1991.

Aguilar, Maria E., and Consuelo S. Mora. "Premisas estructurales de la commercializacion de productes agricoles en Mexico." *Estudios Geograficos*, 53, No. 206 (1992): 45-62.

Pierard, O., et al. "Analyse du risque lie a l'adaption des systemes intensifs de production: Le cas des 'Camellones chantales.'" *Les Cahiers d'Outre Mer*, 46 (1993): 195-208.

Bride, Anne. "Les agriculteurs dans les villes moyennes du Mexique." *L'Information Geographique*, 61, No. 5 (1997): 198-205.

Levy-Tacher, Samuel, and J.R. Aguirre Rivera. "El aprovechamiento agricola intensivo de los Hubches (Acahuales a comunidades secundorias) de Yucatan." *Revista Geografica*, 128 (2000): 79-104.

Salazar, Hilario C., J.R. Aguirre Rivera, and E.O. Saenz. "Proceso de conformacion agricola de la region de Rio Verde, San Luis Potosi, Mexico." *Revista Geografica*, 128 (2000): 105-118.

CROPS AND CROPPING PATTERNS

Foster, Alice. "Sisal Production of Semi-Arid Karst Country of Yucatan." *Journal of Geography*, 29, No. 1 (1930): 16-24.

Dicken, Samuel N. "Cotton Regions of Mexico." *Economic Geography*, 14, No. 4 (1938): 363-371.

Dicken, Samuel N. "Corn and Wheat in Mexico's Changing Economy." *Journal of Geography*, 38, No. 3 (1939): 99-108.

White, C. Langdon, and Edwin J. Foscue. "Henequin: The Green Gold of Yucatan." *Journal of Geography*, 38, No. 4 (1939): 151-155.

Brand, Donald D. "Dividivi and Sesame in Mexico." *Economic Geography*, 17, No. 2 (1941): 141-151.

Bruman, Henry. "Early Coconut Culture in Western Mexico." *Hispanic American Historical Review*, 25, No. 2 (1945): 212-223.

Bruman, Henry. "The Culture History of Mexican Vanilla." *Hispanic American Historical Review*, 28, No. 3 (1948): 360-376.

Aschmann, H. Homer. "The Introduction of Date Palms into Baja California." *Economic Botany*, 11, No. 2 (1957): 174-177.

Pena, Raymond. "A Study of Cotton Production in the Culiacan River Valley, Sinaloa, Mexico." Master's Thesis, University of California, Los Angeles, Calif., 1957.

Chardon, Roland E. *Geographical Aspects of Plantation Agriculture in Yucatan*. Washington, D.C.: National Academy of Science, National Research Council Publication No. 876, 1961.

McDowell, H.G. "Cotton in Mexico." *Journal of Geography*, 63, No. 2 (1964): 67-71.

Mayhew, Barry W. "Lime Cultivation and Processing in Colima, Mexico." Master's Thesis, University of Oregon, Eurgene, Oreg., 1965.

Brown, Charles E. "Viticulture in Northwestern Baja California, Mexico." Master's Thesis, San Diego State University, San Diego, Calif., 1967.

Ugent, Donald. "The Potato in Mexico: Geography and Primitive Culture." *Economic Botany*, 22, No. 2 (1968): 109-123.

Eder, Herbert M. "Palms and Man in Coastal Oaxaca." *Yearbook, Association of Pacific Coast Geographers*, 32 (1970): 41-58.

Williams, Barbara. "Tepetate in the Valley of Mexico." *Annals of the Association of American Geographers*, 62, No. 4 (1972): 618-626.

Edwards, Susan C. "The Almond Industry of Mexico." Master's Thesis, Oregon State University, Corvallis, Oreg., 1975.

Sheldon, Samuel. "Ethnobotany of Agave Lecheguila and Yucca Carnersosona in Mexico's Zona Ixtlera." *Economic Botany*, 34 (1980): 376-390.

Doolittle, William E. "Cabeza de Vaca's Land of Maize: An Assessment of Its Agriculture." *Journal of Historical Geography*, 10, No. 3 (1984): 246-262.

Siemens, Alfred H., et al. "Evidence for a Cultivar and a Chronology from Pattern Wetlands in Central Veracruz, Mexico." *Science*, 242 (1988): 105-107.

ANIMAL HUSBANDRY

Brand, Donald D. "The Early History of the Range Cattle Industry in Northern Mexico." *Agricultural History*, 35, No. 3 (1961): 132-139.

Henderson, David A. "Agriculture and Livestock Raising in the Evolution of the Economy and Culture of the State of Baja California." Ph.D. Dissertation, University of California, Los Angeles, Calif., 1964.

Enjalbert, Henri. "L'elevage au Mexique." *Revista Geografica*, 64 (1966): 53-72.

Sanford, Malcolm T. "A Geography of Apiculture in Yucatan, Mexico." Master's Thesis, University of Georgia, Athens, Ga., 1973.

Sigsworth, Grant W. "The Mexican Epizootic of Foot-and-Mouth Disease: A Study in the Spread, Eradication and Impact of Infectious Livestock Disease and Associated Modernization in the Livestock Industry." Ph.D. Dissertation, University of Illinois, Urbana, Ill., 1975.

Doolittle, William E. "Las Marismas to Panuco to Texas: The Transfer of Open Range Cattle Ranching from Iberia Through Northeastern Mexico." *Yearbook, Latin Americanist Geographers*, 13 (1987): 3-11.

Bouet, Claude. "Irrationalite economique dans le systeme d'elevage bovin de combat Mexicain." *Les Cahiers d'Outre Mer*, 46 (1993): 5-34.

Sluyter, Andrew. "Landscape Change and Livestock in Sixteenth-Century New Spain: The Archival Data Base." *Yearbook, Conference of Latin Americanist Geographers*, 23 (1997): 27-40.

Perramond, Eric P. "Desert Meadows: The Cultural, Political and Ecological Dynamics of Private Cattle Ranching in Sonora, Mexico." Ph.D. Dissertation, University of Texas, Austin, Tex., 1999.

FARM LAND

West, Robert C., and P. Armillas. "Las chinampas de Mexico." *Cuadernos Americanos*, 9, No. 2 (1950): 165-182.

Wilken, G.C. "Drained-Field Agriculture in Southwest Tlaxcala." Ph.D. Dissertation, University of California, Berkeley, Calif., 1967.

Wilken, G.C. "Drained-Field Agriculture: An Intensive Farming System in Tlaxcala, Mexico." *Geographical Review*, 59, No. 2 (1969): 215-241.

Siemens, Alfred H., and D.E. Puleston. "Ridged Fields and Associated Features in Southern Campeche: New Perspectives on the Lowland Maya." *American Antiquity*, 37 (1972): 228-239.

Gunn, Rosemary L. "A Land-Use Study of Commercial Agriculture in Chiapas, Mexico." Ph.D. Dissertation, Michigan State University, East Lansing, Mich., 1976.

Barrett, Ward. "Jugerum and Caballeria in New Spain." *Agricultural History*, 53, No. 2 (1979): 423-437.

Siemens, Alfred H. "Indicas de aprovechamiento agricole Prehispanico de tierras inundales en el centro de Veracruz." *Biotica*, 5 (1980): 83-92.

Siemens, Alfred H. "Aprovechamiento agricole Pre-colombiano de tierras inundales en el norte de Veracruz." *Biotica*, 7 (1982): 343-357.

Siemens, Alfred H. "Oriented Raised Fields in Central Veracruz." *American Antiquity*, 48, No. 1 (1983): 85-102.

Siemens, Alfred H. "Modelling Prehispanic Hydroagriculture on Levee Backslopes in Northern Veracruz, Mexico." Pp. 27-54 in *Drained Field Agriculture in Central and South America*, edited by J.P. Darch. Oxford, England: British Archeological Reports, 1983.

Siemens, Alfred H. "Reducing the Risk: Some Indications Regarding Pre-Hispanic Wetland Agricultural Intensification from Contemporary Use of a Wetland/Terra Firma Boundary Zone in Central Veracruz." Pp. 233-250 in *Agroecology: Researching the Ecological Basis for Sustainable Agriculture*, edited by S.R. Gliessnau. New York, N.Y.: Springer-Verlag, 1990.

Sluyter, Andrew, and Alfred H. Siemens. "Vestiges of Prehispanic Sloping-Field Terraces on the Piedmont of Central Veracruz, Mexico." *Latin American Antiquity*, 3 (1992): 148-160.

AGRICULTURAL POPULATIONS AND SOCIAL ISSUES

Mandell, Paul I. "A Review of Literature Dealing with the Effect of Mexican Agrarian Reform Upon Agricultural Production." Master's Thesis, University of Chicago, Chicago, Ill., 1962.

West, Robert C. "Population Densities and Agricultural Practices in Pre-Columbian Mexico, with Emphasis on Semi-Terracing." Pp. 361-369 in *Proceedings, Thirty-Eighth International Congress of Americanists.* Munich, Germany: Kommissions-verlag Klaus Renner, 1970.

Schroth, Richard E. "Contemporary Agriculture of the Puuc Maya: Study of a Complex Folk System." Master's Thesis, University of Pittsburgh, Pittsburgh, Pa., 1972.

Clawson, David L. "A Comparative Analysis of the Diffusion of Religion and Agricultural Technology Among Peasants of Nealtican, Puebla, Mexico." Master's Thesis, University of Florida, Gainesville, Fla., 1973.

Patrick, Larry L. "A Cultural Geography of the Use of Seasonally Dry, Sloping Terrain: The Metepantli Crop Terraces of Central Mexico." Ph.D. Dissertation, University of Pittsburgh, Pittsburgh, Pa., 1977.

Sander, Hans Jorg. "Rural Skilled Crafts and Agrarian Structure in Mexico." *Rural Systems*, 2, No. 2 (1984): 77-82.

Butzer, Karl W. "Ethno-Agriculture and Cultural Ecology in Mexico: Historical Vistas and Modern Implications." *Yearbook, Conference of Latin Americanist Geographers*, 17-18 (1990): 139-152.

Townsend, J.G., and Bain de Corcuera. "Feminists in the Rainforest of Mexico." *Geoforum*, 24, No. 1 (1993): 45-54.

Biles, James J., and Bruce W. Pigozzi. "The Interaction of Economic Reforms, Social Structure, and Agriculture in Mexico." *Growth and Change*, 31, No. 1 (2000): 3-22.

ENVIRONMENTAL AND CONSERVATION ISSUES

Siemons, Alfred H., and D.E. Puleston. "Ridged Fields and Associated Features in Southern Campeche: New Perspectives on the Lowland Maya." *American Antiquity*, 37 (1972): 228-239.

Gliessman, S.R., et al. "Raised Field Agriculture in the Maya Lowlands of Southwestern Mexico." Pp. 91-110 in *Drained Field Agriculture in Central and South America*, edited by J.P. Darch. Oxford, England: British Archaeological Series, International Series No. 250, 1983.

Turner, Billie L., II. *Once Beneath the Forest: Prehistoric Terracing in the Rio Bec Region of the Maya Lowlands*. Boulder, Colo.: Westview, 1983.

Nabham, G.P. "Ak-cin Arroyo Mouth and the Environmental Settings of the Papago Indian Fields in the Sonoran Desert." *Applied Geography*, 6, No. 1 (1986): 61-75.

Ritter, W., A.M. Naguez, and I. Rosas. "Evaluacion del potencial y la estabilicidad de la producion agricola con base en indices climaticos para algunas localidades de la Republica Mexica." *Geofisica Internacional*, 27, No. 2 (1988): 263-278.

Doolittle, William E. "Arroyos and the Development of Agriculture in Northern Mexico." Pp. 251-269 in *Fragile Lands of Latin America*, edited by John O. Browder. Boulder, Colo.: Westview, 1989.

Livermore, Diana. "Drought Impacts in Mexico: Climate, Agriculture, Technology and Land Tenure in Sonora and Puebla." *Annals of the Association of American Geographers*, 80, No. 1 (1990): 49-72.

Siemens, Alfred H. "Reducing the Risk: Some Implications Regarding Prehispanic Wetland Agricultural Intensification from Contemporary Use of a Wetland/Terra Firma Boundary Zone in Central Veracruz." Pp. 233-250 in

Agro-ecology: Researching the Ecological Basis for Sustainable Agriculture, edited by S.R. Gliessman. New York, N.Y.: Springer-Verlag, 1990.

Goman, Michelle. "Paleoecological Evidence for Prehistoric Agriculture and Tropical Forest Clearance in the Sierra de Los Tuxtlas, Veracruz, Mexico." Master's Thesis, University of California, Berkeley, Calif., 1992.

Richter, M. "Landwirtschaftliche Schaden in verschiedenen Hohenstufen der Sierra Madre de Chiapas/Sudmexiko." *Petermann's Geographische Mitteilungen*, 136, No. 5-6 (1992): 295-308.

Dilley, F.B. "Climate Change and Agricultural Transformation in the Oaxaca Valley, Mexico." Master's Thesis, Pennsylvania State University, University Park, Pa., 1993.

O'Hara, S.L., et al. "Accelerated Soil Erosion Around a Mexican Highland Lake Caused by Pre-Hispanic Agriculture." *Nature*, 362 (1993): 48-51.

Hidalgo-Monroy Wohlgemuth, Neusa. "Organic Agriculture and Indigenous Communities in Chiapas, Mexico: An Alternative to Rural Development." Ph.D. Dissertation, University of California, Berkeley, Calif., 1996.

Heimo, Maija. "Prehistoric Wetland Agriculture South of Laguna Mandinga, Vera Cruz, Mexico: Testing Postulations of Water Management and Agricultural Intensification." Master's Thesis, University of British Columbia, Vancouver, British Columbia, 1998.

Eakin, Hallie. "Smallholder Maize Production and Climatic Risk: A Case Study from Mexico." *Climatic Change*, 45, No. 1 (2000): 19-36.

Rosier, K. "Farming and the Protection of the Tropical Mountain Environment in Chiapas, Mexico: Examples of a System Implemented in Vigurenso" (in French). *Revue de Geographie Alpine*, 88, No. 2 (2000): 59-68.

IRRIGATION

Higbee, Edward C. "Watering Mexico's Deserts." *Agriculture in the Americas*, 1, No. 4 (1941): 11-14.

Thornthwaite, Charles W. "El agua en la agricultura." *Irrigacion en Mexico*, 27, No. 2 (1946): 19-39.

Pederson, J.J. "The Valley of Tecoman: Irrigated Agriculture in the Mexican Tropics." Master's Thesis, University of California, Los Angeles, Calif, 1968.

Clay, James W. "Water-Related Problems and Their Effects on the Spatial Arrangement of Agricultural Production in the Mexicali-San Luis Valley of Northwest Mexico." Ph.D. Dissertation, University of North Carolina, Chapel Hill, N.C., 1969.

Fuentes, Luis. "La productividad de los distritos de riego en la Republica Mexicana." *Informaciones Geograficas* (1970): 69-92.

Gates, Gary R., and Marilyn Gates. "Uncertainty and Developmental Risk in Pequena Irrigation Decisions in Campeche, Mexico." *Economic Geography*, 48, No. 2 (1972): 135-152.

Nabhan, G.P. "Papago Indian Desert Agriculture and Water Control in the Sonoran Desert, 1697–1934." *Applied Geography*, 6, No. 1 (1986): 43-59.

Doolittle, William E. *Canal Irrigation in Prehistoric Mexico: The Sequence of Technological Change.* Austin, Tex.: University of Texas Press, 1990.

McGlade, Michael. "The Efficiency of Irrigation Water Use for Food Production in Mexico." Ph.D. Dissertation, Arizona State University, Tempe, Ariz., 1994.

Crossley, Philip L. "Sub-Irrigation and Temperature Amelioration in Chinampa Agriculture." Ph.D. Dissertation, University of Texas, Austin, Tex., 1999.

AGRICULTURAL MARKETING AND DISTRIBUTION

Manning, E.W. "The Sugar Central as an Agency for the Socio-Economic Development of Properties in the Valley of Tehuacan, Mexico." Master's Thesis, University of British Columbia, Vancouver, British Columbia, 1969.

Kamikihara, Shizue. "Agriculture and Transnational Corporations in the Bajio Region, Mexico." *Geographical Review of Japan*, 66A, No. 1 (1993): 35-51.

Central America as a Whole

GENERAL TEXTS, ARTICLES, AND THESES

Bennett, Hugh H. "Agriculture in Central America." *Annals of the Association of American Geographers*, 16 (1926): 63-84.

Begley, Bryan W. "Agricultural Strategy in a Hot-Wet, Hot-Dry Environment in Tropical Central America and Australia: The Aborigine Tragedy." Master's Thesis, University of Michigan, Ann Arbor, Mich., 1974.

Hardwick, John "A Review of the Problems of Describing and Explaining Agricultural Change in the Maya Lowlands." Master's Thesis, Southern Illinois University, Carbondale, Ill., 1974.

Turner, Billie L., II. "Prehistoric Intensive Agriculture in the Mayan Lowlands." *Science*, 195 (1974): 118-124.

Spielmann, Hans O. "L'agriculture et l'espace agricole de l'Amerique Centrale: Developpement, structure, problemes." *Bulletin de la Societe Geographique de Liege*, 33 (1997): 119-128.

CROPS AND CROPPING PATTERNS

Hearst, Louise. "The Coffee Industry of Central America." *Economic Geography*, 8, No. 1 (1932): 53-66.

Bergman, John F. "The Distribution of Cacao Cultivation in Pre-Columbian America." *Annals of the Association of American Geographers*, 59, No. 1 (1969): 85-96.

Miller, Elbert E. "Small Farms in Central America." *Revista Geografica*, 96 (1983): 38-46.

Maillard, Jean-Claude. "De la Standard Fruit a Castle-and-Cooke, origine et developpement d'une multinationale bananiere." *Les Cahiers d'Outre Mer*, 37 (1984): 371-393.

Whitmore, Thomas M., and Billie L. Turner, II. "Landscapes of Cultivation in Mesoamerica on the Eve of the Conquest." *Annals of the Association of American Geographers*, 82, No. 3 (1992): 402-425.

ANIMAL HUSBANDRY

Suchland, George. "Decisional Influences in Peasant Animal Production: A Central American Comparative Case Study." Ph.D. Dissertation, University of Oklahoma, Norman, Okla., 1972.

Martinson, Tom L. "Rise of the Central American Beef Plantation, 1960–1970." *Geographical Survey*, 3, No. 1 (1974): 1-13.

Miller, Elbert E. "Beef Versus People in Central America and Panama." *Great Plains-Rocky Mountain Geographial Journal*, 4 (1975): 102-108.

Miller, Elbert E. "The Raising of Beef Cattle in Central America and Panama." *Journal of Tropical Geography*, 41 (1975): 59-69.

Miller, Elbert E. "Beef, Food, and Population in Central America and Panama." *B.C. Geographical Series*, 27 (1979): 62-72.

Guatemala

GENERAL TEXTS, ARTICLES, AND THESES

Higbee, Edward C. "The Agricultural Regions of Guatemala." *Geographical Review*, 37, No. 2 (1947): 177-201.

Higbee, Edward C. "Guatemala's Agrarian Problem." Ph.D. Dissertation, Johns Hopkins University, Baltimore, Md., 1949.

Smith, Paul. "Agriculture in Guatemala." Master's Thesis, Eastern Michigan University, Ypsilanti, Mich., 1966.

Wood, Harold A. "The Crop-Livestock Relationship in Guatemala, Central America." *Revista Geografica*, 66 (1967): 95-106.

Hicks, David R. "A Proposed Agricultural/Natural Resources GIS for Guatemala: Problems and Prospects." *Papers and Proceedings of the Applied Geography Conferences*, 17 (1994): 191-200.

CROPS AND CROPPING PATTERNS

Bechtol, Bruce E. "Notes on the Sugar Industry of Guatemala." *Oregon Geographer*, 1, No. 2 (1967): 5-6.

Biechler, Michael J. "The Coffee Industry of Guatemala: A Geographic Analysis." Ph.D. Dissertation, Michigan State University, East Lansing, Mich., 1970.

Biechler, Michael J. "The Regionalization of Coffee Culture in Guatemala." *Revista Geografica*, 77 (1972): 33-56.

Ginsburg, Karen. "Kitchen Gardens in the Peten Region of Guatemala: A Study of the Form, Crops and Contributions of Gardens in Three Villages, 1978–1980." Master's Thesis, University of California, Davis, Calif., 1986.

McCarry, Nancy S. "Variation Among Home Gardens in Guatemala: Reflections of Household Characteristics." Master's Thesis, University of Texas, Austin, Tex., 1990.

ANIMAL HUSBANDRY

Lewis, Robert D. "Animal Domestication in Guatemala." Master's Thesis, University of Oregon, Eugene, Oreg., 1970.

FARM LAND

Pearson, Ross. "Zones of Agricultural Development in Guatemala: An Experiment in Land Reform." *Journal of Geography*, 62, No. 1 (1963): 11-21.

AGRICULTURAL POPULATIONS AND SOCIAL ISSUES

Keys, Eric. "Kakchikel Gardens: Women, Children, and Agricultural Education Among the Maya of Highland Guatemala." *Yearbook, Conference of Latin Americanist Geographers*, 25 (1999): 89-100.

ENVIRONMENTAL AND CONSERVATION ISSUES

Arbana, Sonia J. "Commercial Agriculture and Agrochemicals in Almolonga, Guatemala." *Geographical Review*, 88, No. 1 (1998): 47-63.

IRRIGATION

Mathewson, Kent, ed. *Irrigation Horticulture in Highland Guatemala: The Tablon System of Panajachel.* Boulder, Colo.: Westview, 1984.

AGRICULTURAL MARKETING AND DISTRIBUTION

Biechler, Michael J. "The Movement of Coffee in Guatemala." *Yearbook, Association of Pacific Coast Geographers*, 35 (1973): 115-127.

Lee, Kathryn A. "Illusion of Autonomy: The Impacts of Nontraditional Export Agriculture on Smaller Farmers in Guatemala." Master's Thesis, University of California, Berkeley, Calif., 1993.

Belize

GENERAL TEXTS, ARTICLES, AND THESES

Gouger, James B. "Development of Commercial Agriculture in the Stann Creek Valley, British Honduras." Master's Thesis, University of Florida, Gainesville, Fla., 1968.

Richards, Mary. "The Sibun River Farmers of Belize: A Field Survey and LANDSAT Based Analysis." Master's Thesis, University of Massachusetts, Amherst, Mass., 1983.

Robinson, Guy M., and P.A. Furley, eds. *Resources and Development in Belize: An Account of the University of Edinburgh Expedition to Central America, 1981*. Edinburgh, Scotland: University of Edinburgh, Department of Geography, 1983.

Turner, Billie L., II, and Peter D. Harrison, eds. *Pulltrouser Swamp: Ancient Maya Habitat, Agriculture and Settlement in Northern Belize*. Austin, Tex.: University of Texas Press, 1983.

Furley, P.A., and Guy M. Robinson, eds. *The Agricultural Census of Belize, 1983–4*. Edinburgh, Scotland: University of Edinburgh, Department of Geography, 1984.

Robinson, Guy M. "Agricultural Change in the Belize River Valley." *Caribbean Geography*, 2, No. 1 (1985): 34-44.

Robinson, Guy M., et al. "Developing a Geographical Information System (GIS) for Agricultural Development in Belize, Central America." *Applied Geography*, 9, No. 2 (1989): 81-94.

Whitaker, Victoria A. "Development, Sustainability and Agriculture in Belize." Master's Thesis, University of Texas, Austin, Tex., 1992.

CROPS AND CROPPING PATTERNS

White, Michael G. "The Citrus Industry of the Stann Creek Valley." Master's Thesis, Michigan State University, East Lansing, Mich., 1968.

Bernsten, R.H., and R.W. Herdt. "Towards an Understanding of Milpa Agriculture: The Belize Case." *Journal of Developing Areas*, 11, No. 3 (1977): 373-392.

Robinson, Guy M. "Agricultural Innovation: The Example of Peanut-Growing in the Belize Valley, Central America." *Singapore Journal of Tropical Geography*, 6, No. 2 (1985): 116-126.

FARM LAND

Turner, Billie L., II. "Prehistoric Raised-Field Agriculture in the Maya Lowlands: Pulltrouser Swamp, Northern Belize." *Science*, 213 (1981): 399-405.

Siemens, Alfred H. "Prehistoric Agricultural Use of the Wetlands of Northern Belize." Pp. 205-225 in *Maya Subsistence: Studies in Memory of Dennis E. Puleston*, edited by Kent V. Flannery. New York, N.Y.: Academic Press, 1982.

Lambert, J.D., Alfred H. Siemens, and J.T. Arnason. "Ancient Maya Drained Field Agriculture: Its Possible Application Today in the New River Floodplain, Belize." *Agriculture, Ecosystems, and Environment*, 11 (1984): 67-84.

AGRICULTURAL POPULATIONS AND SOCIAL ISSUES

Hall, Jerry. "Mennonite Agriculture in a Tropical Environment: An Analysis of the Development and Productivity of a Mid-Latitude Agricultural System in British Honduras." Ph.D. Dissertation, Clark University, Worcester, Mass., 1973.

Bliss, Elaine. "Adaptation to Agricultural Change Among Garifuna Women in Hopkins, Belize." *Caribbean Geography*, 3, No. 3 (1992): 160-174.

Nippert, Dale J. "Agricultural Colonization: The Mennonites of Upper Barton Creek, Belize." Master's Thesis, University of Memphis, Memphis, Tenn., 1994.

ENVIRONMENTAL AND CONSERVATION ISSUES

Darch, J.P., and R. Randall. "Maya Drained Field Agriculture at Douglas Village, Orange Walk District, Belize." Pp. 47-52 in *Advances in Environmental and Biogeographical Research in Belize*, edited by P.A. Furley. London, England: Institute of British Geographers Biogeographical Research Group, Monograph No. 3, 1989.

Steinberg, Michael K. "Political Ecology and Cultural Change: Impacts on Swidden-Fallow Agroforestry Practices Among the Mapan Maya in Southern Belize." *Professional Geographer*, 50, No. 4 (1998): 407-417.

Honduras

GENERAL TEXTS, ARTICLES, AND THESES

Martinson, Tom L. "Selected Changes in Agricultural Production and Economic Rent Along the Western Highway of Honduras, 1952–1965." Ph.D. Dissertation, University of Kansas, Lawrence, Kans., 1969.

Martinson, Tom L. "Economic Rent and Selected Changes in Agricultural Production Along the Western Highway of Honduras, 1952–1965." *Revista Geografica*, 71 (1969): 65-74.

Martinson, Tom L. "Copan Counterpoint: Economic Rent and Changing Corn and Tobacco Production in the Agricultural Economy of Western Honduras, 1952–1965." *Proceedings of the Association of American Geographers*, 3 (1971): 114-118.

Martinson, Tom L., et al. "Steps in the Development of a Classification Model for Subsistence Agriculture: An Honduran Example." *Professional Geographer*, 26, No. 1 (1974): 44-47.

Dixon, Clifton V. "Coconuts and Man on the North Coast of Honduras." *Yearbook, Conference of Latin Americanist Geographers*, 11 (1985): 19-24.

CROPS AND CROPPING PATTERNS

Sheldon, Samuel. "The African Oil Palm in Honduras, with Emphasis on the United Brands Operation in San Alejo." Pp. 61-70 in *Field Studies in Central America*, vol. 1. Baton Rouge, La.: Louisiana State University, Department of Geography and Anthropology, 1975.

ANIMAL HUSBANDRY

Alderman, Ralph H. "A Geographic Analysis of the Beef Cattle Industry of Honduras." Master's Thesis, Mississippi State University, Mississippi State, Miss., 1973.

ENVIRONMENTAL AND CONSERVATION ISSUES

Turner, Billie L., II., et al. "Habitat y agricultura en la region de Copan." Pp. 39-149 in *Introduccion a la arqueologia de Copan, Honduras*, tome 1. Tegucigalpa, Honduras: Projects Arqueologico Copan y Secretaria de Estado en el Despacho de Cultura y Turismo, 1983.

Wilvert, Calvin. "Nongovernmental Organizations and Tropical Agroecological Development, with Honduran Examples." *Yearbook, Association of Pacific Coast Geographers*, 57 (1995): 110-133.

AGRICULTURAL MARKETING AND DISTRIBUTION

Martinson, Tom L. "Guatemala City as an Alternative Market for Western Honduras Corn in the Post Miniwar Period." *Proceedings, Indiana Academy of the Social Sciences*, 7 (1972): 73-77.

Nicaragua

ANIMAL HUSBANDRY

Miller, Elbert E. "The Beef Industry in Nicaragua" *B.C. Geographical Series*, 21 (1975): 77-88.

DAIRYING

Patten, George P. "Dairying in Nicaragua." *Annals of the Association of American Geographers*, 61, No. 3 (1971): 303-315.

FARM LAND

Patten, George P. "Agricultural Land Use and Farming Practices in the Managua Region of Nicaragua." Ph.D. Dissertation, Northwestern University, Evanston, Ill., 1954.

AGRICULTURAL POPULATIONS AND SOCIAL ISSUES

Smith, Derek A. "Sumu Settlement and Dooryard Gardens on the Rio Bocay, North Central Nicaragua." Master's Thesis, Louisiana State University, Baton Rouge, La., 1993.

ENVIRONMENTAL AND CONSERVATION ISSUES

Rice, Robert A. "Transforming Agriculture: The Case of Coffee Leaf Rust and Coffee Revolution in Southern Nicaragua." Ph.D. Dissertation, University of California, Berkeley, Calif., 1990.

El Salvador

GENERAL TEXTS, ARTICLES, AND THESES

Masing, U. "The Agricultural Landscape of Eastern El Salvador." Master's Thesis, University of Alberta, Edmonton, Alberta, 1962.

Annegers, John F. "Agricultural Resources and Food Supply in El Salvador." Master's Thesis, Michigan State University, East Lansing, Mich., 1967.

Satterthwaite, Ridgeway. "Campesino Agriculture and Hacienda Modernization in Coastal El Salvador: 1949–1969." Ph.D. Dissertation, University Wisconsin, Madison, Wis., 1971.

ANIMAL HUSBANDRY

Thompson, John. "Production, Marketing and Consumption of Cattle in El Salvador." *Professional Geographer*, 13 (1961):18-22.

ENVIRONMENTAL AND CONSERVATION ISSUES

Laver, Wilhelm. *Vegetation, Landnutzung und Agrarpotential in El Salvador, Zentralamerika*. Kiel, Germany: Kieler Geographische Schriften, Band 16, Heft 1, 1956.

Costa Rica

GENERAL TEXTS, ARTICLES, AND THESES

Thomas, Frank J. "Some Notes on Agriculture in Costa Rica." *Journal of Geography*, 55, No. 3 (1954): 97-103.

Stouse, P.A.D. "Instability of Tropical Agriculture: The Atlantic Lowlands of Costa Rica." *Economic Geography*, 46, No. 1 (1970): 78-97.

Spielmann, Hans O. "La Expancion Ganadera en Costa Rica: Problemes de Desarrollo Agropecuario." *Revista Geografica*, 77 (1972): 57-84.

Spielmann, Hans. "Probleme der agrarwirtschaftlichen Entwicklung in Costa Rica." *Die Erde*, 104 (1973): 27-48.

Arrieta-Chavarria, O. "Desarrollo capitalista y Estructuracion del espacio agricola en Costa Rica." *Revista Geografica de America Central*, 19-20 (1989): 71-88.

CROPS AND CROPPING PATTERNS

Keithan, Elizabeth F. "Cacao in Costa Rica." *Economic Geography*, 16, No. 1 (1940): 79-86.

Jones, Clarence F., and Paul C. Morrison. "Evolution of the Banana Industry of Costa Rica." *Economic Geography*, 28, No. 1 (1952): 1-19.

Pawling, John. "A Geographical Analysis of the Cacao Industy of Costa Rica." Master's Thesis, Pennsylvania State University, University Park, Pa., 1958.

Augelli, John P. "Bananera: A Tropical Plantation on the Pacific Lowlands of Costa Rica." Pp. 31-36 in *Focus on Geographic Activity: A Collection of Original Studies*, edited by Richard S. Thoman and Donald J. Patton. New York, N.Y.: McGraw-Hill, 1964.

Aldridge, Emuel L. "Diffusion of Macadamia Cultivation of Costa Rica." Master's Thesis, University of Georgia, Athens, Ga., 1984.

Hernandez, E. "Sustainable Development in Costa Rica, Between the Theory and the Practice: Insertion of Organic Agriculture into a Traditional System fo Production" (in Spanish). *Revista Geografica Venezuelana*, 36, No. 2 (1995): 137-151.

ANIMAL HUSBANDRY

Smith, Vernon A. "Beef Cattle Production and Marketing in Guanacoste, Costa Rica." Ph.D. Dissertation, University of Florida, Gainesville, Fla., 1970.

Kent, Robert B. "Beekeeping with Honeybees in a Tropical Environment: The Case of Costa Rica." Master's Thesis, University of California, Davis, Calif., 1977.

Hall, Carolyn. "The Tuis Archives: Cattle Ranching on the Frontier of Colonization in Costa Rica, 1873–1876." *Revista Geografica*, 86-87 (1977–78): 101-118.

DAIRYING

Wilvert, Calvin. "Dairying in the Central Highlands of Costa Rica." Master's Thesis, University of California, Berkeley, Calif., 1967.

Fernandez, L.F. "Las condiciones socio-economicas y espaciales de la produccion lechera en Coto Brus Pacifico sur." *Revista Geografica de America Central*, 17-18 (1985): 139-173.

FARM LAND

Stouse, P.A.D. "Effective Agricultural Development of Former Banana Lands: The West Coast of Costa Rica." *Revista Geografica*, 66 (1967): 153-162.

AGRICULTURAL POPULATIONS AND SOCIAL ISSUES

Sandner, Gerhard. *Agrarkolonisation in Costa Rica*. Kiel, Germany: Kieler Geographische Schriften, Heft 3, 1961.

Bergman, Roland. "Shifting Cultivation in the High Rainforests: The Cirripo Indians, Costa Rica." Master's Thesis, University of Wisconsin, Madison, Wis., 1969.

Ulate, Gilbert V. "La Conlonizacion Agricola en la Cuenca del Rio San Lorenzo: Desarrollo y Problemas Ecologicos." *Revista Geografica*, 103 (1986): 69-86.

Dixon, Clifton V., Jr. "Culture in Place: Costa Rica's Italian Agricultural Colony." Pp. 287-310 in *Culture, Form and Place: Essays in Cultural and Historical Geography,* edited by Kent Mathewson. Baton Rouge, La.: Louisiana State University, *Geoscience and Man*, Vol. 32, 1993.

ENVIRONMENTAL AND CONSERVATION ISSUES

Robinson, Timothy. "Fate and Transport of Agricultural Contaminants from Rice Paddies: Impact Sampling Strategies and the Potential Environmental Degradation of Dry Tropical Coastal Wetlands-Guanacoste, Costa Rica." Master's Thesis, University of California, Santa Barbara, Calif., 1993.

Koshear, Jeannine. "Traditional Agricultural Resource Management and Ethnobotany of the Guyami, Coto Brus, Costa Rica." Ph.D. Dissertation, University of California, Berkeley, Calif., 1995.

Northrup, Lisa A. "Precolumbian Agriculture, Fires and Vegetation Dynamics in a Low-Land Rainforest: Paleoecological Evidence from Laguna Bonilla and Bonillita, Costa Rica." Master's Thesis, University of Tennessee, Knoxville, Tenn., 1995.

AGRICULTURAL MARKETING AND DISTRIBUTION

Miranda, Miriam. "The Role of Government in Marketing Fresh Produce in Costa Rica: The Cases of San Jose and San Isidro." Master's Thesis, University of Tennessee, Knoxville, Tenn., 1991.

Panama

GENERAL TEXTS, ARTICLES, AND THESES

Guzman, Louis E. "Farming and Farmlands in Panama." Ph.D. Dissertation, University of Chicago, Chicago, Ill., 1956.

Rubio, Angel. "Las Plantaciones en Panama." *Revista Geografica*, 49 (1959): 5-30.

Freile, Alfonso J. "Agricultural Geography of Panama." Ph.D. Dissertation, Clark University, Worcester, Mass., 1961.

Paganini, Louis A. "The Agricultural Systems of the Chucunaque/Tuira Basin in the Darien Province, Panama." Ph.D. Dissertation, University of Florida, Gainesville, Fla., 1972.

CROPS AND CROPPING PATTERNS

Paganini, Louis A. "Shifting Agriculture and the Staple Rice in Eastern Panama." *Virginia Geographer*, 12, No. 1 (1977): 3-14.

AGRICULTURAL POPULATIONS AND SOCIAL ISSUES

Herlihy, Peter H. "Settlement and Subsistence Change Among the Choco Indians of the Darien Province, Eastern Panama: An Overview." *Yearbook, Conference of Latin Americanist Geographers*, 11 (1985): 13-18.

The Caribbean as a Whole

GENERAL TEXTS, ARTICLES, AND THESES

Shaw, Earl B. "The Food Front in the Greater Antilles." *Economic Geography*, 19, No. 1 (1943): 55-76.

Innes, Frank C. "Approaches to a Study of Small-Scale Agriculture in the Commonwealth Caribbean." Pp. 181-186 in *Proceedings, Canadian Association of Geographers*. Winnipeg, Manitoba: University of Manitoba, Department of Geography, 1970.

Hoy, Don R., and James S. Fisher. "Primary Production and the Measurement of Agricultural Potential in the Caribbean." *Revista Geografica*, 80 (1974): 71-88.

Floyd, Barry. "Problems in the Modernization of Small-Scale Agriculture in Underdeveloped Tropical Countries: A Case Study from the Caribbean." *Proceedings, Moscow: XVII International Geographical Conference*. Moscow, Russia: International Geographical Union, 1976.

Griffith, Anthony D. "A Geographical Perspective on Scale of Farming in a Developing Economy." *Caribbean Quarterly*, 26, No. 3 (1980): 24-35.

Hills, Theo L., and Stanley Iton. "A Reassessment of the Traditional in Caribbean Small-Scale Agriculture." *Caribbean Geography*, 1, No. 1 (1983): 24-35.

Brierley, John S. "Agricultural Policies and Their Impact in the West Indies." *Rural Systems*, 5, No. 1 (1987): 57-72.

Welch, Barbara M. *Survival by Association: Supply Management Landscapes of the Eastern Caribbean*. Montreal, Quebec: McGill-Queen's University Press, 1996.

CROPS AND CROPPING PATTERNS

Shepard, C.Y. "The Sugar Industry of the British West Indies and British Guiana with Special Reference to Trinidad." *Economic Geography*, 5, No. 2 (1929): 149-175.

Landon, Charles E. "Coffee Growing in the Caribbean Countries." *Bulletin of the Geographical Society of Philadelphia*, 29 (1931): 280-297.

Berleant-Schiller, Riva, and Lydia M. Pulsipher. "Subsistence Cultivation in the Caribbean." *Nieuwe West-Indische Gids*, 60 (1986): 1-40.

Brierley, John S. "Kitchen Gardens in the Caribbean, Past and Present: Their Role in Small Farm Development." *Caribbean Geography*, 3, No. 1 (1991): 15-28.

Jaquay, Barbara. "The Caribbean Cotton Production: An Historical Geography of the Region's Mystery Crop." Ph.D. Dissertation, Texas A&M University, College Station, Tex., 1997.

Welch, Barbara M. "The Evolution of Export Banana Cultivation in Dominica and St. Lucia: Implications for Sustainability." *Caribbean Geography*, 7, No. 1 (1997): 62-74.

FARM LAND

Blume, Helmut. "Types of Agricultural Regions and Land Tenure in the West Indies." *Revista Geografica*, 67 (1967): 7-20.

Mills, Beth. "Agriculture and Idle Lands in Carriacou, Grenada, with a Case Study of the Village of Harvey Vole." Master's Thesis, University of New Mexico, Albuquerque, N.Mex., 1987.

Welch, Barbara M. "Institutional Structures as a Factor in Land-Use Decisions: The Impact of Banana Growers' Association in French and Commonwealth Islands in the Eastern Caribbean." Ph.D. Dissertation, University of London, London, England, 1989.

Brierley, John S. "An Analysis of Land Use of Kitchen Gardens, Grenada." Pp. 125-130 in *A Reader in Caribbean Geography*, edited by David Barker, C. Newby, and M. Morrisey. Kingston, Jamaica: Ian Randle Publishers, 1998.

AGRICULTURAL POPULATIONS AND SOCIAL ISSUES

Blume, Helmut. "Wirtschaftslandschaften und Agrarsozialstruktur in West-indien." *Geographische Rundschau*, 19, No. 12 (1967): 449-457.

Grossman, Lawrence S. "Diet, Income and Agriculture in an Eastern Caribbean Village." *Human Ecology*, 26, No. 1 (1998): 21-42.

Cuba

GENERAL TEXTS, ARTICLES, AND THESES

Whitbeck, R.H. "Geographical Relations in the Development of Cuban Agriculture." *Geographical Review*, 12, No. 2 (1922): 223-240.

Tuthill, R.L. "An Independent Farm in Cuba." *Economic Geography*, 25, No. 3 (1949): 201-210.

Semevskiy, B.N. "Social and Natural Conditions of Cuban Agriculture." *Soviet Geography*, 8, No. 1 (1967): 12-24.

Blume, Helmut. "Agrarlandschaft und Agrarreform in Kuba." *Geographische Zeitschrift*, 56 (1968): 295-306.

Dembicz, Andrzej. "Towards an Evolutionary Model of Plantations: A Cuban Typology." *Caribbean Geography*, 2, No. 2 (1986): 110-121.

Gonzalez-Sousa, R. "Algunas consideraciones acerca de desarrollo de la integracion agroindustrial azucaera en la Republica de Cuba." *Acta Universitatis Carolinae Geographica*, 21, No. 2 (1986): 37-43.

Sorzono, Angelina, and Ricardo S. Vernandaz. "La agricultura y e medio ambiente en Cuba." *Acta Universitatis Carolinae Geographica*, 21, No. 2 (1986): 53-64.

Sorzono, Angelina. "La agricultura en Cuba: Cordiciantes sociales de su desarrollo." *Estudios Geograficos*, 57, No. 223 (1996): 291-314.

CROPS AND CROPPING PATTERNS

Platt, Robert S. "Geography of a Sugar District, Mariel, Cuba." *Geographical Review*, 19, No. 4 (1929): 603-612.

Lestina, Mildred L. "Vuelta Abajo, Cuba: A Study of Its Tobacco Industry." *Journal of Geography*, 39, No. 2 (1940): 45-55.

Lestina, Mildred L. "Tobacco Production: San Juan-San Luis, Cuba." *Journal of Geography*, 39, No. 5 (1940): 173-182.

Crist, Raymond. "Some Notes on Recent Trends in Rice Production in Cuba." *Economic Geography*, 28, No. 2 (1956): 126-131.

Dyer, Donald R. "Sugar Regions of Cuba." *Economic Geography*, 32, No. (1956): 177-184.

Winsberg, Morton D. "The Rise and Decline of the American Grapefruit Industry on the Isle of Pines, Cuba." *American Journal of Economics and Sociology*, 20, No. 5 (1961): 543-548.

Gernandez, Maria B.M. "Rice Production in Consolation de Sur, Cuba." Master's Thesis, Clark University, Worcester, Mass., 1964.

Dyer, Donald R. "Cuban Sugar Regions." *Revista Geografica*, 67 (1967): 21-30.

de Rudder, J. "Suiker in Cuba" (in Dutch). *Aardrijkskunde Geographie*, 89 (1971): 95-106.

Honsch, F. "Heraus bildung, Probleme und Tendenzen der Zuckerwirtschaft Kubas." *Geographische Berichte*, 31, No. 2 (1986): 107-123.

Chaplowe, Scott. "Havana's Popular Gardens and the Cuban Food Crisis." Master's Thesis, University of California, Los Angeles, Calif., 1996.

IRRIGATION

Wheatley, Iris V. "The Effects of Irrigation on the Land Use Pattern of the Lajas Valley." Master's Thesis, University of Florida, Gainesville, Fla., 1961.

Borhl, Michael. "Irrigation Farming in the Havana Strath Lowland, 1953–1972." Master's Thesis, Western Illinois University, Macomb, Ill., 1974.

Haiti

GENERAL TEXTS, ARTICLES, AND THESES

Koblinski-Siemens, G. "Zur agrargeographsichen Gliederung von Haiti." *Erdkunde*, 8, No. 3 (1954): 194-197.

Lobeck, E.M. "Haiti: A Brief Survey of Its Past and Present Agricultural Problems." *Journal of Geography*, 53 (1954): 277-290.

FARM LAND

Wood, Harold A. *Northern Haiti: Land, Land Use and Settlement: A Geographical Investigation of the Departement du Nord.* Toronto, Ontario: University of Toronto Press, 1963.

ENVIRONMENTAL AND CONSERVATION ISSUES

Wood, Harold A. "Physical Influences on Peasant Agriculture in Northern Haiti." *Canadian Geographer*, 5, No. 2 (1961): 10-18.

AGRICULTURAL MARKETING AND DISTRIBUTION

Girault, Christian. "Habitants, Speculateurs, et Exporteurs: Le commerce du cafe en Haiti." Ph.D. Dissertation, McGill University, Montreal, Quebec, 1980.

The Dominican Republic

GENERAL TEXTS, ARTICLES, AND THESES

McEachern, J. "The Yuna River Basin, Dominican Republic: Restraints on Agricultural Development." Master's Thesis, McGill University, Montreal, Quebec, 1966.

Antonini, Gustavo A. "Peasant Agriculture in Northwest Dominican Republic." *Journal of Tropical Geography*, 32 (1971): 1-10.

Antonini, Gustavo A. "Evolucion de la agricola tradicional en Santo Domingo." *Eme-Eme Estudios Dominicos*, 9 (1973): 96-122.

Green, Marshall. "Planning Models for Efficient Resource Allocation on Peasant Farms: Some Applications in Northwestern Dominican Republic." Ph.D. Dissertation, University of Florida, Gainesville, Fla., 1974.

O'Ferrall, Andrea M. "Tourism and Agriculture on the North Coast of the Dominican Republic." Master's Thesis, Michigan State University, East Lansing, Mich., 1989.

O'Ferrall, Andrea M. "Tourism and Agriculture on the North Coast of the Dominican Republic." *Revista Geografica*, 113 (1991): 171-192.

Ryder, Roy. "Point Score Analysis of Agricultural Decision Making in the Dominican Republic." *Caribbean Geography*, 4, No. 1 (1993): 2-14.

Ryder, Roy. "La Tecnologia Agricola en la Republica Dominica." *Revista Geografica*, 120 (1994): 79-90.

CROPS AND CROPPING PATTERNS

Girault, Christian. "La Comercializacion de Cafe en la Republica Dominicana: Un Enforque Geografico." *Revista Geografica*, 88 (1978): 49-60.

Chardon, Roland E. "Sugar Plantations in the Dominican Republic." *Geographical Review*, 74, No. 4 (1984): 441-454.

FARM LAND

Zweifler, Mark O. "Land Use Evolution Along the Dominican Republic's Agricultural Frontier: A Comparative Analysis of Two Hill-Land Regions." Master's Thesis, Michigan State University, East Lansing, Mich., 1990.

Ryder, Roy. "Land Evaluation for Steepland Agriculture in the Dominican Republic." *Geographical Journal*, 160, No. 1 (1994): 74-86.

AGRICULTURAL POPULATIONS AND SOCIAL ISSUES

Augelli, John P. "Agricultural Colonization in the Dominican Republic." *Economic Geography*, 38, No. 1 (1962): 15-27.

Augelli, John P. "La Colonizacion Agricola en al Republica Dominica." *Revista Geografica*, 58 (1963): 181-197.

Antonini, Gustavo A. "El Uso de Censas en la Regionalizacion: Agrosocioeconomica de un Pais Latinoamerico." *Revista Geografica*, 90 (1979): 9-32.

Puerto Rico

GENERAL TEXTS, ARTICLES, AND THESES

Haas, William H. "Puerto Rican Agriculture a Century Ago." *Agricultural History*, 10, No. 3 (1936): 97-110.

Imus, Harold R. "The Mayaguez Area of Puerto Rico: A Study in Farm Economy Analysis." Ph.D. Dissertation, Northwestern University, Evanston, Ill., 1951.

Lounsbury, John F. "Farmsteads in Puerto Rico and Their Interpretive Value." *Geographical Review*, 45, No. 3 (1955): 347-358.

Swisher, Carl L. "Agricultural Development Regions and Subregions of Puerto Rico." Ph.D. Dissertation, Northwestern University, Evanston, Ill., 1958.

Leahy, Edward P., and Raymond Crist. "Agricultural Reform in the Humid Tropics: The Example of Las Majaguas." *Professional Geographer*, 21, No. 1 (1969): 8-10.

Huetz de Lemps, Christian. "La place de plantations dans l'economie portoricaine." *Les Cahiers d'Outre Mer*, 24 (1971): 329-393.
Huetz de Lemps, Christian. "La place des plantations dans l'economie portoricaine." *Les Cahiers d'Outre Mer*, 25 (1972): 142-163.
Huetz de Lemps, Christian. "La place des plantations dans l'economie portoricaine." *Les Cahiers d'Outre Mer*, 27 (1972): 57-97.

CROPS AND CROPPING PATTERNS

Whitbeck, R.H. "The Sugar Industry of Porto Rico." *Journal of Geography*, 29, No. 9 (1930): 363-370.
Crist, Raymond. "Sugar Cane and Coffee in Puerto Rico." *American Journal of Economics and Sociology*, 7, No. 2 (1947): 173-184.
Crist, Raymond. "Sugar Cane and Coffee in Puerto Rico, II." *American Journal of Economics and Sociology*, 7, No. 3 (1947): 321-337.
Crist, Raymond. "Sugar Cane and Coffee in Puerto Rico, III." *American Journal of Economics and Sociology*, 7, No. 4 (1947): 469-474.
Campbell, David S. "The Geography of the Coffee Industry of Puerto Rico." Ph.D. Dissertation, University of Chicago, Chicago, Ill., 1948.
Augelli, John P. "Sugar Cane and Tobacco: A Comparison of Agricultural Types in the Highlands of Eastern Puerto Rico." *Economic Geography*, 29, No. 1 (1953): 63-73.
Burchfield, W.W. "Geography of the Pineapple Industry of Puerto Rico." Ph.D. Dissertation, University of Maryland, College Park, Md., 1953.
Burchfield, W.W. "The Pineapple Industry." Pp. 97-128 in *Symposium on the Geography of Puerto Rico,* edited by Clarence F. Jones and Rafael Pico. Rio Piedras, P.R.: University of Puerto Rico Press, 1955.
Imus, Harold R. "The Coffee Industry of the Western Highlands." Pp. 441-472 in *Symposium on the Geography of Puerto Rico*, edited by Clarence F. Jones and Rafael Pico. Rio Piedras, P.R.: University of Puerto Rico Press, 1955.
Cruz-Baez, Angel D. "The Import of Economic Development on an Export Agricultural Sector: A Spatial Analysis of the Sugar Cane Industry in Puerto Rico." Master's Thesis, University of Wisconsin, Madison, Wis., 1971.
Cruz-Baez, Angel D. "Export Agriculture Under Economic Development: A Geographic Analysis of the Decline of Sugar Cane Production in Puerto Rico." Ph.D. Dissertation, University of Wisconsin, Madison, Wis., 1977.
Johnston, Victor W. "The Decline of Sugar in Nineteenth Century Puerto Rico: A Study into the Origins of Underdevelopment." Master's Thesis, Queen's University, Kingston, Ontario, 1983.
Carlo, Louis G. "Changing Crop Production in the Lajas Valley, Puerto Rico, 1976–1991." Master's Thesis, University of Wisconsin, Madison, Wis., 1994.

ANIMAL HUSBANDRY

MacPhail, Donald D. "The Cattle Farms of Southern Puerto Rico." Pp. 213-244 in *Symposium on the Geography of Puerto Rico*, edited by Clarence F.

Jones and Rafael Pico. Rio Piedras, P.R.: University of Puerto Rico Press, 1955.

DAIRYING

Akin, Wallace E. "The Dairy Industry of the San Juan Area of Puerto Rico." Ph.D. Dissertation, Northwestern University, Evanston, Ill., 1952.

Akin, Wallace E. "The Dairy Industry of the San Juan Area." Pp. 47-60 in *Symposium on the Geography of Puerto Rico*, edited by Clarence F. Jones and Rafael Pico. Rio Piedras, P.R.: University of Puerto Rico Press, 1955.

MacPhail, N. "Puerto Rican Dairying: A Revolution in Tropical Agriculture." *Geographical Review,* 53, No. 2 (1963): 224-246.

Morgan, Garrett. "Locational Shifts in the San Juan Dairy Industry, 1935–1969." Master's Thesis, Clark University, Worcester, Mass., 1970.

FARM LAND

Batchelder, Robert B. "Land Use Problems in the Miner Commercial Region of the Hinterland of Isabela." Pp. 129-154 in *Symposium on the Geography of Puerto Rico*, edited by Clarence F. Jones and Rafael Pico. Rio Piedras, P.R.: University of Puerto Rico Press, 1955.

Blount, Stanley F. "A Regional Analysis of Changes from 1950 to 1961 in Rural Land Use in the Isabela Irrigation District, Northwestern Puerto Rico." Ph.D. Dissertation, Northwestern University, Evanston, Ill., 1963.

Wales, Robert W. "Landholdings and Agricultural Development in Puerto Rico, 1508–1970." Ph.D. Dissertation, University of Kansas, Lawrence, Kans., 1973.

Monk, Janice, and Charles Alexander. "Land Abandonment in Western Puerto Rico." *Caribbean Geography*, 2, No. 1 (1985): 2-15.

AGRICULTURAL POPULATIONS AND SOCIAL ISSUES

Augelli, John P. "Geography of Agriculture and Settlement in Interior Puerto Rico: Problems of the Upper Loiza Basin." Ph.D. Dissertation, Harvard University, Boston, Mass., 1950.

ENVIRONMENTAL AND CONSERVATION ISSUES

Netzer, Donald L. "Climate and Agricultural Cropland Use in Puerto Rico." Ph.D. Dissertation, University of Illinois, Urbana, Ill., 1953.

Jamaica

ATLASES AND GRAPHIC PRESENTATIONS

Higman, B.W. "Jamaican Coffee Plantations, 1780–1860: A Cartographic Analysis." *Caribbean Geography*, 2, No. 2 (1986): 73-91.

Higman, B.W. "The Spatial Economy of Jamaican Sugar Plantations: Cartographic Evidence from the Eighteenth and Nineteenth Centuries." *Journal of Historical Geography*, 13, No. 1 (1987): 17-39.

GENERAL TEXTS, ARTICLES, AND THESES

Whitbeck, R.H. "The Agricultural Geography of Jamaica." *Annals of the Association of American Geographers*, 22 (1932): 13-27.

Leeman, Albert. "Die Agrarwirtschaft in den Bauxitgebieten Jamaikas." *Zeitschrift fur Wirtschaftsgeographie*, 13, No. 1 (1969): 20-22.

Iton, S. "The Agronomic Characteristics of Small Scale Jamaican Agriculture." Master's Thesis, McGill University, Montreal, Quebec, 1970.

Thompson, Ronald A. "Small Scale Farming in St. Ann, Jamaica: A Cross Section Analysis of Gross Cash Income on a Selected Number of Farms." Master's Thesis, McMaster University, Hamilton, Ontario, 1970.

Floyd, Barry. "Agriculture in Jamaica." *Geography*, 57, No. 1 (1972): 32-36.

Barker, David. "New Directions for Jamaican Agriculture." *Caribbean Geography*, 2, No. 1 (1985): 56-64.

Found, William C. *Shifting Cultivation in Jamaica: An Invisible Dilemma*. Boston, Mass.: Harvard University Institute for International Development, Development Paper No. 238, 1987.

Barker, David. "Dualism and Disasters on a Tropical Island: Constraints on Agricultural Development in Jamaica." *Tijdschrift voor Economische en Sociale Geographie*, 84, No. 5 (1993): 332-340.

Barker, David, and David J. Miller. "Farming on the Fringe: Small-Scale Agriculture on the Edge of the Cockpit Country." Pp. 271-292 in *Environment and Development in the Caribbean: Geographical Perspectives,* edited by David Barker and Duncan R.M. McGregor. Barbados: University of West Indies Press, 1995.

Spence, Balfour A.B. "Domestic-Food Production and Small Farming in Jamaica." Ph.D. Dissertation, University of Manitoba, Winnipeg, Manitoba, 1995.

Spence, Balfour A.B. "Spatio-Evolutionary Model of Jamaican Small Farming." *Geographical Journal*, 165, No. 3 (1999): 296-305.

CROPS AND CROPPING PATTERNS

Hoy, H.E. "Blue Mountain Coffee in Jamaica." *Economic Geography*, 14, No. 4 (1938): 409-412.

Scott, C.I. "An Analysis of the Supply Characteristics of Jamaican Sugar Cane Farmers." Master's Thesis, McMaster University, Hamilton, Ontario, 1971.

Innis, Donald Q. "Peasant Farming: Banana Intercropping in Jamaica." *Proceedings of the Association of American Geographers*, 5 (1973): 111-117.

Satchell, V. "Pattern of Abandonment of Sugar Estates in Jamaica during the Late Nineteenth Century." *Caribbean Geography*, 2, No. 4 (1989): 251-267.

Meikle, Paulette. "Spatio-Temporal Trends in Root Crop Production and Marketing in Jamaica." *Caribbean Geography*, 3, No. 4 (1992): 223-235.

FARM LAND

Innis, Donald Q. "The Efficiency of Jamaican Peasant Land Use." *Canadian Geographer*, 5, No. 2 (1961): 19-23.

Floyd, Barry. "Agricultural Innovations in Jamaica: The Yallahs Valley Land Authority." *Economic Geography*, 46, No. 1 (1970): 63-77.

Bounds, John H. "Restoration of Mined Lands to Farming in Jamaica." *Revista Geografica*, 80 (1974): 105-110.

McGregor, D.F.M., and David Barker. "Land Degradation and Hillside Farming in the Fall River Basin, Jamaica." *Applied Geography*, 11, No. 2 (1991): 143-156.

Spence, Balfour. "The Influence of Small Farmers' Land Use Decisions on the Status of Domestic Food Security in Jamaica." *Caribbean Geography*, 7, No. 2 (1996): 132-142.

Stirling, Terri. "Alcan Jamaica Company and Small-Scale Agriculture: An Analysis of Competing Land Uses and Changing Land-Use Patterns." Master's Thesis, University of Regina, Regina, Saskatchewan, 1999.

AGRICULTURAL POPULATIONS AND SOCIAL ISSUES

Walters, N.E. "Land Settlement Schemes in Jamaica: A Socio-Agricultural Study." Master's Thesis, McGill University, Montreal, Quebec, 1966.

Barker, David, and Balfour Spence. "Agro-Caribbean Agriculture: A Jamaican Maroon Community in Transition." *Geographical Journal*, 154, No. 2 (1988): 198-208.

Spence, Balfour. "Predicting Traditional Farmers' Responses to Modernization: Case of a Jamaican Maroon Village." *Caribbean Geography*, 2, No. 4 (1989): 217-228.

Thomas-Hope, Elizabeth. "Population Mobility and Land Assets in Hill Farming Area of Jamaica." *Caribbean Geography*, 4, No. 1 (1993): 49-63.

Rashford, John. "Jamaica's Settlement Vegetation, Agroecology and the Origin of Agriculture." *Caribbean Geography*, 5, No. 1 (1994): 32-50.

Schlosser, Tina C. "Local Realities and Structural Constraints of Agricultural Health: Pesticide Poisoning of Jamaican Small-Holders." Master's Thesis, Virginia Polytechnic and State University, Blacksburg, Va., 1999.

ENVIRONMENTAL AND CONSERVATION ISSUES

Ormrod, Richard K. "The Evolution of Soil Management Practices in Early Jamaican Sugar Planting." *Journal of Historical Geography*, 5, No. 2 (1979): 157-170.

IRRIGATION

Found, William C. "Feasibility Analysis for an Irrigation Project: The Case of Clarendon, Jamaica." *Papers and Proceedings of the Applied Geography Conferences*, 6 (1983): 219-224.

AGRICULTURAL MARKETING AND DISTRIBUTION

Belisle, Francois J. "Food Production and Tourism in Jamaica: Obstacles to Increasing Local Food Supplied to Hotels." *Journal of Developing Areas*, 19, No. 1 (1984): 1-20.

TRANSPORTATION AND MECHANIZATION

Anderson, Fred J. "Man Versus Machine in the Sugar Industry in Jamaica." Master's Thesis, Southern Connecticut State College, New Haven, Conn., 1974.

The Lesser Antilles and Virgin Islands

GENERAL TEXTS, ARTICLES, AND THESES

Kopp, A. "L'agriculture a la Guadeloupe." *Annales de Geographie*, 38 (1929): 480-500.

Mbogua, J.P. "Peasant Agriculture in Barbados: A Sample Study." Master's Thesis, McGill University, Montreal, Quebec, 1961.

Brack, D. "Peasant Agriculture in Barbados: A Case Study of Rural Economy." Master's Thesis, McGill University, Montreal, Quebec, 1964.

Henshall, Janet. "The Spatial Structure of Barbadian Peasant Agriculture." Master's Thesis, McGill University, Montreal, Quebec, 1964.

Henshall, Janet, and Leslie King. "Some Structural Characteristics of Peasant Agriculture in Barbados." *Economic Geography*, 42, No. 1 (1966): 74-84.

Watts, David. "Origins of Barbadian Cane Hole Agriculture." *Journal of the Barbados Museum and Historical Society*, 32 (1968): 143-151.

Mills, Frank L. "The Dual Structure of Agriculture in St. Kitts." Master's Thesis, University of Western Ontario, London, Ontario, 1971.

Watts, David. "Dung-Farming: A Seventeenth Century Experiment in Barbadian Agricultural Improvisation." *Journal of the Barbados Museum and Historical Society*, 34 (1972): 58-63.

Griffith, Anthony D. "The Diversification of Peasant Agriculture in Barbados: Problems of Directed Change in a Small Economy." Master's Thesis, McGill University, Montreal, Quebec, 1973.

Mills, Frank L. "The Development of Alternative Farming Systems and Prospects for Change in the Structure of Agriculture in St. Kitts, West Indies." Ph.D. Dissertation, Clark University, Worcester, Mass., 1974.

Gumbs, Howard E. "Problems of Increasing Agricultural Production on St. Thomas, United States' Virgin Islands." Master's Thesis, Illinois State University, Bloomington, Ill., 1981.

Paquette, Romain. *Desengagement paysan et sous-production alimentaire: Martinique, Marie-Galante, Barbade*. Montreal, Quebec: Les Presses de l'Universite de Montreal, 1982.

Berleant-Schiller, Riva. "Grazing and Gardens in Barbuda." Pp. 73-82 in *The Keeping of Animals*, edited by Riva Berleant-Schiller and E. Shanklin. Totowa, N.J.: Rowman & Littlefield, 1983.

De Albuquerque, Klaus, and Jerome L. McElroy. "Agricultural Resurgence in the United States Virgin Islands." *Caribbean Geography*, 1, No. 2 (1983): 121-132.

McQuillan, D. Aidan. "Accessibility and the Development of Export Agriculture in Dominica." *Caribbean Geography*, 1, No. 3 (1984): 149-162.

Rapai, G.R. "Small-Scale Agriculture in St. Lucia." Master's Thesis, University of Western Ontario, London, Ontario, 1984.

Dumaz, Bernard. "Les maraichers de la Cote-sous-le-vent, Guadeloupe: Agriculteurs ou petits patrons?" *Les Cahiers d'Outre Mer*, 34 (1986): 179-204.

Aspinall, R.J., and Janet Momsen. *Small Scale Agriculture in Barbados, 1987*. Newcastle, England: University of Newcastle on Tyne, Department of Geography Paper No. 52, 1987.

Dousson, J., and A.J. Reading. "Small Farm Production in Dominica, West Indies: A Strategy for Survival." *Land Degradation and Rehabilitation*, 1, No. 2 (1989): 153-168.

McElroy, Jerome L., and Klaus De Albuquerque. "Sustainable Small-Scale Agriculture in Small Caribbean Islands." *Society and Natural Resources*, 3, No. 2 (1990): 109-129.

Boulbes, Diane-Colette. "L'agriculture, comme base de developpement economique? Deux cas: Equateur et Guadeloupe." *Acta Geographica*, 88 (1991): 28-37.

CROPS AND CROPPING PATTERNS

James, Preston E. "Geographic Factors in the Trinidad Coconut Industry." *Economic Geography*, 2, No. 2 (1926): 108-125.

Shepard, C.Y. "An Economic Survey of the Cacao Industry of Trinidad, British West Indies." *Economic Geography*, 3, No. 3 (1927): 239-258.

Shaw, Earl B. "St. Croix: A Marginal Sugar-Producing Island." *Geographical Review*, 23, No. 4 (1933): 414-422.

Lasserre, Guy. "Un plantation de canne aux Antilles: La sucrerie Beauport (Guadeloupe)." *Les Cahiers d'Outre Mer*, 5 (1952): 297-329.

Milstead, Harley P. "The Cacao Industry of Grenada." *Economic Geography*, 14, No. 2 (1960): 195-203.

Galloway, J.H. "The Sugar Industry in Barbados during the Seventeenth Century." *Journal of Tropical Geography*, 19 (1964): 35-41.

Oyelese, J.O. "The Cultivation of Food Crops in Barbados." Master's Thesis, McGill University, Montreal, Quebec, 1964.

Kimber, Clarissa. "Dooryard Gardens of Martinique." *Yearbook, Association of Pacific Coast Geographers*, 28 (1966): 97-118.

Sammy, R.V.A. "The Development of Commercial Sugar Cane Agriculture in Trinidad to 1938." Master's Thesis, University of Alberta, Edmonton, Alberta, 1967.

Brierley, John S. "Domestic Food Crop Production, Grenada, West Indies." Master's Thesis, University of Alberta, Edmonton, Alberta, 1968.
Maillard, Jean-Claude. "Un siecle de speculation banniere dans la zone antillaise." *Revista Geografica*, 80 (1974): 89-104.
Manauge, Jacques. *Planteurs et plantations de banane en Guadeloupe*. Talence, France: Travaux et Documents de Geographie Tropicale, CEGET-CNRS, No. 14, 1974.
Burac, Maurice. *L'economie rurale dans le centre nord de la Martinique (Marigot, Sainte-Marie, Gros-Morne et Trinite): Contributions a l'etude de la crise d'une economie de plantation.* Talence, France: Travaux et Documents de Geographie Tropicale, CEGET-CNRS, No. 21, 1975.
Scroggs, Paul. "Problems and Prospect of Tropical Citrus Production: Case Studies from the Eastern Caribbean." Master's Thesis, University of California, Los Angeles, Calif., 1975.
Smith, S.I. "Sugar Production Potential in Barbados." *Journal of Tropical Geography*, 42 (1976): 58-77.
Ferre, Jean-Francoise. "L'economie sucriere et rhumiere martiniquaise en peril (1950–1980)." *Les Cahiers d'Outre Mer*, 34 (1981): 321-360.
Pollard, H.J. "Food Crop Production for Trinidad: Home Market: An Unfulfilled Potential." *Singapore Journal of Tropical Geography*, 2, No. 2 (1981): 91-100.
Brierley, John S. "West Indian Kitchen Gardens: An Historical Perspective with Current Insights from Grenada." *Food and Nutrition Bulletin*, 7, No. 3 (1985): 52-60.
Grossman, Lawrence S. *The Political Economy of Bananas: Contract Farming, Peasants, and Agrarian Change in the Eastern Caribbean*. Chapel Hill, N.C.: University of North Carolina Press, 1988.
Euwema, Jeffrey A. "The Decline of Sugar Production and the Rise of Cocoa Production in Grenada, 1870–1917: The Changing Fortunes of a Cocoa Peasantry." Master's Thesis, Virginia Polytechnic Institute and State University, Blacksburg, Va., 1991.
Thomasson, David A. "Montserrat Kitchen Gardens: Social Functions and Development Potential." *Caribbean Geography*, 5, No. 1 (1994): 20-31.
Drummond, Ian, and Terry K. Marsden. "A Case Study of Unsustainability: The Barbados Sugar Industry." *Geography*, 80, No. 4 (1995): 342-354.
Galloway, J.H. "Botany in the Service of Empire: The Barbados Cane-Breeding Program and the Revival of the Caribbean Sugar Industry, 1802–1930s." *Annals of the Association of American Geographers*, 86, No. 4 (1996): 682-706.
Grossman, Lawrence S. "Food Consumption, Export Bananas, and Local Crops in a Vincentian Village." *Caribbean Geographer*, 9, No. 1 (2000): 1-13.

ANIMAL HUSBANDRY

DeBelle, S.G. "The Cattle Industry of Martinique: A Study in Agricultural Geography." Master's Thesis, McGill University, Montreal, Quebec, 1966.

Berleant-Schiller, Riva. "The Social and Economic Role of Cattle in Barbuda." *Geographical Review*, 67, No. 3 (1977): 299-309.

Berleant-Schiller, Riva. "Grazing and Gardens in Barbuda." Pp. 73-92 in *The Keeping of Animals*, edited by Riva Berleant-Schiller and E. Shanklin. Totowa, N.J.: Rowman & Littlefield, 1983.

DAIRYING

Pollard, H.J., and D.A. Eastwood. "Dairying and Development at Wallerfield, Trinidad." *Tijdschrift voor Economische en Sociale Geographie*, 67, No. 5 (1976): 289-299.

FARM LAND

Hoy, Don R. "Agricultural Land Use of Guadeloupe." Ph.D. Dissertation, University of Illinois, Urbana, Ill., 1960.

Hoy, Don R. *Agricultural Land Use of Guadeloupe*. Washington, D.C.: National Research Council, Publication No. 884, National Academy of Sciences, 1961.

Watts, David. "From Sugar Plantation to Open-Range Grazing: Changes in the Land Use of Nevis, West Indies, 1950–70." *Geography*, 58, No. 1 (1973): 65-68.

Brierley, John S. "Idle Land in Grenada: A Review of Its Causes and the PRG's Approach to Reducing the Problem." *Canadian Geographer*, 29, No. 4 (1985): 298-309.

Brierley, John S. "A Study of Land Redistribution and the Demise of Grenada's Estate Farming System, 1940–1988." *Journal of Rural Studies*, 8, No. 1 (1992): 67-84.

Campbell, J.C., et al. "An Application for Linear Programming and Geographic Information Systems: Cropland Allocation in Antigua." *Environment and Planning A*, 24, No. 4 (1992): 535-550.

Whiteaker, Deanne. "Agricultural Land Use Change and Its Impact on Soil Resources in Southwest Antigua, West Indies from 1964 to 1990." Master's Thesis, University of Oklahoma, Norman, Okla., 1994.

AGRICULTURAL POPULATIONS AND SOCIAL ISSUES

Henshall, D. "The Demographic Factor in the Structure of Agriculture in the Barbados." *Transactions, Institute of British Geographers*, 38 (1966): 183-196.

Bissoon, M. "The Development of Agriculture in South-West Trinidad: A Study of Specialization in a Multi-Racial Society." Master's Thesis, McGill University, Montreal, Quebec, 1968.

Sammy, R.V.A. "Survival Strategies Among Domestic Pure Peasant Farmers in St. Patrick County, Trinidad." Ph.D. Dissertation, University of Western Ontario, London, Ontario, 1977.

Berleant-Schiller, Riva. "The Failure of Agricultural Development in Post-Emancipation Barbuda: A Study of Social and Economic Continuity in a

West Indian Community." *Boletin de Estudios Latinoamericanos y del Caribe*, 25 (1978): 21-36.

Henshall, Janet D. "Women and Small-Scale Farming in the Caribbean." Pp. 28-43 in *Papers in Latin American Geography in Honor of Lucia C. Harrison*, edited by Oscar Horst. Muncie, Ind.: Conference of Latin Americanist Geographers, 1981.

Henshall, Janet D. "Gender Versus Ethnic Pluralism in Caribbean Agriculture." Pp. 173-192 in *Geography and Ethnic Pluralism*, edited by Colin Clarke, David Ley, and Ceri Peach. London, England: Allen and Unwin, 1984.

Brierley, John S. "The Agricultural Strategies and Programs of the People's Revolutionary Government in Grenada, 1979–1983." *Yearbook, Conference of Latin Americanist Geographers*, 11 (1985): 57-62.

Collymore, Jeremy. "Small Farmers as Resource Managers: A Note from St. Vincent." *Caribbean Geography*, 2, No. 2 (1986): 92-99.

Singh, S.P. "Impact of East Indian Immigrants on Trinidad's Agricultural Economy: A Historical Perspective." *National Geographical Journal of India*, 32, No. 2 (1986): 137-163.

Momsen, Janet. "The Feminization of Agriculture in the Caribbean." Pp. 344-347 in *Geography of Gender in the Third World*, edited by Janet Momsen and J.G. Townsend. London, England: Hutchinson, 1987.

Brierley, John S. "A Profile of Grenadian Women Small Farmers." Pp. 194-204 in *Women and Change in the Caribbean: A Pan-Caribbean Perspective*, edited by Janet Momsen. Bloomington, Ind.: Indiana University Press, 1993.

Momsen, Janet. "The Marginalisation of Women Farmers: Gender and Small-Scale Agriculture in Barbados, 1838–1935." *Caribbean Geography*, 6, No. 1 (1995): 52-60.

ENVIRONMENTAL AND CONSERVATION ISSUES

Loslier, Marc. "La Martinique: Essai de determination des regions naturelles, des regions de services et des regions agricoles et analyse de leur integration." Master's Thesis, Universite de Sherbrooke, Sherbrooke, Quebec, 1971.

Frost, David B. "Drag Coefficients over Agricultural Surfaces in Montserrat, West Indies." *Proceedings, New England–St. Lawrence Valley Division of the Association of American Geographers*, 16 (1986): 63-72.

Grossman, Lawrence S. "Pesticides, Caution and Experimentation in St. Vincent, Eastern Caribbean." *Human Ecology*, 20, No. 3 (1993): 315-336.

Singh, B., and M. El Manyor. "Potential Impacts of Greenhouse Gas Climate Change Scenarios on Sugarcane Yields in Trinidad." *Tropical Agriculture*, 75, No. 3 (1998): 348-354.

IRRIGATION

Swanson, W.G. "A Linear Programming Analysis of Agricultural Diversification in Irrigable Areas of Barbados." Master's Thesis, Carleton University, Ottawa, Ontario, 1970.

AQUACULTURE

Considine, James L. "Mariculture and the Turtling Industry of Grand Cayman: Man's Reponse to a Vanishing Resource." Master's Thesis, University of South Carolina, Columbia, S.C., 1973.

AGRICULTURAL MARKETING AND DISTRIBUTION

London, Narrel. "Domestic Agricultural Development Planning and Food Supply in Trinidad, West Indies." Master's Thesis, McGill University, Montreal, Quebec, 1976.

Grossman, Lawrence S. "The Political Ecology of Banana Exports and Local Food Production in St. Vincent, Eastern Caribbean." *Annals of the Association of American Geographers*, 83, No. 2 (1993): 347-367.

Timms, Benjamin. "Linkages Between Domestic Agriculture and the Hotel Sector in St. Lucia." Master's Thesis, Indiana University, Bloomington, Ind., 1999.

TRANSPORTATION AND MECHANIZATION

McQuillan, D. Aidan. "Estate Production and Transportation, Dominica." Master's Thesis, University of Alberta, Edmonton, Alberta, 1968.

AGRICULTURAL LABOR STUDES

Richardson, Bonham C. "Labor Migrants from the Island of Carricou: Workers on the Margins of the Developed World." *Proceedings of the Association of American Geographers*, 6 (1974): 149-152.

South America as a Whole

GENERAL TEXTS, ARTICLES, AND THESES

Jones, Clarence F. "Agricultural Regions of South America." *Economic Geography*: Part I, Vol. 4 (1928): 1-30; Part II, Vol. 4 (1928): 159-186; Part III, Vol. 4 (1928): 267-294; Part IV, Vol. 5 (1929): 109-140; Part V, Vol. 5 (1929): 277-307; Part VI, Vol. 5 (1929): 390-421; and Part VII, Vol. 6 (1930): 1-36.

Platt, Robert S. "Six Farms in the Central Andes." *Geographical Review*, 22 (1932): 245-259.

Enjalbert, Henri. "L'agriculture europeen en Amerique du Sud." *Les Cahiers d'Outre Mer*, 1 (1948): 201-228.

Martinson, Tom L. "Geographers' Contributions to Agricultural Planning in Latin America." *Proceedings of the Indiana Academy of Social Sciences*, 11 (1976): 46-55.

Fonseca Martel, Cesar, and Enrique Mayer. *Comunidad y Produccion en la Agricultura Andina.* Lima, Peru: Fomciencias, 1988.

Arroyo, Monica. "Sobre el Concepto de Estructura Agraria." *Revista Geografica*, 112 (1990): 141-152.

Denevan, William M. "Stone vs. Metal Axes: The Ambiguity of Shifting Cultivation in Prehistoric Amazonia." *Journal of the Steward Anthropological Society*, 20 (1992): 153-165.

Caviedes, Cesar N., and Gregory W. Knapp. "Strategies for Agricultural Subsistence." Pp. 162-183 in *South America*, edited by Cesar N. Caviedes and Gregory W. Knapp. Englewood Cliffs, N.J.: Prentice-Hall, 1995.

Denevan, William M. "Comments on Prehistoric Agriculture in Amazonia." *Culture and Agriculture*, 20, No. 2-3 (1998): 54-59.

Morris, Arthur S. "The Agricultural Base of the Pre-Incan Andean Civilizations." *Geographical Journal*, 165, No. 3 (1999): 286-295.

Sammardi, Marta. "Efectos Ambientales de la Actividad Agricola en el MERCOSUR." *Revista Geografica*, 125 (1999): 187-204.

Zimmerer, Karl S. "Overlapping Patchwork of Mountain Agriculture in Peru and Bolivia: Towards a Regional–Global Landscape Model." *Human Ecology*, 27, No. 1 (1999): 135-166.

CROPS AND CROPPING PATTERNS

McBride, G.M. "Cotton Growing in South America." *Geographical Review*, 9, No. 1 (1920): 35-50.

Jonasson, O. "Potential Areas of Cacao Cultivation in South America: A Review." *Economic Geography*, 27, No. 1 (1951): 90-93.

Eiselon, Elizabeth. "Quinoa: A Potentially Important Food Crop of the Andes." *Journal of Geography*, 55, No. 7 (1956): 330-334.

Dickinson, Joshua C. "The Cultivation and Utilization of the Eucalypt in the Peruvian Sierra and the Industrial Triangle of Brazil." Ph.D. Dissertation, University of Florida, Gainesville, Fla., 1967.

Gade, Daniel W. "Vanishing Crops of Traditional Agriculture: The Case of Tarwi (Lupinus Mutabilis) in the Andes." *Proceedings of the Association of American Geographers*, 1 (1969): 47-51.

Gade, Daniel W. "Ethnobotany of Canihau (Cheropodium pallidicaule), Rustic Seed Crop of the Altiplano." *Economic Botany*, 24 (1974): 55-61.

Bernardes, Nilo. "Sobre Diversificao e Sua Medida Aplicada a Geografia Agraria." *Geografia, Rio Claro*, 3, No. 5 (1978): 15-29.

Breuer, S. "Die Agrarstruktur latinamerikas im Uberblick." *Geographische Berichte*, 35, No. 1 (1990): 39-52.

Works, Martha A. "Continuity and Conversion of House Gardens in Western Amazonia." *Yearbook, Association of Pacific Coast Geographers*, 52 (1990): 31-64.

Zimmerer, Karl S., and D.S. Douches. "Geographical Approaches to Crop Conservation: The Patterning of Genetic Diversity in Andean Potatoes." *Economic Botany*, 45, No. 2 (1991): 176-189.

Preston, David A., and Lorraine Clewer. "Wine Production in a Marginal Area: Tarija Wine in the 1990s." *Journal of Wine Research*, 4, No. 3 (1993): 227-232.

ANIMAL HUSBANDRY

Stewart, Norman R., James Belote, and Linda Belote. "Transhumance in the Central Andes." *Annals of the Association of American Geographers*, 66, No. 3 (1976): 377-397.

Gade, Daniel W. "Leche y civilization andina: Entorno a la ausencia del ordeno de la llama y alpaca." *Yearbook, Conference of Latin Americanist Geographers*, 19 (1993): 3-14.

FARM LAND

Kirby, J. "Agricultural Land Use and the Settlement of Amazonia." *Pacific Viewpoint*, 17 (1976): 105-132.

Wesche, R. "Land Use Alternatives for Small Producers in the Amazon: A Review Article." *Canadian Journal of Development Studies*, 13, No. 2 (1992): 277-285.

AGRICULTURAL POPULATIONS AND SOCIAL ISSUES

Steward, Norman R. "Some Problems in the Development of Agricultural Colonization in the Andean Oriente." *Professional Geographer*, 20, No. 1 (1968): 33-38.

Bromley, R.J. "Agricultural Colonization in the Upper Amazon Basin: The Impact of Oil Discoveries." *Tijdschrift voor Economische en Sociale Geographie*, 63 (1972): 278-294.

Brucher, Wolfgang. "Formen und Effizienzstaatlicher Agrarkolonisation in den ostlichen Regenwaldgebieten der tropischen Andenlander." *Geographische Zeitschrift*, 65 (1977): 3-22.

Turner, Billie L., II, Robert Q. Hanham, and Anthony W. Portaro. "Population Pressure and Agricultural Intensity." *Annals of the Association of American Geographers*, 67, No. 3 (1977): 384-396.

Schuarman, F.J. "From Resource Frontier to Periphery: Agricultural Colonization East of the Andes." *Tijdschrift voor Economische en Sociale Geographie*, 69, No. 1-2 (1978): 95-104.

Gade, Daniel W. "Inca and Colonial Settlement, Coca Cultivation and Endemic Disease in the Tropical Forest." *Journal of Historical Geography*, 5, No. 3 (1979): 263-280.

Bebbington, Anthony. "Los conocimientos tecnicos de los campesinos: Su papel en la elebaracion de una tearin sobre el desarcollo technologico de la agricultura andina." *Documents d'Analisi Geografica*, 17 (1990): 7-26.

Nilsen, A.F., and P. Sjoholt. "Colonization and Adaptation of Agricultural Systems: Some Critical Issues in the Andean Montana." *Norsk Geografisk Tidsskrift*, 47, No. 4 (1993): 229-240.

ENVIRONMENTAL AND CONSERVATION ISSUES

Marbut, C.F., and C.B. Manifold. "The Soils of the Amazon Basin in Relation to Agricultural Possibilities." *Geographical Review*, 16, No. 4 (1926): 414-442.

Field, Chris. "A Reconnaissance of Southern Andean Agricultural Terracing." Ph.D. Dissertation, University of California, Los Angeles, Calif., 1966.

Gade, Daniel. "Ecologia del robo agricola en las tierras altas de los Andes centrales." *America Indigena*, 30 (1970): 3-14.

Fearnside, Philip A. "Effects of Cattle Pasture on Soil Fertility." *Tropical Ecology*, 21 (1980): 125-137.

Blum, W.E.H., and L.M.S. Magalhaes. "Restricoes edaficas de solos na bacia sedimentar Amazonica a utilizacao agraria." Pp. 83-92 in *Hombre y Naturaleza en la Amazonia*, edited by G. Kohlhepp and A. Schrader. Tubingen, Germany: Tubinger Geographische Studien, Heft 95, 1987.

Denevan, William M., Kent Mathewson, and Gregory W. Knapp, eds. *Pre-Hispanic Agricultural Fields in the Andean Region.* Oxford, England: British Archaeological Research International Series, No. 357, 1987.

Fearnside, Philip A. "Rethinking Continuous Cultivation in Amazonia." *Bioscience*, 37, No. 3 (1987): 209-213.

Perez, Francisco L. "Alpine Turf Destruction by Cattle in the High Equatorial Andes." *Mountain Research and Development*, 13, No. 1 (1993): 107-110.

Zimmerer, Karl S. "Agricultural Biodiversity and Peasant Markets in the Central Andes during Inca Rule." *Journal of Historical Geography*, 19, No. 1 (1993): 15-32.

Zimmerer, Karl S. "Transforming Colquepata Wetlands: Landscapes of Knowledge and Practice in Andean Agriculture." Pp. 115-140 in *Irrigation at High Altitudes: The Social Organization of Water Control Systems in the Andes*, edited by W.P. Mitchell and D. Guillet. Washington, D.C.: Society for Latin American Anthropology, 1994.

IRRIGATION

Frederic, Paul B. "Origin and Pre-Columbian Diffusion of Irrigation." *Geographical Perspectives*, 37 (1976): 42-54.

Zimmerer, Karl S. "The Origins of Andean Irrigation." *Nature*, 378 (1995): 481-483.

Argentina

GENERAL TEXTS, ARTICLES, AND THESES

Gooze, Charles. "Northern Mendoza: An Argentine Oasis." Ph.D. Dissertation, Clark University, Worcester, Mass., 1929.

Ryberg, Violet E. "Oasis Agriculture in Tucuman, Argentina." Master's Thesis, University of Washington, Seattle, Wash., 1941.

Eriksen, W., and E. Reisch. "Agrarkrise und Aufforstung in Misiones-Rezessions- und Expansionserscheinungen in Nordostargentinien." *Die Erde*, 105 (1974): 244-264.

Bunstorf, Jurgen. "Die Ackerbauzone im argentinischen Gran Chaco." *Geographische Rundschau*, 28, No. 4 (1976): 144-154.

Zamorano, Gloria L. "La Libertad, Departamento de Rivadavia: Un Tipico Distrito Agrario Mendocino." *Revista Geografica*, 85 (1977): 11-62.

Denis, Paul-Yves. "Espacio Agrario y Centros Urbanos de la Provincia de Cordoba: Conceptualizacion Geografica y Problematica Actual de las Relaciones Ciudad-Campo al Nivel Regional." *Revista Geografica*, 91-92 (1982): 101-140.

Borello, Jose A. "El Desplazamiento de la Frontera Agricola en al Argentina Contemporanea." *Revista Geografica*, 108 (1988): 5-18.

Conte, Analia S. *Geografia Agricola*. Buenos Aires, Argentina: SENCO, Asociacion para la Promocion de Sistemas Educativas no Convencionales, 1988.

Guitierrez de Manchan, Maria J. de Colun, and Silvia L. Monforte de Colun. "Agricultures y Cultivos en un Area del Oasis Norte de Mendoza." *Revista Geografica*, 109 (1989): 199-206.

Daus, Federico A., and Ana del C. Yeannes. "La macroregional para peona agroganada con industrias urban y portuaries." Pp. 476-526 in *La Argentina: Geografia General y Los Macros Regionales*, edited by Juan A. Rocca-Togliata. Buenos Aires, Argentina: Planeta, 1992.

Reboratt, Carlos. "Ambiente, produccion y estructura agraria en el Umbral al Chaco." *Estudios Geograficos*, 53, No. 208 (1992): 503-522.

Arroyo, Monica. "Ajuste economico y sector agrario: Particularidades de esta relacion en la Argentina de los Ochenta." *Revista Geografica*, 118 (1993): 43-50.

Furlani de Civit, Marie, and Maria J. Gutierrez de Manchan. "Dinamica Agraria en un Oasis de Especializacion Viticola." *Revista Geografica*, 115 (1993): 85-137.

Posada, Marcelo G. "Apuntes para una geografia rural del espacio agricola: El caso de Buenos Aires, Argentina." *Revista Geografica*, 120 (1994): 167-201.

Falasca, S.L., A. Ulberich, and M. Bernabe. "Limitantes Fisicas y Quimicas para la Actividad Agricola en el Sudeste Bonaerense, Republica Argentina." *Revista Geografica*, 123 (1997): 107-118.

Valenzuela de Mari, Cristina O. "La Dinamica Productiva de los Paisajes Agricolas en el Nordeste Argentino en la Segunda Mitad del Siglo XX." *Revista Geografica*, 126 (1999): 75-104.

CROPS AND CROPPING PATTERNS

Deasy, George E. "Distribution of Flax Production in Argentina." *Economic Geography*, 19, No. 1 (1943): 45-54.

Zamorano, Mariano. "Le vignoble de Mendoza." *Les Cahiers d'Outre Mer*, 11 (1958): 232-257.

Stewart, Norman R. "Tea: A New Agricultural Industry for Argentina." *Economic Geography*, 36, No. 3 (1960): 267-276.

Gaignard, Romain. "Les plantations de canne a sucre de Tucuman (Argentina)." *Les Cahiers d'Outer Mer*, 17 (1964): 245-291.

Huetz de Lemps, Alain. "Deux vignobles de masse: Algerie et Argentine." *Les Cahiers d'Outre Mer,* 17 (1964): 85-88.

Jones, David M. "Shifting Patterns of Sugarcane Production in Northwest Argentina." Ph.D. Dissertation, Michigan State University, East Lansing, Mich., 1975.

Hansis, Richard A. "Ethnogeography and Science: Viticulture in Argentina." Ph.D. Dissertation, Pennsylvania State University, University Park, Pa., 1976.

Bolsi, A.S. "La Yerba Mate y la Occupacion del Espacio Misiones." *Boletin de Estudios Geograficos*, 21 (1982): 7-65.

Schneider, T. "Weinbau introckenen Westen Argeines: Probleme und entwicklungen in der Oase von Mendoza." *Geographische Rundschau*, 50, No. 11 (1998): 631-635.

Ugolini, M. "Vigne 'bodegas' nel paesaggio mendozino (Argentina)." *Bolletino della Societa Geografica Italiana,* Serie 12, 4, No. 3 (1999): 511-520.

Morris, Arthur S. "Globalization and Regional Differentiation: The Mendoza Wine Region." *Journal of Wine Research*, 11, No. 2 (2000): 145-154.

ANIMAL HUSBANDRY

White, C. Langdon. "The Argentine Meat Question." *Geographical Review*, 35, No. 4 (1945): 634-646.

Snyder, David E. "An Estancia in Argentina: Commercial Livestock Raising." Pp. 53-58 in *Focus on Geographic Activity: A Collection of Original Studies*, edited by Richard S. Thoman and Donald J. Patton. New York, N.Y.: McGraw-Hill, 1964.

Gaignard, Romain. "L'elevage en Argentina: Types d'exploitation pastorale dans la pampa." *Les Cahiers d'Outre Mer*, 18 (1965): 151-193.

Winsberg, Morton D. *Modern Cattle Breeds in Argentina: Origins, Diffusion and Change*. Lawrence, Kans.: University of Kansas Center for Latin American Studies, Occasional Publication No. 13, 1968.

Winsberg, Morton D. "The Introduction and Diffusion of the Aberdeen Angus in Argentina." *Geography*, 55 (1970): 187-195.

Bunstrof, Jurgen. "Formen der Uichwirtschaft im argentinischen Gran Chaco." *Geographische Rundschau*, 23, No. 12 (1971): 462-471.

Miller, Elbert E. "Eight Cows and a Bull: The Influence of Cattle on Argentine Culture." *Great Plains–Rocky Mountain Geographical Journal*, 8 (1979): 20-30.

de Jorge, Carlos A. *Geografiade la Ganaderia Argentina*. Buenos Aires, Argentina: Asociacion para la Promocion de Sistemas Educativas no Convencionales, 1986.

Molinillo, Marcelo F. "Is Traditional Pastoralism the Cause of Erosive Processes in Mountain Environments? The Case of the Cumbres Calchaquies in Argentina." *Mountain Research and Development*, 13, No. 2 (1993): 189-202.

DAIRYING

Morris, Arthur S. "Dairying in Argentina." *Revista Geografica*, 76 (1972): 103-120.

AGRICULTURAL POPULATIONS AND SOCIAL ISSUES

James, Preston E. "The Process of Pastoral and Agricultural Settlement on the Argentine Humid Pampa." *Journal of Geography*, 40, No. 4 (1941): 121-137.

Winsberg, Morton D. *Colonia Baron Hirsch: A Jewish Agricultural Colony in Argentina*. Gainesville, Fla.: University of Florida, Monographs in Social Science, No. 19, 1963.

Gaignard, Romain. "La faillide de l'experience de colonisation agricole des 'Pied-Noirs' en Argentina." *Les Cahiers d'Outre Mer*, 21 (1968): 308-317.

Morris, Arthur S. "The Failure of Small Farmer Settlement in Buenos Aires Province, Argentina." *Revista Geografica*, 81 (1973): 63-78.

Morris, Arthur S. "Diffusion, Deformation or Epigenesis? Growth Processes of the Colonia Agricola in Argentina." Pp. 343-358 in *Perspectives in Agricultural Geography*, vol. 5, edited by Noor Mohammad. New Delhi, India: Concept Publishing Company, 1981.

ENVIRONMENTAL AND CONSERVATION ISSUES

Iglesias de Cuello, A. "Un freia a l'activite pastorale: Le processes de desertification dans les Andes de Patagonia argentine." *Revue de Geographie Alpine*, 76, No. 1 (1988): 83-98.

Falasca, S.L., et al. "Principales caracteristicas agroclimaticas del sudeste bonaerense, Republica de Argentina." *Revista Geografica*, 127 (2000): 91-102.

Degioanni, A., et al. "Evaluacion de los recursos naturales para un uso y gestion sostenible de la producion agraria en una cuenca del centro-sur de Cordoba (Argentina)." *Estudios Geograficos*, 62, No. 2 (2001): 227-256.

TRANSPORTATION AND MECHANIZATION

Miller, Fred. "Highway Improvements and Agricultural Production: An Argentine Case Study." Pp. 104-124 in *Transport and Development*, edited by Brian S. Hoyle. New York, N.Y.: Harper and Row, 1973.

Bolivia

GENERAL TEXTS, ARTICLES, AND THESES

Romecin, Eduardo. "Agricultural Adaptation in Bolivia." *Geographical Review*, 19, No. 3 (1929): 248-255.

Weeks, David. "Bolivia's Agricultural Frontier." *Geographical Review*, 36, No. 4 (1946): 634-646.

Vellard, Jehon. "L'experience agraire en Bolivie." *Les Cahiers d'Outre Mer*, 16 (1963): 201-213.

Grenier, Fernand. "Agriculture et reforme agraire en Bolivie." *Cahiers de Geographie de Quebec*, 9, No. 17 (1964–1965): 25-40.

Borja Berrios, Gover, and Armando Cardozo Gonsalez. *Geographia Agricola de Bolivia*. La Paz, Bolivia: Editorial los Amigos del Libro, 1971.

Henkle, Ray. "The Chapare of Bolivia: A Study of Tropical Agriculture in Transition." Ph.D. Dissertation, University of Wisconsin, Madison, Wis., 1971.

Preston, David A. "L'agriculture dans un desert d'altitude: L'altiplano central de Bolivie." *Les Cahiers d'Outre Mer*, 26 (1973): 113-128.

CROPS AND CROPPING PATTERNS

South, Robert B. "Coca-Cocaine: A Case Crop in Highland Bolivia." *Proceedings of the Association of American Geographers*, 6 (1974): 133-137.

South, Robert B. "Coca in Bolivia." *Geographical Review*, 67, No. 1 (1977): 23-33.

Lawrence, Ivy E. "Coca Production and Use in Bolivia: Prospects for Control." Master's Thesis, Columbia University, New York, N.Y., 1980.

Bonnet, Nicole. "La coca, principal moteur de l'economie bolivienne." *L'Information Geographique*, 51, No. 1 (1987): 18-19.

McGlade, Michael S., Ray Henkle, and Randall S. Cerveny. "The Impact of Rainfall Frequency on Coca (Etythrozyhem coca) Production in the Chopore Region of Bolivia." *Yearbook, Conference of Latin Americanist Geographers*, 20 (1994): 97-106.

ANIMAL HUSBANDRY

Denevan, William D. "Cattle Ranching in the Mojos Savannas of Northeastern Bolivia." *Yearbook, Association of Pacific Coast Geographers*, 25 (1963): 37-44.

FARM LAND

Preston, David A. "Land Tenure and Agricultural Development in the Central Altiplano, Bolivia." Pp. 231-252 in *Spatial Aspects of Development*, edited by Brian S. Hoyle. London, England: John Wiley, 1974.

AGRICULTURAL POPULATIONS AND SOCIAL ISSUES

Schoop, Wolfgang. *Vergleichede Unter suchungen zur Agrarkolonisation der Hochlandindianer am Andenabfall und im Tiefland Ostboliviens*. Aachen, Germany: Aachener Geographische Arbeiten, Heft 4, 1970.

Weil, Connie J. "Prospects of Agricultural Colonization Efforts in Bolivia." Masters' Thesis, Columbia University, New York, N.Y., 1973.

Mahnke, Lothar. "Zur indianischen Landwirtschaft im Siedlungsgebiet der Kallawayas (Bolivien)." *Erdkunde*, 36, No. 4 (1982): 254-265.

Preston, David A. "Post-Peasant Capitalist Graziers: The Twenty-First Century in Southern Bolivia." *Mountain Research and Development*, 18, No. 2 (1998): 151-158.

ENVIRONMENTAL AND CONSERVATION ISSUES

Barber, R.G., and F. Navarro. "The Rehabilitation of Degraded Soils in Eastern Bolivia by Subsoiling and the Incorporation of Cover Crops." *Land Degradation and Rehabilitation*, 5, No. 4 (1994): 247-260.

Brazil

GENERAL TEXTS, ARTICLES, AND THESES

Deffontaines, Pierre. "Une evolution agricole en pays tropical: L'agriculture au Bresil." *Annales de Geographie*, 48 (1939): 385-392.

Monbeig, P. "Les structures agraires dans la frange pionniere de Sao Paulo." *Les Cahiers d'Outre Mer*, 4 (1951): 1-22.

James, Preston E. "Trends in Brazilian Agricultural Development." *Geographical Review*, 43, No. 3 (1953): 301-318.

Sternberg, Hilgard O. "Agriculture and Industry in Brazil." *Geographical Journal*, 121, No. 4 (1955): 488-502.

Faissol, Speridiao. "Problema do Desenvolvimento Agricola do Sudeste do Planalto Central (Brasil)." Ph.D. Dissertation, Syracuse University, Syracuse, N.Y., 1956.

Haskins, Edward C. "An Agricultural Geography of the Reconcavo of Bahia." Ph.D. Dissertation, University of Minnesota, Minneapolis, Minn., 1956.

Faissol, Speridiao. "O problema do desenvolvimento agricole do sudente do Planalto Central do Brasil." *Revista Brasileira de Geografia*, 19 (1957): 3-66.

da Silva, Hilda. "Uma zona agricola do Distrito Federal: O Mendanha." *Revista Brasileira de Geografia*, 20 (1958): 429-462.

Waibel, Leo. *Capitulos de Geografia Tropical e do Brasil*. Rio de Janeiro, Brazil: Servico Grafico do Instituto Brasileiro e Estatistica, 1958.

de Carvalho, Eloisa, and Hilda da Silva. "A producao agricola do Brasil em 1957." *Revista Brasileira de Geografia*, 21 (1959): 195-220.

Webb, Kempton E. "Origins and Developments of a Food Economy in Central Minas Gerais." *Annals of the Association of American Geographers*, 49, No. 3 (1959): 409-419.

Bernardes, Nilo. "Caracteristicas gerais da agricultura brasileira no seculo XX." *Revista Brasileira de Geografia*, 23 (1961): 363-420.

Valverde, Orlando, and Mirian G.C. Mesquita. "Geografia agraria do Baixo Acu." *Revista Brasileira de Geografia*, 23 (1961): 455-494.

Valverde, Orlando. *Geografia Agraria do Brasil.* Rio de Janeiro, Brazil: Centro Brasileiro de Pesquisas Educacionais, Instituto Nacional de Estudoes Pedogogicos, Ministerio da Educacao e Cultura, 1964.

Guerra, Antonio T. "Os solos e a Reforma Agraria no Brasil." *Revista Brasileira de Geografia*, 27, No. 1 (1965): 129-136.

do Carmo Correa Galvao, Maria. "Wandel der Landwirtschaft im Guanabaragebiet." *Heidelberger Geographische Arbeiten*, 15 (1966): 208-218.

Bathelnuss, Heloisa. "Methodolgische Uberlegunen zu einer dynamischen Agrargeographie am Beispiel Paranas." *Geographische Zeitschrift*, 55 (1967): 207-224.

Noble, Allen G. "Geographic Aspects of the Agriculture of Santa Catarina State, Brazil." *Ohio Journal of Science*, 67, No. 5 (1967): 257-273.

Sioli, H. "Entwicklung und Assichten der landwirtschaft im brasilianischen Amazonasgebiet." *Die Erde*, 100 (1969): 307-326.

Berry, Brian J.L., and Gerald F. Pyle. "Grandes Regioes e tipos de Agricultura no Brasil." *Revista Brasileira de Geografia*, 32, No. 4 (1970): 23-40.

Ceron, Antonio O., and Jose A.F. Diniz. "Tipologia da Agricultura: Questoes Metodologicas e Problemas de Aplicacao no Estado de Sao Paulo." *Revista Brasileira de Geografia*, 32, No. 3 (1970): 41-72.

Coelho de Souza Keller, Elza. "Tipos de Agricultura no Parana, uma Analise Fatorial." *Revista Brasileira de Geografia*, 32, No. 4 (1970): 41-86.

Mesquita, Olindina V., and Solange T. Silva. "Regioes Agricolas do Estado do Parana: Uma Definicao Estatistica." *Revista Brasileira de Geografia*, 32, No. 1 (1970): 3-42.

Marques, Jorge S. "Estrutura Agraria do Estado de Pernambuco." *Revista Brasileira de Geografia*, 33, No. 2 (1971): 137-148.

Sternberg, Rolf. "Farms and Farmers in an Estanciero World, 1856–1914." Ph.D. Dissertation, Syracuse University, Syracuse, N.Y., 1971.

Allderdice, William H. "The Expansion of Agriculture Along the Belem-Brasilia Road in Northern Goias, Brazil." Ph.D. Dissertation, Columbia University, New York, N.Y., 1972.

Dewitt, John W. "Food Production and Regional Development in Bahia, Brazil." Ph.D. Dissertation, University of Florida, Gainesville, Fla., 1973.

Pinto de Gusmao, Rivaldo. "Estudo da Organizacao Agraria de Regiao Sul Atraves da uma Analise Fatorial." *Revista Brasileira de Geografia*, 36, No. 1 (1974): 33-52.

Wesche, R. "Planned Rainforest Family Farming on Brazil's Transamazonic Highway." *Revista Geografica*, 81 (1974): 105-114.

Bret, Bernard. "Donnees et reflexions sur l'agriculture bresillienne." *Annales de Geographie*, 84, No. 5 (1975): 557-588.

Galloway, J.H. "Northeast Brazil, 1700–50: The Agricultural Crisis Re-examined." *Journal of Historical Geography*, 1, No. 1 (1975): 21-38.

Geradi, L.H.O. "On Agricultural Typology and Systems Analysis: A Bibliographical Review" (in Portugese). *Boletim Paulista de Geografia*, 52 (1976): 83-92.

Sanders, John H., and Frederick L. Bein. "Agricultural Development in the Brazilian Frontier: Southern Mato Grosso." *Economic Development and Culture Change,* 24, No. 3 (1976): 539-610.

Mesquita, Olinda V., Rivaldo Pinto de Gusmao, and Solange T. Silva. "Modernizacao da Agricultura Brasileira." *Revista Brasileira de Geografia*, 39, No. 4 (1977): 3-65.

Steffan, E.R., and Mario S. Brito. "Applications of Factor Analysis in the Study of Agricultural Organization in Paraibe and Pernambuco" (in Portugese). *Boletim Geografico*, 35, No. 254 (1977): 22-47.

Teixeira, Marlene P.V. "Organizacao Expacial du Agricultura no Estado do Rio de Janeiro." *Revista Brasileira de Geografia*, 39, No. 1 (1977): 41-98.

Valverde, Orlando. "Genese e Evolucao do Problema Agrario Brasileiro." *Finisterra*, 12, No. 24 (1977): 244-279.

Yamamoto, S., and Mario Hiraoka. "Three Farm Types in the Patos Basin of the Inland Paraiba State, Brazilian Northeast" (in Japanese). *Geographical Review of Japan*, 50, No. 9 (1977): 511-529.

de Brito, Sebastiana R., and Nelson de Castro Senra. "Notas Sobre o Pessoal Ocupado no Setar Agro-Pecuario do Parana." *Revista Brasileira de Geografia*, 40, No. 1 (1978): 123-141.

Leo, Ieda R. "Agricultura e Transformacao Estrutural–Estratgias Economicas de Paises em Desenvolvimento." *Revista Brasileira de Geografia*, 40, No. 3-4 (1978): 248-256.

Smith, Nigel J. "Agricultural Productivity Along Brazil's Transamazon Highway." *Agro-Ecosystems*, 4 (1978): 415-432.

van der Pluijm, T. "A Policy Analysis of Brazil's Agriculture." *Tijdschrift voor Economische en Sociale Geographie*, 69, No. 5 (1978): 258-269.

Gerardi, L.H.O., and Antonio O. Ceron. "Disparidades em modernizacao da agricultura no Estado de Sao Paulo: Primeiros resultados." *Geografia, Rio Claro*, 4, No. 8 (1979): 43-53.

Kageyam, Angela A., and Jose F. Graziano. "A proposito da expansao da agricultura Paulista." *Boletim Paulista de Geografia*, 56 (1979): 25-59.

Rocha, Sonia. "As novas medidas de politica economica afetondo o setor agropecuario." *Revista Brasileira de Geografia*, 41, No. 3 (1979): 130-134.

Poltronieri, Ligia C. "Aspectos Sociais, Fundiacios e de Modernizacao da Agricultura Paulista entre 1940 e 1970." *Revista Brasileira de Geografia*, 42, No. 4 (1980): 885-896.

Silva, Solange T. "Os Estudos de Classificacao na Agricultura: Uma Revisao." *Revista Brasileira de Geografia*, 42, No. 1 (1980): 3-30.

Poltronieri, Ligia C. "Adinamica Espacial do Use do Solo Agricola no Estado de Sao Paulo no Periodo 1935–1970." *Revista Brasileira de Geografia*, 43, No. 3 (1981): 405-418.

Romerio, Ademar R., and Fernando J. Abrantes. "Meio Ambiente e Modernizacao Agricola." *Revista Brasileira de Geografia*, 43, No. 1 (1981): 3-46.

Hess, Dora R. "A Apropriacao do Espaco Nas Areas de Fronteira Agricola Brasileiras." *Revista Brasileira de Geografia*, 44, No. 4 (1982): 703-720.

Servolin, Claude. "A Absorcao da Agricultura no Modo de Producao Capitalista." *Revista Brasileira de Geografia*, 45, No. 3-4 (1983): 425-440.

de Figueire, Adma H., Olindina V. Mesquita, and Solange T. Silva. "O Impacto Designal da Modernizacao da Agricultura no Oeste do Parana." *Revista Brasileira de Geografia*, 46, No. 3-4 (1984): 425-550.

Draiz, Jose A.F. *Geografia de Agricultura*. Sao Paulo, Brazil: DIFEL, 1984.

Leao, S.O. "Intensity of Agriculture in the State of Bahia: A Temporal and Multivariate Analysis, 1940–1970." *Rural Systems*, 2, No. 4 (1984): 267-274.

Pebayle, Raymond. "Evolutions rurales et regionalisation au Bresil." Pp. 563-584 in *Geographes d'aujourd'hui*, edited by A. Dogorne and J.P. Paulet. Nice, France: Universite de Nice, 1984.

Ceron, Antonio O., and L.H.O Gerardi. "Modernization of the Brazilian Agriculture: Agrarian Changes in a Developing Country." *Rural Systems*, 3, No. 2 (1985): 121-150.

Ceron, Antonio O., and Lucia H. de O. Gerardi. "Modernizacao da Agricultura Brasileira: Transformacoes Agrarias em Pais em Desenvolvimento." *Revista Geografica*, 101 (1985): 5-28.

Valverde, Orlando. *Estudos de Geografia Agraria Brasileira*. Petropolis, Brazil: Vozes, 1985.

Korp, B. "Das agrare Genossenschaftswesen als raumgestaltender Faktor: Zur Ambivalenz gelenkter Modernisierung am Beispiel von West-Parana (Brasilien)." *Geographische Zeitschrift*, 74, No. 4 (1986): 241-260.

Carmo Carrea Galvao, Maria. "O Espaco Agrario Fluminense–Estrutura e Transformacoes." *Revista Geografica*, 105 (1987): 119-132.

de Azevedo Brito, Maristella. "Questoes Associades a Evolucao Recente da Agricultura Brasileira." *Revista Geografica*, 105 (1987): 83-106.

de Saldanha, Marila, Maristella de Azevedo Brito, and Tereza M. Ramos de Oliveira. "A Evolucao da Agricultura na Regiao Nordeste na Decada de 70." *Revista Brasileira de Geografia*, 49, No. 1 (1987): 47-106.

Hess, Dora R., Maria E. de Paiva Carrea de Sa, and Tereza C. Aguiar. "A evolucao da agricultura na regiao Centro-Oeste na decada de 70." *Revista Brasileira de Geografia*, 49, No. 1 (1987): 197-257.

Mesquita, Olindina V., and Solange T. Silva. "A Evolucao de Agricultura Brasileira na Decada de 70." *Revista Brasileira de Geografia*, 49, No. 1 (1987): 3-10.

Mesquita, Olindina V., and Solange T. Silva. "A Evolucao da Agricultura na Regiao Sul na Decada de 70." *Revista Brasileira de Geografia*, 49, No. 1 (1987): 159-196.

Mueller, Charles C. "Algumas Caracteristicas da Modernizacao Agricola Recente de Regiao Geoeconomica de Brasilia." *Revista Geografica*, 105 (1987): 133-154.

Pieres Guimaraes, Luiz, and Ney R. Innocencio. "A Evolucao da Agricultura na Regiao Sudeste na Decade de 70." *Revista Brasileira de Geografia*, 49, No. 1 (1987): 107-158.

Socorro Brito, Maria, and Mitiko Yanage Une. "A Evolucao da Agricultura na Regiao Nordeste na Decada de 70." *Revista Brasileira de Geografia*, 49, No. 1 (1987): 11-46.

de Azevedo Brito, Maristella, and Ney R. Innocencio. "Organizacao do Espaco Agrario no Estado do Rio de Janeiro." *Revista Brasileira de Geografia*, 50, No. 3 (1988): 85-120.

Hess, Dora R., and Tereza C. Aguiar. "A Apropriacao do Espaco Agricola Pela Pecuaria no Centro-Oeste." *Revista Brasileira de Geografia*, 50, No. 1 (1988): 41-60.

Caldas Lins, Rachel. *As Areas de Execao do Agreste de Pernambuco*. Recife, Brazil: Serie Estudos Regionais, No. 20, 1989.

de Souza Bicalho, Ana Maria, and Scott W. Hoefle. "Urban Capital and the Pseudo-Modernization of Agriculture in Northeast Brazil." *Yearbook, Conference of Latin Americanist Geographers*, 15 (1989): 35-48.

Rheker, Johannes R. *Zur regionalen Entwicklung der Nahrungsmittelproduktion in Pernambuco (Nord ost brasilien).* Bonn, Germany: Bonner Geographische Abhandlungen, Heft 78, 1989.

Rheker, Johannes R. "Untersuchungen zum Ruckgang der Klinbauerlichen Grundnahrungsmittelproduktion in ausgewahlten Munizipien des nordost brasilianischen Agreste." *Erdkunde*, 43, No. 2 (1989): 119-132.

Theilen-Willige, B. "Satelliten-Radaraufnahmen bei agrar- und vegetationsgeographischen Untersuchengen." *Geographische Rundschau*, 41, No. 12 (1989): 706-710.

de Andrade Marinho, Mara, et al. "Bibliographia Brasileira de Levantomento e de Interpretacao de Solos Para Fins Agricolas." *Revista Brasileira de Geografia*, 53, No. 1 (1991): 147-172.

Ceron, Antonio O., and M.C. Sanchez. "Estrutura da producao agricole de regiao de Aracatuba, SP." *Boletim de Geografia Teoretica*, 21, No. 41 (1991): 5-43.

de Souza, J.C., B.C. Nentwig Silva, and S.S. Andrade de Silva. "Combinacoes Agricolas no Estado da Bahia—1970/1980: Uma Contribuicao Metodologica." *Revista Brasileira de Geografia*, 53, No. 2 (1991): 95-112.

Ferreira, Ignez C.B. "A Gestao do Espaco Agrario." *Revista Brasileira de Geografia*, 53, No. 3 (1991): 149-160.

Isenburg, Teresa. "L'agricoltura Nello Stato di San Paolo (Brasile)." *Revista Geografica Italiana*, 98, No. 1 (1991): 33-65.

Kohlhepp, Gerd. "Mundancas Estruturais na Agropecuaria e Mobilidade da Producao Rural no Norte do Parana, Brasil." *Revista Brasileira de Geografia*, 53, No. 2 (1991): 79-94.

Machado, Lia. "A fronteira agricola na Amazonia Brasileira." *Revista Brasileira de Geografia*, 54, No. 2 (1992): 27-56.

Swales, Susan E. "A Case Study of Agricultural Practices in Monte Alegre, Brazil." Master's Thesis, University of Illinois, Chicago, Ill., 1993.

Saito, I., Noritaka Yagasaki, and S. Suyama. "Drought, Canal Irrigation Project and Family Farmers in the Ico Lima Campos Area of Ceara, Northeast Brazil" (in Japanese). *Tsukuba Studies in Human Geography*, 21 (1997): 69-91.

Defouvello, Christope. "Approvisionnement energetique et specialisation agricole des regions pionnieres: L'Amazonie meridionale." *L'Espace Geographique*, 28, No. 3 (1999): 257-268.

Brannstrom, Christian. "Coffee Labor Regimes and Deforestation on a Brazilian Frontier, 1915–1965." *Economic Geography*, 76, No. 4 (2000): 326-346.

CROPS AND CROPPING PATTERNS

Haas, William H. "The Coffee Industry of Brazil." *Journal of Geography*, 28, No. 2 (1929): 41-56.

Cameron, C.R. "Mate: An Important Brazilian Product." *Journal of Geography*, 29, No. 2 (1930): 54-69.

James, Preston E. "The Coffee Lands of Southeastern Brazil." *Geographical Review*, 22, No. 2 (1932): 225-244.

Keithan, Elizabeth F. "The Cacao Industry of Brazil." *Economic Geography*, 15, No. 3 (1939): 195-204.

Russell, Joseph A. "Fordlandia an Belterra, Rubber Plantations on the Tapajos." *Economic Geography*, 18, No. 2 (1942): 125-145.

Strain, Warren. "Carnauba Wax: Brazil's Monopoly Product." *Journal of Geography*, 41, No. 4 (1942): 121-129.

Britton, Charles E. "The Coffee Industry of Sao Paulo, Brazil." Master's Thesis, University of Tennessee, Knoxville, Tenn., 1947.

Kallicott, Signy. "Cotton Production and Potentialities in Northern Brazil." Master's Thesis, University of Maryland, College Park, Md., 1947.

Sullinger, Clara B. "Cotton Production in Sao Paulo State, Brazil." Master's Thesis, Oklahoma State University, Stillwater, Okla., 1950.

Bernardes, L.M.C. "Cultura e producao do arroz no Sud do Brasil." *Revista Brasileira de Geografia*, 16 (1954): 403-438.

Papy, Louis. "Au pays des plantations cafeieres de Sao Paulo." *Les Cahiers d'Outre Mer*, 7 (1954): 195-203.

Dambaugh, Luella N. "Westward Migration of Coffee in Brazil." *Journal of Geography*, 57, No. 4 (1958): 173-176.

Santos, Milton. "La culture du cacao dans l'Etat de Bahia." *Les Cahiers d'Outre Mer*, 16 (1963): 360-378.

da Camara Cascudo, Luis. "A Carnauba." *Revista Brasileira de Geografia*, 26, No. 2 (1964): 159-215.

Adams, J.G.U. "The Banana Industry in the Paulista Litoral." Master's Thesis, University of Western Ontario, London, Ontario, 1967.

Valverde, Orlando. "A Fazenda de Cafe Escravocrata, no Brasil." *Revista Brasileira de Geografia*, 29, No. 1 (1967): 37-81.

Johnson, Dennis V. "The Carnauba Palm (Copernica prunifera) and Its Role as an Economic Plant." Master's Thesis, University of California, Los Angeles, Calif., 1970.

Johnson, Dennis V. "The Cashew of Northeast Brazil: A Geographical Study of a Tropical Tree Crop." Ph.D. Dissertation, University of California, Los Angeles, Calif., 1972.

Pierson, Robert G. "A Geographical Analysis of the Citrus Industry of Sao Paulo, Brazil." Ph.D. Dissertation, University of Tennessee, Knoxville, Tenn., 1972.

Gervaise, Yves. "La culture de la pommer de terre a Ouro Branco (Minas Gerais): Un exemple de specialisation agricole." *Les Cahiers d'Outre Mer*, 26 (1973): 279-313.

Gerardi, L.L., and T.J.L. Ferreira. "O retarno do cafe as antigas areas productoras." *Boletim Geografico*, 33, No. 242 (1974): 103-116.

Kohlhepp, Gerd. *Agrarkolonisation in Nord-Parana: Wirtschafts-und sozialgeographische. Entwicklungsprozesse e. randtrap. Pionierzone Brasiliens under d. Einfluss der koffeeanbaus.* Heidelberg, Germany: Heidelberger Geographische Arbeiten, Heft 41, 1975.

de Araujo Filho, J.R. "O cafe em Sao Paulo." *Boletim Paulista de Geografia*, 50 (1976): 57-82.

Poltronieri, Ligia C. "Difusao Espacial da citricultura no Estado de Sao Paulo analisada at do modelo de Superficie de Tendencia." *Boletim de Geografia Teoretica, Rio Claro*, 6, No. 11-12 (1976): 5-28.

Huetz de Lemps, Alain. *La canne a sucre au Bresil.* Talence, France: Travaux et Documents de Geographie Tropicale, CEGET-CNRS, Vol. 29, 1979.

Bray, Silvio C. "Aspectos da cultura da laranja em Bebedouro." *Boletim Paulista de Geografia*, 57 (1980): 5-24.

Matznetter, Josef. "Der kadao Anbau in Brasilien." *Wirtschaftsgeographische Studien*, 7 (1980): 23-48.

Meaney, Anne M. "The Cashew Industry of Ceara, Brazil: A Study of a Local Agro-Industry." Master's Thesis, University of Calgary, Calgary, Alberta, 1980.

Sanchez, Miguel C. "Expansao das lavouras de soja e trigo no Estado de Sau Paulo." *Geografia, Rio Claro*, 6, No. 11-12 (1981)· 161-177.

Luckner, Reinhold. "Agrarer Strukturwandel unter dem Einflusz des Sojawelmarkets. Das Alto-Uruguai-Bebiet auls Beispiel." *Geographische Rundschau*, 34, No. 8 (1982): 368-373.

Ferreira, E.R., and Silvio C. Bray. "As agroindustrias e a formacao do seter canavieiro em Araras." *Boletim de Geografia Teoretica*, 13, No. 25 (1983): 57-68.

Grigoletto, M.F., and N. de Melo Souza. "Analise de uma agroindustira de tomate na alta Sorocabana: As relacoes industria-productor." *Revista de Geografia*, Vol. 2 (1983): 111-122.

Sanches, Miguel C., and Lucia H. de Oliveira Geradi. "Transformacoes na producao de graos no Brasil Meridional no periodo 1950–1979." *Revista de Geografia*, 2 (1983): 21-56.

Dos Santos, M.J.Z. "Comportamento espacial da cultura canavieira no Estado de Sao Paulo." *Boletim de Geografia Teoretica*, 14, No. 27-28 (1984): 21-34.

Sanchez, Miguel C., and L.H.O. Gerardi. "Transformations in the Grain Yields in Southern Brazil, 1950–79." *Rural Systems*, 2, No. 4 (1984): 245-258.

Sampaio, R.R., and Silvio C. Bray. "O planejamento estatel no processo de modernizacao do setor canaviero de Porto Feliz." *Revista de Geografia*, 4 (1985): 57-68.

Pedrosa, J.M.Y. "O des lo camento da cultura do tomate rasteiro no estado ne Sao Paulo." *Revista de Geografia*, 5 (1986): 55-74.

Graham, Douglas, Howard Gauthier, and Jose R. Mendonca de Barros. "Thirty Years of Agricultural Growth in Brazil: Crop Performance, Regional Profile and Recent Policy Review." *Economic Development and Culture Change*, 36, No. 1 (1987): 1-34.

Meaney-Leckie, Anne M. "The Cashew Industry of Ceara, Brazil: A Development Alternative." *Proceedings of the Middle States Division, Association of American Geographers*, 22 (1989): 35-44.

Dos Santos, M.J.Z. "Comportamento espacial da cultura canavieira no estado de Sao Paulo: 1980." *Revista de Geografia*, 8/9 (1989–90): 31-66.

Meaney-Leckie, Anne M. "The Cashew Industry of Ceara, Brazil: Case Study of a Regional Development Option." *Bulletin of Latin American Research*, 10, No. 3 (1991): 315-324.

Caviedes, Cesar N., and Keith D. Muller. "Fruitculture and Uneven Development in Northeast Brazil." *Geographical Review*, 84, No. 4 (1994): 380-393.

DeBarros, H.M. "Developpement et possibilities de maintien de l'agriculture familiale dans le nordeste bresilien: Le cas de la culture maraicere du Pernambuco." *Travaux, Institut de Geographie / Reims*, 89-90 (1994): 141-156.

Dickinson, John P. "Viticulture in Pre-Independence Brazil." *Journal of Wine Research*, 6, No. 3 (1995): 195-200.

Johns, Norman. "Brazil's Chocolate Forest: Environmental and Economic Roles of Conservation in Bahia's Cocoa Agro-System." Ph.D. Dissertation, University of Texas, Austin, Tex., 1996.

Smith, Nigel J. "Home Gardens as a Springboard for Agroforestry Development in Amazonia." *International Tree Crops Journal*, 9 (1996): 11-30.

ANIMAL HUSBANDRY

Sternberg, Hilgard O. "Die Viehzucht im Careito–Cambixegebiet." *Heidelberger Geographische Arbeiten*, 15 (1966): 171-197.

Correia de Andrade, Manuel. "L'elevage dans le Nord-Est du Bresil." *Les Cahiers d'Outre Mer*, 21 (1968): 56-77.

Becker, Bertha B. "Changing Land Use Patterns in Brazil: The Spread of Cattle Raising in Sao Paulo State." *Revista Geografica*, 71 (1969): 35-44.

Pebayle, Raymond, Orlando Valverde, and Rivaldo Pinto de Gusmao. *Aspects de l'agriculture commerciale et de l'elevage au Bresil.* Talence, France: Travaux et Documents de Geographie Tropicale, No. 11, 1973.

Geiger, Pedro P., et al. "Distribuicao de Atividades Agro-Pastoris em Torno da Metropole de Sao Paulo." *Revista Geografica*, 80 (1974): 35-70.

Kehr, J. "Cattle-Raising in Brazil: An Overview." *Revista Geografica*, 83 (1975): 95-108.

Mamigonian, Armen. "Notas sobre os frigorificos do Brasil central Pecurio." *Boletim Paulista de Geografia*, 51 (1976): 7-14.

Mamigonian, Armen. "Notes sur les frigorifiques de bovins du Bresil Central." *Recherches Geographiques a Strasbourg*, Special Issue (1976): 69-75.

Pereira, Benedito A.S. "Plantas Natives do Cerrado Pastadas par Bovinas na Regiao Geoeconomica do Distrito Federal." *Revista Brasileira de Geografia*, 46, No. 2 (1984): 381-388.

Poelhekke, G.M.N. "Fences in the Jungle: Cattle Raising and the Economic and Social Integration of the Amazon Region in Brazil." *Revista Geografia*, 104 (1986): 33-48.

Sanches, M.C., and Antonio O. Ceron. "A atividade pecuarin na regiao de Aracatuba." *Boletim de Geografia Teoretica*, 19, No. 37-38 (1989): 53-66.

Wolford, Wendy. "The Political Economy of Poultry Production in Brazil." Master's Thesis, University of California at Berkeley, 1997.

Remppis, M. "Brasiliens Rinderweidewirtschaft natur zu fleisch?" *Geographische Rundschau*, 51, No. 5 (1999): 256-262.

FARM LAND

de Oliveira, J.B., and A. Carvalho. "Uso do solo no municipio de Ribeirao Preto, Estado de Sao Paulo." *Boletim Paulista de Geografia*, 43 (1966): 5-58.

Correia de Andrade, M. "Candicoes naturais a sistemas de exploracao da terra no estado de Pernambuco." *Boletim Paulista de Geografia*, 44 (1967): 63-84.

Mandell, Paul I. "The Development of the Southern Goias Brasilia Region: Agricultural Development in a Land-Rich Economy." Ph.D. Dissertation, Columbia University, New York, N.Y., 1969.

Ceron, Antonio O. "Alguns padroes de utilicacao da terra agricola no Planalto Occidental de Sao Paulo." *Boletim Paulista de Geografia*, 47 (1972): 3-29.

Dillman, C. Daniel. "Absentee Landlords and Farm Management in Brazil During the 1960s." *American Journal of Economics and Sociology*, 37, No. 1 (1975): 1-17.

Yamamoto, S. "On the Areal Variation of Land and Its Utilization Patterns in the State of Paraiba, Northeastern Brazil" (in Japanese). *Geographical Review of Japan*, 48, No. 9 (1975): 616-627.

Hicks, David R. "Agricultural Land Use and Related Innovation and Government Assistance in Rio Grande do Sul." Ph.D. Dissertation, Michigan State University, East Lansing, Mich., 1980.

Lucker, Reinhold. *Agrarraumliche Entwicklungsprozesse im Alto-Uruguai-Gebiet (Sudbrasilien)*. Tubingen, Germany: Tubinger Geographische Studien, Heft 94, 1986.

Saito, I., et al. "Agriculture and Land Tenure of the Salgado de Sao Felix in the Agreste of Paraiba, Brazil" (in Japanese). *Tsukuba Studies in Human Geography*, 10 (1986): 77-106.

de Andrade Marinho Weill, Mara. "Metologias de Avaliacao de Terres Para Fins Agricolas." *Revista Brasileira de Geografia*, 52, No. 4 (1990): 127-160.

Brabois, Jose, Mauro J. da Silva, and Caio A.A. Macial. "Reordenacao espacial e evolucao da economia agraria: O caso das Terras Altas da Transicao Agreste-Mata do Norte de Pernambuco." *Revista Brasileira de Geografia*, 54, No. 1 (1992): 121-178.

Donnelly-Morrison, Duane. "Defining Agricultural Land Use in Rondonia, Brazil by Examination of SPOT Multispectral Data." Master's Thesis, Virginia Polytechnic Institute and State University, Blacksburg, Va., 1994.

AGRICULTURAL POPULATIONS AND SOCIAL ISSUES

Pfeifer, Gottfried. "Landwirtschaftliche Betriebssystem und kolonisation serfolg in sudbrasilien auf Grund er Forschungen von Leo Waibel." *Erdkunde*, 7, No. 4 (1953): 241-248.

Roche, G. "L'agriculture des colons allemands dans le Rio Grande do Sul." *Annales de Geographie*, 68 (1959): 224-237.

Galloway, J.H. "The Last Years of Slavery on the Sugar Plantations of Northeast Brazil." *Hispanic American Historical Review*, 51 (1971): 586-605.

Kohlhepp, Gerd. "Gelenkte Agrarkolonisation im Rahmen der Expansion des Koffeeanbaus in Norden Paranas (Brasilien)." Pp. 71-90 in *Czajka-Festschrift*. Gottingen, Germany: Gottinger Geographische Abhandlungen, Heft 66, 1976.

Pebayle, Raymond. *Les gauchos du Bresil: Eleveurs et agriculteurs du Rio Grande do Sul*. Talence, France: Travaux et Documents de Geographie Tropicale, CEGET-CNRS, No. 31, 1977.

Takagi, Hideki. "Socio-Economic Changes of Bastos, a Japanese Colony in Brazil" (in Japanese). *Jimbun Chiri/Human Geography, Kyoto*, 29, No. 1 (1977): 91-112.

Schacht, Siegfried. "Agrarkolonisation in der Zone der Mata Nordbrasiliens am Beispiel der Kolonie Pindorama." *Geographische Zeitschrift*, 68 (1980): 54-76.

Schacht, Siegfried. "Agricultural Colonization of the Zona da Mata of North-East Brazil: The Example of Pindorama." *Applied Geography and Development*, 17 (1981): 71-90.

Brito, Mario S., and Solange T. Silva. "O papel da Pequena producao na agricultura Brasileira." *Revista Brasileira de Geographia*, 44, No. 2 (1982): 191-262.

Poltronieri, Ligia C. "Consequences do modernizacao da agricultura: O uso indiscriminado de praguici das no municipio de Rio Claro (SP)." *Revista de Geografia (Sao Paulo)*, 7 (1988): 35-50.

Muller, Keith D. "Pindorama: European Agricultural Settlement Scheme, Coastal Northeast Brazil." *Papers and Proceedings of Applied Geography Conferences*, 16 (1993): 1-7.

ENVIRONMENTAL AND CONSERVATION ISSUES

Pebayle, Raymond. "Fondements physiques de la geographie agraire et economie de cueillette au Bresil selon Orlando Valverde." *Les Cahiers d'Outre Mer*, 21 (1968): 209-216.

Langsten, M.E. "Drought and Land Tenure in Northeastern Brazil: The Impact of Drought Adjustment Techniques on Farm Size and the Implications of These Techniques for Agrarian Reform." Master's Thesis, University of Michigan, Ann Arbor, Mich., 1974.

Rose, Newton. "Rainfall Reliability and Agriculture in the Semi-Arid Tropics: The Case of the Jaguaribe River Basin, Ceara, North-East Brazil." Pp. 115-128 in *Spatial Aspects of Development*, edited by Brian S. Hoyle. London, England: John Wiley, 1974.

Guitton, Mario C.H. "Influencia dos Fenomenos Climaticos na Producao Agricola Brasileira." *Revista Brasileira de Geografia*, 41, No. 3 (1979): 123-129.

Johnson, Dennis V. "Agroclimatological Zonation of Maize and Sorghum in Northeast Brazil." *Revista Geografica*, 89 (1979): 37-44.

Une, Mitiko Y. "Fatores Climaticos Influenciando a Agricultura em Campo Grande." *Revista Brasileira de Geografia*, 41, No. 1-2 (1979): 3-31.

Une, Mitiko Y. "Interferencias Climaticas nas Atividades Agricoles." *Revista Brasileira de Geografia,* 41, No. 3 (1979): 112-122.

Poltronieri, Ligia C. "Consequencias da moderizacao da agricultura: O uso indiscriminado de pragui idas no munipio de Rio Claro." *Revista de Geografia*, 7 (1988): 35-50.

Cairus, David M. "Spatial Aspects of Witches Broom Disease of Cocoa in Rondonia, Brazil." Master's Thesis, University of Florida, Gainesville, Fla., 1991.

Laroche, Rose C. "Ecossistemes e Impactos Ambientais da Modernizacao Agricola do Vale do Sao Francisco." *Revista Brasileira de Geografia*, 53, No. 2 (1991): 63-78.

Muller, Geraldo. "O agrario verde: Amarelo hoje e amanha." *Revista Brasileira de Geografia*, 54, No. 4 (1992): 29-48.

Mistry, J. "Decision-Making for Fire Use Among Farmers in Savannas: An Exploratory Study in the Distrito Federal, Central Brazil." *Journal of Environmental Management*, 54, No. 4 (1998): 321-334.

IRRIGATION

Andrade, Manuel Correia de Oliveira. *Tradicao e Mundanca: A Organizacao do Espaco Rural e Urbano na Area de Irrigacao do Submedio Sao Francisco*. Rio de Janeiro, Brazil: Zahar Editores, 1983.

Brito, Mario S. "O Programa Nacional de Irrigacao: Unea Avaliacao Previa dos Resultados." *Revista Brasileira de Geografia*, 53, No. 2 (1991): 113-126.

Saito, I., Noritaka Yagasaki, and H. Maruyama. "Irrigation Projects and Corporate Agriculture in the Middle Sao Francisco Valley, Northeast Brazil" (in Japanese). *Tsukuba Studies in Human Geography*, 15 (1991): 269-300.

Broggio, C., M. Droulers, and P. Grandjean. "L'irrigation dans le Nordeste du Bresil." *L'Information Geographique*, 63, 5 (1999): 223-231.

AGRICULTURAL MARKETING AND DISTRIBUTION

Webb, Kempton E. "Food Supply of Belo Horizonte." Ph.D. Dissertation, Syracuse University, Syracuse, N.Y., 1958.

Thompson, Roderick R. "Agricultural Cooperatives: A Means of Promoting Regional Development in Brazil." Master's Thesis, University of Western Ontario, London, Ontario, 1971.

Geiger, Pedro P., Maria Motta Lima, and Myriam E. Abi Abib. "Distribuicao de Atividades Agropastoris em Torno da Metropole da Sao Paulo." *Revista Brasileira de Geografia*, 36, No. 4 (1974): 3-36.

TRANSPORTATION AND MECHANIZATION

Glen, J. "The Technological Structure of Sao Paulo's Agricultural Economy." Master's Thesis, University of Western Ontario, London, Ontario, 1971.

Pinto de Gusmao, Rivaldo, and Olindina V. Mesquita. "Difusao da Infra-Estrutura de Ormazenagem e Suasvinculacoes Com a Atividade Agraria no Noroeste do Rio Grande do Sul." *Revista Brasileira de Geografia*, 40, No. 3-4 (1978): 52-130.

Pinto de Gusmao, Rivaldo, and Olinda V. Mesquita. "Difusao da Intraestrutura de Armazenogem e Suas Vinculcoes Com a Atividade Agraria no Noroeste do Rio Grande do Sul." *Revista Geografica*, 88 (1978): 103-150.

Ceron, Antonio O., and Herbert S.A. Pinho Halsbgut. "Tendences espaciais na difusao dos tratores empregados no agricultura paulista apartir de 1940." *Geografia, Rio Claro*, 3, No. 6 (1979): 75-88.

Guimaraes, Luiz S.P. "Transformacoes Tecnicas e Relacoes de Trabalho na Agricultura Brasileira em Areas de Baixo Nivel de Modernizacao." *Revista Brasileira de Geografia*, 45, No. 2 (1983): 155-204.

Hess, Dora R. "Transformacoes Tecnics e Relacoes de Trabalho na Agricultura Brasileira." *Revista Brasileira de Geografia*, 45, No. 1 (1983): 3-50.

Innocencio, Ney R., and Tereza M. Ramos de Oliveira. "Transformacoes Tecnicas e Relacoes de Trabalho na Agricultura Brasileira em Areas de Nivel de Modernizacao." *Revista Brasileira de Geografia*, 45, No. 3-4 (1983): 263-310.

Kohlhepp, Gerd. "Strukturwandlungen in der Landwirtschaft und Mobilitat der landlichen Bevolkerung in Nord-Parana (Sudbrasilien)." *Geographische Zeitschrift*, 77, No. 1 (1989): 42-61.

Laroche, Rose C. "Ecossistemes e Impactos Ambientais da Modernizacao Agricola do Vale do Sao Francisco." *Revista Brasileira de Geografia*, 53, No. 2 (1991): 63-78.

Muniz, Osvaldo. "Innovation and Diffusion of Electrical Equipment for Agricultural Development in Minas and Goias, Brazil." Ph.D. Dissertation, University of Tennessee, Knoxville, Tenn., 1991.

AGRICULTURAL LABOR STUDIES

Poltronieri, Ligia C. "Agricultural Production and Work Force in the Spatial Agricultural System of Sao Paulo" (in Portugese). *Geografia Teoretica*, 3, No. 5 (1973): 59-65.

Rossini, Rosa E. "Women as a Labor Force in Agriculture: The Case of the State of Sao Paulo, Brazil." *Revista Geografica*, 97 (1983): 91-95.

Pires Guimeraes, Luiz S., Ney R. Innocencio, and Sebastiana Rodriguez de Brito. "O Trabalhdor Volante na Agricultura." *Revista Brasileira de Geografia*, 46, No. 1 (1984): 5-78.

Chase, Jaquelyn. "Controlling Labor Commitments in Brazil's Global Agriculture: The Crisis of Competing Flexibilities." *Environment and Planning D*, 15, No. 5 (1997): 587-610.

Chile

GENERAL TEXTS, ARTICLES, AND THESES

McBride, G.M. "The Agrarian Problem in Chile." *Geographical Review*, 20, No. 4 (1930): 574-586.

Papy, Louis. "Les structures agraires du Chili central." *Les Cahiers d'Outre Mer*, 13 (1960): 461-469.

Smole, William J. "Owner-Cultivatorship in Middle Chile." Ph.D. Dissertation, University of Chicago, Chicago, Ill., 1964.

Santana, R. "Estructura de la ganaderia en Chile central." *Informaciones Geograficas* (1968–69): 57-74.

Kreisman, Arnold. "The Numerical Regionalization of Agricultural Land: A Chilean Example." Ph.D. Dissertation, University of Pittsburgh, Pittsburgh, Pa., 1972.

Wallace, Leilani. "Agrarian Reform and Agricultural Production: Chile, 1970–73." Master's Thesis, University of Windsor, Windsor, Ontario, 1978.

Henriquez, M.E. "Reforma Agrarian en Chile." *Revista de Geografia Norte Grande*, 14 (1987): 61-65.

Bahr, Jurgen, and Antje Fischbock. "Die Veranderungen der Landwirtschaftlichen Betriebsgrosenstruktur durch die Agrarreform in Chile." *Erdkunde*, 41, No. 1 (1988): 52-60.

Rodriguez, Manuel Z. *Geografia Agricola de Chile*. Santiago, Chile: Editorial Universitaria, 1989.

Ortiz-Veliz, J. "Impacto Sociogeografico de la Modernizacion Agricola en dos Valles del Semiarido Chileno." *Revista Geografica de Chile*, 33 (1990): 66-78.

CROPS AND CROPPING PATTERNS

Larrain, P., and T. Vial. "La Producion de Una de Mesa en la Region de Atacama: Requerimiento Actuals y Futuros de Mano de Obra." *Revista Geografica de Chile*, 32 (1990): 73-102.

Gwynne, Robert N. "Globalization, Commodity Chains and Fruit Exporting Regions in Chile." *Tijdschrift voor Economische en Sociale Geographie*, 90, No. 2 (1999): 211-225.

Rosas, Margarita R. "Empleo Temporal y Mercados de Trabajo en la Fruitcultura de Chile Central." *Revista Geografica*, 125 (1999): 139-152.

DAIRYING

Thompson, John. "The Chilean Dairy Industry." *California Geographer*, 1 (1960): 35-40.

FARM LAND

Alaluf, David. *Problemas de la Propiedad Agricola en Chile*. Kiel, Germany: Kieler Geographische Schriften, Band 19, Heft 2, 1961.

Durland, Robert E. "Farm Size and Land Use in the Plan Chillan Area of Chile." Ph.D. Dissertation, Northwestern University, Evanston, Ill., 1963.

AGRICULTURAL POPULATIONS AND SOCIAL ISSUES

Bloom, Reynald J., Jr. "The Influence of Agrarian Reform on the Small Holder Communities in Chile's Central Valley, 1965–1970." Ph.D. Dissertation, University of California, Los Angeles, Calif., 1973.

Bee, Anna E. "The Role of Women in the Grape Economy of the Limari Valley, Chile." Master's Thesis, Miami University, Miami, Ohio, 1991.

Clapp, Roger A. "Regions of Refuge and the Agrarian Question: Peasant Agriculture and Plantation Forestry in Chilean Araucania." *World Development*, 26, No. 4 (1998): 571-589.

Bee, Anna E. "Globalization, Grapes and Gender: Women's Work in Traditional and Agro-Export Production in Northern Chile." *Geographical Journal*, 166, No. 3 (2000): 255-265.

AQUACULTURE

Barton, Jonathan R. "Environment, Sustainability and Regulation in Commercial Aquaculture: The Case of Chilean Salmonid Production." *Geoforum*, 28, No. 3-4 (1997): 313-328.

Barton, Jonathan R. "Revolucion Azul? El Impacto Regional de la Acuicultura del Salmon en Chile." *Revista de Estudios Urbans y Regionales de America*, 22, No. 68 (1997): 57-76.

Barton, Jonathan R. "Salmon Aquaculture and Chile's 'Export-Led' Economy." *Norsk Geografisk Tidsskrift*, 52, No. 1 (1998): 37-47.

Lariau, Sophie, and Jose Jazcano. "Emergence et consolidation d'un pays salmonicole: Le cas du Chili." *Les Cahiers d'Outre Mer*, 51 (1998): 277-304.

AGRICULTURAL MARKETING AND DISTRIBUTION

Glassner, Martin I. "Feeding a Desert City: Antofagasta, Chile." *Economic Geography*, 45, No. 4 (1969): 339-348.

AGRICULTURAL LABOR STUDIES

Polhemus, Deborah L. "The Changing Structure of the Agricultural Labor Force in Central Chile." Master's Thesis, University of California, Berkeley, Calif., 1982.

Colombia

GENERAL TEXTS, ARTICLES, AND THESES

West, Robert C. "Ridge or 'Era' Agriculture in the Colombian Andes." *Actas de XXXIII Congreso Internacional de Americanistas*, 1 (1959): 279-282.

Schier, Friedrich G. "An Application of Von Thunen's Agricultural Model to Bogota, Colombia." Master's Thesis, University of Akron, Akron, Ohio, 1974.

Lazarev, Grigori. "Une analyse regional du probleme agraire en Colombie." *L'Espace Geographique*, 6, No. 2 (1977): 113-123.

Ramsey, Glenn T. "The Impact of Agrarian Reform on the Department of Mela, Colombia." Master's Thesis, University of Florida, Gainesville, Fla., 1980.

CROPS AND CROPPING PATTERNS

Archiniegas, Mercedes. "The Banana Industry in Santa Marta, Colombia." Master's Thesis, Oklahoma State University, Stillwater, Okla., 1951.

Tanner, Hans J. "Der kolumbianische Koffee." *Geografica Helvetica*, 23 (1968): 180-186.

Mertins, Gunter. *Die Bananezone von Santa Marta, Nordkolumbien*. Giessen, Germany: Giessener Geographische Schriften, Heft 17, 1969.

Maillard, Jean-Claude. "Bilon et problemes actuels de l'economie banaiere en Colombie." *Les Cahiers d'Outre Mer*, 25 (1972): 390-417.

Gormsen, Erdmann. "Der Flughafen als Standortfaktor fur den Blumen-Anbau am Beispiel von Kolumbien." *Erdkunde*, 40, No. 4 (1986): 305-317.

ANIMAL HUSBANDRY

Wilhemy, Herbert. "La Cria de Ganagdo en las Costas de Caribe de Colombia." *Revista Geografica*, 68 (1968): 63-82.

FARM LAND

Rau, Herbert L., Jr. "The Agricultural Land Use and Settlement Patterns in the Sabana de Bogota, Colombia." Ph.D. Dissertation, Northwestern University, Evanston, Ill., 1958.

Elbow, Gary S. "Minifundia in Highland Colombia: An Agrarian System." *Proceedings of the Pennsylvania Academy of Science*, 39, No. 2 (1966): 283-288.

Elbow, Gary S. "Agrarian Systems and Land Utilization in Highland Colombia." *Oregon Geographer*, 1, No. 2 (1967): 1-4.

AGRICULTURAL POPULATIONS AND SOCIAL ISSUES

Brucher, Wolfgang. *Die Erschliessung des trapischen Regenwaldes am Ostrand der Kolumbianischen Anden*. Tubingen, Germany: Tubinger Geographische Studien, Heft 28, 1968.

ENVIRONMENTAL AND CONSERVATION ISSUES

Shelmon, Roy J., and L.B. Phelps. "Dredge-Tailing Agriculture on the Rio Nechi, Colombia." *Geographical Review*, 61, No. 3 (1971): 396-414.

Etter, Andres, and L. Albert Villa. "Andean Forests and Farming Systems in Part of the Eastern Cordillera (Colombia)." *Mountain Research and Development*, 20, No. 3 (2000): 236-245.

Ecuador

GENERAL TEXTS, ARTICLES, AND THESES

Miller, E.V. "Agricultural Ecuador." *Geographical Review*, 49, No. 2 (1959): 183-207.

Lopez Cordovez, Luis A. *Zonas Agricolas del Ecuador: Estudio Preliminar*. Quito, Ecuador: Junta Nacional de Planificacion y Coordinacion Economica, 1962.

McGregor, Elizabeth A. "Prediction of Income of Peasant Farms in Andean Ecuador." Master's Thesis, McMaster University, Hamilton, Ontario, 1972.

Bromley, R.J. "El Intercambio de Productas Agricolas entre la Costa y la Sierre Ecuatoriana." *Revista Geografica*, 78 (1973): 15-34.

Antonini, Gustavo A., and Becarios C. Cepeige. "Chambo: Evolusion del Paisaje y su Relacion Con el Potential Productive Agricola." *Revista Geografica*, 96 (1982): 5-24.

Lawson, Victoria. "Government Policy Biases and Ecuadorian Agricultural Change." *Annals of the Association of American Geographers*, 78, No. 3 (1988): 433-452.

Lemky, Kim. "A Comparative Study of Two Agricultural Zones in the Napo Valley, Ecuador." *Geoscope*, Vol. 18, 2 (1988): 50-58.

Breuer, Toni. "Agribusiness als Entwicklungsstimulas? Das Beispiel der Milchwirtschaft in der ecuadorianischen Andean." *Zeitschrift fur Wirtschaftsgeographie*, 36, No. 4 (1992): 193-209.

Lopez, Freddy, ed. *Geografia Agraria: Esbozos de la Problematica Agraria en el Ecuador*. Quito, Ecuador: Corporacion Editora Nacional; Colegio de Geografos del Ecuador, 1993.

CROPS AND CROPPING PATTERNS

Freille, Luis. "Some Notes on the Geography of Sugar Cane in Ecuador." *Proceedings, Oklahoma Academy of Science*, 37 (1957): 136-140.

Parsons, James J. "Bananas in Ecuador: A New Chapter in the History of Tropical Agriculture." *Economic Geography*, 33, No. 2 (1957): 201-216.

Preston, David A. "Changes in the Economic Geography of Banana Production in Ecuador." *Transactions, Institute of British Geographers*, 37 (1965): 484-497.

Ingrun, M. "La Vaca Verde: The Agave Fiber Industry in Riobamba, Ecuador." Master's Thesis, University of Wisconsin, Madison, Wis., 1991.

Bossio, Deborah A., and K.G. Cassman. "Traditional Rainfed Barley Production in the Andean Highlands of Ecuador: Soil Nutrient Limitations and Other Constraints." *Mountain Research and Development*, 11, No. 2 (1991): 115-126.

FARM LAND

Parsons, James J. "Ridged Fields in the Rio Guayas Valley, Ecuador." *American Antiquity*, 34 (1969): 76-80.

Denevan, William M., and Kent Mathewson. "Preliminary Results of the Samborandan Raised-Field Project, Guayas Basin, Ecuador." *British Archaeological Reports, International Series*, 189 (1983): 167-181.

Knapp, Gregory W., and William M. Denevan. "The Use of Wetlands in the Prehistoric Economy of the Northern Ecuadorian Highlands." Pp. 184-207 in *Prehistoric Intensive Agriculture in the Tropics,* edited by I.S. Farrington. Oxford, England: British Archaelogical Reports, No. 232, 1985.

Knapp, Gregory, and David Preston. "Evidence of Prehistoric Ditched Fields on Sloping Land in Northern Highland Ecuador." Pp. 403-424 in *The Ecology and Archeology of Prehistoric Agricultural Fields in the Central Andes*, edited by William M. Denevan, Kent Mathewson, and Gregory W. Knapp. Oxford, England: British Archaeological Reports, No, 359, 1987.

Parsons, James J. "Mapping and Dating the Prehistoric Raised Fields of the Guayas Basin, Ecuador." *British Archaeological Reports, International Series*, 359 (1987): 207-216.

AGRICULTURAL POPULATIONS AND SOCIAL ISSUES

Basile, David G. *Tillers of the Andes: Farmers and Farming in the Quito Basin*. Chapel Hill, N.C.: University of North Carolina Department of Geography, Studies in Geography, No. 8, 1974.

Delavaud, Anne C. "El Papel de la Colonizacion Agricole en la Integration del Espacio Costeno en el Territorio Ecuatoriano." *Revista Geografica*, 84 (1976): 33-44.

Preston, David A., and Gerardo A. Taveras. *Agricultural Change, Land Tenure, and Emigration in Three Metizo Areas of Highland Ecuador*. Leeds, England: University of Leeds, Department of Geography, Working Paper No. 179, 1977.

Preston, David A., and Gerardo A. Taveras. *Emigration, Land Tenure, and Agricultural Change in Areas of Indian Population in Highland Ecuador*. Leeds, England: University of Leeds Department of Geography, Working Paper No. 180, 1977.

Preston, David A., and Gerardo A. Taveras. *Agricultural Change and Emigration in Highland Ecuador*. Leeds, England: University of Leeds, Department of Geography, Working Paper No. 187, 1977.

Preston, Rosemary A., David A. Preston, and Gerardo A. Taveras. "Emigracion Rural y Desarrollo Agricola en al Sierre Ecuatoriana." *Revista Geografica*, 93 (1981): 7-36.

Bebbington, Anthony. "Modernization from Below: An Alternative Indigenous Development?" *Economic Geography*, 69, No. 3 (1993): 274-292.

ENVIRONMENTAL AND CONSERVATION ISSUES

Maggio Pena, Jaime G. *Agrometeorologica de la Provincia de Pichincha*. Mexico, D.F., Mexico: Instituto Panamericano de Geografia e Historia, Seccion de Geografia, 1964.

Knapp, Gregory W. *Andean Ecology: Adaptive Dynamics in Ecuador*. Boulder, Colo.: Westview, 1991.

Smith, B. "Systematic Descriptive Matrix Analysis of the Relationship Between Environment and Agriculture: An Interpretation from Alchipichi-Puellaro, Ecuador." *Singapore Journal of Tropical Geography*, 15, No. 1 (1994): 75-92.

AGRICULTURAL MARKETING AND DISTRIBUTION

Smith, Vernon A. "Marketing Agricultural Commodities in Pichincha Province, Ecuador." *Geographical Review*, 65, No. 3 (1975): 353-363.

TRANSPORTATION AND MECHANIZATION

Huddleston, P. "The Use of Agricultural Machinery in Pichincha Province, Ecuador." Master's Thesis, University of Windsor, Windsor, Ontario, 1974.

AGRICULTURAL LABOR STUDIES

Mathewson, Kent. "Estimating Labor Imputs in Raised Field Complexes of the Guayas Basin, Ecuador." *British Archaelogical Reports, International Series*, 359 (1987): 217-224.

Guyana, Surinam, and French Guiana

GENERAL TEXTS, ARTICLES, AND THESES

Genevois, Louis. "Les recherches agronomiques au Surinam." *Les Cahiers d'Outre Mer*, 16 (1963): 230-232.

Richardson, Bonham C. "Guyana's 'Green Revolution': Social and Ecological Problems in an Agricultural Programme." *Caribbean Quarterly*, 18 (1972): 14-23.

Lall, A., V.C. Lakhan, and D. Ranvana. "Resource Allocation in Agriculture: The Guyana Experience." *Canadian Journal of Development Studies*, 9, No. 2 (1988): 235-248.

Semple, Hugh M. "Agricultural Change and the Decline of Domestic Food Production in Guyana, 1960–1994." Ph.D. Dissertation, University of Manitoba, Winnipeg, Manitoba, 1997.

Semple, Hugh M. "An Overview of Domestic Food Production in Guyana, 1960–1993." *Caribbean Geography*, 9, No. 2 (2000): 20-43.

CROPS AND CROPPING PATTERNS

Richardson, Bonham C. "The Rice Culture of Coastal Guyana: A Study in Location and Livelihood." Ph.D. Dissertation, University of Wisconsin, Madison, Wis., 1970.

Auty, R.M. "Some Problems in Identifying Scale Economies: The Sugar Industry of Demerara, 1930–1965." *Journal of Tropical Geography*, 34 (1972): 8-16.

Richardson, Bonham C. "Distance Regularities in Guyanese Rice Cultivation." *Journal of Developing Areas*, 8, No. 2 (1974): 235-256.

deSouza, Padma D., and V.C. Lakhan. "Long-linear Analysis of Factors Contributing to the Post-Independence Decline of Guyana's Rice Industry." *Applied Geography*, 15, No. 2 (1995): 147-160.

Lakhan, V.C. "Development and Advancement of Guyana's Rice Industry." *Indo-Caribbean Review*, 3 (1997): 81-92.

FARM LAND

Merrill, Gordon C. "Recent Land Development in Coast British Guiana." *Canadian Geographer*, 5, No. 2 (1961): 24-29.

AGRICULTURAL POPULATIONS AND SOCIAL ISSUES

Burrough, J.B. "Ethnicity as a Determinant of Peasant Farming Characteristics: The Canals Polder, Guyana." *Journal of Tropical Geography*, 37 (1973): 1-8.

ENVIRONMENTAL AND CONSERVATION ISSUES

Shaw, Anthony B. "An Agro-Climate Procedure for Rice Cultivation on the Coastal Plain of Guyana." Master's Thesis, University of Western Ontario, London, Ontario, 1977.

Shaw, Anthony B. "An Agroclimatic Procedure for Rice Cultivation on the Coastal Areas of Guyana." *Malaysian Journal of Tropical Geography*, 4 (1982): 67-81.

Williams, Patrick E. "Drainage and Irrigation Projects in Guyana: Environmental Considerations." Pp. 233-248 in *Environment and Development in the Caribbean: Geographical Perspectives*, edited by David Barker and Duncan R.M. McGregor. Barbados: University of West Indies Press, 1995.

Paraguay

GENERAL TEXTS, ARTICLES, AND THESES

Wall, A. "Modernization Patterns of Chulupi and Lengua Farmers in the Paraguayan Chaco." Master's Thesis, University of Western Ontario, London, Ontario, 1985.

Evans, P.T. "Designing Agroforesty Innovations to Increase Their Adaptability: A Case Study from Paraguay." *Journal of Rural Studies*, 4, No. 1 (1988): 45-56.

Peru

GENERAL TEXTS, ARTICLES, AND THESES

Tamayo, Jorge L. "Geografia agricola del Peru." *Revista Geografica*, 4-6 (1942): 101-120.

Dollfus, Olivier. "Quelques donnees sur l'agriculture du Perou." *Les Cahiers d'Outre Mer*, 12 (1959): 112-116.

Denevan, William M. "Campa Subsistence in the Gran Pajonel, Eastern Peru." *Geographical Review*, 61, No. 4 (1971): 496-518.

Mikus, W. "Probleme der regionalen Steuerung der Nahrungswirtschaft an der Costa Perus." *Geographische Zeitschrift*, 62 (1974): 204-232.

Scott, Geoffrey. "Effects of Shifting Cultivation in the Gran Pajoral, Eastern Peru." *Proceedings of the Association of American Geographers*, 6 (1974): 58-61.

Keith, Robert G. *Conquest and Agrarian Change: The Emergence of the Hacienda System on the Peruvian Coast*. Cambridge, Mass.: Harvard University Press, 1976.

Smith, Clifford T. "Location and Agrarian Reform: The Peruvian Experience." Pp. 141-153 in *Latin America: Search for Geographical Explanations, Proceedings of the Fifth Conference of Latin Americanist Geographers*, edited by Robert Tata. Chapel Hill, N.C.: CLAG and University of North Carolina Department of Geography, Vol. 5, 1976.

Denevan, William M., et al. "Indigenous Agroforestry in the Peruvian Amazon: Bora Indian Management of Swidden Fallows." *Interciencia*, 9 (1984): 346-357.

Denevan, William M. "Peru's Agricultural Legacy: Ancient Methods May Be Useful in Reviving Today's Food Production." *Focus*, 35, No. 2 (1985): 16-21.

White, Stuart. "Relations of Subsistence to the Vegetative Mosaic of Vilcabama, Southern Peruvian Andes." *Yearbook, Conference of Latin Americanist Geographers*, 11 (1985): 5-12.

Cardich, Augusto. "Native Agriculture in the Highlands of the Peruvian Andes." *National Geographic Research*, 3, No. 1 (1987): 22-39.

Works, Martha A. "Aguaruna Agriculture in Eastern Peru." *Geographical Review*, 77, No. 3 (1987): 343-358.

Fonseca Martel, Cesar, and Enrique Mayer. *Comunidad y Producion en la Agricultura Andina*. Lima, Peru: Fomciencisas, 1988.

Zimmerer, Karl S. "Wetland Production and Smallholder Persistence: Agricultural Change in a Highland Peruvian Region." *Annals of the Association of American Geographers*, 81, No. 3 (1991): 443-464.

CROPS AND CROPPING PATTERNS

Milstead, Harley P. "Distribution of Crops in Peru." *Economic Geography*, 4 (1928): 88-106.

Sommers, Marjorie S. "Wheat Producing Areas in Peru." *Economic Geography*, 25, No. 1 (1949): 13-16.

Scott, Geoffrey. "Effects of Shifting Cultivation in the Gran Pajonal, Eastern Peru." *Proceedings, Association of American Geographers*, 6 (1974): 58-16.

Hiraoka, Mario. "Cash Cropping, Wage Labor, and Urbanward Migrations: Changing Subsistence in the Peruvian Amazon." *Studies in Third World Societies*, 32 (1985): 199-242.

Kus, James. "The Sugar Cane Industry of the Chicoma Valley, Peru." *Revista Geografica*, 109 (1989): 57-72.

Zimmerer, Karl S. "The Regional Biogeography of Native Potato Cultivation in Highland Peru." *Journal of Biogeography*, 18 (1991): 165-178.

Zimmerer, Karl S. "Managing Diversity in Potato and Maize Fields of the Peruvian Andes." *Journal of Ethnobiology*, 11 (1991): 23-49.

Lerch, Natalie. "Home Gardens, Crop Diversity, and Exchange of Planting Material in the Pacaya-Samiria National Reserve Area, Northeastern Peruvian Amazon." Master's Thesis, McGill University, Montreal, Quebec, 1999.

Delavaud, Collin. "Plantes europeennes introduites au Perou." *Les Cahiers d'Outre Mer*, 53 (2000): 111-128.

ANIMAL HUSBANDRY

Kent, Robert B. "Beekeeping Regions, Technical Assistance and Development Policy in Peru." *Yearbook, Conference of Latin Americanist Geographers*, 12 (1986): 22-33.

Kent, Robert B. "The African Honeybee in Peru: An Insect Invader and Its Impact on Beekeeping." *Applied Geography*, 9, No. 4 (1989): 237-258.

Webber, Ellen R. "Cows in the Colca: Household Cattle Raising in Achoma, Peru." Master's Thesis, University of Wisconsin, Madison, Wis., 1994.

FARM LAND

Collin Delavaud, Claude. *Les Regions Cotieres du Perou Septentrional: Occupation du Sol, Amenagement Regional.* Lima, Peru: Institut francais d'etudes andines, 1968.

Maass, Alfredo. *Entwicklung und Perspektiven der wirtschaftlichen Erschliessung des tropischen Waldlandes von Peru, unter besonderer Berucksichtigung der verkehrsgeographischen Problematik.* Tubingen, Germany: Tubinger Geographische Studien, Heft 31, 1969.

Williams, Lynden S. "Land Use Intensity and Farm Size: Traditional Agriculture in Cuzco, Peru." Ph.D. Dissertation, University of Kansas, Lawrence, Kans., 1973.

Smith, Clifford T. "The Development and Role of Sunken Field Agriculture on the Peruvian Coast." *Geographical Journal*, 145, No. 3 (1979): 387-400.

AGRICULTURAL POPULATIONS AND SOCIAL ISSUES

Smith, Clifford T. "Aspects of Agriculture and Settlement in Peru." *Geographical Journal*, 126 (1960): 397-409.

Anderson, R.G. "Rerene-Satipo-Ene: A Study in Peruvian Agricultural Colonization." Master's Thesis, University of Western Ontario, London, Ontario, 1969.

Monheim, F. "Zur Entwicklung der peruanischen Agrarreform: Beobacktungen auf einer Reisse, 1970." *Geographische Zeitischrift*, 60 (1972): 161-180.

Fauve, Henri. "Le peuplement et la colonisation agricole de la steppe dans le Perou central." *Annales de Geographie*, 84, No. 4 (1975): 415-441.

Julich, Volker. *Die Agrarkolonisation in Regenwald des mittbren Rio Huallaga (Peru).* Marburg, Germany: Marburger Geographische Schriften, Heft 63, 1975.

Hodl, Walter, and Jurg Gaschi. "Indian Agriculture as Exemplified by a Secoya Village on the Rio Yubineto in Peru." *Applied Geography and Development*, 20 (1982): 20-31.

Hiraoka, Mario. "Zonation of Mestizo Riverine Farming Systems in Northeast Peru." *National Geographic Research*, 2, No. 3 (1986): 354-371.

Ackelmire, Ann B. "Agrarian Reform and Internal Migration in Peru." Master's Thesis, University of Arizona, Tucson, Ariz., 1987.

Coomes, Oliver T. "Traditional Peasant Agriculture Along a Blackwater River of the Peruvian Amazon." *Revista Geografica*, 124 (1998): 33-56.

ENVIRONMENTAL AND CONSERVATION ISSUES

Smith, Clifford T., William M. Denevan, and P. Hamilton. "Ancient Ridged Fields in the Region of Lake Titicaca." *Geographical Journal*, 134, No. 3 (1968): 353-366.

Antunez de Mayolo, R., and Erik Santiago. "Fertilizantes agricoles en el antiguo Peru." *Boletin de la Sociedad Geografica de Lima*, 99 (1980): 31-40.

Schroder, Rudolf. "Niederschlagsverhaltnisse und agrarmeteoroligische Bedingungen fur die Landwirkschaft im Einzugsgebiet des Titicacasus." *Erdkunde,* 35, No. 1 (1981): 30-41.

Pozorski, Thomas, et al. "Prehispanic Ridged Fields of the Casma Valley, Peru." *Geographical Review,* 73, No. 4 (1983): 407-416.

Denevan, William M., et al. "Indigenous Agroforestry in the Peruvian Amazon: Bora Indian Management of Swidden Fallow." *Interciencia,* 9 (1984): 346-357.

Hiraoka, Mario. "Floodplain Farming in the Peruvian Amazon." *Geographical Review of Japan,* 58B (1985): 1-23.

Denevan, William M., and C. Padock, eds. *Swidden-Fallow Agroforestry in the Peruvian Amazon.* New York, N.Y.: New York Botanical Gardens, Advances in Economic Botany, Vol. 5, 1987.

Denevan, William M., and John M. Treacy. "Young Managed Fallows at Brillo Nuevo." Pp. 8-46 in *Swidden-Fallow Agroforestry in the Peruvian Amazon,* edited by William M. Denevan and C. Padock. New York, N.Y.: New York Botanical Gardens, Advances in Economic Botany, Vol. 5, 1987.

Treacy, John M. "Building and Rebuilding Agricultural Terraces in the Colca Valley of Peru." *Yearbook, Conference of Latin Americanist Geographers,* 13 (1987): 51-57.

Erickson, Clark L., and Kay L. Chandler. "Raised Fields and Sustainable Agriculture in the Lake Titicaca Basin in Peru." Pp. 230-250 in *Fragile Lands of Latin America: Strategies for Sustainable Development,* edited by John O. Browder. Boulder, Colo.: Westview, 1989.

Hiraoka, Mario. "Agricultural Systems on the Floodplains of the Peruvian Amazon." Pp. 75-101 in *Fragile Lands of Latin America: Strategies for Sustainable Development,* edited by John O. Browder. Boulder, Colo.: Westview, 1989.

Treacy, John M. "Agricultural Terraces in Peru's Colca Valley: Promises and Problems of an Ancient Technology." Pp. 209-229 in *Fragile Lands of Latin America: Strategies for Sustainable Development,* edited by John O. Browder. Boulder, Colo.: Westview, 1989.

Siverton, A., and A. Lundberg. "Farming Practices and Environmental Problems in an Arid Landscape: A Case Study from the Region of Lambayeque, Peru." *Geografiska Annaler B,* 78B, No. 3 (1996): 147-162.

Coomes, Oliver T., F. Corimand, and G.J. Burt. "Tropical Forests and Shifting Cultivation: Secondary Forest Fallow Dynamics Among Traditional Farmers of the Peruvian Amazon." *Ecological Economics,* 32, No. 1 (2000): 109-124.

IRRIGATION

Kus, James S. "A Historical Geography of Irrigated Agriculture in the Chicama Valley, Peru." Master's Thesis, Michigan State University, East Lansing, Mich., 1967.

Amiron, D.H.K. "El Desierto de Suchura, Peru: Problems of Agricultural Use of Deserts." *Revista Geografica*, 72 (1970): 7-12.

Kus, James S. "Selected Aspects of Irrigated Agriculture in the Chima Heartland, Peru." Ph.D. Dissertation, University of California, Los Angeles, Calif., 1972.

Park, Chris C. "Water Resources and Irrigation Agriculture in Pre-Hispanic Peru." *Geographical Journal*, 149, No. 2 (1983): 153-166.

Holmes, Roland C. *Irrigation in Southern Peru: The Chili Basin*. Chicago, Ill.: University of Chicago Department of Geography, Research Paper No. 212, 1986.

Feige, Wolfgang. "Irrigation Projects in the Andean Highlands of Peru." *Applied Geography and Development*, 34 (1989): 23-45.

AGRICULTURAL LABOR STUDIES

Zimmerer, Karl S. "Labor Shortages and Crop Diversity in the Southern Peruvian Sierra." *Geographical Review*, 81, No. 4 (1991): 414-432.

Uruguay

GENERAL TEXTS, ARTICLES, AND THESES

Fitzgibbon, Russel H. "Uruguay's Agricultural Problems." *Economic Geography*, 29 (1953): 251-262.

Griffin, Ernst C. "Testing the Von Thunen Theory in Uruguay." *Geographical Review*, 63, No. 4 (1973): 500-516.

Prost, Gerard. "Structures agraires en Uruguay: L'exemple du departement d'Artigas." *Les Cahiers d'Outre Mer*, 30 (1977): 178-192.

ANIMAL HUSBANDRY

Deffontaines, Pierre. "Histoire du betail dans les pays de la Plata et plus particulierement en Uruguay." *Les Cahiers d'Outre Mer*, 6 (1953): 23-36.

FARM LAND

Griffin, Ernst C. "The Agricultural Land Use Regions in Uruguay." *Revista Geografica*, 76 (1972): 121-148.

Griffin, Ernst C. "Agricultural Land Use in Uruguay." Ph.D. Dissertation, Michigan State University, East Lansing, Mich., 1972.

Griffin, Ernst C. "Causal Factors Influencing Agricultural Land Use Patterns in Uruguay." *Revista Geografica*, 80 (1974): 13-34.

Martinez, Ana M. "Agricultural Land Use Changes in Uruguay: 1956–1981." Master's Thesis, San Diego State University, San Diego, Calif., 1987.

Venezuela

GENERAL TEXTS, ARTICLES, AND THESES

Rudolph, William E. "Agricultural Possibilities in Northwestern Venezuela." *Geographical Review*, 34, No. 1 (1944): 36-56.

Pena, Roberto V. *Estudio Agrologico Semi-Detallado de la Zona Valles del Tuy, Estado Miranda*. Caracas, Venezuela: Estados Unidos de Venezuela, Ministeria de Agricultura y Cria, 1948.

Borcherdt, Christoph. *Geographische Untersurgungen in Venezuela*. Stuttgart, Germany: Stuttgarter Geographische Studien, Band 85, 1973.

Martinson, Tom L. "Oil Profits, Farm to Market Roads, and Corn and Coffee Production in Venezuela." *Proceedings, Indiana Academy of the Social Sciences*, 9 (1974): 75-83.

Machado, Gustavo E. "Nuevastendencias en la reforma agraria de la region Centro-Occidental." *Sintesis Geografica*, 1, No. 2 (1977): 12-17.

Nogueitade Barrios, Sonia. *Cultivos Industriales y Tenencia de la Tierra*. Caracas, Venezuela: Universidad Central de Venezuela, Centro de Estudios de Desarrollo, 1977.

Segnini, Isbelia. "Sobre la crisis de la agricultura venezolana y algunos de sus aspectos geograficos." *Sintesis Geografica*, 1, No. 1 (1977): 9-14.

Waldt, Hans-Otto. *Sonderkulturen in Venezuela: Eine agrargeographische Untersuchung unter besonderer Berucksichtigung der Standortproblematik*. Mainz, Germany: Mainzer Geographische Studien, Heft 15, 1978.

Dufresne, Lucie S. "Procesos de Modernizacion de la Agricultura en los Andes Venezolanos." *Revista Geografica*, 93 (1981): 37-60.

Leddy, George S. "Petroleum Revenue: The State and Agriculture in Venezuela." Master's Thesis, University of California, Berkeley, Calif., 1985.

Rojas-Lopez, Jose J. "La emergencia de Nuevos Sistemas de Producion y la Profundizacaion de los Designaldades Agroespaciales en el Campo Andino Venezolano." *Revista Geografica*, 102 (1985): 107-114.

Sebastiani, M. "La agricultura des de la perspectiva de las sistemas y regiones agricolas: Una linea de investigacion en Venezuela." *Revista Geografica Venezolana*, 32, No. 1 (1991): 65-86.

Ingsbury, Nancy. "Increasing Pressure on Decreasing Resources: A Case Study of Penon Amerindian Shifting Cultivation in the Gran Sabana, Venezuela." Ph.D. Dissertation, York University, North York, Ontario, 1999.

CROPS AND CROPPING PATTERNS

Crist, Raymond. "Wheat Raising in the Venezuelan Andes." *Scientific Monthly*, 56 (1943): 332-338.

Heyman, Arthur. "Physical Parameters in the Development of Peasant Horticulture in the Highland Guayana Region, Venezuela." Ph.D. Dissertation, Columbia University, New York, N.Y., 1967.

Jones, Richard C. "Descriptive and Policy Linear Programming Models for Regional Crop Allocation: The Western Llanos of Venezuela." Ph.D. Dissertation, Ohio State University, Columbus, Ohio, 1973.

Denevan, William M., and R.W. Bergman. "Karinya Indian Swamp Cultivation in the Venezuelan Llanos." *Yearbook, Association of Pacific Coast Geographers*, 37 (1975): 23-37.

Anderson, Thomas D. "Subsystems of Conuco Agriculture in the Basin of Lake Valencia, Venezuela: A Classification and Description." *Ohio Geographer*, 4 (1976): 23-34.

Rojas-Lopez, J.J. "Organizacion espacial de la economia cafetaler andina: La red e Trujillo." *Revista Geografica Venezolana*, 22-23 (1985): 109-130.

ANIMAL HUSBANDRY

James, Preston E. "The Possibilities of Cattle Production in Venezuela." *Bulletin of the Geographical Society of Philadelphia*, 22, No. 2 (1924): 9-20.

Marchand, Bernard. "Les ranchos de Caracas, contribution a l'etude des bidonvilles." *Les Cahiers d'Outre Mer*, 19 (1966): 105-143.

FARM LAND

Denevan, William M., and A. Zucchi. "Ridged Field Excavations in the Central Orinoco Llanos, Venezuela." Pp. 235-246 in *Advances in Andean Archeology*, edited by D.L. Browman. The Hague, Netherlands: Mouton, 1978.

Crist, Raymond. "Development and Agrarian Land Reform in Venezuela's Pioneer Zone." *American Journal of Economics and Sociology*, 43, No. 2 (1984): 149-158.

AGRICULTURAL POPULATIONS AND SOCIAL ISSUES

Aschmann, Homer. "Indian Pastoralists of the Guajira Peninsula." *Annals of the Association of American Geographers*, 50, No. 3 (1960): 408-418.

Miller, E. Willard. "Population Growth and Agricultural Development in the Western Llanos of Venezuela." *Revista Geografica*, 69 (1968): 7-27.

Eidt, Robert C. *Agrarian Reform and the Growth of New Rural Settlements in Venezuela*. Milwaukee, Wis.: University of Wisconsin at Milwaukee, Center for Latin American Studies, Paper No. 25, 1975.

Eidt, Robert C. "Agrarian Reform and the Growth of New Rural Settlements in Venezuela." *Erdkunde*, 29, No. 2 (1975): 118-132.

Waldt, Hans-Otto. "Der Gemusebau als Indikator gesamtwirtschaftlicher Entwicklung-Das Beispiel Venezuela." *Geographische Zeitschrift*, 63 (1975): 123-138.

Ade, Thomas. *Die Agrarreform als Strategie zur Entwicklung des landlichen Raumes in Venezuela: Wirtschafts- und Sozialgeographische*. Stuttgart, Germany: Stuttgarter Geographische Studien, Band 119, 1992.

ENVIRONMENTAL AND CONSERVATION ISSUES

Harris, David R. "The Ecology of Swidden Cultivation in the Upper Orinoco Rainforest, Venezuela." *Geographical Review*, 61, No. 4 (1971): 475-495.

Salinas, Pedro J. "Observaciones ecologicas sobre los modules de Apure, necesidad de una estacion de estudios ecologicos." *Revista Geografica–Merida*, 15, No. 1-2 (1974): 35-54.

Sarmiento, L., M. Monasterio, and M. Montilla. "Ecological Bases, Sustainability and Current Trends in Traditional Agriculture in the Venezuelan High Andes." *Mountain Research and Development*, 13, No. 2 (1993): 167-176.

Maytin, Carlos E., et al. "Potential Effects of Global Climatic Change on the Phenology and Yield of Maize in Venezuela." *Climatic Change*, 29, No. 2 (1995): 189-212.

IRRIGATION

Eden, M.J. "Irrigation Systems and the Development of Peasant Agriculture in Venezuela." *Tijdschrift voor Economische en Sociale Geographie*, 65, No. 1 (1974): 48-54.

Part 4

EUROPE

Europe as a Whole

ATLASES AND GRAPHIC PRESENTATIONS

Thran, P., and Simon Broekhuzer, eds. *Agro-Climatic Atlas of Europe.* New York, N.Y.: American Elsevier Publishers, 1965. 3 volumes.

Kostrowicki, Jerzy. "The Types of Agriculture Maps in Europe." *Geographia Polonica*, 49 (1982): 79-92.

GENERAL TEXTS, ARTICLES, AND THESES

Jonasson, O. "Agricultural Regions of Europe." *Economic Geography*, 1, No. 4 (1925): 277-314.

Jonasson, O. "Agricultural Regions of Europe." *Economic Geography*, 2, No. 1 (1926): 19-48.

Jonasson, O. "Agricultural Regions of Europe." Ph.D. Dissertation, Clark University, Worcester, Mass., 1926.

Hartshorne, Richard, and Samuel N. Dicken. "A Classification of the Agricultural Regions of Europe and North America on a Uniform Statistical Basis." *Annals of the Association of American Geographers*, 25 (1935): 99-120.

Crone, G.R. "Notes on the Agriculture in the Low Countries." *Geographical Journal*, 100, No. 5-6 (1942): 235-244.

Russell, E. John. "Small Farmers and Peasants of Europe Before and After the War." *Geographical Review*, 35, No. 1 (1945): 1-11.

Birchard, Ralph. "Europe's Critical Food Situation." *Economic Geography*, 24, No. 3 (1948): 274-282.

Clout, Hugh D. *Agriculture.* London, England: Macmillan Studies in Contemporary Europe, 1971.

Bronney, Horst M. *Landliche Problemgebiete: Beitrage zur Geographie der Agrarwirtschaft in Europe.* Paderborn, Germany: Bochumer Geographische Arbeiten, Heft 13, 1972.

Clout, Hugh D. "Structural Changes in Agriculture." Pp. 102-117 in *Rural Geography: An Introductory Survey.* Oxford, England: Pergamon, 1972.

Andreae, Bernd. "Strukturzonen und betriebsformen in der Europaischen landwirtschaft: Forschung im Grenzbereich von Agrarokonomie und Agrargeographie." *Geographische Rundschau*, 28, No. 6 (1976): 221-234.

Buchanan, R.H., Robin H. Butlin, and D. McCourt, eds. *Fields, Farms and Settlement in Europe.* Holywood, Northern Ireland: Ulster Folk and Transport Museum, 1976.

Sivini, Giordano. "Some Remarks on the Development of Capitalism and Specific Forms of Part-Time Farming in Europe." Pp. 274-283 in *Part-Time Farming: Problem or Resource in Rural Development. Proceedings of the First Rural Geography Symposium, University of Guelph, June 18–20, 1975*, edited by A.M. Fuller and J.A. Mage. Norwich, England: GeoBooks, 1976.

Stola, Wladyslawa. "Proba zastosowania metod typologicznych do badan porownawczych rozwoju rolnictwa Belgii i Polski." *Przeglad Geographiczny*, 44 (1977): 757-769.

Otremba, Erich. "Agrargeographische Probleme der Europaischen Gemeinschaft." *Geographische Zeitschrift*, 67 (1979): 95-103.

Bonnieux, F., and P. Rainelli. "Les disparities regionales de l'agriculture europeenne." Pp. 319-333 in *L'assiette rurale des departements francais: Actes du colloque de Poitiers 1981*. Poitiers, France: Travaux du Centre de Geographie Humaine et Sociale, Universite de Poitiers, No. 6, 1981.

Rowley, Trevor. *The Origins of Open Field Agriculture*. Totowa, N.J.: Barnes and Noble, 1981.

Fruid, J.P. "Les types regionaux d'agriculture dans les pays de la C.E.E." *L'Information Geographique*, 47, No. 5 (1983): 173-182.

Stola, Wladyslawa. "Essai d'application des methodes typologies a l'etude comparee sur le developpement des agriculture belge et polonaise." *Geographia Polonica*, 46 (1983): 159-174.

Grigg, David B. "The Agricultural Revolution in Western Europe." Pp. 1-17 in *Understanding Green Revolutions: Agrarian Change and Development Planning in South Asia*, edited by Timothy Bayliss-Smith and S. Wanmali. Cambridge, England: Cambridge University Press, 1984.

Kostrowicki, Jerzy. "Types of Agriculture in Europe: A Preliminary Outline." *Geographia Polonica*, 50 (1984): 131-150.

Bowler, Ian R. "Some Consequences of the Industrialization of Agriculture in the European Community." Pp. 75-98 in *The Industrialisation of the Countryside*, edited by Michael J. Healey and Brian W. Ilbery. Norwich, England: Geo Books, 1985.

Mykolenko, L., and E. Calmes. *L'Europe agricole: Une nouvelle geographie des productions*. Paris, France: Ellipses, 1985.

Bowler, Ian R., ed. *Agriculture and the Common Agricultural Policy*. Leicester, England: University of Leicester, Department of Geography, Occasional Paper No. 15, 1986.

Kostrowicki, Jerzy. "Transformations de l'agriculture europeenne a la lumiere de la carte des types agricoles de l'Europe." *Geographia Polonica*, 52 (1986): 191-208.

Von Hecke, E. "Landwirtschaft der Benelux-Staaten." *Geographische Rundschau*, 38, No. 7-8 (1986): 361-368.

Reiners, T.A. "Extensivierung in der Landwirtschaft: Europaische und regionale Anstaze." *Geographische Rundschau*, 41, No. 3 (1989): 156-162.

Meeus, J.H.A., M.P. Wijermans, and M.J. Urom. "Agricultural Landscapes in Europe and Their Transformation." *Landscape and Urban Planning*, 18, No. 3-4 (1990): 289-352.

Chiffue, Jean. "Echanges agro-alimentaire Bourgogne-Europe centrale." *Revue Geographie de l'Est*, 31, No. 1 (1991): 205-218.

Jansen, A.J., and H. Helsen. "Agricultural Development and Spatial Organization in Europe." *Journal of Rural Studies*, 7, No. 3 (1991): 143-152.

Kostrowicki, Jerzy. "Trends in the Transformation of European Agriculture." Pp. 21-47 in *Land-Use Changes in Europe*, edited by F.M. Brouwer, A.J. Thomas, and M.J. Chadwick. Dordrecht, Netherlands: Kluwer, 1991.

Briggs, David, and E. Kerrell. "Patterns and Implications of Policy-Induced Agricultural Adjustments in the European Community." Pp. 85-102 in *Restructuring the Countryside: Environmental Policy in Practice*, edited by Andrew W. Gilg. Aldershot, England: Avebury, 1992.

Robinson, Guy M., and Brian W. Ilbery. "Reforming the CAP: Beyond MacSharey." *Progress in Rural Policy and Planning*, 3 (1993): 197-207.

Cabera, Manuel. "The Common Agricultural Policy and the Development of Agriculture: Problems and Perspectives." Pp. 273-300 in *The European Challenge: Geography and Development in the European Community*, edited by Mark Blacksell and Allan M. Williams. London, England: Oxford University Press, 1994.

Truffelli, Corrado. "Caratteri strutturali delle aziende agricole europee." *Rivista Geografica Italiana*, 101, No. 4 (1994): 525-544.

Whatmore, Sarah J. "Global Agro-Food Complexes and the Refashioning of Rural Europe." Pp. 46-67 in *Globalization, Institutions and Regional Development in Europe*, edited by A. Amin and Nigel J. Thrift. London, England: Oxford University Press, 1994.

Gomez Lopez, Jose D. "El cultivo protegido en la Comunidad Economica Europea y en Espana: De la tecnica informal a la innovacion technologica." *Estudios Geograficos*, 61, No. 218 (1995): 5-28.

Groenendijk, Jan. "Integration and Re-Nationalisation: EU Agricultural Policy with Both Feet on the Ground." *Tijdschrift voor Economische en Sociale Geographie*, 86, No. 4 (1995): 382-389.

Klohn, Werner. "Landwirtschaft in Europa: Strukturen, Problem, Perspektiven." *Praxis Geographie*, 25, No. 5 (1995): 4-10.

Beteille, R. "L'agritourisme dans les espaces euraux europeans." *Annales de Geographie*, 105, No. 6 (1996): 584-602.

Bowler, Ian R., and Brian W. Ilberry. "The Regional Consequences for Agriculture of Changes to the Common Agricultural Policy." Pp. 103-118 in *CAP and the Regions: Building a Multidisciplinary Framework for the Analysis of the EU Agricultural Space*, edited by C. Laurent and Ian R. Bowler. Versailles, France: INRA, 1996.

Clark, J.R.A., et al. "Conceptualizing the Evolution of the European Union's Agri-Environment Policy: A Discourse Approach." *Environment and Planning A*, 20, No. 10 (1997): 1869-1886.

Rey, V. "L'Europe de l'entre-deux: Nouvelles agricultures, nouveaux agriculteurs." *Bulletin de la Societe Geographique de Liege*, 33 (1997): 7-24.

Ilbery, Brian W. "The Challenge of Agricultural Restructuring in the European Union." Pp. 341-358 in *The New Europe: Economy, Society and Environment*, edited by David Pinder. Chichester, England: Wiley, 1998.

Turnock, David. "Private Farming in Eastern Europe and the Former Soviet Union." *GeoJournal*, 45, No. 3 (1998): 229-232.

Unwin, Timothy. "Agricultural Change and Rural Stress in the New Democracies." Pp. 359-380 in *The New Europe: Economy, Society, and Environment*, edited by David Pinder. Chichester, England: Wiley, 1998.

Bowler, Ian R. "Endogenous Agricultural Development in Western Europe." *Tijdschrift voor Economische en Sociale Geographie*, 90, No. 3 (1999): 260-271.

Wilson, Geoff A. "From Productivism to Post-Productivism . . . and Back Again? Exploring the (Un)Changed Natural and Mental Landscapes of European Farming." *Transactions, Institute of British Geographers*, 26, No. 1 (2001): 77-102.

CROPS AND CROPPING PATTERNS

Werle, Otmar. "Das Wein baugebeit der deutsch-luxemburgischen Obermosel." *Geographische Rundschau*, 27, No. 7 (1975): 277-285.

Denecke, D. "Innovation and Diffusion of the Potato in Central Europe in the Seventeenth and Eighteenth Centuries." Pp. 60-96 in *Fields, Farms and Settlement Europe*, edited by R.H. Buchanon, Robin A. Butlin, and D. McCourt. Holywood, Northern Ireland: Ulster Folk and Transport Museum, 1976.

Uhlmann, F. "Regional Angebotsschwerpunkte der pflanzlichen Erzeugung in der EC." *Landbauforschung Volkenrode*, 32, No. 2 (1982): 67-78.

Vaudois, J. "Production et commerce exterieur de l'endive en Europe: Le bilon d'un expansion seculaire." *Hommes et Terres du Nord*, 1 (1984): 8-16.

Unwin, Timothy. "Viticultural Continuity: The Vine in the Landscape 500–1000 A.D." Pp. 105-114 in *The Transformation of Rural Society, Economy and Landscape*, edited by U. Sporrong. Stockholm, Sweden: University of Stockholm, Department of Human Geography, 1990.

Huetz de Lemps, Alain. "La vigne americaise au secours de l'Europe." *Les Cahiers d'Outre Mer*, 45 (1992): 461-478.

Unwin, Timothy. "Continuity in Early Medieval Viticulture: Secular or Ecclesiastical Influences?" Pp. 7-22 in *Viticulture in Geogrpahic Perspective: Proceedings of the Miami AAG Symposium*, edited by Harm deBlij. Miami, Fla.: Miami Geographical Society, 1992.

ANIMAL HUSBANDRY

Evans, E. Estyn. "Transhumance in Europe." *Geography*, 25 (1940): 172-180.

DAIRYING

Naylor, E.L. "EEC Milk Quotas." *Agricultural Administration and Extension*, 24, No. 3 (1987): 177-185.

FARM LAND

Goransson, Solve. "Regular Open-Field Pattern in England and Scandinavian Solskifte." *Geografiska Annaler*, 43 (1961): 80-104.

Timmerman, O.F. "Bedeutung der Wildbeute fur die Entwicklungsstadien der agraren Landnutzung und Parzellierung des Landes in Mitteleuropa." *Geografiska Annaler*, 43 (1961): 277-284.

Uhlig, Harold. "Old Hamlets with Infield and Outfield Systems in Western and Central Europe." *Geografiska Annaler*, 43 (1961): 285-312.

Stamp, L. Dudley. *Land Use Statistics of the Countries of Europe*. Bude, England: World Land Use Survey Occasional Paper No. 3, Geographical Publications Ltd., 1965.

Graul, Hans. "Uber die Brache." *Heidelberger Geographische Arbeiten*, 15 (1966): 37-58.

Belding, Richard. "A Test of the von Thunen Locational Model of Agricultural Land Use with Accountancy Data from the European Economic Community." *Transactions, Institute of British Geographers*, 6, No. 2 (1981): 176-187. (New Series)

Nijkamp, P., and F. Soetema. "Dynamics in Land Use Patterns: Socio-Economic and Environmental Aspects of the Second Agricultural Land Use Revolution." Pp. 153-175 in *Land Use Changes in Europe*, edited by F.M. Brouwer, et al. Dordrecht, Netherlands: Kluwer, GeoJournal Library, No. 18, 1991.

Bolsium, Emmy C.A., Gordon Clark, and Jan Groenendijk, eds. *The Retreat: Rural Land Use and European Agriculture*. Amsterdam, Netherlands: Royal Dutch Geographical Society, No. 172, 1993.

Ilbery, Brian W., and Ian R. Bowler. "Land Diversion and Farm Business Diversification in EC Agriculture." Pp. 15-27 in *The Retreat: Rural Land Use and European Agriculture*, edited by Emmy C.A. Bolsium, Gordon Clark, and Jan Groenendijk. Amsterdam, Netherlands: Royal Netherlands Geographical Society, No. 172, 1993.

Monteaux, D. "Geographes, crise de l'agriculture et degradation des paysages agraires." *La Pensi*, 292 (1993): 23-40.

Inui, T., and Ian R. Bowler. "Agricultural Land Use in the European Union: Past, Present and Future." *Geographical Review of Japan*, 68B, No. 2 (1995): 137-150.

AGRICULTURAL POPULATIONS AND SOCIAL ISSUES

Geddes, A. "The Homestead and Hamlet in Celtic Lands: Their Significance in Agriculture, Pastoralism and Fishing." Pp. 453-471 in *Comptes rendus du Congres International de Geographie, Lisbonne 1949*. Lisbon, Portugal: International Geographical Union, 1951.

Niemeier, Georg. "Agrarlandschaftliche Reliktgebiete und die Morphogenese von Kulturlandschaftten im atlantischen Europa." *Geografiska Annaler*, 43 (1961): 229-235.

Beruzowski, Stanislaw. "Typologie des migrations pastorales en Europe et methodes de leurs etudes." Pp. 165-174 in *L'amenagement de la Montagne, Compte Rendu du III Colloque-franco-polaanis de geographie, Mai 1969*, edited by Jerzy Kostrowicki and Marin Rosciszewski. War-

saw, Poland: Academie Polonaise des Sciences, Centre Scientifique a Paris, Conferences 87, 1971.
Rey, V., and M. Bachvarov. "Rural Settlements in Transition: Agricultural and Countryside Crisis in Central-Eastern Europe." *GeoJournal*, 44, No. 4 (1998): 345-352.

ENVIRONMENTAL AND CONSERVATION ISSUES

Ganser, K. "Eine okonomische und okologische Perspektive fur die Agrarpolitik." *Geographische Rundschau*, 34, No. 3 (1982): 82-87.
Bourke, A. "Impact of Climatic Fluctuations on Agriculture." Pp. 265-313 in *The Climate of Europe: Past, Present and Future*, edited by H. Flohn and R. Fontechi. Dordrecht, Netherlands: Reidel, 1984.
Park, J.R. *Environmental Management in Agriculture: European Perspectives*. London, England: Belhaven, 1988.
Romanova, E.P. "Agricultural Landscapes of Europe: Changes of the Natural Environment Under the Impact of Agriculture" (in Polish). *Przeglad Geograficzny*, 60, No. 4 (1988): 573-584.
Carter, T.R., J.H. Porter, and M.L. Parry. "Climatic Warming and Crop Potential in Europe." *Global Environmental Change*, 1, No. 4 (1991): 291-311.
Carter, T.R., Martin L. Parry, and J.H. Porter. "Climatic Change and Future Agroclimatic Potential in Europe." *International Journal of Climatology*, 11, No. 3 (1991): 251-270.
Buller, Henry J. "Agricultural Change and the Environment in Western Europe." Pp. 68-88 in *Agricultural Change, Environment and Economy: Essays in Honour of W.B. Morgan*, edited by Keith Hoggart. London, England: Mansell, 1993.
Gonin, P., R. Sevestre, and J. Vaudois. "De l'article 19 a l'application des mesures agrienvironnementales: L'integration des fonctions territoires aux politques agricoles europeennes." *Hommes et Terres du Nord*, 4 (1995): 214-223.
Laurila, H.A. "Modelling the Effects of Elevated CO_2 and Temperature on Swedish and German Spring Wheat Varieties with CERES-Wheat and AFRC-Wheat Crop Models." *Journal of Biogeography*, 22, No. 4-5 (1995): 591-596.
Wolf, J., and C.A. Von Dieper. "Effects of Climatic Change on Grain Maize Yield Potential in the European Community." *Climatic Change*, 29, No. 3 (1995): 299-331.
Ahvenjarvi, Hannu, and Matti Hakkila. "Organic Farming in Western Europe" (in Finnish). *Terra*, 109, No. 4 (1997): 198-207.
Jones, Alun, and J.R.A. Clark. "The Agri-Environment Regulation EU 2078/ 92: The Role of the European Commission in Policy Shaping and Setting." *Environment and Planning C*, 16, No. 1 (1998): 51-68.
Wilson, Geoff A., J.E. Petersen, and A. Holt. "EU Member State Responses to Agri-Environmental Regulation 2078/92/EEC: Towards a Conceptual Framework?" *Geoforum*, 30, No. 2 (1999): 185-202.

Barrue-Pastor, Monique. "L'integration de la problematique environnementale dans la politique agrocole europeenne: Analyse comparee en France et en Espagne." *Sud-Ouest Europeen*, 7 (2000): 49-60.

Wilson, Geoff A., and Kaley Hart. "Financial Imperative or Conservation Concern? EU Farmers' Motivations for Participation in Voluntary Agri-Environmental Schemes." *Environment and Planning A*, 32, No. 12 (2000): 2161-2185.

Wilson, Geoff A., and Henry J. Buller. "The Use of Socio-Economic and Environmental Indicators in Assessing the Effectiveness of EU Agri-Environmental Policy." *European Environment*, 11, No. 6 (2001): 297-313.

IRRIGATION

Bohm, Hans. "Die Wiesernbewasserung in Mitteleurope 1937." *Erdkunde*, 44, No. 1 (1990): 1-9.

AGRICULTURAL MARKETING AND DISTRIBUTION

Ilbery, Brian W. and M. Kneafsy. "Niche Markets and Regional Specialty Food Products in Europe: Towards a Research Agenda." *Environment and Planning A*, 31, No. 12 (1999): 2207-2222.

Great Britain

ATLASES AND GRAPHIC PRESENTATIONS

Bowen, E.G. "A Map of the Trehelig Common Fields." *Montgomeryshire Collections*, 41, No. 2 (1930): 163-168.

Wood, Herbert J. *An Agricultural Atlas of Scotland*. London, England: G. Gill, 1931.

Darby, H.C. "The Agrarian Contribution to Surveying in England." *Geographical Journal*, 82, No. 6 (1933): 529-535.

Coppock, J.T. "The Relationship Between Farm and Parish Boundaries." *Geographical Studies*, 1 (1955): 12-25.

Harley, John B. "Uncultivated Fields in the History of British Cartography." *Cartographic Journal*, 4 (1967): 7-11.

Coppock, J.T. *Agricultural Atlas of England and Wales*. London, England: Faber and Faber, 1976.

Coppock, J.T. *An Agricultural Atlas of Scotland*. Edinburgh, Scotland: John Donald, 1976.

Parry, Martin L. "Mapping of Abandoned Farmland in Upland Britain." *Geographical Journal*, 142 (1976): 101-110.

Robinson, Guy M. *Late Victorian Agriculture in the Vale of Evesham*. Oxford, England: Oxford University, School of Geography, No. 16, 1976.

Robinson, Guy M. "Agricultural Depression in the Vale of Hereford, 1870 to 1900." *Transactions, Woolhope Field Nature Club*, 42 (1978): 259-278.

Overton, Mark. "Agriculture." Pp. 34-53 in *Atlas of Industrializing Britain*, edited by J. Langton and R.J. Morris. London, England: Methuen, 1981.

Edwards, Kevin. "The Hunter-Gatherer/Agricultural Transition and the Pollen Record in the British Isles." Pp. 255-268 in *The Cultural Landscape: Past, Present and Future*, edited by Hilary H. Birks, et al. Cambridge, England: Cambridge University Press, 1988.

Campbell, Bruce, and John Power. "Mapping the Agricultural Geography of Medieval England." *Journal of Historical Geography*, 15, No. 1 (1989): 24-39.

Kostrowicki, Jerzy. "Types of Agricultural in Britain in the Light of the Types of Agricultural Map of Europe." *Geographia Polonica*, 56 (1989): 133-154.

Atkinson, P.M. "Selecting the Spatial Resolution of Airborne MSS Imagery for Small-Scale Agricultural Mapping." *International Journal of Remote Sensing*, 18, No. 9 (1997): 1903–1918.

Cobb, Richard, Paul Dolman, and T. O'Riordan. "Interpretations of Sustainable Agriculture in the U.K." *Progress in Human Geography*, 23, No. 2 (1999): 209-235.

Tso, B., and P.M. Mather. "Crop Discrimination Using Multi-Temporal SAR Imagery." *International Journal of Remote Sensing*, 20, No. 12 (1999): 2443-2460.

GENERAL TEXTS, ARTICLES, AND THESES

Roxley, P.M. "The Agricultural Geography of England on a Regional Basis." *Geography Teacher*, 7 (1913–1914): 316-321.

Fussell, G.E. "Agriculture and Economic Geography in the Eighteenth Century." *Geographical Journal*, 43, No. 1 (1927): 170-178.

McClement, James. "The Distribution of Agriculture in Kintyre." *Scottish Geographical Magazine*, 43, No. 1 (1927): 20-30.

Best, S.E.J. *East Yorkshire: A Study in Agricultural Geog*raphy. London, England: Longmans, 1930.

Leppard, Henry M. "Scottish Corse Agriculture: The Corse of Gowrie." *Economic Geography*, 10 (1934): 217-238.

Maxton, J.P., ed. *Regional Types of British Agriculture*. London, England: G. Allen and Unwin, 1936.

Snodgrass, Catherine P. "Lanarkshire: The Agricultural Geography of a Scottish County." *Scottish Geographical Magazine*, 53 (1937): 176-199.

Bernard, Lucie. "Le renouveau agricole d'une region de l'Essex." *Annales de Geographie*, 47 (1938): 397-400.

Hodgson, H.B. "Geographical Aspects of Farming in the Vale of York." *Economic Geography*, 14, No. 1 (1938): 73-78.

Hudson, G. Donald. "Agricultural Patterns of East Lothian, Scotland." *Economic Geography*, 14, No. 1 (1938): 16-22.

Owen, A.E. "Agricultural Divisions of Wales." *Geography*, 26 (1941): 69-76.

Starr, Arthur J. "18th Century Agriculture in Suffolk." *Geography*, 26 (1941): 116-125.

Stamp, L. Dudley. "Wartime Changes in British Agriculture." *Geographical Journal*, 112, No. 1-3 (1947): 39-57.

Borchert, John R. "The Agriculture of England and Wales, 1939–1946." *Agricultural History*, 22, No. 1 (1948): 56-62.

Buchanon, Keith. "Modern Farming in the Vale of Evesham." *Economic Geography*, 24, No. 3 (1948): 235-250.

Fussell, G.E. "High Farming in Southwestern England, 1840–1880." *Economic Geography*, 24, No. 1 (1948): 53-73.

Howe, G. Melvyn. "Note on the Application of Air Photography to the Agricultural Geography of North-West Cardiganshire." *Geography*, 36, No. 1 (1951): 15-20.

Lebon, J.H.G. "The Beginnings of the Agrarian and Industrial Revolutions in Ayrshire." Pp. 175-198 in *London Essays in Geography*, edited by L. Dudley Stamp and S.W. Wooldridge. Cambridge, Massachusetts: Harvard University Press, 1951.

Blume, Helmut. "Die Landwirtschaftsgeographische Strukture von Vier Gebirgstalern der Britischen Inseln." *Erdkunde*, 6 (1952): 213-234.

Henderson, H.C.K. "Agriculture in England and Wales in 1801." *Geographical Journal*, 118, No. 3 (1952): 338-344.

Davies, Ivor. "Urban Farming: A Study of the Agriculture of the City of Birmingham." *Geography*, 38, No. 4 (1953): 296-302.

Birch, J.W. "Observations on the Delimitation of Farming-Type Regions, with Special Reference to the Isle of Man." *Transactions of the Institute of British Geographers*, 20 (1954): 141-158.

Darby, H.C. "Some Early Ideas on the Agricultural Regions of England." *Agricultural History Review*, 2 (1954): 30-47.

Coppock, J.T. "The Statistical Assessment of British Agriculture." *Agricultural History Review*, 4, No. 1 (1956): 4-21.

Coppock, J.T. "The Statistical Assessment of British Agriculture (Continued)." *Agricultural History Review*, 4, No. 2 (1956): 66-79.

Gachon, L., and H. Starr. "Les lecons geographiques de l'experimentation agricole a la premiere des stations agronomiques de l'Angleterre." *Revue de Geographie de Lyon*, 31 (1956): 95-98.

Prince, Hugh C. "The Title Surveys of the Mid-Nineteenth Century." *Agricultural History Review*, 7, No. 1 (1959): 14-26.

Sylvester, Dorothy. "A Note on Medieval Three-Course Arable Systems in Cheshire." *Transactions of the Historic Society of Lancashire and Cheshire for the Year 1958*, 110 (1959): 183-186.

Coppock, J.T. "Agricultural Changes in the Chilterns, 1875–1900." *Agricultural History Review*, 9, No. 1 (1961): 1-16.

Finberg, H.P.R. "Recent Progress in English Agrarian History." *Geografiska Annaler*, 43 (1961): 75-79.

Moindrot, C. "L'espace agricole en Angleterre-Galles." *Norois*, 8 (1961): 5-15.

Grigg, David B. "Changing Regional Values during the Agricultural Revolution in South Lincolnshire." *Transactions, Institute of British Geographers*, 30 (1962): 91-105.

Kampp, A.H. "The Agricultural Geography of Man." *Erdkunde*, 16, No. 3 (1962): 173-190.

Tuck, Gwendolynn M. "Anglesey, Wales: The Economic Geography of an Agricultural Region." Master's Thesis, Columbia University, New York, N.Y., 1962.

Jackson, B.G., C.S. Barnard, and F.G. Sturrock. *The Pattern of Farms in the Eastern Counties: A Report on a Classification of Farms in Eastern England.* Cambridge, England: School of Agriculture, 1963.

Thomas, David. *Agriculture in Wales during the Napoleonic Wars: A Study in the Geographical Interpretation of Historical Sources.* Cardiff, Wales: University of Wales Press, 1963.

Coppock, J.T. "Crop, Livestock and Enterprise Combinations in England and Wales." *Economic Geography*, 40, No. 1 (1964): 65-81.

Coppock, J.T. "Postwar Studies in the Geography of British Agriculture." *Geographical Review*, 54 (1964): 409-426.

Coppock, J.T. "Agricultural Changes in Britain." *Geography*, 49, No. 3 (1964): 322-327.

Hiersemenzel, Sigrid-Elisabeth. *Britische Agrarlandscaften im Rhythmus des Landwirtschaftlichen Arbeitsjahaus.* Berlin, Germany: Berliner Geographische Abhandlungen, Heft 1, 1964.

Simmons, A. Joyce. "An Index of Farm Structure, with a Nottinghamshire Example." *East Midland Geographer*, 3, No. 21 (1964): 255-266.

Grigg, David B. "A Note on Agricultural Rent and Expenditure in Nineteenth Century England." *Agricultural History*, 39, No. 3 (1965): 147-154.

Jones, Glanville. "Agriculture in North-West Wales during the Later Middle Ages." Pp. 47-53 in *Climatic Change with Special Reference to Wales and Its Agriculture: Symposium in Agricultural Meteorology.* Aberystwyth, Wales: University College of Wales, Memorandum No. 8, 1965.

Bennett-Jones, R., and G.E. Jones. "Agriculture." Pp. 253-270 in *Nottingham and Its Region*, edited by K.C. Edwards. Nottingham, England: British Association for the Advancement of Science, 1966.

Grigg, David B. *The Agricultural Revolution in South Lincolnshire.* Cambridge, England: Cambridge University Press, 1966.

de Phanhol, Xavier. "Les structures agraires de l'Europe du Nord-Ouest, au symposium de Bangor-Leicester." *Revue Geographique de l'Est*, 6, No. 1-2 (1966): 109-114.

Shirlaw, D.W.G. *An Agricultural Geography of Great Britain.* Oxford, England: Pergamon Press, 1966.

Thirsk, Joan. "The Farming Regions of England." Pp. 1-160 in *The Agrarian History of England and Wales*, vol. 4, edited by Joan Thirsk. Cambridge, England: Cambridge University Press, 1967.

Barnes, F.A., et al. "Agrarian Landscapes of the East Midlands." Pp. 79-110 in *Agricultural Geography International Geographical Union Symposium 1964*, vol. 2, edited by E.S. Simpson. Liverpool, England: University of Liverpool Department of Geography, Research Paper No. 5, 1968.

Simpson, E.S. "Agriculture in West Cheshire." Pp. 39-59 in *Agricultural Geography International Geographical Union Symposium*, vol. 2, edited by E.S. Simpson. Liverpool, England: University of Liverpool Department of Geography, Research Paper No. 5, 1968.

Taylor, J.A., and I.M. Brightner. "Contract Farming." Pp. 23-25 in *Agricultural Geography International Geographical Union Symposium*, vol. 2, edited by E.S. Simpson. Liverpool, England: University of Liverpool, Department of Geography, Research Paper No. 5, 1968.

Munton, R.J.C., and J.M. Norris. "The Analysis of Farm Organization: An Approach to the Classification of Agricultural Land in Britain." *Geografiska Annaler B*, 51B (1969): 95-103.

Roden, David. "Demesne Farming in the Chiltern Hills." *Agricultural History Review*, 17 (1969): 9-23.

Aitchison, J.W. *The Farming Systems of Wales: A Study of Spatial Variability*. Aberystwyth, Wales: University College of Wales, Department of Geography, 1970.

Moindrot, C. "L'agriculture autaur de Londres." *Acta Geographica*, Serie 3, No. 4 (1970): 205-224.

Armstrong, W.J. "The Definition of Factors Influencing Spatial Variability in Gross Output Per Acre on Farms in South Devon." Pp. 157-168 in *Exeter Essays in Geography in Honour of Arthur Davies*, edited by K.J. Gregory and W.L.D. Ravenhill. Exeter, England: University of Exeter, 1971.

Coppock, J.T. *An Agricultural Geography of Great Britain*. London, England: Bell, 1971.

Henderson, H.J.R. "Agricultural Geography and the Type of Farming Classification in England and Wales." *Swansea Geographer*, 9 (1971): 112-120.

Shirlaw, D.W.G., and Douglas Watson. *An Agricultural Geography of Great Britain*. London, England: Pergamon, 1971.

Bowler, Ian R. "Co-operation: A Note on Governmental Promotion of Change in Agriculture." *Area*, 4 (1972): 169-173.

Frost, L.R. "Recent Trends in Agriculture in Northern Ireland." *Geography*, 57, No. 4 (1972): 337-341.

Perry, Peter J. "Where Was the 'Great Agricultural Depression'? A Geography of Agricultural Bankrupcy in Late Victorian England and Wales." *Agricultural History Review*, 20, No. 1 (1972): 30-45.

Anderson, K.E. *An Agricultural Classification of England and Wales*. Edinburgh, Scotland: University of Edinburgh, Department of Geography, Research Discussion Paper No. 1, 1974.

Edwards, C.J.W. "Farm Enterprise Systems in East County Londonderry." *Irish Geography*, 7 (1974): 29-52.

Henderson, H.J.R., and Brian W. Ilbery. "Factors Affecting the Structure of Agriculture in the Amman and Upper Tawe Valleys." *Swansea Geographer*, 12 (1974): 61-66.

Munton, R.J.C. "Farming on the Urban Fringe." Pp. 201-263 in *Suburban Growth*, edited by James H. Johnson. London, England: John Wiley, 1974.

Perry, Peter J. *British Farming in the Great Depression 1870–1914: An Historical Geography*. Newton Abbot, England: David & Charles, 1974.

Aitchinson, J.W. "Cluster Analysis, Regionalization and the Agricultural Enterprises of Wales." Pp. 17-33 in *Agricultural Typology and Land Utilization*, edited by C. Vanzetti. Verona, Italy: Centre of Agricultural Geography, 1975.

Anderson, K.E. "An Agricultural Classification of England and Wales." *Tijdschrift voor Economische en Sociale Geographie*, 66, No. 3 (1975): 148-158.

Beck, Dietmar. *East Lothian: Die Entwicklung einer schottischen Kulturlandschaft*. Wiesbaden, Germany: Franz Steiner, 1975.

Bowler, Ian R. "Factors Affecting the Trend to Enterprise Specialization in Agriculture: A Case Study in Wales." *Cambria*, 2 (1975): 100-111.

Bowler, Ian R. "Regional Variations in Scottish Agricultural Trends." *Scottish Geographical Magazine*, 91, No. 2 (1975): 114-121.

Bowler, Ian R. "Regional Agricultural Policies: Experience in the United Kingdom." *Economic Geography*, 52 (1976): 267-280.

Coppock, J.T. "Regional Changes in the Agricultural Geography of Great Britain, 1959–1973." Pp. 83-92 in *Regional Development and Planning: British and Hungarian Case Studies*, edited by Paul A. Compton and Martan Pecsi. Budapest, Hungary: Akademiai Kiado, 1976.

Edwards, C.J.W. "Short Term Changes in Farm Enterprise Systems: A Case Study from County Londonderry, 1972–1975." *Irish Journal of Agricultural Economics and Rural Sociology*, 6 (1976): 155-175.

Leonard, Simon. "Von Thunen in British Agriculture." *South Hampshire Geographer*, 8 (1976): 23-37.

McKay, Peter. "Part-Time Farming in Britain: A Study of Statutory Smallholdings in South Staffordshire." Pp. 83-93 in *Part-Time Farming: Problems of Resources in Rural Development. Proceedings of the First Rural Geography Symposium, University of Guelph, 1975*, edited by Anthony M. Fuller and Julius A. Mage. Norwich, England: Geo Abstracts Ltd., 1976.

Robinson, Guy M. *Late Victorian Agriculture in the Vale of Evesham*. Oxford, England: University of Oxford, School of Geography, Research Paper No. 16, 1976.

Layton, Robert L. "The Hobby Farm Issue." *Town and Country Planning*, 49 (1978): 53-54.

Osborne, Brian S. "Glamorgan Agriculture in the 17th and 18th Centuries." *National Libraries of Wales Journal*, 28, No. 4 (1978): 387-407.

Sheail, John. "Rabbits and Agriculture in Post-Medieval England." *Journal of Historical Geography*, 4, No. 4 (1978): 343-356.

Symes, David G., and Terry K. Marsden. "Recent Developments in Agriculture on North Humberside." Pp. 21-31 in *North Humberside: Introductory Essay*, edited by David G. Symes. Hull, England: University of Hull Department of Geography, 1978.

Aitchinson, J.W. "The Agricultural Landscape of Wales, Part 1: The Structure of Agricultural Holdings, 1964–1974." *Cambria*, 6, No. 2 (1979): 32-53.

Bowler, Ian R. *Government and Agriculture: A Spatial Perspective.* London, England: Longman, 1979.

Fox, H.S.A. "Local Farmers' Associations and the Circulation of Agricultural Information in Nineteenth-Century England." Pp. 43-64 in *Change in the Countryside: Essays on Rural England, 1500–1900*, edited by H.S.A. Fox and Robin A. Butlin. London, England: Institute of British Geographers, Special Publications No. 10, 1979.

Ilbery, Brian W. "Decision-Making in Agriculture: A Case Study of North-East Oxfordshire." *Regional Studies*, 13 (1979): 199-210.

Kain, Roger J.P. "Tithe as an Index of Pre-Industrial Agricultural Production." *Agricultural History Review*, 27, No. 2 (1979): 73-91.

MacDonald, Stuart. "The Role of the Individual in Agricultural Change: The Example of George Culley in Fenton, Northumberland." Pp. 5-22 in *Change in the Countryside: Essays on Rural England, 1500–1900*, edited by H.S.A. Fox and Robin A. Butlin. London, England: Institute of British Geographers, Special Publication No. 10, 1979.

Stevenson, Ann, and David G. Symes. *Agricultural Education and Training in North Humberside: A Survey of Experience and Attitudes.* Hull, England: University of Hull, Department of Geography, Paper No. 23, 1979.

Whyte, I.D. *Agriculture and Society in Seventeenth Century Scotland.* Edinburgh, Scotland: Donald, 1979.

Aitchinson, J.W. "The Agricultural Landscape of Wales, Part 2: Patterns of Agricultural Production 1964–1974." *Cambria*, 7, No. 1 (1980): 43-68.

Albarre, Guy, Terry K. Marsden, and David G. Symes. "Intensification de l'agriculture et amenagement du territoire en Grande-Bretagne: L'exemple du North Humberside." *Revue Belge de Geographie*, 104 (1980): 3-12.

Blair, A.M. "Urban Influences on Farming in Essex." *Geoforum*, 11, No. 4 (1980): 371-384.

Caird, J.B. "The Reshaped Agricultural Landscape." Pp. 203-222 in *The Making of the Scottish Countryside*, edited by Martin L. Parry and T.R. Slater. Montreal, Quebec: McGill-Queen's University Press, 1980.

Edwards, C.J.W. "Complexity and Change in Farm Production Systems: A Somerset Case Study." *Transactions, Institute of British Geographers*, 5, No. 1 (1980): 45-52. (New Series)

Giles, A.K., ed. *Farm Business Data.* Reading, England: University of Reading, Department of Geography, Geographical Papers, 1980.

Bowler, Ian R. "Regional Specialisation in the Agricultural Industry." *Journal of Agricultural Economics*, 32 (1981): 43-54.

Campbell, B.M.S. "The Regional Uniqueness of English Field Systems? Some Evidence from Eastern Norfolk." *Agricultural History Review*, 29, No. 1 (1981): 16-28.

Ilbery, Brian W. "Dorset Agriculture: A Classification of Regional Types." *Transactions, Institute of British Geographers*, 6, No. 2 (1981): 214-227. (New Series)

Perry, Peter J. "High Farming in Victorian Britain: Prospect and Retrospect." *Agricultural History*, 55, No. 2 (1981): 156-166.

Robinson, Guy M. "A Statistical Analysis of Agriculture in the Vale of Evesham during the Great Agricultural Depression." *Journal of Historical Geography*, 7, No. 1 (1981): 37-52.

Thomson, K.J. *Farming in the Fringe: An Analysis of Agricultural Census Data Drawn from Parishes Around the Six Metropolitan Counties and London*. Cheltenham, England: County-Side Commission, 1981.

Aitchinson, J.W., and P. Aubrey. "Part-Time Farming in Wales: A Typological Study." *Transactions, Institute of British Geographers*, 7, No. 1 (1982): 88-97. (New Series)

Gasson, R. "Part-Time Farming in Britain: Research in Progress." *GeoJournal*, 6, No. 4 (1982): 355-358.

Clark, Gordon, D.J. Knowles, and H.L. Phillips. "The Accuracy of the Agricultural Census." *Geography*, 68, No. 1 (1983): 115-120.

Gilg, Andrew W. "The Regional Development of Agriculture in Britain." Pp. 99-112 in *Geography and Regional Planning: Proceedings of the 2nd British–Bulgarian Seminar, Sofia*, edited by James H. Johnson. Norwich, England: Geo Books, 1983.

Ilbery, Brian W., and R. Hornby. "Repertory Grids and Agricultural Decision Making: A Mid-Warwickshire Case Study." *Geografiska Annaler B*, 65B (1983): 77-94.

Schroder, Jochem. *Veranderungen in der Agrar- und Sozialstruktur im mittleren Nord-England seit dem Landwirtschaftgesetz von 1947*. Heidelberg, Germany: Heidelberger Geographische Arbeiten, Heft 64, 1983.

Coppock, J.T. "Mapping the Agricultural Returns: A Neglected Tool of Historical Geography." Pp. 8-55 in *Discovering Past Landscapes*, edited by M. Reed. London, England: Croom Helm, 1984.

Overton, Mark. "Agricultural Revolution? Development of the Agrarian Economy in Early Modern England." Pp. 118-139 in *Explorations in Historical Geography*, edited by Alan R.H. Baker and David Gregory. Cambridge, England: Cambridge University Press, 1984.

Overton, Mark. "Probate Inventories and the Reconstruction of Agricultural Landscapes." Pp. 167-194 in *Discovering Past Landscapes*, edited by M. Reed. London, England: Croom Helm, 1984.

Glasson, N.L. "Agricultural Development and Landscape Changes: A Case Study of Eakring and Maplebeck Parishes, Nottinghamshire." *East Midland Geographer*, 8, No. 63 (1985): 206-216.

Hofmeister, B. *Die gemassigten Breiten: Inslusondere die kuhlgemassigten Waldlander*. Braunschweig, Germany: Westermann, 1985.

Wood, B.A., and P.T. Wheeler. "Changing Farm Structure on a Nottinghamshire Landed Estate, 1952–1977." *East Midland Geographer*, 8, No. 63 (1985): 198-205.

Ilbery, Brian W., and M.J. Barrington. "Peri-Urban Farming: A Case Study from Coventry's Urban Fringe." *Journal of the Royal Agricultural Society of England*, 147 (1986): 42-53.

Marsden, T.K., R.J.C. Munton, S. Whatmore, and J. Little. "Towards a Political Economy of Capitalist Agriculture: A British Pespective." *International Journal of Urban and Regional Research*, 10 (1986): 489-521.

Whyte, I.D. "Agriculture in Aberdeenshire in the Seventeenth and Eighteenth Centuries." Pp. 10-31 in *From Lairds to Louns: County and Burgh Life in Aberdeen 1600–1800*, edited by D. Stevenson. Aberdeen, Scotland: Aberdeen University Press, 1986.

Atkins, P. "The Charmed Circle: Von Thunen and Agriculture Around Nineteenth Century London." *Geography*, 72, No. 2 (1987): 129-139.

Bowler, Ian R. "The Geography of Agriculture Under the CAP." *Progress in Human Geography*, 11, No. 1 (1987): 24-40.

Bowler, Ian R. "Locational Aspects of (EEC) Council Regulation No. 1760/87 (UK)." In *Farm Extensification: Implications of EC Regulation 1760/87,* edited by N. Jenkins and M. Bell. Grange-Over-Sands, England: Institute of Terrestrial Ecology, Merlewood Research and Development Paper No. 112, 1987.

Hart, Patrick W.E. "Economic Processes and Spatial Change in Agriculture." Pp. 5-28 in *The Future of the British Rural Landscape*, edited by Douglas G. Lockhart and Brian W. Ilbery. Norwich, England: Geo Books, 1987.

Marsden, Terry K., Sarah Whatmore, and R.J.C. Munton. "Uneven Development and the Restructuring Process in British Agriculture: A Preliminary Exploration." *Journal of Rural Studies*, 3, No. 4 (1987): 297-308.

Thirsk, Joan. *Agricultural Regions and Agrarian History in England, 1500–1750*. Basingstoke, England: Macmillan, 1987.

Whatmore, Sarah, et al. "Towards a Typology of Farm Business in Contemporary British Agriculture." *Sociologica Ruralis*, 27 (1987): 21-37.

Whatmore, Sarah, et al. "Interpreting a Relational Typology of Farm Business in Southern England." *Sociologica Ruralis*, 27 (1987): 103-122.

Campbell, Bruce. "Towards an Agricultural Geography of Medieval England." *Agricultural History Review*, 36, No. 1 (1988): 87-98.

Glennie, Paul. "Continuity and Change in Hertfordshire Agriculture, 1550–1700: I—Patterns of Agricultural Production." *Agricultural History Review*, 36, No. 1 (1988): 55-76.

Grigg, David B. "Types of Farming in England and Wales." *Geography Review*, 1, No. 4 (1988): 20-24.

Munton, R.J.C., Sarah Whatmore, and Terry K. Marsden. "Reconsidering Urban-Fringe Agriculture: A Longitudinal Analysis of Capital Restructuring on Farms in the Metropolitan Green Belt." *Transactions, Institute of British Geographers*, 13, No. 3 (1988): 324-336. (New Series)

Prince, Hugh. "Art and Agrarian Change, 1710–1815." Pp. 98-118 in *The Iconography of Landscape*, edited by D. Cosgrove and S. Daniels. Cambridge, England: Cambridge University Press, 1988.

Robinson, Guy M. *Agricultural Change: Geographical Studies of British Agriculture*. Edinburgh, Scotland: North British Publishing, 1988.

Bowler, Ian R. "The Spatial Restructuring of Farm Types in the English Counties, 1976–1985." *Tijdschrift voor Economische en Sociale Geographie*, 80, No. 5 (1989): 302-311.

Evans, N.J., and Brian W. Ilbery. "A Conceptual Framework for Investigating Farm-Based Accommodation and Tourism in Britain." *Journal of Rural Studies*, 5, No. 3 (1989): 257-266.

Grigg, David B. *English Agriculture: An Historical Perspective*. London, England: Basil Blackwell, 1989.

Munton, R.J.C., Sarah Whatmore, and Terry K. Marsden. "Part-Time Farming and Its Implications for the Rural Landscape: A Preliminary Analysis." *Environment and Planning A*, 21, No. 4 (1989): 523-536.

Overton, Mark. "The Critical Century? The Agrarian History of England and Wales, 1750–1850." *Agricultural History Review*, 38, No. 2 (1990): 185-189.

Walton, J.R. "On Estimating the Extent of Parliamentary Enclosure." *Agricultural History Review*, 38, No. 1 (1990): 79-82.

Ward, N., Terry K. Marsden, and R.J.C. Munton. "Farm Landscape Change: Trends in Upland and Lowland England." *Land Use Policy*, 7, No. 4 (1990): 291-302.

Bowler, Ian R. "The Agricultural Pattern." Pp. 83-114 in *The Changing Geography of the United Kingdom,* edited by R.J. Johnston and Vincent Gardiner. London, England: Routledge, 1991.

Dilley, Robert S. "Agricultural Change and Common Land in Cumberland, 1700–1850." Ph.D. Dissertation, McMaster University, Hamilton, Ontario, 1991.

Ilbery, Brian W. "Farm Diversification as an Adjustment Strategy on the Urban Fringe of the West Midlands." *Journal of Rural Studies*, 7, No. 3 (1991): 207-218.

Ilbery, Brian W., and B. Stiell. "Uptake of the Farm Diversification Grant Scheme in England." *Geography*, 76, No. 3 (1991): 259-262.

Bowler, Ian R. "The Development of Agricultural Resources in Britain." *Nederlandse Geografische Studies,* 153 (1992): 49-59.

Campbell, Bruce. "Cluster Analysis and the Classification of Medieval Demense-Farming Systems." *Transactions, Institute of British Geographers*, 17, No. 2 (1992): 227-245. (New Series)

Dodgshon, Robert A. "Farming Practice in the Western Highlands and Islands Before Crofting: A Study in Cultural Inertia or Opportunity Costs?" *Rural History*, 3 (1992): 173-189.

Evans, N.J., and Brian W. Ilbery. "Farm-Based Accommodation and the Restructuring of Agriculture: Evidence from Three English Counties." *Journal of Rural Studies*, 8, No. 1 (1992): 85-96.

Flynn, Andrew, and T.K. Marsden. "Food Regulation in a Period of Agricultural Retreat: The British Experience." *Geoforum*, 23, No. 1 (1992): 85-94.

Ilbery, Brian W. *Agricultural Change in Great Britain*. Oxford, England: Oxford University Press, 1992.

Ilbery, Brian W. "State-Assisted Farm Diversification in the United Kingdom." Pp. 100-118 in *Contemporary Rural Systems in Transition: Volume 1, Ag-*

riculture and Environment, edited by Ian R. Bowler, Christopher R. Bryant, and M. Duane Nellis. Wallingford, England: C.A.B. International, 1992.

Sheail, John. "Opencast Coal Working and the Perceived Threat to Agriculture in Post-War Scotland." *Scottish Geographical Magazine*, 108, No. 2 (1992): 104-112.

Sheppard, June A. "Small Farms in a Sussex Weald Parish, 1800–1860." *Agricultural History Review*, 40, No. 2 (1992): 127-141.

Shucksmith, Mark. "The Effects of European Agricultural Policies on the Communities of Upland Scotland: The Case of the Grampians." *Revue de Geographie Alpine*, 80, No. 4 (1992): 97-116.

Binns, J.A., D.C. Funnell, and A.G. Stephens. *Agricultural Change in Sussex*. Sussex, England: University of Sussex, Research Paper in Geography No. 10, 1993.

Curry, N., and Caroline Pack. "Planning on Presumption: Strategic Planning for Countryside Recreation in England and Wales." *Land Use Policy*, 10, No. 2 (1993): 140-150.

Dodgshon, Robert A. "Strategies of Farming in the Western Highlands and Islands Prior to Crofting and the Clearances." *Economic History Review*, 46 (1993): 679-701.

Ilbery, Brian W., and Ian R. Bowler. "The Farm Diversification Grant Scheme: Adoption and Nonadoption in England and Wales." *Environment and Planning C*, 11, No. 2 (1993): 161-170.

Robinson, Guy M., and Brian W. Ilbery. "Reforming the CAP: Beyond MacSharry." *Progress in Rural Policy and Planning*, 3 (1993): 197-207.

Marsden, Terry K. "Beyond Agriculture? Regulating the New Rural Spaces." *Journal of Rural Studies*, 11, No. 3 (1995): 285-296.

Sheail, John. "Elements of Sustainable Agriculture: The United Kingdom Experience 1840–1940." *Agricultural History Review*, 43, No. 2 (1995): 178-192.

Evans, N.J. "Evaluating Recent Changes to the Publication of UK Agricultural Census Data." *Geography*, 81, No. 3 (1996): 225-234.

Ilbery, Brian, et al. "Agricultural Adjustment and Business Diversification by Farm Households." *Geography*, 81, No. 4 (1996): 301-310.

Mayfield, Lucy H. "The Local Economic Impact of Small Farms: A Spatial Analysis." *Tijdschrift voor Economische en Sociale Geographie*, 87, No. 5 (1996): 387-398.

Mitchell, Morag, and Chris Doyle. "Spatial Distribution of the Impact of Agricultural Policy Reforms in Rural Areas." *Scottish Geographical Magazine*, 112, No. 2 (1996): 76-82.

Clark, Gordon, et al. "Rural Re-Regulation and Institutional Sustainability: A Case Study of Alternative Farming Systems in England." Pp. 117-134 in *Agricultural Restructuring and Sustainability: A Geographical Perspective*, edited by Brian W. Ilbery, Quentin Chiotti, and Timothy Rickard. Wallingford, England: C.A.B. International, 1997.

Dalton, R.T. "Agricultural Change in Southern Derbyshire 1800 to 1870." *East Midland Geographer*, 20, No. 1-2 (1997): 33-41.

Ilbery, Brian W., and Ian R. Bowler. "From Agricultural Productivism to Post-Productivism." Pp. 57-84 in *The Geography of Rural Change*, edited by Brian W. Ilbery. London, England: Longman, 1998.

Bowler, Ian R. "Modelling Farm Diversification in Regions Using Expert and Decision Support Systems." *Journal of Rural Studies*, 15, No. 3 (1999): 297-305.

Bowler, Ian R. "Agriculture." Pp. 84-107 in *The Changing Geography of the United Kingdom*, 3rd edition, edited by Vincent Gardiner and Hugh Stephens. London, England: Routledge, 2000.

Ilbery, Brian W., and M. Kneafsey. "Producer Constructions of Quality in Regional Specialty Food Production: A Case Study from South West England." *Journal of Rural Studies*, 16, No. 2 (2000): 217-230.

Tsouvalis, Judish, Susanne Seymour, and Charles Watkins. "Exploring Knowledge-Cultures: Precision Farming, Yield Mapping, and the Expert-Farmer Interface." *Environment and Planning A*, 32, No. 5 (2000): 909-924.

Walford, Nigel. "Patterns of Development in Tourist Accommodation Enterprises on Farms in England and Wales." *Applied Geography*, 21, No. 4 (2001): 331-345.

CROPS AND CROPPING PATTERNS

Kendall, M.G. "The Geographical Distribution of Crop Productivity in England." *Journal of the Royal Statistical Society*, 102 (1939): 21-48.

Whyte, R.O. "Crop Geography in Relation to Environment." *Scottish Geographical Magazine*, 62, No. 2 (1946): 49-55.

Taylor, J.A. "The Relation of Crop Distribution to the Drift Pattern in Southwest Lancashire." *Transactions, Institute of British Geographers*, 18 (1952): 77-91.

Leay, Petra. "The Market Garden Industry of the Melbourne District, South Derbyshire." *East Midland Geographer*, 1 (1954): 33-40.

Grigg, David B. "The 1801 Crop Returns for South Lancashire." *East Midland Geographer*, 16 (1961): 43-48.

Harvey, David. "Locational Change in the Kentish Hop Industry and the Analysis of Land-Use Patterns." *Transactions, Institute of British Geographers*, 33 (1963): 123-144.

Pocock, D.C.D. "Some Former Hop-Growing Centres." *Agricultural History Review*, 13, No. 2 (1965): 17-22.

Best, R.H., and R. Gasson. "The Changing Location of Intensive Crops." *Studies in Rural Land Use*. Wye, England: Wye College, Publication No. 6, 1966.

Gasson, R. "The Changing Location of Intensive Crops in England and Wales." *Geography*, 51 (1966): 16-28.

Naylor, E.L. "Market-Gardening Northeast of Nottingham." *East Midland Geographer*, 4, No. 32 (1969): 454-463.

Yelling, James A. "The Combination and Rotation of Crops in East Worcestershire, 1540–1660." *Agricultural History Review*, 17 (1969): 24-43.

Shaw, J.R. "The Location of Main Crop Potato Production in Britain: An Application of Linear Programming." *Journal of Agricultural Economics*, 21 (1970): 267-282.

Brandon, P.F. "Demesne Arable Farming in Coastal Sussex during the Later Middle Ages." *Agricultural History Review*, 19, No. 2 (1971): 113-134.

Dalton, R.T. "Peas for Freezing: A Recent Development in Lincolnshire Agriculture." *East Midland Geographer*, 5, No. 35 (1971): 133-141.

Watts, H.D. "The Location of the Beet-Sugar Industry in England and Wales, 1912–1936." *Transactions, Institute of British Geographers*, 53 (1971): 95-116.

Yelling, James A. "Changes in Crop Production in East Worcestershire, 1540–1867." *Agricultural History Review*, 21, No. 1 (1973): 18-34.

Watts, H.D. "Locational Adjustment in the British Beet Sugar Industry." *Geography*, 59, No. 1 (1974): 10-23.

Beavington, F. "The Development of Market Gardening in Bedfordshire, 1799–1939." *Agricultural History Review*, 23, No. 1 (1975): 23-47.

Hadd, Adrian N.L. "The Cultivation of Orchard Fruits in the Carse of Gowrie, 1750–1900." *Scottish Geographical Magazine*, 91, No. 2 (1975): 79-90.

Hough, M.N. "Mapping Areas of Britain Suitable for Maize on the Basis of Temperature." *ADAS Quarterly Review*, 18 (1975): 64-68.

Emery, Frank. "The Mechanics of Innovation: Clover Cultivation in Wales Before 1750." *Journal of Historical Geography*, 2, No. 1 (1976): 35-48.

Tarrant, J.R. "Maize: A New United Kingdom Agricultural Crop." *Area*, 7, No. 3 (1976): 175-178.

Wrathall, James E. "The Oil-Seed Rape Revolution in England and Wales." *Geography*, 63, No. 1 (1978): 42-45.

Dawson, A.H. "The Great Increase in Barley Growing in Scotland." *Geography*, 65, No. 3 (1980): 213-216.

Briggs, David. "Environmental Influences on the Yield of Spring Barley in England and Wales." *Geoforum*, 12, No. 1 (1981): 99-106.

Turner, M.E. "Arable in England and Wales: Estimates from the 1801 Crop Return." *Journal of Historical Geography*, 7, No. 3 (1981): 291-302.

Griffin, M. "European Challenge for English Hops." *Geographical Magazine*, 54, No. 10 (1982): 564-569.

Ilbery, Brian W. "The Decline of Hop Growing in Hereford and Worchestershire." *Area*, 14, No. 3 (1982): 203-212.

Ilbery, Brian W. "Harvey's Principles Reapplied: A Case Study of the Declining West Midland Hop Industry." *Geoforum*, 14, No. 1 (1983): 111-123.

Ilbery, Brian W. "The Renaissance of Viticulture in England and Wales." *Geography*, 68, No. 4 (1983): 341-344.

Robinson, Guy M. "The Evolution of the Horticulture Industry in the Vale of Evesham." *Scottish Geographical Magazine*, 99, No. 2 (1983): 89-100.

Dury, G.H. "Crop Failures on the Winchester Manors, 1232–1349." *Transactions, Institute of British Geographers*, 9, No. 4 (1984): 401-418. (New Series)

Grigg, David B. "Agricultural Perspectives Under Free Trade." Pp. 24-34 in *The Changing Countryside*, edited by J. Blunden and N. Curry. London, England: Methuen, 1985.

Ilbery, Brian W. "Factors Affecting the Structure of Viticulture in England and Wales." *Area*, 17, No. 2 (1985): 147-154.

Ilbery, Brian W. "Horticultural Decline in the Vale of Evesham, United Kingdom, 1950–1980." *Journal of Rural Studies*, 1, No. 2 (1985): 109-120.

Ilbery, Brian W. "Factors Affecting the Structure of Horticulture in the Vale of Evesham, United Kingdom: A Behavioral Interpretation." *Journal of Rural Studies*, 1, No. 2 (1985): 121-134.

Overton, Mark. "The Diffusion of Agricultural Innovations in Early Modern England: Turnips and Clover in Norfolk and Suffolk, 1580–1740." *Transactions, Institute of British Geographers*, 10, No. 2 (1985): 205-221. (New Series)

Wrathall, James E., and R. Moore. "Oilseed Rape in Great Britain: The End of a Revolution?" *Geography*, 71, No. 4 (1986): 351-354.

Glennie, Paul. "Continuity and Change in Hertfordshire Agriculture, 1550–1700: 2—Trends in Crop Yields and Their Determinants." *Agricultural History Review*, 36, No. 2 (1988): 145-161.

Pitte, J.R. "Le vin des brumes: Note sur la renaissance de la viticulture dans les Iles Brittaniques." Pp. 266-272 in *Campagnes et littoraux d'Europe: Melanges offerts a Pierre Flatres*. Lille, France: Institut de Geographie de Lille: Hommes et Terres du Nord, 1988.

Wrathall, James E. "Recent Changes in Arable Crop Production in England and Wales." *Land Use Policy*, 5, No. 2 (1988): 219-231.

Thery, H. "Le renouveau des vignobles britaniques." *Mappemonde*, 90, No. 2 (1990): 22-26.

Unwin, Timothy. "Saxon and Early Norman Viticulture in England." *Journal of Wine Research*, 1, No. 1 (1990): 61-76.

Whittington, G., and K.J. Edwards. "The Cultivation and Utilisation of Hemp in Scotland." *Scottish Geographical Magazine*, 106, No. 3 (1990): 167-173.

Butterfield, R.E., and J.I.L. Morison. "Modelling the Impact of Climatic Warming on Winter Cereal Development." *Agricultural and Forest Meteorology*, 62 (1992): 241-261.

Pitman, John I. "Changes in Crop Productivity and Water Quality in the United Kingdom." Pp. 89-122 in *Agricultural Change, Environment and Economy: Essays in Honour of W.B. Morgan*, edited by Keith Hoggart. London, England: Mansell, 1993.

Ilbery, Brian W., and Ian R. Bowler. "International Competition in Agriculture and Government Intervention: The Case of Horticulture in England." *Geography*, 79, No. 4 (1994): 361-366.

Unwin, Timothy. "European Wine Sector Policy and the UK Wine Industry." *Journal of Wine Research*, 5, No. 2 (1994): 135-146.

Ilbery, Brian W., and Ian R. Bowler. "Horticultural Change and the Horticultural Buildings and Orchard Replanting Grant Scheme in England." *Environment and Planning C*, 13, No. 1 (1995): 47-66.

Walton, J.R. "Varietel Innovation and the Competitiveness of the British Cereals Sector, 1760–1930." *Agricultural History Review*, 47, No. 1 (1999): 29-57.

Lee, R. "Shelter from the Storm: Geographies of Regard in the Worlds of Horticultural Consumption and Production." *Geoforum*, 31, No. 2 (2000): 137-157.

ANIMAL HUSBANDRY

Walton, C.L. "Some Geographical Aspects of the Sheep Industry." *Scottish Geographical Magazine*, 33, No. 7 (1917): 303-314.

Snodgrass, Catherine P. "Stock Farming in Scotland and Its Relation to Environment." *Scottish Geographical Magazine*, 49, No. 1 (1933): 24-34.

Carrier, E.H. *The Pastoral Heritage of Britain: A Geographical Study*. London, England: Christopher, 1936.

Cumberland, K.B. "Livestock Distribution in Craven." *Scottish Geographical Magazine*, 54 (1938): 75-93.

Hart, John Fraser. "Sheep Farming in Southern Scotland." Ph.D. Dissertation, Northwestern University, Evanston, Ill., 1950.

Hart, John Fraser. "The Changing Distribution of Sheep in Britain." *Economic Geography*, 32, No. 3 (1956): 260-274.

Donkin, R.A. "Cistercian Sheep-Farming and Wool Sales in the Thirteenth Century." *Agricultural History Review*, 6, No. 1 (1958): 2-8.

Donkin, R.A. "Some Aspects of Cistercian Sheep Farming in England and Wales." *Citeaux*, 13 (1962): 296-310.

Hewlett, Geoffrey P. "Geographical Aspects of the Poultry Industry in England and Wales." *Wessex Geographer*, 10 (1969): 11-18.

Edwards, C.J.W. "The Spatial Distribution of Livestock in Northern Ireland." *Irish Geography*, 10 (1977): 58-71.

Job, D.A., and J.A. Taylor. "The Production, Utilization and Management of Upland Grazing on Plynlimon, Wales." *Journal of Biogeography*, 5, No. 2 (1978): 173-191.

Carlyle, William J. "The Changing Distribution of Breeds of Sheep in Scotland, 1795–1965." *Agricultural History Review*, 27, No. 1 (1979): 19-29.

Frost, Pauline. "Yeomen and Metalsmith: Livestock in the Dual Economy in South Staffordshire, 1560–1720." *Agricultural History Review*, 29, No. 1 (1981): 29-41.

Short, Brian. "The Art and Craft of Chicken Cramming: Poultry in the Weald of Sussex, 1850–1950." *Agricultural History Review*, 30, No. 1 (1982): 17-30.

Walton, J.R. "The Diffusion of Improved Sheep Breeds in Eighteenth- and Nineteenth-Century Oxfordshire." *Journal of Historical Geography*, 9, No. 2 (1983): 175-195.

Walton, J.R. "The Diffusion of the Improved Shorthorn Breed of Cattle in Britain during the Eighteenth and Nineteenth Centuries." *Transactions, Institute of British Geographers*, 9, No. 1 (1984): 22-36. (New Series)

Symes, David G., and J. Marsden. "Industrialization of Agriculture: Intensive Livestock Farming in Humberside." Pp. 99-120 in *The Industrialization of the Countryside*, edited by Michael J. Healey and Brian W. Ilbery. Norwich, England: Geo Books, 1985.

Walton, J.R. "Pedigree and the National Cattle Herd, circa 1750–1950." *Agricultural History Review*, 34, No. 2 (1986): 149-170.

Copus, Andrew K. "Changing Markets and the Development of Sheep Breeds in Southern England, 1750–1900." *Agricultural History Review*, 37, No. 1 (1989): 36-51.

Edwards, C.J.W. "Spatial Change in Livestock Production in Northern Ireland: 1974–1986." *Journal of Rural Studies*, 7, No. 3 (1991): 277-284.

Yarwood, Richard, and Vick Evans. "The Changing Geography of Rare Livestock Breeds in Britain." *Geography*, 84, No. 1 (1999): 80-86.

Holloway, L.E. "Pets and Proteins: Placing Domestic Livestock on Hobby-Farms in England and Wales." *Journal of Rural Studies*, 17, No. 3 (2001): 293-307.

DAIRYING

Thompson, Keith W. "The Dairy Industry of England and Wales Since the Establishment of the Milk Marketing Board." Ph.D. Dissertation, University of Washington, Seattle, Wash., 1954.

Barnes, F.A. "Dairying in Anglesey." *Transactions, Institute of British Geographers*, 21 (1955): 137-156.

Simpson, E.S. "The Cheshire Grass-Dairying Region." *Transactions, Institute of British Geographers*, 23 (1957): 141-162.

Barnes, F.A. "The Evolution of the Salient Patterns of Milk Production and Distribution in England and Wales." *Transactions, Institute of British Geographers*, 25 (1958): 167-195.

Boal, F.W., and B.S. McAodha. "The Milk Industry in Northern Ireland." *Economic Geography*, 37, No. 2 (1961): 170-180.

McQueen, J.D.W. "Milk Surpluses in Scotland." *Scottish Geographical Magazine*, 77 (1961): 93-105.

FARM LAND

Stamp, L. Dudley. *Man and the Land*. London, England: Collins, 1955.

Davies, Margaret. "Rhosili Open Fields and Related South Wales Field Patterns." *Agricultural History Review*, 4, No. 2 (1956): 80-96.

Sylvester, Dorothy. "The Common Fields of the Coastlands of Gwent." *Agricultural History Review*, 6, No. 1 (1958): 9-26.

Mills, Dennis R. "Enclosure in Kesteven." *Agricultural History Review*, 7, No. 2 (1959): 82-97.

Coppock, J.T. "Farms and Fields in the Chilterns." *Erdkunde*, 14, No. 2 (1960): 134-145.

Harris, Alan. *The Rural Landscape of the East Riding of Yorkshire, 1700–1850: A Study in Historical Geography*. London, England: Oxford University Press, 1961.

Blake, J.L. "Pasture Improvement Schemes in the Isle of Lewis." *Scottish Studies*, 6 (1962): 108-122.

Coppock, J.T. "Land Use and Land Classification, in Classification of Agricultural Land in Britain." *Agricultural Land Service Technical Report*, 8 (1962): 65-80.

Stamp, L. Dudley. *The Land of Britain: Its Use and Misuse*. London, England: Longman, 1962.

Gailey, R.A. "Agrarian Improvement and the Development of Enclosure in the South West Highlands of Scotland." *Scottish Historical Review*, 42 (1963): 105-125.

Grigg, David B. "Large and Small Farms in England and Wales: Their Size and Distribution." *Geography*, 48 (1963): 268-279.

Johnson, S.A. "Enclosures and Changing Agricultural Landscapes in Lindsay." *Agricultural History Review*, 11, No. 2 (1963): 95-102.

Butlin, Robin A. "Northumberland Field Systems." *Agricultural History Review*, 12, No. 1 (1964): 99-120.

Baker, Alan R.H. "Field Systems in the Vale of Holmesdale." *Agricultural History Review*, 14, No. 1 (1966): 1-24.

Matley, Ian M. "The Origin of Infield–Outfield Agriculture in Scotland: The Linguistic Evidence." *Professional Geographer*, 18 (1966): 275-279.

Wheeler, P.T. "Landownership and the Crofting System in Sutherland Since 1800." *Agricultural History Review*, 14, No. 1 (1966): 45-66.

Wise, Michael. *Statutory and Smallholding Provided by Local Authorities in England and Wales*. London, England: H.M.S.O., 1966.

Butlin, Robin A. "Enclosure and Improvement in Northumberland in the Sixteenth Century." *Archaeologia Aeliana*, 45 (1967): 149-160.

Edwards, C.J.W. "The Assessment of the Quality of Agricultural Land in Central Somerset." Pp. 461-477 in *Liverpool Essays in Geography*, edited by R.W. Steel and R. Lawton. London, England: Longman, 1967.

Aitchinson, J.W. "The Land Factor and Agricultural Production in West Central Lancashire." *Agricultural Geography I.G.U. Symposium*, edited by E.S. Simpson. Liverpool, England: University of Liverpool Department of Geography, Research Paper No. 5, 1968.

Best, R.H., and A.G. Champion. "Regional Conversions of Agricultural Land to Urban Use in England and Wales, 1945–67." *Transactions, Institute of British Geographers*, 49 (1970): 15-32.

Buchanan, R.H. "Common Fields and Enclosure: An Eighteenth Century Example from Lecale, County Down." *Ulster Folklife*, 25-26 (1970): 99-118.

Williams, Michael. "The Enclosure and Reclamation of Wasteland in England and Wales in the Eighteenth and Nineteenth Centuries." *Transactions, Institute of British Geographers*, 51 (1970): 55-70.

Cruickshank, J.G., and W.J. Armstrong. "Soil and Agricultural Land Classification in County Londonderry." *Transactions, Institute of British Geographers*, 53 (1971): 79-94.

Fox, H.S.A. "A Geographical Study of the Field Systems of Devon and Cornwall." Ph.D. Dissertation, University of Cambridge, Cambridge, England, 1971.

Williams, Michael. "The Enclosure and Reclamation of the Mendip Hills, 1770–1870." *Agricultural History Review*, 19, No. 1 (1971): 65-81.

Baker, Alan R.H., and Robin A. Butlin, eds. *Studies of Field Systems in the British Isles*. Cambridge, England: Cambridge University Press, 1973.

Baker, Alan R.H. "A Relatively Neglected Field Form: The Headland Ridge." *Agricultural History Review*, 21, No. 1 (1973): 47-50.

Hewlett, Geoffrey. "Reconstructing a Historical Landscape from Field and Documentary Evidence: Otford in Kent." *Agricultural History Review*, 21, No. 2 (1973): 94-110.

Jones, G.R.J. "Field Systems of North Wales." Pp. 430-479 in *Studies of Field Systems in the British Isles*, edited by Alan R.H. Baker and Robin A. Butlin. Cambridge, England: Cambridge University Press, 1973.

Maida, I. "The Contemporary Regional Structures of British Agriculture: The Characteristics of the Land Utilisation Types from the Viewpoint of Farming System" (in Japanese). *Jimbun Chiri/Human Geography, Kyoto*, 26, No. 5 (1974): 1-29.

Clark, Gordon. *The Amalgamation of Farms in Scotland, 1967–1973: Some Methodological Considerations*. Edinburgh., Scotland: University of Edinburgh Department of Geography, Research Discussion Paper No. 3, 1975.

Dilley, Robert S. "The Customary Acre: An Indeterminate Measure." *Agricultural History Review*, 23, No. 2 (1975): 173-176.

Dodgshon, R.A. "The Landholding Foundations of the Open Field System." *Past and Present*, 57 (1975): 1-29.

Gilg, Andrew. "Agricultural Land Classification in Britain: A Review of the Ministry of Agriculture's New Map Series." *Biological Conservation*, 7, No. 1 (1975): 73-77.

Munton, R.C.J. "The State of the Agricultural Land Market in England, 1971–1973: A Survey of Auctioneers' Property Transactions." *Oxford Agrarian Studies*, 4, No. 2 (1975): 111-130.

Chapman, John. "Parliamentary Enclosure in the Uplands: The Case of the North York Moors." *Agricultural History Review*, 24, No. 1 (1976): 1-17.

Currie, E.A. "Land Tenures, Enclosures and Field Patterns in County Derry in the 18th and 19th Centuries." *Irish Geography*, 9 (1976): 50-62.

Edwards, C.J.W. "The Assessment of the Quality of Agricultural Land in Central Somerset." Pp. 461-477 in *Liverpool Essays in Geography: A Jubilee Collection*, edited by R.W. Steel and R. Lawton. London, England: Longman, 1976.

Lee, J. "Land Resources as a Basic Factor in Livestock Carrying Capacity." *Irish Geography*, 9 (1976): 63-75.

Munton, R.J.C. "An Analysis of Price Trends in the Agricultural Land Market of England and Wales." *Tijdschrift voor Economische en Sociale Geographie*, 67, No. 4 (1976): 202-212.

Parry, Martin L. "A Typology of Cultivation Ridges in Southern Scotland." *Tools and Tillage*, 3 (1976): 3-19.

Phillips, A.D.M. "Agricultural Land Use, Soils and the Nottinghamshire Tithe Surveys, circa 1840." *East Midland Geographer*, 6, No. 46 (1976): 284-301.

Cruickshank, J.G., and M.M. Cruickshank. "A Survey of Neglected Agricultural Land in the Sperrin Mountains, Northern Ireland." *Irish Geography*, 10 (1977): 36-43.

Lindsay, J.M. "Forestry and Agriculture in the Scottish Highlands, 1700–1850: A Problem in Estate Management." *Agricultural History Review*, 25, No. 1 (1977): 23-36.

Munton, R.J.C. "Financial Institutions: Their Ownership of Agricultural Land in Great Britain." *Area*, 9, No. 1 (1977): 29-37.

Chapman, John. "Some Problems in the Interpretation of Enclosure Awards." *Agricultural History Review*, 26, No. 2 (1978): 108-114.

Coleman, Alice. "Agricultural Land Losses: The Evidence from Maps." Pp. 16-36 in *Urban Growth, Farmland Losses and Planning: Symposium Organized by the Rural Geography Study Group of the IBG*, edited by A.W. Rogers. Ashford, England: Wye College, 1978.

Dodgshon, Robert A. "Land Improvement in Scottish Farming: Marl and Lime in Roxburghshire and Berwickshire in the Eighteenth Century." *Agricultural History Review*, 26, No. 1 (1978): 1-14.

Edwards, C.J.W. "The Effects of Changing Farm Size Upon Levels of Farm Fragmentation: A Somerset Case Study." *Journal of Agricultural Economics*, 29, No. 2 (1978): 143-154.

Mathew, Alexander S. "The Alleged Deterioration in Hill Grazing in the Scottish Highlands." *Biological Conservation*, 14, No. 3 (1978): 181-196.

Turnock, David. *The Lochaber Area: A Case Study of Changing Rural Land Use in the West Highlands of Scotland.* Sheffield, England: Regional Studies, Geographical Field Group, No. 22, 1978.

Butlin, Robin A. "The Enclosure of Open Fields and Extinction of Common Rights in England, circa 1600–1750: A Review." Pp. 65-82 in *Change in the Countryside: Essays on Rural England, 1500–1900*, edited by H.S.A. Fox and Robin A. Butlin. London, England: Institute of British Geographers, 1979.

Clark, Gordon. "Farm Amalgamations in Scotland." *Scottish Geographical Magazine*, 95 (1979): 93-107.

Phillips, A.D.M. "The Landlord and Agricultural Improvements: Underdraining on the Lincolnshire Estate of the Earls of Scarborough in the First Half of the Nineteenth Century." *East Midland Geographer*, 7, No. 52 (1979): 168-177.

Whyte, I.D. "Written Leases and Their Impact on Scottish Agriculture in the Seventeenth Century." *Agricultural History Review*, 27, No. 1 (1979): 1-9.

Whyte, I.D. "Infield-Outfield Farming on a Seventeenth-Century Scottish Estate." *Journal of Historical Geography*, 5, No. 4 (1979): 391-402.

Dodgshon, Robert A. *The Origin of British Field Systems: An Interpretation.* London, England: Academic Press, 1980.

Harvey, Mary. "Regular Field and Tenurial Arrangements in Holdernese, Yorkshire." *Journal of Historical Geography*, 6, No. 1 (1980): 3-16.

Laird, J.B. "The Reshaped Agricultural Landscape." Pp. 203-222 in *The Making of the Scottish Landscape*, edited by Martin L. Parry and T.R. Slater. London, England: Croom Helm, 1980.

Turner, M.E. *English Parliamentary Enclosure: Its Historical Geography and Economic History*. Folkestone, England: W. Dawson, 1980.

Bowler, Ian R. "Permanent Grass in Leicestershire: The Decline of a Traditional Land Use and Its Role in the Farming System." *East Midland Geographer*, 7, No. 55 (1981): 272-278.

Coleman, Alice. "Turning the Tide of Agricultural Land Loss." Pp. 133-144 in *Perspectives in Agricultural Geography*, vol. 3, edited by Noor Mohammad. New Delhi, India: Concept, 1981.

Marsden, Terry K. "Farm Organization and Linkages." *Swansea Geographer*, 19 (1981–1982): 10-14.

Chapman, John, and Trevor M. Harris. "The Accuracy of Enclosure Estimates: Some Evidence from Northern England." *Journal of Historical Geography*, 8, No. 3 (1982): 261-264.

Baker, Alan R.H. "Discourses on British Field Systems." *Agricultural History Review*, 31, No. 2 (1983): 149-155.

Harvey, Mary. "Planned Field Systems in Eastern Yorkshire: Some Thoughts on Their Origin." *Agricultural History Review*, 31, No. 2 (1983): 91-103.

Phillips, A.D.M. "Agricultural Land Use on a Nottinghamshire Estate (1849–1899), as Revealed by Cropping Books." *East Midland Geographer*, 8, No. 59 (1983): 70-78.

Harvey, Mary. "Open Field Structure and Landholdings Arrangements in Eastern Yorkshire." *Transactions, Institute of British Geographers*, 9, No. 1 (1984): 60-74. (New Series)

Ilbery, Brian W. "Farm Fragmentation in the Vale of Evesham." *Area*, 16, No. 2 (1984): 159-165.

Harvey, Mary. "The Development of Openfields in the Central Vale of York: A Reconsideration." *Geografiska Annaler B*, 67B, No. 1 (1985): 35-44.

Mather, A.S. "The Classification of the Agricultural Land Resources of Scotland." *Area*, 17, No. 2 (1985): 135-139.

Nash, Alan. "The Size of Open Field Strips: A Reinterpretation." *Agricultural History Review*, 33, No. 1 (1985): 32-40.

Wood, B.A., and P.T. Wheeler. "Changing Farm Structure on a Nottinghamshire Landed Estate, 1952–1977." *East Midland Geographer*, 8 (1985): 198-206.

Baker, Alan R.H. "Evidence in the Norarum Inquisitiones of the Contrasting Arable Lands in England During the Early Fourteenth Century." *Economic History Review*, 19 (1986): 518-532.

Chapman, John. "The Extent and Nature of Parliamentary Enclosure." *Agricultural History Review*, 35, No. 1 (1987): 25-35.

Dawson, A.H. "Farm Size and Land Use in the Urban Fringe of Scotland." Pp. 277-288 in *Demands on Rural Lands: Planning for Resource Use*, edited by Chris Cocklin, Barry E. Smit, and Thomas R.R. Johnston. Boulder, Colo.: Westview, 1987.

Essex, S.J. "Pressures on Farm Woodlands: Present and Future Trends." Pp. 55-70 in *The Future of the British Rural Landscape*, edited by Douglas G. Lockhart and Brian W. Ilbery. Norwich, England: Geo Books, 1987.

Grigg, David B. "Farm Size in England and Wales, from Early Victorian Times to the Present." *Agricultural History Review*, 35, No. 2 (1987): 179-190.

Marsden, Terry K. "Land Ownership and Farm Businesses in the Rural Landscape." Pp. 29-45 in *The Future of the British Rural Landscape*, edited by Douglas Lockhart and Brian W. Ilbery. Norwich, England: Geo Books, 1987.

Grigg, David B. "From Gregory King to Alure: The Changing Amount and Distribution of Arable Land in England and Wales, 1700–1985." *Tijdschrift voor Economische en Sociale Geographie*, 79, No. 3 (1988): 199-209.

Wathern, P., et al. "Recent Upland Land Use Change and Agricultural Policy in Clwyd, North Wales." *Applied Geography*, 8 (1988): 147-163.

Phillips, A.D.M. *The Underdraining of Farmland in England During the Nineteenth Century.* Cambridge, England: Cambridge University Press, 1989.

Ilbery, Brian W. "Adoption of the Arable Set-Aside Scheme in England." *Geography*, 75, No. 1 (1990): 69-72.

Marsden, Terry K., and Jonathan Murdock. *Agriculture in Retreat: Implications for the Changing Control and Development of Rural Land.* Newcastle, England: University of Newcastle, ESRC Countryside Change Initiative, Working Paper No. 9, 1990.

Williams, Michael. "Farm Fragmentation: Modern Variations on an Old Theme." *University of Oxford School of Geography Research Paper*, 45 (1990): 45-53.

Chapman, John. "Confidence Limits and Enclosure Estimates: Some Comments." *Agricultural History Review*, 39, No. 1 (1991): 48-51.

Chapman, John. "The Later Parliamentary Enclosures of South Wales." *Agricultural History Review*, 39, No. 2 (1991): 116-125.

Robinson, Guy M. "EC Agricultural Policy and the Environment: Land Use Implications in the UK." *Land Use Policy*, 8, No. 2 (1991): 95-107.

Walton, J.R. "Parliamentary Enclosure, the Bootstrap and a Red Herring or Two." *Agricultural History Review*, 39, No. 1 (1991): 52-54.

Ilbery, Brian W., and Raymond Riley. "Adoption of the Farm Woodland Scheme in England." *Geography*, 77 (1992): 363-366.

Morris, Joseph. "Agricultural Land Drainage, Land Use Change and Economic Performance: Experience in the UK." *Land Use Policy*, 9, No. 3 (1992): 185-198.

Robinson, Guy M. "The Greening of Agricultural Policy: Scotland's Environmentally Sensitive Areas (ESAs)." *Journal of Environmental Planning and Management*, 37, No. 2 (1994): 215-225.

Boardman, John. "Damage to Property by Runoff from Agricultural Land, South Downs, Southern England, 1976–93." *Geographical Journal*, 161, No. 2 (1995): 177-191.

Evans, R. "Some Methods of Directly Assessing Water Erosion of Cultivated Land: A Comparison of Measurements Made on Plots and in Fields." *Progress in Physical Geography*, 19, No. 1 (1995): 115-129.

Munton, R.J.C. "Sustainable Development: A Critical Review of Rural Land-Use Policy in the UK." Pp. 11-24 in *Agricultural Restructuring and Sus-*

tainability: A Geographical Perspective, edited by Brian W. Ilbery, Quentin Chiotti, and Timothy Rickard. Wallingford, England: C.A.B. International, 1997.

Britt, C.P. "The Cost of Farm Woodland Establishment." *East Midland Geographer*, 21, No. 1 (1998): 49-53.

Pearson, Alastair, and Peter Collier. "The Integration and Analysis of Historical and Environmental Data Using a Geographical Information System: Landownership and Agricultural Productivity in Pembrokeshire, c1850." *Agricultural History Review*, 46, No. 2 (1998): 162-177.

Godfrey, John, and Brian Short. "The Ownership, Occupation and Use of Land on the South Downs, 1840–1940: A Methodological Analysis of Record Linkage over Time." *Agricultural History Review*, 49, No. 1 (2001): 56-78.

Lockhart, Douglas G. "Lotted Lands and Planned Villages in North-East Scotland." *Agricultural History Review*, 49, No. 1 (2001): 17-40.

AGRICULTURAL POPULATIONS AND SOCIAL ISSUES

Gauld, William A. "Agriculture and Populations in Galloway." *Scottish Geographical Magazine*, 38 (1922): 232-242.

Ginier, Jean. "Jersey, ile touristique et agricole." *Norois*, 17 (1970): 371-395.

Fox, H.S.A. "Local Farmers' Associations and the Circulation of Agricultural Information in Nineteenth Century England." Pp. 43-63 in *Change in the Countryside: Essays on Rural England 1500–1900*, edited by H.S.A. Fox and Robin A. Butlin. London, England: Institute of British Geographers, Special Publication No. 10, 1979.

Whyte, I.D. *Agriculture and Society in Seventeenth Century Scotland*. Edinburgh, Scotland: John Donald, 1979.

Dodgshon, Robert A. "Agricultural Change and Its Social Consequences in the Southern Uplands of Scotland, 1600–1780." Pp. 46-59 in *Ireland and Scotland 1600–1850: Parallels and Contrasts in Economic and Social Development*, edited by T.M. Devine and David Dickson. Edinburgh, Scotland: John Donald, 1981.

Ilbery, Brian W. "A Behavioral Analysis of Hop Farming in Hereford and Worchestershire." *Geoforum*, 14, No. 4 (1983): 447-460.

Ilbery, Brian W. "Goals and Values of Hop Farmers." *Transactions, Institute of British Geographers*, 8, No. 3 (1983): 329-342. (New Series)

Schroder, Jochem. *Veranderungen in der Agrar- und Sozialstruktur im mittleren Nord-England seit dem Landwirtschaftsgesetz von 1947 -ein Beitrag zur regionalen Agrargeographie Grossbritanniens, dargestellt anhand eines W-E-Profils von der Irishcen See zur Nordsee*. Heidelberg, Germany: Heidelberger Geographische Arbeiten, Heft 64, 1983.

Symes, David G., and Terry K. Marsden. "Complementary Roles and Asymmetrical Lives: Farmers' Wives in a Large Farm Environment." *Sociologia Ruralis*, 23 (1983): 229-241.

Ilbery, Brian W. "Agricultural Specialization and Farmer Decision Behavior in the West Midlands." *Tijdschrift voor Economische en Sociale Geographie*, 75 (1984): 329-334.

Whyte, I.D., and K.A. Whyte. "Continuity and Change in a Seventeenth-Century Scottish Farming Community." *Agricultural History Review*, 32, No. 2 (1984): 159-169.

Dodgshon, Robert A. "The Scottish Farming Township as Metaphor." Pp. 69-82 in *Perspectives in Scottish Social History: Essays in Honour of Rosalind Mitchison*, edited by L. Leneman. Aberdeen, Scotland: Aberdeen University Press, 1988.

Munton, R.J.C., J. Eldon, and T.K. Marsden. "Farmers' Response to an Uncertain Policy Future." Pp. 19-30 in *Removing Land from Agriculture*, edited by D. Baldock. London, England: Council for the Protection of Rural England, 1988.

Day, G., G. Reese, and Jonathan Murdock. "Social Change, Rural Localities and the State: The Restructuring of Rural Wales." *Journal of Rural Studies*, 5, No. 3 (1989): 227-244.

Marsden, Terry K., et al. "Strategies for Coping in Capitalist Agriculture: An Examination of the Responses of Farm Families in British Agriculture." *Geoforum*, 20, No. 1 (1989): 1-14.

Momsen, Janet. "Genero y agricultura en Inglaterra." *Documents d'Analisi Geografica*, 14 (1989): 115-130.

Shawyer, A.J. "Farm Structure and Farm Families: A Nottinghamshire Field Area." *East Midland Geographer*, 13, No. 1 (1990): 1-18.

Ilbery, Brian W. "Farm Diversification as an Adjustment Strategy on the Urban Fringe of the West Midlands." *Journal of Rural Studies*, 7 (1991).

Munton, R.J.C., and Terry K. Marsden. "Dualism or Diversity in Farmily Farming? Patterns of Occupancy Change in British Agriculture." *Geoforum*, 22, No. 1 (1991): 105-117.

Bowler, Ian R., and Brian W. Ilbery. *Farm Diversification in England and Wales: A Comparison of Participants and Non-Participants in the Farm Diversification Grant Scheme*. Leicester, England: University of Leicester Department of Geography, Occasional Paper No. 21, 1992.

Marsden, Terry K., N. Ward, and R.J.C. Munton. "Farm Businesses in Upland and Lowland Britain: Incorporating Social Trajectories into Uneven Agrarian Developments." *Socologia Ruralis*, 32 (1992): 408-430.

Morris, C., and Clive Potter. "Recruiting the New Conservationists: Farmers' Adoption of Agri-Environmental Schemes in the U.K." *Journal of Rural Studies*, 11, No. 1 (1995): 41-64.

Byron, Richard J. "Field Boundaries in Anglesey, Gweynedd." *Landscape Research*, 21, No. 2 (1996): 189-194.

Wilson, Geoff A. "Farmer Environmental Attitudes and ESA Participation." *Geoforum*, 27, No. 2 (1996): 115-132.

Ilbery, Brian W., Michael Healey, and Julie Higginbottom. "On and Off-Farm Business Diversification by Farm Households in England." Pp. 135-152 in *Agricultural Restructuring and Sustainability: A Geographical Perspective*, edited by Brian W. Ilbery, Quentin Chiotti, and Timothy Rickard. Wallingford, England: C.A.B. International, 1997.

Morris, C., and Charlotte Andres. "The Construction of Environmental Meaning Within Farming Culture in the UK: The Implications for Agri-Environmental Research." Pp. 87-100 in *Agricultural Restructuring and Sustainability: A Geographical Perspective*, edited by Brian W. Ilbery, Quentin Chiotti, and Timothy Rickard. Wallingford, England: C.A.B. International, 1997.

Walsh, M. "The View from the Farm: Farmers and Agri-Environmental Schemes in the Yorkshire Dales." *North West Geographer*, Third Series, 1 (1997): 24-35.

Clark, J.R.A., and A. Jones. "Agricultural Elites, Agrarian Beliefs, and Their Impact on the Evolution of Agri-Environment Policies: An Examination of the British Experience, 1981-92." *Environment and Planning A*, 30, No. 12 (1998): 2227-2242.

Harrison, C.M., J. Burgess, and J.R.A. Clark. "Discounted Knowledges: Farmers' and Residents' Understanding of Nature Conservation Goals and Policies." *Journal of Environmental Management*, 54, No. 4 (1998): 305-320.

ENVIRONMENTAL AND CONSERVATION ISSUES

O'Dell, Andrew C. "Geographical Controls of Agriculture in Orkney and Shetland." *Economic Geography*, 11, No. 1 (1935): 1-11.

Oliver, J. "Problems in Agro-Climatic Relationships in Wales in the Eighteenth Century." Pp. 6-14 in *Climatic Change with Special Reference to the Highland Zone of Britain*, edited by J.A. Taylor. Aberystwyth, Wales: University College of Wales, Memo No. 3, 1965.

Thomas, David. "Climate and Cropping in the Early 19th Century in Wales." Pp. 70-76 in *Climatic Change with Special Reference to Wales and Its Agriculture: Symposium in Agricultural Meteorology*. Aberystwyth, Wales: University College of Wales, Publication No. 8, 1965.

Phillips, A.D.M. "Underdraining and the English Claylands, 1850–1880: A Review." *Agricultural History Review*, 17 (1969): 44-55.

Bunce, Michael F. "The Agricultural Depression in South Yorkshire and North Nottinghamshire, 1875 to 1900: Climatic Hazard and Price Competition." *Canadian Geographer*, 16 (1972): 323-337.

Rathone, A.H. "Environmental Controls in Farming: Scotland, A Case Study." *Pakistan Geographical Review*, 30, No. 1-2 (1975): 24-29.

Milner, J.S. "Weather Variations and Crop Relationships in the East Midlands." *East Midlands Geographer*, 6, No. 48 (1977): 375-381.

Turnock, David. "Stages of Agricultural Improvement in the Uplands of Scotland's Grampian Region." *Journal of Historical Geography*, 3, No. 4 (1977): 327-348.

Mauger, Michelle. "The Shelter Effect of Hedges on Guernsey Dairy Farms." *Brighton Polytechnic Geographical Magazine*, 5 (1979): 13-18.

Munton, R.J.C. "Agriculture and Conservation: What Room for Compromise?" Pp. 353-373 in *Conservation in Perspective*, edited by A. Warren and F.B. Goldsmith. Chichester, England: John Wiley, 1983.

Boardman, John. "Soil Erosion, Climatic Vagary and Agricultural Changes on the Downs Around Lewes and Brighton, Autumn, 1982." *Applied Geography*, 5, No. 3 (1985): 243-258.

Coppock, J.T. "Agriculture and Conservation in the United Kingdom: Some Reflections." *The Geographer, Aligarh*, 32, No. 2 (1985): 21-33.

Tobin, G.A. "Agricultural Policies and Wetlands Management in Wales." *Papers and Proceedings of the Applied Geography Conferences*, 8 (1985): 19-29.

Green, Bryn. "Agriculture and the Environment: A Review of Major Issues in the UK." *Land Use Policy*, 3, No. 3 (1986): 193-204.

Robinson, M. "The Extent of Farm Underdrainage in England and Wales Prior to 1939." *Agricultural History Review*, 34, No. 1 (1986): 79-85.

Munton, R.J.C. "The Conflict Between Conservation and Food Production in Great Britain." Pp. 47-60 in *Demands on Rural Lands: Planning for Resource Use*, edited by Christopher Cocklin, Barry E. Smit, and Thomas R.R. Johnston. Boulder, Colo.: Westview, 1987.

Carter, T.R. "Climatic Change and Cropping Margins in Upland Britain." Ph.D. Dissertation, University of Birmingham, Birmingham, England, 1988.

Dodgshon, Robert A. "The Ecological Basis of Highland Peasant Farming." Pp. 139-151 in *The Cultural Landscape: Past, Present, and Future*, edited by Hilary H. Birks, et al. Cambridge, England: Cambridge University Press, 1988.

Foster, I.D.L., Brian W. Ilbery, and M.A. Hinton. "Agriculture and Water Quality: A Preliminary Examination of the Jersey Nitrate Problem." *Applied Geography*, 9, No. 2 (1989): 95-114.

Overton, Mark. "Weather and Agricultural Change in England, 1660–1739." *Agricultural History*, 63, No. 2 (1989): 77-88.

Robinson, M., M.C. Clayton, and W.C. Henderson. "The Extent of Agricultural Field Drainage in Scotland, 1983–1986." *Scottish Geographical Magazine*, 106, No. 3 (1990): 141-147.

Boardman, John. "Agriculture and Erosion in Britain." *Geography Review*, 6, No. 1 (1992): 15-19.

Buller, Henry J. "Agricultural Change and the Environment in Western Europe." Pp. 68-88 in *Agricultural Change, Environment and Economy: Essays in Honour of W.B. Morgan*, edited by Keith Hoggart. London, England: Mansell, 1992.

Cudjoe, Frank, and Philip Rees. "How Important Is Organic Farming in Great Britain?" *Tijdschrift voor Economische en Sociale Geographie*, 83, No. 1 (1992): 13-24.

Tarrant, J.C., and Richard Cobb. "The Convergence of Agricultural and Environmental Policies: The Case of Extensification in Eastern England." Pp. 153-165 in *Contemporary Rural Systems in Transition: Volume 1, Agriculture and Environment*, edited by Ian R. Bowler, Christopher R. Bryant, and M. Duane Nellis. Wallingford, England: C.A.B. International, 1992.

Ward, N., and R.J.C. Munton. "Conceptualizing Agriculture-Environment Relations: Combining Political Economy and Socio-Cultural Approaches to Pesticide Pollution." *Sociologica Ruralis*, 32, No. 1 (1992): 127-145.

Bolsium, Emmy C.A. "Possible Changes in Land-Use in Rural Areas in the European Community." Pp. 42-51 in *The Retreat: Rural Land-Use and European Agriculture*, edited by Emmy C.A. Bolsium, Gordon Clark, and Jan Groenendijk. Utrecht, Netherlands: Nederlandse Geografische Studies, No. 172, 1993.

Ilbery, Brian W., and Ian R. Bowler. "Land Diversion and Farm Business Diversification in EC Agriculture." Pp. 15-27 in *The Retreat: Rural Land-Use and European Agriculture*, edited by Emmy C.A. Bolsium, Gordon Clark, and Jan Groenendijk. Utrecht, Netherlands: Nederlandse Geografische Studies, No. 172, 1993.

Sheail, John. "The Agricultural Pollution of Water Courses: The Precedents Set by the Beet-Sugar and Milk Industries." *Agricultural History Review*, 41, No. 1 (1993): 31-43.

Robinson, Guy M. "The Greening of Agricultural Policy: Scotland's Environmentally Sensitive Areas (ESAs)." *Journal of Environmental Planning and Management*, 37 (1994): 215-225.

Towers, W. "Towards a Strategic Approach to Sewage Sludge Utilization on Agricultural Land in Scotland." *Journal of Environmental Planning and Management*, 37, No. 4 (1994): 447-460.

Withers, Charles W.J. "On Georgies and Geology: James Hutton's 'Elements of Agriculture' and Agricultural Science in Eighteenth-Century Scotland." *Agricultural History Review*, 42, No. 1 (1994): 38-48.

Mather, A.S., and K.J. Thomson. "The Effects of Afforestation on Agriculture in Scotland." *Journal of Rural Studies*, 11, No. 2 (1995): 187-202.

Sheail, John. "Nature Protection, Ecologists and the Farming Context: A U.K. Historical Context." *Journal of Rural Studies*, 11, No. 1 (1995): 79-88.

Ward, N., et al. "Rural Restructuring and the Regulation of Farm Pollution." *Environment and Planning A*, 27, No. 8 (1995): 1193-1212.

Adger, W.N., and Susan Subale. "Estimating Above-Ground Carbon Fluxes from UK Agricultural Land." *Geographical Journal*, 162, No. 2 (1996): 191-204.

Battershill, Martin R.J., and Andrew Gilg. "Traditional Farming and Agro-Environment Policy in Southwest England: Back to the Future?" *Geoforum*, 27, No. 2 (1996): 133-149.

Battershill, Martin R.J., and Andrew Gilg. "New Approaches to Creative Conservation on Farms in South West England." *Journal of Environmental Planning and Management*, 48 (1996): 321-340.

Battershill, Martin R.J., and Andrew Gilg. "Environmentally Friendly Farming in South West England: An Exploration and Analysis." Pp. 200-224 in *New Directions in Rural Land Use*, edited by N. Curry. Cheltenham, England: Countryside and Community Foundation, 1996.

Goddard, Nicholas. "A Mine of Wealth? The Victorians and the Agricultural Value of Sewage." *Journal of Historical Geography*, 22, No. 3 (1996): 274-290.

Govers, G., et al. "The Relative Contribution of Soil Tillage and Overland Flow Erosion to Soil Redistribution on Agricultural Land." *Earth Surface Processes and Landforms*, 21, No. 10 (1996): 929-946.

Hossell, J.E., P.J. Jones, and R.B. Tranter. "The Likely Effects of Climate Change on Agricultural Land Use in England and Wales." *Geoforum*, 27, No. 2 (1996): 149-158.

Mather, A.S. "The Inter-Relationship of Afforestation and Agriculture in Scotland." *Scottish Geographical Magazine*, 112, No. 2 (1996): 83-91.

Skerratt, Sarah, and J.B. Dent. "The Change of Agri-Environmental Subsidies: The Case of Breadalbane Environmentally Sensitive Area, Scotland." *Scottish Geographical Magazine*, 112, No. 2 (1996): 92-100.

Evans, N.J. "Something Old, New, Borrowed and Blue: The Marriage of Agriculture and Conservation in England." Pp. 233-252 in *Agricultural Restructuring and Sustainability: A Geographical Perspective*, edited by Brian W. Ilbery, Quentin Chiotti, and Timothy Rickard. Wallingford, England: C.A.B. International, 1997.

Gilg, Andrew, and Martin R.J. Battershill. "Farmer Reaction to Agrienvironmental Schemes: A Study in South-West England and the Implications for Research and Policy Development." Pp. 253-266 in *Agricultural Restructuring and Sustainability: A Geographical Perspective*, edited by Brian W. Ilbery, Quentin Chiotti, and Timothy Rickard. Wallingford, England: C.A.B. International, 1997.

Morris, C., and C. Young. Pp. 36-47 in *Never Mind the Environment, Feel the "Quality": A Discussion of the Agri-Environmental Potential of Quality Assuring Schemes.* North West Geographer, Third Series, 1997.

Potter, Clive. "Environmental Change and Farm Restructuring in Britain: The Impact of the Farm Life Cycle." Pp. 73-86 in *Agricultural Restructuring and Sustainability: A Geographical Perspective*, edited by Brian W. Ilbery, Quentin Chiotti, and Timothy Rickard. Wallingford, England: C.A.B. International, 1997.

Lobley, M., and Clive Potter. "Environmental Stewardship in UK Agriculture: A Comparison of the Environmentally Sensitive Area Programme and the Countryside Stewardship Scheme in South East England." *Geoforum*, 29, No. 4 (1998): 413-432.

Ward, N., et al. "Keeping Matter in Its Place: Pollution Regulation and the Reconfiguring of Farms and Farmers." *Environment and Planning A*, 30, No. 7 (1998): 1165-1178.

Bowler, Ian R. "Recycling Urban Waste on Farmland: An Actor-Network Interpretation." *Applied Geography*, 19, No. 1 (1999): 29-43.

Ilbery, Brian W., Lewis Holloway, and Ruth Arber. "The Geography of Organic Farming in England and Wales in the 1990s." *Tijdschrift voor Economische en Sociale Geographie*, 90, No. 3 (1999): 285-295.

Robinson, D.A. "Agricultural Practice, Climatic Change and the Soil Erosion Hazard in Parts of Southeast England." *Applied Geography*, 19, No. 1 (1999): 13-27.

Morgan, K., and Jonathan Murdoch. "Organic vs. Conventional Agriculture: Knowledge, Power and Innovation in the Food Chain." *Geoforum*, 31, No. 2 (2000): 159-173.

IRRIGATION

Knox, J.W., and E.K. Weatherfield. "The Application of GIS to Irrigation Water Resource Management in England and Wales." *Geographical Journal*, 165, No. 1 (1999): 90-98.

AQUACULTURE

Coull, James R. "Fish Farming in the Highlands and Islands: Boom Industry of the 1980s." *Scottish Geographical Magazine*, 104, No. 1 (1988): 4-13.

Ross, L.G., et al. "The Application of Geographical Information Systems to Site Selection for Coastal Aquaculture: An Example Based on Salmonid Cage Culture." *Aquaculture*, 112 (1993): 165-178.

Coull, James R. "Towards a Sustainable Economy for the Shetland Islands: Development and Management Issues in Fishing and Fish Farming." *GeoJournal*, 39, No. 2 (1996): 185-194.

AGRICULTURAL MARKETING AND DISTRIBUTION

Carlyle, William J. "The Marketing and Movement of Scottish Hill Lambs." *Geography*, 57, No. 1 (1972): 10-17.

Peet, J. Richard. "Influences of the British Market on Agriculture and Related Economic Development in Europe Before 1860." *Transactions, Institute of British Geographers*, 56 (1972): 1-20.

Bowler, Ian R. "Spatial Responses to Agricultural Subsidies in England and Wales." *Area*, 8, No. 3 (1976): 225-228.

Bowler, Ian R. "Some Characteristics of an Innovative Form of Agricultural Marketing." *Area*, 13, No. 4 (1981): 307-314.

Bowler, Ian R. *Direct Marketing in British Agriculture: A Survey of Pick-Your-Own Farms in the English Counties*. Leicester, England: University of Leicester Department of Geography, Occasional Paper No. 6, 1981.

Bowler, Ian R. "Direct Marketing in Agriculture: A British Example." *Tijdschrift voor Economische en Sociale Geographie*, 73, No. 1 (1982): 22-31.

Walton, J.R. "The Rise of Agricultural Auctioneering in Eighteenth and Nineteenth Century Britain." *Journal of Historical Geography*, 10, No. 1 (1984): 15-36.

Land, T., and P. Wiggins. "The Industrialization of the UK Food System: From Production to Consumption." Pp. 45-56 in *The Industrialisation of the Countryside*, edited by Michael J. Healey and Brian W. Ilbery. Norwich, England: Geo Books, 1985.

Munton, R.J.C. "Investment in British Agriculture by the Financial Institutions." *Sociologia Ruralis*, 25 (1985): 155-173.

Ilbery, Brian W. "Horticultural Marketing: The Case of the Vale of Evesham." *Transactions, Institute of British Geographers*, 11, No. 4 (1986): 468-478. (New Series)

Marsden, Terry K., Sarah Whatmore, and R.J.C. Munton. "The Role of Banking Capital and Credit Relations in British Food Production." Pp. 36-56 in *Political, Social and Economic Perspectives on the International Food System*, edited by Terry K. Marsden and J. Little. Aldershot, England: Avesbury, 1990.

Cudjoe, F., and P. Reese. *Government Policy Towards Organic Farming: A History of Neglect and Missed Opportunities*. Leeds, England: University of Leeds, School of Geography, Working Paper No. 92-25, 1992.

Campbell, Bruce, J.A. Galloway, and M. Murphy. *A Medieval Capital and Its Grain Supply: Agrarian Production and Distribution in the London Region, c1300*. Cheltenham, England: Historical Geography Research Group, Institute of British Geographers, Research Series No. 30, 1993.

TRANSPORTATION AND MECHANIZATION

Hellen, J.A. "Agricultural Innovation and Detectable Landscape Margins: The Case of Wheelhouses in Northumberland." *Agricultural History Review*, 20, No. 2 (1972): 140-154.

Walton, J.R. *A Study in the Diffusion of Agricultural Machinery in the Nineteenth Century*. Oxford, England: University of Oxford, School of Geography, Research Paper No. 5, 1973.

Walton, J.R. "Mechanization in Agriculture: A Study in the Adoption Process." Pp. 23-42 in *Change in the Countryside: Essays on Rural England, 1500–1900*, edited by H.S.A. Fox and Robin A. Butlin. London, England: IBG Special Publication No. 10, 1979.

Sturrock, F.G., and J. Calthie. *Farm Mechanization and the Countryside*. Cambridge, England: University of Cambridge, School of Geography, Occasional Paper No. 12, 1980.

AGRICULTURAL LABOR STUDIES

Sheppard, June A. "East Yorkshire's Agricultural Labour Force in the Mid-Nineteenth Century." *Agricultural History Review*, 9, No. 1 (1961): 43-54.

Rathone, A.H. "Mobility of Farm Labour in Scotland (1946–65/66): Some Influencing Factors." *Pakistan Geographical Review*, 25, No. 1 (1970): 23-35.

Edwards, C.J.W. "Changes in Agricultural Labour Efficiency in Northern Ireland, 1975–1984." *Irish Geography*, 19, No. 2 (1986): 74-82.

Marsden, Terry K. *Commoditisation of the Labour Process: Farm Households in British Agriculture*. London, England: University College–London, Sociology of Agriculture Working Paper Series, No. 2, 1989.

Bowler, Ian R., et al. "The Development of Alternative Farm Enterprises: A Study of Family Labour Farms in the Northern Pennines of England." *Journal of Rural Studies*, 12, No. 3 (1996): 285-296.

The Irish Republic/Eire

ATLASES AND GRAPHIC PRESENTATIONS

Horner, A.A., James A. Walsh, and J.A. Williams. *Agriculture in Ireland: A Census Atlas*. Dublin, Ireland: University College Dublin, Department of Geography, 1984.

GENERAL TEXTS, ARTICLES, AND THESES

Reid, Courtland. "Geographical Aspects of Ireland's Agriculture." Master's Thesis, University of Illinois, Urbana, Ill., 1941.

Freeman, T.W. "Farming in Irish Life." *Geographical Journal*, 112, No. 1-3 (1947): 38-59.

Carter, Jean. "Farm Types Near Killarney." *Irish Geography*, 3, No. 1 (1954): 40-46.

Freeman, T.W. "The Prospect for Irish Agriculture." *Geographical Journal*, 120, No. 3 (1954): 329-336.

Boal, F.W. "Agriculture in Down and Wexford: A Comparative Study." *Irish Geography*, 4, No. 1 (1959): 55-64.

Flatres, Paul. "Aspects of the Agricultural Geography of Ireland, Wales and Brittany: A Comparative Study." *Irish Geography*, 4, No. 5 (1963): 321-334.

Dooley, William E. "Changing Attitudes in Irish Agriculture." *East Lakes Geographer*, 1 (1964): 53-58.

Gillmor, Desmond A. "The Agricultural Regions of the Republic of Ireland." *Irish Geography*, 5 (1964–1968): 245-261.

Symons, Leslie J. and J.G. Cruickshank. "Agricultural Geography in Ireland." Pp. 462-482 in *Field Studies in the British Isles*, edited by J.A. Steers. London: Nelson, 1965.

Butlin, Robin A. "Agriculture in County Dublin in the Late Eightteenth Century." *Studia Hibernica*, 9 (1969): 93-108.

Tarrant, J.R. "Some Spatial Variations in Irish Agriculture." *Tijdschrift voor Economische en Sociale Geographie*, 60 (1969): 228-237.

Gillmor, Desmond A. "Aspects of Agricultural Change in the Republic of Ireland during the 1960s." *Irish Geography*, 6, No. 4 (1972): 492-498.

Gillmor, Desmond A. *Agriculture in the Republic of Ireland*. Budapest, Hungary: Akademiai Kiado, 1977.

Gillmor, Desmond A. "Agriculture." Pp. 109-136 in *Irish Resources and Land Use*, edited by Desmond A. Gillmor. Dublin, Ireland: Institute of Public Administration, 1979.

Duffy, P.J. "Cooperative Agriculture and Rural Development in Ireland." *Antipode*, 12, No. 1 (1980): 97-107.

Dumortier, B. "Aspects generaux et regionaux de l'agriculture irlandaise." *L'Information Geographique*, 46, No. 4 (1982): 148-151.

Jones-Hughes, T. "The Large Farm in Nineteenth Century Ireland." Pp. 93-100 in *Gold Under the Furze: Studies in Folk Tradition Presented by Caoimhin oDanachair*, edited by A. Gailey and D. O'hOgain. Dublin, Ireland: Glendale Press, 1982.

Walsh, James A. "Agricultural Change and Development." Pp. 11-33 in *Change and Development in Rural Ireland*, edited by Proinnsias Breathnach and M. Cawley. Maynooth, Ireland: Geographical Society of Ireland, Special Publication No. 1, 1986.

Walsh, James A. "Uneven Development of Agriculture in Ireland." *Geographical Viewpoint*, 14 (1986): 37-65.

Gillmor, Desmond A. "Concentration of Enterprises and Spatial Change in the Agriculture of the Republic of Ireland." *Transactions, Institute of British Geographers*, 12, No. 2 (1987): 204-216. (New Series)

Gillmor, Desmond A. "Agricultural Development." Pp. 171-199 in *Ireland: A Contemporary Geographical Perspective*, edited by R. Carter and A.L. Parker. London, England: Routledge, 1989.

Gillmor, Desmond A. "The Political Factor in Agricultural History: Trends in Irish Agriculture, 1922–1985." *Agricultural History Review*, 37, No. 2 (1989): 166-179.

Gillmor, Desmond A. "Butter, Steaks and Schinken: Die irische landwirtschaft." Pp. 138-146 in *Irland: Ein Express Reisenhanbuch*, edited by A. Steinecke. Leer, Germany: Mundo Verlag, 1990.

Walsh, James A. "A Regional Analysis of Enterprise Substitution in Irish Agriculture in the Context of a Changing Common Agricultural Policy." *Irish Geography*, 24, No. 1 (1991): 10-23.

Gillmor, Desmond A. "Les mutations de l'agriculture en Republique d'Irlande: 1960–1990." *Hommes et Terres du Nord*, 2 (1992): 72-78.

Kelly, C.E., Brian W. Ilbery, and Desmond A. Gillmor. "Farm Diversification in Ireland: Evidence from County Wicklow." *Irish Geography*, 25, No. 1 (1992): 23-32.

O'Donnell, Martina. "Farm Clusters in North-West Inishowen, c.1850." *Irish Geography*, 26, No. 2 (1993): 101-119.

Turner, M.E. "Rural Economies in Post-Famine Ireland, c.1850–1914." Pp. 293-337 in *An Historical Geography of Ireland*, edited by B.J. Graham and L.J. Proudfoot. London, England: Academic Press, 1993.

Buttimer, Anne, and Anne-Marie McGauron. "The Impact of EC Agricultural Policies on the Irish Rural Landscape." *Acta Universitatae Carolinae Geographica*, 29, No. 1 (1994): 5-17.

Culleton, Noel, Hubert Tunney, and Brian Coulter. "Sustainability in Irish Agriculture." *Irish Geography*, 27, No. 1 (1994): 36-47.

Gillmor, Desmond A. "The Agriculture of the Republic of Ireland Within the European Union." *Geographia Polonica*, 72, No. 1 (1999): 47-62.

CROPS AND CROPPING PATTERNS

Gillmor, Desmond A. "Spatial Distribution of Tillage and Its Component Crops in the Republic of Ireland." *Irish Journal of Agricultural Economics and Rural Sociology*, 2, No. 2 (1969): 135-170.

Smyth, W.J. "Flax Cultivation in Ireland: The Development and Demise of a Regional Staple." Pp. 234-252 in *Common Ground: Essays on the Historical Geography of Ireland*, edited by W.J. Smyth and Kevin Whelan. Cork, Ireland: Cork University Press, 1988.

Kelly, C.E., Noel Culleton, and V. McCarthy. "The Decline of the Herbage Seed Production Industry in Ireland." *Irish Geography*, 25, No. 1 (1992): 98-101.

ANIMAL HUSBANDRY

Coulter, John W. "Stock Farming in Eire." *Economic Geography*, 25 (1949): 81-93.

Graham, J.M. "Transhumance in Ireland." *Advancement of Science*, 10, No. 37 (1953): 74-79.

Aalen, F.H.A. "Transhumance in the Wicklow Mountains." *Ulster Folklore*, 10 (1964): 65-72.

Gillmor, Desmond A. "Spatial Distributions of Livestock in the Republic of Ireland." *Economic Geography*, 46, No. 4 (1970): 587-597.

Lewis, Colin A., and M.E. McCarthy. "The Horse Breeding Industry in Ireland." *Irish Geography*, 10 (1977): 72-89.

Lewis, Colin A. "Irish Horse Breeding and the Irish Draught Horse, 1917–1978." *Agricultural History Review*, 31, No. 1 (1983): 37-49.

Walsh, James A. "Enterprise Substitution in Irish Agriculture: Sheep Production in the 1980s." *Irish Geography*, 22, No. 2 (1989): 106-109.

Yarwood, Richard, Nick Evans, and Julie Higginbottom. "The Contemporary Geography of Indigenous Irish Livestock." *Irish Geography*, 30, No. 1 (1997): 17-30.

DAIRYING

O'Dwyer, T. "A Comparison of Costs of Milk Assembly in a Cooperative Society Area." *Irish Journal of Agricultural Economics and Rural Sociology*, 1 (1968): 221-243.

Breathnach, Proinnsias. "The Development of the Dairy Industry in County Waterford." Pp. 707-732 in *Waterford: History and Society*, edited by William Nolan and Thomas P. Power. Dublin, Ireland: Geography Publications, 1992.

Jenkins, William. "Restructuring of Irish Dairy Co-operatives Since 1950: An Example from County Tipperary." *Irish Geography*, 29, No. 1 (1996): 38-48.

FARM LAND

Klimm, Lester E. "The Relation Between Field Patterns and Jointing in the Aran Islands." *Geographical Review*, 25, No. 4 (1935): 618-624.

Evans, E. Estyn. "Some Survivals of the Irish Open Fields System." *Geography*, 24 (1939): 24-36.

Aalen, F.H.A. "The Origin of Enclosures in Eastern Ireland." Pp. 209-223 in *Irish Geographical Studies in Honour of E. Estyn Evans*, edited by N. Stephens and R.E. Glasscock. Belfast, Northern Ireland: Queens University, 1970.

Symons, Leslie. "Rural Land Utilisation in Ireland." Pp. 259-273 in *Irish Geographical Studies in Honour of E. Estyn Evans*, edited by N. Stephens and R.E. Glasscock. Belfast, Northern Ireland: Queen's University, 1970.

Buchanan, R.H. "Field Systems of Ireland." Pp. 580-618 in *Studies of Field Systems in the British Isles*, edited by Alan R.H. Baker and Robin A. Butlin. Cambridge, England: Cambridge Univesity Press, 1973.

Leister, Ingeborg. *Peasant Openfield Farming and Its Territorial Organization in County Tipperary*. Marburg, Germany: Marburger Geographische Schriften, Heft 69, 1976.

Grimes, S. "Agricultural Policy and Land Tenure in an Irish Marginal Country." *Irish Geography*, 21 (1988): 33-40.

AGRICULTURAL POPULATIONS AND SOCIAL ISSUES

Hall, D.N., H. Hennessey, and T. O'Keefe. "Medieval Agriculture and Settlement in Oughterard and Castlewarden, County Kildare." *Irish Geography*, 18, No. 1 (1985): 16-24.

Nolan, William. "New Farms and Fields: Migration Policies of State Land Agencies." Pp. 296-319 in *Common Ground: Essays on the Historical Geography of Ireland*, edited by W. Smyth and Kevin Whelan. Cork, Ireland: Cork University Press, 1988.

Gillmor, Desmond A. "The Scheme of Early Retirement from Farming in the Republic of Ireland." *Irish Geography*, 32, No. 2 (1999): 78-86.

ENVIRONMENTAL AND CONSERVATION ISSUES

Gillmor, Desmond A. "Agricultural Aspects and the Irish Environment." *Geographical Viewpoint*, 20 (1991–1992): 5-22.

Willer, Helga, and Desmond A. Gillmor. "Organic Agriculture in the Republic of Ireland." *Irish Geography*, 25, No. 2 (1992): 149-159.

Conry, Michael J., and A. Hegarty. "Effects of Depth Phases of the Clonroche Soil Series and Farming Skill on Spring Barley Yields." *Irish Geography*, 30, No. 1 (1997): 1-8.

Gillmor, Desmond A. "The Implementation of Current European Union Agri-Environmental Policy in the Republic of Ireland." Pp. 19-25 in *Rural Geography* and Environment, edited by J. Munzar and A. Vaishar. Brno, Czech Republic: Institute of Geonics, Academy of Science of the Czech Republic, 1997.

AQUACULTURE

Breathnach, Proinnsias. "The Development of the Irish Fish Farming Industry." *Irish Geography*, 25, No. 2 (1992): 182-187.

AGRICULTURAL MARKETING AND DISTRIBUTION

Gillmor, Desmond A. "Cattle Marketing in Ireland." *Administration*, 15 (1967): 308-327.

TRANSPORTATION AND MECHANIZATION

Walsh, James A. "Adoption and Diffusion Processes in the Mechanization of Irish Agriculture." *Irish Geography*, 25, No. 1 (1992): 33-53.

Scandinavia

GENERAL TEXTS, ARTICLES, AND THESES

Millward, Roy. "The Agricultural Geography of Scandinavia." Pp. 221-259 in *Scandinavian Lands*, edited by Roy Millward. London, England: Macmillan, 1965.

Allefresch, Maurico. "L'agricoltura pioniera nella Fennoscandia subartica." *Bolletin della Societa Geografica Italiana*, Serie 9, No. 9 (1968): 226-236.

Hakkila, Matti. "On the Development and Areal Concentration of Agriculture in Finland and Sweden." *Nordia*, 22, No. 1 (1988): 27-33.

Just, F. "Agriculture and Corporatism in Scandinavia." Pp. 31-52 in *Regulating Agriculture*, edited by Philip Lowe, Terry Marsden, and Sarah Whatmore. London, England: David Fulton, 1994.

Denmark

ATLASES AND GRAPHIC REPRESENTATIONS

Holst, K.A., and H.B. Madsen. "The Elaboration of Drainage Class Maps for Agricultural Planning in Denmark." *Landscape and Urban Planning*, 13 (1986): 199-218.

Jensen, K.M., and A. Reenberg. *Landbrugsatlas Danmark: Atlas over Danmark*, Serie II, Bind 4 Copenhagen, Denmark: C.A. Reitzels Forlug, 1986.

GENERAL TEXTS, ARTICLES, AND THESES

Bergsmark, D.R. "Agriculture in Denmark: A Land of Specialized Farming and Cooperation." *Journal of Geography*, 37, No. 4 (1938): 129-142.

Kampp, A.H. "Les provinces agricoles du Danemark." *Norois*, 5 (1958): 457-463.

Kampp, A.H. *Landbrugsgeografiske studier over Denmark*. Copenhagen, Denmark: C.A. Reitzels Forlug, 1959.

Kampp, A.H. "Some Types of Farming in Denmark." *Oriental Geographer*, 3, No. 1 (1959): 17-32.

Jacobson, N.K. "Agricultural Geography and Regional Planning in a Marine Foreland." *Geografisk Tidsskrift*, 59 (1960): 140-162.

Bunting, B.T. "The Present Re-organization of Agriculture in Denmark." *Geography*, 53, No. 2 (1968): 157-162.

Troughton, Michael J., and F.H. Lummings. "Possibilities for Detailed Agricultural Investigation: An Example Using Danish Farm Data." *Ontario Geography*, 8 (1974): 78-96.

Kampp, A.H. *An Agricultural Geography of Denmark.* Budapest, Hungary: Akademiai Kiado, 1975.

Frandsen, Kark-Erik. *Horndrup ag Skarup: Retrospektiv Analyse af den Agrare-Strukture i Fynske Landsbyeri det 17. Arhundonde.* Odense, Denmark: Odense Universitet, Institet for Historie og Samfunsvienslab, 1976.

Frandsen, Karl-Erik. *Vang og taegt: Studier over dyrkningssystemer og agrarstruktuere: Danmark landsbyer 1682–83.* Esbjerg, Denmark: ByGD, 1983.

Kolb, H.J. "Die danische Landwirtschaft innehalf der EG." *Geographie und Schule*, 11, No. 62 (1989): 6-14.

Kristensen, Soren P. "Gadbjerg and Ginskud Revisited: Changes in Agricultural Structure in Two Parishes in Central Jutland Between 1973 and 1997." *Geografisk Tidsskrift*, 99 (1999): 69-80.

ANIMAL HUSBANDRY

Shaw, Earl B. "The Swine Industry of Denmark." *Economic Geography*, 14, No. 1 (1938): 23-37.

Sundt, H.A. "Sheep and Soil in the Faroe Islands: Examples from an Old Suovray Settlement." *Norsk Geografisk Tidsskrift*, 30, No. 3 (1976): 115-126.

Brandt, J. "Schafzucht auf den Faroern-zur Geivinnung landschaftsokologischer Information durch statische Analyse eines landnutzungssystems." *Petermann's Geographische Mitteilungen*, 136, No. 5-6 (1992): 235-249.

FARM LAND

Bjerre, Arne G. "Geography of Agricultural Land Resource Use in Denmark." Master's Thesis, Oregon State University, Corvallis, Oreg., 1960.

Kampp, A.H. "Danish Agricultural Subdivision and the Majorats." *Geografisk Tidsskrift*, 59 (1960): 132-144.

Gade, Ole. "Land Reclamation and the Cultivation of Reclaimed Waste Land in Denmark." Master's Thesis, Florida State University, Tallahassee, Fla., 1964.

Jensen, Marius. *Opgivne og tilplontede landbrugsarealer i Jylland.* Copenhagen, Denmark: Kongelige Danske Geografiske Selskab, 1976.

Kampp, A.H. "Recent Amalgamation of Agricultural Holdings." *Geografisk Tidsskrift*, 78 (1979): 57-66.

Runberg, Anette. "Agricultural Land Use in Denmark in the 1980s." *Geografisk Tidsskrift*, 88 (1988): 8-13.

Finland

GENERAL TEXTS, ARTICLES, AND THESES

Mead, William R. "Agriculture in Finland, Part I." *Economic Geography*, 15 (1939): 125-134; "Part II," 15 (1939): 217-234.

Mead, William R. *Farming in Finland*. London, England: Athlone, 1953.

Jaatinen, Stig, and William R. Mead. "The Intensification of Finnish Farming." *Economic Geography*, 33, No. 1 (1957): 31-40.

Smeds, Helmer. "Recent Changes in the Agricultural Geography of Finland." *Fennia*, 87 (1962): 1-19.

Fogelberg, Paul. *Regionale Differenzierung in der finnischen Landwirtschaft*. Helsinki, Finland: Einige Beispiele im lichte der Landwirtschaftszahlung, 1965.

Varjo, Uuno. "The Finnish Farm as Seen from the Viewpoint of Geographical Typology of Agriculture." *Fennia*, 92 (1965): 1-18.

Ehlers, Eckert. "Kuparivaara-Puolakkavaara-Jouttiaapa: Beispiele gegenwortiger Agrarkonolisation in Nordfinnland." *Erdkunde*, 21 (1967): 212-226.

Varjo, Uuno. "Uber die regionale Differenzierung der Landwirtschaft im Lan Lappland." *Acta Geographica* (Helsinki), 20, No. 24 (1968): 343-364.

Karkainen, Ulla. "Some Development Trends in Finnish Agriculture, 1959–1969" (in Finnish). *Terra*, 82, No. 4 (1970): 147-154.

Edwards, C.J.W. "Methods of Maintaining Settled Agriculture in Finnish Lapland." *Arctic*, 25, No. 1 (1972): 8-20.

Korva, Timo. "The Prospects of Agriculture in Northern Lapland" (in Finnish). *Terra*, 86, No. 4 (1974): 169-174.

Mead, William R. "The Changing Agricultural Geography of a Finnish Lake-Land Parish." *Norsk Geografisk Tidsskrift*, 30, No. 3 (1976): 93-102.

Yli-Jokipii, Pentti. "Part-Time Farming as a Spatial Phenomenon in South-Western Finland." *Publicationes Instituti Geographici Universitatis Turkuensis*, 70 (1976).

Varjo, Uuno. *Finnish Farming: Typology and Economics*. Budapest, Hungary: Akademiai Kiado, 1977.

Varjo, Uuno. *On the Location of Agriculture in Finland*. Oulu, Finland: Societas Geographica Fenniae Nordicae, 1980.

Cabouret, Michel. "Traits permanents et tendances recentes de l'agriculture finlandaise." *Annales de Geographie*, 91, No. 1 (1982): 87-118.

Varjo, Uuno. "The Location of Finnish Agriculture on Small and Large Farms." *Nordia*, 16, No. 2 (1982): 169-178.

Rikkinen, Kalevi., and T. Vapaaokos. "The Application of World Agricultural Typology to Finland." *Geographia Polonica*, 46 (1983): 93-106.

Varjo, Uuno. "Finnish Farming: Typology and Economics." *Geographia Polonica*, 46 (1983): 107-132.

Varjo, Uuno. "Farming in Finland in the Pressure of Change." *Nordia*, 17, No. 1 (1983): 153-162.

Hakkila, Matti. "The Recent Development of Agriculture with Special Reference to Kainuu, Finland." *Nordia*, 18, No. 2 (1984): 123-133.
Varjo, Uuno. "Changes in Farming in Finland, 1969–1975." *Fennia*, 162, No. 1 (1984): 103-115.
Varjo, Uuno. "Structural Changes in the Economics of Farming in Finland in the Early 1970s." *Rural Systems*, 3, No. 3 (1985): 207-226.
Hakkila, Matti. "Some Regional Trends in Finnish Farming with Special Reference to Agricultural Policy." *Fennia*, 169, No. 1 (1991): 39-56.
Szczesny, R. *Finland's Agriculture: Spatial Differentiation and Transformations* (in Polish). Warsaw, Poland: Polish Academy of Science, No. 27, 1994.

CROPS AND CROPPING PATTERNS

Hult, Juhani. "Areal Variation of the Arable Area of Farms as a Factor Causing Areal Differentiation of Arable Farming" (in Finnish). *Terra*, 80, No. 2 (1968): 41-50.
Varjo, Uuno. "Harvest Yields and the Profitableness of Arable Farming in Finland, 1950–1969" (in Finnish). *Terra*, 84, No. 4 (1972): 233-241.
Varjo, Uuno. "Recent Changes in the Profitableness of Arable Farming in Finland." *Nordia*, 1 (1974): 1-15.
Varjo, Uuno. "Arable Farming in Finland, Types and Economics." *Nordia*, 1 (1976): 1-103.
Varjo, Uuno. "Recent Climatic Trends and the Limits of Crop Cultivation in Finland." *Fennia*, 150 (1978): 45-56.
Varjo, Uuno. "Productivity and Fluctuating Limits of Crop Cultivation in Finland." *Geographia Polonica*, 40 (1979): 225-233.
Varjo, Uuno. "Crop Cultivation in Finland: Recent Temporal and Regional Trends." *Nordia*, 14, No. 2 (1980): 17-47.
Mead, William R. "The Granaries of Finland." *Fennia*, 164, No. 2 (1986): 291-302.
Palomakim Mauri J., and Allen G. Noble. "Greenhouse Horticulture and Economic Transition." *Geographical Review*, 85, No. 2 (1995): 173-184.
Szczesny, R. "Changes in the Spatial Structure of Agriculture in Finland: An Attempt at a Synthesis" (in Polish). *Przeglad Geograficzny*, 67, No. 3-4 (1995): 259-269.
Makela, P., O.W. Archibald, and P. Peltonen-Sainio. "Wild Rice: A Potential New Crop for Finland." *Agriculture and Food Science in Finland*, 7 (1998): 583-597.

ANIMAL HUSBANDRY

Helle, Reijo. *Oulum Yliopistan Maantieteen laitoksen Julkaisuja (An Investigation of Reindeer Husbandry in Finland)*. Oulo, Finland: Publications of the Geographical Institute of the University of Oulo, No. 11, 1966.

Muller-Wille, Ludger, and Olavi Aikio. "Die Auswirkungen der Mechanisierung der Rentierwirtschaft in der Lappischen Gemeide Utsojoki." *Terra*, 83, No. 3 (1971): 179-185.

Helle, Reijo. "Reindeer Husbandry in Finland." *Geographical Journal*, 145, No. 2 (1979): 254-264.

Talman, Paavo. "Areal Livestock Combinations on Finnish Farms." *Fennia*, 152, No. 2 (1979): 155-170.

DAIRYING

Yli-Jokipii, Pentii. "Regional Changes in the Finnish Dairy Industry, 1910–1967." *Fennia*, 100 (1970): 5-65.

Talman, Paavo. *Dairy Farming in Finland: Geographical Aspects of the Development, Typology and Economics of Finnish Dairy Farming*. Oulo, Finland: University of Oulo, 1978.

Hakkila, Matti. "Regional Development of Dairy Farming in Finland." *Nordia*, 21, No. 1 (1987): 27-38.

FARM LAND

Smeds, Helmer. "Post-War Land Clearance and Pioneering Activities in Finland." *Fennia*, 83 (1960): 1-31.

Smeds, Helmer. "Regional Variations in Size Classes of Agricultural Holdings in Finland" (in Finnish). *Terra*, 78, No. 2 (1966): 98-102.

Hult, Juhani. "The Areal Variation of the Arable Size of Farms and the Areal Differentiation of Farming." *Acta Geographica* (Helsinki), 20, No. 11 (1968): 129-153.

Jaatinen, Stig. "The Rank-Size Structure of the Agricultural Holdings in Finland, 1959." *Acta Geographica* (Helsinki), 20, No. 13 (1968): 175-184.

Karkainen, Ulla. "Changes in Farm Size Structure in Finland, 1959–1964." *Acta Geographica* (Helsinki), 20, No. 15 (1968): 201-214.

Jaatinen, Stig, and Mauri Nyard. "Field-Reservation Activity in Finland on a Local Level" (in Finnish). *Terra*, 85, No. 3 (1973): 149-162.

Hakkila, Matti. "Geographical Aspects of the Development and Utilization of Arable Land Area in Finland." *Fennia*, 162, No. 2 (1984): 217-235.

Hakkila, Matti. "Use of Arable Land on Farms in Northern Finland" (in Finnish). *Terra*, 103, No. 2 (1991): 242-249.

Hakkila, Matti. "Geographical Aspects of the Use of Arable Land in Finland." *Nordia Geographical Publications*, 26, No. 1 (1997): 3-16.

AGRICULTURAL POPULATIONS AND SOCIAL ISSUES

Ehlers, Eckert. "Das Ende der Agrarkolonisation in Finnland." *Geographische Rundschau*, 22, No. 1 (1970): 31-33.

Hackman, Patrick. "Attitudes of the Farmers' Unions to Multiple Jobholders in Finland." Pp. 213-219 in *Part-Time Farming: Problem or Resource in Ru-*

ral Development. Proceedings of the First Rural Geography Symposium, University of Guelph, June 18–20, 1975, edited by Anthony M. Fuller and Julius A. Mage. Norwich, England: GeoAbstracts, 1970.
Kampp, A.H., and Kalevi Rikkinen. "Farms in a Finnish Village (Levanto), 1787–1916." *Geografisk Tidsskrift*, 72 (1973): 18-29.

ENVIRONMENTAL AND CONSERVATION ISSUES

Varjo, Uuno. "Productivity of Agricultural Soils in Finland" (in Finnish). *Terra*, 78, No. 2 (1966): 103-110.
Varjo, Uuno. "Soil Productivity and Farm Size Classification in Finland" (in Finnish). *Terra*, 83, No. 3 (1971): 166-171.
Laaksonen, K. "The Effect of Climatic Factors on the Hectare Yields of Barley, Oats, and Spring Wheat in Finland, 1972–1977." *Fennia*, 157 (1979): 199-221.
Varjo, Uuno. "High-Latitude Climate Impacts and Effects." *Nordia*, 20, No. 1 (1986): 113-126.
Kleemola, J., et al. "Modelling the Impact of Climate Change on Growth of Spring Barley in Finland." *Journal of Biogeography*, 22, No. 4-5 (1995): 581-590.
Moto, Miska. "Spatial Analysis of Landscape Ecological Characteristics of Five Agricultural Areas in Finaldn by GIS." *Fennia*, 178, No. 1 (2000): 15-54.

TRANSPORTATION AND MECHANIZATION

Jussila, H. "Mechanization of Farming in Finland." *Nordia*, 20, No. 1 (1986): 41-47.

Iceland

GENERAL TEXTS, ARTICLES, AND THESES

Ashwell, I.Y. "Recent Changes in the Pattern of Farming in Iceland." *Canadian Geographer*, 7 (1963): 174-181.
Gudbergsson, Gylfi. "The Geographical Characteristics of Farm Abandonment in Iceland." Master's Thesis, University of Wisconsin, Madison, Wis., 1966.
White, C. "Current Trends in Icelandic Agriculture." *Swansea Geographer*, 21 (1984): 102-108.

CROPS AND CROPPING PATTERNS

Bryson, R.A. "Heyuppskera: An Heuristic Model for Hay Yield in Iceland." *Research Institute Netri As*, 18. Hveragerti, Iceland, 1974.

AGRICULTURAL POPULATIONS AND SOCIAL ISSUES

Jackson, E.L. "Changes in Settlement and Agriculture in Iceland, 874 A.D. to the Present." Master's Thesis, University of Calgary, Calgary, Alberta, 1970.

ENVIRONMENTAL AND CONSERVATION ISSUES

Bergthrosson, P., et al. "The Effects of Climatic Variations on Agriculture in Iceland." Pp. 381-509 in *The Impact of Climatic Variations on Agriculture, Volume 1: Assessments in Cool Temperate and Cold Regions*, edited by Martin L. Parry, T.R. Carter, and N.T. Konijn. Dordrecht, Netherlands: Kluwer, 1988.

Norway

GENERAL TEXTS, ARTICLES, AND THESES

Thomson, Claudia. "Norway's Agriculture." *Journal of Geography*, 35, No. 5 (1936): 165-178.

Somme, Axel. *Geography of Norwegian Agriculture* (in Norwegian). Bergen, Norway: J.W. Eides, 1949–1954. 2 volumes.

Somme, Axel. "Norwegian Agriculture and Food Supply." *Geography*, 35, No. 4 (1950): 215-227.

Kirk, William, and Francis M. Synge. "Farms of Verdal, Norway." *Scottish Geographical Magazine*, 70, No. 3 (1954): 106-123.

Nordgard, Asbjorn. "Variations in Type of Farming" (in Norwegian). *Norsk Geografisk Tidsskrift*, 22 (1968): 17-38.

Glasser, Ewald. "Jaeren-Norwegens intensivst genutzte Agrarlandschaft." *Geographische Rundschau*, 21, No. 7 (1969): 260-271.

Glasser, Ewald. "Progressive und rezessive Entwicklungstendenzen norwegischer Agrarlandschaftten." *Erdkunde*, 24, No. 4 (1970): 308-326.

Thompson, T. "The Changing Face of Norwegian Fjord Farming." *Wessex Geographer*, 11 (1970): 29-36.

Bronny, Horst M. "Tendencies in the Development of the Agricultural Economy in Sub-Polar Regions of North Europe: A Case Study of Finnmark" (in German). *Bochumer Geographische Arbeiten*, 13 (1972): 181-198.

Longtvet, O. "The Farm-Seter System in Ausdal, Norway: An Analysis of Change." *Norsk Geografisk Tidsskrift*, 28 (1974): 167-180.

Ronneseth, Ottar. *Gard und Ein friedigung: Entwickungsphasen der Agrarlandschaft Jaerens*. Stavanger, Norway: Dryer, 1974.

Nordgard, A. "Orientation and Intensity of Norwegian Agriculture." *Norsk Geografisk Tidsskrift*, 29, No. 4 (1975): 169-220.

Nordgard, A. "Types and Regions of Norwegian Agriculture." *Norsk Geografisk Tidsskrift*, 31, No. 1 (1977): 15-26.

Symes, David G. "Part-Time Farming in Norway." *GeoJournal*, 6, No. 4 (1982): 351-354.

Cabouret, Michel. "Notes sur l'agriculture norvegienne." *Norois*, 30 (1983): 263-278.

Tschudi, A.B., and H. Johansen. "Types of Agriculture in Norway by the Typogram Method: Notes on the Problem of Establishing Commercialism." *Geographia Polonica*, 46 (1983): 83-92.

Sommers, Lawrence S. "Regional Role of Agriculture in the Gross Domestic Product of Norway in 1950–1984." *Nordia*, 21, No. 1 (1987): 77-83.

Glasser, Ewald, and Hakon Hovstad. "Zur Gergbauern problematik in Norwegen: Gezeigt an den Beispielon Fordal (Forbygda) und Budal in Jor-Trondelag." *Erdkunde*, 43, No. 1 (1989): 1-15.

Vorren, K.D., E. Nilssen, and B. Morkved. "Age and Agricultural History of the Stadir Farms of North and Central Norway." *Norsk Geografisk Tidsskrift*, 44, No. 2 (1990): 79-102.

ANIMAL HUSBANDRY

Allefresch, Maurico. "Un exemple de mutation dans la vie pastorale subarctique norvegienne: La ferme de Brattli." *Revue de Geographie de Lyon*, 40, No. 1 (1965): 77-99.

Cabouret, Michel. "La transhumance du mounton dans le sud-ouest de la Norvege." *Norsk Geografisk Tidsskrift*, 21 (1967): 3-38.

Lindemann, Rolf. "Die Norwegische Peltztierzucht." *Zeitschrift fur Wirtschaftsgeographie*, 19, No. 1 (1975): 27-32.

FARM LAND

Rosnes, A.E. "Official Versus Unregistered Change of Agricultural Land in Norway: Extent, Legitimacy, and Regional Distribution Effects." *Norsk Geografisk Tidsskrift*, 39, No. 3 (1985): 125-140.

Moen, A., et al. "Outlying Haymaking Lands at Solendet, Central Norway: Effects of Scything and Grazing." *Norsk Geografisk Tidsskrift*, 53, No. 2-3 (1999): 93-102.

ENVIRONMENTAL AND CONSERVATION ISSUES

Somme, Axel. "The Physical Background of Norwegian Agriculture." *Geography*, 35, No. 3 (1950): 141-154.

Moe, D. "Hustad, Arstad, and Naurstad: A Vegetational Study of Three Farms in Sulten, North Norway." *Norsk Geografisk Tidsskrift*, 45, No. 1 (1991): 11-24.

Hafsten, U. "Vegetation, Climate and Agricultural History on the Extreme South Coast of Norway." *Norsk Geografisk Tidsskrift*, 46, No. 1 (1992): 1-18.

Winkler, S., and N. Haakensen. "Kritische Uber prufung der Moglichkeit zur Prognose des Geitscher-verhaltens auf der Grundlage von Modellierungen-dargestellt anhand von regionalen Beispielen aus Norwegn." *Petermann's Geographische Mitteilungen*, 143, No. 4 (1999): 291-304.

AGRICULTURAL MARKETING AND DISTRIBUTION

Glasser, Ewald. "Spezialisierungsfarmen einer marktorientieten Landwirstschaft in Norwegen: Gezeigt an Beispielen aus Mittel-Jaeren." *Erdkunde*, 31, No. 4 (1980): 245-256.

AQUACULTURE

Jakobsen, K.B. "Political Regulation and Market Allocation: Development Trends in the Norwegian Fish Farming Industry." *Norsk Geografisk Tidsskrift*, 50, No. 3-4 (1996): 182-200.

Sweden

GENERAL TEXTS, ARTICLES, AND THESES

Falkowski, Jan. "Rolnictwo w Szwecji (Agriculture in Sweden)." *Czasopismo Geograficzne*, 44, No. 1 (1973): 33-48.

Lindquist, Sven O. *The Development of the Agrarian Landscape on Gotland during the Early Iron Age*. Stockholm, Sweden: Kulturgeografiska Institutionen vid Stockhoms Universitet, 1975.

Tyszkiewicz, W. "Zastosowanie metod typologieznych dobadan rolnictawa Szwecji." *Przeglad Geograficzny*, 54 (1982): 533-549.

Persson, L.O. "Support for Agriculture—An Appropriate Regional Policy Instrument: The Case of Northern Sweden." *Journal of Rural Studies*, 5, No. 4 (1989): 349-356.

Moller, Jens. "Towards Agrarian Capitalism: The Case of Southern Sweden during the 19th Century." *Geografiska Annaler B*, 72B, No. 2-3 (1990): 59-72.

Peterson, M. "Paradignmatic Shift in Agriculture: Global Effects and the Swedish Response." Pp. 77-100 in *Rural Restructuring*, edited by Terry Marsden, Philip Lowe, and Sarah Whatmore. London, England: David Fulton, 1990.

Ihse, Margarata. "Swedish Agricultural Landscapes: Patterns and Changes During the Last 50 Years, Studies by Aerial Photos." *Landscape and Urban Planning*, 31 (1995): 21-37.

Cabouret, Michel. "Quelques types de paysages agraires du Norrland superieur suidais: Repartition et evolution." *Bulletin de la Societe Geographique de Liege*, 32 (1996): 259-272.

CROPS AND CROPPING PATTERNS

Fullerton, B. "The Northern Margin of Grain Production in Sweden in the 20th Century." *Transactions, Institute of British Geographers*, 20 (1954): 181-191.

ANIMAL HUSBANDRY

Cabouret, Michel. "Etat des recherches sur la vie pastorale en Suede." *Revue Geographique de l'Est*, 6, No. 3-4 (1966): 291-320.

FARM LAND

Enequist, Gerd. "Agricultural Holdings in Sweden 1951–1966 and 1980." *Acta Geographica* (Helsinki), 20, No. 7 (1968): 75-88.

Abrahamson, Hans. "Changes in the Utilisation of Arable Land" (in Swedish). *Svensk Geografisk Arsbok*, 49 (1973): 53-77.

Widgren, Mats. "A Simulation Model of Farming Systems and Land Use in Sweden during the Early Iron Age, c. 500 B.C.-A.D. 500." *Journal of Historical Geography*, 5, No. 1 (1979): 21-32.

AGRICULTURAL POPULATIONS AND SOCIAL ISSUES

Huntington, Ellsworth. "Farms and Farm Villages of Sweden." *Journal of Geography*, 37, No. 3 (1938): 85-90.

Wolpert, Julian. "Decision-Making in Middle Sweden's Farming: A Spatial Behavioral Analysis." Ph.D. Dissertation, University of Wisconsin, Madison, Wis., 1963.

Volkmann, Harmut. "Strukturelle und soziale Wandlungen einer landwirtschaft am Rande der Okumene-augezeist am Beispiel Sweden." *Zietschrift fur Wirtschaftsgeographie*, 21, No. 3 (1977): 77-88.

Sporrang, V. "Agrarian Settlement and Cultivation from a Historical Geographic Perspective: Examples of the Malor District in Central Sweden." *Geografiska Annaler B*, 68B, No. 2 (1986): 69-94.

ENVIRONMENTAL AND CONSERVATION ISSUES

Emanuelsson, Urban, and Jens Moller. "Flooding in Scania: A Method to Overcome the Deficiency of Nutrients in Agriculture in the Nineteenth Century." *Agricultural History Review*, 38, No. 2 (1990): 127-148.

Emanuelson, Urban. "The Debate about a Long-Term Sustainable Cultivation System" (in Swedish). *Svensk Geografisk Arsbok*, 70 (1994): 84-96.

Regnell, G. "Grass Fallow to the People?" (in Swedish). *Svensk Geografisk Arsbok*, 70 (1994): 126-132.

TRANSPORTATION AND MECHANIZATION

Helmfrid, Staffan. "The Mechanisation of Agriculture" (in Swedish). *Svensk Geografisk Arsbok*, 42 (1966): 57-97.

Western Europe

ATLASES AND GRAPHIC PRESENTATIONS

von Hecke, E. "Structure regionale de la production agricole dans la communaute Europeenne: Analyse cartographique." *Bulletin de la Societe Belge d'Etudes Geographie*, 51, No. 2 (1982): 145-190.

GENERAL TEXTS, ARTICLES, AND THESES

Gachon, L. "Les systemes agricoles dans les Alpes, d'upres les travaux recents." *Revue de Geographie de Lyon*, 30 (1955): 153-160.

Pounds, Norman J.G. "Agriculture in the Common Market." *East Lakes Geographer*, 1 (1964): 40-52.

Lendl, Egan. *Die jungsten Wandlungen der europaischen Kulturlandschaft*. Salzburg, Austria: A. Pustet, 1966.

Troll, C. "Die Landnutzunskartierung in den Rheinlanden." *Erdkunde*, 23 (1969): 81-102.

Clout, Hugh. *Agriculture: Studies in Contemporary Europe*. London, England: Macmillan, 1971.

Bowler, Ian R. "Recent Developments in the Agricultural Policy of the European Community." *Geography*, 61, No. 1 (1976): 28-30.

Gillmor, Desmond A. "EEC Scheme for Farming in Less Favoured Areas." *Irish Geography*, 10 (1977): 101-108.

Cuddy, M. "European Agricultural Policy: The Regional Dimension." *Built Environment*, 7 (1981): 200-210.

Bowler, Ian R. "Some Characteristics of the Industrialisation of Agriculture in the European Community." Pp. 75-98 in *The Industrialisation of the Countryside*, edited by Michael J. Healey and Brian W. Ilbery. Norwich, England: Geo Books, 1985.

Bowler, Ian R. "Intensification, Concentration and Specialization in Agriculture: The Case of the European Community." *Geography*, 71, No. 1 (1986): 14-24.

Szczesny, R. "Agricultural Typology of the Alpine Areas: Austria and Switzerland." *Geographia Polonica*, 52 (1986): 209-220.

Leidlmair, Adolph. "Grenzen in der Agrarlandschaft des mittleren Alpenraumes und ihr zeitlicher Wandel." *Geographische Zeitschrift*, 76, No. 2 (1988): 22-41.

Campbell, Bruce, and Mark Overton. *Land, Labour and Livestock: Historical Studies in European Agricultural Productivity*. Manchester, England: Manchester University Press, 1991.

Ronningen, K. "Agricultural Policies and Landscape Management: Some Examples from Norway, Great Britain and Germany." *Norsk Geografisk Tidsskrift*, 47, No. 2 (1993): 93-104.

DAIRYING

Tarrant, J.R. "Production and Disposal of Surplus EEC Milk Products." *Area*, 12, No. 4 (1980): 247-252.

Naylor, E.L. "EEC Dairy Policy." *Geography*, 72, No. 3 (1987): 239-241.

Berenguer, Jeanne. "Quantre ans de quotas laitiers." *Revue Geographique de l'Est*, 29, No. 1 (1989): 49-72.

Downs, Christopher J. "EC Agricultural Policy and Land Use: Milk Quotas and the Need for a New Approach." *Land Use Policy*, 8, No. 3 (1991): 206-210.

FARM LAND

Belding, Richard. "A Test of the Von Thunen Locational Model of Agricultural Land Use with Accountancy Data from the European Economic Community." *Transactions, Institute of British Geographers*, 6, No. 2 (1981): 176-187. (New Series)

ENVIRONMENTAL AND CONSERVATION ISSUES

Huntington, Ellsworth. "Climatic Change and Agricultural Exhaustion as Elements in the Fall of Rome." *Quarterly Journal of Economics*, 31 (1917): 173-208.

Lambert, A. "Farm Consolidation in Western Europe." *Geography*, 48 (1963): 31-48.

Kreisel, W. "Land Utilization in the Alpine Region of Europe." *Rural Systems*, 2, No. 3 (1984): 167-176.

Fanariotu, I.N., and D.G. Skuras. "Environmental Protection Under the EEC's Socio-Structural Policy for Agriculture." *Journal of Environmental Management*, 28, No. 3 (1989): 269-275.

Rohdenburg, H. "Methods for the Analysis of Agro-Ecosystems in Central Europe, with Emphasis on Geoecological Aspects." *Catena*, 16, No. 1 (1989): 1-57.

Robinson, Guy M. "EC Agricultural Policy and the Environment: Land Use Implications in the UK." *Land Use Policy*, 8, No. 2 (1991): 95-107.

AGRICULTURAL MARKETING AND DISTRIBUTION

Bresso, Mercedes. "Les paradoxes de la politique agricole commune." *L'Espace Geographique*, 9, No. 3 (1980): 173-182.

Duchene, Francois, Edward Szczepanik, and Wilfrid Legg. *New Limits on European Agriculture: Politics and the Common Agricultural Policy*. Totowa, N.J.: Rowman & Littlefield, 1985.

Bowler, Ian R. "Direct Supply Control in Agriculture: Experience in Western Europe and North America." *Journal of Rural Studies*, 2, No. 1 (1986): 19-30.

Bonanno, A. "The Agro-Food Sector and the Transnational State: The Case of the EC." *Political Geography*, 12, No. 4 (1993): 341-360.

Austria

GENERAL TEXTS, ARTICLES, AND THESES

Albitreccia, A. "La situation agricole de l'autriche." *Annales de Geographie*, 45 (1936): 529-534.

Demangert, J. "Evolution recente de l'agriculture autrichienne." *Annales de Geographie*, 57 (1948): 61-64.

Dussart, F. "Gennes de vie agricole et paysage rural dans la region de Mayrhafen (Zillertal, Tyrol autrichien)." *Bulletin de la Societe Belge d'Etude Geographique*, 20 (1951): 37-102.

Sehmer, Ingrid. *Studien uber die Differenzierung der Agrarlandschaft im Hochgebirgeim Bereich dreier Staaten*. Kallmunz/Regensburg, Germany: M. Lassleben, 1959.

Lichtenberger, Elisabeth. "Das Bergbauernproblem in den Osterreichischen Alpen Perioden und Typen der Entsiedlung." *Erdkunde*, 19, No. 1 (1965): 39-57.

Fliri, Franz. "Ein betriebswirtscahftlicher Beitrag zur Agrargeographie Osterreich." *Mitteilungen der Osterreich Geographischen Gesellschaft*, 109, No. 1-3 (1967): 257-280.

Krauter, Karl-Gunter. *Die Landwirtschaft im ostlichen Hochpustertal: Sudtirol*. Tubingen, Germany: Tubinger Geographische Studien, Heft 25, 1968.

Roser, Dieter. "Die Almwirtschaft der nordlichen Ortlergruppe- ihre Entwicklung in den vergangenen 100 Jahren." *Zeitschrift fur Wirtschaftsgeographie*, 14, No. 5 (1970): 149-154.

Dohrmann, Wilhelm. *Bonitierung und Tragfahigbeit eines Alpentales: Inners tes Defereggen in Osttirol*. Munster, Germany: Westfaliche Geographische Studien, Heft 24, 1972.

Fischer, Klaus. *Agrargeographie des westlichen Sudtirol, der Vinschgau und seine Nesentaler*. Vienna, Austria: Braumuller, 1974.

Penz, Hugo. "Grunzuge gegenwartiger Veranderungen in der Agrarlandschaft des Bundeslandes Tirol." *Mitteilungen der Osterreichichen Geographischen Gesellschaft*, 117, No. 3 (1975): 334-363.

Hofmayer, A. "Neue Entwicklungen der agrarische Wirtschaftforem im stadtnahen Ackerbangbeit am Beispiel des Marchfeldes." *Geographischer Jahresbericht aus Osterreich*, 35 (1977): 50-88.

Obernberger, Erwin. "Die Entwicklungen der Landwirtschaft im Oberosterreichischen Zentralraum seit 1938." *Wirtschaftsgeographie Studien*, 2 (1977): 1-11.

Szczesny, R. "Przemiang typow rolnictwa Austril w latach 1960–1970." *Przeglad Geograficzny*, 49 (1977): 741-756.

Leidlmair, Adolf. "Tirol auf dem Wege von der Agrar- zur Erholungslandschaft." *Mitteilungen der Osterreich Geographischen Gesellschaft*, 120 (1978): 38-53.

Grief, Franz. "Gedanken zur Alm- und Bergbauern frage." *Osterreich in Geschite und Literatur mit Geographie*, 23, No. 2 (1979): 96-108.

Hestroy, Othmer. "Aspekte des Strukturwandels der osterreichischen Landwirtschaft in den letzten zweit Jahrzchnten." *Osterreich in Geschite un Literatur mit Geographie*, 23, No. 2 (1979): 82-95.

Paschinger, H. "Junge Strukturwandlungen in der Landwirtschaft des osterreichischen Bundeslandes Karut." Pp. 627-641 in *Recherches de Geographie rurale: Hommage au Professeur Frans Dussart*. Liege, Belgium: Bulletin de la Societe de Geographie de Liege, Vol. 1, 1979.

Schroder, P. "Die Almwirtschaft im Osterreich nach den Ergelnissen der Almerheburg 1986." *Geographische Rundschau*, 41, No. 10 (1989): 575-580.

Szczesny, R. "Pastoral Economy and Agriculture of the Alpine Region in Austria" (in Polish). *Przeglad Geograficzny*, 62, No. 3-4 (1990): 311-322.

Hammerl, Reinhold. *Agrargeographische Untersuchungen im Bereich der Gemeince Eidenberg (Ober osterreich)*. Salzburg, Austria: Salzburger Geographische Arbeiten, Band 23, 1993.

Penz, Hugo. "Austrian Agriculture as Challenged by the EU" (in German). *Geographische Rundschau*, 47, No. 1 (1995): 25-29.

Penz, Hugo. "The Competitiveness of Austria's Agricultural Regions." *Mitteilungen der Osterreichischen Geographischen Gesellschaft*, 137 (1995): 417-426.

Grief, Franz. "Austria's Way to Sustainable Agricultural Policy." *Bulletin de la Societe Geographique de Liege*, 32 (1996): 181-184.

Heller, Armin. *Neue Typen der Agrarstruktur Osterreiches*. Innsbruck, Austria: Innsbrucker Geographische Studien, Heft 27, 1997.

Penz, Hugo. "Modernization in Agriculture: Developments in Austria after WWII. Findings of Research into 'Griculture' within the Research Focus of the FWF on Austria—Space and Society" (in German). *Mitteilungen der Osterreichischen Geographischen Gesellschaft*, 139 (1997): 77-100.

ANIMAL HUSBANDRY

Szczesny, R. "Gospodarka posterska i rolnictwo obszarow alpejokich Austrii." *Przeglad Geograficzny*, 62, No. 3-4 (1990): 311-322.

FARM LAND

Henske, Emil. "Dreissig Jahre Zusammenlegung landwirtschaftlicher Grundstucke in Tirol." *Mitteilungen der Osterreichischen Geographischen Gesellschaft*, 119, No. 1 (1977): 66-81.

Nestrog, O. "Flachennutzungsveranderungen in Osterreich." *Marburger Geographische Schriften*, 88 (1982): 103-123.

Zsilincsar, W. "Agricultural Land Use Classification in the Lower Mur Valley, Austria: An Interpretation by Means of a Multitemporal Analysis of Sar-Data." *Moravian Geographical Reports*, 7, No. 2 (1999): 40-46.

ENVIRONMENTAL AND CONSERVATION ISSUES

Strobi, J. "Die Anwendung multivariater statisticher Verfahren auf die Abhangigkeit des landwirtschaftlichen Eruturtrages von Klima am Beispiel von Niederosterreich." *Mitteilungen der Osterreichischen Geographischen Gesellschaft*, 125 (1983): 31-57.

Yamamoto, M. "Transformation of the Mountain Agricultural Areas of Tirol, Austria: An Altitudinal-Zone Analysis" (in Japanese). *Tsukuba Studies in Human Geography*, 14 (1990): 211-230.

TRANSPORTATION AND MECHANIZATION

Stanck, Norbert. "Die Mechanisierung im osterreichischen Weinbau." *Zeitschrift fur Wirtschatsgeographie*, 22, No. 3 (1978): 65-71.

Belgium and Luxembourg

GENERAL TEXTS, ARTICLES, AND THESES

Dussart, F. "Structure agraire et paysages ruraux dans la commune de Bakel (Brabant, septentrionel)." *Bulletin de la Societe Belge d'Etudes Geographique*, 15 (1946): 104-179.

Dendal, R. "Une enquete sur l'agriculture Luxuembourgeoise en 1764." *Bulletin de la Societe Belge d'Etudes Geographique*, 17 (1948): 35-46.

Vince, Stanley W.E. "The Agricultural Regions of Belgium." Pp. 255-288 in *London Essays in Geography*, edited by L. Dudley Stamp and S.W. Wooldridge. Cambridge, Mass.: Harvard University Press, 1951.

Tulippe, Omer. "Le subtrat agraire en Belgique: Etude preliminaire." *Bulletin de la Societe Belge d'Etudes Geographique*, 22 (1953): 19-91.

Christians, Charles. "Un essai de delimitation de regions et de compartiments agricoles dans la Belgique du sud-est." *Bulletin de la Societe Belge d'Etudes Geographique*, 29, No. 2 (1960): 321-361.

Christians, Charles. "Contribution a l'etude geographique de la structure agraire dans la partie wallonne de la Belgique." *Bulletin de la Societe Belge d'Etudes Geographique*, 30, No. 2 (1961): 257-465.

Christians, Charles. "La Belgique: L'evolution de la population active agricole et ses perspectives d'avernir." *Geographie Aardrijkskunde*, 21, No. 3 (1970): 139-150.

Tambuyzer, C. *Analyse regionale de l'agriculture belge. Partie V: Les regions agricoles de la province de Liege*. Brussels, Belgium: Cahiers de l'Institut Economique Agricole, Ministere de l'Agriculture, No. 136/RR/114, 1971.

Tambuyzer, C. *Analyse regionale de l'agriculture belge. Partie VI: Les regions agricoles de la province d'Anvers*. Brussels, Belgium: Cahiers de l'Institut Economique Agricole, Ministere de l'Agriculture, No. 130/RR/110, 1971.

Foutrein, Christian. "Flandre interieure region agricole en voie de specialisation." *Hommes et Terres du Nord*, 2 (1975): 41-59.

Stola, Wladyslawa. "Changes in the Types of Belgian Agriculture in the Years 1950–1970" (in Polish). *Przeglad Geograficzny*, 47, No. 4 (1975): 715-734.

Wieger, Axel. *Das Siedlungs- und Agrargefuge des Condroz und der Famenne in seiner historischen Entwicklung und in der Gegenwort*. Aachen, Germany: Aachener Geographische Arbeiten, Heft 9, 1976.

Christians, Charles. "La typologie de l'agriculture en Belgique: Methodes, problemes, resultats." *Travaux Geographiques de Liege*, 165 (1977): 93-109.

Christians, Charles. "L'agriculture au Grand-Duche de Luxembourg." *Bulletin de la Societe Belge d'Etudes Geographique*, 48, No. 1 (1979): 39-133.

Christians, Charles. "Les resultats de 27 annees de modernisation d'une agriculture avancee: L'exemple belge." *Hommes et Terres du Nord*, 4 (1980): 23-40.

Von Hecke, E. "The Structure of Belgian Agricultual Production." *Tijdschrift voor Economische en Sociale Geographie*, 74, No. 3 (1983): 213-216.

Von Hecke, E. "Les agriculteurs flamands en Wallone: Leur role et leur systeme d'exploitation." *Bulletin de la Societe Belge d'Etudes Geographique*, 55, No. 2 (1986): 315-327.

Christians, Charles. "La croissance de mographique dans les campagnes Belges et la minorisation des agriculteurs." Pp. 139-143 in *Geographes et campagnes*, edited by V. Rey. Paris, France: Ecole Normale Superieure, 1993.

Von Hecke, E. "L'Agriculture dans les regions urbain en belges." *Bulletin de la Societe Geographique de Liege*, 32 (1996): 223-230.

Christians, Charles. "Quarante ans de politique agricole europeenne commune et d'agriculture en Belgique." *Bulletin de la Societe Geographique de Liege,* 35 (1998): 41-56.

CROPS AND CROPPING PATTERNS

de Coster, Paul. "De evolutie van de druivenstukten S.E. von Brussel." Pp. 599-613 in *Recherches de Geographie Rurale: Hommage au Professeur Frans Dussort*. Liege, Belgium: Bulletin, Societe de Geographie de Liege, Vol. 1, 1979.

ANIMAL HUSBANDRY

Fourtrein, Christian. "Elevage et construction subventionnees en Flandre interieure." *Hommes et Terres du Nord,* 1 (1975): 79-90.

FARM LAND

Christians, Charles. "Problemes d'utilisation des espaces agricoles et forestiers en Belgique: Les cas liegeois et ardennais." Pp. 377-393 in *Melanges de geographie physique, humaine, economique applique, offerts a M. Omer Tulippe*, edited by Jose A. Sporck and Bernadette Schoumaker. Gembloux, Belgium: J. Duculot, 1967.

Albarre, Guy. "Le reboisement des terres agricoles en Lorrainebelge." *Revue Belge de Geographie*, 92 (1968): 57-90.

Lootens-de Muynck, M.T. "Het Bedobuen Agrarish Lanschap in de Oest-Vlaamse en Enkele Zeeunes-Vlaamse Polders." *Bulletin de la Societe Belge d'Etudes Geographiques*, 46 (1977): 129-152.

Christians, Charles. "Terroirs Wallons et agriculture wallonne en Belgique." *Bulletin de al Societe Belge d'Etudes Geographiques*, 62, No. 2 (1993): 329-360.

AGRICULTURAL MARKETING AND DISTRIBUTION

Von Hecke, E. "Markten en veilingen: Verschillen in bevoording en afzet." Pp. 581-598 in *Recherches de Geographie rurale: Hommage au Professeur Frans Dussort*. Liege, Belgium: Bulletin de Societe de Geographie de Liege, Vol. 1, 1979.

France

ATLASES AND GRAPHIC PRESENTATIONS

Robert, J., et al. "Quelques observations sur la carte des regions agricoles de la France au 1/500,000a." *Norois*, 6 (1959): 188-198.

Schneider, J.J. "Importance des etudes de geographie agraire sur les photographies aeriennes pour les travaux actuels d'economie agricole." *Bulletin de Societe Francaise de Photogrammetrie*, 28 (1967): 39-53.

Carriere, P. "L'evolution de l'espace agricole languedocien entre 1955 et 1970: Etude cartographique." *Bulletin de la Societe Languedocienne de Geographie*, 11, No. 1 (1977): 45-74.

Cottaz, J.P. "Cartographie des Pentes et du vignoble sur le coteau d'ompus." *Revue de Geographie de Lyon*, 56, No. 4 (1981): 377-390.

Vaudois, J. "L'analyse cartographique de l'evolution des exploitations agricoles dans les zones peri-urbaines: L'exemple de la region dunkerquoise." *Hommes et Terres du Nord*, 1 (1981): 45-51.

GENERAL TEXTS, ARTICLES, AND THESES

Alexandre, A. "L'economie agricole de la Picardie orientale depuis la Guerre." *Annales de Geographie*, 37 (1928): 61-73.

Richard, Marcelle. "L'evolution agricole du plateau de Langres." *Annales de Geographie*, 38 (1929): 133-147.

Lelannou, Maurice. "Le Tregorrois: Etude de geographie agricole." *Annales de Geographie*, 40 (1931): 24-38.

Albitreccia, A. "Une enquete sur la situation agricole de la Corse." *Annales de Geographie*, 46 (1937): 625-629.

Meyer, A. "La structure agraire dans la commune d'Ottersville." *Annales de Geographie*, 46 (1937): 124-136.

Tulippe, Omer. "Structure agraire et paysage rural du Pays de Chimay: La commune de Baileux." *Bulletin de la Societe Belge d'Etudes Geographique*, 13 (1943): 40-105.

Trenard, Louis. "L'evolution de l'economie agraire dans le Nord-Ouest de la Dombes depuis." *Revue de Geographie de Lyon*, 22, No. 1-4 (1947): 1-34.

Charaud, Anne-Marie. "L'habitat et la structure agraire de la Grande Briere et des mareis de Donges." *Annales de Geographie*, 57 (1948): 119-130.

Trenard, Louis, and Louise Trenard. "Structure agraire dans le Bas-Bugey." *Revue de Geographie de Lyon*, 24, No. 4 (1949): 331-340.

Moreau, J.P. "Questions agricoles en Basse-Bourgogne au XVIIIe siecle." *Annales de Geographie*, 59 (1950): 346-353.

Juillard, E. "L'assolement biennal dans l'agriculture septentrionale: Le cas particular de la Basse-Alsace." *Annales de Geographie*, 61 (1952): 34-45.

Duby, Georges. "La revolution agricole medievale." *Revue de Geographie de Lyon*, 29 (1954): 361-366.

Gachon, L. "La crise agricole francaise: Symtomes, manifestations, causes geographiques." *Revue de Geographie de Lyon*, 30 (1955): 35-42.

Juillard, E., and A. Meynier. *Die Agrarlandschaft in Frankreich*. Munich, Germany: Muncherer Geographische, Heft 9, 1955.

Bernard, C. "Contribution e l'etude de la structure agaire de la plaine littorale Montpellieraine." *Bulletin de la Societe Languedocienne de Geographie,* 27 (1957): 175-339.

Becancon, J. "Sur l'agriculture du Moyen-Orient." *Revue de Geographie de Lyon*, 36 (1961): 189-194.

George, Pierre. "Le sous-developpement agricole peril majeur." *Annales de Geographie*, 71 (1962): 630-634.

Gueny, Francois. "Agriculture et industrie dans les Vosges alsaciennes." *Revue Geographique de l'Est*, 2, No. 4 (1962): 327-344.

Flatres, Pierre. "Coup d'oeit sur les institutions agricoles francaises." *L'Information Geographique*, 27, No. 3 (1963): 93-104.

Sanlaville, P. "Les regions agricoles du Libon." *Revue de Geographie de Lyon*, 38 (1963): 47-89.

Price, Homer. "Position and Prospects of Agriculture in French Mediterranea: A Regional Case Study." Ph.D. Dissertation, Columbia University, New York, N.Y., 1964.

Sourdillat, J.M. *Geographie agricole de la France*. Paris, France: Presses Universitaires de France, 1964.

Blache, J. "Exploitations agricoles et amenagements industrials." *Revue de Geographie de Lyon*, 40, No. 3 (1965): 189-220.

Herold, Alfred. *Der zelgenge bundene Anbau im Randgebiet des Frankischen Gaulandes und seine besondere Stellung innerhalb der surwestdeutschen Agrarlandschaft*. Wurzburg, Germany: Wurzburger Geographische Arbeiten, Heft 15, 1965.

Pantard, J. *Les disparites regionales dans la croissance de l'agriculture francaise*. Paris, France: Gauthier-Villars, 1965.

Vernadon, M. "Le bassin de Gouzon (Creuse): Etude de geographie rurale." *Revue d'Auvergne*, 79, No. 1 (1965): 21-30.

Boichard, J. "Perspectives de l'agriculture francaise." *Revue de Geographie de Lyon*, 41, No. 2 (1966): 98-127.

Bras, J. "Problems de contact entre structure agraires en Anjou du Nord-Est." *Norois*, 13 (1966): 473-490.

De Phanhol, X., Francois Houel-Gangloff, and Marie-France Pierquin. "Materiaux pour la geographie historique et agraire de la Lorraine IV: Observations statistiques sur les champs courles en Lorraine." *Revue Geographique de l'Est*, 6, No. 1-2 (1966): 97-100.

Gabel, G. "Les measures agraires sous l'ancien regime en Annis." *Norois*, 13 (1966): 548-551.

Lichtenberger, Elisabeth. "Die Agrarkrise im Franzosischen Zentralmassiv im Spiegel seiner kulturlandschaft." *Mitteilungen der Osterreichischen Geographischen Gesellschaft*, 108, No. 1 (1966): 1-24.

Pijassou, R. "Structures agraires traditionnelles et revolution agricole dans les campagnes perigourdines." *Revue Geographique Pyrenees et du Sud-Ouest*, 37 (1966): 233-262.

Castex, J.M. "Les types de culture sur les hauteurs du Tanneron." *Acta Geographica*, 68 (1967): 6-15.

Demezicq, J. "Note: L'agriculture dans la commune de Dissay (Vienne)." *Norois*, 14 (1967): 601-610.

Flatres, Paul. "Les structures agraires du Nord de la France: Essai d'interpretation." Pp. 309-321 in *Melanges de geographie physique, humaine, economique applique, offerts a M. Omer Tulippe*, Vol. 1, edited by Jose A. Sporck. Gembloux, Belgium: J. Duculot, 1967.

Flechet, J.P. "L'evolution agricole de la Dombes." *Revue de Geographie de Lyon*, 42, No. 1 (1967): 39-79.

Gachon, L. "Pour une France toute entiere, berceau moderne." Pp. 401-407 in *Melanges de geographie physique, humaine, economique applique, offerts a Omer Tulippe*, Vol. 2, edited by Jose A. Sporck and Bernadette Schoumaker. Gembloux, Belgium: J. Duculot, 1967.

LeBerre, Maryvonne. "Paysages agraires et problemes agricoles dans la cluse de Chamberg." *Revue de Geographie Alpine*, 55 (1967): 291-305.

Lequin, Y. "Les grandes traits de l'evolution agricole a Plozevet depuis la fin du XVIIIe siecle." *Revue de Geographie de Lyon*, 42, No. 2 (1967): 165-178.

Lequin-Rimbaud, C. "La commercialisation des produits agricoles a Plozevet." *Revue de Geographie de Lyon*, 42, No. 2 (1967): 215-227.

Lerat, S. "La revolution agricole en Perigard." *Revue Geographique Pyrenees et du Sud-Ouest*, 38 (1967): 369-372.

Mingret, P. "Les exploitations agricoles a Plozevet." *Revue de Geographie de Lyon*, 42, No. 2 (1967): 179-214.

Renard, J. "Problemes agraires du Nord-Est de la Vendee." *Norois*, 14 (1967): 373-397.

Robert, J. "L'economie agricole dans la region de programme Poitou-Charentes." Pp. 5-13 in *Melanges de geographie physique, humaine, economique applique, offerts a M. Omer Tulippe*, Vol. 2, edited by Jose A. Sporck and Bernadette Schoumaker. Gembloux, Belgium: J. Duculot, 1967.

Schneider, S.J. "Importance des etudes de geographie agraire sur les photographies aerinnes pour les travaux actuels d'economie agricole." *Bulletin de la Societe Francaise de Photogrammetrie*, 28 (1967): 39-53.

Calmes, Roger. "Adaptation et modernisation agricole dans une commune du Bas-Segala: Saint-Just-ser-Viaw, Avegron." *Revue Geographique des Pyrenees et du Sud-Ouest*, 39 (1968): 299-332.

Carriere, P. "Agriculture et croissance economique en Languedoc." *Bulletin de la Societe Languedocienne de Geographie*, 2, No. 4 (1968): 285-324.

Clout, Hugh D. "Planned and Unplanned Changes in French Farm Structure." *Geography*, 53, No. 3 (1968): 311-315.

Dietrich, Genevieve. "Structures et problemes de l'exploitation agricole dans le departement des Vosges." *Acta Geographica*, 73 (1968): 8-12.

Henry, J.B. "Quelques aspects des structures agricoles en Bretagne." *Norois*, 15 (1968): 78-96.

Vogt, Jean. "Notes d'histoire agraire rhenone." *Revue Geographique de l'Est*, 8, No. 3-4 (1968): 355-362.

De Phanhol, Xavier, and Andre Perardel. "Materiaux pour la geographie historique et agraire de la Lorraine VI: La repartition geographique des vestiges archeologiques gallo-romains en Lorraine." *Revue Geographique de l'Est*, 9, No. 1-2 (1969): 177-188.

Patin, Jacqueline. "Les transformations de l'agriculture dans la region de Lessay." *Acta Geographica*, 79 (1969): 11-19.

Bernard, Marie Claude. "L'indemnite viagene de depart et la restructuration agricole en Languedoc-Roussillon." *Bulletin de la Societe Languedocienne de Geographie*, 4, No. 2 (1970): 99-114.

Bryant, Christopher R. "Urbanization and Agricultural Change since 1945: A Case Study from the Paris Region." Ph.D. Dissertation, London School of Economics, London, England, 1970.

Houssel, Jean-Pierre. "Villemoyenne, developpement de l'agriculture regionale, une agriculture de banlieue: L'exemple de Roanne." *Revue de Geographie de Lyon*, 45 (1970): 391-408.

Pitte, J.R. "Geographie de l'agriculture francaise." *L'Information Geographique*, 34, No. 3 (1970): 118-127.

Boichard, J. *La vie rural entre Loire et Allier*. Paris, France: Les Belles Lettres, 1971.

Boichard, J. "Chronique comtoise: L'exploitation agricole et ses charges foncieres dans la departement du Doubs." *Revue Geographique de l'Est*, 11, No. 1 (1971): 67-84.

Ginier, Jean. "Jersey, ile touristique et agricole." *Norois*, 17 (1970): 371-395.

Bonnamour, J. "Les systems regionaux d'exploitation agricole en France: Methods d'analyse typologie." *Etudes Rurales*, 43-44 (1971): 78-169.

Bonnamour, J., Yves Guermond, and Chantel Gillette. "Les systemes regionaux d'exploitation agricole en France." *Methodes d'analyse typologique etude rurales*, 43-44 (1971): 78-169.

Bouet, Guy. "La production agricole du Limousin." *Norois*, 18 (1971): 477-489.

Colin, Georges. "Un grand departement agricole: La Marne." *Travaux de l'Institut Geographie de Reims*, 7 (1971): 3-15.

Klatzmann, J. *Geographie agricole de la France*. Paris, France: Presses Universitaires de France, 1971.

Martin, Jacques. "Les G.A.E.C. en France (Groupements agricoles d'exploitative en commun)." *L'Information Geographique*, 35, No. 4 (1971): 163-170.

Montonbaux, G. "La grande agriculture champenoise au sud ouest de Chalons sur Marne." *Travaux de l'Institut de Geographie de Reims*, 7 (1971): 17-35.

Pilleboue, J. "L'evolution recente de l'agriculture dans Midi-Pyrenees d'apres l'enquete agricole de 1967." *Revue Geographique de Pyrenees et du Sud-Ouest*, 42 (1971): 99-132.

Belliard, C. "Les processes de modernisation dans une region de polyculture traditionnelle." *Acta Geographica*, 9 (1972): 33-50.

Bozon, Pierre. "L'agriculture dans un milieu industrialise et urbanise: La vallee du Gier (Loire)." *Revue de Geographie de Lyon*, 47 (1972): 227-257.

Clout, Hugh D. "Part-Time Farming in the Puy de Dome Department." *Geographical Review*, 62, No. 2 (1972): 271-273.

Daude, G. "L'agriculture de Chateau-renard: Tradition, difficultes et perspectives." *Revue de Geographie de Lyon*, 47 (1972): 167-219.

Maury, R. "L'evolution d'une commune agricole proche d'une ville: Saint-Georges-les-Baillargeaux, Vienne." *Norois*, 19, No. 75 (1972): 501-518.

Sceau, R. "La cooperation agricole dans la region Rhone-Alpes." *Revue de Geographie de Lyon*, 47 (1972): 259-296.

Veyret, Paul. "L'agriculture de montagne dans les Alpes Francaise: Le probleme de la survie." *Revue de Geographie Alpine*, 60 (1972): 5-24.

Bryant, Christopher R. "L'agriculture face a la croissance metropolitaine: Le cas des exploitations de grande culture expropriees par l'emprise de l'Aeroport Paris-Nord." *Economie Rurale*, 95 (1973): 23-35.

Bryant, Christopher R. "L'agriculture face a la croissance metropolitaine: Le cas de exploitations fruitiere de Grosley et Deuil-la-Barre dans la grande banlieue nord de Paris." *Economie Rurale*, 98 (1973): 35-55.

Chabaud, M., and A. Champalbert. "Methodes d'analyse de l'espace agricole: Typogramme et matrice ardonnable. Exemples dualiner vois et de la vallee du Rhone." *Bulletin de la Societe Languedocienne de Geographie*, 7, No. 2 (1973): 241-262.

Stola, Wlaydyslawa. "Agriculture in the Voucluse Departement, France: A Typological Study" (in Polish). *Dokumentaija Geografiszna*, 3 (1973): 1-86.

Bonnamour, J. "Regions agricoles et modernization de l'agriculture francaise." *Geographia Polonica*, 29 (1974): 295-306.

Bryant, Christopher R. "An Approach to the Problem of Urbanization and Structural Change in Agriculture: A Case Study from the Paris Region, 1955–1968." *Geografiska Annaler B*, 56B, No. 1 (1974): 1-27.

Erroux, J., and J. Courtin. "Apercu sur l'agriculture prehistorique dans le sud-est de la France." *Bulletin de la Societe Languedocienne de Geographie*, 8, No. 3-4 (1974): 325-336.

Guellec, A. "Les mutations recentes de l'agriculture en Bretagne." *Geographia Polonica*, 29 (1974): 351-362.

Guermond, Yves. "Les inegalites geographiques dans le developpement agricole: L'exemple du Morbihan." *Norois*, 21 (1974): 39-63.

Livet, R. "Quelques aspects de la modernisation agricole dans le sud-est de la France." *Geographia Polonica*, 29 (1974): 343-349.

Phillips, A.D.M., and Hugh D. Clout. "L'evolution agricole dans le departement du Pay-de-Dome pendant la deuxieme moitie du XIXe siecle." *Acta Geographica*, 16-17, Serie 3 (1974): 103-120.

Bozon, P. "Note sur l'evolution recente de l'agriculture ardechoise (Chronique Rhone-Alpes)." *Revue de Geographie de Lyon*, 50 (1975): 177-184.

Bryant, Christopher R. "Metropolitan Development and Agriculture: The SAFER de l'Ile-de-France." *Land Economics*, 51, No. 2 (1975): 158-163.

Clout, Hugh D. "Structural Change in French Farming: The Case of the Puy-de-Dome." *Tijdschrift voor Economische en Sociale Geographie*, 66, No. 4 (1975): 234-246.

Legoy, Pierre. "La diffusion des innovations en agriculture: L'exemple de la stabulation libre en Seine-Maritime." *Cahiers Geographiques de Rouen*, 4 (1975): 1-20.

Loubgery, J. "La mise en valeur agricole des landes du Port Long au nord de Pau." *Revue Geographique des Pyrenees et du Sud-Ouest*, 46 (1975): 313-324.

Pilleboue, J. "La structure des exploitation agricoles dans Midi-Pyrenees." *Revue Geographique des Pyrenees et du Sud-Ouest*, 46 (1975): 5-26.

Berenguer, Jeanne. "Les types d'exploitations agricoles dans le departement des Ardennes." *Travaux de l'Institut de Geographie de Reims*, 27-28 (1976): 15-27.

Gillardot, P. "Diversite des exploitations agricoles en Grande Sologne." *Bulletin de l'Association de Geographes Francais*, 53 (1976): 129-138.

Kroemer, Doris. *Die Agrarlandschaft der Beauce: Historische Entwicklung und heutiges Bild einer hochtechnisierten Getreide baul Landschaft*. Aachen, Germany: Aachener Geographische Arbeiten, Heft 10, 1976.

Pelaton, Jean. "Une region agricole une a travers le recensement general de 1970: Le Perche." *Acta Geographica*, 26 (1976): 39-57.

Pelaton, Jean. "Le boule versement des structures de l'exploitation agricole dans le Perche par la construction de l'Oceane (Autoroute A.11)." *Norois*, 23 (1976): 397-416.

Pletsch, Alfred. *Moderne Wandlungen der Landwirtschaft im Languedoc: Entwicklungstendenzen im einen wirtschaftlichen Schwackeraum Frankreichs im Rahmen der Regionalisation*. Marburg/Lahn, Germany: Selbstverlag des Geographischen Institutes der Universitat Marburg, 1976.

Clout, Hugh D. "Agricultural Change in the Eighteenth and Nineteenth Centuries." Pp. 407-446 in *Themes in the Historical Geography of France*, edited by Hugh D. Clout. London, England: Croom Helm, 1977.

Deffontaines, Jean-Pierre, and Paul-Louis Osty. "De systemes de production agricole aux systemes agraires." *L'Espace Geographique*, 6, No. 3 (1977): 195-199.

MacDonald, Lynn. "Paysan Farming in the Region Auvergne." *South Hampshire Geographer*, 9 (1977): 27-32.

Pletsch, Alfred. "Die Entwicklung des Sonderkulturanbaus im Languedoc/ Sud frankreich nach dem zweiten Weltkrieg." *Erdkunde,* 31, No. 4 (1977): 288-298.

Clinquart, Pierre. *Le Terrier de Rabouillet, Pyrenees-Orientales: Description d'un agrosysteme mediterraneen submontagnard au XVIIIe siecle, interpretation de sa dynamique du XVIe au XIXe siecles.* Poitiers, France: Faculte des Sciences, Services d'ecologie et Biogeographie, 1978.

Allaine, G. "Structures de production et systemes de cultures: Garac, Haute-Garonne." *Revue Geographique des Pyrenees et du Sud-Ouest*, 50, No. 2 (1979): 269-292.

Bonnamour, J., and Chantel Gillette. "Application des normes proposees par J. Kostrowicki a l'agriculture francaise." *Geographia Polonica,* 40 (1979): 151-160.

Bouet, Guy. "Le zonage agriculture: foreten Limousin." *Norois*, 26 (1979): 349-368.

Cohou, M. "Croitre ou disparaitre: Histoire des agriculteurs de la Piege." *Revue Geographique des Pyrenees et du Sud-Ouest*, 52, No. 2 (1979): 251-268.

Durbiano, Claudine, and Andre de Reparaz. "Le foncier agricole en Provences Alpes Cote d'Azur Specificite, defense et restructuration." *Mediterranee,* 37, No. 4 (1979): 55-75.

Guellec, A. "Remembrement, surfaces et orientation des production. L'exemples dans les Cote-du-Nord pour entreprendre l'etude de leurs relations." Pp. 445-458 in *Recherches de Geographie rurale: hommage au Professeur Frans Dussort.* Liege, Belgium: Bulletin de la Societe de Geographie de Liege, Vol. 1, 1979.

Guermond, Yves. *Le systeme de differenciation spatiale en agriculture: La France de l'Ouest de 1950 a 1970.* Paris, France: Librairie Honore Chaupine, 1979.

Maury, R. "Fluctuations de l'espace agricole entre la Loire et le Cher a l'Est de Tours." *Bulletin de l'Association de Geographes Francais*, 56, No. 3 (1979): 125-129.

Renard, J. "Quelques aspects des mutations agricoles dans le sud-est du Massif Armoricain." *Norois,* 26 (1979): 321-347.

Vaudois, J., and A. Barre. "Autoroutes et Agriculture dans le region Nord-Pas-de-Calais." *Bulletin de la Association de Geographes Francais*, 56 (1979): 301-305.

Bernard, Gilles. "Quelques aspects recents de l'agriculture charentaine." *Norois*, 27 (1980): 549-554.

Bernard, Gilles. "Les transformations actuelles de l'agriculture de la plaine du Thouarsais." *Norois,* 28 (1980): 45-54.

Bonnamour, Jacqueline, and Chantel Gillette. *Les types d'agriculture en France, 1970: essai methologique.* Paris, France: Editions du Centre national de la recherche scientifique, 1980.

Calmes, Roger. *Les campagnes des Segalas et du Levezo: Developpement et progres agricole.* Rodez, France: Subervie, 1980.

Clout, Hugh D. *Agriculture in France on the Eve of the Railway Age*. Totowa, N.J.: Barnes and Noble, 1980.

Daumas, M. "L'agriculture peri-urbaine orleanaise." *Norois*, 27 (1980): 61-73.

Vaudois, J. "A propos de l'evolution recente de l'agriculture du Nord-Pas-de-Calais." *Hommes et Terres du Nord*, 1 (1980): 53-57.

Bonneau, Michel. "Etudes d'economie agro-alimentaire." *Hommes et Terres du Nord*, 1 (1981): 56-58.

Bonnieux, F., and P. Rainelli. "Les disparities regionales de l'agriculture europeene." Pp. 319-333 in *L'assiette rurale des departments francais. Actes du colloque de Poitiers 1981*. Poitiers, France: Travaux du Centre de Geographie Humain et Sociale, Universite de Poitiers, No. 6, 1981.

Bryant, Christopher R. "Agriculture in an Urbanizing Environment: A Case Study from the Paris Region, 1968–1975." *Canadian Geographer*, 25, No. 1 (1981): 27-45.

Vaudois, J. "Neuf ans d'evolution dans les campagnes nordistes: les enseiguements des premiers resultats du recensement general de l'agriculture 1979–1980." *Hommes et Terres du Nord*, 1 (1981): 40-44.

Demeulenaere, F. "Evolution recente de l'agriculture dans les zones periurbaines: l'exemple de l'Ouest de l'agglomeration lilloise." *Hommes et Terres du Nord*, 1 (1982): 68-74.

Louault, F. "Espace departemental et regions agricoles: l'Indre-et-Loire." Pp. 117-125 in *L'assiette rurale de departements francais. Actes du colloque de Poitiers 1981*. Poitiers, France: Travaux du Centre de Geographie Humaine et Sociale, Universite de Poitiers, No. 6, 1982.

Maury, R. "La connaissane economique des exploitation agricoles au niveau departemental." Pp. 297-310 in *L'assiette rurale de departments francais: Actes de colloque de Poitiers 1981*. Poitiers, France: Travaux de Centre de Geographie Humaine et Sociale, Universite de Poitiers, No. 6, 1982.

Pitie, J. "L'evolution agricole en Poitou-Charentes dans la dernier demi-siecle." *Norois*, 29 (1982): 141-154.

Cointot, J.C. "Essai de typologie agricole des communes francaises: L'Indre-et-Loire au R.G.A." *Norois*, 30 (1983): 95-101.

Dorel, Gerard, and C. Dumenil. "L'evolution de l'agriculture en Champagne-Ardenne." *Travaux-Institut de Geographie de Reims*, 55-56 (1983): 87-113.

Guermond, Yves. "Les types d'exploitations agricoles en Normandie." *Geographia Polonica*, 46 (1983): 133-148.

Roudie, P. *La France: Agriculture, foret, peche: depuis 1945*. Paris, France: Sirey, 1983.

Dauce, P., and L. Guigeuno. "Aux origines de la modernization agricole et de l'intensification de l'agriculture en Bretagne." *Norois*, 31 (1984): 541-557.

Mercadier, G. "Quelques donnees et remarques sur l'evolution de l'agriculture en Aveyrou." *Revue Geographique des Pyrenees et du Sud-Ouest*, 55, No. 3 (1984): 335-366.

Gillette, C. "Dix ans d'agriculture francaise (1970–1980)." *L'Information Geographique*, 49, No. 1 (1985): 1-16.

Mazataud, P. "La diffusion de la compatibilite de gestion dans les exploitations agricoles du centre-sud-ouest de la France." *Revue Geographique des Pyrenees et du Sud-Ouest*, 56, No. 1 (1985): 39-62.

Oger, O. "L'agriculture du Nord-Pas-de-Calais a la recherche d'un second souffle." *Hommes et Terres du Nord*, 1 (1985): 51-57.

Baker, Alan R.H. "The Infancy of France's First Agricultural Syndicate: The Syndicat des Agriculteurs de Loir-et-Cher, 1881–1914." *Agricultural History Review*, 34, No. 1 (1986): 45-59.

Bonnefaz, S., and J. David. "L'agriculture periurbaine grenobloise: Nouvelle composante de l'amenagement." *Revue Geographique de Alpine*, 74, No. 3 (1986): 265-292.

Vaudois, J. "L'evolution recente de l'agriculture du Nord-Pas-de-Calais: Le poids des facteurs geographiques." *Bulletin de l'Association de Geographes Francais*, 63, No. 2 (1986): 87-95.

Dubesset, P. "Une retambee agricole du nucliaire rhodanien: Les serres chauffees." *Revue de Geograpie de Lyon*, 62 (1987): 192-216.

Guellec, A. "Les perspectives de l'agriculture bretonne et de l'ille-et-Vilaine en particulier: Bilon a grands traits pour un debut de relexion." *L'Information Geographique*, 51, No. 3 (1987): 103-119.

Kayser, B., and D. Romias. "La region, maillon faible dans la congestion de l'agriculture: Le cas de Midi-Pyrenees." *Revue Geographique Pyrenees et du Sud-Ouest*, 58, No. 4 (1987): 399-405.

Rey, V. "Specialisation geographique du systeme agricole francais: Une methode d'evaluation." *Bulletin de l'Association de Geographes Francais*, 64, No. 3 (1987): 211-220.

Solle, Henriette. "Le dynamisme des communes agricoles entre Seine et Loire dans les annees quatre vingts." Pp. 125-154 in *Regions, villes et amenagement*. Paris, France: CREPIF/Societe de Geographie, 1987.

Deslondes, O. "De prise agricole, en resinement et choix actuels: L'exemple des Cavennes." *Revue Geographiques de Pyrenees et du Sud-Ouest*, 59 (1988): 125-132.

Dufour, J. "Crise de l'agriculture et perspectives d'avernir dans la Sarthe." Pp. 132-138 in *Campagnes et littoraux d'Europe: Melanges offers a Pierre Flatres*. Lille, France: Institut de Geographie de Lille, Hommes et Terres du Nord, 1988.

Winchester, Hilary P.M., and Brian W. Ilbery. *Agricultural Change: France and the EEC*. London, England: John Murray, 1988.

Aitchinson, J.W. "Cluster Analysis and Large Data Sets: A Case Study of Farming Systems in France." *Geographia Polonica*, 57 (1989): 13-22.

Castex, J.M., and A. Dagorne. "Les amenagments agricoles en pays grassois: Le cas de Saint-Cezire-sur-Siogne (Alpes Maritimes)." *Analyse Spatiale Quantitative et Applique*, 26 (1989): 45-54.

Jones, Alun. "The Role of the SAFER in Agricultural Restructuring: The Case of the Languedoc-Rousillon, France." *Land Use Policy*, 6, No. 3 (1989): 249-261.

Violier, P. "Urbanisation et dynamique agricole: Les exploitations agricoles face aux expropriations. L'exemple de la Communaute urbaine de Lille." *Hommes et Terres du Nord*, 1-2 (1989): 103-109.

Campagne, P., G. Carrere, and E. Valceschini. "Three Agricultural Regions of France: Three Types of Pluriactivity." *Journal of Rural Studies*, 6, No. 4 (1990): 415-422.

Demonet, Michel. *Tableau de l'agriculture francaise au milieu du 19e siecle: L'enquite de 1852*. Paris, France: Editions de l'Ecole des hautes etudes en sciences sociales, 1990.

Derioz, Pierre, and Pierre Lenormond. "Deprise et resistance de l'activite agricole en basse montagne ardechoise: Region Rhone-Alpes, France." *Geographica Iugoslavica*, 12 (1990): 121-130.

Homeau, J.B. "Les mutations recentes de l'agriculture dans le Baugeois." *Travaux-Universite de Poitiers, Centre de Geographie Humaine et Sociale*, 17 (1990): 479-484.

Limouzin, P. "L'evolution des systemes de culture et des structures d'exploitation (1979–1988) in Picardie." *Hommes et Terres du Nord*, 4 (1990): 239-247.

Meffre, J.C. "Habitats augusteens et a menagements des versants: Seguret (Vaucluse)." *Mediterranee*, 71, No. 3-4 (1990): 17-21.

Mouton, Michel. "L'evolution recente des regions agricoles du Val de Garonne." *Revue Geographique des Pyrenees et du Sud-Ouest*, 61, No. 3 (1990): 335-352.

Derioz, Pierre. "Les consequences spatiales de la deprise agricole en Haut-Languedoc occidental: L'hephemere victoire de la friche." *Revue de Geographie de Lyon*, 66, No. 1 (1991): 47-54.

Mergoil, Guy, and P. Roudie. "Friches et recensements de l'agriculture." *Revue de Geographie de Lyon*, 66, No. 1 (1991): 6-10.

Quaracino, F. "Dynamique spatiale agricole dans l'agglomeration marseillaise: Le modele d'Allauch." *Mediterranee*, 73 (1991): 79-88.

Bernard-Allee, P., and B. Valadas. "L'etat des recherches sur l'archeolgie agraire en Limousin." *Norois*, 39 (1992): 509-518.

Cardi, F. "Decentralisation de l'enseignement agricole et developpement local." *Geographie Sociale*, 12 (1992): 303-310.

Dufour, J. "Le deprise agricole: Methodes d'approche et etudes dans la Sarthe." *Geographie Sociale*, 12 (1992): 227-236.

Gay, J.F. "Agriculture et tourisme en Perigard noir." *Geographie Sociale*, 12 (1992): 123-135.

Maury, R. "Le genese des paysages agraires en France: La Touraine en est-elle une synthese?" *Norois*, 39 (1992): 5-27.

Mignon, C. "La deprise agricole dans les montagnes francais." *Geographie Sociale*, 12 (1992): 261-270.

Piriou, N., and A. Cleach. "La deprise agricole et la recomposition de l'espace littoral sud-Bigouden." *Geographie Sociale*, 12 (1992): 215-226.

Blondeau, P. "Espace agricole et systemes de culture dans la Sarthe." *Norois*, 39 (1993): 253-272.

Millo, C. "Les zones agricoles dans les communes periurbaines." *Mediterranee*, 77, No. 1-2 (1993): 34-36.

Bernard, N. "Restructuration et devenir des exploitations agricoles dans une commune periurbaine: Plomelin (Finisterre)." *Bulletin de l'Association de Geographes Francais*, 71, No. 2 (1994): 196-203.

Charvet, J.P. "Introduction: nouvelles approches et nouvelles questions a propos des agriculture periurbaines." *Bulletin de l'Association de Geographes Francais*, 71, No. 2 (1994): 119-122.

Clout, Hugh D. "Rural Revival in Marne, 1914–1930: French Agriculture." *Agricultural History Review*, 42, No. 2 (1994): 140-155.

Courtot, R. "La huerta de Valence: La fin d'un mythe?" *Bulletin de l'Association de Geographes Francais*, 71, No. 2 (1994): 181-186.

Durbiano, Claudine. "Disparition, marginalisation et intensification de l'agriculture dans l'aire metropolitaine marseillaise." *Bulletin de l'Association de Geographes Francais*, 71, No. 2 (1994): 160-169.

Lees, C., and Pierre Derioz. "Le jardin de la France au peril de la ville: Place et evolution de l'activite agricole dans le 'Grand Avignon.'" *Bulletin de l'Association de Geographes Francais*, 71, No. 2 (1994): 170-180.

Limouzin, P. "Une agriculture periurbaine en voie de banalisation: L'exemple amienois." *Bulletin de l'Association de Geographes Francais*, 71, No. 2 (1994): 156-159.

Pletsch, Alfred. "La transformation de l'agriculture provencale face au marche European." *Revue de Geographie de l'Est*, 34, No. 1 (1994): 31-45.

Prost, B. "L'agriculture periurbaine: Analyse d'une marginalite." *Bulletin de l'Association de Geographes Francais*, 71, No. 2 (1994): 144-151.

Cael, C. "Un plan de developpment agricole durable pour la Montague Nord (departement de Vosges)." *Revue de Geographie Alpine*, 83, No. 3 (1995): 155-160.

Leguilloux, M. "Alimentation et elevage a Marseille au Ve. siecle apres J.C. d'apres les etudes de faunes." *Mediterranee*, 82, No. 3-4 (1995): 85-92.

Rieutart, Ricard L. "Filienes agro-alimentaires et moyennes montagnes francaise." *Revue de Geographie Alpine*, 83, No. 3 (1995): 101-114.

Vitte, P. "Les problemes de l'agritourisme en France." *Bulletin de l'Association de Geographes Francais*, 72, No. 1 (1995): 14-23.

Blondeau, P. "L'agrondisement recent des exploitates agricoles dans la Sarthe." *Norois*, 43 (1996): 375-394.

Karabanick, Edward. "The Var Region: A Microcosm of French Agricultural Modernization." *Yearbook, Association of Pacific Coast Geographers*, 59 (1997): 51-67.

Gilg, Andrew, and Martin R.J. Battershill. "Quality Farm Food in Europe: A Possible Alternative to the Industrialized Food Market and to Current Agri-Environmental Policies: Lessons from France." *Food Policy*, 23 (1998): 25-40.

Mace, Georges. "Les systems d'agriculture dans l'ouest francais." *Acta Geographica*, 116 (1998): 65-84.

Laurens, Lucette. "Le Long parcours de l'Aubrae: Productions de qualite, developpement local et affirmation d'une identite micro-regionale." *Sud-Ouest Europeen*, 6 (1999): 51-60.
Negro, Yves. "L'evolution recente des exploitations agricoles dans le Comminges: L'attende et l'imprevu." *Sud-Ouest Europeen*, 6 (1999): 85-94.
Pilleboue, Jean. "Les produits agro-alimentaires de qualite: Remarques sur leurs liens au territoire." *Sud-Ouest Europeen*, 6 (1999): 69-84.
Bontron, J.C. "L'agriculture francaise a la croisie des chemins: Quatre scenarios pour l'avenir." *Hommes et Terres du Nord*, 2000–4 (2000): 193-201.
Laurens, Lucette. "L'agriculture dans les parcs naturels regionaux, entre multifonctionnalite et concentration." *Hommes et Terres du Nord*, 2000–2 (2000): 238-245.

CROPS AND CROPPING PATTERNS

Zeyl, R. "La culture du houblau en Alsace." *Annales de Geographie*, 39 (1930): 569-578.
Deguil, I. "L'horticulture en Anjou." *Annales de Geographie*, 42 (1933): 601-609.
Marves, P. "L'evolution de la viticulture dans le Bas-Laguedoc." *Bulletin de la Societe Languedocienne de Geographie*, 6 (1935): 26-58.
Perrin, Armand. "Les cooperatives vinicoles en Beaujolais." *Revue de Geographie de Lyon*, 19, No. 1-2 (1944): 1-20.
Agnew, Swanzie. "The Vine in Bas Languedoc." *Geographical Review*, 36, No. 1 (1946): 67-79.
Huser, Simone. "Les vins de Saint-Peray." *Revue de Geographie de Lyon*, 23, No. 3 (1948): 105-120.
Fevret, Maurice. "La sericulture au Libon: Deuxieme partie, son declin actuel." *Revue de Geographie de Lyon*, 29, No. 4 (1949): 341-363.
Gibert, Andre. "Le vignoble francais: D'apres un travail recent." *Revue de Geographie de Lyon*, 24, No. 1 (1949): 59-63.
Weigend, Guido G. "The Basis and Significance of Viticulture in Southwest France." *Annals of the Association of American Geographers*, 44, No. 1 (1954): 75-101.
Dion, Roger. "Le vin d'Arbois au moyen age." *Annales de Geographie*, 64, No. 3 (1955): 162-169.
Menard, A. "Angers: Centre national d'horticulture." *Norois*, 3 (1956): 409-425.
Meynier, A., and E. Juillard. "La geographie agraire en France depuis vingt ans." *Norois*, 3 (1956): 228-232.
Brunet, R. "Les dibuts de la production fruitiere dans le Gers." *Revue Geographique des Pyrenees et du Sud-Ouest*, 31 (1960): 216-221.
Galtice, G. *Le vignoble du Languedoc mediterraneen et du Roussillon: L'etude comparative d'un vignoble de masse.* Montpellier, France: Edition Causse, 1960.
Kayser, B. "Vers une revolution fruitiere en moyenne Garonne." *Revue Geographique des Pyrenees et du Sud-Ouest*, 31 (1960): 205-215.

Bethemont, J. "Le riz et le mise en valeur de la Camarque." *Revue de Geographie de Lyon*, 37 (1962): 153-206.

Gadille, Rolande. "Les exploitations viticoles de la Cote-d'Or." *Revue Geographique de l'Est*, 4, No. 1 (1964): 43-60.

Rousselle-Touchard, M. "Viticulture et agriculture de six communes de la vallee du Layon." *Norois*, 11 (1964): 460-463.

Brunet, R. "Les cereales dans la Gers." *Revue Geographique des Pyrenees et du Sud-Ouest,* 36 (1965): 181-190.

Gautier, M. "La vigne et le vin dans la region pornicaise (Loire-Atlantique)." *Norois*, 12 (1965): 181-199.

Lacaze, R. "La culture de la lavande dans le Quercy." *Revue Geographique des Pyrenees et du Sud-Ouest*, 36 (1965): 39-52.

Laugenie, C. "La culture du noyer en Perigard." *Revue Geographique des Pyrenees et du Sud-Ouest*, 36 (1965): 135-158.

McDonald, James R. "Maize as a Grain Crop in France." *Journal of Geography*, 64, No. 4 (1965): 158-168.

Coustaury, A. "Transformations recentes de l'agriculture fruitiere dans la moyenne vallee du Rhone." *Revue de Geographie de Lyon*, 41, No. 2 (1966): 129-154.

Dufour, J. "Les cultures fruitieres dans le Sud de la Sarthe: Leurs rapports avec les sols et les climats." *Norois*, 13 (1966): 437-452.

Lepagnot, L.J. "Un organisme recent de commercialisation des fruits et legumes en Moyenne-Garonne: Le Marche-gare d'Agen-Boi." *Revue Geographie des Pyrenees et du Sud-Ouest*, 37 (1966): 218-230.

Babonaux, Y. "L'evolution actuelle du vignoble tourangeau." *Norois*, 14 (1967): 64-68.

Godard, Alain. "La culture du mais-grain en Alsace et en Bode du Sud: Reactif du milieu natural out du milieu socio-economique?" *Revue Geographique de l'Est*, 7, No. 1-2 (1967): 3-172.

Colin, Georges. "Le dynamisme du vignoble champenois." *Revue Geographique de l'Est*, 8, No. 3-4 (1968): 337-354.

Durbiano, Claudine. "L'expansion du vignoble des Cotes du Rhone meridionales." *Mediterranee*, 65, No. 3 (1968): 3-11.

Roudie, P. "Les appelations du vignoble bordelais." *Revue Geographiques des Pyrenees et du Sud-Ouest*, 39 (1968): 5-34.

Gadille, Rolande. "Les cultures de petits fruits (cassis, groseilles, framboises) en Cote-d'Or." *Revue Geographique de l'Est*, 9, No. 1-2 (1969): 127-148.

Roux, P. "Groupements agricoles et commercialisation des fruits et legumes en Bas-Languedoc." *Bulletin de la Societe Languedocienne de Geographie*, 4, No. 2 (1970): 213-236.

Brisset, P. "Les marches et les de bouches des legumes en Charente-Maritime." *Acta Geographica*, 6, Serie 3 (1971): 85-132.

Houssel, Jean-Pierre. "L'evolution recente des techniques de culture du ble en France." *L'Information Geographique*, 35, No. 4 (1971): 171-176.

Bozon, Pierre. "L'evolution recente de la production fruitiere en France." *L'Information Geographique*, 36, No. 2 (1972): 84-90.

Desire, Emmanuel. "L'extension recente de la culture du mais en Beauce." *Norois*, 19 (1972): 27-46.

Le Berre, Maryvonne. "La culture des legumes et le developpement urbain autour de Grenoble." *Revue de Geographie Alpine*, 60 (1972): 43-73.

Carrierre, P. "Viticulture et espace rural: L'assiette geographique du vignoble languedocien et son evolution recente." *Bulletin de la Societe Languedocienne de Geographie*, 7, No. 2 (1973): 221-240.

Muller, Bernard. "Le vignoble de Noveant-Sur-Moselle: Histoire d'un declin." *Mosella*, 3, No. 1 (1973): 1-39.

Reitel, Francois. "Les causes du declin du vignoble mosellan: Out plaidoyer pour un renouveau du vignoble mosellan." *Mosella*, 3, No. 1 (1973): 40-71.

Roudie, P. "Evolution recente de la production et du marche des vins de Bordeaux." *Bulletin de l'Association de Geographes Francais*, 50 (1973): 495-500.

Baysong, Jean-Claude. "Croissance et problemes de la viticulture corse." *L'Information Geographique*, 38, No. 4 (1974): 167-173.

Belot, Chaude. "Vers une nouvelle geographie du vignoble charentais." *Norois*, 21 (1974): 263-276.

Fiette, A. "Les legumes de plein-champ en Picardie." *Bulletin de l'Association de Geographes Francais*, 54 (1974): 163-169.

Gilbank, J. "La modernisation de la viticulture francaise." *Geographia Polonica*, 29 (1974): 261-270.

White, Elwood. "A Geographical Analysis of Overdependence on Viticulture in the Bas-Languedoc Region of Southern France, 1941–1973." Master's Thesis, University of Georgia, Athens, Ga., 1974.

Beaue, J.P. "La culture du riz en Camarque: Aspects techniques et commerciaux actuels." *Mediterranee*, 22, No. 3 (1975): 53-68.

Durbiano, Claudine. "Le raison de table: L'element de l'economie maraichere et fruitiere provencale." *Mediterranee*, 23, No. 4 (1975): 11-36.

de Reparz, A., and C. Berard. "Viticulture et cooperation vinicole dans le Sud-Est mediterranee: Le cas du Var." *Mediterranee*, 23, No. 4 (1975): 37-58.

Sceau, R. "Un vignoble dynamique: Le Beaujolais. Quelques aspects de son evolution recente." *Revue de Geographie de Lyon*, 50 (1975): 127-150.

Vaudois, J. "Production legumiere, progres technique et organisation economique des producteurs: L'endive dans la region du nord." *Hommes et Terres du Nord*, 1 (1975): 29-78.

Peyon, Jean-Pierre. "La cooperative viticole en Champagne." *Travaux de l'Institut de Geographie de Rheims*, 27-28 (1976): 29-55.

Enjalbert, Henri. "La naissance des grands vins et la formation du vignoble moderne de Bordeaux." Pp. 59-88 in *Geographie historique des vignobles*. Bordeaux, France: Actes du colloque de Bordeaux, 1977.

Toulze, P. "Le vignoble de Limous." *Revue Geographique des Pyrenees et du Sud-Ouest*, 48 (1977): 321-327.

Bernard, Gilles. "Production et commerce des cereales en Deax-Sevres: Les nouvelles orientations." *Norois*, 25 (1978): 533-549.

Evans, J. "Wine-Growing in France." *Bloomsbury Geographer*, 9 (1978): 6-17.
Gadille, Rolande. "Les viticultures de coteau sur la bardure occidentale du sillon rhodonien." *Revue de Geographie de Lyon*, 53 (1978): 5-6.
Gadille, Rolande. "L'heritage d'une viticulture antique: Vignes et vins de Cote-Rotie et Condrieu." *Revue de Geographie de Lyon*, 53 (1978): 7-21.
Lerat, S. "Collecte et commercialisation des cereales en France." *L'Information Geographique*, 42, No. 1 (1978): 15-27.
Melelli, Alberto, and Renata Perari. "Le developpment du vignoble specialise dans la region ambrienne." *Geographia Polonica*, 38 (1978): 193-206.
Sinn, Peter. "Zur Krise des Weinbaus im Languedoc." *Zeitschrift fur Wirtschaftsgeographie*, 22, No. 8 (1978): 225-232.
Cleach, A., and N. Piriou. "La culture de l'endive dans les Nord-Finistere." *Norois*, 26 (1979): 391-399.
Arnauld, Evelyne. "Le tabac dans l'Est du Bossin Parisien: L'etude de la diffusion d'une culture marginale." *Revue Geographique de l'Est*, 20, No. 1-2 (1980): 15-42.
Cadine, P. "Les producteurs de tabac dans les campagnes du Sud-Ouest de la France." *Revue Geographique des Pyrenees et du Sud-Ouest*, 51, No. 4 (1980): 391-410.
Plet, F. "Localisation et fluctuation de la production de graines de semences en France." *L'Information Geographique*, 45, No. 3 (1981): 104-114.
Cabelguenne, Maurice. "Mais, tournesal, soja, sorghas: Quatre cultures d'ete irriguees dans le sud-ouest de la France." *Revue Geographique des Pyrenees et du Sud-Ouest*, 53, No. 2 (1982): 191-208.
Blondeau, P. "Un vignoble menace: Valencay." *Norois*, 30 (1983): 245-262.
Dorel, Gerard. "Le grande agriculture betteraviere et luzerniere champenoise: Un systeme partaitement adopte au marche commun agricole european." *Travaux-Institut de Geographie de Rheims*, 55-56 (1983): 115-126.
Fourdrignier, M. "Les endiviers du Cambresis." *Hommes et Terres du Nord*, 2 (1983): 69-72.
Vaudois, J. "Mutations et developpment des productions legumieres et maraicheres dans la region Nord-Pas-de-Calais." *Hommes et Terres du Nord*, 2 (1983): 36-64.
Vaudois, J. "Le complexe agro-industrial legumier du Violaine-Lorgies." *Hommes et Terres du Nord*, 2 (1983): 65-68.
Roudie, P. "Le vignoble bordelais: Un monde en mouvement." *Revue Geographique des Pyrenees et du Sud-Ouest*, 55, No. 3 (1984): 337-353.
Sabalcagaray, M.D. "Vin de Cahors et prosperite dans le Lot." *Revue Geographique des Pyrenees et du Sud-Ouest*, 55, No. 3 (1984): 313-335.
Bartoli, P. "L'impact des primes Tarrachage dans le vignoble Languedocien: Resultats recents." *Bulletin de la Societe Languedocienne de Geographie*, 16, No. 1 (1986): 65-80.
Beringuier, C. "La viticulture dans le Midi pyreneen." *Revue Geographique des Pyrenees et du Sud-Ouest*, 57, No. 1 (1986): 73-88.
Blondeau, P. "Vouvray et Montlouis: Deux vignobles tourangeaux dynamiques." *Norois*, 33 (1986): 365-387.

Couderc, Raymond. "Les vignes de la capitale du vin: Viticulture et viticulteurs a Beziers." *Bulletin de la Societe Languedocienne de Geographie*, 16, No. 4 (1986): 487-498.

Astruck, H., et al. "De la cartographie de la vegetation a la noissance d'un enu: Le Cotes de la Maleperre." *Revue Geographique des Pyrenees et du Sud-Ouest,* 58, No. 3 (1987): 259-273.

Baxevanis, John. *The Wines of Champagne, Burgundy, Eastern and Southern France*. Totowa, N.J.: Rowman & Littlefield, 1987.

Deslondes, O. "Vingt ans d'amenagement en Cevenne languedocienne (1967–1987): Bilan et perspectives." *Bulletin de la Societe Languedocienne de Geographie*, 21, No. 3-4 (1987): 405-418.

Durbiano, Claudine, and Andre de Reparaz. "Wein und Weinbau in Frankreich." *Geographische Rundschau*, 39, No. 12 (1987): 688-692.

Durbiano, Claudine. "La vigne et le vin en France: Tendances recentes et transformations." *L'Information Geographique*, 51, No. 5 (1987): 177-189.

Mayberry, Robert. *Wines of the Rhone Valley*. Totowa, N.J.: Rowman & Littlefield, 1987.

Vaudois, J. "Les filieres agro-alimentaires face aux mutations de leur environnement: La dynamique recente des filieres fleurs et legumes de serre aux Pays-Bas." *Revue de Geographie de Lyon*, 62 (1987): 387-410.

De Reparaz, Andre. "Un vignoble mediterraneen francais: L'exemple du Var." *Mediterranee*, 65, No. 3 (1988): 21-27.

Moustier, Philippe. "Le vignoble des Coteaux-d'Aix-en-Provence." *Mediterranee*, 65, No. 3 (1988): 13-20.

Roudie, P. *Vignobles et vignerans du Bordelais, 1850–1980*. Paris, France: Centre National de la Recherches Scientifique, 1988.

Archias, C. "Aspects et perspectives du vignoble provencal." *Bulletin-Societe de Geographie de Marseilles*, 89, No. 18 (1989): 47-52.

Bastie, Jean. "Paris et le vin." *Acta Geographica*, 81 (1990): 43-48.

Bonnaud, Pierre. "Considerations geohistoriques, actuelles et prospectives sur les vignobles d'Auvergne." *Travaux-Universite de Poitiers, Centre de Geographie Humaine et Sociale*, 17 (1990): 497-512.

Fenelon, Pierre. "La culture du tabaco en Perigard." *Travaux-Universite de Poitiers, Centre de Geographie Humaine et Sociale*, 17 (1990): 513-519.

Limouzin, P. "L'evolution des systemes de culture et des structures d'exploitation (1979–1988) en Picardie." *Hommes et Terres du Nord*, 4 (1990): 239-247.

Delcroix, O., and P. Limouzin. "La filiere pomme de terre en Picardie." *Hommes et Terres du Nord*, 4 (1991): 281-283.

Huetz de Lemps, Alain. "Le vin de Benicarlo a Bordeaux au XVIIIe siecle." *Estudios Geograficos*, 51, No. 200 (1991): 479-488.

Blondeau, P. "Lavernir des verges de la Sarthe." *Norois*, 38 (1992): 131-144.

Rouselle, Patrick, Daniel Ellisseche, and Francoise Rousselle. "La pomme de terre." *Les Cahiers d'Outre Mer*, 45 (1992): 303-314.

Crowley, William K. "Changes in the French Winescape." *Geographical Review*, 83, No. 3 (1993): 252-268.

Moran, Warren. "The Wine Appelation as Territory in France and California." *Annals of the Association of American Geographers*, 83, No. 4 (1993): 694-717.

Tezuka, A. "Recent Changes in French Viticulture Areas" (in Japanese). *Tsukuba Studies in Human Geography*, 17 (1993): 33-48.

Rouyres, T. "La cueillete a la ferme: Diversification ou nouvelle activite dans les zones periurbaines: L'exemple de l'ile-de-France." *Bulletin de l'Association de Geographes Francais*, 71, No. 2 (1994): 216-222.

Vaudois, J. "Les zones maraicheres periurbaines: Espaces residuels ou nouvelles formes d'integration des espaces agricoles aux strategies urbaines?" *Bulletin de l'Association de Geographes Francais*, 72, No. 2 (1994): 123-134.

Davey, Charles. "La conchyliculture en France." *L'Information Geographique*, 59 (1995): 98-112.

Bachelier, Eric. "L'emergence des grands domaines du vignoble de Sevre-et-Maine." *Cahiers Nantais*, 46 (1996): 51-68.

Roudie, P. "'Bordeaux': Un modile pour la viti-viniculture mondiale?" *Les Cahiers d'Outre Mer*, 50 (1997): 403-422.

Schirmer, Raphael. "Le vignoble nantais et la naissance d'une qualite." *Acta Geographica*, 114 (1998): 49-64.

Brochot, A. "La qualite, nouvelle arme de la guerre de position dans le vignoble de Champagne." *Sud-Ouest Europeen*, 6 (1999): 21-30.

Hinnewinlal, J.C. "Terroirs et 'qualite des vins': Quels liens dans les vignobles du nord de l'Aquitaine." *Sud-Ouest Europeen*, 6 (1999): 9-20.

Cerveau, M.P. "Le egarements de la region viticole bourguignonne ou les tribulations d'un bateauivre." *Annales de Geographie*, 109, No. 4-5 (2000): 444-458.

Legouy, F. "La renaissance du vignoble des Hautes Cotes de Beaune et des Hautes-Cotes de Nuits." *Annales de Geographie*, 109, No. 4-5 (2000): 459-471.

Rejalot, M. "Bordeaux-Champagne: Propriete ou negou?" *Annales de Geographie*, 109, No. 4-5 (2000): 426-443.

Roudie, P. "Vous a vez di 'chateau'? Essai sur le succes semantique d'un modele viticole venu du Bordelais." *Annales de Geographie*, 109, No. 4-5 (2000): 415-425.

Schirmer, Raphael. "Le regard des geographes francais sur la vigne et le vin (fin du XIXe-XX siecle)." *Annales de Geographie*, 109, No. 4-5 (2000): 345-363.

Unwin, Timothy. "The Viticultural Geography of France in the 17th Century According to John Locke." *Annales de Geographie*, 109, No. 4-5 (2000): 395-414.

ANIMAL HUSBANDRY

Lefebrve, T. "La transhumance dans les Basses-Pyrenees." *Annales de Geographie*, 37 (1928): 35-60.

Wolkowitsch, M. "L'elevage dans le Bazois." *Annales de Geographie*, 55 (1946): 202-210.

Salome, M. "L'elevage au Libon." *Revue de Geographie de Lyon*, 30 (1955): 82-102.

Sivignon, M. "Elevage et embouche en Charolais-Brionnais." *Revue de Geographie de Lyon*, 35 (1960): 357-379.

Bertrand, G. "Vie pastorale et industrie dans le Val d'Ariege." *Revue Geographique des Pyrenees et du Sud-Ouest*, 32 (1961): 63-73.

Perrier, A. "L'elevage du mouton de plein air dans la Basse-Marche Carrondis, de Bellac." *Norois*, 8 (1961): 213-219.

de Falguerolles, G. "Contribution a l'analyse de l'economie agricole en Haut-Languedoc: Le troupeau ovin en metayage au XVIIIe siecle." *Revue Geographique des Pyrenees et du Sud-Ouest*, 34 (1963): 368-386.

Fischer, Andre. "L'elevage bovin en France." *L'Information Geographique*, 29, No. 4 (1965): 164-170.

Gay, F. "Structure agraire et elevage en Champagne du Berry." *Acta Geographica*, 56, No. 2 (1965): 30-33.

Cabouret, Michel. "L'evolution de la vie pastorale dans la vallee de l'Otta." *Revue de Geographie Alpine*, 54 (1966): 631-684.

Bergeron, R. "Problemes de la vie pastorale en Sandaigne." *Revue de Geographie de Lyon*, 42 (1967): 311-328.

Carron, Marie-Antoinette. "Problemes de l'elevage en Creuse." *Acta Geographica*, 68 (1967): 22-26.

Vidal, Christine. "Les aspects recents de l'elevage porcin dans les Bouches-du-Rhone." *Acta Geographica*, 73 (1968): 16-22.

Boichard, J. "La vian de bovine: Sa production et ses problemes en France." *Revue de Geographie de Lyon*, 45 (1970): 325-356.

Gardelle, Charles. "Impressions pastorales corse (ete 1969)." *Revue de Geographie Alpine*, 58 (1970): 383-388.

Le Rhun, P.Y. "La mutation de l'elevage et ses consequences: L'exemple de la region de Lamballe." *Norois*, 17 (1970): 217-234.

Solle, Henriette. "L'aviculture et l'elevage des lapins dans le Gatinais." *Acta Geographica*, 1, Serie 3 (1970): 13-33.

Saussol, Alain. "Les parcours pastoraux en Languedoc oriental." *Bulletin de la Societe Languedocienne de Geographie*, 5 (1971): 429-439.

Dietrich, Genevieve. "L'elevage bovine et le lait dans le departement des Vosges." *Acta Geographica*, 10-11, Serie 3 (1972): 79-114.

Gardelle, Charles. "La vie pastorale dans les Alpes Francaise." *Revue de Geographie Alpine*, 61 (1973): 449-461.

Aubry, M.I. "Les marches du betail en Bretagne." *Geographia Polonica*, 29 (1974): 251-259.

Renucci, J. "L'evolution de l'elevage en Corse depuis la mise en valeur agricole." *Cahiers Geographiques de Rouen*, 5 (1976): 29-39.

Pluvinage, J. "Geographie de l'elevage bovin en evolution." *Comptes Rendes Hebdomadaires des Seances de l'Academic l'Agriculture de France*, 63, No. 13 (1977): 1244-1250.

Gardelle, Charles. "Evolutions recente de la transhumance ovine dans les Alpes de 1964 a 1976." *Revue de Geographie Alpine*, 66, No. 2 (1978): 211-213.

Metaille, J.P. "Les incendies pastoraux dans les Pyrenees centrales." *Revue Geographique des Pyrenees et du Sud-Ouest*, 49, No. 4 (1978): 517-526.

Rinschede, Gisbert. "Die Rindertranshumance in den franzoisichen Alpen." *Mitteilungen der Osterreichischen Geographischen Gesellschaft*, 120, No. 1 (1978): 74-98.

Rinschede, Gisbert. *Die Transhumance in den franzosischen Alpen undin den Pyrenaen*. Munster, Germany: Westfalische Geographische Studien, Heft 32, 1979.

Vergneau, G. "L'elevage bovin et les transformations recentes des systemes de productions agricoles en Deux-Sevres." *Norois*, 26 (1979): 161-180.

Clairacas, J. "Gavage des oies et canards et production de gros dans les exploitations agricoles en polyculture de Chalosse (Landes)." *Revue Geographique des Pyrenees et du Sud-Ouest*, 51, No. 4 (1980): 441-463.

Aubry-Breton, M.L. "Departments et marches du betail en Bretagne." Pp. 159-164 in *L'assiette rurale des departements francais: Actes du colloque de Poitiers, 1981*. Poitiers, France: Travaux du Centre de Geographie Humaine et Sociale, Universite de Poitiers, No. 6, 1981.

Vaudois, J. "La 'crise porcine' dans la region Nord-Pas-de-Calais." *Hommes et Terres du Nord*, 1 (1981): 32-39.

Coquart, D., and L. Mazenc. "L'approvisionnement de l'agglomeration de Toulouse en viande bovine de l'entre-deux-guerres a nos jours." *Revue Geographique des Pyrenees et du Sud-Ouest*, 56, No. 1 (1985): 87-103.

Vivier, M., and C. Douyer. "Evolutions de la geographie pastorale du Pays de Caux." *Etudes Normandes*, 2 (1985): 37-47.

Allaine, G., and F. Bertocchio. "Systemes d'elevage laitier en montagne pyreneenne." *Revue Geographique des Pyrenees et du Sud-Ouest*, 57, No. 1 (1986): 145-161.

Cleary, M.C. "Patterns of Transhumance in Languedoc." *Geography*, 71, No. 1 (1986): 25-33.

Coquart, D., and L. Mazenc. "La production de viande bovine en Midi-Pyrenees et en Bretagne: Un ressort de la differenciation regionale." *Revue Geographique des Pyrenees et du Sud-Ouest*, 58, No. 3 (1987): 245-258.

Maurel, M.C. "Modernisation agro-pastorale et strategies d'integration des eleveurs du Causse Noir." *Bulletin de la Societe Languedocienne de Geographie*, 18, No. 3-4 (1987): 389-404.

Bertocchio, F., A. Cabannes, and J.C. Flamont. "Transformation d'une activite d'elevage en vallee d'Aspe et differenciation de la composition genetique du cheptel bovin." *Revue Geographique des Pyrenees et du Sud-Ouest*, 60, No. 2 (1989): 141-171.

Coquart, D., F. Gary, and L. Mazenc. "La filiene 'broutards' en Midi-Pyrenees: Une articulation effiase entre production et commercialisation de boivins jeuns." *Revue Geographique des Pyrenees et du Sud-Ouest*, 61, No. 1 (1990): 29-45.

Fels, A. "Le paysage agro-pastorale villageois du Massif Central." *Bulletin de l'Association de Geographes Francais*, 68, No. 1 (1991): 37-44.

Chevallier, Denis. "Des vaches, des moutons et des poules ou de quelques usages contemporaires du patrimoine vivant." *Revue de Geographie Alpine*, 86, No. 4 (1998): 61-70.

DAIRYING

Lahardie, R. "L'elevage laitier dans la region d'Aurillac." *Acta Geographica*, 72 (1968): 2-17.

Guermond, Yves. "Deux programmes de simulation." *Cahiers Geographiques de Rouen*, 4 (1975): 21-48.

Charles, J. "La collecte laitiere en Bearn et Pays Basque: Concentration du capital, en cadrement des producteurs." *Revue Geographique des Pyrenees et du Sud-Ouest*, 50, No. 2 (1979): 227-250.

Flamont, J.C. "La production de lait de brebins le modele de Roquefort et son avenir en France meridionale." *Revue Geographiques des Pyrenees et du Sud-Ouest*, 54, No. 3 (1983): 327-340.

Naylor, E.L. "Milk Quotas and the Changing Pattern of Dairying in France." *Journal of Rural Studies*, 9, No. 1 (1993): 53-63.

Lizot, Bernadotte. "Le genie des alpages: Paysage, vacher, fromage en abondance." *Revue de Geographie Alpine*, 86, No. 4 (1998): 35-50.

Delfosse, Clair. "Interactions entre qualites et territoires: L'exemple des bries." *Sud-Ouest Europeen*, 6 (1999): 41-50.

Ricard, Daniel. "Filieres de qualite et ancrage au terroir: La delimitation des zones d'AOC fromageres." *Sud-Ouest Europeen*, 6 (1999): 31-40.

FARM LAND

Biegajlo, Wladislaw. "From the Study of Land Utilization in Provence, France: The Commune of Banou" (in Polish). *Przeglad Geograficzny*, 37, No. 1 (1965): 109-141.

Bell, Timothy A. "The Alsatian Peasant and His Land: A Study in Land Tenure Change." *Yearbook, Association of Pacific Coast Geographers*, 28 (1966): 75-96.

Reitel, Francois. "A propos de l'open field lorrain." *Revue Geographique de l'Est*, 6, No. 1-2 (1966): 29-52.

Perpillou, A. "L'utilisation agricole du sol dans les bocages du Limousin." Pp. 332-343 in *Melanges de geographie physique, humaine, economique, applique, offerts a M. Omer Tulippe*, vol. 1, edited by Jose A. Spork. Gembloux, Belgium: J. Duculot, 1967.

Gay, F. "Economie et formation d'un type de paysage (a propos de la Champagne du Berry)." *Geographische Zeitschrift. Erdkundliches Wissen, Beiheft*, 18 (1968): 38-43.

Sivignon, M. "Les pasteurs du Pinde septentrional." *Revue de Geographie de Lyon*, 43 (1968): 5-43.

Somme, J. "L'utilisation du sols et la pedologie a la bardune orientale de la Thierache." *Acta Geographica*, 71 (1968): 16-26.

Clout, Hugh D., and Keith Sutton. "The Cadastre as a Source for French Rural Studies." *Agricultural History*, 43, No. 2 (1969): 215-224.

Clout, Hugh D. "The Increase in the Grassland Area of the Pays de Bray." *Erdkunde*, 23, No. 1 (1969): 20-29.

Jorre, Georges. *Le terrefort touloiseain et lauragois: Histoire et geographie agraires*. Toulouse, France: E. Privot, 1971.

Fosset, R. "L'evolution de la propriete en Chaouin de 1962 a 1969." Pp. 23-26 in *Les Societes Rurales Mediterraneennes: Actes du Colloque de Geographie Agraire*, edited by Centre Geographique d'Etudes et de Recherches Mediterraneennes. Aix-en-Provence, France: Institut de Geographie, 1972.

Mace, Georges. "La terre et les marchandes de bestiaux dans le departement de la Mayenne." *Norois*, 19 (1972): 481-500.

Vaudois, J. "Un exemple de permance d'un porcellaire agricole en milieu suburbain: L'aire maraichere de Rosendael." *Bulletin de l'Association Geographes Francais*, 49 (1972): 109-115.

Dubosco, J.P. "Une nouvelle agriculture dans les landes de Gascogne." *Revue Geographique des Pyrenees et du Sud-Ouest*, 44 (1973): 185-206.

Guermond, Yves, and Jean-Pierre Massias. "L'utilisation agricole du sol en France: Deux methodes de traitement de l'information." *L'Espace Geographique*, 2, No. 4 (1973): 267-274.

Clout, Hugh D. "Agricultural Plot Consolidation in the Anvergne Region of Central France." *Norsk Geografisk Tidsskrift*, 28 (1974): 181-194.

Guermond, Yves, and Jean-Pierre Massias. "L'evolution de utilisation agricole du sol en Normandie." *Geographia Polonica*, 29 (1974): 193-204.

Baysong, Jean-Claude. "L'utilisation agricole du sol en Corse." *Acta Geographica*, 24 (1975): 47-50.

Flatres, Paul. "L'evolution bocages: La region Bretagne." *Norois*, 26 (1979): 303-320.

Cooper, Sandra H. "Vanishing Farmlands: An Application of Transferable Development Rights to Control Agricultural Land Conversions in the St. Vrain Valley." Master's Thesis, University of Colorado, Boulder, Colo., 1980.

Solle, Henriette. "L'utilisation agricole du sol en France: Les cartes Aime Perpillon." *Acta Geographica*, 45 (1981): 1-25.

Dehaese, H., et al. "Agriculture et bocage a Sains-du-Nord: Donnees et commentaires partiels." *Hommes et Terres du Nord*, 4 (1983): 16-19.

Mace, Georges. "La terre, les agriculteurs, et les problemes fonciers dans une region des marges armoricaines." *Norois*, 30 (1983): 213-226.

Roudie, P. "Le restructuration fonciere en Aquitaine du Nord." *Bulletin de l'Association Geographes Francais*, 60, No. 5 (1983): 269-275.

Bryant, Christopher R. "The Recent Evoluton of Farming Landscapes in Urban-Centred Regions." *Landscape Planning*, 11 (1984): 307-326.

Durbiano, Claudine. "Extension et contraction des terres cultivees en France: Vers une organisation socio-economique dualiste du monde agriculture." *Geographia Polonica*, 49 (1984): 153-160.

Viel, J.M. "Le Sud-Ouest, terre d'accueil pour les agriculteurs: Enquetes dans les Landes et le Gers." *Revue Geographique des Pyrenees et du Sud-Ouest*, 56, No. 1 (1985): 63-85.

Bryant, Christopher R. "Farmland Conservation and Farming Landscapes in Urban-Centred Regions: The Case of the Ile-de-France Region." *Land and Urban Planning*, 13 (1986): 251-276.

Brunet, P. "L'evolution recente des systemes d'utilisation du sol dans la France du Nord-Ouest." Pp. 117-124 in *Regions, villes, et amenagement*. Paris, France: CREPIF/Societe de Geographie de Paris, 1987.

Debarbieux, R. "Propriete des alpagrez structures sociales et organisation de l'espace agraire: Clavous et Besse en Oisans (1830–1986)." *Revue de Geographie Alpine*, 75, No. 1 (1987): 43-62.

Petit, F., C. Cosandey, and T. Maxart. "Defrichement des terres agricoles et risque erosif: Un exemple du Sud du Massif central francais." *Bulletin de la Societe Languedocienne de Geographie*, 18, No. 3-4 (1987): 219-228.

Brun, J.P. "Des terrasses de culture massaliotes sur l'ile de Porquerolles?" *Mediterranee*, 71, No. 3-4 (1990): 13-16.

De Reparaz, A. "La culture en terrasses: L'expression de la petite paysanneurie mediterraneenne tradionnelle." *Mediterranee*, 71, No. 3-4 (1990): 23-30.

Jorda, M., and M. Provansal. "Les terrasses de culture dans le bilon erosif mediterraneen: L'exemple du Vallot de Monsieur (Basse-Provence)." *Mediterranee*, 71, No. 3-4 (1990): 55-61.

Meffre, J.C. "Habitats augusteens et amenagements des versants: Seguret (Vauchuse)." *Mediterranee*, 71, No. 3-4 (1990): 17-21.

Bariou, J.Y., and H. Quere. "L'aide au retrait des terres arables dans l'Ouest de la France." *Geographie Sociale*, 12 (1992): 209-213.

Maury, R. "La genese des paysages agraires en France: La Touraine en est-elle une synthese?" *Norois*, 39 (1992): 5-27.

Rauturier, C. "L'agrandissement des exploitations agricoles dans la region nantaise." *Geographie Sociale*, 12 (1992): 71-80.

Diry, J.P. "Periurbanisation, agriculture et gestion de l'espace: L'exemple des campagnes clermontoises." *Bulletin de l'Association de Geographes Francais*, 71, No. 2 (1994): 204-215.

Piriou, N. "Continuite et ruptures dans l'espace agricole de la communaute urbaine de Brest." *Bulletin de l'Association de Geographes Francais*, 71, No. 2 (1994): 187-195.

Vaudois, J. "Les zones maraicheres periurbaines: Espaces residuels ou novrelles formes d'integration des espaces agricoles aux strategies urbaines?" *Bulletin de l'Association de Geographes Francais*, 71, No. 2 (1994): 123-134.

AGRICULTURAL POPULATIONS AND SOCIAL ISSUES

Meynier, A. "La crise agricole dans le Cantal et ses consequences sociales." *Annales de Geographie*, 40 (1931): 414-419.

Rochefort, Michel. "La penetration des capitaux bourgeois dans la campagne autunoise, ser consequences sur l'habitat et la structure agrairie." *Revue de Geographie de Lyon*, 25, No. 4 (1950): 249-266.

Bertrand, G. "Societe rurale et reamenagement agricole dans le Segala tornais." *Revue Geographiques des Pyrenees et du Sud-Ouest*, 33 (1962): 137-161.

Livet, R. *Habitat rural et structures agraires en Basse-Provence*. Aix-en-Provence, France: Ophrys, 1962.

DeRousseau, B. "La cooperative agricole lauragaise." *Revue Geographique des Pyrenees et du Sud-Ouest*, 35 (1964): 214-220.

Colin, Georges, and Andre Blanc. "Extension et declin de l'assolement triennal au XIXe siecle: L'evolution sociale d'une commune lorraine." *Revue Geographique de l'Est*, 6 (1966): 89-96.

Beguin, Hubert. "Le cout de la densite de population: L'exemple de l'encadrement agricole." Pp. 394-400 in *Melanges de geographie physique, humaine, economique, applique, offerts a M. Omer Tulippe*, Vol. II, edited by Jose A. Sporck and Bernadette Schoumaker. Gembloux, Belgium: J. Duculot, 1967.

Perry, Peter J. "The Structural Revolution in French Agriculture: The Role of 'Societies d'Amenagement foncier et d'Etablissement rural." *Revue de Geographie de Montreal*, 23, No. 2 (1969): 137-151.

Bontron, J.C. "Evolution des systems de production et besoins de main-d'oeuvre dans l'agriculture francaise." *Geographia Polonica*, 29 (1974): 181-191.

Cabouret, Michel. "Apercus nouveaux sur l'agriculture de la Lorraine du Nord-Est: Les repercussions de la division linguistique de departement de la Mosella." *Mosella*, 5, No. 4 (1975): 1-58.

Vitte, P. "Les agriculteurs 'migrants' en Creuse." *Norois*, 23 (1976): 509-527.

Pletsch, Alfred. "Landwirtschaftliche Sanierungs- und Meliorationsmassnahmen im Languedoc unter besonderer Berucksichtigung der Ansiedlung chemaliger Koloniallandwirte." *Dusseldorfer Geographische Schriften*, 7 (1977): 93-112.

Schwab, R. "Les mutations de la societe agricole du grand Ried Alsacien." *Recherches Geographiques a Strasbourg*, 11 (1979): 27-43.

Vergneau, G. "La population agricole, critere de differenciation des espaces ruraux: L'exemple des Deux-Sevres." *Norois*, 26 (1979): 383-390.

Bernard, Marie Claude, and M.C. Maurel. "La pluriactivite de la population agricole francaise: Dimension geographiques et socio-professionnelles." *Bulletin de la Societe Languedocienne de Geographie*, 17, No. 1-2 (1983): 205-240.

Naylor, E.L. *Socio-Structural Policy in French Agriculture*. Aberdeen, Scotland: University of Aberdeen Department of Geography, O'Dell Memorial Monograph No. 18, 1985.

Baker, Alan R.H. "Les syndicats agricoles de la vallee de la Cisse (1883–1914)." *Vallee de la Cisse*, 8 (1986): 59-71.

Massou, F., and P. Roux. "Femmes d'agriculteurs, agricultrices et co-exploitantes en Lomage: Les significations de l'identile." *Revue Geographique des Pyrenees et du Sud-Ouest*, 57, No. 1 (1986): 103-112.

Rieu, A. "Agricultrices et professionnalisation: Une enqueteen Midi-Pyrenees." *Revue Geographique des Pyrenees et du Sud-Ouest*, 57, No. 1 (1986): 89-101.

Stecktman-Labry, G. "Agriculture, amenagement rural: Les etudes ne sont plus ce quelles etaient." *Revue Geographique des Pyrenees et du Sud-Ouest*, 57, No. 1 (1986): 163-169.

Winchester, Hilary P.M. "Agricultural Change and Population Movements in France, 1892–1929." *Agricultural History Review*, 34, No. 1 (1986): 60-78.

Schweyer, Francois-Xavier. *Paysans d'aujourd en Lorraine*. Nancy, France: Presses Universitaires de Nancy, 1987.

Bernard, Gilles. "Population et vignoble dans les Chavertes." *Norois*, 35, No. 4 (1988): 521-541.

Chiffre, J. "Tradition et renouveau dans les agricultures familiales en Bourgogne: La culture des petits fruits en Cote-d'Or." Pp. 319-324 in *Campagnes et littoraux d'Europe: Melanges offerts a Pierre Flatres*. Lille, France: Institut de Geographie de Lille, Hommes et Terres du Nord, 1988.

Berlon, M. "Conocimientos y trajectorias socio-profesionales de las agricultures." *Documents d'Analis: Geografica*, 14 (1989): 53-71.

Cleary, M.C. *Peasants, Politicians and Producers: The Organization of Agriculture in France since 1918*. Cambridge, England: Cambridge University Press, 1989.

Eay, J.F. "Agriculture et tourisme en Perigard noir." *Geographie Sociale*, 12 (1992): 123-135.

Naylor, E.L. "Unionism, Peasant Protest and the Reform of French Agriculture." *Journal of Rural Studies*, 10, No. 3 (1994): 263-274.

ENVIRONMENTAL AND CONSERVATION ISSUES

Charaud, Anne-Marie. "L'habitat et la structure agraire de la Grande Briere et des mareis de Donges." *Annales de Geographie*, 57 (1948): 119-130.

Fels, A. "Le climat agricole et les limites altitudinales de l'occupation du sol dans le Massif Central." *Annales de Geographie*, 64, No. 6 (1955): 401-412.

Suchel, J.B. "L'hydraulique agricole dans le couloir rhodanien entre Vienne et Bollene: Projets et realisations." *Revue de Geographie de Lyon*, 32 (1957): 201-226.

Clout, Hugh D., and A.D.M. Phillips. "Fertilisants mineraux en France au XIXe siecle." *Etudes Rurales*, 45 (1972): 9-28.

Dubesset, P. "Choix agricoles et caracteres du climat dans la region du Rhone mayen." *Revue de Geographie de Lyon*, 47 (1972): 137-165; 297-326; 333-359.

Dujardin, Elisabeth. "Agriculture et pollution industrielle: L'exemple du Plateau de Lannemezan et de la Vallee de la Neste." *Revue Geographique des Pyrenees et du Sud-Ouest*, 49, No. 3 (1978): 403-414.

Callot, Henri-Jacques. *La plaine d'Alsace: Modele agraire et parcellaire*. Nancy, France: Universite de Nancy II, 1980.

Stevenson, Iain. "The Diffusion of Disaster: The Phylloxera Outbreak in the Department of the Herault, 1862–80." *Journal of Historical Geography*, 6, No. 1 (1980): 47-68.

Viel, J.M. "Presentation de l'agriculture biologique dans le Cot-et-Garonne." *Revue Geographique des Pyrenees et du Sud-Ouest*, 51, No. 4 (1980): 465-488.

Robie, M.C., et al. "Accident climatique et fonctionnement de la societe agricole: La secheresse de 1976 chez les eleveurs d'un canton de la Nievre." *L'Espace geographique*, 11, No. 2 (1982): 111-123.

Vivier, M., and C. Douyer. "Les prairies de la plaine alluviale de la Basse-Seine: Un espace agricole menance." *Etudes Normandes*, 3 (1984): 67-78.

Geissert, D., et al. "Systemes de production agricole et contraintes naturelles en Outre-Foret, Alsace du Nord, France." *Recherches de Geographiques a Strasbourgh*, 25-26 (1985): 19-38.

Sick, W.D. "Die alternativ-biologiische landwirtschaft in Elsass." *Recherches de Geographie de Strasbourgh*, 25-26 (1985): 75-81.

Diry, J.P. "Une politique agricole peri-urbaine: L'exemple de Lyon." *Revue de Geographie de Lyon*, 61 (1986): 387-407.

Legagneux, B., A. Veisblat, and A. Peyre. "La foret: Un atout pour les agriculteurs de Midi-Pyrenees." *Revue Geographique des Pyrenees et du Sud-Ouest*, 57, No. 1 (1986): 113-133.

Marchant, J.P., E. Bauquin, and P. Letondaux. "Les calamites agricoles d'origine atmospherique: Le cas de l'Ille-et-Vilaine." *Revue de Geographie de Lyon*, 61 (1986): 319-328.

Auzet, V. "L'erosion des sols cultives en France sous l'action du ruissellement." *Annales de Geographie*, 96 (1987): 529-556.

Wicherek, S. "Viticulture and Soil Erosion in the North of the Parisian Basin. Example: The Mid Aisne Region." *Zeitschrift fur Geomorphologie*, Supplement Band 83 (1991): 115-126.

Arlaud, S. "L'agriculture biologique: Archaisme ou innovation? L'exemple du Poitou-Charentes." *Geographie Sociale*, 12 (1992): 81-90.

Bonnefort, Jean C. "Reflexion sur la pollution agricole en Lorraine." *Revue Geographique de l'Est*, 32, No. 1 (1992): 25-42.

Berenguer, Jeanne. "Les dommages causes a l'agriculture la Vaucluse et dans le sud de la Drome de 22 septembre 1992." *Revue de Geographie de Lyon*, 68, No. 2-3 (1993): 171-174.

Gouin-Leveque, P. "Access to Soils and Training of Farmers in the Perche (Orne Department)." *Norois*, 40 (1993): 431-442.

Josien, E., and M. Tort. "Impact de l'environnement agricole sur un tourbiere de mar: Consequences pour sa protection." *Bulletin de l'Association de Geographes Francais*, 71, No. 3 (1994): 324-336.

Vivier, M., and D. Leconte. "Flore et vegetation: Reflet de la rupture des pratiques agronomiques." *Norois*, 41 (1994): 691-703.

Jauneau, J.C. "Objectifs environnementaux et objectifs de soutien a l'agriculture: Une convergence possible dans les zones de moyenne montagne

menacees de deprise?" *Revue de Geographie Alpine*, 83, No. 3 (1995): 83-100.
Laurens, Lucette. "Les Parces naturals: Du concepts a la pratique d'une agriculture environnmentals." *Annales de Geographie*, 104 (1995): 339-359.
Penven, M.J., and T. Muxant. "Le drainage agricole: Un role fondamental dans les transferts d'eau et sur les transports solides. L'exemple des Pyrenees-Orientales (France)." *Annales de Geographie*, 104 (1995): 88-104.
Demondolx Dedans, J., and G. Bagarre. "La Crau: Du com dur ancien au carrefour des concepts agri-environnementaux." *Mediterranee*, 83, No. 1-2 (1996): 19-26.
Wainwright, J. "Infiltration, Runoff and Erosion Characteristics of Agricultural Land in Extreme Storm Events, Southeastern France." *Catena*, 26, No. 1-2 (1996): 27-48.
Benoit, Guillaume, and Gilles Valero. "Agriculture et environnement dans le parc national des Cevenues: Le concement de l'operation 'agneaux de parcours." *Bulletin de la Societe Languedocienne de Geographie*, 28, No. 3-4 (1997): 101-122.
Salvi, Isabelle. "Les Parcs naturels regionaux et la gestion des espaces agricoles et forestiers: Une logique de projet et un etat d'esprit." *Bulletin de la Societe Languedocienne de Geographie*, 28, No. 3-4 (1997): 81-100.
Battershill, Martin R.J., and Andrew Gilg. "Traditional Low Intensity Farming: Evidence of the Role of Vente Directe in Supporting Such Farms in Northwest France, and Some Implications for Conservation Policy." *Journal of Rural Studies*, 14, No. 4 (1998): 475-486.
Jauneau, Jean-Claude, and Olivier Roque. "Les operations locales agri-environnementales en Rhone-Alpes: Analyse typologique de leur mise en plauen de leur impact." *Revue de Geographie Alpine*, 87, No. 4 (1999): 35-48.

IRRIGATION

Peyre, Marius. "Les irrigations de la basse Durance." *Annales de Geographie*, 36 (1927): 40-57.
Brunet, R. "Ou en est l'irrigation des coteaux de Gascogne?" *Revue Geographique des Pyrenees et du Sud-Ouest*, 31 (1960): 223-228.
Bethemont, J. "Progres techniques et reactions paysannes: L'irrigation dans la plaine de Montelimar." *Revue de Geographie de Lyon*, 36 (1961): 347-377.
Graves, N.J. "Une Californie Francaise: The Languedoc and Lower Rhone Irrigation Project." *Geography*, 50, No. 1 (1965): 71-73.
Martin, Jacques. "L'evolution de l'irrigation en France." *Revue de Geographie Alpine*, 60 (1972): 419-443.
Dubosco, J.P. "La decomposition de la paysannerie et le developpement de l'irrigation en Aquitaine." *Revue Geographique des Pyrenees et du Sud-Ouest*, 50, No. 2 (1979): 199-226.
Vassas, Robert, and Jacques Vigneron. "L'amenagement hydro-agricole de la Petite Camarque et l'assainissement des morais de le Soutegronne." *Bul-*

letin de la Societe Languedocienne de Geographie, 14, No. 2-3 (1980): 267-198.

Faggi, P. "Pour une geographie des grands projets d'irrigation dans les terres seches des pays sous-developpes." *Revue de Geographie de Lyon*, 61 (1986): 5-17.

Popp, Herbert. "L'agriculture irrigee dans la vallee du Sauss: Formes et conflits d'utilisation de l'eau." *Mediterranee*, 59, No. 4 (1986): 33-48.

Martin, G. "Aides et developpement dans les regions defavarisces: La petite irrigation moderne dans l'arriere-pays varois." *Mediterranee*, 67, No. 1 (1989): 33-38.

AQUACULTURE

Doumenge, F. "L'aquaculture francaise: Bilon et perspectives." *Norois*, 31 (1984): 77-96.

van Tilbeurgh, V. "The Oyster-Farms in the Bourgneuf." *Norois*, 34 (1987): 43-50.

Borel, G. "Les conflits d'usage liesaux mutations de l'aquaculture marine sur le littoral atlantique francais: L'exemple du Finistere." *Cahiers Nantais*, 47-48 (1997): 93-101.

AGRICULTURAL MARKETING AND DISTRIBUTION

Galtier, G. "Les cooperatives viticoles du departement de l'Herault." *Bulletin de la Societe Languedocienne de Geographie*, 9 (1938): 33-50.

Delbos, C., and M. Furestier. "Caves cooperatives et commercialisation du vin en Sommierois et Vaunage." *Bulletin de la Societe Languedocienne de Geographie*, 4, No. 2 (1970): 167-182.

Guellec, A. "Les cooperatives de production cerealiere: Laitiere et de vian des bovine et porcine dans les Cotes-du-Nord." *Norois*, 18 (1971): 87-116.

Henniquau, T. "Un exemple de cooperation agricole: La 'Prosperite' fermiere." *Hommes de Terres du Nord*, 1-2 (1972): 27-51.

Canevet, C. "Les modifications dans l'organisation economique de l'agriculture en Bretagne au travers de l'evolution du mouvement cooperatif." *Geographia Polonica*, 29 (1974): 285-294.

Peyon, Jean-Pierre. "La cooperation agricole en France." *Geographia Polonica*, 29 (1974): 271-284.

Pilleboue, Jean. "Tendences recentes dans l'agriculture de Mid-Pyrenees: Vers une nouvelle geographie des revenues." *Revue Geographique des Pyrenees et du Sud-Ouest*, 50, No. 2 (1979): 145-198.

Nicoud, Mireille. "Le Credit Agricole de l'Isere dans le canton de Saint-Marcellin." *Revue de Geographie Alpine*, 71, No. 1 (1983): 95-107.

Crumbach, S. "Les echanges agro-alimentaires franco-saarois." *Mosella*, 17 (1987): 145-155.

Diry, J.P. "Geographie des filienes agro-alimentaires." *Revue de Geographie de Lyon*, 62 (1987): 285-289.

Derruau, Max. "Une union de cooperatives viticoles: Le CEPRO du Haut Biterrois." Pp. 331-334 in *Campagnes et littoraux d'Europe: Melanges offerts a Pierre Flatres*, edited by Jean-Michel Dewailly and Raymond Dion. Lille, France: Institut de Geographie de Lille: Hommes et Terres du Nord, 1988.

Canevet, C. "Les enterprises agro-alimentaires en Bretagne en 1991." *Norois*, 38 (1991): 449-460.

Gilg, Andrew, and Martin R.J. Battershill. "The Role of Household Factors in Direct Selling of Farm Produce in France." *Tijdschrift voor Economische en Sociale Geographie*, 90, No. 3 (1999): 312-319.

Lignon-Darmaillac, S. "Le marche du Xeres en France." *Annales de Geographie*, 109, No. 4-5 (2000): 505-515.

TRANSPORTATION AND MECHANIZATION

Bertrand, Albert. "Problemes d'equipement et de modernisation dans l'agriculture de la region de l'Ornain moyen." *Revue Geographique de l'Est*, 2, No. 1 (1962): 121-158.

Delenne, M., D. Faudrey, and M. Parodi. "Les agriculteurs de Sud-Est face au progres technique: Presentation generale." *Revue de Geographie de Lyon*, 41 (1966): 177-196.

Bonnamour, J., Yves Guermond, and Chantel Gillette. "Typologie des systemes d'exploitation agricole utilises en France: Essai methodologique." *Annales de Geographie*, 80 (1971): 144-166.

Solle, Henriette. "Commercialisation de produits agricoles et transports ruraux dans le Gatinais." *Acta Geographica*, 8, Serie 3 (1971): 257-294.

Baker, Alan R.H. "Adjustment to Distance Between Farmstead and Field: Some Findings from the Southwestern Paris Basin in the Early Nineteenth Century." *Canadian Geographer*, 17 (1973): 259-275.

Sutton, Keith S. "A French Agricultural Canal: The Canal de la Sauldre and the Nineteenth Century Improvement of the Sologne." *Agricultural History Review*, 21, No. 1 (1973): 51-58.

MacArthur, J.A. "Machines agricoles et transformations sociales: Access au material agricole en CUMA et organisations agricoles dans deux cantons de Haute-Garonne." *Revue Geographique des Pyrenees et du Sud-Ouest*, 51, No. 4 (1980): 414-439.

Genevois, Louis. "L'agriculture francaise peut-elle acquerir son autonomie energetique." *Norois*, 28 (1981): 281-287.

Maury, R. "Les exploitations agricoles selon leur orientation technico-economique et leur marge brute standand: L'exemple de l'Indre-et-Loire." *Norois*, 29 (1982): 429-443.

Fels, A. "L'agriculture francaise en mouvement." *Annales de Geographie*, 93 (1984): 303-325.

Dorel, Gerard. "High-Tech Farming Systems in Champagne, France: Change in Response to Agribusiness and International Controls." Pp. 405-424 in *Comparative Farming Systems*, edited by Billie L. Turner II and Stephen B. Brush. New York, N.Y.: Guilford, 1987.

Jones, P.M. "Agricultural Modernization and the French Revolution." *Journal of Historical Geography*, 16, No. 1 (1990): 38-50.

AGRICULTURAL LABOR STUDIES

Bethemont, J. "Sur la rentabilite du travail agricole a Plozevet (Finistere)." *Revue de Geographie de Lyon*, 42, No. 3 (1967): 231-246.

Muller, Konrad. *Arbeitsaufward und Arbeitsrhythmus in den Agrarflandschaften Sud- und Sudost Frankreichs: Les Dombes bis Bouches-du-Rhone.* Berlin, Germany: Berliner Geographische Abhandlungen, Heft 11, 1971.

Bozon, Pierre. "Le pluralisme rural de l'Ardeche." Pp. 13-17 in *L'assiette rurale des departements francais: Actes du colloque de Poitier 1981.* Poitiers: Travaux du Centre de Geographie Humaine et Sociale, Universite de Poitiers, 1981.

Lebleau, J. "L'agriculture de la region Rhone-Alpes vue a travers des travaux recents." *Revue de Geographie de Lyon*, 57 (1982): 173-189.

Germany

ATLASES AND GRAPHIC PRESENTATIONS

Otremba, Erich. *Atlas der deutschen Agrarlandschaft.* Wiesbaden, Germany: Franz Steiner, 1962–1971. 4 Portfolios.

GENERAL TEXTS, ARTICLES, AND THESES

Spering, Fritz. *Agrarlandschaft und Agrarformationen im deutschniederl an ischen Grenzgebiet des Emslaundes und der Provinzen Drenthe/Overijssel.* Gottingen, Germany: E. Goltze, 1881.

Capot-Rey, Robert. "L'agriculture dans le territoire de la Sarre." *Annales de Geographie*, 32 (1923): 97-118.

Niehaus, Heinrich. "Agricultural Conditions and Regions in Germany." *Geographical Review*, 23, No. 1 (1933): 23-47.

Dommermuth, Anton. *Die Landwirtschaftlichen Betriebsformen im Werterwald und im Limburger Becken in ihrer geographischen Bedingtheit.* Frankfurt am Main, Germany: H.L. Bronner, 1940.

Manheim, Felix. "Die Anbauverhaltnisse auf dem Neckarschwemmkegel im Jahre 1669: Ein Beitrag zur historischen Agrargeographie." *Berichte zur deutschen Landeskunde*, 7 (1951): 87-101.

Ketzer, G. "Wandlungen in Agrarlandschaftsbild Rheinbessens." *Erdkunde*, 8, No. 1 (1954): 64-68.

Otremba, Erich. *Die deutsche Agrarlandschaft.* Wiesbaden, Germany: Franz Steiner, 1956.

Jensch, Georg. *Das Indliche Jahr in deutschen Agrarlandscaften.* Berlin, Germany: D. Reimer, 1957.

Linden, Hellmut. *Nature aumliche Kleingliederung und Agrarstruktur an der Grenze des westfalichen Hellwegengengen das Sandmunsterland.* Remagen, Germany: Gundesanstalt fur Landeskunde, 1958.

Bottcher, Gregor. *Die agrargeographische Struktur Westfalens 1890–1950, erlautert an der pflanzlichen Produktion*. Münster, Germany: Selbstverlag der Geographischen Kommission, 1959.

Hannesen, Hans. *Die Agrarlandschaft der schleswig-holsteinischen Geest und ihre neuzeitliche Entwicklung*. Kiel, Germany: Kieler Geographische Schriften, Band 17, Heft 3, 1959.

Roubitschek, Walter. "Die regionale Differenzierung der agraren Bodennutzung 1935 im heutigen Gebiet der Deutschen Demokratischen Republik." *Petermann's Geographische Mitteilungen*, 103 (1959): 190-197.

Otremba, Erich. *Allgemeine Agrar- und Industriegeographie*. Stuttgart, Germany: Erde und Weltwirtschaft, Band 3, 1960.

Jankuhn, Herbert. "Die Entstehung der mittelalterlichen Agrarlandschaft in Angeln." *Geografiska Annaler*, 43 (1961):151-164.

Manheim, Felix. *Agrargeographie des Neckarschwemmkegels: Historische Entwicklung und heutiges Bild einer Kleinraumig differenzierten Agrarlandschaft*. Heidelberg, Germany: Heidelberger Geographische Arbeiten, Heft 5, 1961.

Otremba, Erich. "Die deutsche Agrarlandschaft." *Geographische Zeitschrift Beiheft*, Heft 3 (1961).

Overbeck, H. "Die Geschickte der deutschen Agrarlandschaft im Ubergang vom Mittelalter zur Neu zeit." *Erdkunde*, 15, No. (1961): 126-140.

Panhuysen, Helene. *Die Entwicklung der Agrarlandschaft im Raume Straelen seit 1800 unter besonderer Berucksichtigung des Gemuse- und Blemenanbaus*. Bonn, Germany: F. Dummler, 1961.

Gallusser, Werner A. "Die Landwirtschaftlichen Aussiedlung in der Struckturver besserten Agrarlandschaft." *Erdkunde*, 18 (1964): 311-328.

Hehr, Erich. *Agrargeographische Studien in der Gemarkung Klingenmunster/Sudpfalz*. Speyer am Rhein, Germany: Verlag der Pfalzischen Gesellschaft zur Forderung der Wissenschaften, 1964.

Bottcher, Gregor. "Die Umschichtung der Agrarproduktion im mittleren Ruhrrevier durch den Einfluss der Industrie." Pp. 49-57 in *Bochum und das mittlere Ruhgebiet: Festschrift zum 35. Deutschen Geographentag*, 8-11, 1965.

Kasack, D., and D. Obscherningo. "A Contribution to Agricultural-Geographic Structural Analysis in the Rostock Area, Illustrated by Examples of the VEG (State Farm) Gross-Stieten and the LPG (Cooperative Farm Sildemoau" (in German). *Wissenschaftliche Zeitschrift Universitat Rostock*, 15, No. 7-8 (1966): 1037-1058.

Peltre, Jean. "Quelques en seignements des recherches agraires recentes en Allemagne du Sud." *Revue Geographique de l'Est*, 6, No. 1-2 (1966): 101-108.

Topp, Manfred. *Agrargeographie von Ingelheim*. Bad Godesberg, Germany: Bundesanstalt fur Landeskunde und Raumforschung, 1966.

Vogt, Jean. "Notes agraires rhenanes." *Revue Geographique de l'Est*, 6, No. 1-2 (1966): 53-88.

Giese, Ernst. *Die untere Haseniederung: Eine landlich-bauer liche landschaft im nortwestdeutschen Tiefland.* Munster, Germany: Westfalische Geographische Studien, Heft 20, 1968.

Kessler, Margrit. "Neue Betriebsformen in der Landwirtschaft." *Geographische Rundschau,* 21, No. 12 (1969): 471-475.

Mayhew, Alan. "Structural Reform and the Future of West German Agriculture." *Geographical Review,* 60, No. 1 (1970): 54-68.

Otremba, Erich. *Der Agrarwirtschaftsraum der Bundesrepublik Deutschland.* Wiesbaden, Germany: F. Steiner Verlag, 1970.

Filipp, Karl. "Recherches de geographie agraire et historique dans le Nord-Est du Palatinat." *Revue Geographique de l'Est,* 11, No. 1 (1971): 23-30.

Glasser, Ewald. *Landliche, Siedlung und wirtschaft des Kreises Coesfeld in Vergangenbeit und Gegenwort.* Coesfeld, Germany: Beitrage zur Landes- und Volkskunde des Kreises Coesfeld, Band 12, 1971.

Mayhew, Alan. "Agrarian Reform in West Germany." *Transactions, Institute of British Geographers,* 52 (1971): 61-76.

Vogt, Jean. "Notes agraires rhenanes." *Revue Geographique de l'Est,* 11, No. 1 (1971): 31-36.

Lehmann, Heinrich. *Die Agrarlandschaft in den Nebentalern des oberen Mittelrheins und ihr Strukturwanel.* Bonn-Bad Godesberg, Germany: Selbstverlag der Bundesforschungsanstalt fur landeskunde und Raumordnung, 1972.

Pletsch, Alfred. *Die nordhessische Agrarstruktur unter dem Einfluss der Witrschaftszentralitat Kassels: Die Niederhessische Senke und ihre Randgebiete.* Marburg, Germany: Marburger Geographische Schriften, Heft 56, 1972.

Rupprecht, Alois. "Landwirtschaftsprobleme in der Pfalz." *Geographische Rundschau,* 24, No. 5 (1972): 178-182.

Andreae, Bernd. *Strukturen deutscher Agrarlandschaft: Landgabegiete und Fruchtfolgesysteme in der Bundesrepublik Deutschland.* Bonn-Bad Godesberg, Germany: Forschungen zur deutschen Landeskunde, Band 199, 1973.

Born, Martin. *Die Entwicklung der deutschen Agrarlandschaft.* Darmstadt, Germany: Wissenschaftliche Bunchgesellschaft, 1974.

Stephen, Johannes. *Die Landschaftsentwicklung des Stadtkreises Karlsruhe und seiner Naheren Vangeburg.* Bonn, Germany: Bonner Geographische Abhandlungen, Heft 48, 1974.

Windhorst, Hans-Wilhelm. *Spezialisiche Agrarwirtschaft in Sudoldenburg: Eine agrargeographische Untersuchung.* Leer, Germany: Schuster, Nordwestniedersachsische Regionalforschungen Band 2, 1975.

Borcherdt, Christoph, ed. *Beitrage zur Landeskunde Sudwest deutschlands.* Stuttgart, Germany: Stuttgarter Geographische Studien, Band 90, 1976.

Greeman, Verna D. *Agricultural Development and Rural Change in the German Democratic Republic.* Sheffield, England: Sheffield City Polytechnic, Occasional Papers in Geography and Environmental Science, No. 1, 1979.

Meissner, Horst A. *Beharrung und Wandel in einem nordwestdeutschen Agrarraum: D. Quakenbrucker Becken.* Leer, Germany: Schuster, 1979.

Spering, Fritz. *Agrarlandschaft und Agrarformationen in deutsch niederl and ischen Grenzgebiet des Emslandes und der Provinzen Drenthe/Overijssel.* Gottingen, Germany: E. Goltze, 1981.

Lob, R.E. "Zur Situation der landwirtschaft im sudlichen Niedersachsen: Beispiele aus der Samtgemeinde Bad Essen/Kreisosnabruck." *Essener Geographische Arbeiten*, 1 (1982): 139-162.

Mrohs, E. "Part-Time Farming in the Federal Republic of Germany." *GeoJournal*, 6, No. 4 (1982): 327-330.

Ruppert, Karl. "Die deutschen Alpen-Prozebablanfe spezieller Agrarstrukturen." *Erdkunde*, 36, No. 3 (1982): 176-187.

Thieme, G., and H.D. Laux. "Regional Disparities in West German Agriculture." *Colloquium Geographicum*, 15 (1982): 73-88.

Von Schilling, H. "Regional Schwerpunkte intensiver Landbewirtschaftung." *Geographische Rundschau*, 34, No. 3 (1982): 88-95.

Freeman, V. "Agricultural Reorganization in the German Democratic Republic: 1965–1980." *GeoJournal*, 7, No. 1 (1983): 59-66.

Thieme, G. "Agricultural Change and Its Impact in Rural Areas." Pp. 220-247 in *Urban and Rural Change in West Germany*, edited by Trevor Wild. London, England: Croom Helm, 1983.

Haversath, Johann-Bernhard. *Die Agrarlandschaft in romischen Deutschland der Kaiserzeit.* Passau, Germany: Passauia Universitatsverlag, Passauer Schriften zur Geographie, Heft 2, 1984.

Hohmann, K. "Agrarpolitik und Landwirtschaft in der DDR." *Geographische Rundschau*, 36, No. 12 (1984): 598-604.

Schmitt, Gunther. "Part-Time Farming in the Federal Republic of Germany." *Land Use Policy*, 1, No. 2 (1984): 125-133.

Schroder, Ernst-Jurgen. *Der agrarische Grossbetrieb in Baden-Wurttemberg.* Freiberg, Germany: Freiberger Geographische Hefte, No. 23, 1984.

Eckert, Karl. *Veranderungen der agraren Nutzungsstruktur in den beiden Staaten in Deutschland.* Trier, Germany: Zentralaussehuss fur Deutsche Landeskunde, 1985.

Fasterding, F. "Anpassungsprozesse der Landwirtschaft im Zonenraudgebiet: Dargestellt an sechs ausgewahlten Regionen." *Geographische Rundschau*, 37, No. 8 (1985): 400-407.

Vaudois, J. "Specialisation spatiale et dynamisme de regions d'agriculture intensive." *Hommes et Terres du Nord*, 2 (1985): 80-84.

Sevrin, R. "L'evolution de l'agriculture de la Republique Democratique Allemande depuis 1975." *Bulletin de la Societe Geographique de Liege*, 22-23 (1986–87): 125-148.

Ehlers, Eckert. "Die Agrarlandschaft der Bundesrepublik Deutschland und ihr Wandel seit 1949." *Geographische Rundschau*, 30, No. 1 (1988): 31-40.

Drobig, G. "Die Landwirtschaft der Bundesrepublik Deutschland an der Schwelle des Gemeinsamen Marktes." *Geographie und Schule*, 11, No. 62 (1989): 2-6.

Freund, Bodo. "Landwirtschaft und Ballungsgebieten mit Beispielen aus dem Rhein-Main-Gebiet." *Geographische Rundschau*, 41, No. 3 (1989): 142-148.

Roubitschek, Walter. "Regional Structures and Types of the Agriculture in the GDR." *Geographia Polonica*, 57 (1989): 123-136.

Windhorst, Hans-Wilhelm. *Die Industrialisierung der Agrarwirtschaft: Ein Vergleich ablauforder Prozesse in den USA und der Bundesrepublik Deutschland*. Frankfurt, Germany: Strothe, 1989.

Windhorst, Hans-Wilhelm. "Die 'richtigen' Agrarfabriken." *Der Tierzuchter*, 41 (1989): 191-194.

Sailer-Fliege, U. "Die Landwirtschaft in Baden-Wuttenberg: Gegenwartige Strukturen und Entwicklungstendenzen." *Geographische Zeitschrift*, 78, No. 4 (1990): 227-242.

Windhorst, Hans-Wilhelm. "Die Industrialisierung der Agrarwirtschaft im Olderburger Munsterland." *Westfalische Forschungen*, 40 (1990): 436-463.

Jones, Alun. "The Impact of the EC's Set-Aside Programme: The Response of Farm Businesses in Rendsburg-Eckernforde, Germany." *Land Use Policy*, 8, No. 2 (1991): 108-124.

Miltitz, E. "Zum Tode verurteilt: Landwirtschaft in der neuen Bundeslandern." *Geographie un Schule*, 13, No. 69 (1991): 40-44.

Vock, Hans-Joachim. *Der agrarstrukturelle Wandel im mittleren Niedersachen*. Gottingen, Germany: Gottinger Geographische Abhandlungen, Heft 92, 1991.

Bergman, E. "Raumliche Aspekte des Strukturwandel in der Landwirtschaft." *Geographische Rundschau*, 44, No. 3 (1992): 143-147.

Harle, Josef. "Landwirtschaft und Umwelt in Baden-Wurttemberg." *Geographische Rundschau*, 44, No. 5 (1992): 303-310.

Windhorst, Hans-Wilhelm. "Vom Kombinat Industrielle Mast (KIM) zu vertikal integrierten agrarindustiellen Uterrehmen." *Zeitschrift fur Wirtschaftsgeographie*, 37, No. 3-4 (1993): 190-199.

Thieme, S. "Changes in the Structure of Agriculture in the Land of Brandenburg" (in German). *Zeitschrift fur Wirtschaftsgeogeographie*, 38, No. 1 (1994): 59-65.

Pletsch, Alfred. "East Germany's Agriculture and Rural Society after Unification." Pp. 61-108 in *Espace et culture*, edited by Serge Courville and Normand Seguin. Ste. Foy, Quebec: Les Presses de l'Universite Laval, 1995.

Wollkopt, M., and H.F. Wollkopt. "Changes in the Agricultural Structure in the Five New Federal States since the Reunification of Germany." *Geographical Review of Japan*, 70B, No. 2 (1997): 57-73.

CROPS AND CROPPING PATTERNS

Poggi, E.M. "The German Sugar-Beet Industry." *Economic Geography*, 6, No. 1 (1930): 81-93.

Bruggmann, Gunter. *Die holsteinische Baumschulenlandschaft*. Kiel, Germany: Kieler Geographische Schriften, Band 14, Heft 14, 1953.

Johannes, Egon. *Entwicklung, Funktionswandel und Bedeutung staftischer Kleingarten, dargesstellt am Beispiel der stadte Kiel, Hamburg, und Bremen*. Kiel, Germany: Kieler Geographischen Schriften, Band 15, Heft 3, 1955.

Reitel, Francois. "La viticulture dans la vallee allemande de la Moselle." *Annales de Geographie*, 70 (1961): 287-299.

Reitel, Francois. "La viticulture dans la Republique Federale allemande." *Revue Geographique de l'Est*, 1, No. 2 (1961): 205-222.

Harle, Josef. *Das Obstbaugebiet am Bodensee: Eine agrargeographische Untersuchung*. Tubingen, Germany: Tubinger Geographische Studien, Heft 11, 1964.

Roubitschek, Walter. "Die regionale Struktur des Getreide- und Futterbaues in der Deutschen Demokratischen Republik 1955 und ihre Veranderungen gegenuber 1935." *Petermann's Geographische Mitteilungen*, 108 (1964): 282-291.

Bernrath, Vera. *Landwirtschaftliche Spezial- kulturen im mittleren Saartal*. Saarbrucken, Germany: Universitat des Saarlandes, Arbeiten aus dem Geographischen Institut, Heft 9, 1965.

Jerold, Alfred. *Der zelgengebundene Anbau im Randgabiet des Frankischen Gaulandes und seine besondere Stellung inner halb der sudwestdeutschen Agrarlandschaften*. Wurzburg, Germany: Wurzburger Geographische Arbeiten, Heft 15, 1965.

Hahn, Helmut. "Die Deutschen Weinbaugebiete: Regional Differenzierungen in der Entwicklung der Rebflachen und der Betriebasstruktur." *Erdkunde*, 20 (1966): 128-145.

Schuricht, R. "Multiple-Regressions-analysen als Hilfsmittel der landwirtschaftlichen Standortforschung, dargestellt an der Standortverteilung des Obstbaues im Bezirk Dresden." *Wissenschaft Zeitschrift der Universitat Halle*, 15, No. 3 (1966): 449-458.

Glaser, Gisbert. *Der Sanderkulturanbau zu beiden Seiten des Nordlichen Oberrheins zwischen Karlsrule und Warms: Eine agargeographische Untersuchung unter besonderer Brucksichtigung des Standart-problems*. Heidelberg, Germany: Heidelberger Geographische Arbeiten, Heft 18, 1967.

Blenck, Jurgan. *Die Insel Reichenau: Eine agrargeographische Untersuchung*. Heidelberg, Germany: Heidelberger Geographische Arbeiten, Heft 33, 1971.

Bohm, W. "Der Anbau von Mais in Niedersachsen." *Zeitschrift fur Wirtschaftsgeographie*, 17, No. 5 (1973): 144-149.

Dege, Eckert. *Filsen und Osterspai: Wandlungen d. Sozial- und Agrarstruktur in 2. ehemaligen Weinbaugemeinden am Oberen Mittelrhein*. Bonn, Germany: Dummler, 1973.

Dege, Eckert. "Weinbau, Obstbau und Sozialbrache am Oberen Mittelrhein." *Erdkunde*, 27, No. 1 (1973): 34-54.

Bohm, W. "Die Entwicklung des Haferanbaues in der BRD." *Zeitschrift fur Wirtschaftsgeographie*, 18, No. 7 (1974): 209-214.

Hermann, Heinrich. "Umfangurdf Bedeutung der Landwirtschaft im Raum Nurnberg-aufgezeigt am Beispiel 'Knoblauchsland.'" *Nurnberger Wirschafts- und Sozialgeographische Arbeiten*, 18 (1974): 32-43.

Kolb, Grant. "Viticulture in the Middle Rhine Valley: Its Importance to the Inhabitants and Its Impact on the Landscape." Master's Thesis, University of Michigan, Ann Arbor, Mich., 1975.

Sasaki, Hiroshi. "Vegetable Growing in the Koln-Bonn Hilly Area: The Example of Gielsdorf" (in Japanese). *Geographical Review of Japan*, 49, No. 1 (1976): 1-24.

Hahn, Helmut. "Auswirkungen von Weinbergsflur bereinigungen auf die Sozial- und Wirtschaftstruktur der Weinbaubetriebe am Beispiel der Verfahren Boppard I und Oberwessel 1, Oberes Mittelrheingebiet." *Erdkunde*, 31, No. 3 (1977): 178-192.

Bader, Walter, and J. Frido. "Einige Bemerkungen zu Weinbau und Winzergenossenschaften in Baden-Wurttemberg." *Wirtschaftsgeographie Studien*, 4 (1978): 13-26.

Bohm, Hans. "Das Vorgebirge: Entwicklung und Struktur einer Gortenbaulandschaft am Ronde des Verdichtungstraumes Rhein-Ruhr." *Erdkunde*, 35, No. 3 (1981): 182-193.

Borsdorf, Axel. *Der Weinbau in Unterjesingen: Aufschwung, Niedergang, und Wiederbelebung der Rebkultur an der Peripherie des wurttembergischen Hauptanbaugebietes*. Tubingen, Germany: Tubinger Geographische Studien, Heft 85, 1982.

Eckert, Karl. *Veranderungen der agraren Natzungsstruktur in den beiden Staaden in Deutschland*. Trier, Germany: Forschungen zur deutschen Landeskunde, Band 227, 1985.

Beyer, R. "Aktuelle Prozesse an der Peripherie des deutscehn weinbaus am Beispiel des Main durchbruchstates zwischen Steigerwald und Hassbergen." *Berichte zur Deutschen Landeskunde*, 62, No. 1 (1988): 51-66.

Doll, H. "Struturelle Tendenzen und raumliche Auspragung der Weizen- und Reggenerzengung." *Geographische Rundschau*, 40, No. 2 (1988): 51-57.

Bockmann, M., and J. Mose. "Agrarische Intensivgebiete: Entwicklung, Strukturen und Probleme. Beispiele aus Sudoldenburg und Nord-Limburg." *Vechtaer Arbeiten zur Geographie und Regionalwissenschaft*, 8 (1989): 33-62.

Knod, W. "Wandlungen im deutschen Weinbau am Beispiel Mittelmosel." *Geographie und Scule*, 11, No. 62 (1989): 15-29.

Pez, Peter. *Sonderkulturen im Umland von Hamburg: Eine standortanalytische Untersuchung*. Kiel, Germany: Kieler Geographische Schriften, Heft 71, 1989.

Pez, Peter. "Die Sonderkulteren im Umland von Hamburg in standortanalytischer Sicht." *Zeitschrift fur Wirtschaftsgeographie*, 34, No. 3-4 (1990): 229-240.

Klohn, Werner. "Das raumliche Verbundsystem des Hopfenbaus in der Hallertau." *Zeitschrift fur Wirtschaftsgeographie*, 37, No. 3-4 (1993): 200-207.

Schenk, Winfried. "The Role of Regional Vintners' Cooperatives in the Development of Rural Areas in Germany." *Journal of Wine Research*, 5, No. 3 (1994): 187-204.

ANIMAL HUSBANDRY

Shaw, Earl B. "Potato Fed Swine in Germany." *Economic Geography*, 18, No. 4 (1942): 287-297.

Windhorst, Hans-Wilhelm. "Von der bauerlichen Veredlungswirtschaft zur agrarindustriellen Massentierhaltung: Neue Wege in de Agraren Produktion im Oldenburger Munsterland." *Geographische Rundschau*, 12 (1973): 470-482.

Doll, H. "Bedeutung und Probleme der Tierhaltung." *Geographische Rundschau*, 36, No. 1 (1984): 27-30.

Windhorst, Hans-Wilhelm. "Die Huhner- und Schweinehalturg der B.R. Deutschland im rahmen der E.G." *Deutsche Geflugelwirtschaft und Schweineproduktion*, 39, No. 28 (1987): 853-857.

Achenback, Hermann. "Rinderhaltung in Schleswig-Holstein: Strukturmerkmale und Standortverlagerungen." *Geographische Rundschau*, 42, No. 5 (1990): 256-263.

Windhorst, Hans-Wilhelm. "Prospects for German Intense Livestock Husbandry after 1992." *Feed Magazine*, 1/90 (1990): 4-8.

Achenback, Hermann. "Schafwirtschaft in Schleswig-Holstein: Ruckgag und Come-Bach eines marginalen Betriebszweiges im norddeutschen Kustenruam." *Kieler Geographische Schriften*, 80 (1991): 126-151.

Windhorst, Hans-Wilhelm. *Die sektoralen und regionalen Strukturen der Schweine- und Geflugelhaltung in Nordrhein-Westfalen*. Vechta, Germany: Vechtaer Druckerei und Verlag, 1991.

Gerbaud, F. "L'elevage est-allemand face a la P.A.C." *Revue geographique de l'est*, 35, No. 2 (1995): 137-162.

Kaminske, U., and C. Keipert. "Ostrich Farming in Germany: A Basis for Discussion on a Geoecological and Economic Basis" (in German). *Zeitschrift fur Wirtschaftsgeographie*, 42, No. 1 (1998): 22-34.

DAIRYING

Sauer, Gerhard. *Die Milchwirtschaft des Landes Nordrhein-Westfalen*. Marburg, Germany: Marburger Geographische Schriften, Heft 15, 1961.

Wotha, B. "Die Entwicklung der Milchwirtschaft von der geschlossenen Hauwirtschaft zum genossenschaftlichen Industriebetrieb an einen Beispiel im sudlichen Holstein." *Kieler Geographische Schriften*, 80 (1991): 236-254.

FARM LAND

Marquardt, Gunther. *Die schleswig-holsteinische Knicklandschaft*. Kiel, Germany: Kieler Geographische Schriften, Band 13, Heft 3, 1950.

Jensch, Georg. *Das landliche Jahr in deutschen Agrarlandschaften*. Berlin, Germany: Abhandlungen des Geographischen Institut der Freien Universitat Berlin, Band 3, 1957.

Huttenlocher, Friedrich. "Das Problem der Gewannfluren in Sudwest deutscher Sicht." *Erdkunde*, 17, No. 1-2 (1963): 1-15.

Schroder, K.H. "Der Wandel der Agrarlandschaft im ostelbischen Tiefland seit 1945." *Geographische Zeitschrift*, 52 (1964): 289-316.

Erhart, Rainer R. "Land Consolidation and Its Effects on the Agricultural Landscape of the Vogelsberg, Germany." Ph.D. Dissertation, University of Illinois, Urbana, Ill., 1967.

Roubitschek, Walter. "Entwicklung und regionale Differenzierung der Eigentums formen in der Landwirtschaft der DDR." *Petermann's Geographische Mitteilungen*, 111 (1967): 279-289.

Geyer, Heidemarie. "Verglich landwirtschaftslicher Betriebe aus verschiedensen Teillandscaften der Westfalischen Bucht." *Geographische Rundschau*, 28, No. 11 (1976): 454-464.

Lamping, Heirich. "Die Teillandwirtschaft als Problem der Raumordung in Strukturschwachen Raumen der Bundesrepublik Deutschland." *Frankfurter Wirtschafts- und Sozialgeographische Schriften*, 26 (1977): 275-284.

Kobayashi, Koji. "Regional Disparity of Agricultural Land Use in the Northwestern Fringe of Hamburg" (in Japanese). *Geographical Review of Japan*, 51, No. 9 (1978): 687-703.

Closs, Hans-Martin. *Die nord badische Agrarlandschaft: Aspekte raumlicher Differenzierung*. Trier, Germany: Forschungen zur deutschen Landeskunde, Band 215, 1980.

Kehrer, G., B. Fege, and J. Kurth. "Territorale Aspekte der kombinatsbildung in der GDR." *Geographische Berichte*, 28, No. 1 (1983): 17-29.

AGRICULTURAL POPULATIONS AND SOCIAL ISSUES

Schwald, Mechthild. *Die Entwicklung der bauerlichen kulturlandschaft in Ostriesland und Westoldenburg*. Bonn, Germany: Bonner Geographische Abhandlungen, Heft 12, 1953.

Faigle, Helmut. *Die Zunahme des Davergrunlandes in Wurttenberg und Hohenzollern: Wandel agrar- und sozialgeographischer Verle: Altnisse wahrend der letzten Jahrzehnte*. Tubingen, Germany: Im Selbstvertag des Geographischen Institut der Universitat Tubingen, 1963.

Winter, Heinrich. *Die Entwicklung der Landwirtschaft und Kulturlandschaft des Monschauer Landes unter Berucksichtigung der Rodungen*. Bad Godesberg, Germany: Bundesanstalt fur Landskunde und Raumforschung, 1965.

Mayhew, Alan. *Rural Settlement and Farming in Germany*. London, England: Batsford Historical Geography Series, 1973.

Kohl, Manfred. *Die Dynamik der Kulturlandschaft im Oberen Lahn. Dilldreis: Wandlungen von Haubergswirtschaft und Ackerbaus zu neuen Formen der Landnutzung in der moderen Regionalentwicklung*. Giessen, Germany: Giessenre Geographische Schriften, Heft 45, 1978.

Klohn, Werner. "Kulturlandschaftwandel und Nutzungskonflikte in der Dummerniedurung." *Neus Arkiv fur Niedersachsen*, 39, No. 2/3 (1990): 14-30.

Jones, Alun, F. Fasterding, and Rainer Plankl. "Farm Household Adjustments to the European Community's Setaside Policy: Evidence from Rheinland-Pfalz, Germany." *Journal of Rural Studies*, 9, No. 1 (1993): 65-80.

Schmidt, M. "Women Farmers and the Influences of Ecofeminism on the Greening of German Agriculture." Pp. 102-116 in *Gender and Rurality*, edited by Sarah Whatmore, Terry Marsden, and Philip Lowe. London, England: David Fulton, 1994.

Grabski-Kieran, Ulrike. *Leitzieh der Landschaftspfluge fur die Agrarlandschaft Brandenburgs: Beitrage zur landlichen Entwicklung im Raum Konig Wusterhausen*. Bochum, Germany: Bochumer Geographische Arbeiten, Heft 60, 1995.

Wollkopf, M. "Structural and Social Change in Thuringia's Agriculture" (in German). *Geographische Rundschau*, 48, No. 1 (1996): 18-24.

ENVIRONMENTAL AND CONSERVATION ISSUES

Linden, Hellmut. *Naturraumliche Kleingliederung und Agrarstrukture an der Grenze des westfalischen Hellwegesgegen das Sondmunsterland*. Remagen/Rhein, Germany: Selbstaverlag der Bundesonstalt fur Landeskunde, 1958.

Wiegelmann, Gunter. *Naturliche Gunst und Ungunst im Wandel rheinischer Agrarlandschaften, er ortertanhand von Untersuchungen in der nordlichen Eifel, im Bereich der unteren Saar (Saargau und Hochwold) und im Saarland*. Koln, Germany: Kolner Geographische Schriften, Heft 12, 1958.

Bartel, Jurgen. *Baum und Strauch in der rheinischen Agrarlandschaft: Hecken, Windschutzanlagen und Flurholzon bau am Hohen Venn, in der Nordeifel und in der Jalicher Barde*. Koln, Germany: Kolner Geographischen Arbeiten, Heft 18, 1963.

Hofmeister, B. *Die gemassigten Breiten: Insbesondere die kuhlgemassigten Walderlander*. Braunswig, Germany: Westerman, 1985.

Geyer-Wedel, K., and E. Jordan. "Nitratprobleme in der landschaft: Geookologische Untersuchungen und Modelrechnumgen zur Staffumsatze in Kleinen Einzugsgebieten am Beispiel zweier Teileinzugsgebiete der Visbeker Aue/ Landkreig Vechta." *Vechtaer Arbeiten zur Geographie und Regionalwissenschaft*, 8 (1989): 105-126.

Klohn, Werner. "Die Dummersanierung: Kriegser-Klarung an die Landwirte oder 'Zeichen der Zeit'? Landwirtschaft und okologischen Probleme in der Dummerniederung." *Vechtaer Arbeiten zur Geographie und Regionalwissenschaft*, 8 (1989): 127-154.

Nischewitz, Guido. "Die Entwicklung der deutschen Landwirtschaft im lang fristigen Prozes der evrapaischen Zivilisation." *Geographische Zeitschrift*, 79, No. 4 (1991): 195-211.

Reiners, T. "Zum Verhaltnis von Landwirtschaft und Naturschutz in Schleswid-Holstein." *Kieler Geographische Schriften*, 80 (1991): 57-75.

Wilson, Geoff A. "German Agri-Environmental Schemes—I: A Preliminary Review." *Journal of Rural Studies*, 10, No. 1 (1994): 27-45.

Wilson, Geoff A. "German Agri-Environmental Schemes—II: The MEKA Programme in Baden-Wurttemburg." *Journal of Rural Studies*, 11, No. 2 (1995): 149-160.

AGRICULTURAL MARKETING AND DISTRIBUTION

Nieberger, C. "Die Neuordnung der logistikkette: Raumliche Auswirkungen unternehmens ubergreifender Umstrukturierungsprozesse zwischer Industrie un Handel. Das Beispiel Molkereiprodukte." *Erdkunde*, 51, No. 2 (1997): 131-142.

AGRICULTURAL LABOR STUDIES

Vogt, Jean. "Notes agraires rhenanes: Laboureurs et petites gens a la fin du XVIIIe siecle." *Revue Geographique de l'Est*, 1, No. 1 (1961): 145-150.

Yamamoto, M. "Changes in Agriculture and Rural Employment in Southwestern Thuringia, Germany." *Geographical Review of Japan*, 70B, No. 2 (1997): 83-94.

Knappe, Elke. "The Declining Importance of Agricultural Employment in East Germany." *GeoJournal*, 46, No. 3 (1998): 243-246.

The Netherlands

GENERAL TEXTS, ARTICLES, AND THESES

Davis, Elsie G. "Agriculture in the Netherlands." Master's Thesis, Marshall University, Huntington, W.Va., 1952.

Kleinpenning, J.M.G. *Geographie von de Landbauw*. Utrecht, Netherlands: Het Spectrum, 1968.

Osborne, Richard H. *Rural Studies in the North East Netherlands*. Nottingham, England: Geographical Field Group, Study No. 14, 1969.

Von Welsenes, C. "Farming Problems and Farming Principles in the Dutch Fen-Colonies." *Tijdschrift voor Economische en Sociale Geographie*, 64, No. 4 (1973): 218-230.

Von de Klundert, A.F. "Agricultural Development in the Netherlands: The Spatial Consequences of EEC Agricultural Policy." *Tijdschrift voor Economische en Sociale Geographie*, 78, No. 3 (1978): 232-237.

Spering, Fritz. *Agrarlandschaft und Agrarformationen im deutsch-niederlandischen Grenzgebiet des Emslandes und der Provinzen Drenthe/ Overijssel*. Gottingen, Germany: Gottinger Geographische Abhandlundgen, Heft 76, 1981.

Cardol, G. *Ruimte voor Agribusiness-complexen: Structuur, positieen dynamick von het Noordlimburgse tuinbouw complex vonuit functioneel, geografisch en regional perspectief*. Amsterdam, Netherlands: Netherlands Geographical Studies, No. 57, 1988.

Von Oort, G. "L'agriculture de loisir, une nouvelle forme d'utilisation de l'espace? Le cas de la fringe urbaine de la ville d'Utrecht." *Bulletin de l'Association de Geographes Francais*, 71, No. 2 (1994): 135-143.

Maas, J.H.M. "Toekomst perspectieven voor de Nederlandse agrosector." *Geographie*, 5, No. 1 (1996): 32-37.

CROPS AND CROPPING PATTERNS

Hahn, Arvin. "The South Holland Onion Set Industry: A Study of Specialized Horticulture." Ph.D. Dissertation, Northwestern University, Evanston, Ill., 1952.

Von den Berg, L.M. "Patterns of Harmony and Conflict Between Horticulture and Urban Growth in the Netherlands." *Geography Research Forum*, 13 (1993): 32-45.

DAIRYING

Ginn, E. Margaret. "The Dutch and Dairying." *B.C. Geographical Series*, 9 (1968): 117-138.

Maas, J.H.M., and E. Wever. "Lokationale en innovatieve aspekten von het Nederlandse welk veehouderij complex." *Nederlandse Geografische Studies*, 26 (1986): 34-63.

FARM LAND

Lambert, Audrey M. "Farm Consolidation and Improvement in the Netherlands: An Example from the Land Von Maas en Waal." *Economic Geography*, 37, No. 2 (1961): 115-123.

Gissel, Svend. "Die Dreifelderwirtschaft auf Seeland bis 1700." *Geographische Zeitschrift Erdkundliches Wissen*, Beiheft 18 (1968): 44-49.

Jansen, Marlies E. *Een ouderzoek naar de mogelijkheden voor een landbouwgebied-sindeling met behulip von clusteranalyse: Experiment Noord-Brabant.* Gravenhage, Netherlands: Centraal Bureau voor de Statistik, Hoofdofdeling Statistische Methoden: Staatsuitgeverij, 1983.

Von Uytven, R. "Vroege im breuken op de braak in Brabant in de intensieve landbouw in de Zuidelijke Nederlanden tijdens de dertienc eeuw." *Bulletin de la Societe Belge d'Etude Geographie*, 53, No. 1 (1986): 63-72.

Von Herwijnen, M., R. Janssen, and P. Reitveld. "Herbestemming von landbouwgrond: Een multicriterin benadering." *Nederlandse Geografische Studies*, 107 (1990).

Banski, J. "Transformations in the Agriculture of the Netherlands, with Particular Consideration Given to Land Use in Agriculture" (in Polish). *Przeglad Geograficzny*, 66, No. 3-4 (1994): 357-374.

Van Oort, G. "Agriculture et evolution de la frange urbaine: Le cas de la frange urbaine de la ville d'Utrecht." *Hommes et Terres du Nord*, 4 (2000): 246-252.

ENVIRONMENTAL AND CONSERVATION ISSUES

Dauvellier, Peter L. "Agriculture and Landscape in the Netherlands: The 4th Report on Physical Planning (1988)." *Landscape and Urban Planning*, 18, No. 3-4 (1990): 247-256.

Switzerland and Liechtenstein

GENERAL TEXTS, ARTICLES, AND THESES

Casutt, C.P., and E. Rauch. "Zur Agrargeographie des Vorderrheintals." *Geographica Helvetica*, 16 (1961): 153-161.

Kistler, R. "Die Ausscheidung von Landwirt-schaftszonen bei Orts- und Regional planungen." *Geografica Helvetica*, 18 (1963): 1-86.

Piveteau, J.L. "L'agriculture suisse en question." *Revue de Geographie de Lyon*, 38 (1963): 117-129.

Chiffelle, Frederic. *La Bas-Pays Neuchatelois: Etude de geographie rurale.* Neuchatel, Switzerland: Editions de la Beconniere, Memoires de la Societe neuchateloise de geographie, 1968.

Kreisel, W. "Structures agraires de Waldenfendorf dans le Jura." *Revue de Geographie de Lyon*, 44 (1969): 85-113.

Pfister, C. *Agrarkonjunktur und Witterungsverlauf im Westlichen Schweizer Mittelland.* Bern, Switzerland: Geografica Bernensia, G. 2, 1975.

Bugmann, E. "Neuere Entwicklung der landwirtschaftlichen Nutzung im Walliser Berggebiet." *Geographica Helvetica*, 31 (1976): 65-68.

Cabouret, Michel. "Apercu sur la politique helvetique un faveur de l'agriculture de montagne." *Revue Geographique de l'Est*, 16, No. 1-2 (1976): 3-40.

Bridet, Laurent. "Boltens VD: Le contact entre l'agglomeration Lausannoise et le moyen-pays agricole." *Geographica Helvetica*, 32 (1977): 79-88.

Nuesch, Daniel. "Fernerkundung und Landwirtschaft." *Geographica Helvetica*, 32 (1977): 169-175.

Piot, M.J.C. "Les lignes directrices de la politique agricole suisse." *L'Information Geographique*, 41, No. 3 (1977): 109-112.

Szczesny, R. "Typy rolnictwa Szwajcarii." *Przeglad Geograficzny*, 54 (1982): 511-531.

Chiffelle, Frederic. "Swiss Agricultural Policy in Mountainous Areas." *Nordia*, 17, No. 1 (1983): 27-31.

Dorbellay, Charles. "Terresen friches: Solutions agricoles experimentees en Valais." *Geographica Helvetica*, 38 (1983): 183-186.

Gantner, Urs. "Investitionsverhalten Landwirt-schaftlicher Betriebsleiter." *Geographica Helvetica*, 39 (1984): 189-196.

Schmid, Willy A. "Landwirtschaft und Raumardnung im landlichen Raum (Schweizerisches Mittelland)." *Geographica Helvetica*, 39 (1984): 180-188.

Berthoud, Eric F. "L'evolution de l'agriculture neuchateloise de 1965 a 1980: Problems de comparaison." *Geographica Helvetica*, 40 (1985): 84-90.

Leimgruber, W. "From Plain to Mountain: Zoning in Switzerland. The Example of the Canton of Fribourg." *Nordia*, 20, No. 1 (1986): 49-56.

Muedespacher, A. "Regional Incidence of Agricultural Policy: The Case of Switzerland." *Environment and Planning C*, 4, No. 3 (1986): 343-352.

Bieri, Hans. "Schutz landwirtschaftlicher Boden und bauerlichen Betriebsstandarte im Dorf und im Agglomerationsrandbereich." *Geographica Helvetica*, 43, No. 2 (1988): 69-75.

Klohn, Werner, J. Mose, and Hans-Wilhelm Windhorst. "Landwirtschaft zwischen Uber produktion und Konzentration." *Praxis Geographie*, 18, No. 12 (1988): 42-46.

Chiffelle, Frederic. "L'agriculture suisse dans le dernier quart du XXe siecle." Pp. 337-362 in *Nouvelle geographie de la Suisse et des Suisses*, vol. 2, edited by J.B. Racine and C. Raffestin. Lausanne, Switzerland: Editions Payot, 1990.

Egli, H.R. "Les objectifs de la politque agricole suisse." Pp. 363-375 in *Nouvelle geographie de la Suisse et des Suisses*, vol. 2, edited by J.B. Racine and C. Raffestin. Lausanne, Switzerland: Editions Payot, 1990.

Lentz, S. *Agrargeographie de bundnerischen Sudtaler Val Mustaic und Val Poschiavo*. Mannheim, Germany: Mannheimer Geographische Arbeiten, Heft 28, 1990.

Streckeisen, Willy. "L'agriculture genevoise et son avenir." *Le Globe*, 131 (1991): 13-20.

Lehmann, B., and E.D. Stucki. "L'agriculture et le devenir des espaces ruraux en Suisse." *Revue d'economique regionale et urbaine*, 2 (1996): 424-441.

Gigon, Nathalie, and Olivier Crevoisier. "Internet et la recomposition territoriale des relations dans l'agriculture suisse." *Canadian Geographer*, 43, No. 2 (1999): 152-160.

CROPS AND CROPPING PATTERNS

Daillens, Francoise. "La vigne et levin en Suisse." *Revue Geographique de l'Est*, 2, No. 4 (1962): 345-374.

Chevalier, Michel. "Le vignoble du Jura." *Revue Geographique de l'Est*, 3, No. 1 (1963): 15-36.

Bagutti, Aurelia. "Evoluzione del paesaggio viticola del Mendrisiotto." *Geographica Helvetica*, 42, No. 4 (1987): 249-257.

Raffestin, Claude. "La Suisse dans vingt ans." *Geographica Helvetica*, 49, No. 2 (1994): 81-86.

FARM LAND

Sick, Wolf D. "Flurzusammenlegung und Ausbausiedlung in der Nordostschweiz." *Erdkunde*, 9, No. 3 (1955): 169-187.

Torricelli, Gian Paolo. "Agriculture et territorialite en Valteline." *Geographica Helvetica*, 46, No. 4 (1991): 173-177.

AGRICULTURAL POPULATIONS AND SOCIAL ISSUES

Nageli-Oerthe, Ruedi. "Veranderungen der landwirtschaftlichen Betriebs- und Grundeigentumsstruktur als Abbild des sozialokonomischen Wandels im Berggebiet: Das Beispiel Grindelwald." *Geographica Helvetica*, 43, No. 1 (1988): 3-13.

ENVIRONMENTAL AND CONSERVATION ISSUES

Bugmann, E. "Part-Time Farming and Ecological Aspects in the Swiss Alps." *Geographica Helvetica*, 35, No. 5 (1980): 125-134.

Pfister, C. "An Analysis of the Little Ice Age Climate in Switzerland and Its Consequences for Agricultural Production." Pp. 214-248 in *Climate and History*, edited by T.M.L. Wigley, M.J. Ingram, and G. Farmer. Cambridge, England: Cambridge University Press, 1981.

Siegrist, Sabine. "Experimentelleuntersuchangen uber die Verminderung der Bodenerosian durch biologischen Landbau in einem NW-schweizerischen Lobgebiet." *Die Erde*, 126, No. 2 (1995): 93-107.

IRRIGATION

Crook, Darren S., and Anne M. Jones. "Traditional Irrigation and Its Importance to the Tourist Landscape of Valais, Switzerland." *Landscape Research*, 24, No. 1 (1999): 49-66.

Crook, Darren S., and Anne M. Jones. "Design Principles from Traditional Mountain Irrigation Systems (Bisses) in the Valais, Switzerland." *Mountain Research and Development*, 19, No. 2 (1999): 79-99.

Crook, Darren S. "The Historical Impacts of Hydroelectric Power Development on Traditional Mountain Irrigation in the Valais, Switzerland." *Mountain Research and Development*, 21, No. 1 (2001): 46-53.

AGRICULTURAL MARKETING AND DISTRIBUTION

Laup, J. "Une organisation cooperative agricole helvetique: Provins Valais." *Revue de Geographie Alpine*, 47 (1959): 243-253.

Chiffelle, Frederic. "Agriculteurs-entrepreneurs de plaine et paysons-jardiniers de montagne: Vers une dichotomie croissante. Le cas de la Suisse." Pp. 314-318 in *Campagnes et littoraux d'Europe: Melanges offerts a Pierre Flatres*. Lille, France: Institut de Geographie de Lille, Hommes et Terres du Nord, 1988.

Southern Europe

GENERAL TEXTS, ARTICLES, AND THESES

Semple, Ellen Churchill. "Ancient Mediterranean Agriculture, Part 1." *Agricultural History*, 2, No. 1 (1928): 61-98.

Semple, Ellen Churchill. "Ancient Mediterranean Agriculture, Part 2." *Agricultural History*, 2, No. 3 (1928): 129-156.

Chaire, G. "Problemes agricoles sur les bards septentrionaux du bassin mediterraneen." *Annales de Geographie*, 68 (1959): 318-332.

Brun, Francoise. "La reinstallation des agriculteurs pieds-voirs dans la Midi mediterraneen." *Annales de Geographie*, Vol. 83, 6 (1974): 676-683.

Mykolenko, L., T. deRaymond, and P. Henry. *The Regional Impact of the Common Agricultural Policy in Spain and Portugal.* Luxembourg, Luxemburg: Office for Official Publications of the European Communities, 1987.

Drain, M. "Agricultures et societes rurales." Pp. 141-176 in *Dossiers des images economiques du monde 12: La CEE mediterraneenne*, edited by J. Beaujeu-Garnier and A. Gamblin. Paris, France: Sedes, 1990.

Joannan, M., and Lucien Tirone. "Les systems agricoles en Mediterranie." *Mediterranee*, 70 (1990): 51-60.

CROPS AND CROPPING PATTERNS

Blanchard, W.O., and Elizabeth R. Blanchard. "The Grape Industry of Spain and Portugal." *Economic Geography*, 5, No. 2 (1929): 183-193.

Isaac, Erich. "The Citron in the Mediterranean: A Study in Religious Infuences." *Economic Geography*, 35, No. 1 (1959): 71-78.

Berengi, L. "Les types geographiques de la viticulture en Europe du Sud-East." *Bulletin de la Societe Languedocienne de Geographie*, 8, No. 2 (1974): 107-128.

ANIMAL HUSBANDRY

Gardelle, Charles. "La transhumance ovinee entre les regions mediterraneennes et les Alpes en 1964." *Revue de Geographie Alpine*, 53 (1965): 449-479.

Gardelle, Charles. "La transhumance inverse des bovins vers les regions mediterraneennes." *Revue de Geographie Alpine*, 53 (1965): 503-505.

Zobl, Dorothea. *Die Transhumanz (Wanderschafthaltung) der europaischen Mittel meer lander im Mittelalter in historischer geographische und volkskundlicher Sicht.* Berlin, Germany: Berliner Geographische Studien, Band 10, 1982.

Kayser, B. "Desintegration et integration des relations agriculture: Elevage dans les regions mediterraneennes." *Annales de Geographie*, 100, No. 1 (1991): 18-30.

FARM LAND

Shaw, D.J. "The Problem of Land Fragmentation in the Mediterranean Area: A Case Study." *Geographical Review*, 53, No. 1 (1963): 40-51.

Nicod, J. "Murettes et terrasses de culture dans les regions karstiques mediterraneennes." *Mediterranee*, 71, No. 3-4 (1990): 43-50.

Balm, Roger. "Terrace Condition and Agricultural Change in the Islands of the Mediterranean Region." *Middle States Geographer*, 28 (1995): 234-241.

AGRICULTURAL POPULATIONS AND SOCIAL ISSUES

De Phanhol, Xavier, ed. *Les societes rurales mediterraneennes: Actes du colloque de geographie agraire.* Aix-en-Provence, France: Centre Geographique d'Etudes et de Recherches Mediterraneennes, Institut de Geographie, Aix-en-Provence, 1972.

ENVIRONMENTAL AND CONSERVATION ISSUES

Perez, Manuel R. "Development of Mediterranean Agriculture: An Ecological Approach." *Landscape and Urban Planning*, 18, No. 3-4 (1990): 211-220.

IRRIGATION

Livet, R. "Irrigation et societes dans le monde mediterraneen." *Mediterranee*, 39, No. 2-3 (1980): 3-11.

Greece

GENERAL TEXTS, ARTICLES, AND THESES

Fuldner, Eckart. *Agrargeographische Untersuchungen in der Ebene von Thessaloniki*. Frankfurt am Main, Germany: Frankfurter Geographische, Hefte 44, 1967.

Mistardis, G.G. "L'aire Sud-Egeene: Changements dans le milieu geographique et cycles de developpement economique tres importants." Pp. 253-276 in *Melanges de Geographie physique, humaine, economique, applique offerts a M. Omer Tulippe, Vol. II,* edited by Jose A. Sporck and Bernadette Schoumaker. Gembloux, Belgium: J. Duculot, 1967.

Bonnefort, Jean-Claude. "Recherches sur les paysages agraires de Crete centrale et orientale." *Revue Geographique de l'Est*, 11, No. 1 (1971): 3-22.

Burgal, Guy. "Recherches agraires en Grece." *Memoires et Documents, Services de Documentation et de Cartographie Geographiques, Centre National de la Recherche Scientifique*, 13 (1972): 7-62.

Zentefis, D. "Some Observations of Farming in Macedonia of Greece." *Bulletin, Illinois Geographical Society*, 15, No. 1 (1973): 61-64.

Tank, H. "Wandel und Entwicklungstendenzen der Agrarstuktur Kretas seit 1948." *Die Erde*, 108 (1977): 342-347.

Stanek, Georges. "Politique Agricole Griechenlands." *Wirtschaftsgeographie Studien*, 6 (1980): 45-64.

Oltersdorf, B. "Beginn einer Neuorientierung der Landwirtschaft auf Kreta." *Colloquium Geographicum*, 16 (1983): 319-335.

Tsioubos, Georges. "Politique agricole commune et clientelisme politique en Epire (Grece du Nord-Ouest)." *Revue de Geographie Alpine*, 80, No. 4 (1992): 129-146.

CROPS AND CROPPING PATTERNS

Vouras, Paul P. "The Significance of Cotton Growing in Greece." *Professional Geographer*, 15 (1963): 11-13.

Riccardi, M. "La coltura del cotone in Grecia." *Bolletin della Societa Geografia Italiana*, Serie 9, No. 10 (1969): 23-40.
Pechoux, Pierre-Yves, and Dimitri Katsaros. "Vignes et vins en Grece." *Mediterranee*, 65, No. 3 (1988): 45-54.

ANIMAL HUSBANDRY

Kouraba, P. "Le conflit sylvo-pastoral en Grece et ses aspects historiques, sociaux et legislatifs." *Bulletin de la Societe Languedocienne de Geographie*, 18, No. 3-4 (1987): 263-270.
Tsiboukas, K. "Les formes d'utilization pastorale de la montagne grecque: L'exemple de la Thessalie." *Bulletin de la Societe Languedocienne de Geographie*, 18, No. 3-4 (1987): 247-262.

FARM LAND

Drossos, T. "Farm Fragmentation, Land Consolidation, and Agricultural Development in Greece." Master's Thesis, San Francisco State University, San Francisco, Calif., 1971.
Brunet, M. "Terasses de cultures antiques: L'exemple de Delos, Cyclades." *Mediterranee*, 71, No. 3-4 (1990): 5-12.
Keeler, M.E., and D.G. Skuras. "Land Fragmentation and Consolidation Policies in Greek Agriculture." *Geography*, 75, No. 1 (1990): 73-76.

AGRICULTURAL POPULATIONS AND SOCIAL ISSUES

Lienau, Cay. *Bevolkerungsabwanderung demographische Sturktur und Landwirtschaftsform im W-Peloponnes: Raumliche Ordnung, Entwicklung und Zusammenhange von Wirtschaft und Bevolkerung in einene mediterranean Abwanderunsgebiet*. Giesen, Germany: Giessener Geographische Schriften, Heft 37, 1976.

ENVIRONMENTAL AND CONSERVATION ISSUES

Pepelasis, A.A., and Kenneth Thompson. "Agriculture in a Restrictive Environment: The Case of Greece." *Economic Geography*, 36, No. 2 (1960): 145-157.
Green, A., and M. Lemon. "Perceptual Landscapes in Agrarian Systems: Degredation Processes in North-Western Epirus and the Argolid Valley, Greece." *Ecumene*, 3, No. 2 (1996): 181-199.

IRRIGATION

Knodler, Otto. *Der Bewasserungsfeldbau in Mittelgriechenland und im Peleponnes*. Stuttgart, Germany: Stuttgarter Geographische Studien, Band 81, 1970.

Italy

ATLASES AND GRAPHIC PRESENTATIONS

Ehrlich, Daniele. "Multi-Temporal Satellite Image Analysis for Crop Inventories in the Po River Valley (Italy)." Ph.D. Dissertation, University of California, Santa Barbara, Calif., 1992.

GENERAL TEXTS, ARTICLES, AND THESES

Robertson, J.C. "Agricultural Regions of the North Italian Plain." *Geographical Review*, 28, No. 4 (1938): 573-596.

Basile, David G. "Agricultural Sicily." *Economic Geography*, 17, No. 2 (1941): 109-120.

Van Valkenburg, Samuel. "Structure of Italian Agriculture." *Economic Geography*, 18, No. 2 (1942): 109-124.

McNee, Robert B. "Metaponto: A Geographic Case Study of Agricultural Resource Development in Southern Italy." Ph.D. Dissertation, Syracuse University, Syracuse, N.Y., 1953.

Desplanques, Henri. "La reforme agraire italienne." *Annales de Geographie*, 66 (1957): 310-327.

Bonasera, Francesco. "La Portecipanza agraria di Nonantola (Modena)." *Rivista Geografica Italiana*, 65 (1958): 202-216.

Dongus, Hansjorg. *Die apuanische Kustenebene: Eine agrargeographische Untersuchung*. Stuttgart, Germany: Stuttgarter Geographische Studien, Band 72, 1962.

Peruzzi, Duilio. "Cortana: A Valley-Hill-Mountain Complex of Modified Mediterranean Agriculture in Central Italy." Ph.D. Dissertation, University of Michigan, Ann Arbor, Mich., 1963.

Rochefort, R. "A propos de l'histoire du paysage agraire italien." *Revue de Geographie de Lyon*, 39, No. 3 (1964): 183-195.

Ferrero, Y.F. "Die Agrarwirtschaft Sardiniens." *Zeitschrift fur Wirtschaftsgeographie*, 9, No. 1 (1965): 24-27.

Giuliani, Maria C. *Transformazioni agrarie nel Cesanate del 1500 adoggi*. Milan, Italy: Istituto editoriale Cisalpino, 1965.

Gribaudi, Dino. "The Geographical Aspect of Agriculture" (in Hungarian). *Foldrajzi Ertesito*, 14, No. 2 (1965): 187-199.

Dongus, Hansjorg. *Die Agrarlandschaft der ostlichen Po-Ebene*. Tubingen, Germany: Tubinger Geographische Studien, Heft 2, 1966.

Ferreri, Ferdinando. "Die Agrarwirtschaft in der Lombardei." *Zeitschrift fur Wirtschaftsgeographie*, 10, No. 7 (1966): 211-216.

Cavet, J. "La vieille campagne toscane et la crise agricole: L'evolution rurale du Mugello de puis 1951." *Revue de Geographie de Lyon*, 42 (1967): 329-394.

Ferreri, Ferdinando. "Sudtirols Agrarwirtschaft." *Zeitschrift fur Wirtschaftsgeographie*, 11, No. 6 (1967): 182-187.

Krauter, Karl-Gunter. *Die Landwirtschaft im ostlichen Hochpustertal (Sudtirol)*. Tubingen, Germany: Tubinger Geographische Studien, Heft 25, 1968.

Lago, Luciano. "Rauscedo, Centro di Specializzazion agricole nei 'magredis' del Friuli." *Rivista Geografica Italiana*, 76 (1969): 295-301.

Manheim, Rolf. *Die Agrostadt im Siedlungsgefuge Mittelsizilien*. Bonn, Germany: Bonner Geographische Abhandlungen, Heft 41, 1969.

King, Russell. "Structural and Geographical Problems of South Italian Agriculture." *Norsk Geografisk Tidsskrift*, 24 (1970): 83-95.

Krauter, Karl-Gunter. "Bauerliche Betriebstypen im ostlichen Hochpustertal." *Geographische Rundschau*, 22, No. 7 (1970): 277-282.

Roser, Dieter. "Die nordliche Ortlergruppe-landwirtschaftliche Strukturen eines alpinen Hohentaumes." *Geographische Rundschau*, 22, No. 7 (1970): 270-277.

Colamonico, C. "La geografia agraria della regioni italiane." *Bolletin della Societe Geografia Italiana*, Serie 9, No. 19 (1971): 593-604.

Dumolard, Pierre. "Disparites spatiales de l'intensivite agricole en Italie." *L'Espace Geographique*, 2, No. 1 (1973): 50-61.

Fischer, Klaus. *Agrargeographie des westlichen Sudtirol: Der Vinschgau und Seine Nebentaler*. Vienna, Austria: W. Braumuller Universitats-Verlagsbuchhandlung, 1974.

Ravanelli, S. "Transformaziori agrarie in un settare della piana imolese dal 1500 ad oggi." *Bolletin della Societa Geografica Italiana*, Serie 10, No. 4. (1975): 287-309.

Sabelberg, Elmar. "Kleinbauerntun Mezzadria, Latifundium: Die wichtigsten agraren Betriebssysteme in der kulturlanschaft der Toskana und ihre gegenwartigen Veranderungen." *Geographische Rundschau*, 27, No. 8 (1975): 326-336.

Mannella, S. "Agricultural Reality in Apulia." *Norsk Geografisk Tidsskrift*, 32, No. 4 (1978): 173-180.

Rambotti, Francesco. "La cooperazione agricola nell'alta valle del Tevere, Umbria." Pp. 615-626 in *Recherches de Geographie rurale: Hommage au Professeur Frans Dussart*. Liege, Belgium: Bulletin de la Societe de Geographie de Liege, Vol. 1, 1979.

Tirone, Lucien. "Intensite et productivite dans les agricultures developpees: Les mutations agricoles de la basse Lombardie." *Mediterranee*, 37, No. 4 (1979): 77-93.

Rother, K. "Die agrargeographische Entwicklung und die Wandelbarkiet der Betriebstypen im Kustentiefland von Metapont (Sud italien)." *Dusseldorfer Geographische Schriften*, 15 (1980): 89-104.

Firth, I.J.W. "The Imprint of the Enlightenment: Tuscan Hill Farming in the Eighteenth Century." *Landscape*, 25, No. 1 (1981): 36-41.

Pieroni, O. "Positive Aspects of Part-Time Farming in the Development of a Professional Agriculture: Remarks on the Italian Situation." *GeoJournal*, 6, No. 4 (1982): 331-336.

Pia, Giardine M. "Strutture agrarie in evoluzione: Un esempio della pionura cremonese." *Rivista Geografica Italiana*, 91 (1984): 315-327.

Prost, B., and M. Vandenbroucke. "Le groupe italien Ferruzzi: De l'agro-alimentaire a l'agro-chimie, une strategie audacieuse de l'utilisation des produits agricoles." *Revue de Geographie de Lyon*, 62 (1987): 313-337.

Paloscia, R. "Agriculture and Diffused Manufacturing in the 'Terra Italia': A Tuscan Case Study." Pp. 34-57 in *Rural Enterprise*, edited by Sarah Whatmore, Philip Lowe, and Terry Marsden. London, England: David Fulton, 1991.

Wagner, H., and H.G. Wagner. "Die Veronderung der Bewasserung swirtschaft im nordlichen Suditalien, 1960–1990." *Petermann's Geographische Mitteilungen*, 136, No. 2-3 (1992): 139-154.

Le Meur, P.Y. "Possibilities et limites d'une politique agricole regionale: Le cas de la Vallee d'Aoste." *Revue de Geographie Alpine*, 81, No. 2 (1993): 85-100.

Simoncelli, R. "Dinamismo dell-agricoltura italiana e coinvolgimento del territorio rurale." *Bolletin della Societa Geografica Italiana*, Serie 12, 2, No. 3 (1997): 287-308.

Brogna, Marco. "L'Agriturismo in Italia." *Rivista Geografica Italiana*, 106, No. 4 (1999): 573-594.

DiGiacomo, Maria, ed. *Atlante Tematico dell'Agricoltura Italiana*. Rome, Italy: Societa Geografica Italiana, 2000.

CROPS AND CROPPING PATTERNS

Roletto, Giorgio B. "La culture du chanvre en Italie." *Annales de Geographie*, 32 (1923): 339-348.

Blanchard, W.O. "The Status of Sericulture in Italy." *Annals of the Association of American Geographers*, 19 (1929): 14-20.

Robertson, C.J. "The Italian Beet-Sugar Industry." *Economic Geography*, 14, No. 1 (1938): 1-15.

Desplanques, Henri. "Il paessagio rurale della coltura promiscua in Italia." *Rivista Geografica Italiana*, 66 (1959): 29-64.

Mancosu, F. "Introduzione e coltirazione del riso in Sardegna." *Rivista Geografica Italiana*, 66 (1959): 375-381.

Ronchi, R. "Osservazioni sulla coltivazione della vite in Liguria." *Bolletin della Societa Geografia Italiana*, Serie 8 (1959): 439-451.

Wheeler, David L. "An Experiment in Cooperative Agriculture: The Rice Communities of the Po River Delta." *Journal of Geography*, 62, No. 7 (1963): 310-316.

Rother, K. "Saisonwanderung und Tabakanbau am Golf von Tarento." *Geographische Rundschau*, 20, No. 8 (1968): 296-301.

Beretta, P.L. "Una recente coltura floreale nella pionura dell'Oltrepo Pavene." *Bolletin della Societa Geografica Italiana*, Serie 9, No. 12 (1971): 323-334.

Pipino, Antonio. "La geografia del cedro in Calabria." *Rivista Geografica Italiana*, 78 (1971): 48-66.

Marcarcini, Paolo. "Il limite dell'olivo nella Romagna e in genere nell'Italia continentale." *Rivista Geografica Italiana*, 80 (1973): 155-197.

Tirone, Lucien. "Mutations recentes du vignoble italien." *Mediterranee*, 23, No. 4 (1975): 11-36.
Tirone, Lucien. "Le cadastre viticole italien." *Mediterranee*, 23, No. 4 (1975): 87-92.
Floridi, V. "Recenti suiluppi della carilicoltura nel veterbese Studio di geografia agraria." *Bolletin della Societa Geographica Italiana*, Serie 10, No. 5 (1976): 151-190.
Rother, K. "Ein neues Anbaugebiet fur Erdburen in Suditalien." *Zeitschrift fur Wirtschaftsgeographie*, 24, No. 1 (1980): 33-39.
Tiggesbaumber, Gunter. "Der Agrumen-Anbau am Gardasse: Zur Innovation und Retraktion eines Agrarlandscaftselements." *Die Erde*, 116 (1985): 49-65.
Giuliana, Andreotti G. "La frutticoltura in provincia de Ravenna: Note di geografia agraria." *Rivista Geografica Italiana*, 94 (1987): 185-207.
Marengo, Marina. "Una tipica frangia viticola di qualita: La Cote d'Or." *Rivista Geografica Italiana*, 95 (1988): 559-573.
Azoulay, Raymondo. "Della varieta Spadone alla Bright: L'evoluzione della tabacchio coltura umbra e del relativo paesaggio." *Rivista Geografica Italiana*, 98 (1991): 185-200.
Dagradi, P. "The Development of Crops and the Agricultural Layout of the Po Valley" (in Italian). *Memorie-Societe Geografica Italiana*, 53 (1995): 117-133.
Tirau, L. "Les dynamiques recentes du vignoble italien." *Mediterranee*, 83, No. 1-2 (1996): 87-96.

ANIMAL HUSBANDRY

Bergeron, R. "Problemes de la vie pastorale en Sardigne (deuxieme article)." *Revue de Geographie de Lyon*, 44 (1969): 251-280.
Pagnini, Paola A.M. "La geografia agraria dell'alta Pusteria secondu un recente studio." *Rivista Geografica Italiana*, 77 (1970): 181-189.
Sprengal, Udo. *Die Wanderherdenwirtschaft im mittel- und sudositalien ischen Raum*. Marburg, Germany: Marburger Geographische Schriften, Heft 51, 1971.
Ortoloni, Mario. "Paserizia e agricoltura Sulle Ande Centrali." *Rivista Geografica Italiana*, 81 (1974): 103-106.
Leone, A., and M. Zacagnini. "Evolution de l'elevage des moutons dans la montagne sarde." *Bulletin de la Societe Languedocienne de Geographie*, 18, No. 3-4 (1987): 229-246.
Kennedy, Maya. "Pastoral Identity in the Changing Cultural Landscape of Highland Sardinia." Master's Thesis, University of Wisconsin, Madison, Wis., 1994.

DAIRYING

Maretti, C.L. "Aspetti geografici dell'approvvigio momento di latte a Roma." *Bolletin della Societe Geografica Italiana*, Serie 10, No. 8 (1979): 619-646.

FARM LAND

Despois, J. "Notes sur les culteurs en terrasses de la Sardaigne." *Revue de Geographie de Lyon*, 36 (1961): 7-10.

Rother, K. "Agrargeographische Beobachtungen im Kustentiefland von Metapont." *Erdkunde*, 21 (1967): 26-33.

Desplanques, Henri. "Une nouvelle utilisation de l'espace rural en Italie: L'agritourisme." *Annales de Geographie*, 82, No. 2 (1973): 151-164.

Anon. *Agricultural Typology and Land Utilization Studies Carried out with a Financial Contribution of the Consiglio Nazionale delle Richerche Roma*. Verona, Italy: University of Verona, Center of Agricultural Geography, 1975.

Dziewulska, A. "Kierunki rolniczego uzytkowania ziemi i chowu zwierzat we wloszech." *Przeglad Geograficzny*, 54 (1982): 571-589.

Alexander, David. "The Reclamation of Val-di-Chiana (Tuscany)." *Annals of the Association of American Geographers*, 74, No. 4 (1984): 527-550.

Rother, K. "Agrarian Development and Conflicts of Land Utilization in the Coast Plains of Calabria (South Italy)." *GeoJournal*, 13, No. 1 (1986): 27-36.

King, Russell, and John Killingbeck. "Agricultural and Land Use Change in Central Basilicata: From Carlo Levi to the Communita Montana." *Land Use Policy*, 7, No. 1 (1990): 7-26.

Pungetti, Gloria. "Anthropological Approach to Agricultural Landscape History in Sardinia." *Landscape and Urban Planning*, 31 (1995): 47-56.

AGRICULTURAL POPULATIONS AND SOCIAL ISSUES

King, Russell. "The Human Factor in Agricultural Development: A Case-Study from Sardinia." *Professional Geographer*, 23, No. 3 (1971): 204-207.

Monheim, R. "Die Agrostadt Siziliens: Ein stadtischer Typ agrarischer Grossiedlungen." *Geographische Zeitschrift*, 59 (1971): 204-225.

Schliebe, Klaus. *Die jungere Entwicklung der Kulturlanschaft des Campidano (Sardinien)*. Tubingen, Germany: Im Selbstaverlag des Geographischen Instituts der Universitat, 1972.

Sabelberg, Elmar. *Der Zerfall der Mezzadria in der Toskana urbana: Entstehung Bedeutung und gegenwartige Auflosung eines agraren Betriebssystem in Mittelitalien*. Koln, Germany: Kolner Geographische Schriften, Heft 33, 1975.

Cavazzoni, A. "Social Determinants of Part-Time Farming in a Marginal Region of Italy." Pp. 101-113 in *Part-Time Farming: Problem or Resource in Rural Development. Proceedings First Rural Geography Symposium, University of Guelph, June 18–20, 1975*, edited by A.M. Fuller and Julius A. Mage. Norwich, England: Geo Abstracts Ltd., 1976.

Sciacca, Marisa. *La terre del Sud: La formaziona del paesaggio agrario meridionale moderno*. Cosenza, Italy: Lerici, 1977.

Canosci, Doretta, and Alberto Melelli. "L'evolution recente des structures foncieres et du paysage rural dans un domaine ombrien d'avant-garde: La Tenuta d'Ascognano." Pp. 459-471 in *Recherches de Geographie rurale: Hommage au Professeur Frans Dussart*. Liege, Belgium: Bulletin de la Societe de Geographie de Liege, Vol. 1, 1979.

March, Henning. *Apulkien: Resourcerag kulturlandskab i en syditaliensk egn.* Copenhagen, Denmark: C.A. Reitzel, 1981.

Loose, Rainer. *Agrargeographie des sudwestlichen Trentino: Landwirtschaft und agrarsoziale Verhaltnisse der Valli Giudicarie in der Mitte des 19. Jahrhunderts und in der Gegenwart.* Wiesbaden, Germany: F. Steiner, 1983.

Lutz-Dollinger, Brigitte. *Buchweizenanbau und Buchweizenbauern in Sudtirol: Ein Beitrag zur Agrar- und Sozialgeographie Sudtirols.* Innsbruk, Austria: Universitatsverlag Wagner, 1986.

ENVIRONMENTAL AND CONSERVATION ISSUES

Huntington, Ellsworth. "Climatic Changes and Agricultural Decline as Factors in the Fall of Rome." *Quarterly Journal of Economics*, 31 (1917): 173-208.

Monti, Sabastiano. "La Piana del Sele Ricerca di Geografia Agraria." *Rivista Geografica Italiana*, 81, No. 2 (1974): 145-208.

Ruaro, Paulo, Vittorio Adriano, and Andrea Mezzetti. "The Climate of 1992 in the Alba Wine Region." *Journal of Wine Research*, 4, No. 3 (1993): 233-242.

Clarke, M.L., and H.M. Rendell. "The Impact of the Farming Practice of Remodelling Hillslope Topography on Badland Morphology and Soil Erosion Processes." *Catena*, 40, No. 2 (2000): 229-250.

IRRIGATION

Rother, K. "Bewasserungsgebiete und Bewasserrungsprojekte in-Sudostitaliena." *Erdkunde*, 31, No. 4 (1980): 287-292.

Cusimano, Girolano. "La cultura idraulica nel bacino de Mediterraneo: Territorio e irrigazione in Sicilia." *Geotema*, 4 (1996): 109-122.

AGRICULTURAL MARKETING AND DISTRIBUTION

Merlini, Giovanni. "Il commercio e l'importazione del legnome in Italie." *Rivista Geografica Italiana*, 39 (1932): 109-120.

Malta

GENERAL TEXTS, ARTICLES, AND THESES

Young, B.S. "Agricultural Landscapes of the Maltese Islands." *Journal of Geography*, 63, No. 1 (1964): 23-31.

Short, David E. "Traditional-Style Farming and Values for Sustainable Development in Malta." *Tijdschrift voor Economische en Sociale Geographie*, 88, No. 1 (1997): 41-52.

FARM LAND

Charlton, William, and Brian W. Beeley. "Agriculture, Land Use and Resource Transformation in Malta." *Tijdschrift voor Economische en Sociale Geographie*, 84, No. 5 (1993): 325-331.

Portugal

GENERAL TEXTS, ARTICLES, AND THESES

Weber, Peter. "Alentejo-Bedingungen und Entwicklung smolichkeiten der landwirtschaft im sudlichen Portugal." *Geographische Rundschau*, 21, No. 10 (1969): 389-400.

Freund, Bodo. *Siedlungs- und agrargeographische Studien in der Terra de Barroso/Nordportugal*. Frankfurt am Main, Germany: Frankfurter Geographische Hefte 48, 1970.

Freund, Bodo. *Wirtschaftsgeographie Analyse der portugiesischen Viehhaltung*. Frankfurt, Germany: Frankfurter Wirtschafts- und Sozialgeographische Schriften, Heft 10, 1971.

Lema, Paula B. "Fontes para o estudo da agricultura em Portugal." *Finisterra*, 6 (1971): 131-141.

Freund, Bodo. "Zur Agrarlandschaft-sentwicklung in Transguadianaland (Portugal)." *Erdkunde*, 26, No. 4 (1972): 252-265.

Freund, Bodo. "Struktur und raumliche Ordnung der portugiesischen Landwirtschaft." *Die Erde*, 105 (1974): 151-178.

Moreira da Silva, Rosa F. "Geografia agraria da area das doacoes (de 1928 a 1986)." *Finisterra*, 11, No. 22 (1976): 228-257.

Gaspar, Jorge, and Peter Gould. "The Cova da Beira: An Applied Structural Analysis of Agriculture and Communications." *Lund Studies in Geography*, Series B, 48 (1981): 183-214.

Castro Caldas, Joao, and Manuel Belo Moreira. "Quelle perspective pour les agricultures dans les zones de montagnes au Portugal?" *Revue de Geographie Alpine*, 80, No. 4 (1982): 117-128.

Guichard, D., O. Balabanian, and M. Drain. "Contributions au debort sur l'agriculture au Portugal a l'epoque de Salazar." *Revue Geographique des Pyrenees et du Sud-Ouest*, 54, No. 4 (1983): 403-423.

Poinard, M. "L'agriculture au Portugal a l'epoque de Salazar: Une politique de non-developpement." *Revue Geographique des Pyrenees et du Sud-Ouest*, 54, No. 4 (1983): 389-402.

Rodrigues, A. "Mutations agricoles en cours au Portugal: L'exemple du canton de Silves (Algarve)." *Revue Geographique des Pyrenees et du Sud-Ouest*, 54, No. 4 (1983): 441-458.

Domenech, B., and M. Drain. "La reforme agraire portugaise atelle contreut un espace nouveau?" *Revue Geographique des Pyrenees et du Sud-Ouest*, 55, No. 4 (1984): 457-468.

Wienberg, D. "Information uber die portugiesische Landwirtschaft." *Berichte uber Landwirtschaft*, 62, No. 3 (1984): 422-446.

Cavaco, C. "A Agricultura a tempo porciel como factor de estabilidade e de Paz social em Portugal." *Finisterra*, 20, No. 39 (1985): 47-107.

Cavaco, C. "Lethargie et mutation de l'agriculture portugaise: Quelques aspects structuraux et spatiaux." *Annales de Geographie*, 97, No. 3 (1988): 275-290.

Drain, M. *Portugal: La reforme agrarie*. Paris, France: Universalia, 1985.

Oliveira, Baptista F. "L'agriculture portugaise." *Annales de Geographie*, 97, No. 3 (1988): 259-274.

Raven, G.P. "The Development of the Portugese Agrarian Reform." *Journal of Rural Studies*, 4, No. 1 (1988): 35-44.

Schacht, Siegfried. "Die portugiesische Agrarreform und ihre Auswirkungen auf die landwirtschaftlichen Besitz-und Betriebsverhaltnisse: Dargestelt am Beispiel des kreises Alcazar do Sol." *Erdkunde*, 42, No. 3 (1988): 203-213.

Unwin, Timothy. "The Propagation of Agrarian Change in North-West Portugal." *Journal of Rural Studies*, 4, No. 3 (1988): 223-238.

Sousa Dias, A.J.C.C. "Agricultural Development in the Mondego Valley, Portugal." Ph.D. Dissertation, University of Durham, Durham, England, 1989.

Kummert, C. "Entwicklungen und Entwicklungsperspektiven der landwirtschaft in den portugiesischen Bergregionen." *Geographische Zeitschrift*, 78, No. 1 (1990): 38-47.

CROPS AND CROPPING PATTERNS

Ackerman, Edward A. "The Wine Valley of North Portugal." *Journal of Geography*, 36, No. 10 (1937): 333-353.

Enjalbert, Henri. "L'alto Douro: Les vins de Porto." *Les Cahiers d'Outre Mer*, 2 (1949): 289-342.

Stanislowski, Dan. *Landscapes of Bacchus: The Vine in Portugal*. Austin, Tex.: University of Texas Press, 1970.

Reboul, J., and S. Lerat. "Vignobles portugais." *Revue Geographique des Pyrenees et du Sud-Ouest*, 42 (1971): 233-238.

Guichard, F. "La vigne et le vin au Portugal." *L'Information Geographique*, 44, No. 1 (1980): 12-20.

Guichard, F. "Le vignoble portugais et la communaute economique europeenne." *Mediterranee*, 65, No. 3 (1988): 37-44.

Medeiros, Carlos A. "Contribuicao para o estudo da vinha nos Azores." *Finisterra*, 29, No. 58 (1994): 199-229.

ANIMAL HUSBANDRY

Freund, Bodo. *Wirtschaftsgeographische Analyse der Portugiesischen Viehholtung*. Frankfurt, Germany: Frankfurter Wirtschafts- und Sozialgeographie, Heft 10, 1971.

AGRICULTURAL POPULATIONS AND SOCIAL ISSUES

Unwin, Timothy. "Farmers' Perceptions of Agrarian Change in North-West Portugal." *Journal of Rural Studies*, 1, No. 4 (1985): 339-358.

ENVIRONMENTAL AND CONSERVATION ISSUES

Correia, Teresa P. "Threatened Landscape in Alentejo, Portugal: The Montado and Other Agro-Silvo-Pastoral Systems." *Landscape and Urban Planning*, 24 (1993): 43-48.

IRRIGATION

Bouet, Guy, and O. Balabanian. "Reforme agraire et agriculture irriguee au Portugal: L'evolution du perimetre irrigue du Caia (Haut-Alentejo)." *Revue Geographique des Pyrenees et du Sud-Ouest*, 54, No. 4 (1983): 425-440.

Spain and Andorra

ATLASES AND GRAPHIC PRESENTATIONS

Gonzalvez Perez, Vicente. "El Mapa Actual de Agrios de la Provincia de Alicante." *Estudios Geograficos*, 36 (1975): 509-525.

GENERAL TEXTS, ARTICLES, AND THESES

Jensen, Soren. "Agricultural Methods on the Canaries, Fuerteventura and Lanzarotte." *Economic Geography*, 10, No. 1 (1934): 99-108.

Dobby, E.H.G. "Agrarian Problems in Spain." *Geographical Review*, 26, No. 2 (1936): 177-189.

Garcia-Badell, Gabriel. "Estudio Sobre la Distribucion de la Extension Superficial y de Riqueza de la Propiedad Agricola en Espana Entre las Diferentes Categorias de Fincas." *Estudios Geograficos*, 7 (1946): 171-223.

Fontavella, Vicente. "Los Agrios en el Mundo." *Estudios Geograficos*, 14 (1953): 17-48.

Llobet, Salvador. "De Geografia Agraria de la Comarca del Moresma (Barcelona)." *Estudios Geograficos*, 16 (1955): 23-72.

Llobet, Salvador. "De Geografia Agraria en la Comarca del Moresme." *Estudios Geograficos*, 16 (1955): 215-298.

Pons, Bartolome B. "Evaluacion de la Estructura Agraria de Termino de Ocana." *Estudios Geograficos*, 17 (1956): 185-206.

Matznetter, Josef. *Die Kanarischen Inselen, Wirtschaftgeschichte und Agrargeographie*. Gotha, Germany: Herman Haack, Petermann's Geographische Mitteilungen No. 266, 1958.

Irujo, Carmen. "El Cultivo y Producion del Centeno en Espana." *Estudio Geograficos*, 21 (1960): 287-311.

Mayer, Eberhard. *Moderne Formen der Agrarkolonisation im sommertrocken Spanien: Eine geographische Untersuchung*. Stuttgart, Germany: Stuttgarter Geographische Studien, Band 70, 1960.

Cabo Alonson, Angel. "Fuentes Para la Geografia Agraria de Espana." *Estudios Geograficos*, 22 (1961): 223-249.

Galindo, Jose L.M. "Arcaismo y Modernidad en la Explotacion Agraria de Valdebura (Leon)." *Estudios Geograficos*, 22 (1961): 167-222.

Barbaza-Azema, Y. "Paisaje Rural y Estructura Agraria del Littoral Septentrionel de Cataluna a Principios del Siglo XVIII (Traduccion de S. Llobert)." *Estudios Geograficos*, 24 (1963): 565-590.

Coresa Cortes, C. "Geografia Agraria de Utiel (Valencia)." *Estudios Geograficos*, 24 (1963): 199-256.

Quiros, F. "Sobre Geografia Agraria del Campo de Calatrava y Valle de Alcudia." *Estudios Geograficos*, 26 (1965): 207-230.

Gordon, Marvin. "The Badajoz Plan: A Case Study of Planned Agricultural Development in West Central Spain." Ph.D. Dissertation, Columbia University, New York, N.Y., 1966.

Martin Galindo, Jose L. "La Dehesa Extremena Como Tipo de Explotacion Agraria." *Estudios Geograficos*, 27 (1966): 157-226.

Burriel de Orueta, Eugenio L. "Geografia Agraria de Onda." *Estudios Geograficos*, 29 (1968): 575-640.

Martin Galindo, Jose L. "El Caserio Vasco Como Tipo de Explotacion Agraria." *Estudios Geograficos*, 29 (1968): 205-244.

Zimmermann, Gerd. *Die bauerlich Kulturlandschaft in Sudgalicien: Ein Beitrag zur Geographie eines Ubergangsgebietes auf der Iberishcen Halbinsel*. Heidelberg, Germany: Heidelberger Geographische Arbeiten, Heft 23, 1969.

Burreil e Orueta, Eugenio L. *La Huerta de Valencia: Zona Sur. Estudio de Geografia Agraria*. Valencia, Spain: Instituto de Geografia, Institucion Alfonso el Magnanimo, 1971.

Castell Llacer, Vincente. *El Paisaje Agrario de Algemesi*. Valencia, Spain: Instituto de Geografia, Institucion Alfonso el Magnanimo, 1971.

Gil Olcina, Antonio. *El Campo de Lorca: Estudio de Geografia Agraria*. Valencia, Spain: Departamento de Geografia, Facultad de Filosofia y Letras, 1971.

Odouard, Albert. "Structure foncieres et agriculture speculative dans l'espace canarien." *L'Espace Geographique*, 1, No. 4 (1972): 230-238.

Lopez Gomez, Antonio. "Colectivismo y Sistemas Agrarias en la Serrania de Atienza (Guadalajara)." *Estudios Geograficos*, 35 (1974): 519-578.

Ortega Alba, Francisco. *El Sur de Cordoba: Estudios de Geografia Agraria*. Cordoba, Spain: Trofeo de Investigacion del Monte de Piedad y Caja de Ahorros de Cordoba, 1974. 2 volumes.

Galindo, Jose L.M. "Paisajes Agrarios Moriscos en Almeria." *Estudios Geograficos*, 36 (1975): 673-696.

Castro, Julio V. "Transformaciones Agrarias en le Duero Medio." *Estudios Geograficos*, 37 (1976): 49-82.

Bruser, Gerhard. *Die Landwirtschaftsformationen in Alta Extremadura*. Saarbruken, Germany: Arbeiten aus dem Geographischen Institut der Universitat des Saarlandes, Heft 25, 1977.

Font, Juan G. "Actividades y Estructuras Agrarias en la Sagra (Toledo)." *Estudios Geograficos*, 39 (1978): 49-71.

Fortanete, J.E. "Agricultural Change in Llucerva" (in Spanish). *Guadernos de Geografia*, 22 (1978): 73-92.

Ontiveros, Antonio L. "Algunes Aspectos de la Evolucion Agraria Reciente en Castilla de Nueva." *Estudios Geograficos*, 39 (1978): 5-22.

Teschendorff, Wolfgang. *Der Kustenhof von Valencia: Eine alte Agrarlandschaft im Kraftefeld moderner Wandlungen in Wirtschaft und Gesellschaft*. Regensburg, Germany: Regensburger Geographische Schriften, Heft 10, 1978.

Bosque Maurel, Joaquin. *Andalucia: Estudios de Geografia Agraria*. Granada, Spain: Aljihe-Liberia Galileo, 1979.

Bouhier, Abel. *La Galice: Essai geographique d'analyse et d'interpretation d'un vieux complexe agraire*. La Roche-sur-Yon, France: Yonnaise, 1979. 2 volumes.

Rodriguez, Wladimiro. *Agricultura en Canarias*. Santa Cruz de Tenerife, Spain: Editorial Interinsular Canaria, 1980.

Coli, Enrique M. "Estructuras Agrarias del Alto Vinalepo." *Estudios Geograficos*, 42 (1981): 431-452.

Arnolte, E. "Part-Time Farming in Spain: A Review." *GeoJournal*, 6, No. 4 (1982): 337-342.

Calvo Garcia-Tornel, Francisco. *Continuidad y Cambio en la Huerta de Murcia*. Murcia, Spain: Academia Alphonso X. de Sabio, 1982.

Sancho, Jose. *La Agricultura de Cara al Futuro*. Madrid, Spain: Editorial Cincel, Cuadernos de Estudio. Serie Geografia No. 6, 1982.

Camarero, Concepcion. "La Produccion Agraria en el Siglo XVIII." *Estudios Geograficos*, 45 (1984): 81-107.

Garcia-Ramon, M.D. "Agricultural Change in an Industrializing Area: The Case of the Tarragona Area." Pp. 140-155 in *Uneven Development in Southern Europe: Studies of Accumulation, Class, Migration and the State*, edited by Ray Hudson and James Lewis. London, England: Methuen, 1985.

Roux, M. "Immobilisme des structures et differenciations de la production dans l'agriculture en Andalousie." *Revue Geographique de la Pyranees et du Sud-Ouest*, 56, No. 2 (1985): 151-166.

Valle Buenestado, Bartolome. *Geografia Agraria de los Pedroches*. Cordoba, Spain: Diputacion Provincial de Cordoba, Servicio de Publicaciones, 1985.

Breuer, Toni. "Innovative Entwicklungen der landwirtschaft in Andalusien." *Geographische Rundschau*, 38, No. 1 (1986): 10-17.

Vizoso, Jose M.L. "A Capela: Transformaciones Agrarias en un Municipio de Transicion Hacia la Galicia Interior." *Estudios Geograficos*, 47 (1986): 333-346.

Cruz Villalon, J. "Political and Economic Change in Spanish Agriculture, 1950–1985." *Antipode*, 19 (1987): 119-133.

Guedes, M. "La modernization de l'agriculture en Espagne: Quelques questions." *Revue Geographique des Pyrenees et du Sud-Ouest*, 58, No. 2 (1987): 161-183.

Bachiller-Martinez, Jesus. "El Sector Agrario Soriano Ante et Nuevo Marco Socio-Economico Communitario: Analisis y Perspectives de un Area de Secano." *Estudios Geograficos*, 49 (1988): 477-504.

Butzer, Karl W. "Diffusion, Adaptation and Evolution of the Spanish Agrosystem." Pp. 91-109 in *The Transfer and Transformation of Ideas and Material Culture*, edited by Peter Hugill and D.B. Dickson. College Station, Tex.: Texas A&M University Press, 1988.

Daumas, M. "L'evolution recente des structures agraires en Espagne." *Annales de Geographie*, 97, No. 4 (1988): 419-443.

Garcia, Ramon. "Genero y actividad agraria en Espana: Una aproximacion a partio del Censo Agrario de 1982." *Documents da'Analisi Geografica*, 13 (1988): 16-33.

Juerez-Sanchez-Rubio, Cipriano, and Gregorio C. Martinez. "Transformaciones agrarias Indicativas y Planificacia Urbanistica." *Estudios Geograficos*, 49 (1988): 581-602.

Lora Bajo, Jesus C. "La Modernizacion Agraria en Los Espacios Marginales del Suroeste Espanol: Un Ejemplo en Huelva." *Estudios Geograficos*, 49 (1988): 633-657.

Extezareta, M., and L. Viladomiu. "The Restructuring of Spanish Agriculture and Spain's Accession to the EEC." Pp. 156-182 in *The International Farm Crisis*, edited by D. Goodman and Michael Redclift. Basingstoke, England: Macmillan, 1989.

Knuth, Burkhard. *Agrarstruktur und agrarregionale Differenzierung der Extremadura/Spanien.* Marburg, Germany: Marburger Geographische Schriften, Heft 113, 1989.

Fumega Pineiro, F.X., and R.G. Romani Barrientos. *El Programa de Desarrallo Regional de Galicia 1986–1990 y Su Politica en el Sector Agrario.* Poitiers, France: Travaux-Universite de Poitiers, Centre de Geographie Humaine et Sociale, No. 17, 1990.

Salon, Carrasco J. "La Introduccion de la Agricultura Comercial en el Pais Valenciano y la Terorin de la Difusion de las Innovaciones." *Cuadernos de Geografia*, 49 (1991): 89-120.

Vega, Javier M. "La Estructura Agraria en la Comarca de Los Montes de Toledo." *Estudios Geograficos*, 52 (1991): 673-696.

Delios Espana, E. "La Agricultura en el Area Periurbana de Valencia (L'Harta Nord): Evolucion Reciente de las Estructuras Agrarias." *Mediterranee*, 77, No. 1-2 (1993): 13-16.

Torres-Luna, Maria P., et al. "Deprise agricole en Galice." *Norois*, 40, No. 158 (1993): 217-233.

Bustos Gisbert, Maria Luisa. "La industria agroalimentaria en Espana." *Estudios Geograficos*, 55 (1994): 229-256.

Robinson, Guy M. "Dimensions Medioambientales de la Politica Agricola comun en el Reino Unido." *Agricultura y Sociedad*, 71 (1994): 127-151.

Logron, M.P.A., and M. Luz Hernandez Navarro. "Modificacion del paisage agricola en el regadio de la Ribera del Ebro (Zaragoza)." *Estudios Geograficos*, 57 (1996): 5-30.

Calcerrara, Raul R., and Javier Martinez Vega. "Transformaciones de usos y estructuras agrarias en la Mancha Oriental." *Estudios Geograficos*, 57 (1997): 451-476.

Leco Berrocal, Felipe. "El Envejecimiento del Empresariado Agrario en Extremadura: Una aproximacion mediante al Andies de Componentes Principales." *Estudios Geograficos*, 57 (1997): 625-642.

Navarro, J.G. "Cambios recients en la aplicacion de la nueva politica agraria comun y sus repecusions en Andalucia y Granada." *Cuadernos Geograficos*, 27 (1997): 29-53.

Perez, R.S. "La politica agraria Europea: Oportunidades y amenazas para la sierra norte de Sevilla." *Cuadernos Geograficos*, 27 (1997): 55-75.

Bouhice, A. "Courdeurs du passi et mutations recentes: La situation actuelle de l'agriculture galicienne (1965–1996)." *Annales de Geographie*, 107, No. 1 (1998): 33-58.

Picazo, M.T. "Fuentas fiscales e historia agraria: El debate en torno de las posibilidades heuristicas de las amillaramientos." *Estudios Geograficas*, 59 (1998): 285-310.

Mazarra, Angel P. "Toma de decision e implantacion en la polita agrambiental espanola." *Estudios Geograficos*, 61 (2000): 325-352.

Mazarra, Angel P. "Agri-Environmental Policy in Spain: The Agenda of Socio-Political Developments at the National, Regional and Local Levels." *Journal of Rural Studies*, 17, No. 1 (2001): 81-97.

CROPS AND CROPPING PATTERNS

Bull, William F. "The Olive Industry of Spain." *Economic Geography*, 12, No. 2 (1936): 136-154.

Lopez-Gomez, Antonio. "Riegos y cultivos en la Huerta de Alicante." *Estudios Geograficos*, 12 (1951): 701-771.

Fontavella, Vicente. "El Azafron: Su cultivo y comercio en Espana." *Estudios Geograficos*, 13 (1952): 239-256.

Crist, Raymond. "Rice Culture in Spain." *Scientific Monthly*, 84 (1957): 66-74.

Benito Arrang Cesteros, J. "El Cultivo y la Industrializacion de la Achicaria en Expana." *Estudios Geograficos*, 19 (1958): 135-140.

Frey, H. "Vegetacion y Cultivos de Akocebre (Castellon) y Sus Alvededores." *Estudios Geograficos*, 20 (1959): 269-272.

Huetz de Lemps, Alain. "El Vinedo de la 'Tierra de Medina' en los Siglos XVII y XVIII." *Estudios Geograficos*, 20 (1959): 111-126.

Llobet, S. "La Industria del Vino Espumoso Espanol." *Estudios Geograficos*, 20 (1959): 459-482.

Lauer, Wilhelm. "Tipos de Cultivo en la Espana Semiarida: Illustrades con el Ejemplo de la Mancha." *Estudios Geograficos*, 21 (1960): 509-527.

Castell, Llacer V. "Transformacion de cultivos en la ribera del Jucar: La sustitucion del arroz par el maiz." *Estudios Geograficos*, 23 (1962): 251-258.

Cardo, J. "Los Cultivos del Siglo XIX en el 'Alt Camp de Tarrogona.'" *Estudios Geograficos*, 24 (1963): 5-38.

Hofmeister, B. "The Impact of State Regulations on Spain's Major Olive Growing Area." Pp. 11-24 in *Agricultural Geography: IGU Symposium*, edited by E.S. Simpson. Liverpool, England: University of Liverpool, Department of Geography, 1964.

Quadra-Salcedo y Gayorre, S. "Limite Meridional del Haja y Septentrional del Olivo en Navarra." *Estudios Geograficos*, 26 (1965): 41-84.

Huetz de Lemps, Alain. *Vignobles et vins du Nord-Ouest de l'Espagne*. Bordeaux, France: Editeurs, 1967. 2 volumes.

Geiger, Folkwin. *Die Aridit at in Sudostspanien: Ursachen und Kuswirkunge im Landschaftsbild*. Stuttgart, Germany: Stuttgarter Geographische Studien, Band 77, 1970.

Fernandez, Jesus G. "El Cultivo del Arroz y su Expansion en el Siglo XVIII en los Llanos Litorales del Golfo de Valencia." *Estudios Geograficos*, 32 (1971): 163-187.

Hofmeister, B. "Four Types of Agriculture with Predominant Olive Growing in Southern Spain." *Geoforum*, 8 (1971): 15-30.

Regio, Jose M.R. "Le Economia del Vino Jerzano, Segun Garcia de Quevedo." *Geographica*, 13, No. 4 (1971): 249-250.

Schacht, Siegfried. *Drei ausgewahlte Reisbaulanschaften im westlichen Mittelmurgebeit: Kustenhof von Valencia, Sudobecken Camargue*. Kohn, Germany: Kolner Geographische Schriften, Heft 25, 1971.

Lopez Gomez, Antonio. "El Cultivo del Platano en Canarias." *Estudios Geograficos*, 33 (1972): 5-68.

Garcia Mamrique, E. "E Vinedo en la Costa Alpujarrena." *Estudios Geograficos*, 34 (1973): 501-538.

Parsons, James J. "Sand-Bed Agriculture: A Remarkable System for Early Truck Crop Production on the Arid Southeast Coast of Spain." *Proceedings, Association of American Geographers*, 5 (1973): 216-220.

Fourveau, F. "La riziculture dans les Marismas del Guadalquivire Isla Minima." *Revue Geographique des Pyrenees et du Sud-Ouest*, 45 (1974): 317-320.

Breuer, Toni. "Der Hopfenanbau in der Provinz Leon (Spanien): Eine Diffusionsanalyse." *Erdkunde*, 33, No. 1 (1979): 23-35.

Breuer, Toni. "Untersuchungen zur Adaption des Sonnenblumenanbaus in Niederandalusien." *Dusseldorfer Geographische Schriften*, 15 (1980): 69-87.

Breuer, Toni. "Speiseol-Produktion und Sonnenblumenanbau in Spanien." *Zeitschrift fur Wirtschaftsgeographie*, 22, No. 6 (1980): 177-180.

Piqueras, Juan. *La Vid y el Vino en el Pais Valenciano: Geografia Economica*, 1564–1980. Valencia, Spain: Institucion Alfonso el Magnanimo, 1981.

Daumas, M. "La culture du ble en Espagne." Pp. 429-451 in *Geographes d'aujaurd'hui*, edited by A. Dagarne and J.P. Paulet. Nice, France: Universite de Nice, 1984.

Olmedo, Victoriano G. "El Tabaco en las Islas Canarias (Tenerife) y su Problematica." *Estudios Geograficos*, 45 (1984): 35-57.

Huetz de Lemps, Alain. "Les vins espagnals et la CEF." *Travau-Institut de Geographie de Reims*, 67-68 (1986–1987): 35-57.

Justicia Segovia, Agustin. *Especializacion Agricole y Desorticulacion del Espacio: La Viticultura en Malaga Durante en Siglo XIX.* Malaga, Spain: Servicio de Publicaciones, Disputacion Provincial de Malaga, 1987.

Piqueras, Juan. "Evolution recente et dynamique spatiale du vignoble espagnol." *Mediterranee*, 65, No. 3 (1988): 29-36.

Lasanta-Martinez, T., et al. "Evolucion superficial del espacio cultivado en Cameros Viejo (Sistema Iberico) y su relacion con algunos factores geoecologicos." *Estudios Geograficos*, 50 (1989): 533-572.

Tout, D. "The Horticultural Industry of Almeria Province, Spain." *Geographical Journal*, 156, No. 3 (1990): 304-312.

Angles, Stephane. "Declin et renouveau de l'oleiculture dans la province de Seville (Espagne)." *Mediterranee*, 76, No. 3-4 (1992): 27-34.

Huetz de Lemps, Alain. *Vignobles et vins d'Espagne.* Talence, France: Presses Universitaires de Bordeaux, 1993.

Garcia-Martinez, P. "An Approximation of the Current State of the Main Crops in Alta Alpujarra, Western Granada." *Cuadernos Geograficos*, 24-25 (1995): 73-102.

Gomez Lopez, Jose D. "El cultivo protegido en la Communidad Economica Europa y en Espana: De la temia informal a la innovacion tecnologica." *Estudios Geograficos*, 56 (1995): 5-28.

Voth, A. "La difusion del cultivo del chirimoyo en la Costa del Sol." *Cuadernos Geograficos*, 27 (1997): 77-98.

Huetz de Lemps, Alain. "La resurrection des vignobles du Duaro en Espagan." *Annales de Geographie*, 109, No. 4-5 (2000): 488-504.

Menor Toribio, Jose, and Mari Jose Cuesta Aguilar. "L'erosion et les mesures agri-environnementales dans l'oliveraie andalouse." *Sud-Ouest Europeen*, 7 (2000): 61-68.

ANIMAL HUSBANDRY

Sorre, Max. "Nomadisme agricole et transhumance dans la Sierra Nevada." *Annales de Geographie*, 41 (1932): 301-305.

Aitken, R. "Routes of Transhumance on the Spanish Meseta." *Geographical Journal*, 106, No. 1 (1945): 59-69.

Lopez Gomez, Antonio. "La Transhumancia en Valdelaguna (Burgos)." *Estudios Geograficos*, 16 (1955): 163-166.

Gil Crespo, Adela. "Hoyos de Espino." *Boletin de la Real Sociedad Geografica*, 97 (1961): 176-205.

Torres Luna, Maria P. *La Navarra Humeda del Noroeste: Estudio Geografico de la Ganaderia.* Madrid, Spain: Instituto de Geografia Aplicada del Patronato Alonso de Herrera, 1971.

Fernandez, Julia A. "Un Modelo Geografico Sobre el Potencial Ganadero de Territorio." *Geographica*, 14, No. 2 (1972): 129-132.

Garcia, Antonio A., and Ann O. Poll. "La Transhumancia par Ferrocarril en Espana." *Estudios Geograficos*, 40 (1979): 385-413.

Rinschede, Gisbert. "Die Transhumance der Rinder in du spanischen Pyrenaen." *Zeitschrift fur Wirtschaftsgeographie*, 27 (1983): 92-103.

May, T. "Geographische Aspekte der Bienenhaltung an der ostlichen Costa del Sol un ihrem Hinterland." *Die Erde*, 121 (1990): 191-206.

Piqueras, Juan, and Carmen Sanchis-Deusa. "La Transhumancia Iberico-Valenciana en la Edad Moderna." *Cuadernos de Geografia*, 49 (1991): 35-47.

Jimenez, Paulino N. "Explotaciones de ovino en los montes de Ciudad Real: Viabilidad economica e impactes de la reforma de la P.A.C." *Estudios Geograficos*, 55 (1994): 109-134.

DAIRYING

Albert, P.C. "Studies on Geographical Prices for Milk" (in Spanish). *Revista de Estudios Agro-Sociales*, 83 (1973): 1-115.

FARM LAND

Naylor, J. "Land Consolidation in Spain." *Annals of the Association of American Geographers*, 49, No. 3 (1959): 361-373.

Quiros Linares, F. "La Desamartizacion, Factor Condicionante de la Estructura de la Propiedad Agraria en el Valle de Alcudia y Compos de Catatrava." *Estudios Geograficos*, 25 (1964): 367-408.

Garcia Fernandez, Jesus. "La Comarca de Tera: Datos para su Estructura Agraria." *Estudios Geograficos*, 27 (1966): 227-248.

Engass, Peter M. "A Method for Measuring and Comparing Agricultural Land Use: An Andalusian Experiment." *Professional Geographer*, 19, No. 3 (1967): 125-129.

Manuel, Joaquin B. "Latifundio y Minifundio en Andalucia Oriental." *Estudios Geograficos*, 34 (1973): 457-500.

Giner, J. C., and J. Romero Gonzalez. "La Estructura de la Propiedad y los Cultivos en Cullera (Valencia) Durante el Siclo XX." *Estudios Geograficos*, 39 (1978): 347-361.

Vega, Miguel P. "La Gran Propiedad Agraria en la Provincia de Valencia." *Estudios Geograficos*, 39 (1978): 73-92.

Lopez Andian, J.M. *Estructura y Morfologia Agraria en la Terra Cha*. Santiago de Compostela, Spain: Monografias de la Universidad de Santiago de Compostela, No. 55, 1979.

Mata Olmo, Rafael. *La Gran Propiedad en Los Llanos de Atequera*. Madrid, Spain: Instituto Juan Sebastian Elcano, Universidad Autonoma, 1979.

Bouhier, Abel. "Les phenomines recents de degradation des farmes d'organisation agraire en open field de l'extreme Sud-Est galacien." *Revue Geographique de l'Est*, 23, No. 3-4 (1983): 277-290.

Rodriguez, Vicente. *La Tierra en la Sagra Toledara: Su Evolucion de los Siglos XIX a XX.* Toledo, Spain: Caja de Ahorro Provincial de Toledo, 1984.

Lasanta-Martinez, T. "Organisation spatiale et dynamique recente de l'utilisation agraire du sols dans les Pyrenees centrales espagnoles." *Revue Geographique des Pyrenees et du Sud-Ouest*, 60, No. 2 (1989): 173-198.

Lasanta-Martinez, T. "L'agriculture en terrasses dans les Pyrenees centrales espagnoles." *Mediterranee*, 71, No. 3-4 (1990): 37-41.

Undillo, J.A., et al. "L'abandon de l'espace agricole dans la montagne sub-mediterraneenne en Espagne." *Revue Geographique des Pyrenees et du Sud-Ouest*, 61, No. 2 (1990): 237-253.

Douglas, T.D., S.J. Kirkby, and G.J. Park. "Agricultural Terrace Abandonment in the Alpujarra, Andalusia, Spain." *Land Degradation and Rehabilitation*, 5, No. 4 (1994): 281-294.

AGRICULTURAL POPULATIONS AND SOCIAL ISSUES

Mayer, Eberhard. *Moderne Formen der Agrarkolonisation im sommertrock-enen Spanien: Eine geographische Untersuchung.* Stuttgart, Germany: Selbstverlag des Geographischen Institut de T.H. Stuttgart, 1960.

Rodrigues Alfageme, M. "Castres y Luces: Estudio Comparado de la Estructura Social y Economica de un Pueblo Pesquero y Otro Agricola del Littoral Asturiano." *Estudios Geograficos*, 23 (1962): 553-582.

Garcia Mamrique, E. "Un Ejemplo de Paisaje Agrario de la Vega Alta del Segura: Abavan." *Estudios Geograficos*, 36 (1975): 417-451.

Maas, J.H.M. "The Behavior of Landowners as an Explanation of Regional Differences in Agriculture: Latifunists in Sevilla and Cordoba, Spain." *Tijdschrift voor Economische en Sociale Geographie*, 74 (1983): 87-95.

Corbera, M. "El Papel de la Politica Agraria Comun en la Evolucion de la Agricultura Familiar." *Anales de Geografia de la Universidad Computense*, 7 (1987): 241-251.

Daumas, M. "Structures foncieres agricoles et milieu geographique dans les Pyrenees espagnoles." *Revue de Geographique Alpine*, 75, No. 3 (1987): 213-232.

Mazarra, Angel P. "Reforma y Colonization Agraria: La Intervencion Oficial en la Comarca de Valdepus, 1932–1987." *Estudios Geograficos*, 49 (1988): 357-380.

Garcia Ramon, M.D. "Actividad Agraria y Genero en Espana: Une Aproximacion a Partir del Censo Agrario de 1982." *Documents d'Analise Geografica*, 14 (1989): 89-114.

Tulla, A.F. "La Mujer e las Explotaciones Agrarias del Piriveo Catalan." *Documents d'Analise Geografica*, 14 (1989): 171-201.

Painague, Angel, Eduardo Moyano, and Cecilia Diez. "Articulacion de Intereses y Conflicto entre Actones Sociales Debido a la Introduccion de Medidas de Regulacion Autsientel en la Agricultura Espanola." *Estudios Geograficos*, 59, No. 4 (1998): 661-688.

Caballe, Alba. "Farm Tourism in Spain: A Gender Perspective." *GeoJournal*, 48, No. 3 (1999): 245-252.

ENVIRONMENTAL AND CONSERVATION ISSUES

Gaussen, Henri. "A View from Canigou: Nature and Man in the Eastern Pyrenees." *Geographical Review*, 26, No. 2 (1936): 190-204.

Lauer, Wilhelm. *Formen des Feldbaus im semiariden Spanien: Oargestelit am Beispiel der Mancha*. Kiel, Germany: Kieler Geographische Schriften, Band 15, Heft 1, 1954.

Rossello Verger, V.M. "El Paisaje Natural y la Vida Agraria en el Sur y Sureste de Mallorca." *Estudios Geograficos*, 24 (1963): 311-368; Part II, 475-546.

Vaudor, J. "Sols et structures agraires dans la region de Madrid." Pp. 51-62 in *Les Societes Rurales Mediterraneennes: Actes du Colloque de Geographie Agraire*, edited by Centre Geographique et de Recherches Mediterraneennes. Aix-en-Provence, France: Institut de Geographie, 1972.

Ebert, Charles H.V. "Cinder Mulch Farming in Lonzarotte." *Ecumene*, 8 (1976): 18-26.

Lodez Ontiveros, A. "Chasse et activite agricole en Espagne et en Andalousie: Evolution recente." *Revue Geographique de Pyrenees et du Sud-Ouest*, 56, No. 2 (1985): 203-223.

Pardo, Fernando, and Luis Gil. "La transformacion del paisaje en la Sierra Pobre de Madrid: Influencia de la agricultura y ganaderia en la extinction local de los pinares." *Estudios Geograficos*, 57 (1997): 397-424.

Domingo, C. "Fuentes Agrarias Pre-Catastrales Valencianas: Su Utilidad Geografica." *Estudios Geograficos*, 59 (1998): 225-244.

Palet-Martinez, J.M., and J.M. Gurt Esparraguerra. "Amenagement et drainage des zones humides du littoral emporitain (Catalogue): Une lecture diachronique des structures agraires antiques." *Mediterranee*, 90, No. 4 (1998): 41-48.

Paniagua Mazorra, Angel. "Toma de decision e implantacion en la politica agroambiental espanola." *Estudios Geograficos*, 61, No. 2 (2000): 325-352.

IRRIGATION

Houston, J.M. "Irrigation as a Solution to Agrarian Problems in Modern Spain." *Geographical Journal*, 116, No. 1-3 (1950): 55-63.

Frohling, Margret. *Die Bewasserungslandschaften an der spanischen Mittelmurkuste*. Munster, Germany: Westfalische Geographische Studien, Heft 17, 1965.

Naylon, J. "An Appraisement of Spanish Irrigation and Land Settlement Policies Since 1939." *Iberian Studies*, 2 (1973): 12-17.

Courtot, R. "Teledetection et evolution des cultures irriguees dans la basse vallee du Jucar (Espagne)." *Mediterranee*, 37, No. 4 (1979): 95-182.

Butzer, Karl W., Juan F. Matea, Elizabeth K. Butzer, and Pavel Kraus. "Irrigation Agrosystems in Eastern Spain: Roman or Islamic Origins?" *Annals of the Association of American Geographers*, 75, No. 4 (1985): 479-509.

Bazzana, Andre. "Irrigation et maitrise de l'eau dans l'Espagne medievale: Quelques remarques methodologiques." *Revue Geographique de l'Est*, 26, No. 3-4 (1986): 199-204.

Feijoo, M. Luisa, Elena Calvo, and Jose Albinae. "Economic and Environmental Policy Analysis of the Flumen-Monegros Irrigation System in Huesca, Spain." *Geographical Analysis*, 32, No. 3 (2000): 187-204.

AQUACULTURE

Pinieiro-Antelo, A., and X. Santos-Solla. "Culture de la moule en Galcie: Aspects socio-economiques et spatiaux." *Cahiers Nantais*, 47-48 (1997): 125-132.

AGRICULTURAL LABOR STUDIES

Exterzareta, M. "Transformation of the Labour System and Work Processes in a Rapidly Modernizing Agriculture: The Evolving Case of Spain." Pp. 44-67 in *Labour and Locality*, edited by Terry Marsden, Philip Lowe, and Sarah Whatmore. London, England: David Fulton, 1992.

Eastern Europe

GENERAL TEXTS, ARTICLES, AND THESES

McDonald, John W. "The Spread of Socialized Agriculture in Eastern Europe." Master's Thesis, University of California, Los Angeles, Calif., 1965.

Enyedi, Gyorgy. "The Changing Face of Agriculture in Eastern Europe." *Geographical Review*, 57, No. 3 (1967): 358-372.

Kostrowicki, Jerzy. "Agricultural Types and Regions in East Central Europe." Pp. 229-248 in *Studies in Applied and Regional Geography*, edited by M. Shafi and Mehdi Raza. Aligarh, India: Aligarh Muslim University, Department of Geography, 1971.

Romanowski, Jacek I. "Geographic Research and Methodology on East-Central and Southeast European Agriculture." Pp. 127-159 in *Eastern Europe: Essays in Geographical Problems*, edited by George Hoffman. London, England: Methuen, 1971.

Tyszkiewicz, Wieslawa. "Problemes agricoles de la Cujavie et du pays de Chelmno." *Geographia Polonica*, 29 (1974): 399-418.

Szyrmer, Jacek. "The Degree of Agricultural Specialization" (in Polish). *Przeglad Geograficzny*, 47, No. 1 (1975): 117-135.

Dietl, J., and B. Gregor. "Dysproporcje w rozmieszczeiu produkcyjno-handlowej obslugi rolnictwa." *Przeglad Geograficzny*, 49 (1977): 657-676.

Kozysa, Jozef. "Agriculture in the European Countries of Council for Mutual Economic Co-operation (CMEC): The Present State and Development" (in Czech). *Czasopismo Geograficzne*, 49, No. 1 (1978): 53-65.

Spisiak, Peter. "The Analysis of Agro-Space of the Chosen CMEA Countries" (in Czech). *Acta Facultatis Rerum Naturalium Universitatis Comenianae Geographica*, 17 (1979): 241-253.

Istok, M.N. "An Economic Geographical Analysis of Food Economy in the Lower Danube Region" (in Serbian). *Foldrajzi Kozlemenyek*, 29, No. 3 (1981): 241-250.

Spisiak, Peter. "The Integration Process in the Agriculture of the COMECON Countries and This Impact upon the Spatial Structure of Slovakia" (in Slovak). *Acta Facultatis Rerum Naturalium-Universitatis Comenianae Geographica*, 20 (1982): 115-137.

Olszewski, Tadeusz. "Landwirtschaft in Mittelosteuropa: Geographische Anschaung." *Frankfurter Wirtschafts- und Sozialgeographische Schriften*, 47 (1985): 89-105.

Maurel, M.C. "A l'est des agricultures en transition: La question de la reversibilite." *Bulletin de l'Association Geographes Francais*, 69, No. 1 (1992): 3-12.

Kondras, Janet E. "Geopolitical Transitions in Agriculture: The Changing Role of American Food Policy in Eastern Europe, 1955–91." Pp. 53-70 in *The New Political Geography of Eastern Europe*, edited by John O'Loughlin and Herman van der Wusten. London, England: Belhaven, 1993.

Turnock, David. *Agricultural Change in the New Eastern Europe*. Cheltenham, England: Stanley Thomas, 1993.

Young, Craig. "A Bitter Harvest? Problems of Restructuring East-Central European Agriculture." *Geography*, 78, No. 1 (1993): 69-72.

Turnock, David. "Agriculture in Eastern Europe: The Transition and the Future." *GeoJournal*, 38, No. 2 (1996): 137-149.

Turnock, David. "Private Farming in Eastern Europe and the Former Soivet Union." *GeoJournal*, 45, No. 3 (1998): 229-232.

CROPS AND CROPPING PATTERNS

Strauss, David L. "Specialty Crop Agriculture in Eastern Europe and the Soviet Union. Hops: A Case Study." Ph.D. Dissertation, University of Washington, Seattle, Wash., 1968.

Biegajlo, Wladislaw, and Roman Kulikowski. "Crop Combination: A Comparative Case Study of Poland, Czechoslovakia and Hungary" (in Polish). *Przeglad Geograficzny*, 44, No. 3 (1972): 540-547.

Roudie, P. "Le role des techniques dans l'evolution des paysages viticoles d'Europe occidentale." *Geographia Polonica*, 38 (1978): 253-256.

ENVIRONMENTAL AND CONSERVATION ISSUES

Loczy, D., and L. Szalai. "Assessment of Loess as Parent Material for Agroecological Potential." *GeoJournal*, 36, No. 2-3 (1995): 275-280.

IRRIGATION

Carriere, P. "Les progres de l'irrigation des terres dans la vallee moyenne du Syr-Daria, Plaine de Ferghana: Steppe de la Faim." *Bulletin de l'Association de Geographes Francais*, 348 (1966): 15-33.

Albania

GENERAL TEXTS, ARTICLES, AND THESES

Blanc, Andre. "Naissance et evolution des paysages agraires en Albanie." *Geografiska Annaler*, 43, No. 1-2 (1961): 8-16.

Carriere, P. "L'agriculture albanaise: Un modile socialiste originale." *Bulletin de la Societe Languedocienne de Geographie*, 14, No. 1 (1980): 97-112.

Toepfer, Helmuth. "Zur Entwicklung der Landwirtschaft in Albanien." *Zeitschrift fur Agrargeographie*, 3, No. 2 (1985): 136-157.

Sjoberg, O. *The Agrarian Sector in Albania During the 1980s: A Changing Regional Focus*. Stockholm, Sweden: School of Economics, Studies in International Economics and Geography, 1990.

Toepfer, Helmuth. "Kultivierungsmassnahamen in der Albanischen Landwirtschat." *Erdkunde*, 45, No. 4 (1991): 308-312.

ANIMAL HUSBANDRY

Blanc, Andre. "L'evolution contemporaine de la vie pastorale en Albanie meridionale." *Revue de Geographie Alpine*, 51, No. 3 (1963): 429-461.

AGRICULTURAL POPULATIONS AND SOCIAL ISSUES

Blanc, Andre. "Recherches sur les communautes patriarcales et les structures agraires en Albanie du nord." *Bulletin de l'Association de Geographes Francais*, 292-293 (1960): 117-128.

ENVIRONMENTAL AND CONSERVATION ISSUES

Nuttonson, Michael Y. *Ecological Plant Geography of Albania: Its Agricultural Crops and Some North American Climatic Analogues*. Washington, D.C.: American Institute of Crop Ecology, 1947.

Bulgaria

GENERAL TEXTS, ARTICLES, AND THESES

Beshkov, Anastov. "The Upper Thracian Plain in Bulgarian Agriculture." *Economic Geography*, 15, No. 2 (1939): 179-184.

Billaut, Micheline. "La collectivisation agraire en Bulgarie: L'exemple du village de Petartch." *Annales de Geographie*, 69 (1960): 484-492.

Borisov, Zdravko. "Economico–Geographic Study of Agriculture in the Blagoevgrad Micro-Zone" (in Bulgarian). *Godisnik na Sofijskina Universitet. Geologo-Geografski fakultet, Geografija*, 59, No. 2 (1964/1965): 303-325.

Jordanov, T. "Utilization of the Comparative Method in Studying the Geography of Farming in Bulgaria" (in Bulgarian). *Izvestija-Balgarskoto Geografsko Druzestvo*, 6, No. 16 (1966): 121-130.

Ilieva, Margarita. "Zonality of the Rural Economy in Northwest Bulgaria" (in Bulgarian). *Problemi na Geografijata*, 2, No. 2 (1976): 44-53.

Tyszkiewicz, Wieslawa. "Typologia rolnictwa uspolexznionego kotling Trackiej (Bulgaria)" (in Polish). *Przeglad Geograficzny*, 52 (1980): 61-80.

Kolev, B. "Determining the Agro-Industrial Potential of the Southern Central Region with Respect to Major Industries" (in Bulgarian). *Izvestija-Balgarskoto Geografsko Druzestvo*, 19 (1981): 31-40.

Galczynka, B. "Typologia rolnictwa Bulgarii" (in Polish). *Przeglad Geograficzny*, 54 (1982): 551-570.

Jordanov, T. "Influence of the Natural Conditions and Resources on the Specialization of Agriculture in the Rhodope Mountains" (in Bulgarian). *Problemi na Geografijata*, 3 (1983): 15-28.

Lenormond, Pierre. "L'agriculture bulgare a l'age des complexes agro-industriels." *Annales de Geographie*, 92, No. 2 (1983): 129-151.

Galczynska, B. "Agricultural Typology of Bulgaria." *Geographia Polonica*, 50 (1984): 169-178.

CROPS AND CROPPING PATTERNS

Beshkov, Anastov. "Tobacco in Bulgaria." *Economic Geography*, 14, No. 2 (1938): 188-194.

Batakliev, Ivan. "Viticulture in Bulgaria." *Geography*, 24, No. 1 (1939): 85-94.

Demerdziev, Z. "Development of Tobacco Production and Creation of Tobacco Producing Regions in Bulgaria" (in Bulgarian). *Godisnik na Sofijskija universitet, Geologo-geografski fakultet*, 60, No. 2 (1967): 303-333.

Borisov, Zdravko. "Geography of Hot-House Production of Vegetables in Bulgaria" (in Bulgarian). *Godisnik na Sofijskija universitet, Geologo-geografski fakultet*, 62, No. 2 (1967/1968/1969): 327-335.

Demerdziev, Z. "Problems of the Prospective Development and Territorial Distribution of Tobacco Farming in the South-Western Tobacco Region in Bulgaria" (in Bulgarian). *Godisnik na Sofijskija universitet, Geologo-geografski fakultet*, 67, No. 2 (1975): 87-102.

Gergov, Georgi. "The Territorial Distribution of Grain Production in Bulgaria" (in Bulgarian). *Geografija*, 4 (1975): 1-4.

Garabedova, V. "Flax Production in Bulgaria" (in Bulgarian). *Geografija*, 26, No. 7 (1976): 1-4.

Dermerdziev, Z. "Viticulture in Bulgaria" (in Bulgarian). *Geografija*, 4 (1978): 1-5.

Jordanov, T. "Main Changes in the Geographic Distribution of Sunflower in Bulgaria and Their Influence on the Economic Effectiveness of Its Production" (in Bulgarian). *Problemi na Geografijata*, 3 (1978): 24-36.

Savov, Jordan. "Tobacco Complex in Bulgaria and Its Mapping" (in Bulgarian). *Problemi na Geografijata*, 2 (1978): 32-41.

Dermerdziev, Z. "Hothouse Vegetable Production in Bulgaria" (in Bulgarian). *Geografija*, 3 (1979): 1-4.

Jordanov, T. "Changes in the Structure of Plant-Growing and Their Influence on the Economic Effectiveness of the Utilization of Arable Land in the People's Republic of Bulgaria" (in Bulgarian). *Problemi na Geografijata*, 3 (1979): 13-22.

Atanasov, Nikolai. "The Culture of Etheric Essential Oils in Bulgaria" (in Bulgarian). *Geografija*, 6 (1980): 1-4.

Tankova, S. "The Production Relations of the Sugar Industry in the Districts Along the Danube River and Possibilities for Their Improvement" (in Bulgarian). *Problemi na Geografijata*, 4 (1981): 60-68.

Dzhambazo, I. "The Production of Flax and Hemp in Bulgaria" (in Bulgarian). *Geografija*, 5 (1982): 5-7.

Ilieva, Margarita. "Use of Forage Areas in Rita, Pirin, and the Western Borderland Mountains (Bulgaria)" (in Bulgarian). *Izvestija-Balgarskoto Geografsko Druzestvo*, 22 (1984): 63-72.

Cekov, N. "Classification of the Agro-Industrial Complexes: Producers of Grain Corn in the South Central Economic Region (Bulgaria)" (in Bulgarian). *Problemi na Geografijata*, 2 (1985): 59-68.

Popova, S. "Soya: An Important Reserve for Bulgaria's Cereal Production" (in Bulgarian). *Geografija*, 40 (1985): 4-8.

Jordanov, T. "Basin Changes in the Geography of Grain Produce in Bulgaria during the 40-Year Period of Socialist Construction" (in Bulgarian). *Problemi na Geografijata*, 1 (1986): 14-25.

ANIMAL HUSBANDRY

Gesev, Geso, and Anton Dinev. "All-Round Geographical Estimate of the Possibilities for Sheep-Breeding in the Fore-Balkan and the Highest Balkan Range Areas Based on Data for the Districts of Veliko Tarnovo and Gabrovo" (in Bulgarian). *Problemi na Geografijata*, 4 (1977): 39-46.

Demerdziev, Z. "Livestock Production in Bulgaria" (in Bulgarian). *Geografija,* 4 (1980): 1-5.

Gesev, Geso, and Anton Dinev. "Complex Geographical Assessment of the Hilly and the Mountainous Belt of Bulgaria with a View to the Development of Sheep-Breeding" (in Bulgarian). *Izvestija-Balgarskoto Geografsko Druzestvo*, 19 (1981): 41-47.

Kolev, B. "Applying Extrapolation in Geography of Stock Breeding in Order to Forecast the Number of Farm Animals" (in Bulgarian). *Izvestija-Balgarskoto Geografsko Druzestvo*, 20 (1982): 151-160.

Petrov, T. "Golemite problemi na 'Malkite' zhivitnov'dni podotraski v B'lgarija." *Geografija*, 42 (1988): 5-9.

FARM LAND

Gadomski, Wladyslaw, and Roman Kulikowski. "Land Utilization in the Tobacco Zone of Eastern Rhodope Mountains: The Case of the Collective Farm Shiroko Polje" (in Polish). *Dokumentacja Geografiszna*, 5 (1969): 73-90.

Lakov, Lazar. "Mapping the Land Stock of Agro-Industrial Complexes (on the Example of the Agro-Industrial Complex 'Septemvriiska Slava' in the District of Pazardzik)" (in Bulgarian). *Godisnik na Sofijskija universitet, Geologo-geografski fakultet*, 64, No. 2 (1969/1970/1972): 125-145.

Daneva, Marija, and Kiril Misev. "The Cadastre of Farming Land in the People's Republic of Bulgaria and Possibilities of Its Use in Geographical Studies" (in Bulgaria). *Problemi na Geografijata*, 1 (1978): 57-65.

Yarnel, Brent. "Decollectivization of Bulgarian Agriculture." *Land Use Policy*, 11, No. 1 (1994): 67-70.

ENVIRONMENTAL AND CONSERVATION ISSUES

Jordanov, T. "The Effect of Precipitation in the Growth Period on the Average Yields of Certain Farm Crops in the Eastern Upper Thracian Subregion" (in Bulgarian). *Problemi na Geografijata*, 1, No. 2 (1975): 34-48.

Yarnal, Brent. "Agricultural Decollectivization and Vulnerability to Environmental Change: A Bulgarian Case Study." *Global Environmental Change*, 4, No. 3 (1994): 229-244.

IRRIGATION

Petkov, P., and S. Dimitrov. "Irrigation in Bulgaria." *GeoJournal*, 40, No. 4 (1996): 405-411.

Petrov, K.V., and N. Kathijotes. "Non-Convential Water for Irrigation During Drought in Bulgaria." *GeoJournal*, 40, No. 4 (1996): 413-419.

TRANSPORTATION AND MECHANIZATION

Dzarova, E. "Economic Geographical Analysis of the Transportation of Farming Produce in the Southeastern Economic Region" (in Bulgarian). *Problemi na Geografijata*, 4 (1978): 54-64.

Czechoslovakia, The Czech Republic, and The Slovak Republic

GENERAL TEXTS, ARTICLES, AND THESES

Horbaly, William. "Agricultural Conditions in Czechoslovakia, 1950." Ph.D. Dissertation, University of Chicago, Chicago, Ill., 1951.

Isnard, Hildebert. "Notes sur l'agriculture tchecoslovaque." *Revue Geographique de l'Est*, 3, No. 2 (1963): 117-130.

Ivanicka, Koloman. "The Systems Analysis of Agriculture in the Hinterland of Bratislava" (in Slovak). *Acta Geographica Universitatis Comenianae, Economico-Geographica*, 10 (1971): 9-51.

Fialova, Ludmila. "Origin and Present Day State of Prague's Suburban Farmsteads" (in Czech). *Acta Universitatis Carolinae Geographica*, 2 (1973): 47-72.

Ivanicka, Koloman. "Facteurs dynamiques de transformation des structures agricoles en Slovaquie occidentale." *Bulletin de la Societe Languedocienne de Geographie*, 8, No. 2 (1974): 143-154.

Gotz, Antonin. "Geographical Differences in the Increase of Agricultural Production in Czechoslovakia" (in Slovak). *Studia Geographia Brno*, 52 (1975): 35-42.

Zelensky, Konstantine. "Types of Agricultural Production in Slovakia" (in Slovak). *Geograficky Casopis*, 31, No. 1 (1979): 92-99.

Zelensky, Konstantine. "Influence of the Geographic Environment in Slovakia on the Development of Agricultural Production" (in Slovak). *Geograficky Casopis*, 32, No. 2-3 (1980): 140-147.

Loboda, Jan. "The Problem of Diffusion Innovation Processes in a Farming Area of Lower Silesia." *Acta Universitatis Carolinae Geographica*, 23, No. 2 (1988): 20-31.

Spisiak, Peter. "Priestorove vztaly v agrokomplexe." *Geograficky Casopis*, 42, No. 4 (1990): 401-409.

Bockova, I. "Conditions et difficultes du passage de l'agriculture tchecoslovaque a l'economie de marche." *Bulletin de l'Association de Geographes Francais*, 69, No. 1 (1992): 39-44.

Baticle, Y. "Impact du developpement de Bongue (R.C.A.) sur le paysage periurbaine." *Bulletin de l'Association de Geographes Francais*, 71, No. 2 (1994): 152-155.

Gotz, Antonin. "Regional Differences in Transformation of Czech Agriculture After 1989" (in Czech). *Sbornic Cesta Geograficke Spolecnosti*, 99, No. 2 (1994): 93-100.

Gotz, Antonin. "Regional Inequalities in Agricultural Transformation in the Czech Republic." *Acta Universitatis Carolinae Geographica*, 29, No. 1 (1994): 19-29.

Ptacek, J. "Czech Agriculture in Transition" (in Czech). *Sbornik Ceske Geograficke Spolesnosti*, 101-102 (1996): 110-127.

Rialland, C. "Decollectivisation et paysages agraires dans une region peripheriques de la Republique Tcheque: La Valachie Morave." *Cahiers Nantais*, 46 (1996): 69-80.

Spisiak, Peter. "Agriculture in Slovakia and Its Future Prospects." *Acta Universitatis Carolinae, Geographica*, 32 (1997): 403-409.

CROPS AND CROPPING PATTERNS

Sabaka, Jan. "The Geography of Potato-Growing in Slovakia" (in Slovak). *Acta Universitatis Carolinae Geographia*, 2 (1968): 51-66.

Gotz, Antonin. *Rostlinna vyroba vychodnich Cech: Geograficka studie o rozmisteni zemedelske vyroby*. Brno, Czech Republic: Geograficky ustav CSAV, 1974.

Mocko, Zdenko. "Viticulture in Slovakia and Its Development" (in Slovak). *Studia Geographica Brno*, 52 (1975): 139-148.

Ungerman, J. "Correlating the Increase of Crop Production and Hazard for Drinking Water Resources in the Area of the Svratecka hornatina" (in Czech). *Zpravy-Geografickeho Ustavu CSAV*, 17, No. 4 (1980): 179-185.

Ungerman, J. "Intensification de la production vegetale en evolution: Methode de la differentiation des regions agricoles de la Republique Socialiste Tcheque." *Zpravy-Geografickeho Ustavu CSAV*, 17, No. 2 (1980): 79-87.

ANIMAL HUSBANDRY

Gotz, Antonin. *Animal Production in Eastern Bohemia* (in Czech). Brno: Studia Geographica Brno, No. 60, 1977.

Zrincak, G. "Le systeme d'elevage bovin bitier en Republique Tcheque." *Revue Geographique de l'Est*, 35, No. 2 (1995): 163-184.

DAIRYING

Zrincak, G. "Le systeme d'elevage bovin laitier en Republique Tcheque." *Revue Geographique de l'est*, 35, No. 2 (1995): 163-184.

FARM LAND

Gadomsky, Wladyslaw. "Land Utilization in the Upper Nitra Valley, Slovakia" (in Slovak). *Dokumentacja Geograficzna*, 1 (1967): 5-26.

Dorsz, B., and Wladyslawa Stola. "Agricultural Land Utilization in the Market-Gardening Zone of Litomeric, Bohemia" (in Czech). *Dokumentacja Geograficzna*, 5 (1969): 91-104.

Ivanickova, Alzbeta. "Problems of Transformation of Land Use in West Slovakia as Exemplified in the Village of Horne Oresang." *Acta Geographica Universitatis Comenianae, Economic-Geographica*, 10 (1971): 175-189.

Ivanicka, Koloman. "The Development of Agriculture and the Map of Utilization of the Land of the Slovak Socialist Republic" (in Slovak). *Studia Geographica Brno*, 61 (1977): 41-55.

Bucek, A., and J. Ungerman. "Complete Land Reformations in the Territory of the Productional-Organisation Unit Merin and Agricultural Utilization of the Landscape" (in Czech). *Zpravy-Geografickehlo Ustavu CSAV*, 15, No. 6-7 (1978): 93-104.

Kostrubiec, B. "Zagadnienia pomiaru stopnia komasaeji parcel w gos podartswack wiejskiek." *Acta Universitatis Wratislaviensis Prace Instytutu Geograficznego Seria B*, 438 (1979): 87-92.

Ungerman, J. "Hodnoceni krajiny z hlediska zemedelske produkcni funkce." *Zpravy-Geografickeho Ustavu CSAV*, 20, No. 1 (1983): 35-61.

Zelensky, Konstantine. "Landscape Types of Slovakia from the Agricultural Potential Stand Point." *Geograficky Casopis*, 36, No. 4 (1984): 378-391.

Zelensky, Konstantine. "Types of Arable Landscapes of the Slovak Socialist Republic." *Geografia Polonica*, 57 (1989): 175-185.

AGRICULTURAL POPULATIONS AND SOCIAL ISSUES

Spisiak, Peter. "Economic Classification and Typology of Settlements in a High-Production Agricultural Area" (in Czech). *Acta Facultetis Reum Naturalium Universitatis Comenianae, Geographica*, 19 (1981): 85-101.

ENVIRONMENTAL AND CONSERVATION ISSUES

Kurpelova, Margita. "Agricoclimatic Conditions of the Cultivation and Yields of Some Agricultural Cultures in Slovakia" (in Czech). *Geograficky Casopis*, 28, No. 1 (1976): 58-75.

Drgova, V. "Ecological Problems Arising from Intensive Agriculture in Western Slovakia." *GeoJournal*, 38, No. 2 (1996): 213-218.

AQUACULTURE

Lenhard, Franz. "The South Bohemian District—An Economic—Geographic Study with Special Regard to Fish Farming." *Geographische Berichte*, 36, No. 3 (1965): 161-182.

Hungary

GENERAL TEXTS, ARTICLES, AND THESES

Enyedi, Gyorgy. *A Delkelet-Alford mezogazdasagi foldrajza*. Budapest, Hungary: Akademiai Kiado, 1964.

Enyedi, Gyorgy. "Geographical Types of Agriculture in Hungary." Pp. 1-6 in *Agricultural Geography, International Geographical Union Symposium*, edited by E.S. Simpson. Liverpool, England: University of Liverpool Department of Geography, Research Paper No. 3, 1965.

Asztalos, Istvan, Gyorgy Enyedi, Bela Sarfalvi, and Laszlo Simon. *Geographical Types of Hungarian Agriculture*. Budapest, Hungary: Hungarian Academy of Sciences, Geographical Research Institute, Studies in Geography No. 3, 1966.

Enyedi, Gyorgy. "Regional Deviations of Agricultural Incomes in the Great Plains of Hungary" (in Hungarian). *Foldrajzi Ertesito*, 15, No. 3 (1966): 297-307.

Enyedi, Gyorgy. "Trends and Results of Work in Hungarian Agricultural Geography" (in German). *Wissenschaftliche Zeitschrift der Universitat Halle*, 15, No. 3 (1966): 473-483.

Kovacs, Csaba. "The Main Factors and Rules of the Spatial Arrangement of Subsistence and Commodity Producing Agriculture" (in Hungarian). *Foldrajzi Ertesito*, 15, No. 2 (1966): 211-234.

Enyedi, Gyorgy. "La collectivisation de l'agriculture hongroise: Son contenu humain et economique." Pp. 14-24 in *Melanges geographie physique, humaine, economique, applique, offerts a M. Omer Tulippe*, vol. 2, edited by Jose A. Sporck and Bernadette Schoumaker. Gembloux, Belgium: J. Duculot, 1967.

Bernat, Tivador, and Gyorgy Enyedi. "Some Questions of the Regional Development of Hungarian Agriculture" (in Hungarian). *Foldrajzi Ertesito*, 17, No. 4 (1968): 407-427.

Kristof, J. "The Predominant Production-Types and Regions of Agriculture in Hungary" (in Hungarian). *Foldrajzi Ertesito*, 17, No. 2 (1968): 229-248.

Markos, G. "Strukturwandlungen und Spezialisierung der Landwirtschaft Ungarns." *Geographische Zeitschrift*, 56 (1968): 18-57.

Bernat, Tivador, and Gyorgy Enyedi. "La localisation optionale de l'agriculture hongroise." *Bulletin de la Societe Languedocienne de Geographie*, 8, No. 2 (1974): 155-170.

Enyedi, Gyorgy. "A Regional Subdivision of the Agriculture of Hungary." Pp. 93-102 in *Regional Development and Planning, British and Hungarian Case Studies*, edited by Paul A. Compton and Martan Pecsi. Budapest, Hungary: Akademiai Kiado, 1976.

Enyedi, Gyorgy. "Rozwoj rolnictwa wegierskiego." *Przeglad Geograficzny*, 44 (1977): 773-780.

Sipos, A. "Az agrar-ipari integracio (the Agro-Industrial Integration)." *Kozgazdasagi Szemle*, 24, No. 6 (1977): 613-627.

Enyedi, Gyorgy. "Die Entwicklung der ungarischen Landwirtschaft nach dem Zeveiten Weltkrieg." *Mitteilungen der Osterreichischeen Geographischen Gesellschaft*, 120, No. 2 (1978): 265-279.

Makkai, Laszlo. *Agrarian Landscapes of Historical Hungary in Feudal Times.* Budapest, Hungary: Akademiai Kiado, 1980.

Enyedi, Gyorgy. "Part-Time Farming in Hungary." *GeoJournal*, 6, No. 4 (1982): 323-326.

Enyedi, Gyorgy. "Der Einflus der modernen Landwirtschaft auf die rurale Siedlungsstruktur in Ungarn." *Petermann's Geographische Mitteilungen*, 126, No. 1 (1982): 1-4.

Varga, Z.S. "Changes in Hungary's Agrarian Sector with Special Reference to the Northern Hill Region." *GeoJournal*, 38, No. 2 (1996): 181-184.

CROPS AND CROPPING PATTERNS

Bernat, Tivador. *A magyar mzogazdasag termelesi korzetei (irtak).* Budapest, Hungary: Mezagazdasagi Kiado, 1961.

Dudas, G. "Neue Moglichkeiten der Entwicklung des Weinbaus in Ungarn." *Annales Universitas Scientiarum Budapestiensis de Rolardo Eotuos Nominatae, Sectio Geographica*, 1 (1965): 69-80.

Boros, J. "Viticulture and Physio-Geographical Endowments of Tokaj-Hegyalja." *Foldrajzi Ertesito*, 20, No. 3 (1971): 343-358.
Birkas, Maria. "An Historical Geography of Paprika in Hungary." Master's Thesis, University of California, Los Angeles, Calif., 1976.
Charvet, J.P. "Mise au point sur le role des 'systemes de production industrialisee des cereales' dans l'evolution recente de la culture du mais en Hongrie." *L'Information Geographique*, 44, No. 1 (1980): 21-26.
Plet, F. "La viti-viniculture hongroise post communiste a la fin du Xxe siecle." *Annales de Geographie*, 109, No. 4-5 (2000): 473-487.

ANIMAL HUSBANDRY

Asztalos, Istvan. "A Geographical Study of Stockbreeding in Hungary." Pp. 183-203 in *Research Problems in Hungarian Applied Geography*, edited by Bela Sarfalvi. Budapest, Hungary: Akademiai Kiado, 1969.

FARM LAND

Stola, Wladyslawa, and W. Jankowski. "Land Utilization and Farming in the Badacsony Uraeyard District: The Case of Badacsonyto Maj" (in Polish). *Dokumentacja Geograficzna*, 1 (1967): 51-70.
Enyedi, Gyorgy. "Die Landwirtschaftliche Bodennut zung in Ungarn." *Petermann's Geographische Mitteilungen*, 112 (1968): 81-91.
Enyedi, Gyorgy. "Land Utilization Mapping Methods" (in Hungarian). *Foldrajzi Kozlemenyek*, 17, No. 4 (1970): 309-323.
Enyedi, Gyorgy. "The Land of Hungary and Its Agricultural Utilization." Pp. 249-270 in *Studies in Applied and Regional Geography*, edited by M. Shafi and Mehdi Raza. Aligarh, India: Aligarh Muslim University, Department of Geography, 1971.
Enyedi, Gyorgy. "Distribution and Regionalization of the Agricultural Area in Hungary" (in Hungarian). *Foldrajzi Ertesito*, 24, No. 1 (1975): 33-53.
Berenyi, Istvan. "Main Tendencies of Changes in Hungarian Land Use Structure." Pp. 74-84 in *Rural Transformations in Hungary and Poland. Polish-Hungarian Seminar*, edited by Gyorgy Barta. Budapest, Hungary: Hungarian Academy of Sciences, Geographical Research Institute, 1979.
Berenyi, Istvan. "Die geographischen Typen der Brache in Ungarn." *Erdkunde*, 34, No. 1 (1980): 36-45.

AGRICULTURAL POPULATIONS AND SOCIAL ISSUES

Enyedi, Gyorgy. "Le village hongrois et la grande exploitation agricole." *Annales de Geographie*, 73 (1964): 687-700.
Schultz, J. "Habitat disperse et agriculture collective: Le probleme des tonyas dans l'Alford." *Bulletin de la Societe Languedocienne de Geographie*, 3, No. 1 (1969): 85-114.
Enyedi, Gyorgy. "Modern Agriculture and Rural Settlement: Network in Hungary." Pp. 105-118 in *Perspectives in Agricultural Geography*, vol. 5, ed-

ited by Noor Mohammad. New Delhi, India: Concept Publishing Company, 1981.

ENVIRONMENTAL AND CONSERVATION ISSUES

Dohres, Fred E. "Nature Versus Ideology in Hungarian Agriculture: Problems of Intensification." Pp. 271-295 in *Eastern Europe: Essays in Geographical Problems,* edited by George Hoffman. London, England: Methuen, 1971.

Berenyi, Istvan. "The Environmental Problems of Agricultural Land Use in Rural Areas." Pp. 111-116 in *Rural Transformations in Hungary*, edited by Gyorgy Enyedi. Budapest, Hungary: Akademiai Kiado, Studies in Geography in Hungary, No. 13, 1976.

Molnar, K. "Agroecologial Microregionalization in Hungary." Pp. 145-160 in *Contemporary Essays in Austrian and Hungarian Geography*, edited by Elisabeth Lichtenberger and Martan Pecsi. Budapest, Hungary: Akademiai Kiado, Studies in Geography in Hungary, No. 22, 1988.

Engel-Di Maruo, Salvatore. "Farming Under a Regime of Harder Work: Ecological Maintenance and Degradation in the Ormanysag, Southwest Hungary." *Middle States Geographer*, 31 (1998): 45-53.

IRRIGATION

Simon, L. *Possibilities of Irrigation Agriculture at the Sandridges of Our Plains (in Hungarian).* Budapest, Hungary: Akademiai Kiado, Foldrajzi Tanulmanyok, Geographical Study No. 8, 1968.

AGRICULTURAL LABOR STUDIES

Sarfalvi, Bela. "Social Restratification and Changes in the Geographical Distribution of Agricultural Labour" (in Hungarian). *Foldrajzi Ertesito*, 14, No. 2 (1965): 243-254.

Poland

GENERAL TEXTS, ARTICLES, AND THESES

Kostrowicki, Jerzy. "Trends in the Development of Agricultural Geography and Its Status in Poland" (in Polish). *Przeglad Geograficzny*, 29, No. 1 (1957): 3-16.

Straszewicz, L. "Agricultural Problems of the Suburban Zone of Lodz" (in Polish). *Przeglad Geograficzny*, 29, No. 1 (1957): 87-108.

Tobjasz, J. "Review and Assessment of the Achievements of Polish Geography of Agriculture" (in Polish). *Przeglad Geograficzny*, 29, No. 1 (1957): 67-86.

Olszewski, Tadeusz. *Rejony rolnicze polnocnego Oszaru wojewodztwa lodzkiego*. Lodz, Poland: Acta Geographica Lodziensia, No. 15, 1962.

Kostrowicki, Jerzy. "Contemporary Problems and Methods of Polish Agricultural Geography" (in Polish). *Czasopismo Geograficzne*, 37, No. 3 (1966): 289-304.

Kostrowicki, Jerzy. "Die Agrargeographie in Poland: Entwicklung, Forschungsgebiete und Ergebnise." *Wissenschaftliche Zeitschrift der Universitat Halle, Math-Natural Reihe*, 16 (1967): 165-177.

Kostrowicki, Jerzy. "La geographie polonaise d'agriculture: Ses problemes ses methodes et son application pratique." Pp. 361-373 in *Melanges de geographie physique, humaine, economique, applique, offerts a M. Omer Tulippe*, vol. 1, edited by Jose A. Sporck. Gembloux, Belgium: J. Duculot, 1967.

Kostrowicki, Jerzy. "Problems of Agricultural Geography" (in Polish). Pp. 147-182 in *Zarys geografii ekonomiczney Polski*. Warsaw, Poland: Panstwowe Wydaurnictwo Naukowe, 1967.

Szczesny, R. "Problems and Methods of the Macroscale Studies in Agricultural Geography" (in Polish). *Dokumentacja Geograficzna*, 4 (1968): 56-67.

Jedut, R. "The Regions of Agricultural Production in the Lublin Province, Poland" (in Polish). *Annales Universitatis Mariae Curie-Sklodowska,* 26 (1971): 349-390.

Kostrowicki, Jerzy, and R. Szczesny. *Polish Agriculture: Characteristics, Types and Regions*. Budapest, Hungary: Akademiai Kiado, 1972.

Kostrowicki, Jerzy. *Zarys geografii rolnictwa.* Warsaw, Poland: Panstonowe Wydawn. Naukowe, 1973.

Matusik, Marian. *Probatypologii i regionalizacji rolnictwa na obszarze Dolnego Powisla*. Wroclaw, Poland: Zaklad Narodowy im. Ossolinskich, 1973.

Matusik, Marian. *An Attempt at an Agricultural Typology and Regionalization of Lower Powisle, Northern Poland* (in Polish). Prace Geograficzne, No. 102, 1973.

Szyrmer, Jacek. "The Proposal of the New Taxonomic Method to be Applied in Agricultural Typology." *Przeglad Geograficzny*, 45, No. 4 (1973): 739-756.

Kostrowicki, Jerzy. "Les transformations dans le repartition spatiale des types d'agriculture en Pologne et essai du pronostic de l'evolution ulterieure." *Geographia Polonica*, 29 (1974): 307-329.

Peter, W. "Problemes de la campagne et de l'agriculture dans la voivodie de Bialystok." *Geographia Polonica*, 29 (1974): 375-378.

Stola, Wladyslawa. "Etudes regionales sur la typologie de l'agriculture polonaise." *Geographia Polonica*, 29 (1974): 331-341.

Stola, Wladyslawa, and Roman Kulikowski. "Les transformations de l'agriculture de la voivodie de Bialystok au cours des vingt dermieres annees." *Geographia Polonica*, 29 (1974): 363-374.

Szemberg, A. "Les tendences actuelles des changements dans la structure des exploitations de l'agriculture polonaise." *Geographia Polonica*, 29 (1974): 157-168.

Kostrowicki, Jerzy. "Essai preliminaire de la typologie de l'agriculture Polonaise." *Geographica Slovenica*, 4 (1975): 43-55.

Stola, Wladyslawa. "Procede typologique de l'agriculture a l'exemple du bassin de nida, Pologne meridionale." *Geografica Slovenica*, 4 (1975): 62-71.

Galaj, Dyzma. "Part-Time Farmers in Poland." Pp. 258-266 in *Part-Time Farming: Problem or Resource in Rural Development. Proceedings, First Rural Geography Symposium, University of Guelph, June 18–20, 1975*, edited by Anthony M. Fuller and Julius A. Mage. Norwich, England: Geo Abstracts Ltd., 1976.

Muszynski, Marek. "Part-Time Farming in Poland: Conclusions and Prospects." Pp. 130-133 in *Part-Time Farming: Problem or Resource in Rural Development. Proceedings, First Rural Geography Symposium, University of Guelph, June 18–20, 1975*, edited by Anthony M. Fuller and Julius A. Mage. Norwich, England: Geo Abstracts Ltd., 1976.

Stola, Wladyslawa, and R. Szczesny. *Geografia rolnictwa Polski*. Warsaw, Poland: Wydawnictwa Szkolne i Pedagogiczne, 1976.

Szyrmer, Jacek. "The Spatial Changes of the Productivity and Specialization of Individual Farming in Poland in 1960–1970" (in Polish). *Dokumentacja Geograficzna*, 4-5 (1976).

Falkowski, Jan. "Proba typologii i regionalizacji rolnictwa aglomeracji bydgosko-torunskiej." *Przeglad Geograficzny*, 49 (1977): 713-730.

Kulikowski, Roman. "Przemiany kierundow produkcji globalnej w rolnictwie indywidualnym w Polsce (1960/1965/1975)." *Przeglad Geograficzny*, 49 (1977): 689-702.

Manteuffel, R. "Vjecie regionalne relacji obszar-produckeja w indywidulanych gospodarstwach rolniczych." *Przeglad Geograficzny*, 49 (1977): 633-656.

Szyrmer, Jacek. "Zmiany przestrzennego zroznicowania wazniejszch cech produkcyjnych polskiego rolnictwa indywidualnego 1960–1970." *Przeglad Geograficzny*, 49 (1977): 703-712.

Kostrowicki, Jerzy, Roman Kulikowski, and Jacek Szyrmer. "Components of Agricultural Production" (in Polish). *Prace Geograficzne Instytuta Geografii i Przestrzennego Zagospodarowanie PAN*, 127 (1978): 181-359.

Kulikowski, Roman, and Jacek Szyrmer. "Agricultural Productivity and Commercialization Structure of Agricultural Production" (in Polish). *Prace Geograficzne Instytuta Geografii i Przestrzennego Zagospodarowania PAN*, 127 (1978): 360-427.

Tyszkiewicz, Wieslawa. "Changes in the Agrarian Structure in Poland 1945–1975: A Spatio-Temporal Analysis" (in Polish). *Dokumentacja Geograficzna*, 1 (1978).

Bromek, Karol, and C. Guzik. "Rolnictwo." *Folia Geographica: Series Geographical Oeconomica*, 12 (1979): 79-92.

Durbiano, Claudine. "Les transformations recentes de l'agriculture polonaise: Une voie nouvelle pour la socialisation de l'agriculture privie." *Annales de Geographie*, 88, No. 3 (1979): 351-368.

Poland. Glowny Urzad Statystyeny. *Przestrzenne zroznicowanie zasieurow i peglowia zwierzat*. Warsaw, Poland: GUS, 1979.

Szczesny, R. "Transformation of Different Types of Agriculture in Poland Between 1970–1976." Pp. 48-61 in *Rural Transformations in Hungary and*

Poland, edited by Gyorgy Barta. Budapest, Hungary: Hungarian Academy of Sciences, Geographical Research Institute, 1979.

Szyrmer, Jacek. "Changes in Spatial Structure of the Commercial Output of Individual Farming in Poland in 1960–1970" (in Polish). *Prace Geograficzne, Instytuta Geografii, Przestrzennego Zagospodorowania PAN*, 134 (1979).

Tyszkiewicz, Wieslawa. "Changes in the Agrarian Structure of Poland." Pp. 62-73 in *Rural Transformations in Hungary and Poland*, edited by Gyorgy Barta. Budapest, Hungary: Hungarian Academy of Sciences, Geographical Research Institute, 1979.

Falkowski, Jan. "Selected Problems of Suburban Agriculture in the Central Part of the Lower Vistula Region." *Geographica Slovenica*, 11 (1980): 103-116.

Gatczynska, B., and Roman Kulikowski. "Agricultural Suburban Zone of Warsaw in the Lights of Studies of the Spatial Differentiation of Agriculture in the Warsaw Voivodship." *Geografica Slovenica*, 11 (1980): 295-214.

Bielecka, K., M. Paprzycki, and Z. Pinsecki. "Applicability of Quantitative Methods in Spatial Analysis of Agriculture." Pp. 389-416 in *Perspectives in Agricultural Geography*, vol. 1, edited by Noor Mohammad. New Delhi, India: Concept Publishing Company, 1981.

Haase, Alfons. *Schlesiens Landwirtschaft: Ein Gangdurch die Geschichte der schlesischens Landwirtschaft von den ersten Anfongen bis zum Liestungsstand bei Beginn des Zweiten Weltkrieges: Eine agrarhistorische und agrargeographische Darstellung*. Wolfenbuttel, Germany: Grenzland-Verlag Rock, 1981.

Szczesny, Roman. "Transformations of Agricultural Types in Poland in 1970–1976." Pp. 209-228 in *Perspectives in Agricultural Geography*, vol. 1, edited by Noor Mohammad. New Delhi, India: Concept Publishing Company, 1981.

Dawson, A.H. "An Assessment of Poland's Agricultural Resources." *Geography*, 67, No. 4 (1982): 297-309.

Galczynska, B., and Roman Kulikowski. "Struktura przestrzenna rolnictwa indywidualnego w woje wodztevie stoleczaym warszawskin." *Dokumentacja Geograficzna*, 5-6 (1982).

Kulikowski, Roman. "Przestrzenne zroznicowanie produckcji rolniczez w Polsce." *Przeglad Geograficzny*, 54 (1982): 475-497.

Gorz, B. "Problemy rozwoju rolnictwa drobnoobszorowego." *Folia Geographica: Series Geographica-Oeconomica*, 15 (1983): 97-117.

Kosinski, J., and A. Wielonski. "Selected Problems of Agricultural Economy in the Capital Voivodship of Warsaw" (in Polish). *Prace i Studia Geograficzne*, 3 (1983): 183-202.

Kostrowicki, Jerzy. "Polska geografia rolnictwa Mechaizny rozwoja." *Przeglad Geograficzny*, 55 (1983): 597-631.

Kurek, W. "Gospodoarka rolna-ston obecny i kierunki rozwoju: Warunki przyrodnicze." *Folia Geographica: Series Geographica-Oeconomica*, 16 (1984): 67-85.

Tyszkiewicz, Wieslawa. "Agrarian Structure in Poland, 1950–1983." *Geographia Polonica*, 52 (1986): 221-234.

Kulikowski, Roman. "Production Types in Polish Non-socialized Agriculture." *Rural Systems*, 5, No. 1 (1987): 11-16.

Morgan, W.B., B. Galczynska, and Roman Kulikowski. "Productivity and Location in Private Agriculture Around Warsaw, Poland." *GeoJournal*, 15, No. 4 (1987): 375-386.

Szczesny, R. "Types of Individual Farming in Poland in the Years 1960–1983" (in Polish). *Przeglad Geograficzny*, 59, No. 4 (1987): 511-526.

Gemeinsome Deutsche-Polnische Schulbuchkommission, Sympozjum Geografow. *Geograficzne problemy rolnictwa w Polsce.* Poznan, Poland: VI Sympozjum Geografow Wspolnej Komisji Padrecznikowej PRL-RFN, Instytut Zachedni, 1988.

Szczesny, R. "Transformations in the Spatial Structure of Individual Agriculture in Poland" (in Polish). *Dokumentacja Geograficzna*, 2-3 (1988): 15-24.

Szczesny, R. *Changes of the Spatial Structure of Polish Agriculture in the Years 1970–1980.* Warsaw, Poland: Instytut Geografi i Przestrzennego Zagospodarowania, 1988.

Guzik, C. "Problems of Agriculture in the Urbanizing Zone of Upper Silesia-Cracow Agglomeration." *Bochumer Geographische Arbeiten*, 51 (1989): 163-173.

Kulikowski, Roman. "Agricultural Geography in the Current Work on the Plan of the Country's Spatial Organization up to 2000." *Geografia Polonica*, 56 (1989): 115-120.

Kulikowski, Roman. "Changing Spatial Patterns of Polish Agriculture." *Rural Systems*, 7, No. 1 (1989): 39-56.

Morgan, W.B. "Individual Farming and Spatial Policy in Poland." *Geografia Polonica*, 56 (1989): 121-132.

Morgan, W.B. "Some Aspects of Recent Improvements in the Productivity of Private Agriculture in Poland." *Geographica Polonia*, 57 (1989): 99-110.

Morgan, W.B. "Recent Government Policy and Private Agriculture in Poland." *Resource Management and Optimization*, 6 (1989): 291-305.

Szczesny, R. "The Place of the Polish Agriculture in European Agriculture: An Attempt at a Comparison of Chosen Areas." *Przeglad Geograficzny*, 61, No. 4 (1989): 441-456.

Morgan, W.B., Roman Kulikowski, and B. Galczynska. "Typology of Individual Agriculture in the Suburban Zone of Warsaw." *Geographica Slovenia,* 21 (1990): 133-151.

Christians, C., and P. Dabrowski. "Les nouvelles donnees de l'agriculture polonaise au ferme de 45 ans de systeme planifie." *Revue Belge de Geographie*, 115, No. 1-3 (1991): 209-218.

Busek, B., and A. Doninik. "Czynniki warunkujace funke jonowanimie wojewodztwie katowickim." *Studia Geograficzne* (Wroclaw), 52 (1992): 187-203.

Maurel, M.C. "L'agriculture polonaise et le choc du marche." *Bulletin de l'Association de Geographes Francais*, 69, No. 1 (1992): 13-26.

Morgan, W.B. "Economic Reform, the Free Market and Agriculture in Poland." *Geographical Journal*, 158, No. 2 (1992): 145-156.

Szczesny, R. "The Spatial Structure of Individual Farming in Poland: An Attempt of a Synthesis" (in Polish). *Przeglad Geograficzny*, 63, No. 1-2 (1992): 37-52.

Klodzinski, Marek. "Processes of Agricultural Change in Eastern Europe: The Example of Poland." Pp. 123-137 in *Agricultural Change, Environment and Economy: Essays in Honour of W.B. Morgan*, edited by Keith Hoggart. London, England: Mansell, 1993.

Zglinski, W. *Development of the Food-Zone of Warsaw Agglomeration* (in Polish). Prace Geografiszne, PAN 162, 1994.

Dabrowski, P.H. "Le developpement regional et les organisations professionnelles agricoles en Pologne: Perspectives historique." *Bulletin de la Societe Geographique de Liege*, 32 (1996): 171-176.

Kurek, W. "Agriculture Versus Tourism in Rural Areas of the Polish Carpathians." *GeoJournal*, 38, No. 2 (1996): 191-196.

Szczesny, R. *Agriculture and Food Supply During Restructuring, 1989–1994* (in Polish). Warsaw, Poland: Polish Academy of Science, No. 40, 1996.

Szczesny, R. *Spatial Structure of Poland's Agriculture in 1995* (in Polish). Warsaw, Poland: Polish Academy of Science, No. 50, 1997.

Grykien, Stanislaw. "Tourist Farms in Lower Silesia, Poland." *GeoJournal*, 46, No. 3 (1998): 279-281.

Ingham, Hilary, Mike Ingham, and Grzegorz Weclawowicz. "Agricultural Reform in Post-Transition Poland." *Tidjschrift voor Economische en Sociale Geographie*, 89, No. 2 (1998): 150-160.

Janzen, J. "Changes in the Agriculture Structure in Western Poland" (in German). *Geografische Rundschau*, 50, No. 1 (1998): 42-48.

Bellows, Anne. "The Praxis and Production of Food Security: Urban Agriculture in Silesia, Poland." Ph.D. Dissertation, Rutgers University, New Brunswick, N.J., 1999.

CROPS AND CROPPING PATTERNS

Kulikowski, Roman. "Changes in Crop Combination Patterns in Poland Between 1958–1965" (in Polish). *Przeglad Geograficzny*, 41, No. 2 (1969): 281-286.

Romanowski, Jacek J. "Factors of Location of Fresh Vegetable Production in Poland." Ph.D. Dissertation, University of Washington, Seattle, Wash., 1970.

Dubel, Krystyna. "Possibilities and Economic Effects of Changes in the Structure of the Four Cereals Conditioned by the Complexes of Soils Suitable for Agricultural Use: A Case Study of the Opole Voivodship" (in Polish). *Przeglad Geograficzny*, 47, No. 2 (1975): 393-405.

Kwiecien, Ireneusz. "Agriculture and Horticulture" (in Polish). Pp. 85-94 in *Rozwoj i wspolczesna struktura spolecznoekonomiczna miasta krakowa*. Krakow, Poland: Folia Geographica, Series Geographica-Oeconomica, No, 8, 1975.

Kwiecien, Ireneusz. "Spatial Aspects of Truck Farming in Poland" (in Polish). *Prace Geograficzne*, 47 (1979): 117-127.

Nowalkowski, Marian. "Horticulture in Cracow" (in Polish). *Prace Geograficzne*, 50 (1980): 87-105.

Guzik, C. "Problemy rozwojn sadownictwa w karpackim rejorie sadowniczym." *Folia Geographica: Series Geographica-Oeconomica*, 14 (1981): 129-141.

Antoszek, J. "Regionalizm uprawy warzyw gruntowyeh w Polsce no podstowie metody koncentracji." *Czasopismo Geograficzne*, 53, No. 2 (1982): 179-186.

Guzik, C. "Systemy gospodorowania w sadach karpackiego rejonce sadowniczego (Systems of Orchard Farming in the Carpathians)." *Folia Geographica: Series Geographica-Oeconomica*, 18 (1985): 63-78.

Antoszek, J. "Uprawa warzyw gruntorych na rolniczym obszorze hebelskiego Zaglebia Weglowego w latach 1978–1984 (Vegetable Cultivation in the Agricultural Area of the Lublin Coal District)." *Czasopismo Geograficzne*, 58, No. 1 (1987): 57-66.

Swic, Halina. "Zroznicowanie przestrzenne i strukuralne upraw roslin przemyslowych w Polsce." *Czasopismo Geograficzne*, 63, No. 1 (1992): 51-65.

ANIMAL HUSBANDRY

Berezowski, S. "Les migrations pastorales et l'amenagement des alpages dans les Carpathes Polonaises." *Geographica Polonica*, 2 (1964): 115-122.

Szczesny, R. "Essay on Methods of Determining Trends in Livestock Breeding" (in Polish). *Przeglad Geograficzny*, 41, No. 4 (1969): 683-699.

Dramowitz, E. "Chow zwierzat gospodarskich w Panstwowych gospodarstwach Rolnych w Polsce (Livestock Breeding in State Farms in Poland)." *Prace i Studia Geograficzne*, 3 (1983): 163-182.

Dziewulska, A. "Transformations of the Spatial Management of Livestock Breeding in Poland, 1970–1988." *Dokumentacja Geograficzna*, 3-4 (1999).

FARM LAND

Aretowski, Henryk. "Agriculture and Landownership in Poland." *Geographical Review*, 11 (1921): 161-171.

Dobrowalska, Maria. "The Morphogenesis of the Agrarian Landscape of Southern Poland." *Geografiska Annaler*, 43 (1961): 26-45.

Gadomski, Wladyslaw. "Land Use and Agriculture in the Kosice Basin: The Example of the Drienovec Collective Farm" (in Polish). *Dokumentacja Geograficzna*, 2-3 (1963): 25-47.

Macznik-Stola, W. "The Orientation of Arable Land Utilization in the Voivodship of Bialystok" (in Polish). *Przeglad Geograficzny*, 36, No. 1 (1964): 143-156.

Hanzer, Stefan. "Land Use and Rural Economy in Lomianki Coummunity" (in Polish). *Dokumentacja Geograficzna*, 5 (1965): 39-63.

Kostrowicki, Jerzy. "Methods of Presenting Material Collected by the Land Utilization Survey" (in Polish). *Dokumentacja Geograficzna*, 2-3 (1966): 1-23.

Dubel, Krystyna. "Natural Conditions and the Land Use: The Case Study of the Opole, Poviat" (in Polish). *Przeglad Geograficzny*, 42, No. 3 (1970): 529-537.

Mydel, R. "Land Use in the 14th Cadastral District of the City of Cracow-Czarna Wies" (in Polish). *Folia Geographica: Series Geographica-Oeconomica*, 4 (1971): 137-166.

Tkocz, Jan. "General Types of Field Patterns in Agricultural Typology" (in Polish). *Przeglad Geograficzny*, 45, No. 3 (1973): 631-635.

Tyszkiewicz, Wieslawa. "Land Utilization in Relation to the Forms of Land Tenure and the Size of Farms in Kujawy, Central Poland" (in Polish). *Prace Geografizne*, 107 (1974).

Tyszkiewicz, Wieslawa. "The Structure of Cropland Use in the Kujawy Region in Relation to the Forms of Land Tenure" (in Polish). *Przeglad Geograficszny*, 46, No. 1 (1974): 83-98.

Wilamowski, B. "La renouvellement de l'agriculture et de l'amenagement spatial sur les territoire du nord-est de la Pologne." *Geographia Polonica*, 29 (1974): 379-398.

Kulinskowski, Roman. "An Attempt to Determine Land Use Orientations by the Method of Successive Quotients" (in Polish). *Przeglad Geograficzny*, 47, No. 2 (1975): 383-391.

Stola, Wladyslawa. "Land Utilization" (in Polish). *Prace Geograficzne Instytuta Geografii i Przestrzennego Zagospodarowania PAN*, 127 (1978): 110-180.

Tyszkiewicz, Wieslawa. *Struktura agrarna Polski 1945–1975: Analiza przestrzenno-czasowa*. Wroclaw, Poland: Zaklad Narodowy im. Ossolinskich, 1978.

Januszewski, J., and F. Grzesiowski. "The Size of the Land Expanse Commitment in Individual Farms in Poland" (in Polish). *Acta Universitatis Wratislaviensis: Prace Instytuta Geograficznego*, Seria B, 438 (1979): 93-103.

Kulikowski, Roman. "Technique of Successive Quotients to Determine Agricultural Land Use Orientation." Pp. 417-428 in *Perspectives in Agricultural Geography*, vol. 1, edited by Noor Mohammad. New Delhi, India: Concept Publishing Company, 1981.

Stola, W. "Transformations in Agricultural Land Use in Poland, 1946–1978." Pp. 231-244 in *Perspectives in Agricultural Geography*, vol. 3, edited by Noor Mohammad. New Delhi, India: Concept Publishing Company, 1981.

Tyszkiewicz, Wieslawa. "A Method of Identification of the Size Structure of Agricultural Holdings." Pp. 429-438 in *Perspectives in Agricultural Geography*, vol. 1, edited by Noor Mohammad. New Delhi, India: Concept Publishing Company, 1981.

Antoniak, W. "Komasacja gruntowjako czynnik rozwoju rolnictwa na przykladzie wojewodztwa Bialostockiego." *Dokumentacja Geograficzna*, 6 (1983).

Falkowski, J. "Geographical Environment and Land Use in the Structure of Agricultural Ownership in Poland" (in Polish). *Acta Universitatis Nicolai Copernici Geografia*, 19 (1986): 89-108.

Falkowski, Jan. "Types of Agricultural Landscape in Poland" (in Polish). *Acta Universitatis Nicolai Copernici Geografia*, 20 (1987): 127-141.

Galczynska, B., and Roman Kulikowski. "The Size of Holdings Versus Productive Effects in Polish Individual Farming" (in Polish). *Dokumentacja Geograficzna*, 2-3 (1988): 39-45.

Mazurski, K.R. "Natural Recultivation of Land in Poland" (in Polish). *Przeglad Geograficzny*, 60, No. 4 (1988): 599-612.

Tyszkiewicz, Wieslawa. "Changes in the Structure of Individual Agricultural Holdings in Poland in the Period 1950–1985" (in Polish). *Dokumentacja Geograficzna*, 2-3 (1988): 25-37.

Banski, J. "Changes of Agricultural Land Use in Poland in the Period 1975–1988" (in Polish). *Prace Geograficzne, PAN*, 168, 1997.

Wielonski, A., H. Dziakowska, and M. Durydiwka. "Contemporary Ownership Transformation in Polish Agriculture." *Acta Universitatae Carolinae, Geografica*, 32 (1997): 411-417.

AGRICULTURAL POPULATIONS AND SOCIAL ISSUES

Muszynski, Marek. "The Economic Costs of Transforming Dual Occupational People into Single Occupational." Pp. 267-273 in *Part-Time Farming: Problem or Resource in Rural Development. Proceedings of the First Rural Geography Symposium, University of Guelph, 1975*, edited by Anthony M. Fuller and Julius A. Mage. Norwich, England: GeoAbstracts, 1976.

Stasiak, A. "Przemiany w strukturze zawodowej ludnosci wsi w Polsce." *Przeglad Geograficzny*, 49 (1977): 677-688.

Saniech, Z. "Indywidualne gospodarstwa rone w Polsce nalezace do vzytkownikow w wieku ponad 55 lat." *Czasopismo Geograficzne*, 53, No. 1 (1982): 67-74.

Schirmer, J.H. "La 'socialisation' de l'agriculture en Pologne: Origins, evolution, et faillite." *Bulletin de la Societe Geographique de Liege*, 32 (1996): 215-218.

ENVIRONMENTAL AND CONSERVATION ISSUES

Siuta, J. "Typology of Cultivated Soils of Mazury Lake District" (in Polish). *Przeglad Geograficzny*, 31 (1959): 327-344.

Galon, R. "The Physiographical Base of Agriculture in the Bydgoszcz Voivodship" (in Polish). *Przeglad Geograficzny*, 36, No. 1 (1964): 50-54.

Deja, W. *Appraisal of Geographic Environment for Rural Needs and for Agricultural Planning, with Selected Regions of Central Part of the Great Poland Lowlands as a Basis* (in Polish). Poznan, Poland: Poznanskie Towarzystwo Przyjaciol Nauk. Wydzial Nauk Matematyczno-Przrodniczck. Prace Komisji Geograficzno-Geologicznez, 9, No. 2, 1968.

Kozminski, Czeslaw. "Hailstorm Damage to Tobacco Plantations in Poland's Territory" (in Polish). *Przeglad Geograficzny*, 46, No. 2 (1974): 317-332.

Koreleski, K., and G. Magiera-Bras. "The Natural Conditions of Agricultural Production in the South-East Macroregion" (in Polish). *Czasopismo Geograficzne*, 50, No. 1-2 (1979): 19-28.

Guzik, C. "Biogeographical Agents of the Integrated Protection of Orchards in the Carpathian Fruit-Growing Region." *Folia Geografica: Series Geographica-Oeconomica*, 13 (1980): 93-102.

Kurek, W. "Problems and Research Methods of Agriculture of Mountain Regions as Exemplified by a Chosen Area in the Polish Carpathians." *Folia Geografica: Series Geographica-Oeconomica*, 13 (1980): 103-113.

Walewski, Andrejs. "Relations entre la production agricole et le milieu naturel en Pologne: Approche quantitative." *Bulletin de la Societe Languedocienne de Geographie*, 17, No. 1-2 (1983): 179-186.

Stryjakiewicz, Tadeus. "The Relationships Between the Location of the Agricultural Processing Industry and the Agricultural Produce Base in the Poznan Region." *Geographia Polonica*, 51 (1985): 189-198.

Grykien, Stanislaw. "Ekologiczne gapo dorswa rolne w Polse." *Czasopismo Geograficzne*, 66, No. 2 (1995): 175-189.

Solon, J. "Anthropogenic Disturbance and Vegetation Diversity in Agricultural Landscapes." *Landscape and Urban Planning*, 31 (1995): 171-180.

Durydiwka, M. "Ecological Aspects of Polish Agriculture" (in Polish). *Prace i Studia Geograficzne*, 18 (1996): 49-64.

Jedut, R. "Ecological Agriculture in Poland: Barriers and Perspective of Development" (in Polish). *Przeglad Geograficzny*, 69, No. 3-4 (1997): 239-260.

Grykien, Stanislaw. "Ecofarming in Poland" (in Polish). *Acta Universitatis Wratislaviensis, Studia Geograficzne*, No. 68, 1997.

Durydiwka, M. "Ecological Farming in Poland as an Example of Implementation of the Sustainable Development Concept" (in Polish). *Prace i Studia Geograficzne*, 23 (1998): 79-105.

Nalborczyk, Emil, et al. "The Emission, Absorption, and Retention of Greenhouse Gases (GHG) in Polish Agriculture." *Geographia Polonica*, 72, No. 2 (1999): 89-98.

IRRIGATION

Somorowski, C. "The Influence of Climate Changes on Demand for Irrigation Water in Poland." *Geografia Polonica*, 71 (1997): 121-126.

Romania

GENERAL TEXTS, ARTICLES, AND THESES

Roubitschek, Walter. "Zur Bevolkerungs und Agrarstruktur Rumaniens." *Petermann's Geographische Mittelungen*, 104 (1960): 23-32.

Poncet, Jean. "Les transformations de l'agriculture roumaine." *Annales de Geographie*, 73 (1964): 540-567.

Iacob, G. "Contributions to the Geography of the Giurgeu Depression Agriculture" (in German). *Revue roumaine de geologie, geophisque et geographie, Serie geographie.* 10, No. 1 (1966): 85-96.

Radulescu, N., Ion Velcea, and N. Petrescu. *Geografia Agriculturii Romanici.* Bucharest, Romania: Editura Stiintifica, 1968.

Hartia, S. "Distribution by Area of Gross Output in Agricultural Cooperations" (in Romanian). *Revista de Statistica*, 18, No. 7 (1969): 19-33.

Turnock, David. "Geographical Aspects of Romanian Agriculture." *Geography*, 55, No. 2 (1970): 169-186.

Marian, Teodor. *Romania: Agriculture—Figures, Facts, Pictures.* Bucharest, Romania: Redactia Revistelor Agricole, 1971.

Chiriac, D. "Contributions of Industry to the Development of Agriculture" (in Romanian). *Terra*, 4, 24, No. 2 (1972): 14-19.

Mihail, M. "Typization of Agriculture in the Transylvania Plain, Romania" (in Romanian). *Studie Universitatis Babes-Bolyai, Geographia*, 17, No. 1 (1972): 57-64.

Molnar, E. "Agriculture et regiones agricoles en Roumanie." *Bulletin de la Societe Languedocienne de Geographie*, 8, No. 2 (1974): 129-142.

Velcea, Ion. "Agro-Geographical Differentiations in the Romanian Region of the Danube Valley." *Geografica Slovenica*, 4 (1975): 72-80.

Dragomirescu, D. "Note sur l'agriculture roumaine en 1990." *Bulletin de l'Association de Geographes Francais*, 69, No. 1 (1992): 55-56.

Ianos, I., et al. "Changements recents dans l'agriculture roumanie." *Revue Roumanie: Geographie*, 36 (1992): 9-15.

Turnock David. *Agricultural Change in the Romanian Carpathians.* Leicester, England: University of Leicester, Papers in Geography, 93/2, 1993.

CROPS AND CROPPING PATTERNS

Velcea, I. "The Vine-Growing Regions of Romania." *Revue Roumaine de geologie, geophysique et geographie. Serie de Geographie*, 12, No. 1-2 (1968): 175-180.

Tomoroga, P., and I. Lupu. "Agrogeographical Considerations Regarding the Distribution of Wheat Growing in Dobrudja, Romania. Studii si cercetoride geografie aplic" (in Romanian). *Dobrojei* (1971): 315-324.

Baranovsky-Ban, N. "Vine Growing" (in Romanian). *Studii si cercetori de geologie, geofizica, geografie: Serie Geografie*, 19, No. 1 (1972): 41-47.

Lacob, G. "Some Geographical Aspects Regarding the Present Territorial Distribution of Sugar Beet and Supply Areas of the Processing Units in Romania" (in Romanian). *Studii si cercetori de geologie, geofizica, geografie: Serie Geografie*, 19, No. 1 (1972): 49-55.

Lobotka, Vojtech. "Vine Growing Districts in Romania" (in Slovak). *Acta Geographica Universitatis Camenianae, Economico-Geografia*, 11 (1972): 215-222.

Dando, William A. "Wheat in Romania." *Annals of the Association of American Geographers*, 64, No. 2 (1974): 241-257.

Iordan, Ian. "The Viticultural Zone of Panciu-Odobesti-Otesti" (in Romanian). *Studii si cercetori de geologie, geofizica, geografie: Serie Geografie*, 23 (1976): 61-69.

Stefanescu, I. "Romania's Fruit Growing Potential" (in Romanian). *Studii si cercetori de geologie, geofizica, geografie: Serie Geografie*, 28 (1981): 77-86.

ANIMAL HUSBANDRY

Apavaloaei, M. "Pastoral Economy in the Trotus Basin" (in Romanian). *Analele stiintifice Universitatii Al. I. Cuzean Iasi, Geologie-Geografie*, 14 (1968): 179-183.

Matley, Ian M. "Traditional Pastoral Life in Romania." *Professional Geographer*, 22, No. 6 (1970): 311-316.

Schneider, Karl S. "Fernweidwirtschaft in den Sudost-Karpaten" (in Romanian). *Geographische Rundschau,* 7 (1973): 282-289.

FARM LAND

Biegajlo, Wladislaw, and Wieslawa Tyszkiewicz. "Studies of Land Utilization in Rumania" (in Polish). *Przeglad Geograficzny*, 39, No. 3 (1967): 635-639.

Iordan, Ian, and D. Cancea. "Amenagements du territoire agricole de la zone preurbaine Galati-Braila." Pp. 362-376 in *Melanges de geographie physique, humaine, economique, applique, offerts a M. Omer Tulippe*, vol. 2, edited by Jose A. Sporck and Bernadette Schoumaker. Gembloux, Belgium: J. Duculot, 1967.

Iacob, G. "Geographical Characterization on Agricultural Use of the Slope of North Dobrudja, the Bugeac-Ceatalul Tukei Sector" (in Romanian). *Studii si cercetori geografie aplic. Dobrogii.* (1971): 331-336.

Boeru, S. "Operations of Land Improvement: Achievements and Prospects" (in Romanian). *Terra*, 4, 24, No. 2 (1972): 20-29.

Bordanc, Florica D. "Spatial Variations in the Process of Agricultural Land Privatisation." Pp. 39-47 in *Rural Change in Romania*, edited by David Turnock. Leicester, England: University of Leicester, Department of Geography Occasional Paper No. 33, 1995.

AGRICULTURAL POPULATIONS AND SOCIAL ISSUES

Roucek, Joseph S. "Romanian Peasants and Agriculture." *Journal of Geography*, 31, No. 8 (1932): 279-287.

Muica, Nicolae, Lesley Roberts, and David Turnock. "Transformation of a Border Region: Dispersed Agricultural Communities in Brasov County, Romania." *GeoJournal*, 46, No. 3 (1998): 305-317.

ENVIRONMENTAL AND CONSERVATION ISSUES

Muica, C. "The Ecological Consequences of Privatisation in Romanian Agriculture." *GeoJournal*, 38, No. 2 (1996): 207-212.

Ploaic, G. "The Impact of Tourism and Conservation on Agriculture in the Mountains of Valcea County, Romania." *GeoJournal*, 38, No. 2 (1996): 219-227.

AGRICULTURAL MARKETING AND DISTRIBUTION

Radulescu, N.A., and S. Vlad-Radulescu. "Agriculture Co-operization in Romania and Its Geographical Implications" (in Romanian). *Terra*, 4, 24, No. 3 (1972): 52-61.

TRANSPORTATION AND MECHANIZATION

Bordanc, Florica D. "Geographical Considerations Concerning the Mechanisation of Agriculture in Romania" (in Romanian). *Studii si cercetori de geologie, geofizica, geografie: Serie Geografie*, 25 (1978): 97-104.

Yugoslavia, Bosnia, Croatia, Kosovo, Macedonia, Montenegro, Serbia, and Slovenia

GENERAL TEXTS, ARTICLES, AND THESES

Peyre, Marius. "La reforme agraire en Yugoslavie et ses effets dans les provinces de Backka et de Srem." *Annales de Geographie*, 44 (1935): 58-64.

Johnston, W.B., and Ivo Crkvencic. "Changing Peasant Agriculture in Northwestern Hrvatsko Primorje, Yugoslavia." *Geographical Review*, 44, No. 3 (1954): 352-372.

Johnston, W.B., and Ivo Crkvencic. "Examples of Changing Peasant Agriculture in Croatia, Yugoslavia." *Economic Geography*, 33, No. 1 (1957): 50-71.

Hoffmann, George W. "Changes in the Agricultural Geography of Yugoslavia." Pp. 101-140 in *Geographical Essays on Eastern Europe*, edited by Norman J.G. Pounds. Bloomington, Ind.: Indiana University Publications, Russian and Eastern European Series, No. 24, 1961.

Kostrowicki, Jerzy, and Wladislaw Biegojlo. "Investigations Carried Out in Yugoslavia in 1962–1964 by the Department of Agricultural Geography in the Institute of Geography, Polish Academy of Sciences" (in Polish). *Przeglad Geograficzny*, 37, No. 4 (1965): 697-702.

Sore, Anton. *Geografija neka terih delov celjske makroregije*. Celje, Yugoslavia: Svetza kulturo in znanost Skupscine obcine, 1969.

Markovic, Petar J. *Poljoprivredna geografija*. Zagreb, Croatia: Informatov, 1970.

Degenhardt, Bodo. "Zum Strukturwandel in der Jugoslawischen Agrarlandschaft." *Geographische Rundschau*, 7 (1973): 266-271.

Crkvencic, Ivo, and Vladimir Klemencic. "Agrarian-Geographical Research Work in Yugoslavia Since the Conference in Budapest in 1964." *Geografica Slovenica*, 4 (1975): 9-23.

Ingolic, B. "The Developmental Tendencies of the Farming Economy in the Cadastral Parish of Krajna vas and in Kras in General." *Geografica Slovenica*, 4 (1975): 119-121.

Jordanov, T. "Factors and Criteria Determining the Zonal Distribution of Agriculture." *Geographica Slovenica*, 4 (1975): 24-27.

Frank, Walter. "Wandel der landwirtschaftlichen Betriebsformen in Jugoslawien." *Zeitschrift fur Wirtschaftsgeographie*, 21, No. 3 (1977): 85-89.

Tyszkiewicz, Wieslawa. "Typy rolnictwa Macedonii jako przyklad typologii rolnictwa swiata." *Przeglad Geograficzny*, 44 (1977): 781-806.

Gramatnikovski, V. "Transformation of Agrarian Space in the Suburban Zone of Skopje." *Geographica Slovenica*, 11 (1980): 45-49.

Klemencic, Vladimir. "The Problems of the Semi-Agrarian Structure in the Semi-Urban Zones of Slovenia." *Geographica Slovenica*, 11 (1980): 9-15.

Durie, V. "Poljoprivreda SAP Kosovo." *Zbornik Radova Institutaza Geografiju*, 28 (1981): 37-46.

Feletar, D. "Utjecaj industrije na promjene u agrarnoj proizvodnji i raspored kooperanata-na primjeru 'Podravke.'" *Geografski Glasnik*, 43 (1981): 109-125.

Meze, D. "The Mountain Farms in the Kokra Valley" (in Croatian). *Geografski Zbornik*, 21 (1981): 73-114.

Milojevic, Miroslav. "On the Development of the Agrarian Complex in the Mladerovac Commune in the Post-War Period" (in Serbo-Croat). *Posebna Izdanje-Geografski Institut "Jovan Cvijic,"* No. 34 (1984).

Buschenfeld, H. "Entwicklungstendenzen der jugoslawischen Landwirtschaft." *Zeitschrift fur Wirtschaftsgeographie*, 33, No. 1-2 (1989): 68-78.

Buschenfeld, H. "Die jugoslawische Landwirtschaft an der Schwelle der neunziger Jahre: Perspektiven eines binomen Agrarsystems." *Geographische Rundschau*, 41, No. 3 (1989): 169-176.

Malic, A. "Regional Characteristics of Agriculture in the Privatization Process in the Republic of Croatia" (in Croatian). *Annales-Universitatis Scientiarum Budapestinensis de Rolando Eotvos Nominatau, Sectio Geograhia*, 22-23 (1993): 149-155.

Vriser, I. "Agrarian Economy in Slovenia." *GeoJournal*, 31, No. 4 (1993): 373-378.

Brancelj, Irena R. *Agricultural Geography of the Koper Littoral from the Standpoint of Environmental Protection* (in Slovenian). Ljubljova, Slovenia: Institut za Geografijo Universe v Ljubljoni, 1994.

Vriser, I. "The Development, Present Significance and Future Prospects of Socialist Agricultural Enterprises in Slovenia." *GeoJournal*, 38, No. 2 (1996): 151-156.

Pak, Mirko, and Valentina Brecko. "Problems of Agriculture in Slovenia with Special Reference to Cirkovce." *GeoJournal*, 46, No. 3 (1998): 257-261.

CROPS AND CROPPING PATTERNS

Kostic, Mihajlo, and Zivormir Gajic. "Specialized Market-Gardening in the Southern Morava Region" (in Serbian). *Zbornic Radova Geografski Institut, "Jovan Cvijic,"* 20 (1965): 103-145.

Medved, J. "The Development of the Growing of Non-Cereal Field Crops in the Socialist Republic of Slovenia from 1961 Until 1974." *Geographica Slovenica,* 11 (1980): 175-180.

Grcic, M. "Vloga mesto vinogradarskva u prostornoj organizaciji poljoprivredes Zupe Alekson drovake (The Role of Viticulture)." *Zbornic Radova Institut za Geografski,* 28 (1981): 53-62.

Durik, V. "The Grocka Danube Basin: Specialized Region for Fruit Growing and Viticulture" (in Serbian). *Zbornik Radova-Prirodno-Matematicki Facultera Univerziteta u Beogradu, Geografski Insitut,* 31 (1984): 49-60.

ANIMAL HUSBANDRY

Matley, Ian M. "Transhumance in Bosnia and Herzegovina." *Geographical Review,* 58, No. 2 (1968): 231-261.

Lutovac, M.V. "Livestock Rearing and the Pastoral Life on the Pester and Sjenica Plateau" (in Serbian). *Zbornik Radova, Geografski Institut, "Jovan Cvijic,"* 29 (1977): 157-170.

Milojevic, Miroslav. "Post War Livestock Rearing in Western Serbia: A Research Monograph in Agricultural Geography" (in Serbian). *Posebna Izdonje, Geografski Institut, "Jovan Cvijeic,"* 30 (1977).

FARM LAND

Crkvencic, Ivo. "Kulturlandschaftliche Veranderungen in Hrvatsko Zagorje, (Kroatisch Zagorien), Jugoslawien." *Erdkunde,* 16, No. 3 (1962): 161-173.

Susic, M.B. "The Natural Conditions Affecting Land Use on Bar Polje" (in Serbian). *Zbornik Radova Geografski Institut, "Jovan Cvijie,"* 20 (1965): 181-225.

Biegajlo, Wladyslaw. "Land Utilization in the Eastern Slovenia Village of Belski" (in Polish). *Dokumentacja Geograficzna,* 2-3 (1966): 49-79.

Biegajlo, Wladyslaw. "Land Utilization in Central Dalmatia: The Cadastral Unit of Jesenice in the Commune of Omis" (in Polish). *Dokumentacja Geograficzna,* 2-3 (1966): 81-103.

Hauzer, Stefan. "Land Use and Agricultural Economy in the Villages of Zvad and Runec, Slovenia, Yugoslavia" (in Polish). *Dokumentacja Geograficzna,* 2-3 (1966): 105-127.

Tyszkiewicz, Wieslawa. "Land Utilization in the Eastern Slovenian Village of Goricak" (in Polish). *Dokumentacja Geograficzna,* 2-3 (1966): 81-103.

Biegajlo, Wladyslaw. "Land Utilization and Farming in Central Macedonia: The Case of Elovets Village, District Titov Veles" (in Polish). *Dokumentacja Geograficzna,* 5 (1969): 5-32.

Tyskiewicz, Wieslawa. "Land Utilization in Southwestern Macedonia: The Case of Asamati Village, Resen District" (in Polish). *Dokumentacja Geograficzna*, 5 (1969): 33-50.

Ilesic, S., I. Vriser, and J. Medved. "Recherches typologiques sur l'utilisation agricole du sol en Slovenie." *Bulletin de la Societe Languedocienne de Geographie*, 8, No. 2 (1974): 171-190.

Belec, Borut. "Agrarian Economy and Land Use in Eastern Slovenske Gorice." *Geografica Slovenica,* 4 (1975): 104-108.

Rakitnikov, Andre. "On Methods of Studies of Agricultural Utilization of Land." *Geografica Slovenica*, 4 (1975): 81-94.

Kladnik, Drago. "Problems of Agrarian Land-Use with Special Reference to the Commune of Ptuj (in Slovenian)." *Geographica Slovenia*, 13 (1982): 105-115.

Tyskiewicz, Wieslawa, and B. Jacimovic. "Changes in the Structure of Size of Private Farms in the Republic of Serbia in the Period 1961–1981" (in Polish). *Przeglad Geograficzny*, 60, No. 1-2 (1988): 49-70.

Pavlin, Branko. "Contemporary Changes in the Agricultural Land Use as an Indicator of Discrepancies in the Regional Development of Littoral Sloveni" (in Slovenian). *Geographica Slovenica*, 21 (1990): 283-294.

Pavlin, Branko. *Contemporary Changes in the Agricultural Use of Land in the Border Landscape Units of the Slovene Littoral* (in Slovenian). Ljubljana, Slovenia: Institut za Geografijo Univerze v Ljubljani, 1991.

AGRICULTURAL POPULATIONS AND SOCIAL ISSUES

Vresk, M. "Tendencies in the Present Evolution of an Overpopulated Farming Region: The Case of the Varazdin District" (in Croatian). *Geografski Glasnik*, 30 (1968): 143-155.

Belec, Borut. "Soziookonomische Gliederung von Agrarraum West Sloveniens aus der Sicht der Sonderkulturen." *Geografica Slovenica*, 8 (1978): 172-179.

Crkvenic, Ivo. "Socijalnogeografski aspekti pojave ugara odnosno neobradenile aranica." *Geografski Glasnik*, 43 (1981): 95-107.

Grcic, M. "The Areal Structure of Agriculture in the Community of Sabac (Yugoslavia)" (in Serbian). *Glasnik-Srpskog Geografskog Drustva*, 64, No. 2 (1984): 33-43.

Stombuk, M. "Agricultural Depopulation in Croatia." *Sociologia Ruralis*, 31, No. 4 (1991): 281-289.

ENVIRONMENTAL AND CONSERVATION ISSUES

Gams, Ivon. "Systems of Adapting the Littoral Dinaric Karst to Agrarian Land Use." *Geografski Zbornik* (Ljubljana), 31 (1991): 5-106.

Lovrencak, F. "Soils as a Basis of Farming in Slovenia." *GeoJournal*, 30, No. 3 (1993): 349-353.

Lampie, Barbara. "Agricultural Pollution of the Environment in Slovenia from the Aspect of Energy and Nitrate Consumption." *Moravian Geographical Reports*, 8, No. 2 (2000): 2-13.

Estonia, Latvia, and Lithuania

GENERAL TEXTS, ARTICLES, AND THESES

Smits, Visvoldis. "Impact of Collectivization on Latvian Agriculture." Master's Thesis, University of Washington, Seattle, Wash., 1961.

Stadelbauer, Jorg. "Landwirtschaft und Landlicher Raum in den Baltischen Republiken." *Geographische Rundschau*, 40, No. 9 (1988): 44-51.

Sando, Paul. "The Privatization of Latvian Agriculture: Problems and Prospects." Master's Thesis, Indiana State University, Terre Haute, Ind., 1992.

Sando, Paul. "Recovery from Food Shortage: Current Privatization and Past Resislience of Latvian Agriculture." *Professional Paper*, Department of Geography and Geology, Indiana State University, 20 (1994): 27-36.

Unwin, Timothy. "Agrarian Change in Estonia: Historical Context and Contemporary Restructuring." *Eastern European Contryside*, 1 (1995): 37-51.

Sando, Paul. "Latvian Agriculture after Communism: Restructuring and Privatization." Ph.D. Dissertation, Indiana State University, Terre Haute, Ind., 1996.

Maciulyte, Jurgita, and M.C. Maurel. "Le retour de l'agriculture familiale en Lituanie." *Bulletin de la Societe Languedocienne de Geographie*, 28, 1-2 (1997): 153-201.

FARM LAND

French, R.A. "Field Patterns and the Three Field System: The Case of Sixteenth Century Lithuania." *Transactions, Institute of British Geographers*, 48 (1969): 121-134.

Part 5

RUSSIA, UKRAINE, TRANSCAUCASIA, AND THE SOUTHERN FRINGE

Russia, Ukraine, Transcaucasia, and the Southern Fringe

ATLASES AND GRAPHIC PRESENTATIONS

Simonova, L.P. "Thematic Maps in Solving the Problems Involved in Developing Adaptive Systems of Land Cultivation" (in Russian). *Vestnik Moskovskogo Universiteta, Seriya Geografiya*, 3 (1984): 26-30.

Vostokova, A.V., et al. "New Types of Maps for Managing the Agriculture of a Region" (in Russian). *Vestnik Moskovskogo Universiteta, Seriya Geografiya*, 2 (1986): 43-50.

Komedchikov, N.N. "Mathematic–Cartographic Optimisation of the Utilisation of Alpine Pastures of the Kabardino-Balkar Autonomous Republic" (in Russian). *Vestnik Moskovskogo Universiteta, Seriya Geografiya*, 6 (1990): 55-61.

GENERAL TEXTS, ARTICLES, AND THESES

Tulaikov, N.M. "Agriculture in the Dry Region of the U.S.S.R." *Economic Geography*, 6, No. 1 (1930): 54-80.

Hopkins, I.M. "Sovietizing Agriculture." *Journal of Geography*, 30, No. 8 (1931): 279-290.

George, Pierre. "Les methods de reconstruction agricole en U.R.S.S." *Annales de Geographie*, 55 (1946): 247-258.

Cressy, George. "U.S.S.R.: The Geographic Base for Agricultural Planning." *Land Economics*, 25 (1949): 334-336.

Harris, Chauncy D. "Growing Food by Decree in the Soviet Union." *Foreign Affairs*, 33 (1955): 1-14.

Harris, Chauncy D. "Soviet Agricultural Resources Reappraised." *Journal of Farm Economics*, 38 (1956): 258-273.

Anderson, Jeremy H. "The Agricultural Development of Yakutia." Master's Thesis, University of Washington, Seattle, Wash., 1959.

Mensheha, Mark P. "Geographical Distribution of Soviet Agriculture." Master's Thesis, University of Minnesota, Minneapolis, Minn., 1962.

Zimmerman, Robert C. "Subtropical Agriculture in Western Georgian S.S.R." Master's Thesis, Syracuse University, Syracuse, N.Y., 1962.

Mikheyeva, U.S. "An Economic–Mathematical Model of the Location of Farm Production by Regions of the Soviet Union." *Soviet Geography*, 4, No. 3 (1963): 24-29.

Jensen, Robert G. "Soviet Agricultural Regionalization and Price Zonation." Ph.D. Dissertation, University of Washington, Seattle, Wash., 1964.

Jensen, Robert G. "Soviet Subtropical Agriculture: A Microcosm." *Geographical Review*, 54, No. 2 (1964): 185-202.

Rakitnikov, Andre N., and V.G. Krychkov. "Agricultural Regionalization." *Soviet Geography*, 7, No. 5 (1966): 48-57.

Symons, Leslie J. "Agricultural Production and the Changing Roles of State and Collective Farms in the U.S.S.R." *Pacific Viewpoint*, 7, No. 1 (1966): 54-66.

Anderson, Jeremy H. "Fodder and Livestock Production in the Ukraine: A Case Study of Soviet Agricultural Policy." *East Lakes Geographer*, 3 (1967): 29-46.

Jensen, Robert G. "The Soviet Concept of Agricultural Regionalization and Its Development." Pp. 77-98 in *Soviet and Eastern European Agriculture*, edited by J.F. Karcz. Berkeley, Calif.: University of California Press, 1967.

Backus, G.B. "Some Recent Developments in the Evolution of the Soviet Collective Farm." *Oregon Geographer*, 2, No. 2 (1968): 2-7.

Kalustyan, L.K. "The Odessa Suburban Agricultural Zone" (in Ukrainian). *Economichna Geografiya, Mizhvidomchyi Naukovyi Zbirbyk, Kiev*, 4 (1968): 199-204.

Kjeld, Brackhus. "Some Geographic Aspects of Soviet Agriculture." *Norsk Geografisk Tijdsskrift*, 22 (1968): 39-55.

Stepanenko, A.V. "Determining Territorial Specialisation of Agriculture and Methods for Evaluating Its Economic Effectiveness" (in Ukrainian). *Ekonomichna Geografiya, Mezhvidomchey Nankovyi Zbirbyk, Kiev*, 5 (1968): 94-101.

Harris, Chauncy D., and W.A.D. Jackson. "U.S.S.R.: Resources for Agriculture." *Focus*, 4 (1969): 1-11.

Troitskaya, A.V. "An Attempt to Differentiate Types of Circum-Town Farm Units in the Amur Region" (in Russian). Pp. 155-157 in *Sbornik: Vopr. Geogr. Sibiri i Dal'n Collection: Problems of Geography of Siberia and the Far East*. Irkutsk, 1969.

Giese, Ernst. "Agrarwirtschaft des Siebenstromlandes in Sudkazachstan." *Die Erde*, 101 (1970): 92-122.

Preuss, H. "Agrargeographische Veranderungen in der Sovietunion." *Zeitschrift fur Wirtschaftsgeographie*, 14, No. 5 (1970): 137-141.

de Vries, J. "Changes in Soviet Agriculture: Intensification Versus Growing State Influence" (in Dutch). *Geografisch Tijdschrift*, 6, No. 4 (1972): 314-326.

Kryuchkov, V.G. "The Territorial Organization and the Efficiency of Agriculture in Landscapes of Different Types" (in Russian). *Voprosy Geografii*, 90 (1972): 127-145.

Symons, Leslie J. *Russian Agriculture: A Geographic Survey*. New York, N.Y.: Wiley, 1972.

Zhukovskaya, V.M., V.G. Kryuchkov, and I.B. Muchnik. "Utilization of Methods of Images Recognition in the Classification of Productive Forms in Agriculture" (in Russian). *Voprosy Geografii*, 90 (1972): 146-157.

Giese, Ernst. *Sovchoz, kolchoz und personliche Nebernerwerbswirtschaft in Sowjet-Mittelasien: Eine Analyse der raumlichen Verteilungs- und Verflechtungsstysteme*. Munster, Germany: Westfaliche Geographische Studien, Heft 27, 1973.

Laykin, V.I. "Some Methods for an Agricultural Typology of the South of Krasnoyorsk Kray." *Soviet Geography*, 14, No. 9 (1973): 564-571.

Towber, Richard I. "The Locational Responses of Soviet Agriculture to Central Decision Making." Ph.D. Dissertation, University of Washington, Seattle, Wash., 1974.

Mejendorf, A.L. "Measuring the Differential Income Ascribable to Locational Advantages in Agricultural Production" (in Russian). *Vestnik Moskovskogo Universiteta, Seriya Economicka*, 30, No. 1 (1975): 55-65.

Stetsenko, S.V. "The Zonal Peculiarities of Concentration and Specialization of Agricultural Production on the State-Owned and Collective Peasant Farms in the Ukraine" (in Russian). *Vestnik Moskovskogo Universiteta, Seriya Geografiya*, 5 (1975): 11-16.

Maurel, M.C. "Differenciations spatiales des unites de production agricole dans la Russie d'Europe." *Annales de Geographie*, 85, No. 6 (1976): 694-724.

Bakiev, S.G. "The Problems of Regional Agro-Industrial Unit Formation" (in Russian). *Vestnik Leningradskogo Universiteta, Geologiya-Geografiya*, 6, No. 1 (1977): 134-137.

Kriuchkov, V.G. *Territorial'naia organizatsiia sel'skogo Khoziaist va: Problemy i metody ekonomiko–geograficheskogo issledovaniia*. Moscow, Russia: Mysl', 1978.

Carriere, P. "Le restructuration recente des exploitations agricoles en Moldavie." *Annales de Geographie*, 88, No. 5 (1979): 581-604.

Carriere, P. "La restructuration de l'agriculture sovietique." *L'Information Geographique*, 43, No. 2 (1979): 59-70.

Gorbunova, L.I., et al. "A Typology of Soviet Agriculture Using the Method of the IGU Commission on Agricultural Typology." *Soviet Geography*, 20, No. 1 (1979): 41-48.

Matlock, Thomas S. "The Agro-Industrial Complex: A New Stage in the Soviet Union's Agrarian Program." Master's Thesis, Kent State University, Kent, Ohio, 1979.

Shotski, V.P. *Agro-Industrial Complexes and Types of Agriculture in Eastern Siberia*. Budapest, Hungary: Akademiai Kiado, 1979.

Stadelbauer, Jorg. "Horizontale und vertikale Kooperation in der sowjetischen Agrarwirtschaft Entwicklungsprobleme des nahrungswirtschaftlichen Agrarkomplexes, auf gezeigt und Beispielen aus der RSFSR." *Geographische Zeitschrift*, 67 (1979): 217-239.

Zaltsman, L.M., and S.J. Polovenko. "Les mutations progressives dans la concentration et typisation de l'agriculture de l'URSS." *Geographia Polonica*, 40 (1979): 75-82.

Graeme, W.D., and Walter Roubitschek. "Zur zweiglichen und territorialen Entwicklung der Landwirtschaft der USSR." *Geographische Berichte*, 97 (1980): 225-248.

Kozlovsky, F.I. "The Method of Belt Farming in an Arid Zone" (in Russian). *Izvestiya Akademii Nauk SSSR, Seriya Geograficheskaya*, 5 (1981): 50-65.

Oak, B. "Reviews of the Studies in Agro-Productivity in the Soviet Union." *Geographical Observer*, 17 (1981): 40-44.

Shunkov, V.I. "Geographical Distribution of Siberian Agriculture in the 17th Century, Part 1." *Soviet Geography*, 22, No. 6 (1981): 381-392.

Volf, Mark B. *Geografia mirovogo sel'skogo khoziaistva*. Moscow, Russia: Mysl', 1981.

Laird, Roy D. "Soviet Meat and Grain 1981–1985: Output Projections." *Soviet Geography*, 23, No. 2 (1982): 65-72.

Mochalov, V.D. "Agricultural Geography of Transcaucasia in the Last Third of the 19th Century." *Soviet Geography*, 23, No. 9 (1982): 654-664.

Monbrod, Timothy P. "The Problem of Soviet Agriculture: Constraints and Potentialities." Master's Thesis, University of Chicago, Chicago, Illinois, 1982.

Yatsunskiy, V.K. "Changes in the Distribution of Agriculture in European Russia from the End of the 18th Century Until World War I, Part 1." *Soviet Geography*, 23, No. 4 (1982): 251-269; "Part 2." 23, No. 5 (1982): 326-345.

Araslanova, A.N. "Objective Factors of the Formation of the Agrarian-Industrial Relationships and Their Classification" (in Russian). *Vestnik Leningradskogo Universiteta, Geologiya-Geografiya*, 12 (1983): 110-115.

Giese, Ernst. "Der Private Produktionssektor in der sowjetischen Landwirtschaft." *Geographische Rundschau*, 35, No. 11 (1983): 554-564.

Pronshteyn, A.P. "Development of Agriculture on the Don in the 18th Century." *Soviet Geography*, 24, No. 10 (1983): 739-755.

Stadelbauer, Jorg. *Studien zur Agrargoegraphie Transkaukasiens: Subtropische Landwirtschaft im gesamt sowjetischen Rahmen*. Berlin. Germany: Duncker and Humblot, 1983.

Ioffe, G.V. "On Methodology of the Geography of Agriculture within the Zone of Influence of the Largest City: A Case Study of the Moscow Region" (in Russian). *Izvestiya-Akademiya Nauk SSSR, Seriya Geograficheskaya*, 1 (1984): 63-72.

Wein, Norbert. "Agriculture in the Pioneering Regions of Siberia and the Far East: Present Status, Problems and Prospects." *Soviet Geography*, 25, No. 8 (1984): 592-625.

Palamarchuk, M.M., and A.M. Palamarchuk. "Typology of Agro-Industrial Territorial Complexes" (in Russian). *Izvestiya-Vsesoyuznogo Geograficheskogo Obshchestva*, 117, No. 5 (1985): 404-413.

Stebelsky, Ihor. "Agriculture of the Soviet Union and Eastern Europe." *Journal of Geography*, 84, No. 6 (1985): 264-273.

Zhikharevich, B.S. "Some Features in the Formation of an Agro-Industrial Complex in the Zone of Influence of a Large City" (in Russian). *Izvestiya-Vsesoyuznogo Geograficheskogo Obshchestva,* 117, No. 5 (1985): 413-420.

Stadelbauer, Jorg. "Agricultural Cooperation and Agribusiness in Soviet Transcaucasia." *Soviet Geography*, 27, No. 8 (1986): 553-566.

Ioffe, G.V., T.G. Nefedova, and D.G. Pines. "Intensification of Agriculture in the European USSR: Theoretical and Methodological Approaches" (in

Russian). *Izvestiya-Akademiya Nauk SSSR, Seriya Geograficheskaya*, 1 (1988): 64-77.

Ioffe, G.V., et al. "Intensification of Agriculture within the European USSR" (in Russian). *Izvestiya-Akademiya Nauk SSSR, Seriya Geografischeskaya*, 5 (1988): 5-15.

Burakov, V.I. "Agro-Landscapes for Agricultural Intensification" (in Russian). *Izvestiya-Vsesoyuznogo Geograficheskogo Obshchestva*, 122, No. 1 (1990): 80-86.

Chistobayev, A.I. "A State Farm in the Depths of the Nonchernozem Zone." *Soviet Geography*, 31, No. 3 (1990): 151-159.

Ioffe, G.V. "The Level of Economic Development and Agriculture in the European USSR." *Izvestiya-Akademiya Nauk SSSR, Seriya Geograficheskaya*, 2 (1990): 63-70.

Stebelsky, Ihor. "Restructuring Soviet Agriculture." *Ontario Geography*, 34 (1990): 7-14.

Manasyan, V.G. "Spatial Organization of Individual Subsidiary Farms in Mountain Territories of Armenia SSR" (in Russian). *Izvestiya- Vsesoyuznogo Geograficheskogo Obshchestva*, 123, No. 4 (1991): 377-382.

Nefedova, T.G. "Natural and Socio-Economic Factors of Regional Development of Agriculture of the USSR European Territory." *Revue Belge de Geographie*, 115, No. 1-3 (1991): 219-236.

Wegren, Stephen K. "Agricultural Reform in the Nonchernozem Zone: The Case of Kostroma Oblast." *Post-Soviet Geography*, 33, No. 10 (1992): 645-685.

Rose, Richard, and Yevgeniy Tikhomirov. "Who Grows Food in Russia and Eastern Europe." *Post-Soviet Geography*, 34, No. 2 (1993): 111-126.

Van Atta, Dan. "The Current State of Agrarian Reform in Uzbekistan." *Post-Soviet Geography*, 34, No. 9 (1993): 598-606.

Wegren, Stephen K. "New Perspectives on Spatial Patterns of Agrarian Reform: A Comparison of Two Russian Oblasts." *Post-Soviet Geography*, 35, No. 8 (1994): 455-481.

Pallot, Judith. "Agriculture and Rural Development." Pp. 101-116 in *The Post-Soviet Republics: A Systematic Geography*, edited by Denis J.B. Shaw. New York, N.Y.: Longman, 1995.

Wegren, Stephen K. "Regional Development of Russian Private Farms: A Comment." *Post-Soviet Geography*, 36, No. 3 (1995): 176-182.

Nefedova, T.G. "Agriculture in Russia in the 1990s: Geography of Crisis and Recent Trends" (in Russian). *Izvestiya-Akademiya Nauk, Seriya Geograficheskaya*, 2 (1996): 85-100.

Stebelsky, Ihor. "The Promise Unfulfilled: Ukrainian Agriculture in Kuchma's First Year." *Ukrainian Economic Review*, 2, No. 3 (1996): 129-136.

Kegel, H. "Georgian Agriculture in Transition" (in German). *Zeitschrift fur Wirtschaftsgeographie*, 41, No. 2-3 (1997): 143-152.

Leonard, C. Scott, and J. Szyrmer. "Agrarian Transition in the Former Soviet Union." *Bulletin de la Societe Geographique de Liege*, 33 (1997): 31-26.

Marchard, P. "L'agriculture post-sovietique: La crise sons mutation?" *Annales de Geographie*, 106, No. 4 (1997): 459-478.

Stebelsky, Ihor. "Ukrainian Agriculture and Kuchma's First Two Years of Reforms." *Choices*, 12, No. 2 (1997): 21-27.

Stein, C. "The State of the Transformation of the Agrarian Sector in Uzbekistan" (in German). *Zeitschrift fur Wirtschaftsgeographie*, 41, No. 2-3 (1997): 162-173.

Szyrmer, Jacek, and T.A. Reiners. "Ukraine's Agriculture, 1990–1996: An Incomplete Transition." *Bulletin de la Societe Geographique de Liege*, 33 (1997): 25-30.

O'Hara, Sarah L., and Robert Hudson. "Agricultural Decline in Uzbekistan: The Case of Yazyavan Region." *Post-Soviet Geography and Economics*, 40, No. 6 (1999): 440-452.

Ioffe, Grigory, and T.G. Nefedova. "Areas of Crisis in Russian Agriculture: A Geographic Perspective." *Post-Soviet Geography and Economics*, 41, No. 4 (2000): 288-305.

CROPS AND CROPPING PATTERNS

Hinrichs, Carl G. "Limits of the Spring Wheat Belt in European Russian." Master's Thesis, University of Iowa, Iowa City, Iowa, 1952.

Bone, Robert M. "The Development and Significance of Tea Cultivation in the Soviet Union." *Yearbook, Association of Pacific Coast Geographers*, 18 (1956): 63-73.

Bone, Robert M. "The Development and Significance of Tea Cultivation in the Soviet Union." Master's Thesis, University of Washington, Seattle, Wash., 1957.

Jackson, W.A.D. "The Russian Non-Chernozem Wheat Base." *Annals of the Association of American Geographers*, 49 (1959): 97-109.

Edwards, Richard L. "A Survey of Cotton Production on the Irrigated Lands of Soviet Central Asia." Master's Thesis, University of Washington, Seattle, Wash., 1960.

Anderson, Jeremy H. "The Soviet Corn Program: A Study in Crop Geography." Ph.D. Dissertation, University of Washington, Seattle, Wash., 1964.

Brereton, Barbara. "Viticulture and Viniculture in the U.S.S.R." Master's Thesis, University of Washington, Seattle, Wash., 1967.

Birdsall, Stephen S. "The Effect of Management on Crop Yields in Soviet Agriculture." *Journal of Geography*, 67, No. 1 (1968): 95-103.

Hultquist, Warren E. "The Geographic Structure of the Soviet Sugar Industry." Ph.D. Dissertation, University of Washington, Seattle, Wash., 1968.

Polosa, O.I. "The Most Important Changes in the Structure and Distribution of Major Field Crops in the Southern Ukraine" (in Ukrainian). *Ekonomichna Geografiya, Mizhvidomchyi Naukovyi Zbirbyk, Kiev*, 4 (1968): 163-168.

Shapiro, L.L. "The Problem of Distribution of Planting of Winter Crops in the Ukraine" (in Ukrainian). *Ekonomichna Geografiya, Mizhvidomchyi Naukovyi Zbirbyk, Kiev*, 4 (1968): 169-177.

Turner, Robert L. "Growing Crops By Decree: Khrushchev's Corn Campaign in the Ukraine." Master's Thesis, Wayne State University, Detroit, Mich., 1970.

Karger, Adolf. "Probleme der sowjetischen Getreideerzeugung." *Geographische Rundschau*, 28, No. 7 (1976): 265-269.

Turdibekov, Z.K. "The Questions of Development and Specialization of Vine-Growing in Uzbekistan" (in Russian). *Vestnik Leningradskogo Universiteta, Geologiya-Geografiya*, 24 (1978): 141-143.

Rostonkaowski, Peter. *Agrarraum und Getreiddeanbau in der Sowjetunion 1948–1985: Eine agrargeographische Studie.* Berlin, Germany: Duncker and Humblot, 1979.

Oak, B. "Geography of Crop Productivity with Respect to Climatic and Soil Variations: A Methodological Study with Reference to the European Part of the USSR." *Geographical Observer*, 16 (1980): 21-27.

Rakovetskaya, L.I. "The Trends of Changes in the Yields of Cereal Crops in the Main Natural Zones" (in Russian). *Vestnik Moskovskogo Universiteta, Seriya Geografiya*, 6 (1980): 42-46.

Berentsen, W.H. "Spatial Variability in Soviet Grain Productivitiy, 1955–1974." *Soviet Geography*, 23, No. 9 (1982): 630-639.

Rakovetskaya, L.I. "Grain-Yield Changes in the Main Natural Zones." *Soviet Geography*, 23, No. 8 (1982): 608-612.

Kogon, Felix. "Soviet Grain Production: Resources and Prospects." *Soviet Geography*, 24, No. 9 (1983): 631-661.

Rauner, Y.L. "Harvest of Grain Crops and the Factors of Agrotechnics" (in Russian). *Izvestiya Akademii Nauk SSSR, Seriya Geograficheskaya*, 5 (1983): 46-58.

Sinnott, Peter. "The Soviet Sugar Beet Industry: Its Diffusion and Systematic Articulation." Master's Thesis, Columbia University, New York, N.Y., 1983.

Zhitirchuk, Y.V., and L.E. Strashnenko. "The Optimization of the Structure of Areas Under Agricultural Crops with Regard to Climatological Information" (in Russian). *Trudy-Glavnaya Geofizicheskaya Observatoriya*, 466 (1983): 104-109.

Tarrant, J.R. "Predicting U.S.S.R. Wheat Production." *Applied Geography*, 4, No. 1 (1984): 47-58.

Tarrant, J.R. "The Significance of Variability in Soviet Cereal Production." *Transactions, Institute of British Geographers*, 9, No. 4 (1984): 387-400. (New Series)

Zolotokrylin, A.N. "Wheat Yield Variability in the European USSR in a Quasi-Two-Year Cycle of Atmospheric Processes." *Soviet Geography*, 26, No. 9 (1985): 680-690.

Zaitsev, G.A., and E.P. Titov. "Recultivation of Disturbed Lands Along the Line of the Berkakit-Tommot-Yakutsk" (in Russian). *Vestnik Moskovskogo Universiteta, Seriya Geografiya*, 5 (1988): 74-80.

Rakovetskaya, L.I. "Territorial Aspects of Development of Grain Production in the USSR" (in Russian). *Vestnik Moskovskogo Universiteta, Seriya Geografiya*, 2 (1989): 37-44.

Zolotogrylin, A.N., and L.A. Lozovskaya. "Obtaining Stable Grain Yields in Siberia and Kazakhstan." *Soviet Geography*, 30, No. 10 (1989): 743-751.

Rakovetskaya, L.I. "Spatial Aspects of Grain Production in the USSR." *Soviet Geography*, 31, No. 2 (1990): 108-115.

Danshin, A.I., and M.A. Kaz'min. "The Role of the Grain Economy of the Moscow Region in the Organization of Stock-Breeding" (in Russian). *Vestnik Moskovskogo Universiteta, Seriya Geografiya*, 6 (1991): 48-54.

Severin, Barbara S. "The 1992 Grain Harvest in Russian and the Republics." *Post-Soviet Geography*, 33, No. 9 (1992): 557-568.

ANIMAL HUSBANDRY

Albano, H.K. "Livestock Organization in the Formerly Nomadic Areas of the Soviet Union." *Yearbook, Association of Pacific Coast Geographers*, 18 (1956): 57-62.

Anon. "Distribution of Branches of Livestock-Rearing in the Volgo–Vyatka Zone" (in Russian). Pp. 23-35 in *Distribution and Specialisation of Farming and Animal Farming in the Volgo-Vyatskaya Area*. Moscow, Russia: 1968.

Kolomiets, M.F. "Some Problems of Developing Livestock Rearing in the Kirovograd Oblast" (in Russian). *Materialy Khar'kovsk Std. Geogra. o-vs Ukrainy*, 5 (1968): 73-78.

Avagyan, G.E. *Sovetakan Hayastani gyughatntesutyne: Tutese-ashkharhagrakan usumnasirutyun*. Erevan, Armenia: Ervani Hamalsarani Hratarkchutyun, 1970.

Fry, Virginia K. "Reindeer Ranching in Northern Russia." *Professional Geographer*, 23, No. 2 (1971): 146-152.

Balland, Daniel. "L'elevage du moutan karakul en Asie centrale sovietique." *Revue Geographique de l'Est*, 17, No. 1-2 (1977): 73-88.

Potakhin, S.B. "The Effect of the Seasonal Cyclicity of Stock-Breeding on the Landscapes of the Tuva Basin" (in Russian). *Vestnik Leningradskogo Universiteta, Seriya Geologiya-Geografiya*, 2 (1989): 103-105.

DAIRYING

Fondahl, Gail. "Reindeer Dairying in the Soviet Union." *Polar Record*, 25, No. 155 (1989): 285-294.

Kaz'min, M.A. "The Territorial Organization of the Dairy Industry in the Moscow Region and Ways of Improving It" (in Russian). *Vestnik Moskovskogo Universiteta, Seriya Geografiya*, 1 (1989): 44-51.

FARM LAND

Kuznetsov, G.A. "An Economic Appraisal of Lands in Connection with the Organization of Lands on State Farms of the Virgin Lands." *Soviet Geography*, 4, No. 9 (1963): 10-15.

Stebelsky, Ihor. "An Analysis of Agricultural Production and Land Productivity Gradients in the Soviet Union in 1953." Master's Thesis, University of Toronto, Toronto, Ontario, 1964.

Cleary, J.W. "The Virgin Lands." *Survey*, 56 (1965): 95-105.

Stebelsky, Ihor. "Land Tenure and Farm Holding in European Russia on the Eve of Collectivization." Ph.D. Dissertation, University of Washington, Seattle, Wash., 1967.

Bel'skii, N.N. "Improving the Use of Land in Collective Farms and State Farms of the Bobrov and Liski Regions in the Voronezk Oblast." *Nauchnye Zapiski, Geograficheskoe Obshchestva SSSR, Varonezhskii Otdel* (1968): 85-91.

Shukeikin, A.D., and M.V. Andriishin. "Land Stocks of the Western Regions of the Ukrainian SSR and the Effectiveness of Their Utilization" (in Russian). Pp. 9-18 in *Problems of Rational Utilisation, Organisation and Evaluation of Land*. Lvov, Ukraine: 1968.

Matiiv, I.O., and O.G. Topchev. "Natural-Economic Conditions and Specialization of Agricultural Production in the Mountain-Carpathian Zone" (in Ukrainian). *Geografichnyi Zbirnik, Livivskyi Viddil, Geografichne Zovarystvo URSR*, 9 (1969): 16-24.

Medvedev, A.G., and N.I. Zhuravskaya. "Effect of Quality of Land Resources on Specialization and Effectiveness of Agricultural Production (in the Case of the Vitebskaya Oblast)" (in Russian). *Vestnik Belarussk, Universita Seriya*, 2, No. 1 (1969): 59-64.

Hahn, Roland. *Jungere Veranderungen der landlichen Siedlungen im europaischen Teil der Sowjetunion*. Stuttgart, Germany: Stuttgarter Geographische Studien, Band 79, 1970.

Silvestrov, S.I. "Efforts to Combat the Process of Erosion and Deflation of Agricultural Land." Pp.161-186 in *Natural Resources of the Soviet Union: Their Use and Renewal*, edited by I.P. Gerasimov, D.L. Armand, and K.M. Yefron. San Francisco, Calif.: W.H. Freeman, 1971.

Sobelev, L.N. "Natural Fodderlands: Their Use and Improvement." Pp. 217-250 in *Natural Resources of the Soviet Union: Their Use and Renewal*, edited by I.P. Gerasimov, D.L. Armand, and K.M. Yefron. San Francisco, Calif.: W.H. Freeman, 1971.

Wadekin, Karl-Eugen. "Kolkhoz, Sovkhoz and Private Production in Soviet Agriculture." Pp. 106-137 in *Agrarian Policies and Problems in Communist and Non-Communist Countries*, edited by W.A.D Jackson. Seattle, Wash.: University of Washington Press, 1971.

Zimina, R.P., et al. "The Plowing Up of the Virgin Lands and the Babac Marmot Resources of Central Kazakhstan." *Soviet Geography*, 13, No. 4 (1972): 246-255.

Gurin, I.I., N.G. Lysenko, and N.K. Zadoya. "The Recultivation of Shoals of the Dnieper Reservoirs" (in Russian). *Voprosy Geografii*, 99 (1975): 157-161.

Kryuchkov, V.G. "Ways of Rationalization of Land Utilization in Different Types of Natural Environment of the Altai Territory" (in Russian). *Voprosy Geografii*, 99 (1975): 96-130.

Zamkov, O.K. and G.I. Valeshko. "Land Evaluation by the Major Agricultural Zones and Regions of the U.S.S.R." (in Russian). *Voprosy Geografii*, 99 (1975): 41-63.

Zhuravskaya, N.I. "Changes in the Agricultural Utilization of Lands in Connection with Their Drainage in the Byelorussian Polesya" (in Russian). *Voprosy Geografii*, 99 (1975): 133-146.

Amelchenko, V.I. "The Agricultural Use of Lands in the North of the Nonchernozem Zone of the Russian Federation Republic" (in Russian). *Vestnik Moskovskogo Universiteta, Seriya Geografiya*, 1 (1981): 8-91.

Oak, B. "Land Ownership and Crop Productivity in Pre-Revolutionary Russia." Pp. 249-258 in *Perspectives in Agricultural Geography*, vol. 2, edited by Noor Mohammad. New Delhi, India: Concept Publishing Company, 1981.

Puchkova, E.I., and Y.N. Tseselchuk. "Some Features in the Agricultural Use of Lands of the Northwestern and the Northern Territories of the Non-Chernozem Zones" (in Russian). *Vestnik Moskovskogo Universiteta, Seriya Geografiya*, 6 (1981): 36-41.

Paallot, J. "Open Fields and Individual Farms: Land Reform in Pre-Revolutionary Russia." *Tijdschrift voor Economische en Sociale Geographie*, 65 (1984): 46-60.

Bozhyeva, T.G. "Types of Agricultural Utilization of Land of the Moscow Region (USSR)" (in Russian). *Vestnik Moskovskogo Universiteta, Seriya Geografiya*, 5 (1987): 23-30.

Lerman, Zvi, et al. "Self-Sustainability of Subsidiary Household Plots: Lessons for Privatization of Agriculture in Former Socialist Countries." *Post-Soviet Geography*, 35, No. 9 (1994): 526-542.

Wegren, Stephen K. "Yel'tsin's Decree on Land Relations: Implications for Agrarian Reform." *Post-Soviet Geography*, 35, No. 3 (1994): 166-177.

Delehanty, James, and Kathryn Rasmussen. "Land Reform and Farm Restructuring in the Kyrgyz Republic." *Post-Soviet Geography*, 36, No. 9 (1995): 565-586.

Heron, Lena. "Wandering the Wilderness Between Plan and Market: Contemporary Land Reform and Agricultural Restructuring in Russia." Master's Thesis, University of Washington, Seattle, Wash., 1995.

Wegren, Stephen K. "The Development of Market Relations in Agricultural Lands: The Case of Kostroma Oblast." *Post-Soviet Geography*, 36, No. 8 (1995): 496-512.

O'Hara, Sarah L. "Agriculture and Land Reform in Turkmenistan Since Independence." *Post-Soviet Geography and Economics*, 38, No. 7 (1997): 430-444.

AGRICULTURAL POPULATIONS AND SOCIAL ISSUES

Cox, Kevin R., and George J. Demko. "Agrarian Structure and Peasant Discontent in the Russian Revolution of 1905." *East Lakes Geographer*, 3 (1967): 3-20.

Pifer, Joseph. "Agricultural Development and Resultant Rural Population Decline in the Central Chernozem Region of the U.S.S.R." *Pennsylvania Geographer*, 11 (1973): 12-20.

Dando, William A. "The Soviet Famine of 1946–1947." *Great Plains–Rocky Mountain Geographical Journal*, 5 (1976): 15-21.

Kuritsin, I.I. "Density of Population as a Factor and Consequence of Agricultural Location" (in Russian). *Izvestiya-Vsesoyuznogo Geograficheskogo Obshchestva*, 116, No. 4 (1984): 304-311.

Ioffe, G.V. "Shifts in Rural Population Patterns and Changes in the Agriculture of the Central Regions of the European USSR." *Geoforum*, 22, No. 1 (1991): 91-98.

Craumer, Peter R. "Regional Patterns of Agricultural Reform in Russia." *Post-Soviet Geography*, 35, No. 6 (1994): 329-351.

ENVIRONMENTAL AND CONSERVATION ISSUES

Stevens, A. "On Russian Climates and Agriculture." *Geography*, 28 (1943): 6-11.

Nuttonson, Michael Y. *Ecological Crop Geography of the Ukraine and the Ukrainian Agro-Climatic Analogues in North America*. Washington, D.C.: American Institute of Crop Ecology, 1947.

Gerasimov, I.P. "Reducing the Dependence of Soviet Agriculture on Natural Elements to a Minimum." *Soviet Geography*, 4, No. 2 (1963): 3-11.

Stankowna, Z. "The Northern Limit of Forest and Its Significance for Agriculture in the Soviet Union" (in Polish). *Przeglad Geograficzny*, 39, No. 4 (1967): 791-808.

Dizengof, G.I., and L.F. Dizengof. "Converting Mountain Relief by Planning for Agricultural Utilization" (in Russian). *Doklady Sochinsk, Otdel, Geograficheskoe Obshchestva SSSR*, 1 (1968): 143-161.

Davitaya, F.F., and G.A. Sapozhnikova. "Agroclimatic Resources." Pp. 55-70 in *Natural Resources of the Soviet Union: Their Use and Renewal*, edited by I.P. Gerasimov, D.L. Armand, and K.M. Yefron. San Francisco, Calif.: W.H. Freeman, 1971.

Babushkiu, L.N., and N.A. Kogy. "Principles of the Methods of Evaluation of Natural Resources for the Purposes of Agriculture" (in Russian). *Voprosy Geografii*, 99 (1975): 64-77.

Bridges, E.M. "Agriculture and Soils in Tajikistan." *Swansea Geographer*, 13 (1975): 10-14.

Rakitnikov, Andre N. "Regional Types of Agriculture as Related to the Types of Natural Environment" (in Russian). *Voprosy Geografii*, 99 (1975): 25-40.

Kosmakhov, K.P., and V.P. Zhotskij. "Bezichungen zwischen naturlichen Zonen und landwirtschaftlichen Produktiostypen." *Geographische Berichte*, 79 (1976): 100-110.

Rauner, Y.L. "The Periodicity of Droughts in the Grain-Growing Areas of the USSR." *Soviet Geography*, 18, No. 9 (1977): 625-645.

Maslov, E.P. "Problems of Developing Agricultural Resources of Mountainous Regions of the USSR" (in Russian). *Izvestiya Akademii Nauk SSSR, Seriya Geograficheskaya*, 5 (1979): 60-71.

Rakhmatov, J. "The Natural Conditions of Agriculture of the Karakulskaja Delta of the Zerevshan River" (in Russian). *Vestnik Leningradskogo Universiteta, Seriya Geologiya-Geografiya*, 6 (1979): 96-103.

Wein, Norbert. "Die oistsibirische Steppenlandwirtschaft-Neulandgewinnung und ihn okologische Problematik." *Erdkunde*, 35, No. 4 (1981): 263-273.

Runova, T.G., and A.D. Akhaminov. "The Study of the Impact of Agriculture on the Natural Environment: A Case Study of Kursk Region" (in Russian). *Izvestiya-Akademiya Nauk SSSR, Seriya Geograficheskaya*, 3 (1982): 30-41.

Runova, T.G., and A.D. Akhaminov. "The Impact of Agriculture on the Natural Environment (with Particular Reference to Kurst Oblast)." *Soviet Geography*, 25, No. 10 (1984): 733-748.

Rauner, Y.L. "Clusters of Drought Years in the Grain Zone of the USSR." *Soviet Geography*, 26, No. 2 (1985): 73-90.

Zonn, S.V., et al. "The Study of the Agro-Industrial Potential of Ecosystems in the South of the Russian Plain and of Its Efficient Use" (in Russian). *Izvestiya-Akademiya Nauk SSSR, Seriya Geograficheskaya*, 3 (1988): 61-71.

Bayeva, R.I., A.D. Gorbovskaya, and L.L. Dvornikova. "On the Possibility of Utilization of Industrial Waste When Recultivating the Outworn Areas" (in Russian). *Vestnik Leningradskogo Universiteta, Seriya Geologiya-Geografiya*, 4 (1989): 88-91.

Bozhyeva, T.G., and Y.G. Zharkova. "Specialization of Agriculture and Anti-Erosion Capability of Agrocoenoes (as Shown by the Moscow Region)" (in Russian). *Vestnik Moskovskogo Universiteta, Seriya Geografiya*, 5 (1989): 67-72.

Tulokhonov, A.K. "Historic-Geographical Approach for the Estimation of Agrarian Nature Management in the Baikal Region" (in Russian). *Izvestiya Vsesoyuznogo Geograficheskogo Obshchestva*, 121, No. 4 (1989): 354-360.

Chernyshev, E.P., and I.Z. Shaidulin. "Methods of Increasing the Agronomic Efficiency of Field-Protecting Belts" (in Russian). *Izvestiya-Akademiya Nauk SSSR, Seriya Geograficheskaya*, 2 (1990): 57-62.

Grin, G.B., and S.A. Lapin. "On the Efficiency of Flooding and Agricultural Use of the Irtysh Floodplain" (in Russian). *Izvestiya-Akademiya Nauk SSSR, Seriya Geografisheskaya*, 2 (1990): 75-82.

Kalutskov, V.N. "Ecological Aspects of the Use of Agricultural Lands in the Zones of Industrial Contamination" (in Russian). *Vestnik Moskovskogo Universiteta, Seriya Geografiya*, 1 (1990): 68-73.

Pruzhin, M.K. "On Interactions Between Land Cultivation Systems and Natural Environmental Conditions and Resources" (in Russian). *Izvestiya-Vsesoyusznogo Geograficheskogo Obshchestva*, 122, No. 1 (1990): 75-80.

Sautieva, T.B. "Ecological Problems of Agricultural Production in the Northern Osetin ASSR" (in Russian). *Vestnik Leningradskogo Universiteta, Seriya Geologiya-Geografiya*, 3 (1990): 103-107.

Sirotenko, O.D., et al. "An Assessment of Agroclimatic Resources of the Russian Plain as Related to Global Warming" (in Russian). *Izvestiya-Akademiya Nauk SSSR, Seriya Geograficheskaya*, 6 (1990): 29-38.

Tulokhonov, A.K. "Historico-Geographical Aspects of the Interrelationship of the Baikal Region's Agriculture with the Natural Environment" (in Russian). *Izvestiya-Akademiya Nauk* SSSR, *Seriya Geograficheskaya*, 1 (1990): 38-45.

Veretenova, N.N., et al. "Dynamics of Cultivation Improvement and Productivity of Soils of the Moscow Region" (in Russian). *Vestnik Moskovskogo Universiteta, Seriya Geografiya*, 1 (1990): 30-35.

Nefedova, T.G. "Natural and Socio-Economic Factors of Regional Development of Agriculture of the USSR European Territory." *Revue Belge de Geographie*, 115, No. 1-3 (1991): 219-236.

Rybintsev, V.A., and A.G. Amirjanov. "Evaluation of Bioclimatic Resources of a Region for the Prediction of Vineyard Productivity." *Journal of Wine Research*, 4, No. 2 (1993): 109-118.

IRRIGATION

Field, N.C. "The Role of Irrigation in the South European U.S.S.R. in Soviet Agricultural Growth: An Appraisal of the Resource Base and Development Problem." Ph.D. Dissertation, University of Washington, Seattle, Wash., 1957.

Lewis, Robert A. "A Geographic Appraisal of Irrigation Agriculture in Soviet Central Asia." Master's Thesis, George Washington University, Washington, D.C., 1959.

Kunin, V.N. "Local Water Supplied in the Desert and Agricultural Development." *Soviet Geography*, 4, No. 2 (1963): 39-45.

Lewis, Robert A. "Early Irrigation in West Turkestan." Ph.D. Dissertation, University of Washington, Seattle, Wash., 1964.

Carriere, P. "Irrigations et cultures irriguees en Turkmenie sovietique." *Bulletin de la Societe Languedocienne de Geographie*, 2, No. 2 (1968): 105-140.

Avaryanov, S.F., E.N. Minayeva, and U.A. Timost-Kina. "Increasing Agricultural Productivity Through Irrigation and Drainage." Pp. 131-160 in *Natural Resources of the Soviet Union: Their Use and Renewal*, edited by I.P. Gerasimov, D.L. Armand, and K.M. Yefron. San Francisco, Calif.: W.H. Freeman, 1971.

Spath, Hans-Joachim. "Das Konya-Cumra-Projekt, Ein Beitrag zur Problematik des Bewasserungsfeldbaus in winterkolten Trockensteppen." *Geographische Zeitschrift*, 62 (1974): 81-105.

Matley, Ian M. "The Murgab Oasis: The Modernization of an Ancient Irrigation System." *Canadian Slavonic Papers*, 17 (1975): 417-435.

L'vovich, M.I., et al. "The Water Aspect of the Geography of Agriculture in the USSR." *Soviet Geography*, 20, No. 9 (1979): 515-533.

Berengner, A.S. "Specificity of Forecasting Water Consumption by Irrigation Agriculture for a Remote Perspective" (in Russian). *Izvestiya Akademii Nauk SSSR, Seriya Geograficheskaya*, 1 (1982): 44-52.

Berengner, A.S. "Long-Term Projection of Water Needs for Irrigation in the USSR." *Soviet Geography*, 24, No. 5 (1983): 333-343.
Ermakov, Y.G., and T.J. Kondratyena. "The Influence of Irrigated Agriculture on the Environment of the West of the USSR" (in Russian). *Vestnik Moskovskogo Universiteta, Seriya Geografiya*, 3 (1986): 59-66.
Micklin, Philip P. "Irrigation and Its Future in Soviet Central Asia: A Preliminary Analysis." Pp. 229-262 in *Soviet Geographical Studies in Our Time*, edited by Lutz Holzner and Jeane M. Knapp. Milwaukee, Wis.: American Geographical Society and University of Wisconsin-Milwaukee, 1987.
O'Hara, Sarah L. "Irrigation and Land Degradation: Implications for Agriculture in Turkmenistan, Central Asia." *Journal of Arid Environments*, 37 (1997): 165-179.
Saiko, Tatyana, and Igor S. Zonn. "Irrigation Expansion and Dynamics of Desertification in the Circum-Aral Region of Central Asia." *Applied Geography*, 20, No. 4 (2000): 349-367.

AGRICULTURAL MARKETING AND DISTRIBUTION

Davitaya, F.F. "A Natural Transmission Belt for the Year-Round Supply of Fresh Farm Products." *Soviet Geography*, 4, No. 2 (1963): 25-31.
Gumerov, R.M. "Basic Principles of a Zonal Differentiation of Farm Procurement Prices." *Soviet Geography*, 9, No. 3 (1968): 194-202.
Leversedge, Francis M., and Robert C. Stuart. "Soviet Agricultural Restructure and Urban Markets." *Canadian Geographer*, 19, No. 1 (1975): 73-93.
Lasis, Y.V. "Questions of the USSR Regions Self-Provision with Food-Stuffs in a Comprehensive Economico-Geographical Analysis" (in Russian). *Izvestiya-Akademii Nauk SSSR, Seriya Geograficheskaya*, 6 (1982): 100-112.
Ingram, Michael C. "The Ukrainian Agricultural Sector: Foreign Export Trade Potential in the Intermediate Term." Ph.D. Dissertation, University of Kansas, Lawrence, Kans., 1996.

TRANSPORTATION AND MECHANIZATION

Earle, Frances M. "Mechanization of Agriculture in the U.S.S.R." *Economic Geography*, 7, No. 3 (1931): 297-307.
Pechoux, Pierre-Yves. "L'agriculture russe et les machines." *Annales de Geographie*, 50 (1941): 94-113.
Anon. "Mechanization of Agriculture in the Volga-Vyatka Zone" (in Russian). Pp. 36-64 in *Distribution and Specialization of Farming and Animal Farming in the Volga–Vyatka Zone*. Moscow, Russia: 1968.

AGRICULTURAL LABOR STUDIES

Khinchuk, Ksenya. "The Agricultural Labor Force in the Soviet Union." *Soviet Geography*, 28, No. 2 (1987): 90-115.

Part 6

SOUTHWEST ASIA AND NORTH AFRICA

Southwest Asia as a Whole

GENERAL TEXTS, ARTICLES, AND THESES

Butzer, Karl W. "Agricultural Origins in the Near East as a Geographical Problem." Pp. 416-437 in *Environment and Archeology: An Introduction to Pleistocene Geography*, edited by Karl W. Butzer. Chicago, Ill.: Aldine, 1964.

Hannemann, Manfred A. "A Culture-Historical Geography of the Plow-Cattle Complex in the Near East." Master's Thesis, University of California at Los Angeles, 1967.

Asadi, Fawzi. "Some Socio-Economic Institutions Inhibiting Economic Progress in the Agrarian Sector of the Arab East." *Geographical Perspectives*, 26 (1969): 17-31.

Stevens, J.H. "Changing Agricultural Practice in an Arabian Oasis." *Geographical Journal*, 136, No. 3 (1970): 410-418.

Clawson, Marion, Hans H. Landsberg, and Lyle T. Alexander. *The Agricultural Potential of the Middle East.* New York, N.Y.: Elsevier, 1971.

Stevens, John H. "Oasis Agriculture in the Central and Eastern Arabian Peninsula." *Geography*, 54, No. 4 (1972): 321-326.

McLachlan, Keith S. "The Agricultural Potential of the Arab Gulf States." Pp. 103-137 in *The Impact of Oil Revenues on Arab Gulf Development*, edited by M.S. Al-Azhary. London, England: Croom Helm, 1984.

Beaumont, Peter. "Trends in Middle Eastern Agriculture." Pp. 305-322 in *Agricultural Development in the Middle East*, edited by Peter Beaumont and Keith S. McLachlan. New York, N.Y.: John Wiley, 1985.

Beaumont, Peter, and Keith S. McLachlan, eds. *Agricultural Development in the Middle East.* New York, N.Y.: John Wiley, 1985.

Lawless, Richard I. "The Agricultural Sector in Development Policy." Pp. 107-122 in *Agricultural Development in the Middle East*, edited by Peter Beaumont and Keith S. McLachlan. New York, N.Y.: John Wiley, 1985.

McLachlan, Keith S. "The Agricultural Development of the Middle East: An Overview." Pp. 27-50 in *Agricultural Development in the Middle East*, edited by Peter Beaumont and Keith S. McLachlan. New York, N.Y.: John Wiley, 1985.

Shafi, M. "Contribution of Arabs to Agriculture: A Spatial Analysis in Methods and Techniques." *The Geographer, Aligarh*, 35, No. 2 (1988): 1-7.

CROPS AND CROPPING PATTERNS

Beaumont, Peter. "Wheat Production and the Growing Food Crisis in the Middle East." *Food Policy*, 14, No. 4 (1989): 378-384.

Hashmi, S.N.J. "Growth and Production of Horticultural Crops and Level of Self Sufficiency in West Asia." *Asian Profile*, 24, No. 1 (1996): 69-82.

Schirmer, J.H. "L'Agriculture des pays arabes: Entre le desert et le petrole." *Bulletin de la Societe Geographique de Liege*, 33 (1997): 73-92.

ANIMAL HUSBANDRY

Mikesell, Marvin W. "Notes on the Dispersal of the Dromedary." *Southwestern Journal of Anthropology*, 11 (1955): 231-245.

De Phanhol, Xavier. "Nomades et pasteurs, V." *Revue Geographie de l'Est*, 5 (1965): 355-374.

De Phanhol, Xavier. "Nomades et pasteurs, VII." *Revue Geographique de l'Est*, 7 (1967): 373-374.

Johnson, Douglas L. *The Nature of Nomadism: A Comparative Study of Pastoral Migrations in Southwestern Asia and Northern Africa.* Chicago, Ill.: University of Chicago Department of Geography, Research Paper No. 118, 1969.

Chatterton, Brian, and Lynne Chatterton. "Increasing Livestock Production in Dry Zones: Policy Options for the Middle East and North Africa." *Land Use Policy*, 4, No. 2 (1987): 121-132.

ENVIRONMENTAL AND CONSERVATION ISSUES

Beaumont, Peter. "The Agricultural Environment: An Overview." Pp. 3-26 in *Agricultural Development in the Middle East*, edited by Peter Beaumont and Keith S. McLachlan. New York, N.Y.: John Wiley, 1985.

Butzer, Karl W. "The Islamic Traditions of Agroecology: Crosscultural Experience, Ideas and Innovations." *Ecumene*, 1, No. 1 (1994): 7-50.

Beaumont, Peter. "Agricultural and Environmental Changes in the Upper Euphrates Catchment of Turkey and Syria and Their Political and Economic Implications." *Applied Geography*, 16, No. 2 (1996): 137-158.

Meir, Avinoam, and Haim Tsoar. "International Borders and Range Ecology: The Case of Bedouin Transborder Grazing." *Human Ecology*, 24, No. 1 (1996): 39-64.

IRRIGATION

Shafi, Ali. "Problems of Irrigation in Arid Zones." *Bulletin de la Societe de Geographie d'Egypte*, 32 (1959): 143-164.

Bederman, Sanford H. "Koranic Laws Pertaining to Water, Land and Agricultural Practices." *Southeastern Geographer*, 3 (1963): 9-16.

English, Paul W. "The Origin and Spread of Qanats in the Old World." *Proceedings of the American Philosophical Society*, 112, No. 3 (1968): 170-181.

Stevens, John H. "Irrigation in the Arab Countries of the Middle East." Pp. 73-81 in *Change and Development in the Middle East: Essays in Honour of W.B. Fisher*, edited by John Clarke and Howard Bowen-Jones. London, England: Methuen, 1981.

Allan, J.A. "Irrigated Agriculture in the Middle East: The Future." Pp. 51-62 in *Agricultural Development in the Middle East*, edited by Peter Beaumont and Keith S. McLachlan. New York, N.Y.: John Wiley, 1985.

Perennes, J.J. "Les politiques de l'eau au Maghreb: D'une hydraulique miniere a nune gestion sociale de la rarete." *Revue de Geographie de Lyon*, 65 (1990): 9-20.

Schopke, H. "Agrarwirtschaftliche Nutzung von Staudammen in Nordafrika." *Geographie und Schule*, 15 (1993): 17-23.

Beaumont, Peter. "The Myth of Water Wars and the Future of Irrigated Agriculture in the Middle East." *International Journal of Water Resources Development*, 10, No. 1 (1994): 9-21.

Lightfoot, Dale R. "The Origin and Diffusion of Qanats in Arabia: New Evidence from the Northern and Southern Peninsula." *Geographical Journal*, 166, No. 3 (2000): 215-226.

Afghanistan

GENERAL TEXTS, ARTICLES, AND THESES

Grotzbach, E. "Formen des zelgengebundenen Feldbaus und dessen Auflosungserscheinungen im Hindukush (Nordost-Afghanistan)." *Die Erde*, 101 (1970): 23-40.

Amin, Hamid A. *Nazari bih Jughrafiagayi ziraati-i Afghanistan*. Kabul, Afghanistan: Pohanhi Aadabiyat va Ulum Bashari, 1974.

CROPS AND CROPPING PATTERNS

Balland, Daniel. "Le coton en Afghanistan: Essai d'analyse geographique et economique d'un culture industrielle dans un pays sous-industrialise." *Revue Geographique de l'Est*, 13, No. 1-2 (1973): 17-76.

De Phanhol, Xavier. "Le vin de l'Afghanistan et de l'Himalya occidental." *Revue Geographique de l'Est*, 17, No. 1-2 (1977): 3-26.

FARM LAND

Jentsch, Christoph. *Zur Geographie der landlichen Siedlungen in Afghanistan: Zweit Beitrage*. Mannheim, Germany: Mannheimer Geographische Arbeiten, Heft 6, 1980.

ENVIRONMENTAL AND CONSERVATION ISSUES

Mikeladze, L.B. "The Influence of Natural Environment on the Development and Location of Agriculture in Afghanistan" (in Russian). *Izvestiya-Akademiya Nauk SSSR, Seriya Geografiya*, 2 (1985): 77-86.

IRRIGATION

Humlum, Johannes. "L'Agriculture par irrigation en Afghanistan." *Comptes rendus du Congres International de Geographie, Lisbon 1949*, 3 (1951): 318-328.

Cyprus

GENERAL TEXTS, ARTICLES, AND THESES

Upton, M., C. Bishop, and R. Pearu. "Part-Time Farming: The Cyprus Case." *GeoJournal*, 6, No. 4 (1982): 343-350.

IRRIGATION

Andre, F., and A. Robert. "Maitrise de l'eau et developpement agricole dans le Sud Chypriote." *Mediterranee*, 55, No. 4 (1985): 3-12.

Iraq

GENERAL TEXTS, ARTICLES, AND THESES

Lennie, A.B. "Agriculture in Mesopotamia in Ancient and Modern Times." *Scottish Geographical Magazine*, 52, No. 1 (1936): 33-46.

Fast, Nathan. "Agricultural and Agrarian Problems in Alluvial Iraq." Master's Thesis, Johns Hopkins University, Baltimore, Md., 1955.

Al-Mayad, Ali M. "Analysis of the Spatial Relationships Among Agricultural Phenomena in Iraq, 1953." Ph.D. Dissertation, University Iowa, Iowa City, Iowa, 1958.

Najim, Hassan T. "Agriculture in Iraq: Problems and Prospects." Ph.D. Dissertation, Syracuse University, Syracuse, N.Y., 1960.

Wirth, Eugen. *Agrargeographie des Irak*. Hamburg, Germany: Hamburger Geographische Studien, Heft 13, 1962.

Sethom, H. "Reforme agraire, petrole et urbanisation en Irak." *Annales de Geographie*, 92, No. 5 (1983): 548-573.

CROPS AND CROPPING PATTERNS

Al-Awway, Hammadi M. "The Crops of Iraq." Master's Thesis, University of Texas, Austin, Tex., 1955.

Kerim Ali, Ali A. "The Date Palms on the Shores of the Shat-el-Arab Bay" (in Russian). *Vestnik Moskovskogo Universiteta, Seriya Geografiya*, 6 (1981): 90-94.

AGRICULTURAL POPULATIONS AND SOCIAL ISSUES

Lebon, J.H.G. "Population Distribution and the Agricultural Regions of Iraq." *Geographical Review*, 43, No. 2 (1953): 223-228.

IRRIGATION

Al-Hadeed, Abbas A. "Irrigation in Iraq." Master's Thesis, University of Texas, Austin, Tex., 1954.

Barazi, Nuri K. *The Geography of Agriculture in Irrigated Areas of the Middle Euphrates Valley*. Baghdad, Iraq: Al-Aani Press, 1961–63. 2 volumes.

Al-Kasab, Nafi N. "Agrarreform und Bewasserungsfeldbau in Irak." *Geographische Rundschau*, 29, No. 4 (1977): 121-133.

Iran

GENERAL TEXTS, ARTICLES, AND THESES

Beckett, Philip H.T. "Agriculture in Central Persia." *Tropical Agriculture*, 34 (1957): 9-28.

Flower, David J. "The Agricultural Geography of the Mashad Region." Master's Thesis, University of Durham, Durham, England, 1966.

Malek, Hossein. "Apris la reforme agraire iranienne." *Annales de Geographie*, 75 (1966): 268-285.

Bowen-Jones, Howard. "Agriculture." Pp. 565-598. In *The Cambridge History of Iran: Volume 1—The Land of Iran*, edited by W.B. Fisher. Cambridge, England: Cambridge University Press, 1968.

Goodell, G. "Agricultural Production in a Traditional Village of Northern Khuzestan." Pp. 243-289 in *Tradtionelle und moderne Formen der Landwirtschaft in Iran*, edited by Eckert Ehlers and G. Goodell. Marburg, Germany: Marburger Geographische Schriften, Heft 64, 1975.

Planck, Ulrich. "Die Reintegrationphase der iranischen Agrarreform." *Erdkunde*, 29, No. 1 (1975): 1-9.

Mojtahedi, Ahmad. "Iranian Agriculture and Its Development." *Pakistan Geographical Review*, 32, No. 1 (1977): 43-50.

Ehlers, Eckert, and J. Safi-Nejad. "Formen Kollektiver Landwirtschaft in Iran." Pp. 55-82 in *Beitrage zur kulturgeographie des islanischen Orients*, edited by Eckert Ehlers. Marburg, Germany: Marburger Geographische Schriften, Heft 78, 1979.

Tavana, M.H.Z. *Die Agrarlandschaft Iranish-Sistans: Aspekte des Strukturwandels im 20 Jahrhundurt*. Marburg, Germany: Marburger Geographische Schriften, Heft 91, 1983.

Suzuki, Toshio. *Land Reform, Technology, and Small-Scale Farming: The Ecology and Economy of Gilaki-Rashtio Rice Cultivators, Northern Iran*. Ann Arbor, Mich.: University of Michigan Department of Geography, Michigan Geographical Publication No. 29, 1985.

McLachlan, Keith S. *The Neglected Garden: The Politics and Ecology of Agriculture in Iran*. New York, N.Y.: St. Martin's, 1990.

CROPS AND CROPPING PATTERNS

Beaumont, Peter, and J.H. Neville. "Rice Cultivation in Iran's Caspian Lowlands." *World Crops*, 20 (1968): 70-73.

Mojtahedi, Ahmad. *Rice Growing in Northern Iran*. Durham, England: University of Durham Department of Geography, Occasional Paper in Geography No. 15, 1980.

Khan, Aziz W.R. "An Investigation of Crop Yield Index Method and Its Application to Iran." *The Geographer, Aligarh*, 41, No. 1 (1994): 51-63.

ANIMAL HUSBANDRY

Kessler, Margrit. "Die Viehwirtschaft im Intensitatsprofil des Agrarraumes Iran-Anatolian-Balkan-Mitteleuropa." *Geographische Rundschau*, 21, No. 2 (1969): 51-59.

FARM LAND

Ahghari, Mohammad R. "Land Reform and Agricultural Development in Iran." Master's Thesis, Florida State University, Tallahassee, Fla., 1980.

Sheikh-Hassani, G.H. "Agricultural Land Use Change in Gilan, Iran." Ph.D. Dissertation, University of Southampton, Southampton, England, 1987.

IRRIGATION

Beckett, Philip H.T. "Qanats in Persia." *Journal of the Iran Society*, 1 (1952).

Beckett, Philip H.T. "Qanats, Aran and Kirman." *Journal of the Royal Central Asian Society*, 40 (1953): 47-58.

Bemont, F. "L'irrigation en Iran." *Annales de Geographie*, 70 (1961): 597-620.

Cordonnier, J.C. "Les tendances nouvelles de l'agriculture irriguee dans l'oasis d'Isfahan (Iran)." *Revue Geographique de l'Est*, 4, No. 4 (1964): 387-392.

Basset, Jacques. "Irrigation et cultures irriguees dans l'oasis d'Isphahan." *Bulletin de la Societe Languedocienne de Geographie*, 3, No. 3 (1969): 341-366.

Bavis, Myrna. "Some Observations on a Unique Irrigation System Found in Persia." *Oregon Geographer*, 4, No. 1 (1969): 9-10.

Beaumont, Peter. "Qanat Systems in Iran." *Bulletin of the International Association of Scientific Hydrology*, 16, No. 1 (1971): 39-50.

Kortum, Gerhard. *Die Marvdasht-Ebene in Fars: Grundlagen und Entwicklung einer alten iranischen Bewasserungslandschaft*. Kiel, Germany: Kieler Geographische Schriften, Band 44, 1976.

Ehlers, Eckert. "Social and Economic Consequences of Large-Scale Irrigation Developments: The Dez Irrigation Project, Khuzestan, Iran." Pp. 85-97 in *Arid Land Irrigation in Developing Countries*, edited by E.B. Worthington. Oxford, England: Pergamon, 1977.

Bonine, Michael. "Qanats, Field Systems and Morphology: Rectangularity on the Iranian Plateau." Pp. 35-57 in *Qanat, Kariz, and Khettara*, edited by Peter Beaumont, Michael Bonine, and Keith S. McLachlan. Wisbech, Germany: Menan, 1989.

Abbunoari, Abbas A. "Irrigation and Agricultural Development in Drought Prone Areas of Iran: A Water Balance Approach." *Transactions, Institute of Indian Geographers*, 12, No. 1 (1990): 39-56.

Bonine, Michael. "Qanats and Rural Socieites: Sustainable Agriculture and Irrigation Cultures in Contemporary Iran." Pp. 183-209 in *Canals and Communities: Small-Scale Irrigation Systems*, edited by J.B. Makay. Tucson, Ariz.: University of Arizona Press, 1996.

Israel

GENERAL TEXTS, ARTICLES, AND THESES

Hubbard, G.D. "Agriculture in Palestine." *American Journal of Economics and Sociology*, 10, No. 3 (1951): 247-268.

Kedar, Y. "The Problem of the Mounds of Tuleilat et Anab and Their Relation to Ancient Agriculture in the Central Negev." *Bulletin of the Israel Exploration Society*, 20 (1956): 31-43.

Halperin, Haim. *Changing Patterns in Israeli Agriculture*. London, England: Routledge and Kegan Paul, 1957.

Evenari, M., et al. "The Ancient Desert Agriculture of the Negev. III. Early Beginnings." *Israel Exploration Journal*, 8 (1958): 231-268.

Amiron, David H.K., and Y. Kedar. "Techniques of Ancient Agriculture in the Negev of Israel." *Comptes rendus, XVIII Congres de Institute de Geographie, Rio de Janeiro, 1956*, 1 (1959): 207-217.

Aharani, Y., et al. "The Ancient Desert Agriculture of the Negev: An Israelite Settlement at Ramat Matred." *Israel Exploration Journal*, 10 (1960): 23-36, 97-111.

Evenari, M., et al. "Ancient Agriculture in the Negev." *Science*, 133 (1961): 979-996.

Criscuolo, Anthony. "Agriculture in Israel." Master's Thesis, Southern Connecticut State University, New Haven, Conn., 1962.

Orni, Efraim. *L'Agriculture en Israel: Formes de la vie rurale*. Paris, France: Centre israelian de Documentation, 1963.

Kedar, Y. *Ancient Agriculture in the Negev* (in Hebrew). Jerusalem, Israel: Bialik Institute, 1967.

Richter, Werner. *Historische Entwicklung und junger Wandel der Agrarlandschaft Israels, dargestelit insbesandere am Beispiel Nordgalilans*. Cologne, Germany: Kolner Geographische Schriften, Heft 21, 1969.

Bar-Gal, Y. "On the Agricultural Character of the Various Regions in Israel" (in Hebrew). *Merhavim*, 2 (1975): 89-97.

Haupert, John S. "Prospects for an Agro-Industrial Complex on the Gaza-Sinai Sea Coast." *Ecumene*, 7 (1975): 21-24.

Feldman, Shelden. "Increasing Agricultural Productivity and Regional Development." Pp. 211-226 in *Perspectives in Agricultural Geography*, vol. 4, edited by Noor Mohammad. New Delhi, India: Concept Publishing Company, 1981.

Kellerman, A. "The Pertinence of the Macro-Thunian Analysis: The Case of Israel." *Geography Research Forum*, 4 (1981): 46-54.

Asadi, Fawzi. "How Viable Will the Agricultural Economy Be in the New State of Palestine." *GeoJournal*, 21, No. 4 (1990): 375-383.

Rokach, A. and R. Weitz. "Agriculture and Rural Planning During the First Decade of Israel's Independence." Pp. 184-192 in *Eretz Israel: Archeological, Historical and Geographical Studies*, vol. 22, edited by J. Aviram and A. Shachur. Jerusulam, Israel: Israel Exploration Society, 1991.

Amit, I. "Private Capital Investment in Agricultural Enterprise in the Sharon Plain During the Latter 1920s." *Studies in the Geography of Israel*, 15 (1993): 122-141.

CROPS AND CROPPING PATTERNS

Crickenberger, Jeanne F. "Pattern and Prospects of the Israeli Orange Industry." Master's Thesis, Columbia University, New York, N.Y., 1967.

Orni, Efraim. "Eshkolbezirk und Aravatal: Zwei Beispiele landwirtschaft Intensivbetriebe in Wustgengebieten Israels." *Geographische Rundschau*, 4 (1974): 125-131.

Avitsur, S. "Melon Cultivation in the Land of Israel: 1948–1973" (in Hebrew). *Merhavim*, 2 (1975): 82-88.

Reinhold, Horst G. *Citruswirtschaft in Israel: Eine geographische Untersuchung des Agrumenbaus, seiner Varaussetzungen, Formen und Moglichekeiten*. Heidelberg, Germany: Heidelberger Geographische Arbeiten, Heft 43, 1975.

Clancy, S. "The Geography of Food Crops in the Gaza Strip." Master's Thesis, California State University, Northridge, Calif., 1996.

ANIMAL HUSBANDRY

Meir, Avinoam and Haim Tsoar. "International Borders and Range Ecology: The Case of Bedouis Transborder Grazing." *Human Ecology*, 24, No. 1 (1996): 39-64.

FARM LAND

Ben-Arieh, Y. "A Comparison of Agricultural Land Use in a Semi-Arid Region: Ashquelon-Bet Guvrin Area, 1946–1968." *Jerusalem Studies in Geography*, 1 (1970): 107-127.

Amiron, David H.K. "Changes in the Assessment of Land in Israel as an Agricultural Resource" (in Hebrew). *Karka*, 23 (1983): 11-32.

Amiron, David H.K. "Preservation of Land for Future Use in Agriculture." Pp. 107-112 in *The Mosaic of Israeli Geography*, edited by Yehuda Gradus and Gabriel Lipshitz. Beersheba, Israel: Ben-Gurion University of the Negev Press, 1996.

AGRICULTURAL POPULATIONS AND SOCIAL ISSUES

Warr, E.C.R. "The Kibbutz: An Agricultural Entity." Pp. 39-64 in *Proceedings, Second New Zealand Geography Conference*. Palmerston North, New Zealand: New Zealand Geographical Society, 1961.

Kark, Ruth. "The Agricultural Character of Jewish Settlement in the Negev (1939–1947)." *Jewish Social Studies*, 45 (1983): 157-174.

Kark, Ruth. "Millenarism and Agricultural Settlements in the Holy Land in the Nineteenth Century." *Journal of Historical Geography*, 9, No. 1 (1983): 47-62.

Grossman, David. "Agricultural Intensification and Population in Northern Samaria: Changes in the Olive-Grain Ratio." *Geografiska Annaler B*, 66B, No. 2 (1984): 135-144.

Thalmann, Naftali. "Introducting Modern Agriculture into Nineteenth-Century Palestine: The German Templers." Pp. 90-104 in *The Land That Became Israel: Studies in Historical Geography*, edited by Ruth Kark. New Haven, Conn.: Yale University Press, 1990.

Katz, Yossi, and Shoshana Narmon. "Women's Quest for Occupational Equality: The Case of Jewish Female Agricultural Workers in Pre-State Israel." *Rural History*, 7, No. 1 (1996): 33-52.

Sofer, M. "Pluriactivity in the Moshav-Family Farming in Israel." *Journal of Rural Studies*, 17, No. 3 (2001): 363-375.

ENVIRONMENTAL AND CONSERVATION ISSUES

Kedar, Y. "Water and Soil from the Desert: Some Ancient Agricultural Achievements in the Central Negev." *Geographical Journal*, 123 (1957): 179-187.

Bar-Gal, Y. "Decision Environment and Agricultural Activity: The Israeli Case." *Asian Geographer*, 1, No. 2 (1982): 25-33.

Bruins, H.J. "Desert Environment and Agriculture in the Central Negev and Kadesh-Barnea During Historical Times." Ph.D. Dissertation, Agricultural University of Wageningen, Netherlands, 1986.

Amiron, David H.K. "Arid Zone Agriculture in the Aravah in Israel: Unconventional Agroclimatic Resources and Risks." *Geography Research Forum*, 13 (1993): 109-115.

IRRIGATION

Gottman, Jean. "L'irrigation en Palestine." *Annales de Geographie*, 44 (1935): 143-161.

Igbozurike, Matthias. "Irrigation Expansion in Israel." Master's Thesis, University of Wisconsin, Milwaukee, Wis., 1969.

Yair, A. "Hillslope Hydrology, Water Harvesting, and Areal Distribution of Some Ancient Agricultural Systems, Northern Negev, Israel." *Journal of Arid Environments*, 13 (1983): 9-24.

Rowley, Gwyn. "Irrigation Systems in the Holy Land: A Comment." *Transactions, Institute of British Geographers*, 11, No. 3 (1986): 356-359. (New Series)

Shanon, Leslie. "Irrigation Development: Proactive Planning and Interactive Management." Pp. 251-276 in *The Arid Frontier: Interactive Management of Environment and Development*, edited by H.J. Bruins and Harvey Lithwick. Dordrecht, Netherlands: Kluwer Academic, 1998.

Beaumont, Peter. "Water for Peace in the Middle East: The Sacrifice of Irrigated Agriculture in Israel?" *The Arab World Geographer*, 3, No. 2 (2000): 97-112.

Jordan

GENERAL TEXTS, ARTICLES, AND THESES

Dufourge, J. "Notes et reflexions sur l'agriculture jordanienne." *Revue de Geographie de Lyon*, 34 (1959): 255-270.

Birch, Brian P. "Recent Developments in Agriculture, Land, and Water Use in Jordan." *World Crops*, 25 (1973): 66-76.

Borham, Nasim, and Horst G. Mensching. "Pressure on the Pastureland Environment in Jordan." *Applied Geography and Development*, 31 (1988): 37-53.

Borham, Nasim, and Horst G. Mensching. "Agrargeographische Folgen der Urbanisierung von Ammon/Jordanien." *Erdkunde*, 41, No. 3 (1988): 241-245.

Borham, Nasim, and E. Grotzbach. "Landwirtschaftliche rekultivierung und Siedlungsausbau in Bergland von Ajlun (Nordjordanien)." *Petermann's Geographische Mitteilungen*, 144 (1999): 203-218.

CROPS AND CROPPING PATTERNS

Khan, Chhutton. "Fluctuations in Foodgrain Production in Jordan." *Rural Systems*, 4, No. 1 (1986): 45-56.

Lines, Lee. "Bronze Age Orchard Cultivation and Urbanization in the Jordan River Valley." Ph.D. Dissertation, Arizona State University, Tempe, Ariz., 1995.

FARM LAND

Edwards, M., J. Wellens, and D. Al-Eisawi. "Monitoring the Grazing Resources of the Badia Region, Jordan, Using Remote Sensing." *Applied Geography*, 19, No. 4 (1999): 385-398.

ENVIRONMENTAL AND CONSERVATION ISSUES

Mahmoud, Ali A. "Climate and Agriculture in Jordan." Master's Thesis, South Dakota State University, Brookings, S.Dak., 1983.

IRRIGATION

Smith, R.A., and Brian P. Birch. "The East Ghor Irrigation Project in the Jordan Valley." *Geography*, 48, No. 4 (1963): 406-409.

Sahawneh, Fouzi E. "Irrigation and Irrigated Agriculture in Jordan: An Economic Geographic Appraisal." Ph.D. Dissertation, Michigan State University, East Lansing, Mich., 1970.

Manners, Ian R. "Problems of Water Resource Management in a Semi-Arid Environment: The Case for Irrigation Agriculture in the Central Jordan Rift Valley." Pp. 95-115 in *Spatial Aspects of Development*, edited by Brian S. Hoyle. New York, N.Y.: John Wiley, 1974.

Lebanon

GENERAL TEXTS, ARTICLES, AND THESES

Klaer, Wendelin. *Eine Landnutzungs Karte von Libanon*. Heidelberg, Germany: Heidelberger Geographische Arbeiten, Heft 10, 1952.

Sanlaville, P. "Les regions agricoles du Liban." *Revue de Geographie de Lyon*, 38 (1963): 47-90.

CROPS AND CROPPING PATTERNS

Jones, David R.W. "Apple Production in the Lebanon: A Study of Agricultural Development in an Under-Developed Area." *Economic Geography*, 39, No. 3 (1963): 245-257.

FARM LAND

Irby, Joseph K. "Aspects of Musha Land Tenure in Lebanon." *Yearbook, Association of Pacific Coast Geographers*, 33 (1971): 153-160.

Saudi Arabia and Kuwait

GENERAL TEXTS, ARTICLES, AND THESES

Crary, Douglas D. "Recent Agricultural Developments in Saudi Arabia." *Geographical Review*, 41, No. 3 (1951): 366-383.

Stevens, John H. "The Changing Agricultural Practice of an Arabian Oasis." *Geographical Journal*, 136 (1970): 410-418.

Al-Zokair, Abdulazuz. "A Study of the Potential for Agricultural Development in Al-Kharuj, Saudi Arabia." Master's Thesis, Michigan State University, East Lansing, Mich., 1981.

Hall, James R. "Potential Agricultural Development in the Hijaz Region of Saudi Arabia." Master's Thesis, University of Texas, Austin, Tex., 1982.

Ashwon, Majed S. "Agricultural Patterns Around Al-Kharj, Saudi Arabia." Master's Thesis, University of Cincinnati, Cincinnati, Ohio, 1984.

Al-Feel, Muhammad R. "Kuwait: Small-Scale Agriculture in an Oil Economy." Pp. 279-288 in *Agricultural Development in the Middle East*, edited by Peter Beaumont and Keith S. McLachlan. New York, N.Y.: John Wiley, 1985.

Al-Turki, Abdul. "A Geographical Analysis of the Al-Kharg Agricultural Region of Saudi Arabia." Master's Thesis, University of South Florida, St. Petersburg, Fla., 1986.

Searight, S. "Farmers of the Desert." *Geographical Magazine*, 58, No. 3 (1986): 27-31.

Akkad, A.A. "Socio-Economic Factors Affecting Agricultural Development in Al-Hassa Oasis, Saudi Arabia." *GeoJournal*, 20, No. 3 (1990): 259-270.

FARM LAND

Al-Shomrony, Salih. "Types, Distribution and Significance of Agricultural Terraces in Assarah, Saudi Arabia." Master's Thesis, Michigan State University, East Lansing, Mich., 1980.

AGRICULTURAL POPULATIONS AND SOCIAL ISSUES

Abdulfattah, K. *Mountain Farmer and Fellah in Asir, South-West Saudi Arabia*. Erlanger, Germany: Erlanger Geographische Arbeiten, Heft 12, 1981.

El-Hadad, Awad. "The Feasibility of Agricultural Settlement Projects in Jabal Al-Akhdar: An Example of Wadi Al-Kharroubah Pastoral Settlement Scheme." Master's Thesis, Michigan State University, East Lansing, Mich., 1985.

ENVIRONMENTAL AND CONSERVATION ISSUES

Al-Zaidy, Abdulrahman. "Food and Agriculture in Saudi Arabia Under Aridity Environment and Rapid Economic Change." Master's Thesis, Michigan State University, East Lansing, Mich., 1980.

Wildenhahn, Eberhard. "Hydrological and Agricultural Conditions in South-West Saudi Arabia." *Applied Geography and Development*, 17 (1981): 58-70.

El-Sharif, A.A. "Climatic Constraints and Potential Corn Production in Saudi Arabia: A Study in Agroclimate." *GeoJournal*, 13, No. 2 (1986): 119-127.

Al Mass, Ahmad. "Agriculture and Sand Incursion in the Al-Ahsa Oasis, Saudi Arabia." Master's Thesis, Indiana State University, Terre Haute, Ind., 1999.

IRRIGATION

Al-Thubaity, K. "Impact of the Malaki Dam on Agricultural Development in the Wadi Jizan, Saudi Arabia." Master's Thesis, Michigan State University, East Lansing, Mich., 1978.

AGRICULTURAL MARKETING AND DISTRIBUTION

Mohammed, Alfair. "The Agricultural Marketing System in Taif, Saudi Arabia." Master's Thesis, Michigan State University, East Lansing, Mich., 1972.

Syria

GENERAL TEXTS, ARTICLES, AND THESES

Rosciszewski, Marcin. "Quelques remarques sur la geographie agraire de la Syrie." *Mediterranee*, 6 (1965): 171-184.

Bianquis, A.M. "Les cooperatives agricoles en Syrie: L'exemple de l'oasis de Damas." *Revue de Geographie de Lyon*, 54 (1979): 289-303.

Metral, J., and P. Sanlaville. "Problemes agraires en Syrie." *Revue de Geographie de Lyon*, 54 (1979): 229-325.

Manners, Ian R., and Tazi Sagafi-Nejad. "Agricultural Development in Syria." Pp. 255-278 in *Agricultural Development in the Middle East*, edited by Peter Beaumont and Keith S. McLachlan. New York, N.Y.: John Wiley, 1985.

Hopfinger, Hans. "Kapitalistische Agro-Business in einer sozialistischen Land? Syrien versucht neue Wege in der Landwirtschaft." *Die Erde*, 121 (1990): 157-176.

Hopfinger, Hans. "Offentliche Landwirtschaft in Syrien: Staatsformen als Beispiele fur die Wirtsamkeit staatlicher Wirtschaftstatigkeit im Agrarsektor des Landes." *Erdkunde*, 44, No. 2 (1990): 123-135.

CROPS AND CROPPING PATTERNS

Isnard, Hildebert. "La culture de la vigne dans la Fruska Gora." *Annales de Geographie*, 67 (1958): 404-429.

Shah, Abdul R. "Role of Fluctuation in Foodgrain Production in Syria." *Rural Systems*, 4, No. 2 (1986): 99-106.

ENVIRONMENTAL AND CONSERVATION ISSUES

Wirth, Eugen. "Die Ackerebene Nordost syriens." *Geographische Zeitschrift*, 52 (1964): 7-42.

AGRICULTURAL POPULATIONS AND SOCIAL ISSUES

Rosciszewski, Marcin. "Changes in Social Relations and Ownership Structures in Agriculture in Syria" (in Polish). *Prace Geograficzne*, 59 (1967): 29-61.

IRRIGATION

Lightfoot, Dale R. "Syrian Qanat Romani: History, Ecology, Abandonment." *Journal of Arid Environments*, 33, No. 3 (1996): 321-336.

Lightfoot, Dale R. "Qanats in the Levant: Hydraulic Technology at the Periphery of Early Empires." *Technology and Culture*, 38 (1997): 432-451.

Turkey

GENERAL TEXTS, ARTICLES, AND THESES

Ering, Siari, and Necdet Tuncoilek. "The Agricultural Regions of Turkey." *Geographical Review*, 42, No. 2 (1952): 179-203.

Helburn, Nicholas. "A Stereotype of Agriculture in Semiarid Turkey." *Geographical Review*, 45, No. 3 (1955): 375-384.

Hiltner, John, Jr. "An Analysis of the Contemporary Agricultural Geography of Cukurova." Ph.D. Dissertation, Northwestern University, Evanston, Ill., 1958.

Tumertekin, Erol. "La dispersion des ouvrienes dans l'agriculture de la Turquie." *Review of the Geographical Institute of the University of Istambul*, 5 (1958–1959): 159-161.

Tumertekin, Erol. "The Structure of Agriculture in Turkey." *Review of the Geographical Institute of the University of Istambul*, 5 (1958–1959): 77-93

Hutteroth, Wolf D. *Bergnomaden und Yaylabauer im mittleren kurdischen Tauras*. Marburg, Germany: Marburger Geographische Schriften, Heft 11, 1959.

De Phanhol, Xavier. "Expansion et problemes de l'agriculture turque." *Revue de Geographie de Lyon*, 35 (1960): 91-103.

Denker, Bedriye. "Die heutige Agrarwirtschaft der Bursa-Ebene (Turkei)." *Review of the Geographical Institute of the University of Istambul*, 9-10 (1963–1964): 116-132.

Kolars, John F. "Bahtili: A Turkish Village in Transition from Subsistence to Commercial Agriculture." Pp. 12-18 in *Focus on Geographic Activity: A Collection of Original Studies*, edited by Richard S. Thoman and Donald J. Patton. New York, N.Y.: McGraw-Hill, 1964.

Kolars, John F. "Decision and Commitment in Turkish Agriculture." *Review of the Geographical Institute of the University of Istambul*, 11 (1966–1968): 37-46.

Atanasiu, Nicolae. "Wandlungen und Wandlungsmoglichkeiten der turkischen Ararstruktur." *Geographische Rundschau*, 22, No. 1 (1970): 19-22.

Kolars, John F. "Systems of Change in Turkish Village Agriculture." Pp. 204-233 in *Turkey: Geographic and Social Perspectives*, edited by Peter Benedict, Erol Tumertekin, and F. Mansur. Leiden, Belgium: E.J. Brill, 1974.

Weiss, Walter. "Die landwirtschaft der Turkei." *Zeitschrift fur Wirtschaftsgeographie*, 23, No. 4 (1979): 110-112.

Dewdney, John C. "Agricultural Development in Turkey." Pp. 213-223 in *Change and Development in the Middle East*, edited by John Clarke and Howard Bowen-Jones. London, England: Methuen, 1981.

Schweiser, Gunter. "The Changing Structure of Mountain Farming in the Pontus Mountains (Eastern Black Sea Region, Turkey)." *Applied Geography and Development*, 20 (1982): 46-62.

Tumertekin, Erol. "Les structures de l'agriculture Turque: Tendences novelles." *Travaux-Institut de Geographie de Reims*, 65-66 (1986): 31-48.

Manegold, D. "Agrarwirtschatliche Aspekte eines Beitrits der Turkei zur Europaischen Gemeinschaft." *Geographie un Schule*, 11, No. 62 (1989): 32-38.

Toepfer, Helmuth. "Die Diffusion von Innovation in Gebieten intensiver landwirtschaftlicher Nutzung in der Westturkei." *Erdkunde*, 44, No. 3 (1990): 171-179.

CROPS AND CROPPING PATTERNS

Stratil-Sauer, G. "Cereal Production in Turkey." *Economic Geography*, 9, No. 4 (1933): 325-336.

Tumertekin, Erol. "Time Relationship Between the Wheat Growing Period and Dry Months in Turkey." *Review of the Geographical Institute of the University of Istambul*, 2 (1952–1953): 73-84.

Tuncoilek, Necdet. "Tea Growing in Turkey." *Review of the Geographical Institute of the University of Istambul*, 7 (1961–1962): 53-74.

Hutteroth, Wolf D. "Getreidekonjunktur und Jungerer Siedlungsausbau im Sudlichen Inneranatolien." *Erdkunde*, 16 (1962): 249-271.

Wajtkowiak, Georg. *Die Zitruskulturen in der Kustennahen Agrarlandschaft der Turkei: Wirtschaftsgeographische Betrachtung eines Produktionszweiges*. Hamburg, Germany: Mitteilungen der Geographischen Gesellschaft in Hamburg, Band 8, 1971.

Beeley, Brian W. "A Fieldscape in Transition: The Case of a Turkish Orchard." Pp. 224-232 in *Change and Development in the Middle East: Essays in Honour of W.B. Fisher*, edited by John Clarke and Howard Bowen-Jones. London, England: Methuen, 1981.

Kortum, Gerhard. "Zuckerrubenanbau und regionale Agrarentwicklung in der Turkei." *Die Erde*, 113 (1982): 21-42.

Kortum, Gerhard. *Zuckerrubenanbau und Entwicklung landlicher Wirtschaftsraume in der Turkei: Ausbreitung und Auswirkung einer Industriepflanze unter besonderer Berucksichtigung des Bezirks Beypazan (Provinz Ankana)*. Kiel, Germany: Kieler Geographische Schriften, Band 63, 1986.

Struck, Ernst. "The Innovation of Greenhouse Crops in the Mediterranean Area: The Turkish Case." *GeoJournal*, 13, No. 1 (1986): 37-46.

Struck, Ernst. "Die Treibhauskulturen der turkischen-Sudkuste: Eine Diffusionsanalyse." *Erdkunde*, 44, No. 3 (1990): 161-170.

FARM LAND

Tumertekin, Erol. "Some Maps of Agricultural Land Holdings in Turkey." *Professional Geographer*, 12 (1962): 21-22.

AGRICULTURAL POPULATIONS AND SOCIAL ISSUES

Tumertekin, Erol. "Changing Picture of Female Participation in Turkish Agriculture." *Professional Geographer*, 16 (1964): 17-20.

Mitchell, W.A. "Turkish Villages in Interim: Anatolia and Von Thunen's Isolated States." *Middle East Journal*, 25 (1971): 355-369.

ENVIRONMENTAL AND CONSERVATION ISSUES

Tumertekin, Erol. "Some Observations Concerning Dry Farming in Arid Regions of Turkey." *Review of the Geographical Institute of the University of Istambul*, 3 (1954–1955): 19-30.

Hummer, P. "Die Schadlingskatastrophe im Baumwolfanbangebiet der Cukurova/Turkei: Ihre geographischen, wirtschaftlichen und sozialen Konsequenzen." *Zeitschrift fur Auslandischelandwirtschaft*, 16, No. 4 (1977): 372-381.

IRRIGATION

Spath, Hans-Joachim. "The Konya-Cumra Project in Turkey: A Study of the Problems of Irrigated Agriculture in Dry Steppe Climates with Cold Winters." *Applied Sciences and Development*, 10 (1977): 122-145.

Kora, H. "Irrigation et culture du coton en Turquie." *Revue Geographique de l'Est*, 30, No. 2 (1990): 169-192.

TRANSPORTATION AND MECHANIZATION

Tanoglu, Ali. "La medanisation de l'agriculture en Turquie." *Review of the Geographical Institute of the University of Istambul*, 8 (1963–1964): 33-46.

The United Arab Emirates, Oman, Qatar, and Bahrain

GENERAL TEXTS, ARTICLES, AND THESES

Stevens, John H. "Arid Zone Agricultural Development in the Trucial States." *Journal of Soil and Water Conservation*, 24, No. 5 (1969): 181-183.

Dutton, R.W. "The Agricultural Potential of Oman." Pp. 170-184 in *Issues in Development: The Arab Gulf States*, edited by M. Ziwar Daftari. London, England: MD Services, 1980.

Ahmed, J.H., and John C. Doornkamp. "The Farmer's Perception of Soil for Agriculture in Bahrain." *Third World Planning Review*, 3, No. 3 (1981): 275-288.

Babikir, Ahmed A.A. "Farms Abandonment in Qatar." *Annals of the National Association of Geographers, India*, 3, No. 1 (1983): 19-27.

Unwin, Timothy. "Agriculture and Water Resources in the United Arab Emirates." *Arab Gulf Journal*, 3, No. 1 (1983): 75-85.

ENVIRONMENTAL AND CONSERVATION ISSUES

Bowen-Jones, Howard, et al. *Survey of Soils and Agricultural Potential in the Trucial States*. Durham, England: University of Durham, Department of Geography, 1967.

Pike, J.G. "Water Resources and Agriculture in Qatar." *Arabian Studies*, 5 (1979): 67-86.

IRRIGATION

Wilkinson, J.C. *The Organisation of the Falaj Irrigation System in Oman*. Oxford, England: Oxford University, School of Geography, No. 10, 1974.

Yemen, North and South

GENERAL TEXTS, ARTICLES, AND THESES

Kopp, Horst. *Agrargeographie der Arabischen Republik Jemen: Landnutzung und agrarsoziale Verhaltnisse in einen islamisch-orientalischen Entwicklungsland mit alter bauerlicher Kultur*. Erlangen, Germany: Palm and Enke, 1981.

Banski, J. "Types of Agriculture in the South Yemen" (in Polish). *Przeglad Geograficzny*, 63, No. 3-4 (1991): 307-324.

Schmitz, Charles. "Privatization in Agriculture: A Case Study from South Yemen." Ph.D. Dissertation, University of California, Berkeley, Calif., 1997.

ENVIRONMENTAL AND CONSERVATION ISSUES

Brunner, V., and H. Haefner. "The Successful Floodwater Farming System of the Sabeans, Yemen Arab Republic." *Applied Geography*, 6, No. 1 (1986): 77-86.

Schorlemer, Dietmar. *Die Al Mahwit Provinz/Jemen: Das naturliche Entwicklungspotential einer randtropischen Gebirgsregion: Eine okologische landklassifizierung auf der Grundlage einer physiographischen Bodenkartierung*. Giessen, Germany: Giessener Geographische Schriften, Heft 65, 1990.

IRRIGATION

Caton-Thompson, G., and E.W. Gardner. "Climate, Irrigation, and Early Man in the Hadh-Ramant." *Geographical Journal*, 93, No. 1 (1939): 18-38.

Serjeant, R.B. "Some Irrigation Systems in Hadramawt." *Bulletin of the School of Oriental and African Studies*, 27 (1964): 33-76.

North Africa

GENERAL TEXTS, ARTICLES, AND THESES

White, K.D. "Roman Agriculture in North Africa." *Nigerian Geographical Journal*, 6 (1963): 39-49.

Depois, Jean. "Les paysages agraires traditionnels du Maghreb et du Sahara septentrional." *Annales de Geographie*, 73, No. 2 (1964): 129-171.

Rosciszewski, Marcin. "The Directions of the Evolution of Agriculture in the Countries of the Maghreb" (in Polish). *Prace Geograficzne*, 88 (1970).

Lawless, Richard I. "Progress and Problems of Maghreb Agriculture." *Maghreb Review*, 1, No. 2 (1976): 6-11.

Achenback, Hermann. *Agrargeographie-Nordafrika (Tunesien, Algerien)*. Berlin, Germany: Borntraeger, 1983.

Kassah, Ahmed. "Le probleme alimentaire et la crise de l'agriculture au Maghreb." *Erdkunde*, 39, No. 1 (1985): 42-49.

Kassah, Ahmed. "Le probleme alimentaire et la crise de l'agriculture au Maghreb." *Revue Tunisienne de Geographie*, 16 (1989): 9-44.

Maures, G. "Les dynamiques agraires dans les montagnes rifaines et telliennes au Maghreb." *Bulletin de l'Association de Geographes Francais*, 68, No. 4 (1991): 267-280.

Maures, G. "Agriculture in the Rif and Tell Mountains of North Africa." *Mountain Research and Development*, 12, No. 4 (1992): 337-347.

FARM LAND

Ballais, J.L. "Terrasses de culture et jessours du Maghreb oriental." *Mediterranee*, 71, No. 3-4 (1990): 51-53.

AGRICULTURAL POPULATIONS AND SOCIAL ISSUES

Lawless, Richard I. "L'evolution du peuplement de l'habitat et des paysages agraires du Maghreb." *Annales de Geographie*, 81, No. 4 (1972): 451-464.

ENVIRONMENTAL AND CONSERVATION ISSUES

Manners, Ian R. "The Desert and the Sown: An Ecological Appraisal of the Middle East." *Focus*, 22, No. 2 (1971): 1-8.

Achenback, Hermann. "Klimagebundene Risikostufen der Ert ragsbildung und raumliche Standart differenzierung der landwirtschaft im Maghreb." *Erdkunde*, 33, No. 4 (1979): 275-281.

Anhuf, D. "Climate and Crop Yields in Northern Africa (Senegal, Sudan, Tunisia)." *GeoJournal*, 20, No. 3 (1990): 221-230.

Algeria

GENERAL TEXTS, ARTICLES, AND THESES

Karabenick, Edward. "An Agricultural Comparison of the Toulon Region and the European-Settled Area of the Algiers Region." Master's Thesis, Wayne State University, Detroit, Mich., 1956.

Isnard, Hildebert. "Agriculture europeenne et agriculture indigenine en Algerie." *Les Cahiers d'Outre Mer*, 12 (1959): 147-159.

Prenant, Andre. "La propriete fonciere des citadins dans les regions de Tlemcen et Sidi-Bel-Abbis." *Annales Algeriennes de Geographie*, 2, No. 3 (1967): 2-94.

Boumaza, Nadir. "Un domaine de la vallee le Sersou: Chebbah Mohamed. (daira de Tissemsilt)." *Annales Algeriennes de Geographie*, 4, No. 8 (1969): 123-128.

Brule, Jean-Claude. "Une comperative d'Anciens Moujahidine dans le bassin de Djidjelli: Khelalef Mohamed Tahar." *Annales Algeriennes de Geographie*, 4, No. 8 (1969): 43-50.

Castevert, Claude. "Un domaine autogere de la region d'Annaba: Le domaine 'Boucetta Ali'." *Annales Algeriennes de Geographie*, 8 (1969): 21-42.

Mutin, G. "L'Algerie et ses agromes." *Revue de Geographie de Lyon*, 44 (1969): 5-36.

Mutin, G. "Un domaine intensif de la Mitidju: Le domaine 113 a Oued el Alleug." *Annales Algeriennes de Geographie*, 8 (1969): 7-20.

Nesson, Claude. "Un domaine autogere au Sahara: le domaine Hamlaoui Brahim (Ourir)." *Annales Algeriennes de Geographie*, 8 (1969): 149-164.

Sari, Djilali. "Un domaine de la vallee du Chelif: Le domaine Sidi Oqba (El Asnam)." *Annales Algeriennes de Geographie*, 8 (1969): 107-122.

Tomas, Francois. "Annaba et sa region agricole." *Revue de Geographie de Lyon*, 44 (1969): 37-74.

Mutin, Georges. "Agriculture et agriculteurs prives en Mitidja." *Annales Algeriennes de Geographie*, 5, No. 10 (1970): 63-87.

Sari, Djilali. "La vie rurale dans l'Ouer-Senis." *Annales Algeriennes de Geographie*, 5, No. 10 (1970): 25-62.

Zimmermann, Bernard. "Les jardins de Kristel." *Annales Algeriennes de Geographie*, 5, No. 10 (1970): 9-24.

Beauvillain, Jean. "Aspects agraires de la region de Sidi Aissa." *Annales Algeriennes de Geographie*, 6, No. 11 (1971): 10-23.

Isnard, Hildebert. "Les structures de l'agriculture privee en Algerie." Pp. 61-73 in *Les Societes Rurales Mediterraneennes: Actes du Colloque de Geographie Agraire*. Aix-en-Provence, France: Centre Geographique d'Etudes et de Recherches Mediterraneennes, 1972.

Marthelot, P. "Le remodelage des exploitations agricoles en Algerie." *Geographia Polonica*, 29 (1974): 419-425.

Couderc, Raymond, and Guy Desire. "Croissance urbaine et milieu rural: Le d'organisation de l'agriculture autogeree entre Oran et Arzew (Algerie)." *L'Espace Geographique*, 4, No. 1 (1975): 17-30.

Cote, Marc. "Revolution agraire et espace rural: Le cas hautes plaines de l'Est algerien." *Bulletin de la Societe Languedocienne de Geographie*, 10, No. 1 (1976): 75-94.

Le Coz, Jean. "Agriculture, developpment et organisations spatiales: Le modile algerien." *Bulletin de la Societe Languedocienne de Geographie*, 10, No. 1 (1976): 5-32.

Sutton, Keith S. "The Progress of Algeria's Agrarian Reform and Its Settlement Implications." *Maghreb Review*, 3, No. 5-6 (1978): 10-16.

Cote, Marc. "Le produit agricole algerien." *L'Espace Geographique*, 8, No. 2 (1979): 143-152.

Trautman, Wolfgang. "Entwicklung und Probleme der Agrarreform in Algerien." *Erdkunde*, 33, No. 3 (1979): 215-225.

Schnetzler, J. "Ou en est l'agriculture algerienne?" *L'Information Geographique*, 44, No. 2 (1980): 59-72.

Brule, J.-C. "Geographie regionale de la Revolution Agraire Algerienne." *Bulletin de l'Association Geographes Francais*, 62, No. 1 (1985): 5-20.

Szrmer, Jacek. "Types of Algerian Socialist Agriculture" (in Polish). *Dokumentacja Geograficzna*, 2, No. 3 (1988): 47-73.

CROPS AND CROPPING PATTERNS

Barrere, Pierre. "Le vignoble algerien et le milieu natural." *Les Cahiers d'Outre Mer*, 6 (1953): 187-191.

Barrere, Pierre. "Une histoire du vignoble algerien." *Les Cahiers d'Outre Mer*, 10 (1957): 373-378.

Isnard, Hildebert. "Les caracteres originaux du vignoble oranais." *Les Cahiers d'Outre Mer*, 10 (1957): 97-106.

Grandet-Vache, M. Claude. "La riziculture dans les terraine sales d'Oranie." *Les Cahiers d'Outre Mer*, 11 (1958): 87-90.

Cote, Marc. "Un bel exemple de reconversion du vignoble: Un domaine Bouglouf Braiek (Bassin de Skikda)." *Annales Algeriennes de Geographie*, 8 (1969): 73-88.

Cote, Marc. "Un domaine cerealier de 3000 hectares dans le Setifois: Rouba Tayeb." *Annales Algeriennes de Geographie*, 8 (1969): 129-148.

Lenormond, Pierre. "Un domaine viticole de l'Oranie." *Annales Algeriennes de Geographie*, 8 (1969): 51-72.

Brule, Jean-Claude, Marc Cote, and Claude Nesson. "Les palmeraies de l'Oued Righ." *Annales Algeriennes de Geographie*, 5, No. 9 (1970): 93-106.

Isnard, Hildebert. "La viticulture algerienne: Colonisation et decolonisation." *Mediterranee*, 23, No. 4 (1975): 3-10.

Sutton, Keith S. "Algeria's Vineyards: A Problem of Decolonization." *Mediterranee*, 65, No. 3 (1988): 55-66.

Sutton, Keith S. "Algeria's Vineyards: An Islamic Dilemma and a Problem of Decolonization." *Journal of Wine Research*, 1, No. 2 (1990): 101-120.

FARM LAND

Mutin, G. "Quelques exemples de morcellement foncier en terres socialistes: Les domaines de la Mitidja." *Annales Algeriennes de Geographie*, 8 (1969): 165-168.

Brule, Jean-Claude. "Propriete et exploitation dans les montagnes de Petite Kaylie." *Annales Algeriennes de Geographie*, 6, No. 11 (1971): 3-9.

Couderc, Raymond. "Perspectives d'amenagement de la steppe occidentale." *Annales Algeriennes de Geographie*, 6, No. 11 (1971): 65-72.

AGRICULTURAL POPULATIONS AND SOCIAL ISSUES

Cote, Marc. "Ain Oulmene, une paysannerie dynamique." *Annales Algeriennes de Geographie*, 5, No. 10 (1970): 88-115.

Cote, Marc. "Terraguelt, une paysannerie ecrosie." *Annales Algeriennes de Geographie*, 6, No. 11 (1971): 24-64.

Sutton, Keith S. "The Progress of Algeria's Agrarian Reform and Its Settlement Implications." *Maghreb Review*, 2 (1978): 10-16.

Trautman, Wolfgang. "The Impact of the Agrarian Revolution on Nomadism of the Algerian Steppe." *Nomadic Peoples*, 17 (1985): 23-33.

ENVIRONMENTAL AND CONSERVATION ISSUES

Prinz, D., D. Gomer, and S. Belz. "Studies of the Causes of Soil Erosion on Marl Soils in Nothern Algeria: The Role of Traditional Soil Tillage." *Land Degradation and Rehabilitation*, 5, No. 4 (1994): 271-280.

IRRIGATION

Dupois, Jean. "Types d'irrigation et modes de portage et de possession des eaux dans les plaines du Hodra (Algerie)." *Bulletin de la Societe de Geographie d'Egypte*, 25 (1953): 137-145.

Sanlaville, P. "Les centres de cultures de l'Ahaggar." *Revue de Geographie de Lyon*, 30 (1955): 333-342.

Mingasson, Christian. "Le perimetre d'irrigation du Tadla." *Annales Algeriennes de Geographie*, 1, No. 1 (1966): 41-65.

Mingasson, Christian. "Le perimetre d'irrigation du Tadla (2e et 3e partie)." *Annales Algeriennes de Geographie*, 1, No. 2 (1966): 82-119.

Castevert, Claude. "Mise au point sur: Le pirimetre irrigable de l'Oued Bou Namoussa." *Annales Algeriennes de Geographie*, 3, No. 5 (1968): 93-100.

Tomas, Francois. "Les debuts d'un perimetre irrigue: Le domaine MAIZ Bachir (plaine d'Annaba)." *Annales Algeriennes de Geographie*, 8 (1969): 169-172.

AGRICULTURAL MARKETING AND DISTRIBUTION

Brule, Jean-Claude. "Une cooperative d'Anciens Moujahidine dans le bassin de Djidjelli: Khelalef Mohamed Tahar." *Annales Algeriennes de Geographie*, 8 (1969): 43-50.

Casabo-Berna, Suzel. "Un groupe de domaines socialistes maraichers: Ain Taya." *Annales Algeriennes de Geographie*, 8 (1969): 89-106.

Egypt

GENERAL TEXTS, ARTICLES, AND THESES

Finch, Vernar C. "Agriculture on the Nile Delta." *Journal of Geography*, 16 (1917–1918): 241-244.

Guadia, J.E. "The Agricultural Nile Delta." *Bulletin of the Geographical Society of Philadelphia*, 25 (1927): 163-186.

Rosciszewski, Marcin. "Agricultural Geography of Egypt." *Africana Bulletin* (Warsaw), 5 (1966): 33-52.

Hance, William A. "The Gezira Scheme: A Study in Agricultural Development." Pp. 31-53 in *African Economic Development*, edited by William A. Hance. New York, N.Y.: Praeger, 1967.

Rosciszewski, Marcin. "Development Problems of Egyptian Agriculture" (in Polish). *Prace Geograficzne*, 59 (1967): 63-109.

Simons, Peter. *Die Entwicklung des Anbaus und die Verbreitung der Nutzfianzen in der agyptischen Nilstromoase von 1800 bis zur Gegenart: Eine agrargeographische Untersuchung*. Cologne, Germany: Kolner Geographische Arbeiten, Heft 20, 1968.

Nasr, N. *Jughra fiyet Misr al-ziraiyah*. Cairo, Egypt: Maktabat Saie Rafat, 1988.

Ishay, K., et al. "Farm Location and Farmers' Performance on the Hamza Minor Canal." *Water Resources Development*, 7, No. 1 (1991): 2-15.

Schmitz, Charles. "Agrarian Transformation in Egypt." Master's Thesis, University of California, Berkeley, Calif., 1992.

Meyer, Gunter. "Liberalisierung und Privatisierung der agyptischen Landwirtschaft." *Erdkunde*, 49, No. 1 (1995): 17-31.

Bethemont, J., and P. Faggi. "Originalite, potentialites, et limites de l'agriculture dans le nord-Sinai." *Revue de Geographie de Lyon*, 73, No. 3 (1998): 239-246.

CROPS AND CROPPING PATTERNS

n.a. "Cotton Cultivation in the British Empire and Egypt." *Scottish Geographical Magazine*, 20, No. 7 (1904): 354-370.

Harris, Ralph. "Geographic Aspects of Egyptian Cotton Production." Master's Thesis, University of Illinois, Urbana, Ill., 1937.

Weheba, A.F. "Rice Culture in Egypt: A Study in Historical Geography." *Bulletin de la Societe de Geographie d'Egypte*, 40 (1967): 213-236.

Allan, J.A. "Some Phases in Extending the Cultivated Area in the Nineteenth and Twentieth Centuries in Egypt." *Middle Eastern Studies*, 19, No. 4 (1983): 470-481.

Abou El-Ela, Mohamed. "Le sucre en Egypte: L'equilibre entre la production et la consommation." *Bulletin de la Societe de Geographie d'Egypte*, 70 (1997): 87-108.

Hobbs, Joseph J. "Troubling Fields: The Opium Poppy in Egypt." *Geographical Review*, 88, No. 1 (1998): 64-85.

ANIMAL HUSBANDRY

Kazmi, Hasam A. "Livestock Industry and Crop Responses in Egypt." *The Geographer, Aligarh*, 34, No. 2 (1987): 67-86.

FARM LAND

Hassan, Mohamed I. "Physical Elements of Agricultural Land Use in the Nile Delta." *Bulletin de la Societe de Geographie d'Egypte*, 26 (1953): 227-241.

Hassan, Mohamed I. "Physical Elements of Agricultural Land Use in the Fayum Depression." *Bulletin de la Societe de Geographie d'Egypte*, 27 (1954): 51-64.

Hassan, Mohamed I. "Agricultural Land Use in the Fayum Depression." *Bulletin de la Societe de Geographie d'Egypte*, 28 (1955): 99-112.

Asadi, Fawzi. "Agricultural Land in Egypt: The Quest for Rationale." *Geographical Survey*, 1 (1972): 3-24.

Born, Martin. "Landerschliessung durch bauerliche Gruppen und Nomaden zwischen Nil und Rotem Meer." Pp. 211-219 in *Czajka-Festschrift*. Gottingen, Germany: Gottinger Geographische Abhandlungen, Heft 66, 1976.

El-Einein, Salah M.A. "Dune Encroachment on the Cultivated Land in Egypt." *Bulletin de la Societe de Geographie d'Egypte*, 64 (1991): 107-124.

Lenney, Mary Helen. "Monitoring Agricultural Lands in Egypt with Multitemporal Vegetation Indices Derived from Satellite Imagery." Ph.D. Dissertation, Boston University, Boston, Mass., 1996.

ENVIRONMENTAL AND CONSERVATION ISSUES

Ahmed, G., and Abd El-Samie. "Soil Survey, Classification and Management of the Mariut Agricultural Project." *Bulletin de la Societe de Geographie d'Egypte*, 33 (1960): 147-176.

Wagner, Leo F. "The Impact of the Aswan High Dam on Egyptian Agriculture." Master's Thesis, San Francisco State University, San Francisco, Calif., 1970.

IRRIGATION

Joannides, P.N. "Egyptian Agriculture, with Special Reference to Irrigation." *Scottish Geographical Magazine*, 20, No. 11 (1904): 561-568.

Shafei, A. "Fayoum Irrigation as Described by Nabulsi in 1245 AD with a Description of the Present System of Irrigation and a Note on Lake Moeris." *Bulletin de la Societe de Geographie d'Egypte*, 20, No. 3 (1940): 283-327.

Shafei, A. "Irrigation of the Wilderness of Etham." *Bulletin de la Societe de Geographie d'Egypte*, 22, No. 3-4 (1948): 151-155.

Hamdon, G. "Evolution of Irrigation Agriculture in Egypt." Pp. 119-142 in *A History of Land Use in Arid Regions*, edited by L. Dudley Stamp. Paris, France: UNESCO, Arid Zone Research No. 17, 1961.

Samie, Ahmed. "Water Requirements for Irrigated Areas in the Fayyoum Province." *Bulletin de la Societe de Geographie d'Egypte*, 37 (1964): 35-56.

Whittington, Dale, and Kingsley E. Haynes. "Valuing Water in the Agricultural Environment of Egypt: Some Estimations and Policy Considerations." *Regional Science Perspectives*, 10 (1980): 109-126.

Whittington, Dale, and Kingsley E. Haynes. "Nile Water for When? Emerging Conflicts in Water Allocation for Agricultural Expansion in Egypt and Sudan." Pp. 125-150 in *Agricultural Development in the Middle East*, edited by Peter Beaumont and Keith S. McLachlan. New York, N.Y.: John Wiley, 1985.

Brookes, Ian A. "Above the Salt: Sediment Accretion and Irrigation Agriculture in an Egyptian Oasis." *Journal of Arid Environments*, 17 (1989): 335-348.

Radwan, L.S. "Farmer Responses to Inefficiencies in the Supply and Distribution of Irrigation Requirements in Delta Egypt." *Geographical Journal*, 163, No. 1 (1997): 78-92.

Libya

GENERAL TEXTS, ARTICLES, AND THESES

Ahmad, Naji Abbus. *Die Landlichenlebensformen und die Agrarentwicklung in Tripolitanien.* Heidelberg, Germany: Heidelberger Geographische Arbeiten, Heft 25, 1967.

Allan, J.A., Keith S. McLachlan, and Edith T. Penrose, eds. *Libya: Agriculture and Economic Development.* London, England: Frank Cass, 1973.

El-Meheski, Abdelkader M. "Ezzera Ah Al-Libiy Ah: A Study of Libyan Agriculture." Master's Thesis, San Jose State University, San Jose, California, 1974.

Alawar, Mohamed A. "An Examination of the Agricultural Development Projects in the Fezzan Region, Libya." D.A. Dissertation, Northern Colorado University, Greeley, Colo., 1979.

El-Meheshi, Abdelkader M. "The Transformation of Libyan Agriculture: The Impact of Oil Revenues on Agriculture of a Developing Country." Ph.D. Dissertation, University of California, Riverside, Calif., 1980.

Schliephake, Konrad. "Libyan Agriculture: Natural Constraints and Aspects of Development." *Maghreb Review*, 5, No. 2-4 (1980): 51-56.

McLachlan, Keith S. "Strategies for Agricultural Development in Libya." Pp. 9-24 in *Libya Since Independence*, edited by J.A. Allan. London, England: Croom Helm, 1982.

CROPS AND CROPPING PATTERNS

Taylor, A.R. "Regional Variations in Olive Cultivation in Northern Tripolitania." Pp. 80-100 in *Field Studies in Libya*, edited by S.G. Willimott and John Clarke. Durham, England: University of Durham, Department of Geography, Research Paper No. 4, 1960.

Taylor, A.R. "The Olive in Northern Tripolitania: Some Aspects of Agrarian Geography." Master's Thesis, University of Durham, Durham, England, 1961.

FARM LAND

Willimott, S.G., and John Clarke, eds. *Field Studies in Libya*. Durham, England: University of Durham, Department of Geography Research Series, No. 4, 1960.

ENVIRONMENTAL AND CONSERVATION ISSUES

El-Meheshi, Abdelkader M. "Sand Fixation and Its Implications for Agricultural Development in Libya." *California Geographer*, 20 (1980): 63-80.

IRRIGATION

Hill, R.W. "Some Problems of Economic Geography in Northern Tripolitania: A Study of Agriculture and Irrigation on the Jefara Plain." Ph.D. Dissertation, University of Durham, Durham, England, 1960.

Eldblam, L. "Quelques points de vue comparatifs sur le problemes d'irrigation dans les trois oases Libyennes de Brak, Ghadanes, et particulierment Mourzouk." *Svensk Geografisk Arsbok*, 37 (1961): 124-145.

Morocco

GENERAL TEXTS, ARTICLES, AND THESES

Amphoux, Marcel. "L'evolution de l'agriculture europeenne au Maroc." *Annales de Geographie*, 42 (1933): 175-185.

Raynal, R. "Problemes et bilan de l'agriculture marocaine." *Les Cahiers d'Outre Mer*, 4 (1951): 342-362.

Gadille, J. "L'agriculture europeenne au Maroc: Etude humain et economique." *Annales de Geographie*, 66 (1957): 144-158.

Naciri, M. "Les experiences de modernisation de l'agriculture au Maroc." *Revue de Geographie du Maroc*, 11 (1967): 102-114.

Le Coz, Jean. "Le troisieme age agraire du Maroc." *Annales de Geographie*, 77 (1968): 385-413.

Delannoy, H. "Le code des investissements agricoles." *Revue de Geographie du Maroc*, 16 (1969): 151-157.

Legras, J.C. "Aspects agricoles de la plaine de Taza-Oujda (Maroc Oriental)." *L'Information Geographique*, 33, No. 4 (1969): 180-186.

Bouderbala, N. "Quelques donnees elementaires sur l'evolution des structures agraires dans la plaine du Rharb." *Revue de Geographie du Maroc*, 20 (1971): 119-124.

Flouriot, J. "Le perimetre agricole du Loukhos." *Revue de Geographie du Maroc*, 21 (1972): 31-42.

Riser, J. "Le barrage de Monsour Eddahbi et les amenagement agricoles du Dra Moyen." *Revue de Geographie du Maroc*, 23-24 (1973): 167-178.

Karabenick, Edward. "Morocco: A Microcosm of Evolving Agriculture in the Arab World." *Journal of Geography*, 75, No. 9 (1976): 531-540.

Fadloullah, A., and M. Belfquih. "Colonisation rurale et espace agricole dans l'arriere: Pays de Rabat-Sale." *Revue de Geographie du Maroc*, 2 (1977): 95-116.

Pletsch, Alfred. "Traditionelle landwirtschaft in Marokko." *Geographische Rundschau*, 29, No. 4 (1977): 107-114.

Bencherifa, Abdellatif. *Chtouka et Massa: Une region du sous en cours de transformation: Etude de geographie agraire*. Rabat, Morocco: Faculte des Lettres et des Sciences Humaines, Universite Mohamed V, 1980.

Popp, Herbert. "Entkolonialisierung und Agrarreform in Marokko: Das Beispiel des Gharb." *Erdkunde*, 31, No. 4 (1980): 257-268.

Bencherifa, Abdellatif. "Centres de collecte du lait et evolution des systems de production agricole au Maroc: Cas du Rharb, Doukkala et Souss." *Revue de Geographie du Maroc*, 5 (1981): 135-150.

Graul, Franz. *Tarhzout: Grundlagen und Strukturen des Wirtschaftslebens einer Talschaft im Zentralen Rif Morokko*. Hamburg, Germany: Hamburger Geographischen Studien, Heft 38, 1982.

Labouesse, F. "L'agriculture marocaine au debut des annees 80: Situation et perspectives." *Mediterranee*, 59, No. 4 (1986): 93-102.

Badidi, Brahim. "Agriculture et modernite en zone de moyenne montagne: Le cas du Moyen-Atlas central (Maroc)." *Acta Geographica*, 110, No. 2 (1997): 24-38.

CROPS AND CROPPING PATTERNS

Le Coz, Jean. "La riziculture marocaine." *Les Cahiers d'Outre Mer*, 8 (1955): 257-282.

Charrie, J.P. "La culture du coton au Maroc." *Revue de Geographie du Maroc*, 22 (1972): 103-118.

Sebillotte, M. "Les cultures de cerales en sec dans le Maroc oriental." *Revue de Geographie du Maroc*, 23-24 (1973): 51-78.

Ladwig, Rolf. *Staatliche Forderungsmassnahmen in der landwirtschaft Marokkos: Am Beispiel des Zuckerrubenanbaus im Bewasserungsgebiet Tadla.* Frankfurt am Main, Germany: Frankfurter Geographische Hefte 52, 1978.

Papy, F., and F. Lefievre. "Les pratiques de cerealiculture dans une region a climat aride de type mediterraneen la plaine de Benguerir." *Revue de Geographie du Maroc*, 3 (1979): 23-44.

Jouve, A.M. "Demographie et cerealiculture: Evolution comparee de la demographie et la cerealiculture au Maroc depuis de debut du siecle." *Revue de Geographie du Maroc*, 5, No. 2 (1981): 5-20.

Jenare, L. "Oleiculture et paysannerie dans la region de Fes." *Revue de Georaphie du Maroc*, 9 (1985): 3-19.

Ait Hamzsa, M. "Les cereales dans le Maroc du Centre-ouest." *Mediterranee*, 88, No. 1 (1998): 27-32.

Dougnedroit, A., and A. Messaoudi. "Essai de prediction des rendemonts des cereales dans le Maroc de Centre-ouest." *Mediterranee*, 88, No. 1 (1998): 45-50.

ANIMAL HUSBANDRY

Couvrer, G. "La vie pastorale dans le Haut Atlas central." *Revue de Geographie du Maroc*, 13 (1968): 3-52.

Hudowicz, S. "L'elevage au Maroc: Situation actuelle." *Revue de Geographie du Maroc*, 17 (1970): 116-120.

Martin, J.E. "Les potentialites agro-pastorales du Moyen Atlas central: Premicre approche cartographique." *Revue de Geographie du Maroc*, 10, No. 1-2 (1986): 217-234.

AGRICULTURAL POPULATIONS AND SOCIAL ISSUES

De Farcy, Henri. "Techniques indigenes et techniques des colons dans l'agriculture marocaine." *Revue de Geographie de Lyon*, 25, No. 2 (1950): 99-105.

Le Coz, Jean. "Les tribes guichs au Maroc." *Revue de Geographie du Maroc*, 7 (1972): 1-52.

Fadloullah, A., and M. Belfquih. "Colonisation rurale et espace agricoles dans l'arriere, pays de Rabat-Sale." *Revue de Geographie du Maroc*, 2 (1978): 95-116.

Marcarcini, Paolo. "Esperienze di riforma agraria e problemi sociali nell'area mediterranea: Il caso de Marocco." *Bolletin della Societa Geografia Italiana*, Serie 10, 10 (1981): 247-306.

ENVIRONMENTAL AND CONSERVATION ISSUES

Coulter, John W. "Aspects of Morocco Today: Climate and Agriculture." *Journal of Geography*, 63, No. 9 (1964): 402-412.

Naciri, M. "Conditions climatiques, recoltes cerealienes et situation des campagnes traditionnelles marocaines." *Revue de Geographie du Maroc*, 16 (1969): 35-70.

Papy, F., M. Oussible, and P.H. Jouve. "Les contraintes pedoclimatiques a l'exploitation agricole des zones semi-aride et aride du Maroc-Oriental." *Revue de Geographie du Maroc*, 5 (1981): 121-134.

Kutsch, Horst. *Principal Features of a Form of Water-Concentrating Culture on Small-Holdings with Special Reference to the Anti Atlas: A Contribution to the Debate on Aridity Thresholds and the Question of Transferring Cultivation Methods to Areas of the World with Similar Climates*. Trier, Germany: Trierer Geographische Studien, Heft 5, 1982.

Jouve, P.H., and F. Papy. "Les systemes de culture dans les zones semi-aride et aride du Maroc-Occidental." *Revue de Geographie du Maroc*, 7 (1983): 3-20.

Fay, G. "Un projet agro-sylvo-pastoral pur le Rif Occidential." *Revue de Geographie du Maroc*, 8 (1984): 3-23.

IRRIGATION

Fenelon, Pierre. "L'Irrigation dans le Haouz de Marrakech." *Bulletin de l'Association de Geographes Francais*, 18 (1941): 67-70.

Mercier, M. "L'irrigation au Fezzan." *Annales de Geographie*, 55 (1946): 282-289.

Poupart, J.M. "Les problems de l'eau a Marrakech." *Les Cahiers d'Outre Mer*, 2 (1949): 38-53.

Lahlimi, A. "Les terres irriguees et le monde rural de la Tessouf moyennes." *Revue de Geographie du Maroc*, 11 (1967): 3-38.

Awad, H. "Les grandperimetres irrigues de type moderne et leur incidence sur les modes de vie traditionnels. Deux grandes realisations: La Gezira au Soudan et le Tadla au Maroc." *Revue de Geographie du Maroc*, 15 (1969): 81-92.

Couvreur-Laraichi, F. "Les perimetres irrigues du Rharb." *Revue de Geographie du Maroc*, 22 (1972): 75-82.

Swearingen, W.D. "Not a Drop of Water to the Sea: The Colonial Origins of Morocco's Present Irrigation Programme." *Maghreb Review*, 9, No. 1-2 (1984): 26-38.

Lightfoot, Dale R. "Moroccan Khettara: Traditional Irrigation and Progressive Dessication." *Geoforum*, 27, No. 2 (1996): 261-273.

Kales, Matthew. "Evaluating Agricultural Sustainability: A Case Study of Large-Scale Irrigation in Southwestern Morocco." Master's Thesis, University of Georgia, Athens, Ga., 1998.

Tunisia

GENERAL TEXTS, ARTICLES, AND THESES

Poncet, Jean. "Les champs et l'evolution du paysage agraireen Tunisie." *Annales de Geographie*, 71 (1962): 620-629.

Achenback, Hermann. "Die Agrarlandschaft der Tunisischen Nordkuste um Bizerte." *Erdkunde*, 21 (1967): 132-146.

Makhlouf, Ezzedine. *Structures agraires et modernisation de l'agriculture dans les plaines du Kef: Les unites cooperatives de production*. Tunis, Tunisia: Centre d'etudes et de recherches economiques et sociales, 1968.

Achenback, Hermann. *Agrargeographische Entwicklungsprobleme Tunesians und Ostalgerians: Exemplarische Strukturanalyse aus gewahlter Reform- und Traditionsraume zwischen Mittelmeerkuste und Nordsahara*. Hannover, Germany: Jahrbuch der Geographische Gesellschaft zur Hannover, 1970.

Kassab, A. "La crise des grandes cultures dans la Tunisie tellienne: causes naturelles et structurelles." *Revue tunisienne de geographie*, 1 (1978): 11-39.

Sethom, H. *L'agriculture de la presqu'ile du Cap Bon (Tunisie): Structures sociales et economie rurale*. Tunis, Tunisia: Publications de l'Universite de Tunis, Faculte des lettres et sciences humaines de Tunis, Deuxieme serie, Geographie, Vol. 5, 1978.

Kassab, A. "L'agriculture tunisienne sur la voie de l'intensification." *Annales de Geographie*, 90, No. 1 (1981): 55-86.

Cherif, A. "Un agriculture stabilisee pour la Tunisie au XXIe siecle 'tetes du collque tuniso-americain: Tunis 25-27 October 1988.'" *Revue Tunisienne de Geographie*, 19-20 (1991): 225-230.

Krapiwinicka, K. "Typology of Agriculture in Tunisia" (in Polish). *Przeglad Geograficzny*, 63, No. 1-2 (1992): 53-66.

CROPS AND CROPPING PATTERNS

Laitman, L. "Le marche et la production de l'huile ville de d'olive en Tunisie." *Annales de Geographie*, 62 (1953): 271-286.

Winkler, Erhardt. "Wandlungen der Halfagraswirtschaft in Tunesien." *Mitteilungen der Osterreichischen Geographischen Gesellschaft*, 110, No. 2-3 (1968): 159-174.

ANIMAL HUSBANDRY

Kassab, A. "Types d'elevage bobin en Tunisie." *Mediterranee*, 29, No. 2 (1977): 11-21.

AGRICULTURAL POPULATIONS AND SOCIAL ISSUES

Poncet, Jean. *La colonisation et l'agriculture europeennes en Tunis depuis 1881: Eetude de geographie historique et economique*. Paris, France: Mouton, 1961.

Lambooy, J.G. *De agrarische hervormingin Tunesie: Proeve von een sociaal-geografische onderzoek*. Assen, Netherlands: Von Goraim, 1969.

ENVIRONMENTAL AND CONSERVATION ISSUES

May, Detlef. *Untersuchungen zur geo-okologischen Situation der nordlichen Nefzaoua-Oasen Tunisien*. Stuttgart, Germany: Stuttgarter Geographische Studien, Band 102, 1984.

Meuere, M. "Okologische und okonomische Aspekte der Ziegenhaltung in NW-Tunesien: Ein Beitrag zum Problem der Nutzungskonflikte zwischen Tierhalturn, Pflanzenproduktion und Landschaftsschutz." *Geographische Zeitschrift*, 73, No. 3 (1985): 162-173.

IRRIGATION

Atkinson, K. "Saline Water Irrigation." *Geography*, 57, No. 3 (1972): 244-246.

Winkler, Erhard. "Die Bewasserungswirtschaft im tunesischen Medjerdatal." *Mitteilungen der Osterreichischen Geographischen Gesellschaft*, 110, No. 1 (1968): 79-84.

Part 7

AFRICA SOUTH OF THE SAHARA

Africa South of the Sahara

ATLASES AND GRAPHIC PRESENTATIONS

Bennett, Merrill K. "An Agroclimatic Mapping of Africa." *Food Research Institute Studies*, 3, No. 3 (1962): 195-216.

Atlas of African Agriculture: African Agriculture—The Next 25 Years. Rome, Italy: FAO, 1986.

GENERAL TEXTS, ARTICLES, AND THESES

Light, Richard, and Mary Upjohn. "Contrasts in African Farming." *Geographical Review*, 28, No. 4 (1938): 529-555.

Schantz, Homer L. "Agricultural Regions of Africa." *Economic Geography*; Part I, 16, No. 1 (1940): 1-47; continued, 16, No. 2 (1940): 122-161; Part II, 16, No. 4 (1940): 341-389; continued, 17, No. 3 (1941): 217-249; Part III, 17, No. 4 (1941): 353-379; continued 18, No. 3 (1942): 229-246; Part IV, 19, No. 1 (1943): 77-109; continued 19, No. 2 (1943): 217-269.

Robiero, Orlando. "Agriculture in West Africa." *Indian Geographical Journal*, 26 (1951): 65-72.

Gourou, Pierre. "Les enseignments d'une experience agricole en pays equatorial: Aubeville, 1952." *Les Cahiers d'Outre Mer*, 6 (1953): 232-243.

Morgan, W.B. "Some Comments on Shifting Cultivation in Africa." *Research Notes*, 9 (1957): 1-10. Ibadan, Nigeria: University College Ibadan, Department of Geography.

Pelissier, P. "Sur une synthise nouvelle de la vie agricole en Afrique noire." *Les Cahiers d'Outre Mer*, 11 (1958): 197-199.

Morgan, W.B. "The Forest and Agriculture in West Africa." *Journal of African History*, 3 (1962): 235-239.

Isnard, Hildebert. "Agriculture et developpement en Afrique Occidentale." *Les Cahiers d'Outre Mer*, 16 (1963): 253-262.

Wilmut, J. "Contributions recentes a la connaissance de l'agriculture itinerate en Afrique occidentale et centrale." *Bulletin de la Societe Belge d'Etudes Geographiques*, 32, No. 1 (1963): 51-63.

Tulippe, Omer, and J. Wilmut. "Geographie de l'agriculture en Afrique Centrale: Essai de synthese." *Bulletin de la Societe Belge d'Etudes Geographiques*, 33, No. 2 (1964): 303-374.

Tulippe, Omer, and J. Wilmut. "Geographie de l'agriculture en Afrique Centrale: Essai de synthese." *Travaux Geographique de Liege*, 150 (1965): 303-374.

White, H.P. "Terre de Barre: Basis of a West African Agricultural Region." *Bulletin de l'Institut fond. Africains noire*, 27, Series B (1965): 169-182.

Lemaire, M.E. "Agriculture in West Africa." *Journal of Geography*, 66, No. (1967): 380-384.

Coppock, J.T. "Agricultural Geography in Tropical Africa." Pp. 9-20 in *Environment and Land Use in Africa*, edited by M.F. Thomas and G.W. Whittington. London, England: Methuen, 1969.

Hellen, J.A. "Colonial Administration Policies and Agricultural Patterns in Tropical Africa." Pp. 321-354 in *Environment and Land Use in Africa*, edited by M.F. Thomas and G.W. Whittington. London, England: Methuen, 1969.

Morgan, W.B. "Peasant Agriculture in Tropical Africa." Pp. 241-272 in *Environment and Land Use in Africa*, edited by M.F. Thomas and G.W. Whittington. London, England: Methuen, 1969.

Guinard, Andre. "Le developpement de l'agriculture en Afrique tropicale." *Les Cahiers d'Outre Mer*, 23 (1970): 113-139.

Baker, P. Randall. "Agricultural Changes in Bunyaro, 1954–1968." Pp. 123-136 in *Studies in East African Geography and Development*, edited by S.H. Ominde. Berkeley, Calif.: University of California Press, 1971.

Benneh, George. "Systems of Agriculture in Tropical Africa." *Economic Geography*, 48, No. 3 (1972): 244-257.

Gleave, M.B., and H.P. White. *Agricultural Systems and Pastoralism in Tropical Africa.* Wellington, New Zealand: Hicks and Smith and Sons, 1972.

Jackson, R.T. "A Vicious Circle? The Consequences of Von Thunen in Tropical Africa." *Area*, 4, No. 4 (1973): 258-261.

Carol, Hans. "The Calculation of Theoretical Feeding Capacity for Tropical Africa." *Geographische Zeitschrift*, 62 (1974): 81-94.

Allan, J.A. "The Kufrah Agricultural Scheme." *Geographical Journal*, 142, No. 1 (1976): 48-56.

Knight, C. Gregory. "Prospects for Peasant Agriculture." Pp. 195-211 in *Contemporary Africa: Geography and Change*, edited by C. Gregory Knight and James L. Newman. Englewood Cliffs, N.J.: Prentice-Hall, 1976.

Tibahikaho, Vivina. "Problems of Inducing Agricultural Change in East Africa." Master's Thesis, Wayne State University, Detroit, Mich., 1976.

Jeske, I. *Botswana–Lesotho–Swaziland: Agrargeographische Struktur und wirtschaftliche Verflachtung im sudlichen Afrika.* Afrika-Studien, IFO Institut fur Wirtschaftsforschung, No. 97, 1977.

Hecklau, Hans. *Agrargeographie, Ostafrika (Kenya, Uganda, Tanzania): Landwirstschaftliche Flachennatzungsstile.* Berlin, Germany: Gebruder Borntraeger, 1978.

Hunter, John M., and George K. Ntiri. "Speculations on the Future of Shifting Agriculture in Africa." *Journal of Developing Areas*, 12, No. 2 (1978): 183-208.

Stevens, Richard, and Yuk Lee. "A Spatial Analysis of Agricultural Intensity in a Basotho Village of Southern Africa." *Professional Geographer*, 31, No. 2 (1979): 177-183.

Bahr, J. "Die neure Entwicklung der sudwestafrikanischen Farmwirtschaft unter dem Einefluss Veranderter politishcer Bedingungen." *Afrika-Information, Forschung berichte Deutscher Geographen*, 3 (1980): 15-26.

Coppock, J.T. "Agricultural Geography in Developing Countries, with Special Reference to Tropical Africa." Pp. 311-328 in *Perspectives in Agricultural*

Geography, vol. 1, edited by Noor Mohammad. New Delhi, India: Concept Publishing Company, 1981.

Cech, Diethard, et al. *Agrargeographie, Sudafrika (Mocambique, Swaziland, Republik Sudafrika).* Berlin, Germany: Gehruder Borntraeger, 1982.

Richards, Paul. "Farming Systems and Agrarian Change in West Africa." *Progress in Human Geography*, 7, No. 1 (1983): 1-39.

Vermeer, Donald. "Food Sufficiency and Farming in the Future of West Africa: Resurgence of Traditional Agriculture?" *Journal of African Studies*, 10, No. 3 (1983): 74-83.

Achtwich, W. "Agrarproduktion im Sahel: Moglichkeiten und Grenzen." *Die Erde*, 116 (1985): 159-168.

Chaleard, Jean-Louise. "Intensification agricole et ravitaillement urbain: Le cas des perimetres vivriers du departement d'Agboville." *Les Cahiers d'Outre Mer*, 34 (1986): 157-178.

Sibanda, Backson M.C. "Impacts of Agricultural Microprojects on Rural Development: Lessons from Two Projects in the Zambezi Valley." *Land Use Policy*, 3, No. 4 (1986): 311-329.

Funnell, D.C. "Crisis in Africa: The Agrarian Dimension." *Geography*, 73, No. 1 (1988): 54-59.

Villien, Francois. "L'agriculture dans la ville: L'exemple de Bangui." *Les Cahiers d'Outre Mer*, 41 (1988): 283-302.

Watts, Michael J. "The Agrarian Crisis in Africa: Debating the Crisis." *Progress in Human Geography*, 13, No. 1 (1989): 1-41.

Weiner, Daniel. "Agricultural Restructuring in Zimbabwe and South Africa." *Development and Change*, 20, No. 3 (1989): 401-428.

Gyasi, Edwin A. "Nucleus Agricultural Estates and Food Security Lessons from Two Case Studies in Ghana and Nigeria." *Rural Systems*, 8, No. 2 (1990): 61-68.

Unruh, Jon D. "Integration of Transhumant Pastoralism and Irrigated Agriculture in Semi-Arid East Africa." *Human Ecology*, 18, No. 3 (1990): 223-246.

Binns, J.A. "Traditional Agriculture, Pastoralism and Fishing." Pp. 153-191 in *Tropical African Development: Geographical Perspectives*, edited by M.B. Gleave. London, England: Longman, 1992.

Combin, Eric F., et al. "Agricultural Production Monitoring in the Sahel Using Remote Sensing: Present Possibilities and Research Needs." *Journal of Environmental Management*, 38, No. 4 (1992): 301-322.

Krings, T. "Die Bedeutung autochthanen Agrarwissens fur die Ernabrungsicherung in den Landern Tropische Afrikas." *Geographische Rundschau*, 44, No. 2 (1992): 88-93.

McMaster, D.N. "Agricultural Development." Pp. 192-222 in *Tropical African Development: Geographical Perspectives*, edited by M.B. Gleave. London, England: Longman, 1992.

Gleave, M.B. "Urban Bias in Agricultural Development in West Africa." Pp. 138-158 in *Agricultural Change, Environment and Economy: Essays in Honour of W.B. Morgan*, edited by Keith Hoggart. London, England: Mansell, 1993.

Wegulo, Francis. "Farm and Nonfarm Activities: A Study of Their Linages and Impact upon Rural Development in Kakamega District." Ph.D. Dissertation, Clark University, Worcester, Mass., 1993.

Clapp, Roger A. "The Moral Economy of the Contract." Pp. 78-94 in *Peasants Under Contract: Contract Farming and Agrarian Transformation in Sub-Saharan Africa*, edited by Michael J. Watts and Peter D. Little. Madison, Wis.: University of Wisconsin Press, 1994.

Little, Peter D., and Michael J. Watts, eds. *Peasants Under Contract: Contract Farming and Agrarian Transformation in Sub-Saharan Africa.* Madison, Wis.: University of Wisconsin Press, 1994.

Morgan, W.B., and Jerzy A. Solarz. "Agricultural Crisis in Sub-Saharan Africa: Development Constraints and Policy Problems." *Geographical Journal*, 160, No. 1 (1994): 57-73.

Goldman, Abe. "Threats to Sustainability in African Agriculture: Searching for Appropriate Paradigms." *Human Ecology*, 23, No. 3 (1995): 291-334.

Chaleard, Jean-Louise. "Les mutations de l'agriculture commerciale en Afrique de l'Ouest." *Annales de Geographie*, 105, No. 6 (1996): 563-583.

Nicolai, H. "Reflexions sur les caracteres originaux des paysages agraires de l'Afrique tropical." *Bulletin de la Societe Geographique de Liege*, 32 (1996): 311-320.

Gunnell, Y. "Comparative Regional Geography in India and West Africa: Soils, Landforms, and Economy Theory in Agricultural Development Strategies." *Geographical Journal*, 163, No. 1 (1997): 38-46.

Obio, G.C. "Agricultural Development in Sub-Saharan Africa." Pp. 286-324 in *Geography of Sub-Saharan Africa*, edited by Samuel Aryeetey-Attoh. Upper Saddle River, N.J.: Prentice-Hall, 1997.

Wammali, Sudhir, and Yassir Islam. "Rural Infrastructure and Agricultural Development in Southern Africa: A Centre-Periphery Perspective." *Geographical Journal*, 163, No. 3 (1997): 259-269.

CROPS AND CROPPING PATTERNS

Le Coeur, C. "Le commerce de la noix de kola en Africa occidentale." *Annales de Geographie*, 36 (1927): 143-149.

Porterers, Roland. "Geographie alimentaire, bereaux agricoles et migration des plants cultives en Afrique intertropical." *Compte Rendu des Seances Societes de Biogeographie*, 239 (1951): 16-21.

Miege, J. "Les cultures vivrieres en Afrique occidentale." *Les Cahiers d'Outre Mer*, 7 (1954): 25-50.

Paulme, Denise. "Des riziculteurs africaines: Les Baga." *Les Cahiers d'Outre Mer*, 10 (1957): 257-278.

Jarrett, H.R. "The Present Setting of the Oil-Palm Industry, with Special Reference to West Africa." *Journal of Tropical Geography*, 11 (1958): 59-69.

Murdock, George P. "Staple Subsistence Crops of Africa." *Geographical Review*, 50, No. 4 (1960): 523-540.

Jatzold, Ralph. "Die Nachwirkungen des Fehlgeschlagenen Erdnuss-Projekts in Ostafrika." *Erdkunde*, 19 (1965): 210-233.

Poole, Peter. "Wildlife Cropping in East Africa: Present Problems and Future Prospects." Master's Thesis, Columbia University, New York, N.Y., 1968.

Harrison-Church, R.J. "The Firestone Rubber Plantations in Tropical Africa." *Geography*, 54 (1969): 430-437.

Engelhard, Karl, and Cay Kienan. "Der Zuckerrohraubau in Ostafrika." *Geographische Rundschau*, 22, No. 2 (1970): 55-63.

Randiga, Henry. "The Banana Regions of East Africa: The Regional Distribution and Cultural Significance of a Traditional Food Crop." Master's Thesis, Western Kentucky University, Bowling Green, Ky., 1971.

Hankins, Thomas D. "So You Get Out Early to Plant, Who Are You Ahead Of? The Role of Cotton Planting Time in Sukuma Agriculture." Ph.D. Dissertation, Clark University, Worcester, Mass., 1974.

Harris, David R. "Traditional Systems of Plant Food Production and the Origins of Agriculture in West Africa." Pp. 311-356 in *Origins of African Plant Domestication*, edited by Jack R. Harlan, Jan M. J. deWet, and Ann B.L. Stemler. The Hague, Netherlands: Mouton, 1976.

Oyendo, R.B., and J.C.A. Obiero. "The East African Sugar Industry." *GeoJournal*, 2, No. 4 (1978): 343-354.

Stromgaard, P. "Potential Crop Production, Illustrated by an Example from West Africa." *Geografisk Tidsskrift*, 77 (1978): 6-10.

Obara, Dunston A. "Mechanization of Small-Scale Cotton Farms in East Africa: A Study of Problems and Opportunities." Pp. 445-462 in *Perspectives in Agricultural Geography*, vol. 2, edited by Noor Mohammad. New Delhi, India: Concept Publishing Company, 1981.

Moliki, Mbulyo. "La culture du cafe et le developpement de l'Vele." *Les Cahiers d'Outre Mer*, 34 (1986): 143-156.

Richards, Paul. *Coping with Hunger: Hazard and Experiment in an African Rice-Farming System*. Boston, Mass.: Allen & Unwin, 1986.

Hervouet, J.P., and C. Laveissiere. "Cash Crop Development and Sleeping Sickness in the Forest Belt of West Africa." Pp. 373-382 in *Health and Disease in Tropical Africa: Geographical and Medical Viewpoints*, edited by Rais Akhtar. London, England: Harwood, 1987.

Seavoy, Ronald E. "Hoe Shifting Cultivation in East African Subsistence Culture." *Singapore Journal of Tropical Geography*, 8, No. 1 (1987): 60-71.

Bogh, Eva. "Water Use and Millet Production in the Sahel." *Geografisk Tidsskrift*, 93 (1993): 80-89.

Carter, J.E., and P.G. Jones. "A Model of the Distribution of Cassava in Africa." *Applied Geography*, 13, No. 4 (1993): 353-372.

Carney, Judith. "Landscapes of Technology Transfer: Rice Cultivation and African Continuity." *Technology and Culture*, 37 (1996): 5-35.

ANIMAL HUSBANDRY

Blache, J. "La question pastorale en Afrique occidentale." *Annales de Geographie*, 51 (1942): 26-44.

Veyret, Paul. "L'elevage dans la zone tropicale." *Les Cahiers d'Outre Mer*, 5 (1952): 70-83.

Deshler, Walter. "Cattle in Africa: Distribution, Types and Problems." *Geographical Review*, 53, No. 1 (1963): 52-58.

Turner, B. "Ecological Problems of Cattle Ranching in Bunyaro." *East African Geographical Review*, 5 (1967): 9-20.

Santoir, C., et al., eds. *Strategies pastorales et agricoles des saheliens durant la secheresse, 1969–1974*. Talence, France: Travaux et documents de geographie tropicale, No. 30, 1977.

Bock, Wilfred. "Pensionsvichhaltung in Berchtesgadener Land Ausmasz, Herkunft, und Probleme au Beispiel de Berechtigungsolmen des Fors tamtes Berchtesgadenerland." *Zeitschrift fur Wirtschaftsgeographie*, 23, No. 2 (1979): 52-58.

Gallais, Jean. "La situation de l'elevage bovin et le probleme des elevesusen Afrique occidentale en centrale." *Les Cahiers d'Outre Mer*, 32 (1979): 113-144.

Wilson, R.T. "Distribution and Importance of the Domestic Donkey in Circumsaharan Africa." *Singapore Journal of Tropical Geography*, 2, No. 2 (1981): 136-143.

Illgner, Peter M., E.L. Nel, and Mark P. Robertson. "Beekeeping and Local Self-Reliance in Rural Southern Africa." *Geographical Review*, 88, No. 3 (1998): 349-362.

DAIRYING

Kivikkokangas, Ritva. "Milk Cattle as an Innovation Problem in East Africa" (in Finnish). *Terra*, 87, No. 2 (1975): 123-131.

FARM LAND

Born, Martin. "Anbauformen an der agronomischen Trockengrenze Nordostafrikas." *Geographische Zeitschrift*, 55 (1967): 243-278.

Floyd, Barry N. *Some Spatial Aspects of Rural Land Use in Tropical Africa: Typologies, Models and Case Studies*. Mona, Jamaica: University of the West Indies, Department of Geography, 1972.

Mensching, Horst G. "Nomads and Farmers in the West African Sahel: Problems of Competing Land Use." *Applied Geography and Development*, 20 (1982): 7-19.

Mascarenhas, Adolfo, et al. "Land Use Policies and Farming Systems: Kenya, Tanzania, Zambia and Mozambique." *Land Use Policy*, 3, No. 4 (1986): 286-303.

Adewole-Osunade, M.A. "A Viable Method of Land Capability Classification for Small Farmers." *Journal of Environmental Management*, 25, No. 1 (1987): 81-94.

Biswas, A.K., et al. "Land Use and Farming Systems in the Horn of Africa." *Land Use Policy*, 4, No. 4 (1987): 419-443.

Bassett, Thomas J. "Introduction: The Land Question and Agricultural Transformation in Sub-Saharan Africa." Pp. 3-31 in *Land in African Agrarian*

Systems, edited by Thomas J. Bassett and Donald E. Crummey. Madison, Wis.: University of Wisconsin Press, 1993.

Bassett, Thomas J., and Donald E. Crummey, eds. *Land in African Agrarian Systems*. Madison, Wis.: University of Wisconsin Press, 1993.

Moss, R.P. "Land Evaluation for Agricultural Development in Tropical Africa." Pp. 223-250 in *Agricultural Change, Environment and Economy: Essays in Honour of W.B. Morgan*, edited by Keith Hoggart. London, England: Mansell, 1993.

Watts, Michael J. "Idioms of Land and Labor: Producing Politics and Rice in Senegambia." Pp. 157-193 in *Land in African Agrarian Systems*, edited by Thomas J. Bassett and Donald E. Crummey. Madison, Wis.: University of Wisconsin Press, 1993.

Chokor, B.A., and F.O. Odemerho. "Land Degradation Assessment by Small-Scale Traditional African Farmers and Implications for Sustainable Conservation Management." *Geoforum*, 25, No. 2 (1994): 145-154.

Reenberg, A. "Agricultural Land Use Dynamics in a Sahelian Environment: Does Reality Match Hypothesized Trends?" *Geografisk Tidsskrift*, 94 (1994): 58-68.

Turner, Matthew D. "Merging Local and Regional Analysis of Land-Use Change: The Case of Livestock Change in the Sahel." *Annals of the Association of American Geographers*, 89, No. 2 (1999): 191-219.

AGRICULTURAL POPULATIONS AND SOCIAL ISSUES

Gleave, M.B., and H.P. White. "Population Density and Agricultural Systems in West Africa." Pp. 273-300 in *Environment and Land Use in* Africa, edited by M.F. Thomas and G.W. Whittington. London, England: Methuen, 1969.

Gengenbach, Inge. "Farmleben in Sudwestafrika." *Zeitschrift fur Wirtschaftsgeographie*, 16, No. 2 (1972): 51-54.

Dupre, Herb, and Wolf Roder. "Coping with Drought in a Preindustrial, Preliterate Farming Society." Pp. 115-119 in *Natural Hazards: Local, National, Global*, edited by Gilbert F. White. New York, N.Y.: Oxford University Press, 1974.

Udo, Reuben K. "Rural Migrations and the Problems of Agricultural Labour in Western Tropical Africa." Pp. 189-208 in *Spatial Aspects of Development* London, edited by Brian S. Hoyle. New York, N.Y.: John Wiley, 1974.

Gleave, M.B. "Some Further Thoughts on Population Density and Agricultural Systems in West Africa." Pp. 59-74 in *The Rural-Agricultural Sector*, edited by E.S. Simpson. Newcastle, England: Institute of British Geographers, Developing Areas Study Group, 1980.

Kebbede, Girma. "Subsaharan Africa: The Political Economy of Food Shortage." *Geographical Perspectives*, 54 (1984): 21-33.

Fuchs, P. "Agrarsoziale Situation im Sahel." *Die Erde*, 116 (1985): 169-184.

McCall, M.K. "Accessibility and Mobility in Peasant Agriculture in Tropical Africa." Pp. 42-63 in *Rural Accessibility and Mobility*, edited by Paul

Cloke. Lampeter, England: Institute of British Geographers, Rural Geography Study Group, 1985.

McCall, M.K. "The Significance of Distance Constraints in Peasant Farming Systems with Special Reference to Sub-Saharan Africa." *Applied Geography*, 5 (1985): 325-345.

Grivetti, L.E., et al. "Bush Foods and Edible Weeds of Agriculture: Perspectives on Dietary Use of Wild Plants in Africa, Their Role in Maintaining Human Nutritional Status and Implications for Agricultural Development." Pp. 51-82 in *Health and Disease in Tropical Africa: Geographical and Medical Viewpoints*, edited by Rais Akhtar. London, England: Harwood, 1987.

Mortimore, Michael J. *Adapting to Drought: Farmers, Famines, and Desertification in West Africa.* Cambridge, England: Cambridge University Press, 1989.

Schaaf, Thomas, and Walther Manshard. "The Growth of Spontaneous Agricultural Colonization in the Border Area of Ghana and the Ivory Coast." *Applied Geography and Development*, 34 (1989): 7-22.

Malkhazova, S.M., and T.Y. Karimova. "Assessment of the Medico-Geographical Consequences of Farming: The Case Study of Natural Endemic Diseases in Africa." *Geographica Medica*, 20 (1990): 51-60.

Krings, T. "Kulturbaumparke in den Agrarlandschaften Westafrikas: Eine Form autochthaner Agroforstwirtschaft." *Die Erde*, 122 (1991): 117-130.

Drummond, James, and Andy Manson. "The Rise and Demise of African Agricultural Production in Dinokana Village, Bophuthatswana." *Canadian Journal of African Studies*, 27, No. 3 (1993): 462-479.

Freeman, Donald B. "Survival Strategy or Business Training Ground? The Significance of Urban Agriculture for the Advancement of Women in African Cities." *African Studies Review*, 36, No. 3 (1993): 1-22.

Hyden, Goram, Robert W. Kates, and Billie Lee Turner II. "Beyond Intensification." Pp. 401-439 in *Population Growth and Agricultural Change in Africa*, edited by Robert W. Kates, Goram Hyden, and Billie Lee Turner II. Gainesville, Fla.: University Press of Florida, 1993.

Kates, Robert W., Goram Hyden, and Billie Lee Turner II. "Theory, Evidence, Study Design." Pp. 1-40 in *Population Growth and Agricultural Change in Africa*, edited by Robert W. Kates, Goram Hyden, and Billie Lee Turner II. Gainesville, Fla.: University Press of Florida, 1993.

Turner, Billie Lee, II, Goram Hyden, and Robert W. Kates, eds. *Population Growth and Agricultural Change in Africa.* Gainesville, Fla.: University of Florida Press, 1993.

ENVIRONMENTAL AND CONSERVATION ISSUES

Knight, Chester G. "Unyiha: The Ecology of Agriculture and Change in an East African Community." Ph.D. Dissertation, University of Minnesota, Minneapolis, Minn., 1969.

Morgan, W.B., and R.P. Moss. "Soils, Plants and Farmers in West Africa." Pp. 1-31 in *Human Ecology in the Tropics*, edited by J.P. Garlich. Oxford, England: Pergamon, 1970.

Andreae, Bernd. *Die Farmwirtschaft an den agronomischen Trockengrenzen: Uber den Wettbewerb Okologischer Varianten in der okonomischen Evolution: Betriebs- und standartsokonomische Studien en der Farmzone des sudlichen Afrika und der westlichen USA.* Wiesbaden, Germany: F. Steiner, 1974.

Kabwoto, E. "Problems of Gilgaied Soils for Agricultural Use in East Africa." *Journal of the Geographical Association of Tanzania*, 11 (1974): 45-55.

Chamard, Philippe C. "Environnement, agriculture et nutrition dans la zone sahelienne." *L'Information Geographique*, 39, No. 3 (1975): 124-129.

Porter, Philip W. "Climate and Agriculture in East Africa." Pp. 112-139 in *Contemporary Africa: Geography and Change*, edited by C. Gregory Knight and James L. Newman. Englewood Cliffs, N.J.: Prentice-Hall, 1976.

Kowal, J.M., and A.H. Kassam. *Agricultural Ecology of Savanna: West Africa.* Oxford, England: Clarendon Press, 1978.

Lema, A.J. "A Note on the Lysimeter Method of Measuring Crop-Water Consumptive Use in East Africa." *Journal of the Geographical Association of Tanzania*, 17 (1978): 94-119.

Nieuwalt, S. "The East African Monsoons and Their Effects on Agriculture." *GeoJournal*, 3, No. 2 (1979): 193-200.

Bein, Frederick L. "Response to Drought in the Sahel." *Journal of Soil and Water Conservation*, 35, No. 3 (1980): 121-124.

Dennett, M.D., J. Elston, and C.B. Speed. "Rainfall and Crop Yields in Seasonally Arid West Africa." *Geoforum*, 12 (1981): 203-209.

Oguntoyinbo, J. "Climatic Variability and Food Crop Production in West Africa." *GeoJournal*, 5 (1981): 139-149.

Moss, R.P. "Ecological Constraints on Agricultural Development in Africa." *African Research and Documentation*, 25 (1982): 1-12.

Bruins, H.J., et al. "Rainwater-Harvesting Agriculture for Food Production in Arid Zones: The Challenge of the African Famine." *Applied Geography*, 6, No. 1 (1986): 13-32.

Stromgaard, P. "Soil and Vegetational Changes Under Shifting Cultivation in the Miombo of East Africa." *Geografiska Annaler B*, 70B, No. 3 (1988): 363-374.

Ezaza, W.P. "The Main Environmental Issues and Options for the Sustainable Development of Agriculture Associated with the Use of Mineral Fertilizer and Agro-Chemicals." *Journal of Eastern African Research and Development*, 19 (1989): 73-82.

Anhuf, D. "Niederschlagsschwankungen und Anbau un sicherheit in der Sahelzone." *Geographische Rundschau*, 42, No. 3 (1990): 152-158.

Sarpong, Ofori. "Moisture Availability and Agricultural Potentialities of Mali and Niger." *Indonesian Journal of Geography*, 20, No. 60 (1990): 57-70.

Agnew, C.T. "Evaluation of a Soil Water Balance Model for the Analysis of Agricultural Drought in the Sahel." Pp. 583-592 in *Proceedings of a Workshop on Soil Water Balance in the Sudano Sahelian Zone, Feb. 18-23*, edited by M. Sivakumar, et al. Wallingford, England: CAHS Publications No. 194, 1991.

Shisanya, Chris A. "A Note on the Problems Experienced in Agro-Ecological Research Work in Developing Countries." *Journal of Eastern African Research and Development*, 21 (1991): 128-132.

Turner, Matthew D. "Overstocking the Range: A Critical Analysis of the Environmental Science of Sahelian Pastoralism." *Economic Geography*, 69, No. 4 (1993): 402-421.

Baker, Kathleen M. "Drought, Agriculture and Environment: A Case Study from the Gambia, West Africa." *African Affairs*, 94, No. 1 (1995): 67-86.

Reenberg, A., and B. Fog. "The Spatial Pattern and Dynamics of a Sahelian Agroecosystem." *GeoJournal*, 37, No. 4 (1995): 489-500.

Salam, Ademola T. "Agrarian Crisis, Ecology and the Challenge of Sustainable Development in Africa." *Nigerian Geographical Journal*, 2 (1995): 1-18.

Schulze, R.E., and R.P. Kunz. "Potential Shifts in Optimum Growing Areas of Selected Commercial Tree Species and Subtropical Crops in Southern Africa Due to Global Warming." *Journal of Biogeography*, 22, No. 4-5 (1995): 679-688.

Warren, Andrew. "Changing Understandings of African Pastoralism and the Nature of Environmental Paradigms." *Transactions, Institute of British Geographers*, 20, No. 2 (1995): 193-203. (New Series)

Rasmussen, Kathryn, et al. "Agricultural Systems and Transnational Water Management in the Senegal River Basin." *Geografisk Tidsskrift*, 99 (1999): 59-68.

Bohn, L.E. "The Use of Climate Information in Commercial Agriculture in Southeast Africa." *Physical Geography*, 21, No. 1 (2000): 57-67.

IRRIGATION

Robequain, Charles. "L'irrigation dans l'Afrique noire francaise." *Bulletin de la Societe de Geographie d'Egypte*, 25 (1953): 147-163.

Gallais, Jean. "Options prise ou ignorees dans les amenagements hydro-agricoles en Afrique Sahelienne." *Cahiers Geographiques de Rouen*, 6 (1976): 77-95.

Adams, W.M., and A.T. Grove. *Irrigation and Tropical Africa: Problems and Problem Solving*. Cambridge, England: Cambridge University African Studies Centre, Monograph No. 3, 1984.

Adams, W.M., and R.C. Carter. "Small-Scale Irrigation in Sub-Saharan Africa." *Progress in Human Geography*, 11, No. 1 (1987): 1-27.

Biswas, A.K. "Water for Food Production in Sub-Saharan Africa." *GeoJournal*, 15, No. 3 (1987): 233-242.

Schliephake, Konrad. "Irrigation and Food Production: Experiences from North Africa and Application to East Africa." *Applied Geography and Development*, 30 (1987): 30-45.

Funnell, D.C. "Intervention and Indigenous Management: The Geography of Small-Scale Irrigation Development in Morocco and Swaziland." *Land Use Policy*, 11, No. 1 (1994): 45-54.

TRANSPORTATION AND MECHANIZATION

Obara, Dunston A. "Mechanization of Small-Scale Farms in East Africa: A Study of Problems and Opportunties." Pp. 445-462 in *Perspectives in Agricultural Geography*, vol. 2, edited by Noor Mohammad. New Delhi, India: Concept Publishing Company, 1981.

Deichmann, Uwe. "A Spatial Analysis of Technology Adoption in Sub-Saharan Farming Systems." Ph.D. Dissertation, University of California, Santa Barbara, Calif., 1995.

AGRICULTURAL LABOR STUDIES

Bickel, Samuel J. "The Role of Human Labor in Rice Growing West African Peasant Societies." Master's Thesis, Syracuse University, Syracuse, N.Y., 1983.

Turner, Matthew D. "Labor Process and the Environment: The Effects of Labor Availability and Compensation on the Quality of Herding in the Sahel." *Human Ecology*, 27, No. 2 (1999): 267-296.

Angola

GENERAL TEXTS, ARTICLES, AND THESES

Boleo, Jose de Oliveira. *En saio sobre geografia agraria: Angola, a terra e o homen*. Lisbon, Portugal: Agencia-Geral do Uttfamar, 1958.

CROPS AND CROPPING PATTERNS

Van Dongen, I.S. "Coffee Trade, Coffee Regions, and Coffee Ports in Angola." *Economic Geography*, 37, No. 4 (1961): 320-346.

AGRICULTURAL POPULATIONS AND SOCIAL ISSUES

Slezak, F. "Angola: Agrarsiedlungen." *Mitteilungen der Osterreichischen Geographischen Gesellschaft*, 108, No. 1 (1966): 196-198.

ENVIRONMENTAL AND CONSERVATION ISSUES

Diniz, A.C. *Caracteristicas mesologicas de Angola: Descricao e carrelacao dos aspectos fisiograficos, dos solos e da vegetacao das zonas agricolas angolanas*. Nova Lisboa, Angola: Missao de Inqueritos Agricolas de Angola, 1973.

Benin/Dahomey

GENERAL TEXTS, ARTICLES, AND THESES

N'Bessa, Benoit. "Les exploitations agricoles des citadins en milieu rural: L'exemple beninois." *Les Cahiers d'Outre Mer*, 52 (1999): 275-292.

FARM LAND

Swoboda, Jan, and Hans-Jurgen Sturm. "Traditionelle Bodennutzurgssysteme und Bodenbewertung bei Ackerbauern und Tierhaltern in Nord-Benin." *Die Erde*, 126, No. 1 (1995): 53-71.

ENVIRONMENTAL AND CONSERVATION ISSUES

Kamuanga, Mulumba, and Venuste M. Murinda. "The Value of Agro-Climatic Conditions and Farmers' Resources Level in Determining Farm Typologies: The Case of Northern Benin." *Singapore Journal of Tropical Geography*, 14, No. 1 (1993): 58-71.

Botswana

GENERAL TEXTS, ARTICLES, AND THESES

Ingersall, Jeanne. "The Prospects for Agricultural Development in Eastern Botswana." Master's Thesis, Columbia University, New York, N.Y., 1967.

Arntzen, J.W. "Crop Production, Cattle Rearing, and Land Use in Kgatleng District, Botswana." *Norsk Geografisk Tidskrift*, 38, No. 2 (1984): 95-108.

ANIMAL HUSBANDRY

Storey, A. "The Botswana Livestock Industry." *Geography*, 53, No. 1 (1968): 87-89.

Isaksen, Jon. "Some Aspects of the Cattle Industry in the Economy of Botswana." *Norsk Geografisk Tidskrift*, 38, No. 2 (1984): 85-94.

Chi, Z., et al. "Analyzing Rangeland Grazing Intensity Using Landsat Imagery: A Case Study in the Southern District of Botswana." *Papers and Proceedings of the Applied Geography Conferences*, 20 (1997): 334-342.

FARM LAND

Lado, Cleophas. "Agricultural Production and Land-Use Pattern in Kgatleng District, Botswana." *Indonesian Journal of Geography*, 28, No. 72 (1996): 27-44.

Burkina Faso/Upper Volta

GENERAL TEXTS, ARTICLES, AND THESES

Savonnet, Georges. *Les Birifor de Diepla et sa region: Insulaires du rameau lobi: Houte-Volta*. Paris, France: Mouton, 1976.

Marchal, Monique. *Les paysages agraires de Haute-Volta: Analyse structurale par la methode graphique*. Paris, France: Editions de l'Office de la recherche scientifique et technique outre-mer, 1983.

Nebie, Ousmane. "Evolution des systemes agraires Beva et Pougouli de Po-Ouest, Burkina Faso." *Les Cahiers d'Outre Mer*, 41 (1988): 259-282.

Braun, Michaela. *Subsistenzsicherung und Marktpartizipation: Eine agrargeographie Untersuchung zu klein bauerlichen Produktionsstrategien in der Province de la Comoe, Burkina Faso*. Heidelberg, Germany: Heidelberger Geographische Arbeiten, Heft 105, 1996.

CROPS AND CROPPING PATTERNS

Pale, Frederic O. "Introduction d'une culture de rapport dans une agriculture traditionnelle le coton en Pays Lobi (Haute-Volta)." *Recherches Geographiques a Strasbourg*, 5 (1978): 163-168.

Hartog, Thierry. "Le perimetre sucrier de Banfora (Haute-Volta): Du pouvoir technocratique aux deboins paysons." *Les Cahiers d'Outre Mer*, 36 (1983): 119-135.

Schwarz, A. "Des temps anciens a la devaluation du franc C.F.A., les tribulations de la culture du coton au Burkina Faso." *Annales de Geographie*, 106, No. 2 (1997): 288-312.

ANIMAL HUSBANDRY

Piquemal, D. "L'agro-pastoralisme du departement de Sebba (Burkina Faso)." *Les Cahiers d'Outre Mer*, 50 (1997): 293-314.

FARM LAND

Hansen, Tina S., and A. Reenberg. "Approaching Local Limits to Field Expansion: Land Use Pattern Dynamics in Semi-Arid Burkina Faso." *Geografisk Tidsskrift*, 98 (1998): 56-85.

AGRICULTURAL POPULATIONS AND SOCIAL ISSUES

Freidberg, Susanne E. "Making a Living: A Social History of Market-Garden Work in the Regional Economy of Bobo-Dioulasso, Burkina Faso." Ph.D. Dissertation, University of California, Berkeley, Calif., 1996.

Thorsen, Dorte, and A. Reenberg. "Marginal Producers or Breadwinners: Women's Cropping Strategies and Access to Agricultural Key Resources in Boulgou Province, Burkina Faso." *Geografisk Tidsskrift*, 100 (2000): 47-60.

IRRIGATION

Nebie, Ousmane. "Le perimetre irrigue de la vallee du Kou (Burkina Faso): Limites d'une operation." *Les Cahiers d'Outre Mer*, 45 (1996): 273-296.

AGRICULTURAL MARKETING AND DISTRIBUTION

Nebie, Ousmane. "Approvisionnement urbain et developpement des cultures maraicheres dans la region de Ouagdougou (Burkina Faso)." *Les Cahiers d'Outre Mer*, 52 (1999): 257-274.

Cameroon

GENERAL TEXTS, ARTICLES, AND THESES

Guillaume, Geoffrey M.D., and Sanford H. Bederman. *Subsistence Activity in Five Villages in and Around Mount Cameroun in Victoria Division, West Cameroon*. Atlanta, Ga.: Georgia State College, School of Arts and Sciences Research Paper No. 14, 1967.

Dongmo, J.L. "Typologie de l'agriculture camerounaise: Essai d'application a l'Afrique noire de la methode statistico-geographique du professeur Kostrowicki." *Annales de la Faculte des Lettres et Sciences Humains, Universite de Yaounde*, 5 (1973): 19-40.

Dongmo, J.L. "Typologie de l'agriculture camerounaise." Pp. 115-145 in *Agricultural Typology and Land Utilisation*, edited by C. Vanzetti. Verona, Italy: Center for Agricultural Geography, 1975.

Ejedepang-Koge, S.N. "Etude sur les cultures vivrieres destinees a la consommation urbaine dans la province du Sud-Ouest du Cameroun." *Travaux et Documents de Geographie Tropicale, CEGET-CNRS*, 28 (1977): 49-82.

Floyd, Barry N., and L. Tandap. "Intensification of Agriculture in the Republique Unie du Cameroun." *Geography*, 65, No. 4 (1980): 324-327.

Riddell, J.C., and D. Campbell. "Agricultural Intensification and Rural Development: The Mandora Mountains of North Cameroon." *African Studies Review*, 29, No. 3 (1986): 89-106.

Von der Ruhren, N. "Kamerun-Agrarstrukturen in ein em tropischen Entwicklungsland." *Geographie und Schule*, 11, No. 60 (1989): 14-25.

CROPS AND CROPPING PATTERNS

Bederman, Stanley H. "Plantation Agriculture in Victoria District, West Cameroon." *Geography*, 51 (1966): 349-360.

Fossung, E.Y.W. "Plantation Agriculture: A Case Study of the Cameroun Development Corporation in Cameroun." Master's Thesis, University of Waterloo, Waterloo, Ontario, 1980.

Levrat, Regine. "La place du coton dans la vie des paysons du Nord-Cameroun." *Les Cahiers d'Outre Mer*, 37 (1984): 33-62.

FARM LAND

Champaud, Jacques. *Mom, terroir, bassa (Cameroun).* Paris, France: Mouton, 1973.

Boulet, Jean. *Magomaz, pays Mafa (Nord Cameroun): Etude d'un terroir de montagne.* Paris, France: ORSTROM, 1975.

Ngwa, Nebasina E. "Time and Land Space Utilization Within an Urban Confine: The Case of Buea Town Gardeners in the Republic of Cameroon." *GeoJournal,* 15, No. 1 (1987): 77-82.

Morin, Serge. "Colonisation agraire, espaces pastoraux et degradation des milieux dans les hautes terres de l'Ouest Cameroun." *Les Cahiers d'Outre Mer,* 47 (1994): 79-104.

IRRIGATION

Schroder, R. "Effectiver Regenfall und Wasseranspruch fur die Landwirtschaft, dargestellt am Beispiel Kameruns." *Petermann's Geographische Mitteilungen,* 125, No. 4 (1981): 253-262.

AGRICULTURAL MARKETING AND DISTRIBUTION

Fodouop, Kengne. "Demande urbaine et evolution de l'agriculture vivriere et maraichine dans le sud de Cameroun." *Les Cahiers d'Outre Mer,* 52 (1999): 293-322.

Central African Republic

GENERAL TEXTS, ARTICLES, AND THESES

Prioul, Christian. "Le reseau routier et l'agriculture commerciale en Republique Centrafricaine." *Les Cahiers d'Outre Mer,* 21 (1968): 434-437.

ANIMAL HUSBANDRY

Suchel, J.B. "L'elevage des bovins en Republique Centraficaines." *Les Cahiers d'Outre Mer,* 20 (1967): 137-149.

AGRICULTURAL POPULATIONS AND SOCIAL ISSUES

Prioul, Christian. "Villes et agriculture vivriere en Republique Central Africaine." *Travaux et Documents de Geographie Tropicale,* 7 (1972): 83-118.

Chad

CROPS AND CROPPING PATTERNS

Cabot, Jean. "La culture du coton au Tchad." *Annales de Geographie,* 66 (1957): 499-508.

Cabot, Jean. "Un domaine nouveau de riziculture inondee: Les plaines du Moyen-Logane." *Les Cahiers d'Outre Mer*, 10 (1957): 158-173.

ANIMAL HUSBANDRY

Bouquet, Christian. "La transformation industrielle des products de l'agriculture et de l'elevage au Tchad." *L'Information Geographique*, 34, No. 1 (1970): 13-24.

ENVIRONMENTAL AND CONSERVATION ISSUES

Magrin, G. "Crise climatique et mutations de l'agriculture: L'emergence d'un grenier cerealier entre Tchad et Chari." *Annales de Geographie*, 105, No. 6 (1996): 620-644.

Sarch, Marie-Therese, and Charon Birkett. "Fishing and Farming at Lake Chad: Responses to Lake-Level Fluctuations." *Geographical Journal*, 166, No. 2 (2000): 156-172.

IRRIGATION

Adamu, Manu. "The Lake Chad Basin Development Scheme: Farmer Participation in the South Chad Irrigation Project." Master's Thesis, Ohio University, Athens, Ohio, 1984.

Congo

GENERAL TEXTS, ARTICLES, AND THESES

Vennetier, Pierre. "Le vie agricole urbaine a Pointe-Noire (Congo)." *Les Cahiers d'Outre Mer*, 14 (1961): 60-84.

Mengho, B.M. "L'agriculture traditionelle chez les Bakouele et les Djem du Congo." *Les Cahiers d'Outre Mer*, 31 (1978): 48-84.

Mengho, B.M. "Les transformations de l'agriculture vivriere la long de la RN2 au nord de Brazzaville." *Les Cahiers d'Outre Mer*, 52 (1999): 233-256.

FARM LAND

Guillot, Bernard. *La terre Enkou: Recherches sur les structures agraires du plateu Koukouya (Congo)*. La Haye, France: Mouton, 1973.

Ethiopia

GENERAL TEXTS, ARTICLES, AND THESES

Reta, Taye. "Alem Gena: A Study in Local Geography." *Ethiopian Geographical Journal*, 3, No. 1 (1965): 21-34.

Horvath, Ronald J. "Von Thunen's Isolated State and the Area Around Addis Ababa, Ethiopia." *Annals of the Association of American Geographers*, 59 (1969): 308-323.

Jackson, R.T. "Some Observations on the Thunen Method of Analysis with Reference to Southern Ethiopia." *East African Geographical Review*, 8 (1971): 39-46.

Aklilu, B., and Lakshman Yapa. "Diffusion of Agricultural Modernization in Ethiopia." *Proceedings, New England–St. Lawrence Valley Division, Association of American Geographers*, 6 (1976): 81-85.

Brune, S. "Agrarian Change and Agricultural Policy in Post-Imperial Ethiopia (1974–1988)." *Zeitschrift fur Wirtschaftsgeographie*, 33, No. 4 (1989): 208-218.

Engdawork, Desta. "Agricultural Producer Cooperatives: Some Lessons of Experience from Ethiopia." *GeoJournal*, 36, No. 4 (1995): 353-360.

CROPS AND CROPPING PATTERNS

Simoons, Frederick J. "Two Ethiopian Gardens." *Landscape*, 14, No. 3 (1965): 15-20.

Stanley, S. "Ensete in the Ethiopian Economy." *Ethiopian Geographical Journal*, 4, No. 1 (1966): 30-37.

Schumacher, Gunther. "Der Koffee in Landschaft und Wirtschaft Aethiopians." *Geografia Helvetica*, 21 (1966): 13-19.

Winidowa, Jadwiga. "Geographic Aspects of Coffee Cultivation in Ethiopia" (in Polish). *Czasopismo Geograficzne*, 43 (1972): 19-31.

FARM LAND

Wubneh, Mulatu. "Communal Land Ownership and Its Effects on the Current State of Agriculture in the Province of Begimdir and Simen, Ethiopia." Master's Thesis, Florida State University, Tallahassee, Fla., 1972.

AGRICULTURAL POPULATIONS AND SOCIAL ISSUES

Smeds, Helmer. "The Ensete Planting Culture of Eastern Sidamo, Ethiopia." *Acta Geographica*, Finland, 13 (1955): 1-39.

Wood, Adrian P. "Spontaneous Agricultural Settlement in Ethiopia, 1950–1974." Pp. 157-165 in *Redistribution of Population in Africa*, edited by John Clarke and L.A. Kosinski. London, England: Heinemann, 1982.

Abate, Alula. "Peasant Association and Collective Agriculture in Ethiopia: Promise and Performance." *Erdkunde*, 37, No. 2 (1983): 118-126.

ENVIRONMENTAL AND CONSERVATION ISSUES

Malesani, Emilio. "Agricoltura e Insediamento Umano in Etiopia." *Rivista Geografica Italiana*, 57 (1950): 100-106.

Dyer, J.A., A. Teshance and J.K. Torrance. "Agroclimatic Profiles for Uniform Productivity Areas in Ethiopia." *Water International*, 18 (1993): 189-199.

IRRIGATION

Kloos, Helmut. "Peasant Irrigation Development and Food Production in Ethiopia." *Geographical Journal*, 157, No. 3 (1991): 295-307.

Fernando Poo

GENERAL TEXTS, ARTICLES, AND THESES

Teran, M. "El Bosque: La Agricultura Indigena y la Agricultura de Plantacion de Fernando Poo." *Estudios Geograficos*, 23 (1962): 487-516.

Gambia

GENERAL TEXTS, ARTICLES, AND THESES

Jarrett, H.R. "The Strangers-Farmers of the Gambia." *Geographical Review*, 39, No. 4 (1949): 649-657.

CROPS AND CROPPING PATTERNS

Barrett, Hazel R., and Angela W. Browne. "Export Horticultural Production in Sub-Saharan Africa: The Incorporation of the Gambia." *Geography*, 81, No. 1 (1996): 47-56.

FARM LAND

Eastman, Clyde. "Gambian Usufruct Tenure: Help or Hindrance to Agricultural Production?" *Land Use Policy*, 7, No. 1 (1990): 27-40.

AGRICULTURAL POPULATIONS AND SOCIAL ISSUES

Barrett, Hazel R., and Angela W. Browne. "Women's Horticulture in the Peri-Zone, The Gambia." *Geography*, 73, No. 2 (1988): 158-160.

Barrett, Hazel R., and Angela W. Browne. "Environmental and Economic Sustainability: Women's Horticultural Production in the Gambia." *Geography*, 76, No. 3 (1991): 241-248.

Carney, Judith. "Converting the Wetlands, Engendering the Environment: The Intersection of Gender with Agrarian Change in the Gambia." *Economic Geography*, 69, No. 4 (1993): 329-348.

Schroeder, Richard A. "Shady Practice: Gender and the Political Ecology of Resource Stabilization in Gambian Gardens/Orchards." Ph.D. Dissertation, University of California, Berkeley, Calif., 1993.

Schroeder, Richard A. "Shady Practice: Gender and the Political Ecology of Resource Stabilization in Gambian Gardens/Orchards." *Economic Geography*, 69, No. 4 (1993): 349-365.

Swindell, Kenneth. "From Migrant Farmer to Permanent Settler: The Strange Farmers of the Gambia." Pp. 96-101 in *Redistribution of Population in Africa*, edited by John Clarke and L.A. Kosinski. London, England: Heinemann, 1982.

ENVIRONMENTAL AND CONSERVATION ISSUES

Baker, Kathleen M. "Traditional Farming Practices and Environmental Decline, with Special Reference to the Gambia." Pp. 180-202 in *Agricultural Change, Environment and Economy: Essays in Honour of W.B. Morgan*, edited by Keith Hoggart. London, England: Mansell, 1993.

AGRICULTURAL MARKETING AND DISTRIBUTION

Swindell, Kenneth. "African Food Imports and Agricultural Development: Peanut Basins and Rice Bowls in the Gambia, 1843–1933." Pp. 159-179 in *Agricultural Change, Environment and Economy: Essays in Honour of W.B. Morgan*, edited by Keith Hoggart. London, England: Mansell, 1993.

Ghana

GENERAL TEXTS, ARTICLES, AND THESES

White, H.P. "Provisional Agricultural Regions of Ghana." *Journal of Tropical Geography*, 11 (1958): 90-99.

Ball, Mary S. "Future Developments in the Utilization of Agricultural Resources in Ghana." Master's Thesis, University of Cincinnati, Cincinnati, Ohio, 1962.

Dickson, K.B. "The Agricultural Landscape of Southern Ghana and Ashanti-Brong Ahafo: 1800 to 1850." *Bulletin, Ghana Geographical Association*, 9, No. 1 (1964): 25-35.

Benneh, George. "The Role of Agricultural Geography in the Economic Development of Ghana." *Bulletin, Ghana Geographical Association*, 12 (1967): 51-64.

Benneh, George. "Small-Scale Farming Systems in Ghana." *Africa*, 43 (1973): 134-146.

Bigelow, Ross. "Impact of Selected Agricultural Schemes on the Innovation Patterns Among Traditional Farmers in the Southern Savanna of Ghana." Ph.D. Dissertation, Michigan State University, East Lansing, Mich., 1973.

Boaten, K. "Commercial Agriculture in Asante." *Bulletin Ghana Geographical Association*, 15 (1973): 30-49.

Webber, Paul. "Agrarian Change in Kusasi, North-East Ghana." *Africa*, 66, No. 3 (1996): 437-457.

Awongo, Louis. "Unveiling the 'Hidden' Agrarian Processes Under a Structural Adjustment Program in Ghana: A Study of Agricultural Production in Buckum, 1985–1995." Ph.D. Dissertation, Queen's University, Kingston, Ontario, 1999.

Awongo, Louis. "Labor, Ecology, and a Failed Agenda of Market Incentives: The Political Ecology of Agrarian Reforms in Ghana." *Annals of the Association of American Geographers*, 91, No. 1 (2001): 92-121.

CROPS AND CROPPING PATTERNS

Ballwanz, Howard R. "The Cocoa Industry of Ghana." Master's Thesis, Northern Illinois University, DeKalb, Ill., 1965.

Gyasi, Edwin A. "The Spatial Diffusion of Rubber in Western Ghana." Master's Thesis, University of Wisconsin, Madison, Wis., 1971.

Brand, R. "The Role of Cocoa in the Early Economic Growth of Accra, Ghana." *African Studies Review*, 15 (1972): 271-284.

Gyasi, Edwin A. "A Comparative Geographical Analysis of Maize Production in Ghana." Ph.D. Dissertation, University of Wisconsin, Madison, Wis., 1975.

Dotse, J.M. *Shallot Farming in Anloga, Ghana: A Case-Study in Agricultural Geography*. Cape Coast, Ghana: University of Cape Coast, 1980.

Gyasi, Edwin A. "The Oil Palm Industry and Its Implications for Rural Development in Ghana." *Malaysian Journal of Tropical Geography*, 18 (1988): 1-9.

Gyasi, Edwin A. "Emergence of a New Oil Palm Belt in Ghana." *Tijschrift voor Economische en Sociale Geographie*, 83, No. 1 (1992): 39-49.

Porter, G., E. Young, and A. Dzietror. "Pressures on an Intensive Irrigated Cash-Crop System in Coastal Ghana." *Geoforum*, 28, No. 3-4 (1997): 329-340.

FARM LAND

Manshard, Walther. "Land Use Patterns, Settlement and Agricultural Migration in Central Ghana." *Bulletin, Ghana Geographical Association*, 5, No. 2 (1960): 23-24.

Gyasi, Edwin A. "The Adaptability of African Communal Land Tenure to Economic Opportunity: The Example of Land Acquisition for Oil Palm Farming in Ghana." *Africa*, 64, No. 3 (1994): 391-405.

AGRICULTURAL POPULATIONS AND SOCIAL ISSUES

Hill, Polly. "The Migrant Cocoa Farmers of Southern Ghana." *Africa*, 31 (1961): 209-230.

Djeagbo, Grace, and Adeline Mate. "Settlement and Farming in the Manya Krobo District." *Bulletin, Ghana Geographical Assocication*, 7, No. 1-2 (1962): 54-58.

Benneh, George. "The Response of Farmers in Northern Ghana to the Introduction of Mixed Farming: A Case Study." *Geografiska Annaler B*, 54B, No. 2 (1972): 95-103.

Hunter, John M. "Ascertaining Population Carrying Capacity Under Traditional Systems of Agriculture in Developing Countries." Pp. 299-302 in *People and Land in Africa South of the Sahara*, edited by R.M. Prothero. Oxford, England: Oxford University Press, 1972.

Doran-Adjobu, C. "The Impact of Migrant Ewe Cocoa Farmers in Buen, the Volta Region of Ghana." *Bulletin, Ghana Geographical Association*, 16 (1974): 45-53.

Hill, Polly. "Food-Farming and Migration from Fonte Villages." *Africa*, 48 (1978): 220-230.

Schaaf, Thomas, and Walter Manshard. "Die Entwicklung der ungelenkten Agrarkolonisation im Grenzgebiet von Ghana und der Elfenbeinkuste." *Erdkunde*, 42, No. 1 (1988): 26-36.

Kusi-Appiah, Andrew. "Population Distribution and Agricultural Production in Northern Ghana." Master's Thesis, Carleton University, Ottawa, Ontario, 1994.

Awongo, Louis. "Culture, Markets, and Agricultural Production: A Comparative Study of the Investment Patterns of Migrant and Citizen Cocoa Farmers in the Western Region of Ghana." *Professional Geographer*, 50, No. 4 (1998): 516-530.

ENVIRONMENTAL AND CONSERVATION ISSUES

Dedoo, Robert. "Agricultural Conservation: Eastern Region of Ghana." Master's Thesis, University of California, Los Angeles, Calif., 1967.

Benneh, George. "The Importance of Rainfall and Soil Moisture in the Agricultural Development of the Closed Forest Zone of Ghana." *Bulletin, Ghana Geographical Association*, 13 (1968): 37-45.

Sarpong, E. "Water Budget Characteristics, Availability of Water Periods, and Crop Production in Ghana." *Journal of the Geographical Association of Tanzania*, 27 (1992): 1-22.

Mikkelsen, Jari H., and Roger Langoho. "Comparison of International, National and Farmers' Classification Systems, Applied to Soils of the Western Dagumb District (Northern Ghana)." *Geografisk Tidsskrift*, 97 (1997): 47-57.

AGRICULTURAL MARKETS AND DISTRIBUTION

Manshard, Walther. "Agrarische Organisations-formen der fur den Binuenmarkt bestionten Kulturen in Waldgurtel Ghanas." *Erdkunde*, 11, No. 3 (1957): 215-223.

Guinea

CROPS AND CROPPING PATTERNS

Gallais, Jean. "La riziculture de plains en Haute-Guinee." *Annales de Geographie*, 68 (1959): 207-223.

Ivory Coast

GENERAL TEXTS, ARTICLES, AND THESES

Genevois, Louis. "Le premier congres international des industries agricoles et alimentaires des zones tropicales et subtropicales (Abidjan, 14-20 decembre 1964)." *Les Cahiers d'Outre Mer*, 18 (1965): 194-198.

Sawadogo, A. "Strategy of Development in Ivory Coast Agriculture." Bulletin de l'Association de Geographes Francais, 51 (1974): 87-103.

CROPS AND CROPPING PATTERNS

Hiernaux, C. "Les aspects geographiques de la production bananiere de la Cote d'Ivoire." *Les Cahiers d'Outre Mer*, 1 (1948): 68-84.

Frechou, Hubert. "Les plantations europeennes en Cote d'Ivoire." *Les Cahiers d'Outre Mer*, 8 (1955): 56-83.

Tricart, Jean. "Le cafe en Cote d'Ivoire." *Les Cahiers d'Outre Mer*, 10 (1957): 209-233.

Boni, Dion. "Situation actuelle de la culture bananiere en Cote d'Ivoire." *Les Cahiers d'Outre Mer*, 26 (1973): 225-249.

Boni, Dion. "Plantations paysannes et developpement dans le Sud-Est Forestier: L'exemple du Departement D'adzope." *Bulletin de l'Association Geographes Francais*, 51 (1974): 105-115.

Saint-Vil, Jean. "La riziculture intra-urbaine a Gagnon (Cote d'Ivoire)." *Travaux et Documents de Geographie Tropicale, CEGET-CNRS*, 28 (1977): 231-257.

Hauhouot, A. "Une premiere approche geographique de l'integration regionale des complexes sucriers de Cote d'Ivoire: L'exemple de Ferke I et de Borotou la Boa." *Annales de l'Universite de Abidjan*, Serie G, 8 (1979): 5-32.

Camara, C. "Les cultures vivrieres en Republique de Cote d'Ivoire." *Annales de Geographie*, 93, No. 4 (1984): 432-453.

Bassett, Thomas J. "Breaking Up the Bottlenecks in Food Crop and Cotton Cultivation in Northern Ivory Coast." *Africa*, 52, No. 2 (1988): 147-174.

Bassett, Thomas J. "The Development of Cotton in Northern Ivory Coast, 1910–1965." *Journal of African History*, 29, No. 2 (1988): 267-284.

Becker, Laurence, and Roger Diallo. "The Cultural Diffusion of Rice Cropping in Cote d'Ivoire." *Geographical Review*, 86, No. 4 (1996): 505-528.

Bassett, Thomas J. *The Peasant Cotton Revolution in West Africa: Cote d'Ivoire, 1880–1995*. Cambridge, England: Cambridge University Press, 2001.

ANIMAL HUSBANDRY

Bassett, Thomas J. "Hired Herders and Herd Management in Fulani Pastoralism (Northern Cote d'Ivoire)." *Cahiers d'Etudes Africaines*, 34 (1994): 147-173.

FARM LAND

Wurtz, Jacqueline. *Adiamprikofikro-Douakankro: Etude geographique d'un terroir baoule de Cote d'Ivoire.* La Haye, France: Mouton, 1971.

Bassett, Thomas J. "Land Use Conflicts in Pastoral Development in Northern Cote d'Ivoire." Pp. 139-156 in *Land in African Agrarian Systems*, edited by Thomas J. Bassett and Donald E. Crummey. Madison, Wis.: University of Wisconsin Press, 1993.

AGRICULTURAL POPULATIONS AND SOCIAL ISSUES

Koffi, K. "Agricultural paysanne, element strategique d'une problematique de developpement rural integre: Cas de la Cote d'Ivoire." Master's Thesis, Universite Laval, Ste. Foy, Quebec, 1987.

Bassett, Thomas J. "The Political Ecology of Peasant–Herder Conflicts in the Northern Ivory Coast." *Annals of the Association of American Geographers*, 78, No. 3 (1988): 453-472.

ENVIRONMENTAL AND CONSERVATION ISSUES

Bouet, Claude. *Bettie et Akiekrou: Etude comparee de deux terrois en zone forestiere ivoirienne.* Paris, France: ORSTOM, 1972.

IRRIGATION

Hauhouot, A., and K. Atta. "Un exemple d'amenagement hydro-agricole en Cote d'Ivoire: Le perimetre agricole de Koubi-Tiebissou." *Les Cahiers d'Outre Mer*, 34 (1981): 111-137.

Kenya

GENERAL TEXTS, ARTICLES, AND THESES

Case, Earl C. "The Pastoral and Agricultural Industry of Kenya Colony Protectorate." *Economic Geography*, 6, No. 3 (1930): 243-256.

Lehrer, Paul L. "The Agricultural Potentialities of Kenya." Master's Thesis, Ohio State University, Columbus, Ohio, 1951.

Kearn, H. "The Diverse Pattern of African Agriculture in the Nyanga Province of Kenya." Pp. 21-28 in *Natural Resources, Food and Population in Inter-Tropical Africa: International Geographical Union Symposium, Makerere*, edited by L. Dudley Stamp. Bude, England: Geo Publications, 1955.

Lehrer, Paul L. "African Agriculture in Kenya: A Study of a Changing System of Subsistence Agriculture." *Nigerian Geographical Journal*, 7, No. 1 (1964): 24-33.

Senkowski, Margie. "The Agrarian Revolution in Kenya." Master's Thesis, Western Michigan University, Kalamazoo, Mich., 1966.

Bernard, Frank E. "Recent Agricultural Change East of Mount Kenya." *Papers in International Studies*. Athens, Ohio: Ohio University African Series, No. 4, 1968.

Fliedner, Hanfried. "Die Wandlung der Agrarstruktur in Kenia." *Geographische Rundschau*, 20, No. 3 (1968): 81-86.

Taylor, D.R.F. "Agricultural Change in Kikuyuland." Pp. 463-493 in *Environment and Land Use in Africa*, edited by M.F. Thomas and G.W. Whittington. London, England: Methuen, 1969.

Bernard, Frank E. *East of Mount Kenya: Meru Agriculture in Transition*. New York, N.Y.: Humanities Press, 1972.

Andreae, Bernd. "Planungsstafen von Kleinfarmen im Generationswechsel: Das Hockland Kenias als Modell." *Geographische Rundschau*, 30, No. 8 (1978): 304-308.

Freeman, Donald B. "The Importance of Being First: Preemption by Early Adopters of Farming Innovations in Kenya." *Annals of the Association of American Geographers*, 75, No. 1 (1985): 17-28.

Hecklau, H. "Kenyas Landwirtschaft auf dem Weg in die Moderne: Probleme der agrarer Wandels." *Geographische Rundschau*, 41, No. 11 (1989): 620-626.

Burnett, G.W., and K.M. Rowntree. "Agriculture, Research and Tourism in the Landscape of Lake Baringo, Kenya." *Landscape and Urban Planning*, 19, No. 2 (1990): 159-172.

Freeman, Donald B. *A City of Farmers: Informal Urban Agriculture in the Open Spaces of Nairobi, Kenya*. Montreal, Quebec: McGill-Queen's University Press, 1991.

Bernard, Frank E. "Increasing Variability in Agricultural Production: Meru District, Kenya, in the Twentieth Century." Pp. 80-113 in *Population Growth and Agricultural Change in Africa*, edited by Robert W. Kates, Goram Hyden, and Billie Lee Turner II. Gainesville, Fla.: University Press of Florida, 1993.

Goldman, Abe. "Agricultural Innovation in Three Areas of Kenya: Neo-Boserupian Theories and Regional Characterization." *Economic Geography*, 69, No. 1 (1993): 44-71.

Memon, Pyar Ali, and Diane Lee-Smith. "Urban Farming in Kenya." *Canadian Journal of African Studies*, 17, No. 1 (1993): 25-42.

Lado, Cleophas. "The State and Peasant Farmers' Leverage in Kenya." *GeoJournal*, 41, No. 1 (1997): 33-41.

Tiffen, Mary. "Comment favoriser l'investissement paysan dans l'amelioration de l'agriculture." *Les Cahiers d'Outre Mer*, 50 (1997): 87-104.

CROPS AND CROPPING PATTERNS

Munger, Edwin S. "African Coffee on Kilimanjaro: A Chaggo Kihamba." *Economic Geography*, 28, No. 2 (1952): 181-185.

Taylor, D.R.F. "New Tea-Growing Areas in Kenya." *Geography*, 50 (1965): 373-375.

Arvo, Waino E. "The Geography of the Tea Industry in Kenya." Master's Thesis, Syracuse University, Syracuse, N.Y., 1968.

Morgon, William T. "The Role of Temperate Crops in the Kenya Highlands." *Acta Geographica*, 20, No. 19 (1968): 273-278.

Shisanya, Chris A., and Hoger Thunenson. "The Use of Dynamic Crop Production: Simulation Model in Quantifying Agricultural Resources: An Example from Machakos District, Kenya." *Journal of Eastern African Research and Development*, 23 (1993): 176-191.

Lado, Cleophas. "Capitalism and Rural Change in Africa: An Historical Case Study of the Gusii Coffee Industry in Kisii District, Kenya." *Malaysian Journal of Tropical Geography*, 28, No. 1 (1997): 21-32.

MacKenzie, Fiona. "Betterment and the Gendered Politics of Maize Production, Murang's District, Central Province, Kenya, 1880–1952." *Canadian Journal of African Studies*, 33, No. 1 (1999): 64-97.

DAIRYING

Ogendo, R.B. "Kenya Dairy Industry: Part 1. The Historical Background." *Journal of Eastern African Research and Development*, 1, No. 2 (1971): 161-166.

Ogendo, R.B. "Kenya Dairy Industry: Part 2. Location and Structure of the Industry." *Journal of Eastern African Research and Development*, 2, No. 1 (1972): 41-68.

FARM LAND

Maina, J.W. "Agriculture and Land Tenure in Kenya." Pp. 177-188 in *East Africa: Its People and Resources*, edited by William T. Morgon. Nairobi, Kenya: Oxford University Press, 1969, London, 1971.

Odingo, Richard S. *The Kenya Highlands: Land Use and Agricultural Development*. Nairobi, Kenya: East African Publishing House, 1971.

Bullock, Ronald A. *Ndeiya, Kikuyu Frontier: The Kenya Land Problem in Microcosm*. Waterloo, Ontario: University of Waterloo Department of Geography, Research Paper No. 6, 1975.

Holmes, R.S. "Land Tenure and Agricultural Productivity in the Trans-Nzoia, Kenya." *Geography*, 60, No. 2 (1975): 137-138.

Hecklau, Hans. "Ursachen und Wirkungen Konkurrierender Raumanspruche zwischen Hirten und Bauern bei der Expansion der Anbauflachen in Kenya." Pp. 221-228 in *Landerschliessung und Kulturlandscahftswandel an den Siedlungsgrenzen der Erde*, edited by Hans Jurgen Nitz. Gottingen, Germany: Gottingen Geographische Abhandlungen, Heft 66, 1976.

Stein, Christoph, and Christa Schulze. "Landnutzung und Entwicklungsmoglichkeiten in der Trockengebieten Kenias." *Geographische Rundschau*, 28, No. 3 (1976): 86-98.

Thomas, D.B., R.G. Barker, and T.R. Moore. "Terracing of Cropland in Low Rainfall Areas of Machakos District, Kenya." *Journal of Agricultural Engineering Research*, 25 (1981): 57-63.

Flury, Manuel. "Small-Scale Farming and Changes of Land Use in the Highland of Laikipia, Kenya." *Mountain Research and Development*, 8, No. 4 (1988): 265-272.

Lado, Cleophas. "Informal Urban Agriculture in Nairobi, Kenya: Problem or Resource in Development and Land Use Planning?" *Land Use Policy*, 7, No. 3 (1990): 257-266.

Mackenzie, Fiona. "A Piece of Land Never Shrinks: Reconceptualizing Land Tenure in a Smallholding District, Kenya." Pp. 194-221 in *Land in African Agrarian Systems*, edited by Thomas J. Bassett and Donald E. Crummey. Madison, Wis.: University of Wisconsin Press, 1993.

AGRICULTURAL POPULATIONS AND SOCIAL ISSUES

Mortimore, Gretchen. "Women Agricultural Producers in Western Kenya." Master's Thesis, University of Iowa, Iowa City, Iowa, 1990.

Mutisya, Dorothy N., and Cleophas Lado. "Some Socio-Economic Factors Behind Roadside Farming in Kiambu District, Kenya." *Journal of Eastern African Research and Development*, 21 (1991): 107-127.

Nyakaana, J.B. "Feminization of Agriculture Without Loss of Productivity and Income: A Case Study from Nyeri District, Kenya." *Indonesian Journal of Geography*, 22, No. 63 (1992): 57-71.

Nyakaana, J.B. "Feminization of Agriculture Without Loss of Productivity and Income: A Case Study of Nyeri District, Kenya." *Journal of the Geographical Association of Tanzania*, 27 (1992): 73-86.

Hyma, B., and Brez Nuamwange. "Women's Role and Participation in Farm and Community Tree-Growing Activities in Kiambu District, Kenya." Pp. 30-45 in *Different Places, Different Voices: Gender and Development in Africa, Asia and Latin America*, edited by Janet Momsen and Vivian Kinnaird. London, England: Routledge, 1993.

Barbina, Guido. "La riforma agraria del Kenya: Il fallimento di cena politica di decolonizzazione." *Rivista Geografica Italiana*, 104, No. 3 (1997): 303-337.

Geheb, Kim, and Tony Binns. "Fishing Farmers or Farming Fisherman? The Quest for Household Income and Nutritional Security on the Kenyan Shores of Lake Victoria." *African Affairs*, 96 (1997): 73-93.

Mutoro, Basilida A. "Agricultural Policy, Marginalisation of Women and Their Effect on Agricultural Production in Vihiga District, Kenya." *Malaysian Journal of Tropical Geography*, 28, No. 2 (1997): 103-114.

Thompson, John J. "Cooperative in the Commons: The Emergence and Persistence of Reciprocal Altruism and Collective Action in Farmer-Managed Irrigation Systems in Kenya." Ph.D. Dissertation, Clark University, Worcester, Mass., 1997.

Campbell, David J. "Response to Drought Among Farmers and Herders: A Comparison of 1972–1976 and 1994–1995." *Human Ecology*, 27, No. 3 (1999): 377-416.

ENVIRONMENTAL AND CONSERVATION ISSUES

Lehrer, Paul L. "Effects of African Agricultural Practices on Soil Productivity and Nutrient Levels in Kenya." Ph.D. Dissertation, University of Nebraska, Lincoln, Nebr., 1962.

Wisner, Ben, and P. Mbithi. "Drought in Eastern Kenya: Nutritional Status and Farmer Activity." Pp. 87-97 in *Natural Hazards*, edited by Gilbert F. White. New York, N.Y.: Oxford University Press, 1974.

Morgon, William T. "Altitude and Crop Zone: A Transect in the Kenya Highlands." *Die Erde*, 108 (1977): 124-128.

Abok, Wilson O. "The Impact of Paddy Rice Cultivation on the Ecology of Malarial Mosquitos in the Ahero Irrigation Scheme, Kenya." Master's Thesis, Ohio University, Athens, Ohio, 1982.

Kalff, J., J.A. Downing, and T.T. Smith. "Rainfall, Agriculture, Livestock and Human Density in the Dry Regions of Kenya." *Journal of Arid Environments*, 9, No. 2 (1985): 173-183.

Lewis, John E., and I.J. Ndolo. "Estimating Moisture Availability for Rice Cultivation in Mwea, Kenya." *Agricultural and Forest Meteorology*, 41 (1987): 87-95.

Finkel, Moshe, and M.B.K. Darkoh. "Sustaining the Arid and Semi-Arid Environment in Kenya Through Improved Pastoralism and Agriculture." *Journal of Eastern African Research and Development*, 21 (1991): 1-20.

Ofari-Sarpong, E. "Implications of Rainfall for Agricultural and Urban Development of Eldaret, Kenya." *Indonesian Journal of Geography*, 21, No. 62 (1991): 55-69.

IRRIGATION

Hornby, W.F. "The Mwea Irrigation Settlement." *Geography*, 58, No. 3 (1973): 255-259.

Watson, Elizabeth E. "Indigenous Irrigation, Agriculture and Development, Marakwet, Kenya." *Geographical Journal*, 164, No. 1 (1998): 67-84.

AGRICULTURAL MARKETING AND DISTRIBUTION

Gyllstrom, B. "Administered Interdependence Versus Spatial Integration: The Case of Agricultural Service Cooperatives in Kenya." *Lund Studies in Geography*, Series B, 53 (1989): 5-56.

Barrett, Hazel R., et al. "Globalization and the Changing Networks of Food Supply: The Imporation of Fresh Horticultural Produce from Kenya to the UK." *Transactions, Institute of British Geographers*, 24, No. 2 (1999): 259-274.

TRANSPORTATION AND MECHANIZATION

Birdsall, Stephen S. "Transportation and Agricultural Development in Western Kenya: An Approach to the Problem of Investment Priorities." Ph.D. Dissertation, Michigan State University, East Lansing, Mich., 1968.

AGRICULTURAL LABOR STUDIES

Levesque, J., and Fiona Mackenzie. "A Farm Is Like a Child Who Cannot Be Left Unguarded: Gender, Land and Labour in Central Province, Kenya." *IDS Bulletin*, 26, No. 1 (1995): 17-23.

Liberia

CROPS AND CROPPING PATTERNS

Church, R.J.H. "The Firestone Rubber Plantations in Liberia." *Geography*, 54 (1957): 430-437.

Koch, Peter. "Brandrodungsfeldbau in Liberia." *Geographica Helvetica*, 51, No. 1 (1996): 31-36.

AGRICULTURAL MARKETING AND DISTRIBUTION

Amegashie, Austin K. "Liberian Agricultural Programs: The Theory and Practice of Marketing and Food Distribution Systems." Ph.D. Dissertation, University of Florida, Gainesville, Florida, 1978.

Madagascar

GENERAL TEXTS, ARTICLES, AND THESES

Jackson, R.T. "Agricultural Development in the Malagasay Republic." *East African Geographical Review*, 9 (1971): 69-78.

Carlson, Roger A. "Agricultual Development in the Lake Alaotoa Basin, Malagasay Republic." Ph.D. Dissertation, University of Nebraska, Lincoln, Nebr., 1973.

Douessin, Rene. "Geographie agraire des plaines de Tananarive (lere partie)." *Madagascar Revue de Geographie*, 25 (1974): 11-148.

Douessin, Rene. "Geographie agraire des plaines de Tananarive (2nd article)." *Madagascar Revue de Geographie*, 26 (1975): 9-91.

Portais, M. "Les cultures commercials dans un milieu geographique original: La cuvette d'Andopa, Madagascar." *Travaux et Documents de Geographie Tropical, CEGET-CNRS*, 20 (1975): 325-356.

Douessin, Rene. *Geographie agraire des plaines de Tananarive*. Tananarive, Madagascar: Association des Geographes de Madagascar, 1976.

Ramamonjisoa, J. "Les operations de developpement agricole dans le Unkinonkaratra." *Madagascar Revue de Geographie*, 46 (1985): 23-45.

Neuvy, Guy. "Le developpement agricole du bassin superieur de la Lokoho, a Madagascar." *Les Cahiers d'Outre Mer*, 42 (1989): 133-154.

Underwood, Emma. "An Application of Remote Sensing and Geographic Information Systems to a Study of Shifting Cultivation and Sustainable Development in Eastern Madagascar." Master's Thesis, Ohio University, Athens, Ohio, 1995.

CROPS AND CROPPING PATTERNS

Defos du Rau, J. "Le sisal dans le Sug Malgache." *Les Cahiers d'Outre Mer*, 7 (1954): 51-83.

Guilcher, Andre. "Un cas d'economie tropicale de plantation: Les iles Radama (Madagascar)." *Les Cahiers d'Outre Mer*, 10 (1957): 23-33.

Le Bourdiec, F. "Le sisal a Madagascar." *Madagascar Revue de Geographie*, 13 (1968): 57-89.

Dewailly, Jean-Michel. "La culture de la Banane d'exportation a Madagascar." *Madagascar Revue de Geographie*, 16 (1970): 7-43.

Douessen, Rene. "Le sucre a Madagascar: Etude geographique." *Madagascar Revue de Geographie*, 22 (1973): 95-153.

Le Bourdiec, F. "L'evolution de la riziculture dans l'Ouest Malgache." *Madagascar Revue de Geographie*, 30 (1977): 9-32.

Le Bourdiec, F. "Geographie historique de la riziculture malgache." *Madagascar Revue de Geographie*, 31 (1977): 11-69.

Ramamonjisoa, J. "Le sisal a Madagascar." *Madagascar Revue de Geographie*, 31 (1977): 87-122.

Salomon, Jean-Noel. "Les vignobles et les vins de Madagascar." *Les Cahiers d'Outre Mer*, 33 (1980): 335-362.

Guebourg, Jean-Louis. "Analyse de la filiere tabac malgache: De l'etatisation pure a la societe mixte." *Les Cahiers d'Outre Mer*, 50 (1997): 245-270.

ANIMAL HUSBANDRY

Raison, J.P. "Mouvements et commerce des bovins dans la region de Mandoto (Moyen Ouest de Madagascar)." *Madagascar Revue de Geographie*, 12 (1968): 7-57.

Petit, M. "L'elevage du mouton astrakan en Afrique australe." *Madagascar Revue de Geographie*, 18 (1971): 103-129.

Donque, Gerald. "L'agriculture et l'elevage a Madagascar." *Les Cahiers d'Outre Mer*, 27 (1974): 105-127.

Randrianarison, J. "Le bouef dans l'economie rurale de Madagascar." *Madagascar Revue de Geographie*, 28 (1976): 9-122.

Randrianarison, J. "Le bouef dans l'economie rurale de Madagascar, 2." *Madagascar Revue de Geographie*, 29 (1976): 9-81.

Cori, Gilles. "Deux types d'elevage bovin a Madagascar: L'elevage extensif de l'ouest and l'elevage des paysons de hauts-plateaux." Pp. 5-120 in *Types d'elevage et de vie rurale a Madagascar*, edited by Gilles Cori and Pierre Trama. Talence, France: Travaux et Documents de Geographie Tropicale, CEGET-CNRS, No. 37, 1979.

Cori, Gilles, and Pierre Trama, ed. *Types d'elevage et de vie rurale a Madagascar*. Talence, France: Travaux et Documents de Geographie Tropicale, CEGET-CNRS, No. 37, 1979.

Hoerner, J.M. "Les vols de bouef dans le sud Malgache." *Madagascar Revue de Geographie*, 41 (1982): 85-103.

FARM LAND

Kull, Christian A. "Leimavo Revisited: Agrarian Land-Use Change in the Highlands of Madagascar." *Professional Geographer*, 50, No. 2 (1998): 163-175.

AGRICULTURAL POPULATIONS AND SOCIAL ISSUES

Sarremejean, Marguerite. "La vie agricole et la calendrier du payson malgache dans les plaines de Tananarive." *Les Cahiers d'Outre Mer*, 14 (1961): 349-371.

Koerner, Francois. "Les types de societes agricoles privees a Madagascar." *Les Cahiers d'Outre Mer*, 21 (1968): 276-297.

Jarosz, L. "Rice on Shares: Agrarian Change and the Development of Sharecropping in Alaotru, Madagascar." Ph.D. Dissertation, University of California, Berkeley, Calif., 1990.

Jarosz, L. "Defining and Explaining Tropical Deforestation: Shifting Cultivation and Population Growth in Colonial Madagascar, 1896–1940." *Economic Geography*, 69, No. 4 (1993): 366-379.

Rakoto Ramiarandtoa, Herve. "Paysanniers merina des hautes terres centrales malgaches: Une ethno-agronomie de phasie." *Les Cahiers d'Outre Mer*, 50 (1997): 43-56.

IRRIGATION

Hoerner, J.M. "L'eau et l'agriculture dans le sud-ouest de Madagascar." *Madagascar Revue de Geographie*, 30 (1977): 63-101.

Neuvy, Guy. "Les amenagements hydro-agricoles pour le developpement a Madagascar." *Revue Geographie de l'Est*, 28, No. 1 (1988): 19-38.

Malawi

GENERAL TEXTS, ARTICLES, AND THESES

Lienau, Cay. "Agrarlandschaften in Malawi." *Die Erde*, 100 (1969): 5-29.

Erhard, Andres. *Malawi: Agrarstruktur und Unterentwicklung*. Innsbruk, Austria: Innsbrucker Geographische Studien, Heft 22, 1994.

CROPS AND CROPPING PATTERNS

Burfaud, Pierre. "La riziculture dans la vallee du Limphasa (Malawi)." *Les Cahiers d'Outre Mer*, 30 (1977): 111-130.

Mali

GENERAL TEXTS, ARTICLES, AND THESES

Barth, Hans Karl. *Der Geokomplex Sahel: Unters zur landschaftsokologie im Sahel Malis als Grundlage agrar- und weidewirtschaftlungen Entwicklungsplanung*. Tubingen, Germany: Tubinger Geographische Studien, Heft 7, 1977.

Krings, T. "The Advantages and Risks of Ox-Ploughing and Monoculture in the Central and Southern Savannas of the Republic of Mali." *Applied Geography and Development*, 30 (1987): 46-63.

Thoms, Derrick, and John Wells. "Farming Systems in the Niger Inland Delta, Mali." *Geographical Review*, 77, No. 3 (1987): 302-342.

Fournier, I. "Tentatives Tomacheq pour la mise en place d'un nouveau systeme de production: Region de Gao (Mali)." *Cahiers Geographiques de Rouen*, 35 (1991): 29-39.

CROPS AND CROPPING PATTERNS

Krings, T. "Die Vorteile und Risitan von Pflugbau und Monokultur in den Zentralen und sudlichen Savana en der Republik Mali." *Die Erde*, 117 (1986): 201-216.

AGRICULTURAL POPULATIONS AND SOCIAL ISSUES

Raoult, B. "Evolution des systems de production dans la region de Gao (Mali): Un fructueux malentendu?" *Cahiers Geographiques de Rouen*, 35 (1991): 41-60.

ENVIRONMENTAL AND CONSERVATION ISSUES

Krings, T. "Periodic Fluctuations in the Lake Level and Their Effect upon the Agricultural Economy in Lake Faguibim (Republic of Mali)." *Applied Geography and Development*, 27 (1986): 19-38.

Ofori-Sarpong, E. "Moisture Availability and Agricultural Potentialities of Mali and Niger." *Indonesian Journal of Geography*, 20, No. 60 (1990): 57-70.

IRRIGATION

Krings, T. "Die Bedentung des Bewasserungsfeldbaus fur die Subsistenzsicherung bei den Dogou, Republik Mali." *Die Erde*, 119 (1988): 185-192.

AGRICULTURAL MARKETING AND DISTRIBUTION

Schmidt-Wulffen, Wulf. "Ernahrungssicherung und technischer Fortschrift durch Marktliberalisierung? Zur Akzeptonz des Freien Getreidemarkets durch Bauern in Mali." *Geographische Zeitschrift*, 79, No. 3 (1991):168-180.

Mauritania

ENVIRONMENTAL AND CONSERVATION ISSUES

Vermeer, Donald E. "Collison of Climate, Cattle and Culture in Mauritania during the 1970s." *Geographical Review*, 71, No. 2 (1981): 281-297.

Mozambique

GENERAL TEXTS, ARTICLES, AND THESES

de Carvalho, M. *A agricultura tradicional de Mocambique: 1. Distribucao geografica dos culturas e sua relacao com o meio*. Lorenzo Marques, Mozambique: Missao de Inquerito Agricola de Mozambique, 1969.

Weber, Peter. "Agrargeographische Skizze des Makonde-Gebietes (Mozambique)." *Die Erde*, 103 (1972): 60-76.

AGRICULTURAL POPULATIONS AND SOCIAL ISSUES

Weber, Peter. *Die agrargeographische Struktur von Mittel-Mocambique: Natur- und sozial raumliche Grundlagen der Bantu-land-wirtschaft*. Marburg, Germany: Marburger Geographische Schriften, Heft 48, 1971.

ENVIRONMENTAL AND CONSERVATION ISSUES

Vache-Grandet, Claude. "Probleme agraire en Mauritanie: L'amenagement du Lac Rkiz." *Les Cahiers d'Outre Mer*, 12 (1959): 251-264.

Wisner, Ben. "Ecodevelopment and Ecofarming in Mozambique." Pp. 157-168 in *Ecodevelopment: Concepts, Projects, Strategies*, edited by B. Glaeser. Oxford, England: Pergamon, 1984.

Namibia/Southwest Africa

GENERAL TEXTS, ARTICLES, AND THESES

Leser, Hartmut. "Weidewirtschaft und Regenfeldbau in Sandveld: Westliche Kalahari von Schwarzen Nossob und Epukiro, Ostliches Sudwestafrika." *Geographische Rundschau*, 27, No. 3 (1975): 108-122.

Bahr, Jurgen. "Veranderungen in der Farmwirtschaft sudwestafrikas/ Namibias zwischen 1965 und 1980." *Erdkunde*, 35, No. 4 (1981): 274-289.

Bahr, Jurgen. "Das zentralartliche System in der Farmzone Sudwestafrikas / Namibias: Entwicklung, Struktur und jungere Veranderungen." *Die Erde*, 115 (1984): 209-234.

Schneider, M.B. "Agriculture in Namibia." *Frankfurter Wirtschafts- und Sozial geographische Schriften*, 56 (1991): 141-160.

AGRICULTURAL POPULATIONS AND SOCIAL ISSUES

Eggers, Heinz. "Das Ovamboland-Sonderstellung und Probleme eines Dichtegebietes in Sudwestafrikas." *Geographische Rundschau*, 18, No. 12 (1966): 459-468.

Niger

GENERAL TEXTS, ARTICLES, AND THESES

Sergent, H. "Au Niger dans la vallee du fleuve et sa region: Quelles mutations des systems de production depuis les secheresses?" *Cahiers Geographiques de Rouen*, 35 (1991): 5-27.

CROPS AND CROPPING PATTERNS

Arnold, Wolfgang. "Die Republik Niger lebt von der Erdmess." *Zeitschrift fur Wirtschaftsgeographie*, 10, No. 5 (1966): 152-155.

ANIMAL HUSBANDRY

Mainet, Guy. "L'elevage dans la region de Maradi (Republique du Niger)." *Les Cahiers d'Outre Mer*, 18 (1965): 32-72.

Beauvilain, A. "Eleveurs et amenagments hydro-pastoraux dans le Nord Kourfeye (Niger) au tours de la recente secheresse." *Cahiers Geographiques de Rouen*, 20 (1976): 97-107.

Tetaille, Denis. "Interventions externes et reponses des pasteurs: L'exemple de la zone des forages nord Goure." *Les Cahiers d'Outre Mer*, 38 (1985): 103-120.

FARM LAND

Weicker, M. "Vieh Statt Hirse over Vieh gegen Hirse: Probleme der Resourcenutzung und der Nahrungsmittelversorgung beig den Iforas-Tuareg." *Die Erde*, 119 (1988): 243-252.

ENVIRONMENTAL AND CONSERVATION ISSUES

Agnew, C.T. "The Water Balance Approach to the Development of Rainfed Agriculture in South West Niger." Ph.D. Dissertation, University of East Anglia, Ipswich, England, 1980.

IRRIGATION

Chudeau, R. "Les irrigation du Niger et la culture du coton." *Annales de Geographie*, 31 (1922): 155-163.

Nigeria

ATLASES AND GRAPHIC PRESENTATIONS

Agboola, S.A. *An Agricultural Atlas of Nigeria*. Oxford, England: Oxford University Press, 1979.

GENERAL TEXTS, ARTICLES, AND THESES

Morgan, W.B. *The Change from Shifting Agriculture to Fixed Settlement in Southern Nigeria*. Ibadan, Nigeria: University of Ibadan Department of Geography, Research Note No. 7, 1955.

Morgan, W.B. "Some Comments on Shifting Cultivation in Africa." *Research Notes*, University College Ibadan, Department of Geography, 9 (1957): 1-10.

Morgan, W.B. "Agriculture in Southern Nigeria." *Economic Geography*, 35, No. 2 (1959): 138-150.

Higgins, G.M. "Experimental Farm Coverage in Northern Nigeria." *Nigerian Geographical Journal*, 4, No. 2 (1961): 11-25.

Coppock, J.T. "Agricultural Geography in Nigeria." *Nigerian Geographical Journal*, 7, No. 2 (1964): 67-90.

Goddard, S. "Town-Far Relationships in Yorubaland: A Case Study from Oyo." *Africa*, 35 (1965): 21-29.

Coppock, J.T. "Agricultural Development in Nigeria." *Journal of Tropical Geography*, 23 (1966): 1-18.

Norman, D.W. "Rural Economy with Special Reference to Agriculture." Pp. 39-148 in *Zaria and Its Region*, edited by Michael J. Mortimore. Nigeria: Ahmadu Bello University, Department of Geography, Occasional Paper No. 4, 1970.

Mbagwu, T.C. "Agricultural Enterprise Combination in Ngwa-Land." *Nigerian Geographical Journal*, 17, No. 2 (1974): 127-142.

Adisa, Joseph. "An Analysis of Recent Agricultural Planning in Nigeria: A Geographic Approach." Master's Thesis, University of Akron, Akron, Ohio, 1977.

Ekwere, Sunday D. "An Agricultural Development Program for Nigeria." Master's Thesis, Mankato State University, Mankato, Minn., 1978.

Iliya, Mohammad A. "Diffusion of Agricultural Innovations in the Sokota Rima Basin of Nigeria: A Study of the Bakolori Project." Master's Thesis, Ohio University, Athens, Ohio, 1981.

Floyd, Barry N. "Agricultural Planning in Nigeria." *Geography*, 67, No. 4 (1982): 345-348.

Okafor, F.C. "Diffusion of Agricultural Practices and Information in Bendel State, Nigeria." *Geographical Perspectives*, 52 (1983): 35-44.

Udo, Reuben K. "Assessment of Recent Government Efforts to Increase Food Production in Nigeria." *Rural Systems*, 2, No. 4 (1984): 259-265.

Davis, J. Tait. "Agricultural Development Projects in Nigeria: Experiments in Induced Diffusion." *Canadian Geographer*, 31, No. 2 (1987): 127-139.

Okafor, F.C., and E.C.M. Fernandes. "Compound Farms of Southeastern Nigeria." *Agroforestry Systems*, 5 (1987): 153-168.

Sokari-George, E. "Planning in Nigeria: The Agricultural Base 1962–1985." *GeoJournal*, 14, No. 1 (1987): 97-108.

Oyelese, J.O. "Agricultural Development in Nigeria: A Case Study of the Utility of Agro-Service System." *Nigerian Geographical Journal*, 1 (1994): 165-186. (New Series)

Adams, W.M., and Michael J. Mortimore. "Agricultural Intensification and Flexibility in the Nigerian Sahel." *Geographical Journal*, 163, No. 2 (1997): 150-160.

Jagtap, S., and A. Amissale-Arthur. "Stratification and Synthesis of Crop-Livestock Production System Using GIS." *GeoJournal*, 47, No. 4 (1999): 573-582.

Swindell, Kenneth, M.A. Jliya, and A.B. Mamman. "Making a Profit, Making a Living: Commercial Food Farming and Urban Hinterlands in North-West Nigeria." *Africa,* 69, No. 3 (1999): 386-403.

Lynch, K., Tony Binns, and E. Olofin. "Urban Agriculture Under Threat: The Land Security Question in Kano, Nigeria." *Cities*, 18, No. 3 (2001): 159-171.

CROPS AND CROPPING PATTERNS

Morgan, W.B. "The Nigerian Oil Palm Industry." *Scottish Geographical Magazine*, 71 (1955): 174-177.

Prothero, R.M. "Recent Developments in Nigerian Export Crop Production." *Geography*, 40 (1955): 18-27.

Mabogunje, A.L. "Rice Cultivation in Southern Nigeria." *Nigerian Geographical Journal*, 2, No. 2 (1959): 59-69.

Johnsrud, Robert. "A Decade of Nigerian Cotton, 1949–1958." *Nigerian Geographical Journal*, 3, No. 2 (1960): 1-14.

Coppock, J.T. "Tobacco Growing in Nigeria." *Erdkunde*, 19 (1965): 297-306.

Udo, Reuben K. "Sixty Years of Plantation Agriculture in Southern Nigeria: 1902–1962." *Economic Geography*, 41, No. 4 (1965): 356-368.

Moisley, H.A. "The Establishment of a Sugar Industry in Nigeria." *Geography*, 51, No. 1 (1966): 142-147.

Udo, Reuben K. "British Policy and the Development of Export Crops in Nigeria." *Nigerian Journal of Economic and Social Studies*, 9, No. 3 (1967): 299-314.

Agboola, S.A. "The Introduction and Spread of Cassava in Western Nigeria." *Nigerian Journal of Economics and Social Studies*, 10, No. 3 (1968): 369-385.

Ajaegbu, H.I. "Food Crop Farming in the Coastal Area of Southwestern Nigeria." *Journal of Tropical Geography*, 31 (1970): 1-9.

Oyeleye, D.A. "Tobacco Cultivation in Oyo Division, Western Nigeria." *Nigerian Geographical Journal*, 14, No. 2 (1971): 165-184.

Uzozie, L.C. "Patterns of Crop Combination in the Three Eastern States of Nigeria." *Journal of Tropical Geography*, 33 (1971): 62-72.

Adejuwon, O. "Agricultural Areal Differentiation in the Cocoa Producing Areas of Western Nigeria." *Journal of Tropical Geography*, 35 (1972): 1-10.

Kassam, A.H., and J.M. Kowal. "Productivity of Crops in the Savanna and Rain Forests Zones in Nigeria." *Savanna*, 2, No. 1 (1973): 39-49.

Akinbode, Ade. "The Spatial Diffusion of Kolanut Production in South-Western Nigeria." *Nigerian Geographical Journal*, 17, No. 2 (1974): 111-125.

Umachi, U.N. "The Nigerian Cocoa Industry: An Analysis of Spatial Diffusion and Production Patterns." Master's Thesis, Wayne State University, Detroit, Mich., 1974.

Vermeer, Donald E. "The Tradition of Experimentation in Swidden Cultivation Among the Tiv of Nigeria." *Papers and Proceedings of the Applied Geography Conference*, 2 (1979): 244-257.

Filoni, M.O. "Rice in the Nigerian Economy: A Geo-Economic Analysis." *Nigerian Journal of Economic and Social Studies*, 22, No. 1 (1980): 145-176.

Ekanale, Olusegun. "Small-Scale Cocoa Farmers and Environmental Change in the Tropical Rain Forest Regions of South-Western Nigeria." *Journal of Environmental Management*, 25, No. 1 (1987): 61-70.

Adeniyi, J.P. "Farm Size and Resource Efficiency in Peasant Production of Rice." *Nigerian Geographical Journal*, 30-31 (1987–1988): 93-102.

Afolobi, Falola J. "The Changing Organisation of Tobacco Production in Oyo State: Lessons for the Restructuring of Peasant Agricultural Production in Nigeria." *Rural Systems*, 8, No. 1 (1990): 15-24.

Edanade, Olusegun, and F.A. Adesina. "Sustaining Tree Crop Production Under Intensive Land Use: An Investigation into Soil Quality Differentiation Under Varying Cropping Patterns in Westen Nigeria." *Journal of Environmental Management*, 32, No. 2 (1991): 105-113.

Kimmage, K. "The Evolution of the 'Wheat' Trap: The Nigerian Wheat Boom." *Africa*, 60, No. 4 (1991): 471-500.

Odemerho, F.O., and A. Avurunadiogba. "The Effects of Changing Cassava Management Practices on Soil Loss: A Nigerian Example." *Geographical Journal*, 159, No. 1 (1993): 63-69.

Porter, G., and Kevin Phillips-Howard. "Bitter Brew? Barley Growing in Nigeria." *Geography Review*, 6, No. 5 (1993): 34-37.

Porter, G., and Kevin Phillips-Howard. "Outgrower Barley Cultivation on the Jos Plateau, Nigeria: A Case Study of Agricultural Innovation." *Geographical Journal*, 160, No. 3 (1994): 319-327.

Bello, N.J. "Climatic Aspects of Cane Sugar (Saccharum L.) Yield at Bacita, Nigeria." *Malaysian Journal of Tropical Geography*, 27, No. 1 (1996): 29-34.

Adebayo, A.A., and E.F. Adebayo. "Forecasting the Yield of Upland Rice in Adamawa State, Nigeria on the Basin of Agroclimatic Factors." *Malaysian Journal of Tropical Geography*, 28, No. 2 (1997): 97-102.

ANIMAL HUSBANDRY

Fricke, Werner. "Herdenzahheng mit Hilfe von Luftbibdern im Gebiet des kundfligen Niger-Stausees (Nordnigeria)." *Die Erde*, 96 (1966): 208-223.

Burchardt, Helmut H. "Die Fulbe und ihre Rinderzucht in Nord Nigeria." *Geographische Rundschau*, 20, No. 3 (1968): 98-108.

Dunbar, Gary S. "African Ranches Ltd., 1914–1931: An Ill-Fated Stock-Raising Enterprise in Northern Nigeria." *Annals of the Association of American Geographers*, 60, No. 1 (1970): 102-123.

Van Raay, J.G.T. "Animal Husbandry." Pp. 149-156 in *Zaria and Its Region,* edited by Michael J. Mortimore. Nigeria: Ahmado Bello University Department of Geography, Paper No. 4, 1970.

Van Raay, J.G.T., and P.N. de Leeuw. "The Importance of Crop Residues as Fodder: A Resource Analysis in Katsina Province, Nigeria." *Tijdschrift voor Economische en Sociale Geographie,* 61, No. 3 (1970): 137-147.

Fricke, Werner. *Cattle Husbandry in Nigeria: A Study of the Ecological Conditions and Social-Geographical Differentiations* (in German). Heidelberg, Germany: Heidelberger Geographische Arbeiten, Heft 52, 1979.

Bassett, Thomas J. "Fulani Herd Movements." *Geographical Review,* 76, No. 2 (1986): 233-248.

Falela, J.A. "Factors of Poultry Farming Adoption in Oyo State, Nigeria: A Study in Aging and Infrastructure-Based Diffusion." *GeoJournal,* 15, No. 1 (1987): 83-90.

FARM LAND

Morgan, W.B. "The Strip Fields of Southern Nigeria." Pp. 33-37 in *Natural Resources, Food and Population in Inter-Tropical Africa,* edited by L. Dudley Stamp. London, England: Geographical Publications, 1956.

Floyd, Barry N. "Terrace Agriculture in Eastern Nigeria: The Case of Makee." *Nigerian Geographical Journal,* 7, No. 2 (1964): 91-108.

Adeniyi, E.O. "Land Tenure and Agricultural Development in Nupeland." *Nigerian Geographical Journal,* 15, No. 1 (1972): 49-58.

Oyeleye, D.A. "Agricultural Land Use in Oyo Division Viewed Against the Background of von Thunen's Theory of Land Utilization." *Journal of Tropical Geography,* 37 (1973): 39-52.

Wunglueck, Letty L. "Agriculture and Land Tenure in Transition: The Case of Western Nigeria." Master's Thesis, University of Michigan, Ann Arbor, Mich., 1973.

Atteh, O.D. "The Effect of Land Tenure Systems on Agricultural Production: A Case Study of the Land Tenancy System in the Okun Area of Kwara State, Nigeria." *Geoforum,* 16, No. 3 (1985): 277-286.

Afolayan, A.A. "Land Policy, Cash Cropping and In-Migration into Ife, Southern Area of Nigeria." *GeoJournal,* 14, No. 1 (1987): 87-96.

Francis, Paul. "Land Tenure Systems and Agricultural Innovation: The Case of Alley Farming in Nigeria." *Land Use Policy,* 4, No. 3 (1987): 305-320.

Gyasi, Edwin A. "The Rison Palm Nucleas Estate of Nigeria: An Agricultural Plantation and Its Effect in the Surrounding Area." *Malaysian Journal of Tropical Geography,* 15 (1987): 26-38.

Bdliya, H.H. "Complementary Land Evaluation for Small-Scale Farming in Northern Nigeria." *Journal of Environmental Management,* 33, No. 2 (1991): 105-116.

Ariyo, J.A., and D.O. Ogbonna. "The Effects of Land Speculation on Agricultural Production Among Peasants in Kachia Local Government Area of Kaduna State, Nigeria." *Applied Geography,* 12, No. 1 (1992): 31-46.

Odemerho, F.O. "Land Degradation Impacts on Tropical Agricultural Basins and Their Management: The Nigerian Case." *Geoforum*, 23, No. 4 (1992): 499-506.

Osemeobo, G.J. "Agricultural Land Use in Nigerian Forestry Reserves: Towards a Solution to Problems or Conflict in Biotic Conservation?" *Land Use Policy*, 10, No. 1 (1993): 44-48.

Alexander, M.J., and A.D. Kidd. "Farmers' Capability and Institutional Incapacity in Reclaiming Disturbed Land on the Jos Plateau, Nigeria." *Journal of Environmental Management*, 59, No. 2 (2000): 151-155.

Tukur, A.L., and H.H. Ray. "Farming and Land Degradation in Michika-Madagali Area of Adamawa State, Nigeria." *Nigerian Geographical Journal*, 3-4 (2000): 64-71.

AGRICULTURAL POPULATIONS AND SOCIAL ISSUES

Morgan, W.B. "Farming Practice, Settlement Pattern and Population Density in Southeastern Nigeria." *Geographical Journal*, 121, No. 3 (1955): 320-333.

Grove, A.T. "Population Densities and Agriculture in Northern Nigeria." Pp. 115-136 in *Essays in African Population*, edited by Kenneth M. Barbour and R.M. Prothero. London, England: Routledge and Kegan Paul, 1962.

Ojo, G.J.A. "The Changing Patterns of Group Farming in Ekiti, Northeastern Yoruba Country." *Nigerian Geographical Journal*, 6, No. 1 (1963): 31-38.

Udo, Reuben K. "The Migrant Tenant Farmer of Eastern Nigeria." *Africa*, 34 (1964): 326-338.

Grossman, David. "Migrating Tenant Farming in Northern Iboland in Relation to Resource Use." Ph.D. Dissertation, Columbia University, New York, N.Y., 1968.

Vermeer, Donald E. "Population Pressure and Crop Rotational Changes Among the Tiv of Nigeria." *Annals of the Association of American Geographers*, 60, No. 2 (1970): 299-314.

Adejuwon, O. "Agricultural Colonization in Twentieth Century Western Nigeria." *Journal of Tropical Geography*, 33 (1971): 1-8.

Udo, Reuben K. "Food-Deficit Areas of Nigeria." *Geographical Review*, 61, No. 3 (1971): 415-430.

Grossman, David. "The Effect of Migratory Tenant Farming on Food Production in Northern Iboland, Nigeria." *National Geographical Journal of India*, 20, No. 4 (1974): 201-215.

Udo, Reuben K. *Migrant Tenant Farmers of Nigeria*. Lagos, Nigeria: African University Press, 1975.

Ballot, J.M. "Ke Tamasheq de Gourma Nigerien et Peul du Torodi: Societes agropastorales en mutation." Ph.D. Dissertation, Universite de Bordeaux, Bordeaux, France, 1980.

Okafor, F.C. "The Adoption Strategies of Smallholders in Parts of Southeastern Nigeria: Implications for Agricultural Planning." *Journal of Rural Studies*, 2, No. 2 (1986): 117-126.

Spiro, Heather M. "Women Farmers and Traders in Oyo State, Nigeria: A Case Study of Their Changing Roles." Pp. 173-191 in *Geography of Gender in the Third World*, edited by Janet Momsen and J.G. Townsend. London, England: Hutchinson, 1987.

Goldman, Abe. "Population Growth and Agricultural Change in Imo State, Southeastern Nigeria." Pp. 250-301 in *Population Growth and Agricultural Change in Africa*, edited by Robert W. Kates, Goram Hyden, and Billie Lee Turner II. Gainesville, Fla.: University Press of Florida, 1993.

Martin, Susan. "From Agricultural Growth to Stagnation: The Case of the Ngwa, Nigeria, 1900–1980." Pp. 302-323 in *Population Growth and Agricultural Change in Africa*, edited by Robert W. Kates, Goram Hyden, and Billie Lee Turner II. Gainesville, Fla.: University Press of Florida, 1993.

Mortimore, Michael J. "The Intensification of Peri-Urban Agriculture: The Kano Close-Settled Zone, 1964–1986." Pp. 358-400 in *Population Growth and Agricultural Change in Africa*, edited by Robert W. Kates, Goram Hyden, and Billie Lee Turner II. Gainesville, Fla.: University Press of Florida, 1993.

Okafor, F.C. "Agricultural Stagnation and Economic Diversification: Awka-Nnewi Region, Nigeria, 1930–1980." Pp. 324-357 in *Population Growth and Agricultural Change in Africa*, edited by Robert W. Kates, Goram Hyden, and Billie Lee Turner II. Gainesville, Fla.: University Press of Florida, 1993.

Shimoda, S. "Migration and Change in Agricultural Production Systems in Rural Nigeria: A Case Study." *Science Reports, Tohoku University, Seventh Series, Geography*, 43, No. 2 (1993): 63-90.

ENVIRONMENTAL AND CONSERVATION ISSUES

Adejuwon, O. "Crop-Climate Relationship: The Example of Cocoa in Western Nigeria." *Nigerian Geographial Journal*, 5, No. 1 (1962): 21-33.

Morgan, W.B., and R.P. Moss. "Farming, Forest and Savanna in Western Nigeria." *Erdkunde*, 24, No. 1 (1970): 71-80.

Ashaye, T.I. "Soils and Land Capability of Some Areas in South West Nigeria in Connection with Sugarcane Cultivation." *Nigerian Journal of Science*, 9, No. 1 (1975): 131-173.

Moss, R.P. and W.B. Morgan. "Soils, Plants, and Farmers in West Africa, I: A Consideration of Some Aspects of Their Relationships, with Special Reference to Continguous Areas of Forest and Savana in South-West Nigeria." Pp. 27-57 in *Human Ecology in the Tropics*, edited by J.P. Garlich and R.W. J. Keay. London, England: Taylor and Francis, 1977.

Moss, R.P. "Soils, Plants, and Farmers in West Africa, II: Further Examination of the Role of Edaphic Factors and Cropping Patterns in South-West Nigeria." Pp. 59-77 in *Human Ecology in the Tropics*, edited by J.P. Garlich and R.W.J. Keay. London, England: Taylor and Francis, 1977.

Ogbuji, R.O., and C.T. Uwakah. "Outlook on Farming Systems and Root-Knot Damage to Crops in Eastern Nigeria." *GeoJournal*, 3, No. 6 (1979): 567-570.

Okafor, F.C. "Environmental Constraints and Part-Time Farming in Southeastern Nigeria." *GeoJournal*, 6, No. 4 (1982): 359-366.

Ogbuji, R.O. "Nematode Damage to Crops Affects Decision Making of the Nigerian Farmer and Consumer." *GeoJournal*, 7, No. 3 (1983): 279-282.

Odemerho, F.O. "The Effects of Shifting Cultivation on Stream Channel Size and Hydraulic Geometry in Small Headwater Basins of Southwestern Nigeria." *Geografiska Annaler A*, 66A, No. 4 (1984): 327-340.

Adefolalu, B.O. "On Rainfall Distribution and Agricultural Planning in Nigeria." *Malaysian Journal of Tropical Geography*, 11 (1985): 1-11.

Okafor, F.C. "River Basin Management and Food Crisis in Nigeria." *Geoforum*, 16, No. 4 (1985): 413-421.

Omiunu, F.G.I. "Effects of Drought on Food Production in Nigeria." *Malaysian Journal of Tropical Geography*, 11 (1985): 54-62.

Adams, W.M. "Traditional Agriculture and Water Use in the Sokoto Valley, Nigeria." *Geographical Journal*, 152, No. 1 (1986): 30-43.

Gbadegesin, A. "A Method for Identifying Soil Properties Influencing Crop Yield: The Example of Maize in the Savanna Belt of Southwestern Nigeria." *Geoforum*, 17, No. 1 (1986): 109-118.

Ojo, S.O., M. Ijioma, and A.O. Ojo. "Consumptive Water Use of Rainfed Maize in Nigeria." *Singapore Journal of Tropical Geography*, 7, No. 2 (1986): 128-144.

Gbadegesin, A. "Soil Properties and Maize Yield in the Savanna Belt of Southwest Nigeria." *Malaysian Journal of Tropical Geography*, 16 (1987): 37-42.

Olaniran, O.J., and J.S. Babatohu. "The Effect of Climate on the Growth of Early Maize at Kabba, Nigeria." *GeoJournal*, 14, No. 1 (1987): 71-75.

Kolawok, A. "Cultivation of the Floor of Lake Chad: A Response to Environmental Hazard in Eastern Barno, Nigeria." *Geographical Journal*, 154, No. 2 (1988): 243-258.

Adejuwan, J.O., F.A. Adesina, and E. Onijingin. "Some Aspects of Soil and Vegetation Recovery in Forest and Savanna Fallows in Western Nigeria and Their Implications for Agricultural Development." *Malaysian Journal of Tropical Geography*, 20 (1989): 1-11.

Ekanade, Olusegan. "Degradation of the Physical Elements of the Rural Environment Resulting from Tree Crops Cultivation in the Nigerian Cocoa Belt." *Singapore Journal of Tropical Geography*, 12, No. 2 (1991): 82-94.

Adewole-Osunade, M.A. "Identification of Crop Soils by Small Farmers of Southwestern Nigeria." *Journal of Environmental Management*, 35, No. 3 (1992): 193-203.

Kimmage, K., and W.M. Adams. "Wetland Agricultural Production and River Basin Development in the Hadejia-Jame'are Valley, Nigeria." *Geographical Journal*, 158, No. 1 (1992): 1-12.

Adewole-Osunade, M.A. "Indigenous Climate Knowledge and Agricultural Practice in Southwestern Nigeria." *Malaysian Journal of Tropical Geography*, 25, No. 1 (1994): 21-28.

Phillips-Howard, Kevin, and Fergus Lyon. "Agricultural Intensification and the Threat to Soil Fertility in Africa: Evidence from Jos Plateau, Nigeria." *Geographical Journal*, 160, No. 3 (1994): 252-265.

Adediji, A., O. Ogunkoya, and L. Jeje. "Soil Loss from Maize on Ultisols, Ife Area, Southwestern Nigeria." *Nigerian Geographical Journal*, 2 (1995): 30-43.

Adesola, Ogidiolu. "Effects of Oil Palm (Elaeis quineesis) and Rubber (Hevea brasiliensis) Plantations on Forest Soil in South-Western Nigeria." *Nigerian Geographical Journal*, 2 (1995): 74-80.

Ariyo, J.A., and M.A. Gashur. "Indigenous On-Farm Techniques of Coping with Droughts in the Sahel Region of Nigeria." *Malaysian Journal of Tropical Geography*, 27, No. 1 (1996): 19-28.

Ite, Uwem E. "Small Farmers and Forest Loss in Cross River National Park, Nigeria." *Geographical Journal*, 163, No. 1 (1997): 45-56.

Oche, C.Y. "Agroclimatic Zonation for Wheat Production in the Savanna Region of Nigeria." *Singapore Journal of Tropical Geography*, 19, No. 1 (1998): 39-50.

Harris, Frances. "Nutrient Management Strategies of Small-Holder Farmers in a Short-Fallow Farming System in North-East Nigeria." *Geographical Journal*, 165, No. 3 (1999): 275-285.

Adebayo, A.A. "Agro-Climatic Classification of Adamawa State, Nigeria for Upland Rice Production." *Nigerian Geographical Journal*, 3-4 (2000): 83-98.

IRRIGATION

Roder, Wolf. *The Irrigation Farmers of the Kainji Lake Area*. Rome, Italy: FAO, 1970.

Ajao, Florence A. "Nigeria's Yoke Irrigation Scheme: 1959 to 1978." Ph.D. Dissertation, University of Pittsburgh, Pittsburgh, Pa., 1980.

Bello, N.J. "An Assessment of Water Supply for Agriculture in the Niger River Basin Development Authority Area, Nigeria." *Agricultural and Forest Meteorology*, 40 (1987): 109-121.

Kimmage, K., and W.M. Adams. "Smallscale Farmer-Managed Irrigation in Northern Nigeria." *Geoforum*, 21, No. 4 (1990): 435-446.

Pratt, N.D., et al. "Estimating Areas of Land Under Small-Scale Irrigation Using Satellite Imagery and Ground Data for a Study Area in North East Nigeria." *Geographical Journal*, 163, No. 1 (1997): 65-77.

AGRICULTURAL MARKETING AND DISTRIBUTION

Essong, S.M. "Impact of the Marketing Board on the Distribution of Cocoa Income in Western Nigeria." *Nigerian Geographical Journal*, 15, No. 1 (1972): 35-48.

Scott, Earl P. "The Spatial Structure of Rural Northern Nigeria: Farmers, Periodic Markets, and Villages." *Economic Geography*, 48, No. 3 (1972): 316-332.

Oyeleye, D.A. "The Diffusion of Agricultural Cooperatives in Western Nigeria." *Singapore Journal of Tropical Geography*, 3, No. 1 (1982): 69-85.

TRANSPORTATION AND MECHANIZATION

Ojo, G.J.A. "Trends Towards Mechanized Agriculture in Yorubaland." *Nigerian Geographical Journal*, 6, No. 2 (1963): 116-129.

Ijere, M.O. "The Impact of Nigerian Railways on Agricultural Development." *Nigerian Geographical Journal*, 16, No. 2 (1973): 137-144.

Onyewuenyi, Remy N. "Railway Development and the Growth of Export Agriculture in Nigeria During the 1900–1995 Period." Master's Thesis, University of Ottawa, Ottawa, Ontario, 1982.

Salau, A.T., and J.M. Baba. "Spatial Impact of the Relocation of a Section of the Zaria-Kano Road: A Study of Change and Development in Rural Zaria." *Applied Geography*, 4, No. 4 (1984): 283-292.

AGRICULTURAL LABOR STUDIES

Ojo, G.J.A. "Some Observations on Journey to Agricultural Work in Yorubaland, Southwestern Nigeria." *Economic Geography*, 46, No. 4 (1970): 459-471.

Ojo, G.J.A. "Journey to Agricultural Work in Yorubaland." *Annals of the Association of American Geographers*, 63, No. 1 (1973): 85-96.

Okafor, F.C. "Labour Shortage in Nigerian Agriculture." *Labor, Capital and Society*, 12 (1979): 108-122.

Okoji, Maurice A. "A Regional Survey of Labor Problems in Nigerian Peasant Agriculture." *Philippine Geographical Journal*, 34, No. 4 (1990): 181-194.

Rwanda and Burundi

GENERAL TEXTS, ARTICLES, AND THESES

Vennetier, Pierre. "Une micro-realisation de developpement agro-artisanal au Rwanda." *Les Cahiers d'Outre Mer*, 31 (1978): 209-224.

Rossi, G. "Evolution des versants et mise en valeur agricole au Rwanda." *Annales de Geographie*, 93, No. 1 (1984): 23-43.

Nkurunziza, Francois. "Structures agraires et intensification agricole sur une colline du Burundi." *Bulletin de le Societe de Geographique de Belge*, 56 (1987): 385-404.

Wakana, Mathias. *Diagnostic de la region naturelle du Buyenzi*. Ngozi, Burundi: Institut des sciences agronomiques du Burundi, 1989.

Delaine, Bernard, et al. *Diagnostic de la region de l'Imbo-Nord*. Cibitoke, Burundi: Institut des sciences agronomiques du Burundi, 1990.

CROPS AND CROPPING PATTERNS

Baker, P. Randall. "The Introduction of Rice in Rwanda." *Journal of Tropical Geography*, 31 (1970): 27-32.

Bart, Francois. "Le cafe dans l'agriculture rwandaise: L'exemple de Kidahire." *Les Cahiers d'Outre Mer*, 33 (1980): 301-317.

Gotanegre, Jean-Francoise. "La banane au Rwanda." *Les Cahiers d'Outre Mer*, 36 (1983): 303-310.

Chapuis, Odile. "La thei culture rwandaise." *Les Cahiers d'Outre Mer*, 39 (1986): 117-142.

FARM LAND

Nwafor, J.C. "Agricultural Land Use and Associated Problems in Rwanda." *Journal of Tropical Geography,* 48 (1979): 58-65.

de la Masselliere, B.C. "Le resserrement de l'espace agraire au Rwanda: Les paysans dans la crise." *Etudes Rurales*, 125-126 (1992): 99-116.

Ford, Robert E. "Marginal Coping in Extreme Land Pressures: Ruhengeri, Rwanda." Pp. 145-186 in *Population Growth and Agricultural Change in Africa*, edited by Robert W. Kates, Goram Hyden, and Billie Lee Turner II. Gainesville, Fla.: University Press of Florida, 1993.

ENVIRONMENTAL AND CONSERVATION ISSUES

Ntibashirwa, Sylvestre, and Bernard Delaine. *Diagnostic de la region naturelle du Kirimiro*. Bitega, Burundi: Institut des sciences agronomiques du Burundi, 1989.

Storz, S. "Ecofarming in Ruanda" (in German). *Geographie und Schule*, 11, No. 60 (1989): 8-13.

Lewis, L.A. "Relations Between Crops, Topography and Degradation in Westen Rwanda: Some Strategies for Sustainable Agriculture." *Zeitschrift fur Geomorphologie*, Supplement, Band 83 (1991): 23-28.

Stache, N., and P. Drechsel. "Okologischer Landbau in de Landlichen Entwicklung Ruandas." *Geographische Rundschau*, 49, No. 1 (1997): 32-38.

Stache, N., and A. Wirthmann. "Die Misere in der Landwirtschaft im tropischen Afrika am Beispiel einer geookologischen Studie in Rwanda." *Petermann's Geographische Mitteilungen*, 142, No. 5-6 (1998): 339-354.

São Tomé and Príncipe

CROPS AND CROPPING PATTERNS

Matznetter, Josef. "Die Guineainseln San Tome und Principe und ilne Plantagen." *Geographische Zeitschrift*, 51 (1963): 268-301.

Senegal

GENERAL TEXTS, ARTICLES, AND THESES

Chamard, Philippe C. "L'agriculture senegalaise." *L'Information Geographique*, 43, No. 4 (1979): 180-186.

CROPS AND CROPPING PATTERNS

Pelissier, P. "Les Diola, etude sur l'habitat des riziculteurs de Basse-Casamance." *Les Cahiers d'Outre Mer*, 11 (1958): 334-388.

AGRICULTURAL POPULATIONS AND SOCIAL ISSUES

Sy, Eliane. "Cayar, village des pecheurs: Cultivateurs au Senegal." *Les Cahiers d'Outre Mer*, 18 (1965): 342-368.

ENVIRONMENTAL AND CONSERVATION ISSUES

Lericollais, Andre. *Etude geographique d'un terroir serer (Senegal).* Paris, France: Mouton, 1972.

Anhuf, D. "Climate and Crop Yields in Northern Africa (Senegal, Sudan, Tunisia)." *GeoJournal*, 20, No. 3 (1990): 221-230.

Sall, Mamadau. "Transformation de l'agriculture et crises de l'environnement dans la Saloum meridional (Senegal)." *L'Espace Geographique*, 21, No. 4 (1992): 343-346.

IRRIGATION

Mathieu, P. "Irrigation, reforme fonciere et reactions paysonnes dans l'amenagement de la vallee du fleure Senegal." *Revue de Geographie de Lyon*, 65 (1990): 46-51.

Seck, S.M. "Gestion et organisation des amenagements irrigation en exploitation paysonne dans la vallee du Senegal." *Revue de Geographie de Lyon*, 65 (1990): 38-45.

Sierre Leone

GENERAL TEXTS, ARTICLES, AND THESES

Binns, J.A. *Integrated Agricultural Development: A Case-Study from Sierre Leone*. Oxford, England: Discussion Paper in Geography, Oxford Polytechnic Institute, No. 6, 1977.

Gleave, M.B. *Agricultural Development in Sierre Leone: Some Basic Patterns.* Salford, England: Salford University, Department of Geography, Discussion Paper in Geography No. 6, 1978.

Kaloko, F.R. "Rural Development and Agriculture in Sierre Leone." *Philippine Geographical Journal*, 28, No. 1-2 (1984): 72-84.

Binns, J.A. "Agricultural Change in Sierre Leone." *Geography*, 67, No. 2 (1985): 113-125.

CROPS AND CROPPING PATTERNS

Jarrett, H.R. "Rice Production in Sierre Leone." *Journal of Tropical Geography*, 8 (1956): 73-81.

Jarrett, H.R. "The Oil Palm and Its Changing Place in the Economy of Sierre Leone." *Geography,* 42, No. 1 (1957): 50-59.

Gwynne-Jones, D.R.G. "Rubber Production in Sierra Leone." *Bulletin of the Sierra Leone Geographical Association*, 9 (1965): 23-26.

Lappia, Joseph. "Comparative Analysis of Two Efforts Designed to Expand Rice Production in Sierre Leone." Master's Thesis, University of Wisconsin, Madison, Wis., 1972.

Hewapathirane, D.U. "The Use of Swamplands for Rice Farming in Sierra Leone." *Eastern and Southern Africa Geographical Journal*, 2, No. 1 (1991): 111-128.

ANIMAL HUSBANDRY

Gwynne-Jones, D.R.G. "Cattle in a Mende Chiefdom: Aspects of the Cattle Trade in the Eastern Province." *Sierra Leone Geographical Journal*, 13 (1969): 28-31.

ENVIRONMENTAL AND CONSERVATION ISSUES

Gleave, M.B. "The Length of the Fallow Period in Tropical Fallow Farming Systems: A Discussion with Evidence from Sierra Leone." *Geographical Journal*, 162, No. 1 (1996): 14-24.

TRANSPORTATION AND MECHANIZATION

Gleave, M.B. *Mechanisation of Peasant Farming: Experience in Sierre Leone.* Salford, England: University of Salford, Department of Geography, Discussion Paper No. 3, 1977.

Somalia

GENERAL TEXTS, ARTICLES, AND THESES

Dainelli, Giotto. "The Agricultural Possibilities of Italian Somalia." *Geographical Review*, 21, No. 1 (1931): 56-69.

Salisbury, Lance. "The State and the Crisis of Agrarian Production in the Lower Shabelle of Somalia." Master's Thesis, University of Iowa, Iowa City, Iowa, 1991.

Unruh, Jon D. "Nomadic Pastoralism and Irrigated Agriculture in Somalia." *GeoJournal*, 25, No. 1 (1991): 91-108.

CROPS AND CROPPING PATTERNS

Somatar, Abdi I. "Structural Adjustment as Development Strategy? Bananas, Boom and Poverty in Somalia." *Economic Geography*, 69, No. 1 (1993): 25-43.

ANIMAL HUSBANDRY

Al-Najim, M., and John A. Briggs. "Livestock Development in Somalia: A Critical Review." *GeoJournal,* 26, No. 3 (1992): 357-362.

AGRICULTURAL POPULATIONS AND SOCIAL ISSUES

Janzen, J. "Anthropogeographische Forschung in der Bewasserungsregion Sud-Somalis-Ziele, Methodik, Ergebnisse." *Geomethodica,* 15 (1990): 71-106.

ENVIRONMENTAL AND CONSERVATION ISSUES

Scholz, M. "Problems of Rainfed Agriculture in Southern Somalia." *Geomethodica,* 15 (1990): 47-70.

IRRIGATION

Janzen, J. "Staudamme und grobflachiger Bewasserrungs feldbau contra mobile Viehwirtschaft? Das Baardhure-Staudammproject in Sud-Somalia und mogliche Folgen fur die mobile Viehwirtschaft." *Die Erde,* 121, No. 1 (1990): 25-38.

South Africa, Lesotho, and Swaziland

GENERAL TEXTS, ARTICLES, AND THESES

MacDonald, William. "The Agricultural Potentialities of the Transvaal." *Scottish Geographical Magazine,* 25, No. 1 (1909): 12-17.

Wellington, John H. "Some Geographical Factors Affecting Agriculture in South Africa." *South African Geographical Journal,* 6 (1923): 41-66.

Carol, Hans. "Das agrargeographische Betrachtungssystem: Beitrag zur landschaftskindlichen Methodik, dargelegt am Beispiel der Kauru." *Geographica Helvetica,* 7, No. 1 (1952): 17-67.

McDaniel, J.B. "Some Government Measures to Improve African Agriculture in Swaziland." *Geographical Journal,* 132, No. 4 (1966): 506-514.

Pollock, N. "Problems of Agricultural Diffusion in Natal, South Africa." *Geography,* 59, No. 4 (1974): 257-259.

Christopher, A.J. "The Variability of the Southern African Standard Farm." *South African Geographical Journal,* 58, No. 2 (1976): 107-117.

Lageat, Yannick. "Structures agraires et developpement agricole au Swaziland." *Les Cahiers d'Outre Mer,* 30 (1977): 246-277.

Thailer, Donald F. "Factors Affecting Relative Agricultural Production Amongst Land Holders on the Leribe Development Project in Lesotho, Africa." Ph.D. Dissertation, University of Wisconsin, Madison, Wis., 1977.

Sterkenberg, Jan. "Agricultural Commercialization in Africa South of the Sahara: The Cases of Lesotho and Swaziland." *Development and Change,* 11 (1980): 573-606.

Daniel, J.B.M. "Agricultural Development in the Ciski: Review and Assessment." *South African Geographical Journal*, 63, No. 1 (1981): 3-23.

Rogerson, C.M. "Urban Agriculture in South Africa: Policy Issues from the International Experience." *Development South Africa*, 10, No. 1 (1993): 33-44.

Rogerson, C.M. "Urban Agriculture in South Africa: Scope, Issues, and Potential." *GeoJournal*, 30, No. 1 (1993): 21-28.

Mather, Charles. "Towards Sustainable Agriculture in Post-Apartheid South Africa." *GeoJournal*, 39, No. 1 (1996): 41-49.

Rogerson, C.M. "Urban Cultivation and Urban Reconstruction in South Africa." *Geographical Journal of Zimbabwe*, 27 (1996): 11-20.

Porter, G., and Kevin Phillips-Howard. "Contract Farming in South Africa: A Case Study from KwaZulu-Natal." *Geography*, 82, No. 1 (1997): 38-44.

Nel, E., and J. Davies. "Farming Against the Odds: An Examination of the Challenges Facing Farming and Rural Development in the Eastern Cape Province of South Africa." *Applied Geography*, 19, No. 3 (1999): 253-274.

CROPS AND CROPPING PATTERNS

Cole, Monica M. "The Growth and Development of the South African Citrus Industry." *Geography*, 39, No. 2 (1954): 102-112.

Williams, Owen. "Sugar Growing and Processing in the Union of South Africa." *Economic Geography*, 35, No. 4 (1959): 356-366.

Griffiths, I.L. "Changes in the South African Pineapple Industry." *Geography*, 46, No. 4 (1961): 360-363.

Impey, L.H. "The Olive Industry in South Africa." *South African Geographical Journal*, 44 (1962): 34-49.

Benseler, Arthur. "Weinbau in Sudafrika: Kapwein—Ein neuer konkurrent auf dem Weltmarkt." *Zeitschrift fur Wirtschaftsgeographie*, 9, No. 8 (1965): 242-247.

Wiese, B. "Die Kultur tropischer Fruchtbaum in Sudafrika." *Erdkunde*, 25, No. 2 (1971): 135-147.

Sternkenberg, Jan, A.J. Diphoorn, and K.P. Spoorman. *The Production and Marketing of Vegetables in Swaziland*. Utrecht, Netherlands: University of Utrecht, Department of Geography of Developing Nations, 1978.

Gillooly, J.F., and T. Dyer. "On Spatial and Temporal Variations of Maize Yields over South Africa." *South African Geographical Journal*, 61 (1979): 111-118.

Lewis, Colin A. "The South African Sugar Industry." *Geographical Journal*, 156, No. 1 (1990): 70-79.

Makhanya, F.M. "Factors Influencing the Viability and Sustainability of Smallholder Sugar Cane Production in Umbumbulu." *South African Geographical Journal*, 79, No. 1 (1997): 19-26.

Terry, Alan K. "Extending Participation in the Swaziland Sugar Industry to Small-Scale Growers: Patterns and Prospects." *Singapore Journal of Tropical Geography*, 18, No. 2 (1997): 196-209.

ANIMAL HUSBANDRY

Baticle, Y. "Les differents types d'elevage ovin dans la province du Cap." *Les Cahiers d'Outre Mer*, 23 (1970): 233-258.

FARM LAND

Andreae, Bernd. *Weidewirtschaft im Sudlichen Afrika*. Wiesbaden, Germany: F. Steiner, 1966.

Weiner, Daniel, and R. Levin. "Land and Agrarian Transition in South Africa." *Antipode*, 23, No. 1 (1991): 92-120.

AGRICULTURAL POPULATIONS AND SOCIAL ISSUES

Bob, Urmilla. "Gender Geographies and Plantation Agriculture: A Case Study of a Coffee Plantation in Lebowa, South Africa." Master's Thesis, West Virginia University, Morgantown, W.Va., 1994.

Myburgh, D.W. "The Response of Farmers to the Drought Hazard in the Karoo Environment." *GeoJournal*, 33, No. 4 (1994): 401-410.

Kruger, Darrell P. "Pastoral Strategies and Settlement Systems in Colonial South Africa." *Historical Geography*, 24, No. 1-2 (1995): 57-70.

May, J., and C. Rogerson. "Poverty and Sustainable Cities in South Africa: The Role of Urban Cultivation." *Habitat International*, 19, No. 2 (1995): 165-182.

Crush, Jonathan. "The Culture of Failure: Racism, Violence, and White Farming in Colonial Swaziland." *Journal of Historical Geography*, 22, No. 2 (1996): 177-197.

Watson, H.K. "Short and Long Term Influence on Soil Erosion of Settlement by Peasant Farmers in Kavazulu-Natal." *South African Geographical Journal*, 78, No. 1 (1996): 1-6.

Mather, Charles. "Populist Agrarian Visions and Rural Livelihoods in KwaZulu-Natal." *Canadian Journal of African Studies*, 32, No. 1 (1998): 139-145.

ENVIRONMENTAL AND CONSERVATION ISSUES

Wilken, G.C. *Agroclimatic Hazard Perception, Prediction and Risk-Avoidance Strategies in Lesotho*. Boulder, Colorado: University of Colorado, Natural Hazard Research Working Paper No. 44, 1982.

Smit, H., and J. Olivier. "Agro-environmental Analogues for Jojoba Production in South Africa." *South African Geographical Journal*, 73, No. 1 (1991): 3-9.

Vogel, C. "Usable Science: An Assessment of Long-Term Seasonal Forecasts Amongst Farmers in Rural Areas of South Africa." *South African Geographical Journal*, 82, No. 2 (2000): 107-116.

IRRIGATION

Phillips-Howard, Kevin, and G. Porter. "Small-Scale Irrigation and the Reconstruction and Development of Transkei, South Africa." *Area*, 28, No. 3 (1996): 373-383.

AGRICULTURAL MARKETING AND DISTRIBUTION

Mather, Charles. "Agro-Commodity Chains, Market Power and Territory: Re-Regulating South African Citrus Exports in the 1990s." *Geoforum*, 30, No. 1 (1999): 61-70.

AGRICULTURAL LABOR STUDIES

Borchert, Gunter. *Agricultural Efficiency in South Africa: A Survey Based on Value Added Per Labour Day*. Hamburg, Germany: Hamburger Geographische Studien, Heft 39, 1983.

Crush, Jonathan. "The Long-Averted Clash: Farm Labour Competition in the South African Countryside." *Canadian Journal of African Studies*, 27, No. 3 (1993): 404-423.

Mather, Charles. "Foreign Migrants in Export Agriculture: Mozambican Labour in the Mpumalanga Low Veld, South Africa." *Tijdschrift voor Economische en Sociale Geographie*, 91, No. 4 (2000): 426-436.

Sudan

GENERAL TEXTS, ARTICLES, AND THESES

Lamberty, Mary A. "The Agricultural and Pastoral Resources of Kordofan Province, Sudan." Master's Thesis, Columbia University, New York, N.Y., 1963.

Wood, Roland G. "Agricultural Systems in the Guba Mountains, Sudan." Ph.D. Dissertation, University of California, Los Angeles, Calif., 1971.

Mohamed, Y.A. "Diffusion of Agricultural Innovations Among Traditional Farmers of Western Sudan: The Case of East Kordofan and El Fasher Districts." *GeoJournal*, 6, No. 1 (1982): 31-40.

Dettman, K. "Rainfed Agriculture in the Republic of the Sudan: Its Role in Development Planning." *Rural Systems*, 1, No. 4 (1983): 283-288.

El Arifi, S.A. "Problems in Planning Extensive Agricultural Projects: The Case of New Halfa, Sudan." *Applied Geography*, 8, No. 1 (1988): 37-52.

Lavergne, Marc E. "L'agriculture dans la Republique du Soudan." *Les Cahiers d'Outre Mer*, 42 (1989): 293-317.

Wallach, Bret. "Improving Traditional Grassland Agriculture in Sudan." *Geographical Review*, 79, No. 2 (1989): 143-160.

CROPS AND CROPPING PATTERNS

Hoyle, S.G. "Developments in Sugar Cane Production in the Sudan." *Geography*, 64, No. 2 (1979): 123-125.

Briggs, John A. "An Exploratory Study of Farmers' Choice of Crops in Central Sudan." *Transactions, Institute of British Geographers*, 10, No. 2 (1985): 170-180. (New Series)

Lado, Cleophas. "Some Aspects of Cotton as a Cash Crop Development in an Historical Perspective in Maridi District, Southern Sudan." *Journal of Eastern African Research and Development*, 18 (1988): 24-43.

ANIMAL HUSBANDRY

Davies, H.R.J. "Nomadism in the Sudan: Aspects of the Problem and Suggested Lines for Its Solution." *Tijdschrift voor Economische en Sociale Geographie*, 57 (1966): 193-202.

Baumer, Michel C. "Pastoralism au Kordofan (Republique du Soudan): 1. Les hommes et les paturages." *Bulletin de la Societe de Geographie d'Egypte*, 47-48 (1974–1975): 1-60.

El Arifi, S.A. "Some Recent Changes in Pastoral Nomadism in the Sudan." *Bulletin de la Societe de Geographie d'Egypte*, 47-48 (1974–1975): 79-94.

El Arifi, S.A. "Pastoral Nomadism in the Sudan." *East African Geographical Review*, 13 (1975): 89-104.

Baumer, Michel C. "Pastoralism au Kordafon." *Bulletin de la Societe de Geographie d'Egypte*, 57-58 (1984–1985): 3-200.

Khogali, M.M. "An Assessment of Gereih Al Sarfha Pilot Scheme for the Livestock Sector in the Sudan." *GeoJournal*, 14, No. 1 (1987): 53-62.

Ibrahim, F.N. "Viehholtung bei den Hirsebauern der Sahelzone des Sudan: Eine Uberle bens strategie." *Die Erde*, 119 (1988): 219-226.

FARM LAND

Willimott, S.G. "Cultivable Land and Land Use in Equatoria Province, Sudan." *Journal of Tropical Geography*, 8 (1956): 40-50.

Hale, G.A. "Cultivation Terraces in Western Darfur, Sudan: A Study in an Agricultural Form and Practice." Ph.D. Dissertation, University of California, Los Angeles, Calif., 1966.

Lado, Cleophas. "A Note on Traditional Agricultural Land Use: A Case Study from Maridi District, Southern Sudan." *Malaysian Journal of Tropical Geography*, 18 (1988): 25-34.

Lado, Cleophas. "Traditional Agricultural Land Use and Some Changing Trends in Maridi District, Southern Sudan." *Land Use Policy*, 6, No. 4 (1989): 324-340.

Reenberg, A. "Agricultural Land Use Pattern Dynamics in the Sudan-Sahel: Towards an Event-Driven Framework." *Land Use Policy*, 18, No. 4 (2001): 309-319.

AGRICULTURAL POPULATIONS AND SOCIAL ISSUES

Renner, G.T. "A Famine Zone in Africa: The Sudan." *Geographical Review*, 16, No. 4 (1926): 583-596.

Briggs, John A. "Farmers' Responses to Planned Agricultural Development in the Sudan." *Transactions, Institute of British Geographers*, 3, No. 4 (1978): 464-475. (New Series)

Ebene, R.T. "Health Risks to the Population of Gezira, Sudan Following Irrigation of Cotton Fields." *GeoJournal*, 7, No. 3 (1983): 283-290.

Lado, Cleophas. "Agricultural and Environmental Knowledge: A Case Study of Peasant Farming in Maridi District, Southern Sudan." *Malaysian Journal of Tropical Geography*, 13 (1986): 7-36.

Ebene, R.T. "Health Risks to the Population of Gezira, Sudan Following Irrigation of Cotton Fields." Pp. 293-304 in *Health and Disease in Tropical Africa: Geographical and Medical Viewpoints*, edited by Rais Akhtar. London, England: Harwood, 1987.

Ibrahim, F.N. "Combating Famine by Grain Storage in Western Sudan." *GeoJournal*, 14, No. 1 (1987): 29-35.

Ibrahim, Mohamed B. "Drought, Attitude and Crop Diffusion Among the Peasant Farmers of Western Sudan." *Geography*, 73, No. 1 (1988): 71-73.

Lado, Cleophas. "Peasant Farmers' Rationality and the Diffusion of Agricultural Innovations in Maridi District, Southern Sudan." *Malaysian Journal of Tropical Geography*, 20 (1989): 20-29.

Asfana, Shoa. "Un espace agraire Loti au Soudan: Les sites d'installation des refugies erythreens." *L'Espace Geographique*, 2, No. 4 (1992): 347-356.

ENVIRONMENTAL AND CONSERVATION ISSUES

Ismail, A.M.A. "Soil Properties and Moisture Characteristics and Their Relationship with Crop-Mid-Day Stress in the Sudan Gezira." *GeoJournal*, 23, No. 3 (1991): 233-240.

IRRIGATION

Briggs, John A. "The Development of Irrigated Agriculture in the Sudan." *Journal of the Geographical Association of Tanzania*, 16 (1978): 89-96.

Heinritz, Gunter. "Des Rahad-Scheme: Ein neues Bewasserungsprojekt in der Republik Sudan." *Zeitschrift fur Wirtschaftsgeographie*, 24, No. 3 (1980): 81-87.

Ibrahim, F.N. "Water Harvesting und Water Spreading fur Flutanbau in der Sahelzone des Sudan." *Die Erde*, 119 (1988): 179-184.

Birciato, F., and P. Faggi. "Gezira Schemes Between State and Market: Some Remarks on Privatization of Irrigation in Sudan." *GeoJournal*, 37, No. 1 (1995): 101-104.

AGRICULTURAL MARKETING AND DISTRIBUTION

Trilsbach, A., and S. Wood. "Livestock Markets and the Semi-Arid Environment: A Case Study from Sudan." *Geografiska Annaler B*, 68B, No. 1 (1986): 51-58.

Lado, Cleophas. "Some Aspects of Rural Marketing Systems and Peasant Farming in Maridi District, Southern Sudan." *Transactions, Institute of British Geographers*, 13, No. 3 (1988): 361-374. (New Series)

Speece, M. "Market Performance of Agricultural Commodities in Semi-Arid South Kordofan, Sudan." *Geoforum*, 20, No. 4 (1989): 409-426.

Tanzania

GENERAL TEXTS, ARTICLES, AND THESES

Changoni, A.S. "Seasonal Food Shortages and Agricultural Development." *Journal of the Geographical Association of Tanzania*, 8 (1973): 40-62.

Hoza, R.B. "The Factors of Change in the Evolution of an Agrarian Landscape." *Journal of the Geographical Society of Tanzania*, 10 (1974): 43-65.

Hathout, Salah A. "Some Aspects of Agricultural Planning in Tanzania." *Journal of the Geographical Society of Tanzania*, 13 (1975): 1-10.

Reineberg, Zindra J. "A Spatial Analysis of Colonial Agricultural Development in German East Africa, 1884–1914." Master's Thesis, Arizona State University, Tempe, Ariz., 1975.

Lema, A.J. "Decision Making in Agriculture: A Study of Farmers in Kibaka District, Tanzania." *Journal of the Geographical Association of Tanzania*, 25 (1986): 42-57.

Schultz, J. "Agrarwirtschaftsliche Entwicklung in Tanzania: Diskrepanz zwicken naturlichen Agrarpotential und dessen Inwertsetzung." *Geographische Rundschau*, 41, No. 11 (1989): 613-619.

Ezaza, W.P. "Towards a Strategy for Agricultural Development in the United Republic of Tanzania." *Eastern and Southern Africa Geographical Journal*, 2, No. 1 (1991): 23-36.

Birch-Thomsen, Torben, and B. Fog. "Changes Within Small-Scale Agriculture: A Case Study from Southwestern Tanzania." *Geografisk Tidsskrift*, 96 (1996): 60-69.

CROPS AND CROPPING PATTERNS

Gould, Peter. "Wheat on Kilimanjaro: The Perception of Choice Within Game and Learning Model Frameworks." *General Systems Yearbook*, 10 (1965): 157-166.

Daya, M.F. "A Micro-Study of a Banana-Coffee Format in Tengeru Arusha District, Tanzania." *Journal of the Geographical Association of Tanzania*, 2 (1968): 39-49.

Waller, C. "A Case Study of a Sisal Estate: Kingolwira and Pangawe Sisal Estates, Morogoro, Tanzania." *Journal of the Geographical Association of Tanzania*, 3 (1968): 5-24.

Adamu, E.S.S. "Small-Hold Cashew Farmers in Mtwara Region: A Case Study." *Journal of the Geographical Association of Tanzania*, 5 (1969): 71-93.

Mascarenhas, Adolfo. "Resistance and Change in the Sisal Plantation System of Tanzania." Ph.D. Dissertation, University of California, Los Angeles, Calif., 1970.

Mmbando, J.S. "The Production of Vegetables in Lushoto." *Journal of the Geographical Association of Tanzania*, 6 (1970): 5-28.

Malema, E.C. "Small Holder Wheat Cultivation in Matai Subdivision, Sumbawonga District." *Journal of the Geographical Association of Tanzania*, 9 (1973): 5-24.

Heijnen, J.D. "National Policy, Foreign Aid and Rural Development: A Case Study of LIDEP's Vegetable Component in Lushoto District (Tanzania)." *Bulletin, Social Geographie Ontwikkelingslanden*, Series II, No. 3, 1974.

Okere, John. "The Impact of the Mumias Sugar Scheme." *Journal of the Geographical Association of Tanzania*, 10 (1974): 4-23.

Lomoy, J. "Explaining Agricultural Decline: A Case Study of Castor Seeds in Dodoma Region, Tanzania." *Norsk Geografisk Tidsskrift*, 40, No. 1 (1986): 15-26.

Briggs, John A. *Agriculture and Farmers' Choice of Crops in the Peri-Urban Zone of Dar es Salaam, Tanzania.* Glasgow, Scotland: University of Glasgow, Department of Geography, Occasional Paper No. 27, 1989.

Lynch, K. "Commercial Horticulture in Rural Tanzania: An Analysis of Key Influences." *Geoforum*, 30, No. 2 (1999): 171-183.

ANIMAL HUSBANDRY

Moule, G.R. "Kangwa Ranching." *Journal of the Geographical Society of Tanzania*, 1 (1967): 327-331.

Kivikkokangon, Ritva. "Pastoralism and National Development in Tanzania, as Illustrated by Baraguyu" (in Finnish). *Terra*, 90, No. 4 (1978): 189-192.

Kjaerby, Finn. *The Development of Agro-Pastoralism Among the Borbaig in Honong District.* Dar es Salaam, Tanzania: Bureau of Resource Assessment and Land Use Planning, University of Dar es Salaam, 1979.

DAIRYING

Williamson, Helen. "Experiences with Grade Dairy Cattle in the Peasant Economy on Mount Meru, North Tanzania." *Nurnberger Wirtschafts- und Sozialgeographische Arbeiten*, 8 (1968): 372-376.

FARM LAND

Hamersley, A. "Agriculture and Land Tenure in Tanzania." Pp. 189-197 in *East Africa: Its People and Resources*, edited by W.T.W. Morgon. Nairobi, Kenya: Oxford University Press, 1969.

Pitblado, J.R. "Land Capability and Land Tenure: Problems and Prospects for Agricultural Development in the North Mkata Plain, Tanzania." Ph.D. Dissertation, University of Toronto, Toronto, Ontario, 1975.

Hathout, Salah and D. Hiebert. "Quantitative Methods in Land Evaluation for Wheat Production in West Kilimanjaro, Tanzania." *Singapore Journal of Tropical Geography*, 1, No. 2 (1980): 47-54.

Maro, Paul. "Agricultural Land Management Under Population Pressure: The Kilimanjaro Experience, Tanzania." *Mountain Research and Development*, 8, No. 4 (1988): 273-282.

Briggs, John A. "The Peri-Urban Zone of Dar es Salaam, Tanzania: Recent Trends and Change in Agricultural Land Use." *Transactions, Institute of British Geographers*, 16, No. 3 (1991): 319-331. (New Series)

Sawio, Camillus. "Feeding the Urban Masses? Towards an Understanding of the Dynamics of Urban Agriculture and Land-Use Change in Dar es Salaam, Tanzania." Ph.D. Dissertation, Clark University, Worcester, Mass., 1993.

AGRICULTURAL POPULATIONS AND SOCIAL ISSUES

Brooke, Clarke. "Types of Food Shortages in Tanzania." *Geographical Review*, 67, No. 3 (1967): 333-357.

Berry, L., et al. *Human Adjustments to Agricultural Drought in Tanzania: Pilot Investigations*. Toronto, Ontario: University of Toronto Department of Geography, Natural Hazards Research Working Paper No. 19, 1971.

Datoo, B.A. *Population Density and Agricultural Systems in the Uluguru Mountains, Marogoro District*. Dar-es-Salaam, Tanzania: Bureau of Resources Assessment and Land Use Planning, University of Dar-es-Salaam, Research Paper No. 26, 1973.

Rand, Jorgen, and Karen Rand. "Rural Organization in Bukoba District, Tanzania." *Folia Geographica Danica*, 14 (1975).

Datoo, B.A. "Relationship Between Population Density and Agricultural Systems in the Uluguru Mountains, Tanzania." *Journal of Tropical Geography*, 42 (1976): 1-12.

Ibrahim, B., and F.N. Ibrahim. "Pastoralists in Transition: A Case Study from Lenijape, Massai Steppe." *GeoJournal*, 36, No. 1 (1995): 27-48.

ENVIRONMENTAL AND CONSERVATION ISSUES

Mascarenhas, A. "Agricultural Vermin in Tanzania." Pp. 259-268 in *Studies in East African Geography and Development*, edited by S.H. Ominde. London, England: Heinemann, 1971.

Kaduma, J.D. "A Soil Survey for Irrigation at Upper Ngerengere Valley: Morogoro." *Journal of the Geographical Association of Tanzania*, 13 (1975): 87-103.

Luk, S.H. "Soil Erosion with Special Reference to Agriculture in Tanzania." *Journal of the Geographical Society of Tanzania*, 13 (1975): 64-86.

Lema, A.J. "An Assessment of Water Availability for Maize in Lower Moshi, Moshi District." *Journal of the Geographical Association of Tanzania*, 16 (1978): 108-139.

Mielke, Howard W., and Paul W. Mielke, Jr. "Termite Mounds and Chitemene Agriculture: A Statistical Analysis of Their Association in Southwestern Tanzania." *Journal of Biogeography*, 9, No. 6 (1982): 499-504.

Jones, Samantha. "Farming Systems and Nutrient Flow: A Case of Degradation?" *Geography*, 81, No. 4 (1996): 289-300.

IRRIGATION

Clore, Francis. "The Mborali Irrigation Scheme." *Journal of the Geographical Association of Tanzania*, 2 (1968): 18-34.

Lema, A.J. "Lake Victoria Waters in the Context of Irrigation Development in Tanzania." *Journal of Eastern African Research and Development*, 23 (1993): 192-209.

Adams, W.M, T. Potkanski, and J.E.G. Sutton. "Indigenous Farmer-Managed Irrigation in Sonjo, Tanzania." *Geographical Journal*, 160, No. 1 (1994): 17-32.

AGRICULTURAL MARKETING AND DISTRIBUTION

Clare, Frances. "Mbarali Irrigation Scheme." *Journal of the Geographical Association of Tanzania*, 2 (1968): 18-38.

Sporrek, A. "The Marketing of Fruits and Vegetables in Dar es Salaam" (in Swedish). *Svensk Geografisk Arbok*, 61 (1985): 203-217.

Togo

GENERAL TEXTS, ARTICLES, AND THESES

Akibode, Ayechorok K. "Modernisation de l'agriculture et modification des relations de production chez les Moba de l'Ertreme-Nord Togo." *Bulletin de la Societe Belge d'Etudes Geographiques*, 57, No. 2 (1988): 273-284.

CROPS AND CROPPING PATTERNS

Akibode, Ayechorok K. "Structures agraires et evolution des milieux ruraux: Le cas de la region cotonniere de l'extreme-nord Togo." *Bulletin de la Societe Belge d'Etudes Geographiques*, 63, No. 1 (1994): 87-112.

Akibode, Ayechorok K. "L'evolution recente du marche des produits vivriers au Togo." *Les Cahiers d'Outre Mer*, 48 (1995): 477-500.

ENVIRONMENTAL AND CONSERVATION ISSUES

Antheaume, Benoit. *Agbetiko: Terroir de la basse vallee du Mono (Sud-Togo)*. Paris, France: ORSTOM, 1978.

Uganda

GENERAL TEXTS, ARTICLES, AND THESES

Sentango, David S. "Uganda: A Geographical Study in Agriculture." Master's Thesis, Ohio State University, Columbus, Ohio, 1961.

Manshard, Walther. "Kigezi (Sudwest-Uganda): Die agrargeographische Struktur eines ostafrikanischen Berglandes." *Erdkunde*, 19 (1965): 192-210.

McMaster, D.N. "Uganda: Initiatives in Agriculture, 1888–1966." Pp. 241-260 in *Land Use and Resources: Studies in Applied Geography*, edited by

C. Embleton and J.T. Coppock. London, England: Institute of British Geographers, Special Publication No. 1, 1968.

Naughton-Treves, Lisa. "Farming the Forest Edge: Vulnerable Places and People Around Kibale National Park, Uganda." *Geographical Review*, 87, No. 1 (1997): 27-46.

CROPS AND CROPPING PATTERNS

Deshler, Walter. "Livestock Trypanosomiasis and Human Settlement in Northeast Uganda." *Geographical Review*, 50, No. 4 (1960): 541-554.

McMaster, D.N. "Subsistence Crop Geography of Uganda." Ph.D. Dissertation, University of London, London, England, 1960.

McMaster, D.N. "Speculations on the Coming of the Banana to Uganda." *Journal of Tropical Geography*, 16 (1962): 57-69.

McMaster, D.N. *A Subsistence Crop Geography of Uganda*. Bude, England: International Geographical Union, 1962.

O'Connor, A.M. "The Geography of Tea and Sugar Production in Uganda: Some Comparisons and Contrasts." *East African Geographical Review*, 3 (1965): 27-36.

McMaster, D.N. "Changes in the Extent and Distribution of Cultivation in Uganda, 1952–1958." *Uganda Journal*, 30, No. 1 (1966): 63-74.

Baker, P. Randall. "The Distribution of Cattle in Uganda." *East African Geographical Review*, 6 (1968): 63-74.

Baker, P. Randall. "Cattle, Cash and Condemnation: A Study of the Supply of Slaughter Cattle to the Kampala-Mukono Area." Pp. 139-150 in *Geography Papers: University of East Africa, Makerere Institute of Social Research*. Kampala, Uganda: Social Science Council Conference, 1969.

MacKenzie, M.K. "The Geography of Tobacco Production in West Nile District, Uganda." *East African Geographical Review*, 8 (1970): 31-38.

Smith, J.A. "The Production of Sugar Cane by Outgrowers in East Mengo District, Uganda." *East African Geographical Review*, 8 (1970): 47-54.

Harkema, R.C. "The Production of Tobacco in Uganda." *East African Geographical Review*, 10 (1972): 51-64.

Hyde, R.J., and B.W. Langlands. *Patterns of Food Crop Production and Nutrition in Uganda*. Kampala, Uganda: Makerere University, Department of Geography, Occasional Paper No. 58, 1974.

Hyde, R.J. *The Analysis of Crop Distribution in Uganda*. Kampala, Uganda: Makerere University, Department of Geography, Occasional Paper No. 65, 1975.

DAIRYING

Baker, P. Randall. "Stages in the Development of a Dairy Industry in Bunyaro, Western Uganda." *Transactions, Institute of British Geographers*, 53 (1971):
43-54.

FARM LAND

Anderson, G.W. "Agriculture and Land Tenure in Uganda." Pp. 199-207 in *East Africa: Its Peoples and Resources*, edited by William T. Morgon. Nairobi, Kenya: Oxford University Press, 1969.

Grenzeback, Klaus. "Methods of Regional Land-Use Analysis in South-West Uganda: A Precondition for Agricultural Regional Planning." *Applied Geography and Development*, 16 (1980): 28-52.

Mwaka, Victoria M. "The Agricultural Utility of Wetlands of the Lake Kyoga Basin." *Eastern and Southern Africa Geographical Journal*, 1, No. 1 (1990): 10-25.

Hunter, S.S., E. Bulirwa, and E. Kisseka. "AIDS and Agricultural Production: Report of a Land Utilization Survey, Masaka and Rakai Districts of Uganda." *Land Use Policy*, 10, No. 3 (1993): 241-258.

AGRICULTURAL POPULATIONS AND SOCIAL ISSUES

Mwaka, Victoria M. "Agricultural Production and Women's Time Budgets in Uganda." Pp. 46-51 in *Different Places, Different Voices: Gender and Development in Africa, Asia, and Latin America*, edited by Janet Momsen and Vivian Kinnaird. London, England: Routledge, 1993.

ENVIRONMENTAL AND CONSERVATION ISSUES

Hanna, Lewis W. "Climate and Crop-Potential in Uganda." Pp. 99-112 in *Studies in East African Geography and Development*, edited by S.H. Ominde. Berkeley, Calif.: University of California Press, 1971.

Hanna, Lewis W. "Climatic Influences on Yields of Sugarcane in Uganda." *Transactions, Institute of British Geographers*, 52 (1971): 41-59.

Drennon, Christine. "Agricultural Encroachment in Two of Uganda's Forest and Game Reserves." Master's Thesis, University of Wisconsin, Madison, Wis., 1990.

IRRIGATION

Carruthers, I.D. "Irrigation Development in Uganda: Past Experiences and Future Prospects." *East African Geographical Review*, 8 (1970): 11-22.

Western Sahara

ANIMAL HUSBANDRY

Arkell, T. "The Decline of Pastoral Nomadism in the Western Sahara." *Geography*, 76, No. 2 (1991): 162-166.

Zaire

GENERAL TEXTS, ARTICLES, AND THESES

Gassiat, Anne. "Enclavement et developpement rural: Le cas du project contonnier de la zone de Bambesa (Haut-Zaire)." *Les Cahiers d'Outre Mer*, 44 (1991): 243-258.

ANIMAL HUSBANDRY

Nzeka Kabu Zex-Kongo, Jean-Pierre. "Le probleme de l'approvisionne de Kinshasa en viande bovine locale." *Les Cahiers d'Outre Mer*, 52 (1999): 169-196.

ENVIRONMENTAL AND CONSERVATION ISSUES

Mpasi, M. "Connaissances empiriques des mecanismes meteorologique, rites agraires et techniques agricoles au Bas-Zaire." *Revue Belge de Geographie*, 119, No. 1-2 (1995): 85-94.

AGRICULTURAL MARKETING AND DISTRIBUTION

Kayser, B., Abdel Kader Fahein, and Marc Pain. "Cherte du manioc et pauvrete paysanne dans le Bas-Zaire." *Les Cahiers d'Outre Mer*, 34 (1981): 97-110.

Zambia

GENERAL TEXTS, ARTICLES, AND THESES

Kay, G. "Agricultural Progress in Zambia." Pp. 495-524 in *Environment and Land Use in Africa*, edited by M.F. Thomas and G.W. Whittington. London, England: Methuen, 1969.

Martin, P. "An Example of Co-operative Mixed Farming in the Zambian Copperbelt." *Geography*, 59, No. 3 (1974): 264-266.

Woods, J.P. "The History of a Commercial Farm in the Periurban Fringe of Lusaka: Waterfall Farm." *Zambian Geographical Journal*, 31-32 (1976–1977): 53-66.

Woods, J.P., Ian Condliffe, and Adrian P. Wood. "Mukushi Farm Block: The Development of a Commercial Farming Area in Zambia." *Zambian Geographical Journal*, 33-34 (1978–1979): 1-16.

Smith, W. "Crisis and Response: Agricultural Development and Zambia's Third National Development Plan, 1979–1983." *Geography*, 66, No. 2 (1981): 134-136.

Wood, Adrian P., and W. Smith. "Zambia Up for Grabs? The Regionalization of Agricultural Aid." *Area*, 16, No. 1 (1984): 3-8.

Scott, Earl P. "Development Through Self-Reliance in Zambia." *Journal of Geography*, 84, No. 6 (1985): 282-290.

Stocking, Michael. "Agricultural Development for the Small Farmer: An Evaluation of the Lima Programme in Zambia." *Zambian Geographical Journal*, 35 (1985): 45-54.

Chanda, R. "The Usefulness of Principal Components Analysis in Development Planning and Geographic Research with an Example Based on Zambian Agricultural Data." *Zambian Geographical Journal*, 36 (1986): 49-56.

Stromgaard, Peter. "Traditional Agriculture in Zambia." *Geography Review*, 5, No. 3 (1992): 18-20.

Mulerga, Mark C. "Peri-Urban Agriculture: A Case of Small-Scale Peasant Cultivation in and Around Zambia Towns and Cities, with Special Reference to Lusaka." *Eastern and Southern Africa Geographical Journal*, 6, No. 1 (1995): 1-16.

CROPS AND CROPPING PATTERNS

Whittington, G.W. "Tobacco Production in the Eastern Province of Zambia." *Erdkunde*, 21 (1967): 297-309.

FARM LAND

Priestley, J.R. "The Transfer of Agricultural Land in the Eastern Province of Zambia." *Journal of Tropical Geography*, 46 (1978): 47-60.

Lado, Cleophas. "Sustainable Agricultural Land Use and Farmers' Participation in Agroforestry Technology in Zambia." *Eastern and Southern Africa Geographical Journal*, 5, No. 1 (1994): 18-27.

AGRICULTURAL POPULATIONS AND SOCIAL ISSUES

Stromgaard, Peter. "Peasant Household Economy in Rural Zambia: The Dilemma of Small-Scale Farmers in Transition." *Fennia*, 168, No. 2 (1990): 201-210.

ENVIRONMENTAL AND CONSERVATION ISSUES

Pullan, R.A. "Biogeographical Studies and Agricultural Development in Zambia." *Geography*, 59, No. 4 (1974): 309-321.

Salter, Leo F. "A Study of the Run-Off of Fertilizers from Agricultural Land on the Kafue Flats, Zambia." *Zambian Geographical Journal*, 33-34 (1978–1979): 95-104.

Stocking, Michael. "Farming and Environmental Degradation in Zambia: The Human Dimension." *Applied Geography*, 3, No. 1 (1983): 63-77.

Chidumago, E.N. "A Study of the Impact of Agricultural Development on the Natural Resources of Chipangali, Eastern Zambia." *Zambian Geographical Journal*, 35 (1985): 31-44.

Chipanshi, A.C. "Analysis of Rainfall Probabilities to Determine Maize Species Suitability: An Agro-Meteorological Study of Zambia." *Singapore Journal of Tropical Geography*, 10, No. 2 (1989): 110-118.

Johnson, P.C. "Ecology and Change in the Agricultural System of the Kaonde of Northwestern Zambia." *Singapore Journal of Tropical Geography*, 15, No. 1 (1994): 1-16.

Sumida, H., H. Tanahashi, and K. Hanzawa. "Soil Fertility of Dombo Area in Zambia." *Science Reports, Tohoku University, Seventh Series, Geography*, 44, No. 1 (1994): 65-70.

Chipanshi, A.C. "Relationships of Non-Rainfall Events and Total Rainfall to Maize Yields at Selected Climatological Stations in Zambia." *South African Geographical Journal*, 82, No. 2 (2000): 117-121.

Zimbabwe/Rhodesia

GENERAL TEXTS, ARTICLES, AND THESES

Richards, Audrey. "A Changing Pattern of Agriculture in East Africa: The Bemba of Northern Rhodesia." *Geographical Journal*, 124, No. 3 (1958): 302-314.

Gwarada, Nason. "Historical Development and Future Aspects of Agriculture in Zimbabwe." Master's Thesis, University of Washington, Seattle, Wash., 1979.

Zinyama, L.M. "Agricultural Development Policies in the African Farming Areas of Zimbabwe." *Geography*, 71, No. 2 (1986): 105-115.

Zinyama, L.M. "Commercialization of Small-Scale Agriculture in Zimbabwe: Some Emerging Patterns of Spatial Differentiation." *Singapore Journal of Tropical Geography*, 9, No. 2 (1988): 151-162.

Weiner, Daniel. "Socialist Transition in the Capitalist Periphery: A Case Study of Agriculture in Zimbabwe." *Political Geography Quarterly*, 10, No. 1 (1991): 54-75.

Weiner, Daniel, Barry Munslow, and S. Mogo. "Energy for Sustainable Agricultural Development in Zimbabwe." *Growth and Change*, 23, No. 3 (1992): 335-381.

CROPS AND CROPPING PATTERNS

Locke, R. "Cotton in Rhodesia." *Geographical Association of Rhodesia Proceedings*, 4 (1971): 42-45.

Smith, Jose. "Zimbabwe's Horticultural Export Trade." *Geography*, 75, No. 2 (1990): 160-162.

ANIMAL HUSBANDRY

Nel, E., et al. "Rural Self-Reliance in Bordolfi, Zimbabwe: The Role of Beekeeping." *Geographical Journal*, 166, No. 1 (2000): 26-34.

FARM LAND

Hellen, Anthony. "A Note on Development of the Kafue Flats in Northern Rhodesia with Particular Reference to Older Methods in Agriculture." *Erdkunde*, 18 (1964): 149-157.

Hamilton, P. "The Changing Pattern of African Land Use in Rhodesia." Pp. 247-271 in *Essays in Geography for Austin Miller*, edited by J.B. Whittaw and P.D. Wood. Reading, England: University of Reading Department of Geography, 1965.

M'Tukudzi, Bonet. "The Relation of Land Tenure to Change and Development in the System of Subsistence Agriculture in the Umzingwane District of Rhodesia." Master's Thesis, University of Wisconsin, Madison, Wis., 1974.

Rambanapasi, C.O. "Agricultural Land-Use Policies and Rural Spatial Differentiation in Colonial and Postcolonial Zimbabwe." *Environment and Planning C*, 9, No. 2 (1991): 225-236.

AGRICULTURAL POPULATIONS AND SOCIAL ISSUES

Drakakis-Smith, David W., T. Bowyer-Bower, and D. Tevera. "Urban Poverty and Urban Agriculture: An Overview of the Linkage in Harare." *Habitat International*, 19, No. 2 (1995): 183-193.

Mudimu, Godfrey D. "Urban Agricultural Activities and Women's Strategies in Sustaining Family Livelihoods in Harare, Zimbabwe." *Singapore Journal of Tropical Geography*, 17, No. 2 (1996): 179-194.

ENIVRONMENTAL AND CONSERVATION ISSUES

Vogel, H. "Effects of Conservation Tillage on Sheet Erosion from Sandy Soils at Two Experimental Sites in Zimbabwe." *Applied Geography*, 12, No. 3 (1992): 229-242.

Weiner, Daniel, and Marilyn Murphy. "Differential Vulnerability to Severe Agricultural Drought in Zimbabwe." *Eastern and Southern Africa Geographical Journal*, 3, No. 1 (1992): 13-33.

IRRIGATION

Kuwana, Samuel M. "Sabi-limpopo Irrigation Development Scheme of Rhodesia." Master's Thesis, Indiana University, Bloomington, Ind., 1967.

Williams, Owen. "Irrigation Farming in the Southeast Lowveldt of Zimbabwe: Retrospect and Prospect." *Geography*, 66, No. 3 (1981): 228-231.

AGRICULTURAL MARKETING AND DISTRIBUTION

Smith, J.A. "The Transport and Marketing of Horticultural Crops by Communal Farmers into Harare." *Geographical Journal of Zimbabwe*, 20 (1989): 1-13.

AGRICULTURAL LABOR STUDIES

Zinyama, L.M. "Rural Household Structure, Absenteeism, and Agricultural Labour: A Case Study of Two Subsistence Farming Areas in Zimbabwe." *Singapore Journal of Tropical Geography*, 7, No. 2 (1986): 163-173.

Joseph, Tarek. "Cyclical Labor Migration and Peasant Agriculture in Zimbabwe." Master's Thesis, West Virginia University, Morgantown, W.Va., 1991.

Part 8

ASIA

Asia as a Whole

ATLASES AND GRAPHIC PRESENTATIONS

Huke, Robert E. *Agroclimate and Dry-Season Maps of South, Southeast and East Asia.* Los Banos, Philippines: International Rice Research Institute, 1982.

GENERAL TEXTS, ARTICLES, AND THESES

Noble, William A. "The Earliest Agriculture: Some Theoretical Speculations." *Asian Profile*, 4, No. 1 (1976): 73-80.

Sum, Kongsut. "Japanese and Chinese Agriculture: Comparison and Appraisal" (in Japanese). *Science Reports, Tohoku University, Seventh Series, Geography*, 28, No. 2 (1978): 379-385.

Dhillon, S.S. "The Origins of Agriculture." *Asian Profile*, 9, No. 4 (1981): 371-382.

Yeung, Yueman. *Urban Agriculture in Asia: A Substantive and Policy Review.* Hong Kong, China: Chinese University of Hong Kong, Department of Geography Occasional Paper No. 81, 1985.

CROPS AND CROPPING PATTERNS

Bassin, Mark. "Cotton Cultivation in Central Asia, 1880–1940, as Part of an Economic and Social Network." Unpublished Master's Thesis, University of Toronto, Toronto, Ontario, 1977.

Huke, Robert E. *Rice Area by Type of Culture: South, Southeast and East Asia.* Los Banos, Philippines: International Rice Research Institute, 1982.

Hossain, M., and Klaus Fischer. "Rice Research for Food Security and Sustainable Agricultural Development in Asia: Achievements and Future Challenges." *GeoJournal*, 35, No. 3 (1995): 286-298.

Zeigler, R.S., and D.W. Puckridge. "Improving Sustainable Productivity in Rice-Based Rainfed Lowland Systems of South and Southeast Asia." *GeoJournal*, 35, No. 3 (1995): 307-324.

Smil, Vaclav. "Food, Energy, and the Environment: Implications for Asia's Rice Agriculture." Pp. 321-331 in *Sustainability of Rice in the Global Food System*, edited by N.G. Dawling, S.M. Greenfield, and Klaus Fischer. Los Banos, Philippines: International Rice Research Institute, 1998.

FARM LAND

Takaya, Yoshikazu, and Y. Itoh. "Vertical Zoning of Monsoon Asia by Agricultural Land Use" (in Japanese). *Southeast Asian Studies, Kyoto*, 14, No. 1 (1976): 99-122.

Rostotsky, S.B. "Use of Land Resources and the Way to Increase Production of Food Stuffs in the Delta Regions of Southern and South-East Asia" (in Russian). *Izvestiya Akademii Nauk SSSR, Seriya Geograficheskaya*, 5 (1978): 63-71.
Pannell, Clifton W. "Farmland and Agricultural Development in India and China." *Rural Systems*, 4, No. 1 (1986): 1-18.
Yeung, Yue-man. "Agricultural Land Use in Asian Cities." *Land Use Policy*, 5, No. 1 (1988): 79-82.

DAIRYING

Simoons, Frederick J. "Dairying, Milk Use and Lactose Malabsorption in Eurasia: A Problem in Culture History." *Anthropos*, 74 (1979): 61-80.

AGRICULTURAL POPULATIONS AND SOCIAL ISSUES

Farmer, Bertram H. *Agricultural Colonization in South and South-East Asia*. Hull, England: Hull University Press, 1969.

ENVIRONMENTAL AND CONSERVATION ISSUES

Sirinanda, K. "A Survey of the Variability of Water Supply as a Physical Constraint on the Optimization of Paddy Production in South and Southeast Asia." *Geographica*, 12 (1978): 45-55.
Richards, John F. "Agricultural Impacts in Tropical Wetlands: Rice Paddies for Mangroves in South and Southeast Asia." Pp. 217-233 in *Wetlands: A Threatened Landscape*, edited by Michael Williams. London, England: Basil Blackwell, 1990.
Woodhouse, P. "Farming a Wetland Ecosystem: Rice Cultivation in Asia." Pp. 85-112 in *Environment, Population and Development*, edited by P. Sarre. London, England: Hodder and Stoughton, 1991.
Blumler, M.A. "Ecology, Evolutionary Theory, and Agricultural Origins." Pp. 25-50 in *The Origins and Spread of Agriculture and Pastoralism in Eurasia*, edited by David R. Harris. London, England: University College London Press, 1996.

South Asia as a Whole

CROPS AND CROPPING PATTERNS

Kingsbury, Robert C. "The Jute Industry of India and Pakistan, with Emphasis on the Changes Resulting from the Partition of the Indian Subcontinent." Master's Thesis, University of Kansas, Lawrence, Kans., 1951.
Rathjens, Carl. *Vergleichende Kultureographie der Hochgebirge des sudlichen Asien*. Wiesbaden, Germany: F. Steiner Verlag, 1973.

Farmer, Bertram H., ed. *Green Revolution? Technology and Change in Rice-Growing Areas of Tamil Nadu and Sri Lanka.* Boulder, Colo.: Westview, 1977.
Allen, Nigel J.R. "Men and Crops in the Central Hindu Kush." Ph.D. Dissertation, Syracuse University, Syracuse, N.Y., 1978.
Farmer, Bertram H. "Perspectives on the Green Revolution in South Asia." *Modern Asian Studies,* 20 (1986): 175-199.
Huke, Robert E. *Human Geography of Rice in South Asia.* Manila, Philippines: International Rice Research Institute, 1987.

ANIMAL HUSBANDRY

Villegas, Valente. "Zebu Cattle of India and West Pakistan." *Philippine Geographical Journal,* 17, No. 2 (1973): 64-73.

DAIRYING

Simoons, Frederick J. "The Traditional Limits of Milking and Milk Use in Southern Asia." *Anthropos,* 65 (1970): 547-593.

FARM LAND

Pulyarkin, V.A. "New Trends in Land Utilization in the Developing Countries of South Asia" (in Russian). *Voprosy Geografii,* 99 (1975): 174-190.

ENVIRONMENTAL AND CONSERVATION ISSUES

Farmer, Bertram H. "The Green Revolution in South Asian Rice Fields: Environment and Production." *Journal of Development Studies,* 15 (1979): 304-319.
Farmer, Bertram H. "The Green Revolution in South Asia." *Geography,* 66, No. 3 (1981): 202-207.
Baylis-Smith, Timothy, and S. Wanmali, eds. *Understanding Green Revolutions: Agrarian Change and Development Planning in South Asia.* Cambridge, England: Cambridge University Press, 1984.
Giambelluca, T.W., D. Nullet, and M.A. Nullet. "Agricultural Drought in South-Central Asia." *Professional Geographer,* 40, No. 4 (1988): 404-415.

Bangladesh/East Pakistan

ATLASES AND GRAPHIC PRESENTATIONS

Huke, Eleanor H., Robert E. Huke, and T. Woodhead. *Rice-Wheat Atlas of Bangladesh.* Manila, Philippines: International Rice Research Institute, 1993.

GENERAL TEXTS, ARTICLES, AND THESES

Khan, Fazle K. "Comparative Agricultural Geography of East Pakistan and Louisiana." Ph.D. Dissertation, Louisiana State University, Baton Rouge, La., 1958.

Khan, Fazle K. "Agricultural Origins in East Pakistan and Adjoining Areas." *Oriental Geographer*, 4, No. 1 (1960): 47-56.

Shajaat Ali, Abu M. "Part-Time Farming in Bangladesh." *Labor, Capital and Society*, 13, No. 2 (1980): 23-38.

Shajaat Ali, Abu M. "Urban Expansion and Agricultural Change: A Comparative Analysis of Metropolitan Detroit, U.S.A. and Metropolitan Dakka, Bangladesh." Master's Thesis, Wayne State University, Detroit, Mich., 1983.

Ahmed, Mushtaq. *Bangladesh Agriculture: Towards Self Sufficiency*. Dhaka, Bangladesh: External Publicity Wing, Ministry of Information-Bangladesh, 1988.

Shajaat Ali, Abu M. "Part-Time Farming in Bangladesh: A Village Case Study of Its Causes." *Contemporary South Asia*, 4, No. 3 (1995): 309-330.

Mayeed, Maliha, and Nusha Y. Choudhury. "Agriculture in Urban Context: A Case Study of Dhaka City." *Oriental Geographer*, 40, No. 1-2 (1996): 22-35.

Turner, Billie L., II, and Abu M. Shajaat Ali. "Induced Intensification: Agricultural Change in Bangladesh with Implications for Malthus and Boserup." *Proceedings of the National Academy of Science*, 93 (1996): 14984-14991.

CROPS AND CROPPING PATTERNS

Ahad, Nafis, and Fazle K. Khan. "Betel Vine Cultivation in East Pakistan." *Oriental Geographer*, 2, No. 1 (1958): 1-10.

Johnson, B.L.C. "Crop-Association Regions in East Pakistan." *Geography*, 43, No. 2 (1958): 86-103.

Islam, Mohammed A. "Crop Combination Regions in East Pakistan." Ph.D. Dissertation, Clark University, Worcester, Mass., 1964.

Sopher, David E. "The Swidden/Wet-Rice Transition Zone in the Chittagong Hills." *Annals of the Association of American Geographers*, 54, No. 1 (1964): 107-126.

Islam, Mohammed A. "Crop Combination Regions in East Pakistan." *Oriental Geographer*, 9 (1965): 1-16.

Islam, Mohammed A. "Changes in Cropping Patterns in East Pakistan." *Oriental Geographer*, 12, No. 1 (1968): 1-23.

Khan, Fazle K., and Amarendra L. Khisha. "Shifting Cultivation in East Pakistan." *Oriental Geographer*, 14, No. 2 (1970): 22-43.

Huq, Lutful. "Sericulture in Bangladesh." *Oriental Geographer*, 16, No. 2 (1972): 71-90.

Huq, Lutful. "Low Rice Productivity in Bangladesh." Master's Thesis, Carleton University, Ottawa, Ontario, 1977.

Paul, Bimal. "Adoption Period of High Yielding Varieties of Rice Among the Farmers of Bangladesh." *Oriental Geographer*, 21, No. 1-2 (1977): 44-51.

Shajaat Ali, Abu M. "Intensive Paddy Agriculture in Shyampur, Bangladesh." Pp. 276-306 in *Comparative Farming Systems*, edited by Billie L. Turner II and Stephen B. Brush. New York, N.Y.: Guilford, 1987.

Ahmed, Rafique. "Changes in Rice Productivity in Bangladesh from 1974–1975 Through 1988–1989." *Indian Geographical Journal*, 67, No. 2 (1992): 17-28.

Mahmood, Rezaul, and John T. Hayes. "A Model-Based Assessment of Impacts of Climate Change on Boro Rice Yield in Bangladesh." *Physical Geography*, 16, No. 6 (1995): 463-486.

ANIMAL HUSBANDRY

Schmidt, Oscar. "Daulatjanj: The Major Egg-Exporting Center of East Pakistan." *Professional Geographer*, 15 (1963): 5-7.

FARM LAND

Uddin, Borban. "Cropland Productivity and Its Spatial Patterns in Bangladesh." *Oriental Geographer*, 25, No. 1-2 (1981): 22-36; 37-53.

Shajaat Ali, Abu M. "Agricultural Land in an Urbanizing Environment: A Case Study in the Dhaka Metropolitan Area." *Rural Systems*, 1, No. 3 (1983): 213-224.

Ahson, Rosie M., and Ekramul Ahson. "Socio-Economic Considerations of Land Use at Farm Level." *Oriental Geographer*, 29-30 (1985–1986): 5-24.

Dayal, Edison. "Land and Labor Productivity in Bangladesh Agriculture." *Asian Profile*, 17, No. 4 (1989): 337-350.

AGRICULTURAL POPULATIONS AND SOCIAL ISSUES

Patel, Ahmad M. "Population, Food and Agriculture in East Pakistan." *Pakistan Geographical Review*, 23, No. 2 (1968): 61-77.

Islam, Mohammed A. "The Future Relationshps Between Rural Population and Agriculture in East Pakistan." *Pakistan Geographical Review*, 21, No. 2 (1968): 39-55.

Noble, William A., Ashok K. Dutt, and Abu Rahman. "Food Production and Population Growth in Bangladesh." *Asian Profile*, 8, No. 1 (1980): 52-77.

Paul, Bimal K. "Perception of Agricultural Adjustments to Floods in Jamuna Floodplain, Bangladesh." *Human Ecology*, 12, No. 1 (1984): 3-20.

Baqee, Abdul. "Violence and Agricultural Seasonality in Char-Lands of Bangladesh." *Oriental Geographer*, 29-30 (1985–1986): 25-36.

Elahi, M.K. "Farm Size, Women and Fertility Differential in Rural Bangladesh." *Oriental Geographer*, 35, No. 1-2 (1991): 32-42.

Paul, Bimal K. *Farmers' and Public Responses to the 1994–1995 Drought in Bangladesh: A Case Study*. Boulder, Colo.: University of Colorado, Natural Hazards Research and Applications Information Center, QR76, 1995.

Paul, Bimal K. "Farmers' Responses to the Flood Action Plan (FAP) of Bangladesh: An Empirical Study." *World Development*, 23, No. 2 (1995): 299-310.

ENVIRONMENTAL AND CONSERVATION ISSUES

Huq, Obaedul. "Dry Season Fallow in East Pakistan." *Oriental Geographer*, 10, No. 2 (1966): 92-102.

Islam, Mohammed A. "Agricultural Adjustments to Flooding in Bangladesh: A Preliminary Report." *National Geographical Journal of India*, 26, 1-2 (1980): 50-59.

Dara Shamsuddin S., and M. Shamsul Alam. "Monsoon Seasonality and Aspect of Hydrological Regime with Reference to Agriculture in Bangladesh." *Malaysian Journal of Tropical Geography*, 21, No. 1 (1990): 25-34.

Paul, Bimal K., and Harun Rasid. "Geographical Aspects of Food Crop Damage in Bangladesh Due to River Flooding." *Asian Affairs*, 13, No. 2 (1991): 33-55.

Lein, H. "Floods and Agricultural Changes: Some Observations from Bangladesh, 1986–1990." *Norsk Geografisk Tidsskrift*, 47, No. 4 (1993): 211-228.

Paul, Bimal K., and Harun Rasid. "Flood Damage to Rice Crop in Bangladesh." *Geographical Review*, 83, No. 2 (1993): 150-159.

Rasid, Harun. "Flood Regulation by Compartmentalization: Is It Compatible with Indigenous Agricultural Adjustments to Floods in Bangladesh?" Pp. 147-161 in *Proceedings, International Conference on Floods in Bangladesh: An Interdisciplinary Analysis of Alternative Solution Strategies*, edited by M.M. Ali. Urbana, Ill.: University of Illinois, Center for International Business Education and Research, Working Paper No. 93-007, 1993.

Rasid, Harun, and Azim Mallik. "Flood Adaptation in Bangladesh: Is the Compartmentalization Scheme Compatible with Indigenous Adjustments of Rice Cropping to Flood Regions?" *Applied Geography*, 15, No. 1 (1995): 3-17.

Choudhury, Nusha Y. "Impact of Poldering on Agriculture and Urbanization: DND Scenario." *Oriental Geographer*, 43, No. 1 (1999): 41-50.

Gafur, Abdul, et al. "Changes in Soil Nutrient Content Under Shifting Cultivation in the Chittagong Hill Tracts of Bangladesh." *Geografisk Tidsskrift*, 100 (2000): 37-46.

AGRICULTURAL LABOR STUDIES

Dayal, Edison. "Sources of Labor Productivity in Bangladesh Agriculture." *GeoJournal*, 20, No. 3 (1990): 293-300.

Shajaat Ali, Abu M. "Unemployment in Agriculture and Opportunities for and Contributions of Off-Farm Employment to Rural Economy: A Case Study from Southwestern Bangladesh." *Human Ecology*, 21, No. 4 (1993): 431-445.

Shajaat Ali, Abu M. "Toward an Ecological Explanation of Agricultural Unemployment in Bangladesh." *Professional Geographer*, 50, No. 2 (1998): 176-191.

India

ATLASES AND GRAPHIC PRESENTATIONS

Jones, Wellington D. "An Isopleth Map of Land Under Crops in India." *Geographical Review*, 19 (1929): 495-496.

Sen, Amal K. "Crop Yield Map of Rajasthan." *Deccan Geographer*, 6, No. 1 (1966): 77-88.

Tiwari, P.S. *Agricultural Atlas of Uttar Pradesh*. Pantnagar, India: B.B. Pant University of Agriculture and Technology, 1973.

Singh, Jasbir. *An Agricultural Atlas of India: A Geographical Analysis*. Kurukshetra, India: Vishal Publications, 1974.

Agro-Climatic Atlas of India. New Delhi, India: Indian Meteorological Department, 1978.

Allan, J.A. "Mapping Agricultural Resources by Remote Sensing Techniques: A North Indian Example." *Remote Sensing of the Electromagnetic Spectrum*, 5, No. 2 (1978): 40-51.

Allan, J.A. "Mapping Agricultural Resources by Remote Sensing Techniques: A North Indian Example." Pp. 353-364 in *Perspectives in Agricultural Geography*, vol. 1, edited by Noor Mohammad. New Delhi, India: Concept Publishing Company, 1981.

Ghosh, Sanjib K. "Quantitative Aspects of Remote Sensing: A Tool for Agricultural Mapping." Pp. 365-380 in *Perspectives in Agricultural Geography,* vol. 1, edited by Noor Mohammad. New Delhi, India: Concept Publishing Company, 1981.

Hehar-Homji, V.M. "Multi-Disciplinary Land Use Maps: Gaussen's Approach and Its Application to India." Pp. 381-388 in *Perspectives in Agricultural Geography*, vol. 1, edited by Noor Mohammad. New Delhi, India: Concept Publishing Company, 1981.

Woodland, T., Robert E. Huke, Eleanor H. Huke, and L. Balababa. *Rice-Wheat Atlas of India*. Manila, Philippines: International Rice Research Institute, 1994.

Nandi, P. "Remote Sensing: A Great Potential in Observing Agricultural Activies." *Geographical Perspective*, 10, No. 2 (1996): 5-9.

GENERAL TEXTS, ARTICLES, AND THESES

Simkins, Ethel. "The Agricultural Geography of the Deccan Plateau of India." *Geographical Teacher*, Supplement No. 2. London, England: George Philip, 1927.

Van Valkenburg, Samuel. "Agricultural Regions of Asia, Part 3." *Economic Geography*, 9 (1933): 1-18; Part 4, 9 (1933): 109-135; Part 5, 10 (1934): 14-34.

Bergsmark, D.R. "Agricultural Production in the Major Regions of India." Pp. 188-225 in *Economic Geography of Asia*. New York, N.Y.: Prentice-Hall, 1940.

Ghosh, Suprakash. "Geography in the Food and Agricultural Organization and Regional Planning in India." *Professional Geographer*, 5 (1947): 5-8.

Dayal, P. "The Agricultural Regions of Bihar." *Indian Geographical Journal*, 25, No. 3-4 (1950): 14-26.

Kuriyon, G. "The Food Problem in India." *Indian Geographical Journal*, 26 (1951): 203-216.

Spate, O.H.K. "The Measurement of Agricultural Efficiency with Special Reference to India." *Indian Geographical Journal*, 26 (1951): 117-180.

Kulkarni, L.Y. "Agricultural Patterns of Dharwar District." *Indian Geographical Journal*, 30, No. 3 (1955): 88-94.

Kuriyon, G. "Agricultural Planning in India." *Indian Geographical Journal*, 31, No. 1-2 (1956): 26-38.

Sen, Probhat K. "Agriculture in the Andamans." *Geographical Review of India*, 19, No. 4 (1957): 48-58.

Singh, R.L., and S. Pannu. "The Soryu-Par Plain: A Study in Agro-Industrial Relationships." *National Geographical Journal of India*, 3, No. 3-4 (1957): 109-116.

Bhatia, Shyam S. "Analysis of the Regional Agricultural Patterns of India." Ph.D. Dissertation, University of Kansas, Lawrence, Kans., 1958.

Shafi, M. "Measurement of Agricultural Efficiency in Uttar Pradesh." *Economic Geography*, 36, No. 3 (1960): 296-305.

Pande, Y.D. "Agriculture in Nainital Tarai and Bhabar." *Geographical Review of India*, 23, No. 2 (1961): 19-39.

Chaterjee, S.P. "Planning for Agricultural Development in India (With Particular Reference to Eastern India)." *National Geographer*, 5 (1962): 48-56.

Kanshie, S.D. "Agricultural in the Himalayan Ganga Basin." *National Geographical Journal of India*, 8, No. 3-4 (1962): 270-283.

Kayastha, S.L. "Suratgarh Farm: A Geographical Appraisal." *National Geographical Journal of India*, 8, No. 2 (1962): 136-144.

Mukerji, A.B. "Agricultural Regions and Geographical Planning for Indian Agriculture." *National Geographer*, 5 (1962): 40-47.

Bajwa, Rajinder S. "Agricultural Changes in the Sutlej Basin Since Independence." Master's Thesis, Ohio State University, Columbus, Ohio, 1963.

Chaturvedi, B.N., and K.S. Reddy. "Regional Setting and Changing Agricultural Practices in Hyderbad District." *Deccan Geographer*, 1, No. 2 (1963): 141-167.

Misra, S.D. "Agricultural Geography of Himachal Pradesh." *Oriental Geographer*, 7, No. 1 (1963): 46-58.

Devi, M.N.V. "Some Aspects of the Agricultural Geography of South India." *Indian Geographical Journal*, 39, No. 1-2 (1964): 1-41.

Devi, M.N.V. "Some Aspects of the Agricultural Geography of South India." *Indian Geographical Journal*, 39, No. 3-4 (1964): 59-122.

Gopalkrishnan, M.D., and T.R. Rao. "Regional Variations in Agricultural Productivity in Andhra Pradesh." *Indian Journal of Agricultural Economics*, 19, No. 1 (1964): 227-236.

Sopre, S.G., and V.D. Deshpande. "Inter-District Variations in Agricultural Efficiency in Maharashtra." *Indian Journal of Agricultural Economics*, 19, No. 1 (1964): 242-252.

Anderson, A.G. "Problems and Prospects for Indian Agriculture." *New Zealand Geographical Society Record*, 39 (1965): 11-15.

Kuriyan, G. "Food Problems in India: A Continuing Crisis." *Indian Geographical Journal*, 40, No. 3-4 (1965): 51-74.

Thanbad, S.B. "Spatial and Temporal Variation in Agricultural Productivity in Mysore State." *Indian Journal of Agricultural Economics*, 20, No. 4 (1965): 39-45.

Ahmad, Zamir. "The Measurement of Productivity and Farming Efficiency in Village of Manupura." *Geographical Observer*, 2 (1966): 68-82.

Bhatia, Shyam S. "A New Measure of Agricultural Efficiency in Uttar Pradesh, India." *Economic Geography*, 43, No. 3 (1967): 244-260.

Chandra, Gopal. "Agricultural Zones in the Ghar (Cis-Siwalik) Tract of the District of Saharandpur." *Geographical Observer*, 3 (1967): 21-28.

Mukerjee, Bibha, J.N. Singh, and K.L. Mukherjee. "Changes in Agricultural Landscape in Varanasi District During the Past Fifty Years." *National Geographical Journal of India*, 13, No. 4 (1967): 187-193.

Mukerjee, Sudershan. "Agricultural Production in Vidarbha." *Deccan Geographer*, 5, No. 1-2 (1967): 86-97.

Noble, William A. "Agricultural Classification, Extreme Southern India." *Professional Geographer*, 19, No. 5 (1967): 247-250.

Batt, Dattatraya R. "Agricultural Patterns in the Konkan Karnatak." Master's Thesis, Wayne State University, Detroit, Mich., 1968.

Heston, Alan. "Variations in Agricultural Growth and Output Between and Within Regions of India." *Asian Survey*, 8, No. 3 (1968): 174-187.

Whyte, R.O. *Land, Livestock and Human Nutrition in India*. New York, N.Y.: Praeger, 1968.

Banerjee, Bireswar, ed. *Essays on Agricultural Geography: A Memorial Volume to Dr. B.N. Mukherji*. Calcutta, India: Banerjee, 1969.

Dutt, Ashok K., and R.S. Gupta. "An Assessment of Agricultural Development in West Bengal." *Journal of Tropical Geography*, 28 (1969): 17-22.

Nath, V. "The Growth of Indian Agriculture: A Regional Analysis." *Geographical Review*, 59, No. 3 (1969): 348-372.

Bhattacharyya, S.R., and S.G. Saba. "Changes in Agricultural Pattern in Bongaon Sub-District." *Geographical Review of India*, 32, No. 4 (1970): 223-242.

Negi, B.S. "Agricultural Regions of Garhwal District." *Geographical Observer*, 6 (1970): 27-32.

Blaikie, P.M. "Spatial Organization of Agriculture in Some North Indian Villages." *Transactions, Institute of British Geographers*, 52 (1971): 1-40.

Blaikie, P.M. "Spatial Organization of Agriculture in Some North Indian Villages, Part II." *Transactions, Institute of British Geographers*, 53 (1971): 15-30.

Nitz, Hans Jurgen. *Formen der Landwirtschaft und ihre raumliche Ordung in der oberen Gangesebene*. Wiesbaden, Germany: Heidelberger Geographische Arbeiten, Heft 28, 1971.

Sharma, Rajendra P. "Three Orders of Agriculture in Madhya Pradesh." *Geographical Viewpoint*, 2, No. 1 (1971): 13-28.

Duggal, H.S. "Morpho-Agricultural Regions of the Lower Chambal Valley, India." *Geographical Review of India*, 34, No. 2 (1972): 109-122.

Gupta, N.L., and S.L. Hiran. "Surplus/Deficit Areas of Food Production in Rajasthan." *Indian Geographical Journal*, 47, No. 3-4 (1972): 19-30.

Shafi, M. "Agricultural Geography: A Trend Report." Pp. 3-18 in *A Survey of Research in Geography*. New Delhi, India: Indian Council of Social Science Research, 1972.

Sharma, Harisanka. "Morpho-Agricultural Regions of the Lower Chambal Valley, India." *Geographical Review of India*, 34, No. 2 (1972): 109-122.

Tripathi, R.R. "Agricultural Dynamicity and Efficiency Levels: A Case Study of the Upper Ganga Gomati Doab." *Indian Geographical Journal*, 47, No. 1-2 (1972): 12-19.

Tyagi, B.S. "Agricultural Intensity in Chunar Tahsil District, Mirzapur, Uttar Pradesh." *National Geographical Journal of India*, 18, No. 1 (1972): 42-48.

Bergman, T. "Resource Mobilization and Obstacles to Development in Indian Farming." *Pacific Viewpoint*, 14, No. 1 (1973): 1-22.

Chakravarti, A.K. "India's Agriculture." *Focus*, 23, No. 5 (1973): 1-3.

Dikshit, K.R. "Agricultural Regions of Maharashtra." *Geographical Review of India*, 35, No. 4 (1973): 384-396.

Gupta, N.L., and S.L. Hiran. "Agricultural Regions of Rajasthan." *Deccan Geographer*, 11, No. 1-2 (1973): 57-74.

Kayastha, S.L., and S.N. Mishra. "Agricultural Innovations Among Farmers of Mahasu, Mandi and Kinnaur (Uttar Pradesh): Correlates and Levels." *National Geographical Journal of India*, 19, No. 1 (1973): 1-14.

Shinde, Suresh D. "The Panchgonga Basin: An Appraisal of Some Aspects of Its Agricultural Geography." *Geographical Review of India*, 35, No. 3 (1973): 263-276.

Agarwal, P.C., and M.P. Gupta. "Changing Patterns of Agricultural Density in Madhya Pradesh." Pp. 119-124 in *Proceedings, Eighth New Zealand Geography Conference*, edited by William Brockie, Richard LeHeron, and Evelyn Stokes. Christchurch, New Zealand: New Zealand Geographical Society, 1974.

Bhattacharya, Anima. "Farm Activity and Its Place in a Planned Economy." *North Eastern Geographer*, 6, No. 1-2 (1974): 20-25.

Dagli, V., ed. *A Regional Profile of Indian Agriculture*. Bombay, India: Vora and Company, 1974.

Mitra, Champa. "Some Aspects of Indian Agriculture in the Context of Changing International Patterns." *Geographical Review of India*, 36, No. 1 (1974): 50-61.

Mohammad, Noor. "The Diffusion of an Agricultural Innovation in Haryana." *The Geographer, Aligarh*, 21, No. 1 (1974): 11-31.

Raagard, S. "The Tilling in a Village in Southern India." *Geografisk Tidsskrift*, 73 (1974): 66-81.

Shafi, M. "Perspective on the Measurement of Agricultural Productivity." *The Geographer, Aligarh*, 21, No. 1 (1974): 1-10.

Singh, L.R. "Indian Contribution to Agricultural Typology, Agricultural Systems and Agricultural Regionalization: A Bibliography." *National Geographer*, 9 (1974): 85-94.

Ishida, Hiroshi, ed. *Dynamic Regional Geographical Studies of the Punjab, India* (in Japanese). Hiroshima, Japan: University of Hiroshima, Centre for Research on World Regional Geography, 1975.

Mamoria, C.B. *Geography of India: Agricultural Geography*. Agra, India: Shiva Lal Agarwala, 1975.

Murton, Brian J. "Agrarian System Dynamics in Interior Tamil Nadu Before 1800 AD." *National Geographical Journal of India*, 21, No. 3-4 (1975): 151-165.

Singh, Abha L. "New Measures of Agricultural Efficiency." *The Geographer, Aligarh*, 22, No. 1 (1975): 25-35.

Taher, M. "Regional Basis of Agricultural Planning in the Brahmaputra Valley." *North Eastern Geographer*, 7, No. 1-2 (1975): 10-18.

Farmer, Bertram H. "Geography and Agrarian Research: Experience from Tamil Nadu and Sri Lanka." Pp. 198-206 in *Indian Geographical Society Jubilee Volume*, edited by V.L.S. Prakasa Rao. Madras, India: Indian Geographical Society, 1976.

Husain, Majid. "A New Approach to the Agricultural Productivity of the Sutlej Ganga Plains of India." *Geographical Review of India*, 38, No. 3 (1976): 230-236.

Husain, Majid. "Agricultural Productivity of India: An Exploratory Analysis." *National Geographical Journal of India*, 22, No. 3-4 (1976): 113-123.

Panda, B.P. "Agricultural Types in Madhya Pradesh." *The Geographer, Aligarh*, 23, No. 2 (1976): 11-29.

Shafi, M. "Approaches to the Measurement of Agricultural Efficiency: A Study in Agricultural Typology." Pp. 67-77 in *Essays in Applied Geography*, edited by V.C. Mishra, N.P. Ayyar, and P. Kumar. Sagar, India: University of Sagar, Department of Geography, 1976.

Sharma, M.K. "The Agricultural Regions of the Brahmaputra Valley." *North Eastern Geographer*, 8, No. 1-2 (1976): 21-29.

Sharma, Tara C., and O. Coutinho. "Aspects of Agricultural Productivity in Karnataka, 1961–1972." *Deccan Geographer*, 14, No. 1 (1976): 10-21.

Singh, G.B. *Transformation of Agriculture: A Case Study of Punjab*. Kurukshetra, India: Vishal, 1976.

Singh, Jasbir. *An Agricultural Geography of Haryana*. Kurukshetra, India: Vishal Publications, 1976.

Chauhan, V.S., and Surendra Singh. "Agricultural Production and Sufficiency in Rohilkhand, Uttar Pradesh." *Geographical Observer*, 13 (1977): 30-38.

Gowda, P.B., and K.P.D. Mahadev. "Changing Nature of Agriculture in the Rural-Urban Fringe: A Case Study of Bangalore." *Indian Geographical Journal*, 52, No. 1 (1977): 1-9.

Izhar, Nilofar. "Measurement of Agricultural Productivity in Western Uttar Pradesh." *Geographical Observer*, 13 (1977): 27-29.

Negi, B.S. "Food Production Studies in Eleven Villages of Garhwal District." *Geographical Observer*, 13 (1977): 51-54.

Singh, S., and V.S. Chauhon. "Measurement of Agricultural Productivity: A Case Study of Uttar Pradesh, India." *Geographical Review of India*, 39, No. 3 (1977): 222-231.

Bhagirath, J.S. "Jalanzi Region: A Study of Agricultural Growth 1855–1945." *Geographical Review of India*, 40, No. 2 (1978): 182-194.

Das, Manamohan. "Problems of Agricultural Growth in Assam." *Geographical Review of India,* 40, No. 2 (1978): 144-154.

Izhar, Nilofar. "Agricultural Productivity and Standard Nutrition Units in Uttar Pradesh: A Study in Spatial Correlation." *National Geographical Journal of India*, 24, No. 1-2 (1978): 47-49.

Mohammad, Ali, and Shah Alam. "Spatial and Temporal Variations in Levels and Growth of Agricultural Output: A Case of Assam." *The Geographer, Aligarh*, 25, No. 2 (1978): 28-41.

Rehman, Hifzur, and Mazhar Ali Khan. "Resource Allocation and Variations in Agricultural Productivity in Aligarh District." *The Geographer, Aligarh*, 25, No. 2 (1978): 6-13.

Sadhukkan, S.K. "A Geo-economic Dissection of South Indian Agriculture." *Geographical Review of India*, 40, No. 1 (1978): 51-56.

Shinde, Suresh D., M.G. Jadhov, and C.T. Pawar. "Agricultural Productivity in Maharashtra: A Geographical Analysis." *National Geographer*, 13, No. 1 (1978): 35-42.

Siddiqi, Farasat Ali. "Agricultural Density and Its Changing Pattern in Uttar Pradesh." *The Geographer, Aligarh*, 25, No. 1 (1978): 50-56.

Singh, V.R., and R.S. Singh. "Spatial Regularity of Agricultural Location in Von Thunen Concept: A Case." *Indian Geographical Journal*, 53, No. 2 (1978): 53-58.

Husain, Majid. *Agricultural Geography*. Delhi, India: Inter-India Publications, 1979.

Panda, B.P. "Agricultural Types in Madhya Pradesh." *Geographia Polonica*, 40 (1979): 133-150.

Saxena, J.P., and S.N. Mehrotra. "Regional Imbalances in the Agricultural Development of Madhya Pradesh, India." *Geographia Polonica*, 40 (1979): 201-208.

Singh, Abha L. "Agricultural Efficiency in Uttar Pradesh." *National Geographical Journal of India*, 25, No. 2 (1979): 129-135.

Singh, B.B. "Trends and Concepts in Agricultural Geography: A Review." *National Geographical Journal of India*, 25, No. 3-4 (1979): 269-280.

Singh, G.B. *Transformation of Agriculture: A Case Study of Punjab*. Kurukshetra, India: Vishal, 1979.

Singh, Jhujar. "Commercialization of Agriculture in Punjab: 1971–1972: A Spatial Analysis." *Geographical Review of India*, 41, No. 1 (1979): 1-8.

Singh, K. "Factors Determining Agricultural Productivity in Rajasthan." *Geographical Observer*, 15 (1979): 55-70.

Singh, V.R. "Agricultural Typology of India." *Geographia Polonica*, 40 (1979): 113-131.

Thakur, Rameshwar. "Patterns of Agricultural Growth in Bihar." Master's Thesis, University of Delhi, Delhi, India, 1979.

Bhagabati, A.K. "Spatial Analysis of Small-Scale Agriculture in Assam: A Case Study of Nalbari District." Ph.D. Dissertation, Gauhati University, India, 1980.

Chattopadhyay, R.N. "Relative Contributions of Agricultural Output Growth Components in Some Selected Districts of Eastern India." *Geographical Review of India*, 42, No. 1 (1980): 16-21.

Mitra, Manju. *Agricultural Geography of Chhattisgarh Basin*. Kanpur, India: Sahitya Ratnalaya, 1980.

Shinde, Suresh D. *Agriculture in an Underdeveloped Region: A Geographical Survey*. Bombay, India: Himalaya Publications, 1980.

Subbiah, S., and A. Ahmad. "Determinants of Agricultural Productivity in Tamil Nadu, India." *Transactions, Institute of British Geographers*, 2, No. 1 (1980): 19-31. (New Series)

Tawde, M.D., and B.N. Gophane. "Assessment of Agricultural Efficiency for Better Rural Economy: A Case Study of the Manjra Plateau (Maharashtra)." *National Geographer*, 15 (1980): 39-44.

Chauchon, V.S., and Surinder Singh. "Measurement of Regional Disparities in Agricultural Output in Rohilkand, Uttar Pradesh." Pp. 271-280 in *Perspectives in Agricultural Geography*, vol. 4, edited by Noor Mohammad. New Delhi, India: Concept Publishing Company, 1981.

Das, M.M. "Spatio-Temporal Variations of Agricultural Efficiency in Assam: A Quantitative Analysis." Pp. 257-270 in *Perspectives in Agricultural Geography*, vol. 4, edited by Noor Mohammad. New Delhi, India: Concept Publishing Company, 1981.

Ghosh, Dhira. "Agricultural Trends in Bankura District—1950–1974: A Review." Pp. 351-378 in *Perspectives in Agricultural Geography*, vol. 4, edited by Noor Mohammad. New Delhi, India: Concept Publishing Company, 1981.

Husain, Majid. "Shifting Cultivation in North East India: Prospects and Problems." Pp. 241-258 in *Perspectives in Agricultural Geography*, vol. 1, edited by Noor Mohammad. New Delhi, India: Concept Publishing Company, 1981.

Jana, M.M. "Programmes for Agricultural Development in India: A Case Study of Lower Silabati Basin." Pp. 289-304 in *Perspectives in Agricultural Geography*, vol. 4, edited by Noor Mohammad. New Delhi, India: Concept Publishing Company, 1981.

Kelly, Kenneth. "Agricultural Change in Hooghly, 1850–1910." *Annals of the Association of American Geographers*, 71, No. 2 (1981): 237-252.

Mathur, R.N., Kamla Prasad, and R.S. Singh. "Spatio-Temporal Variation in Agricultural Productivity in Mirzapur District." Pp. 281-288 in *Perspectives in Agricultural Geography*, vol. 4, edited by Noor Mohammad. New Delhi, India: Concept Publishing Company, 1981.

Mohammad, Ali. "Regional Imbalances in Levels of Agricultural Productivity." Pp. 227-256 in *Perspectives in Agricultural Geography*, vol. 4, edited by Noor Mohammad. New Delhi, India: Concept Publishing Company, 1981.

Mohammad, Ali. "Sources of Variations in Levels of Agricultural Productivity." Pp. 437-454 in *Perspectives in Agricultural Geography*, vol. 4, edited by Noor Mohammad. New Delhi, India: Concept Publishing Company, 1981.

Mohammad, Noor, and Rameshwar Thakur. "Spatial Variations and Determinants of Agricultural Productivity in Bihar." Pp. 415-436 in *Perspectives in Agricultural Geography*, vol. 4, edited by Noor Mohammad. New Delhi, India: Concept Publishing Company, 1981.

Mukhopadhyay, J. "Concept of Agricultural Region." Pp. 45-60 in *Perspectives in Agricultural Geography*, vol. 4, edited by Noor Mohammad. New Delhi, India: Concept Publishing Company, 1981.

Murton, Brian J. "Traditional Systems of Agricultural Knowledge and Contemporary Agricultural Development in India." Pp. 401-414 in *Perspectives in Agricultural Geography*, vol. 4, edited by Noor Mohammad. New Delhi, India: Concept Publishing Company, 1981.

Nagpol, Usha. "Agricultural Regionalism of Uttar Pradesh." Pp. 305-316 in *Perspectives in Agricultural Geography*, vol. 4, edited by Noor Mohammad. New Delhi, India: Concept Publishing Company, 1981.

Qureshi, M.H., and G. Parimala. "Components of Agricultural Growth: A Case Study of Tamil Nadu (1960–1963 to 1972–1975)." *National Geographer*, 16, No. 1 (1981): 43-50.

Ramachandran, R. "Correlates of Innovativeness Among Farmers in the Ganga Plains." Pp. 383-402 in *Perspectives in Agricultural Geography*, vol. 5, edited by Noor Mohammad. New Delhi, India: Concept Publishing Company, 1981.

Raza, Moonis. "Regional Disparities in India: A Preliminary Exploration of the Regional Dimension of Agricultural Development." Pp. 103-144 in *Perspectives in Agricultural Geography*, vol. 4, edited by Noor Mohammad. New Delhi, India: Concept Publishers, 1981.

Schoen, Meera G. "Changing Attitudes to Agriculture in Urbanizing Areas." Pp. 251-266 in *Pespectives in Agricultural Geography*, vol. 5, edited by Noor Mohammad. New Delhi, India: Concept Publishing Company, 1981.

Singh, B.B. "Trends and Concepts in Agricultural Geography." Pp. 293-310 in *Perspectives in Agricultural Geography*, vol. 1, edited by Noor Mohammad. New Delhi, India: Concept Publishing Company, 1981.

Singh, Jasbir. "A New Technique for Measurement of Agricultural Productivity." Pp. 169-182 in *Perspectives in Agricultural Geography*, vol. 4, edited by Noor Mohammad. New Delhi, India: Concept Publishing Company, 1981.

Singh, Surendra. "Levels of Input-Output Relationship in the Agriculture of Uttar Pradesh." *Deccan Geographer*, 19, No. 2-3 (1981): 119-130.
Sinha, B.N. "Agricultural Efficiency in India." Pp. 183-210 in *Perspectives in Agricultural Geography*, vol. 4, edited by Noor Mohammad. New Delhi, India: Concept Publishing Company, 1981.
Thakur, Rameshwar. "Patterns of Agricultural Growth in Bihar." Pp. 317-350 in *Perspectives in Agricultural Geography*, vol. 4, edited by Noor Mohammad. New Delhi, India: Concept Publishing Company, 1981.
Tyagi, S.C. "Agricultural Development in Meerut City Region" (in Hindi). *Geographical Observer*, 17 (1981): 58-64.
Singh, Minati. "Agricultural Characteristics of Lower Ganga-Ghahgra Doab." *Deccan Geographer*, 20, No. 1 (1982): 165-182.
Singh, Surendra. "Identification and Phasing of Agricultural Development in Rohilkhand, Uttar Pradesh." *National Geographical Journal of India*, 28, No. 1-2 (1982): 28-36.
Yadov, J.R. "Adoption of Agricultural Innovations in Spatial Framework: A Case of Kanpur District." *National Geographer*, 17, No. 1 (1982): 75-80.
Agarwal, P.C. "Spatial Analysis of the Level of Agricultural Development in Madhya Pradesh." *National Geographer*, 18, No. 2 (1983): 205-214.
Bagi, F.S. "Relationship Between Farm Size, Productivity, and Returns to Scale in Haryana Agriculture." *Rural Systems*, 1, No. 3 (1983): 189-212.
Banerjee, Swapna, and S. Barman. "Input Availability and Modern Agriculture in Burdwan District, West Bengal." *Transactions, Institute of Indian Geographers*, 5, No. 2 (1983): 109-114.
Joshi, Y.G. "Agricultural Development Around a City: A Case Study of Indore Tahsil, Madhya Pradesh." *Deccan Geographer*, 21, No. 3 (1983): 427-440.
Lal, Munnu. "Agricultural Production in Central India: A Study." *Rural Systems*, 1, No. 1 (1983): 61-72.
Parimada, G., and M.H. Qureshi. "Levels of Agricultural Development in Tamil Nadu." *Indian Geographical Journal*, 58, No. 2 (1983): 119-126.
Pollard, H.J. "Development from an Agricultural Base in Punjab." *Geography*, 68, No. 1 (1983): 16-24.
Sawant, S.B., and Alka Gadgil. "Factors Affecting Agricultural Efficiency in India." *National Geographer*, 18, No. 1 (1983): 107-114.
Sharma, B.L. "Agricultural Typology: A Case Study of the Ajmer District, Rajasthan, India." *Geographia Polonica*, 46 (1983): 79-82.
Sharma, B.L. "A Typological Analysis of Agriculture in the Rajasthan State." *Geographia Polonica*, 46 (1983): 71-78.
Singh, J. "A Typology of Agriculture: The Indian Experience." *Geographia Polonica*, 46 (1983): 49-70.
Singh, Jasbir, and Sher S. Dhillon. "Statistical Technique for Agricultural Regionalization—A New Approach: India—A Case Study." *Asian Profile*, 11, No. 2 (1983): 167-174.
Singh, Jhujar. "Spatial Organization of Agriculture in a North Indian Village: Deductive Model." *Transactions, Institute of Indian Geographers*, 5, No. 2 (1983): 115-126.

Bagi, F.S. "Economic Evaluation of Some Agrarian Policies: An Analysis of Farm Level Data from Haryana, India." *Rural Systems*, 2, No. 1 (1984): 27-40.

Bhagabati, A.K. "Levels of Agricultural Productivity in the Brahmaputra Valley: A Regional Analysis." *North Eastern Geographer*, 15, No. 1-2, and 16, No. 1-2 (1984): 31-38.

Dayal, Edison. "Agricultural Productivity in India: A Spatial Analysis." *Annals of the Association of American Geographers*, 74, No. 1 (1984): 98-123.

Dayal, Edison. "Factors Affecting Modernization of Agriculture in India." *Asian Profile*, 12, No. 3 (1984): 243-256.

Kelly, Kenneth. "Commercial Agriculture in the Region North and Northeast of Calcutta During the Nineteenth Century." Pp. 83-107 in *Canadian Papers in Rural History*, vol. 4, edited by Donald Akenson. Gananoque, Ontario: Langdale, 1984.

Ramanaiah, Y.V., and N.B.K. Reddy. "Regionalisation of Agricultural Productivity in Andhra Pradesh." *Transactions, Institute of Indian Geographers*, 6, No. 2 (1984): 1-18.

Shafi, M. "New Sources of Food: A Study in Food Systems." *The Geographer, Aligarh*, 31, No. 1 (1984): 10-13.

Shukla, S.P. "Planning and the Agricultural Development in Arunachal Pradesh: The Task Ahead." *National Geographical Journal of India*, 30, No. 3 (1984): 191-198.

Siddiqi, Shamsul H., Hifzur Rehman, and M. Farooq Siddiqi. "Regional Analysis of Agricultural Productivity in Bihar." *The Geographer, Aligarh*, 31, No. 1 (1984): 77-85.

Singh, Jhujar. "Distance and Spatial Organization of Agriculture: A Case Study of a North India Village." *Geographical Review of India*, 46, No. 1 (1984): 29-35.

Thakur, M.C., and R.C.S. Taragi. "Spatial Variations in the Levels and Growth of Agricultural Productivity in Western Uttar Pradesh." *Rural Systems*, 2, No. 2 (1984): 109-118.

Das, M.M. "Agricultural Regionalisation of Assam: A Quantitive Analysis." *North Eastern Geographer*, 17, No. 1-2 (1985): 29-34.

Komoguchi, Yoshimi. *Agricultural Systems in Tamil Nadu*. Chicago, Ill.: University of Chicago, Department of Geography Research Paper No. 219, 1985.

Qureshi, Azimuddin, and Masood Ahsan. "Levels of Agricultural Productivity in Eastern Uttar Pradesh." *The Geographer, Aligarh*, 32, No. 1 (1985): 14-19.

Shafi, M. "Perspectives on Agricultural Development in Tribal Areas with Special Reference to Chhatanagpur Plateau." *The Geographer, Aligarh*, 32, No. 2 (1985): 6-12.

Sharma, S.K., and C.K. Jain. "Changes in Agricultural Productivity in Madhya Pradesh: Spatial and Temporal Dimensions." *The Geographer, Aligarh*, 32, No. 2 (1985): 34-55.

Singh, Abha L., and Arun P. Saxena. "Agricultural Development in Uttar Pradesh." *The Geographer, Aligarh*, 32, No. 1 (1985): 20-28.
Vidyanath, V. "Agricultural Productivity in Andhra Pradesh." *National Geographical Journal of India*, 31, No. 3 (1985): 204-207.
Banerjee, Smriti. *Regional Imbalances in the Agricultural Development in Uttar Pradesh*. Vananasi, India: Sudha Publications, 1986.
Bano, Hanifa. "Economic Exploitation by Way of Agriculture in Dal Lake (Srinagar)." *Annals of the National Association of Geographers, India*, 6, No. 1 (1986): 38-42.
Datye, V.S. "Agriculture in the Vedic and Later Vedic India." *Transactions, Institute of Indian Geographers*, 8, No. 2 (1986): 5-12.
Kuniyal, J.C., and R.L. Singh. "Measurement of Agricultural Productivity in Nainital District, Uttar Pradesh." *National Geographer*, 21, No. 2 (1986): 193-201.
Lal, Munnu. *Agro-Types in Central India*. Varanasi, India: Star, 1986.
Pannell, Clifton W. "Farmland and Agricultural Development in India and China." *Rural Systems*, 4, No. 1 (1986): 1-18.
Pandekor, V., and V.S. Phadke. "Agricultural Changes in the Rural Periphery of a Metropolis: A Case Study of Chinchani." *Rural Systems*, 4, No. 2 (1986): 73-84.
Raina, Jawahar L. "Analysis of the Application of the Von Thunen Hypothesis in Jommu." *Deccan Geographer*, 24, 1-2 (1986): 22-28.
Singh, B.B., and Rajputi Ram. "Food Production System and Efficiency in Azamgarh District." *National Geographical Journal of India*, 32, No. 2 (1986): 127-136.
Singh, B.B., and Rajputi Ram. "Factor Analysis of Regional Disparities in Agricultural Production in Eastern Uttar Pradesh." *Rural Systems*, 4, No. 1 (1986): 19-28.
Sujan, D.K., and A. Dimri. "Agricultural Broadcast: Impact, Needs, and Problems Amongst Farmers." *National Geographer*, 21, No. 2 (1986): 187-192.
Thakur, M.C., and Rajhubir Chand. "Agricultural Variations within Uttar Pradesh: A Temporal Analysis." *National Geographer*, 21, No. 2 (1986): 153-160.
Tiwari, P.D. "Agricultural Products and Nutritional Availability in the Normada Basin." *Deccan Geographer*, 24, No. 1-2 (1986): 51-66.
Tiwari, P.S., ed. *Contributions to Indian Geography, Vol. VIII: Agricultural Geography*. New Delhi, India: Heritage, 1986.
Vidyanath, V. "Agricultural Regionalization of Andhra Pradesh." *Annals of the National Association of Geographers, India*, 6, No. 2 (1986): 49-56.
Bhaskar, K.S., et al. "Agricultural Efficiency of Vidarbha Region (Maharashtra)." *National Geographical Journal of India*, 33, No. 2 (1987): 154-159.
Chouhon, T.S. *Agricultural Geography: A Study of Rajasthan State*. Jaipur, India: Academic, 1987.
Dube, R.S. *Agricultural Geography: Issues and Applications*. Delhi, India: Gian Publishers, 1987.

Ramanaiah, Y.V., and Naveen K. Charles. "Agricultural Geography of Anantopur District, Andhra Pradesh." *Indian Geographical Journal*, 62, No. 2 (1987): 106-118.

Siddiqi, Shamsul H. "Agricultural Productivity in North Bihar Plain." *Indian Geographical Journal*, 62, No. 2 (1987): 130-136.

Singh, Abha L., and S.N.I. Hashmi. "Technological Determinants of Food Productivity in Aligarh District." *Annals of the National Association of Geographers, India*, 7, No. 1 (1987): 9-14.

Singh, V.R. "Agricultural Change in Budelkhand Region, Uttar Pradesh." *Rural Systems*, 5, No. 4 (1987): 225-236.

Sinha, R.N.P. "The Choice Before the Indian Small Farmer." *Geographical Perspective*, 1, No. 2 (1987): 1-8.

Vaid, Varsha, and V.S. Datye. "Influence of Some Selected Variables on Agricultural Productivity of Maharashtra." *National Geographical Journal of India*, 33, No. 2 (1987): 175-183.

Datye, V.S., and N.S. Pawar. "Agricultural Productivity: A Micro-Level Study of Two Villages (Maharashtra)." *Transactions, Institute of Indian Geographers*, 10, No. 1 (1988): 1-7.

Jain, C.K. *Patterns of Agricultural Development in Madhya Pradesh*. New Delhi, India: Northern Book Centre, 1988.

Mishra, S.P. "Agricultural Productivity and Its Spatial Variation in Eastern Uttar Pradesh." *National Geographical Journal of India*, 34, No. 2 (1988): 113-125.

Munir, Abdul. "Agricultural Productivity and Regional Development: A Case Study of the Sub-Himalayan East Region of Uttar Pradesh." *The Geographer, Aligarh*, 35, No. 2 (1988): 45-59.

Prasad, Har. "Agricultural Geography of the Dehradum: A Study in Methodology." *Geographical Viewpoint*, 17-18 (1988–1989): 31-38.

Singh, D.N. "Infrastructural Framework and Agricultural Development: An Analysis of Existing Structure and the Future Strategy with Particular Reference to India." *Philippine Geographical Journal*, 32, No. 2-3 (1988): 83-91.

Sujan, D.K., and D. Sujan. "Effects of Rumors in Agricultural Practices." *National Geographer*, 23, No. 1 (1988): 55-64.

Tiwari, P.D. "Levels of Agricultural Development in the Rewa Plateau." *National Geographer*, 23, No. 1 (1988): 65-74.

Tiwari, P.D. "Levels of Agricultural Development in the Rewa Plateau." *Deccan Geographer*, 26, No. 2-3 (1988): 377-388.

Chand, Rajhubir, and Kailosh C. Joshi. "Productivity Constraints in the Himalayan Agriculture." *Rural Systems*, 7, No. 4 (1989): 227-240.

Chandel, Rajendon S. "Agricultural Productivity of Konch Block: A Case Study." *Rural Systems*, 7, No. 3 (1989): 193-206.

Chidonanda, Sach. *Shifting Cultivation in India*. New Delhi, India: Concept Publishing Company, 1989.

Gopalakrishnan, R. "The State of Agricultural Development in Assam." *Geographical Review of India*, 51, No. 2 (1989): 62-67.

Gupta, N.L., and L.C. Khatri. "Agricultural Typological Analysis: A Case Study of the Tribal Region of Southern Rajasthan." *Annals of the National Association of Geographers, India*, 9, No. 1 (1989): 41-48.

Krooneman, P. "Commercialisering Overheidsbeleid in de agrarische sector in ontwik-kelingslanden: Het voor beeld von de droge gebieden in India." *Geografisch Tijdschrift*, 23, No. 5 (1989): 312-322.

Misra, B.N., and S.N. Tripathi. "Level of Agricultural Development and Its Contribution to Regional Economy in Basti District, Uttar Pradesh." *National Geographer*, 24, No. 2 (1989): 97-108.

Munir, Abdul, M.F. Khan, and Qazi M. Ahmad. "Agricultural Productivity in Azangarh District, Uttar Pradesh." *Geographical Review of India*, 51, No. 1 (1989): 78-89.

Naranind, N. "Agricultural Typology of Southern Rajasthan." *Geographical Review of India*, 51, No. 2 (1989): 101-103.

Raina, Jawahar L. *Locational Analysis in Agricultural Geography*. Jaipur, India: Pointer Publishers, 1989.

Rawat, D.S. "Initial Agrarian Industrial Structure: A Case Study of the Indian District of Muzaffornagar." *Geographical Observer*, 25 (1989): 17-28.

Singh, V.R. "Need for Agro-Typological Studies in India." *Annals of the National Association of Geographers, India*, 9, No. 1 (1989): 61-65.

Singh, V.R. "Microlevel Typological Classification of Indian Agriculture: The Case of Uttar Pradesh." *Geografia Polonica*, 57 (1989): 149-166.

Thakur, Rameshwar. "Agricultural Productivity in South Bihar Plain: A Spatial Analysis." *Transactions, Institute of Indian Geographers*, 11, No. 2 (1989): 27-42.

Tiwari, P.D., and C.K. Jain. "Levels of Agricultural Innovations and Agricultural Development in Reva Plateau, Mahdya Pradesh." *National Geographer*, 24, No. 2 (1989): 129-138.

Bhagabati, A.K. "Spatial Pattern of Agricultural Development in Brahmaputra Valley, Assam." *Rural Systems*, 8, No. 1 (1990): 69-76.

Chand, Rajhubir, and Kailosh C. Joshi. "Determinants of Agricultural Productivity in the Village of Patalgaon, Kumaun Himalaya." *Geographical Viewpoint*, 19, No. 1-2 (1990): 5-14.

Kaur, Dhian. "Influence of Distance in the Location of Agriculture in Bist Doab (Punjab)." *The Geographer, Aligarh*, 37, No. 1 (1990): 41-49.

Khatri, L.C. *Agricultural Typology: A Tribal Region of Southern Rajasthan*. Udaipur, India: Himanshu Publications, 1990.

Krishnakumari, A., and E. Swaminathan. "Changing Farming Scene in Nellore District, India." *Indian Geographical Journal*, 65, No. 2 (1990): 123-129.

Ramanaiah, Y.V. "Changing Pattern of Agricultural Landscape in the Drought Prone Area, Anantapur District (M.L.)." *Rural Systems*, 8, No. 2 (1990): 99-112.

Shafi, M. "Agricultural and Rural Transformation in India." *The Geographer, Aligarh*, 37, No. 1 (1990): 1-8.

Siddiqi, M. Farooq. "Agriculture in Bundelkhand (Uttar Pradesh): A Study of Post-Independence Changes." *The Geographer, Aligarh*, 37, No. 1 (1990): 50-56.

Siddiqi, Shamsul H. "Trends of Agricultural Development in the North Bihar Plain." *The Geographer, Aligarh*, 37, No. 2 (1990): 46-57.

Singh, J. "Regional Agricultural Disparities in Uttar Pradesh." *National Geographical Journal of India*, 36, No. 3 (1990): 199-210.

Vaid, Varsha, and V.S. Datye. "Agricultural Productivity Regions of Maharashtra." *National Geographical Journal of India*, 36, No. 3 (1990): 153-161.

Ahmed, Jameel. "Spatial Pattern of Agriculture in Vaishali District (Bihar)." *Geographical Perspective*, 5, No. 1-2 (1991): 1-10.

Bhagabati, A.K., and M.M. Das. "Structural Constraints of Agricultural Development in Assam." *Annals of the National Association of Geographers, India*, 11, No. 2 (1991): 68-71.

Chandel, Rajendon S. *Agricultural Change in Bundelkhand Region*. Varanasi, India: Star Distributors, 1991.

Chand, Rajhubir, and Kailosh C. Joshi. "Measurement of Agricultural Productivity in a Himalayan Village." *Indian Geographical Journal*, 66, No. 1 (1991): 14-19.

Shukla, Lakshmi, ed. *Readings in Agricultural Geography*. Jodhpur, India: Scientific Publishers, 1991.

Thakur, Rameshwar. "Regional Imbalances in Agricultural Productivity in South Bihar Plain, India." *Asian Profile*, 19, No. 4 (1991): 341-358.

Datta, Archana, and R. Bhattacharyya. "Level of Agricultural Development in Serampore Subdivision, Hooghly District, West Bengal." *Geographical Review of India*, 54, No. 3 (1992): 95-103.

Khan, Zafar I., and Shafat Ali Khan. "Levels of Agricultural Productivity in Jammu and Kashmir State." *Geographical Review of India*, 54, No. 4 (1992): 41-47.

Khan, Zafar I., and Shafat Ali Khan. "Agricultural Productivity of the Jammu and Kashmir." *Indian Geographical Journal*, 67, No. 2 (1992): 72-74.

Khan, Zafar I., and Shafat Ali Khan. "Regional Variations of Agricultural Productivity in Jammu and Kashmir." *Philippine Geographical Journal*, 36, No. 3-4 (1992): 108-126.

Khanna, Sulbha. "Sustainable Agricultural Development: A Comparison Between Commercial and Traditional Agricultural Systems." *National Geographer*, 27, No. 2 (1992): 147-158.

Kumar, Ashok. *Development of Indian Agriculture*. New Delhi, India: Anmal Publishers, 1992.

Marius-Gnonan, K. "Le revolution verte et le maitrise de l'espace agricole: Le cas de la region de Pondichery (Inde meridionale)." *Les Cahiers d'Outre Mer*, 45 (1992): 189-212.

Misra, B.N., and Vandana Pandy. "Role of Agriculture in Rural Development: A Case of Mirzapur District, Uttar Pradesh." *Geographical Review of India*, 54, No. 1 (1992): 37-49.

Misra, B.N. "Indian Agriculture: The Progress and the Predicament." *National Geographer*, 27, No. 2 (1992): 85-100.

Ojha, Brahm S., Jasbir Singh, and Jai P. Gupta. *Technological and Socio-economic Changes in Agriculture in India: A Case Study: A Study at Spatio-Temporal Scale*. New Delhi, India: Oxford and IBH Publications Company, 1992.

Schar, P. "Capitalisme agraire et industrialisation rurale en Inde meridionale (Etude de la region de Coimbatore–Tamil Nadu)." *Les Cahiers d'Outre Mer,* 45 (1992): 125-158.

Shafi, Mohammad. "Changing Role of Agriculture in the Economic Development of India." *The Geographer, Aligarh*, 39, No. 1 (1992): 1-7.

Singh, S., and Neera Rastogi. "Morpho-Agricultural Regionalisation of Belan-Son Interstream Region of the Eastern Rewa Plateau." *National Geographer*, 27, No. 2 (1992): 159-167.

Vaid, Varsha, and V.S. Datye. "Agricultural Productivity in Western Maharashtra and Konkan." *Transactions, Institute of Indian Geographers*, 14, No. 2 (1992): 141-148.

Bhagwati, Aboni K. "Agricultural Development in Assam." *Annals, National Association of Geographers, India*, 13, No. 2 (1993): 19-28.

Desai, A., and Sulbha Khanna. "Problems and Prospects of Achieving Sustainability in Indian Agriculture: A Case Study of Gujarat." *National Geographer*, 28, No. 2 (1993): 157-164.

Khanna, Sulbha. "Diffusion of Agricultural Information: A Case Study of Gujarat." *The Geographer, Aligarh*, 40, No. 1 (1993): 35-42.

Krishnakumari, A., and E. Swaminathan. "Levels of Agricultural Modernization in Nellore District, Andhra Pradesh." *The Geographer, Aligarh*, 15, No. 2 (1993): 21-26.

Reddy, G.J., and N.B.K. Reddy. "Agricultural Development in the Districts of Andhra Pradesh." *Indian Geographical Journal*, 68, No. 2 (1993): 41-47.

Tiwari, R.C., and S. Tripathi. "Agricultural Development: Retrospects and Prospects—A Case of Gorakhpur District." *National Geographer*, 28, No. 2 (1993): 97-114.

Chand, Rajhubir, and Chandra Mehra. "Regional Imbalances in the Levels of Agricultural Development and Nutrition in Kumaun." *Geographical Review of India*, 56, No, 4 (1994): 33-42.

Chendrayuda, N., and Y.V. Ramanaiah. "Typification of Agriculture and Typological Regions: A Case Study of Chittoor District, Andhra Pradesh." *National Geographer*, 29, No. 2 (1994): 99-106.

Munir, Abdul, Shamsul H. Siddiqui, and Abu Sufiyan. "Agricultural Modernization in Uttar Pradesh." *The Geographer, Aligarh,* 41, No. 1 (1994): 13-19.

Prasad, Sri Ram Lagna. "Agro-Based Industries of Patna District: A Geographical Study." *Geographical Perspective*, 9, No. 1-2 (1995): 12-14.

Prasad, Sri Ram Lagna. "The Ribbon Development of Agricultural Belt in the Rural-Urban Fringe of Patna." *Geographical Perspective*, 9, No. 1-2 (1995): 18-20.

Raina, Jawahar L. *Dryland Farming in India: Constraints and Challenges.* Jaipur, India: Pointer Publications, 1995.

Shafi, M. "Agricultural Policy of India: A Geographical Perspective." *The Geographer, Aligarh*, 42, No. 2 (1995): 72-75.

Sharma, P.K. "Agricultural Development in the State of Punjab During the Last Two Decades." *Transactions, Institute of Indian Geographers*, 17, No. 2 (1995): 107-118.

Siddiqui, A.M.A. "Computing the Agricultural Productivity in the Absence of Genuine Village Data." *The Geographer, Aligarh*, 42, No. 1 (1995): 51-65.

Singh, Shiv Kumar, and M. Tajuddin. "Agricultural Resources of Buxar District (Bihar): A Geographical Study." *Geographical Perspective*, 9, No. 1-2 (1995): 21-26.

Vaidya, Balakrishna C. "Agricultural Regions of Yashoda Basin, Maharashtra." *Indian Geographical Journal*, 70, No. 1 (1995): 35-46.

Abbasi, S.A., and S. Vinithon. "Impact of Industrialization on Agriculture in an Industrial Complex at Pondicherry." *Indian Geographical Journal*, 71, No. 2 (1996): 102-105.

Bhaskar, B.P., O. Challa, and R.S. Reddy. "Spatial Variations in Determinants of Agricultural Efficiency in Andhra Pradesh." *Geograhical Review of India*, 59, No. 1 (1997): 54-61.

Jadhav, M.G. "Agricultural Development in Maharashtra: A Spatial Interpretation." *Transactions, Institute of Indian Geographers*, 19, No. 1 (1997): 39-46.

Jeffrey, C. "Richer Farmers and Agrarian Change in Meerut District, Uttar Pradesh, India." *Environment and Planning A*, 29, No. 12 (1997): 2113-2128.

Khan, W. "Agricultural Development in Western Uttar Pradesh, India." *Indian Geographical Journal*, 72, No. 2 (1997): 146-149.

Krishnakumari, A. "Agricultural Modernization in Nellore District, India." *Indian Geographical Journal*, 72, No. 1 (1997): 71-73.

Landy, F. "Une revolution vert pale: Les limites de l'intensification agricole en Inde semi-arid." *Annales de Geographie*, 106 (1997): 571-591.

Prakash, Jai, and Noor Mohammad. "Impact of Energy Input on Agricultural Productivity in Sonepat District." *Geographical Review of India*, 59, No. 4 (1997): 313-319.

Reddy, V. "L'Agriculture de l'Inde: Tendences et problems." *Bulletin de la Societe Geographique de Liege*, 33 (1997): 67-72.

Misra, K.K. "Diffusion of Agricultural Innovations: A Case Study of Atarra Tehsil, Banda District, Uttar Pradesh." *Geographical Review of India*, 61, No. 3 (1999): 220-230.

Shafiqullah. "Levels of Agricultural Development in Gonda District, Uttar Pradesh." *Geographical Review of India*, 61, No. 4 (1999): 361-371.

CROPS AND CROPPING PATTERNS

Rajamanikhan, A.S. "Crops of the Trichinopoly District." *Journal of the Madras Geographical Association*, 8 (1933): 173-178.

Garu, G.J.R. "Distribution of Crops in Vizagapatnam District." *Journal of the Madras Geographical Association*, 9 (1934): 24-31.

Ayyar, S.N.C. "Major Cultivated Crops of the Coimbatore District." *Journal of the Madras Geographical Association*, 14 (1939): 117-126.

Banerjee, Bireswar. "Relief as a Factor in the Location of the West Bengal Tea Gardens." *Geographical Review of India*, 16, No. 2 (1954): 1-9.

Banerjee, Bireswar. "A Study of Jute Cultivation in West Bengal." *Indian Geographical Journal*, 30, No. 3 (1955): 65-78.

Gupta, R.N. "Jute Problem in India." *Geographial Review of India*, 18, No. 4 (1955): 43-50.

Banerjee, Bireswar. "Tobacco in West Bengal." *Geographical Review of India*, 18, No. 3 (1956): 18-39.

Kirk, William. "The Cotton and Jute Industries of India." *Scottish Geographical Magazine*, 72, No. 1 (1956): 38-52.

Bhat, L.S. "Geography of the Spice Gardens." *Bombay Geographical Magazine*, 4 (1957): 21-28.

Banerjee, Bireswar. "Betel-Vine in West Bengal." *National Geographical Journal of India*, 4, No. 2 (1958): 79-88.

Banerjee, Bireswar. "Mulberry Cultivation and Sericulture in West Bengal." *Oriental Geographer*, 3, No. 1 (1959): 33-46.

Sharma, S.S. "Tea Culture in Assam." *National Geographical Journal of India*, 5, No. 2 (1959): 91-108.

Banerjee, Bireswar. "Cinchona in West Bengal." *Oriental Geographer*, 4, No. 1 (1960): 21-36.

Krishnon, V. "Self-Sufficiency of Tamiland in Food Grains." *Indian Geographical Journal*, 35, 1-2 (1960): 37-51.

Sengupta, Phulrani. "Some Aspects of Jute Production in India." *National Geographical Journal of India*, 7, No. 2 (1961): 124-133.

Reddy, R.S. "Viticulture in Hyderbad City." *Oriental Geographer*, 6, No. 1 (1962): 23-38.

Ramachandran, R. "Crop Regions in India." *Indian Geographical Journal*, 38, No. 2 (1963): 58-64.

Varma, K.N. "Orange Cultivation in Vidarbha." *Deccan Geographer*, 2, No. 1 (1963): 63-70.

Nasuti, James J. "The Geography of Jute in India from Partition in 1947 to 1963." Master's Thesis, University of Hawaii, Manoa, Hawaii, 1964.

Bhatia, Shyam S. "Patterns of Crop Concentration and Diversification in India." *Economic Geography*, 41, No. 1 (1965): 39-56.

Singh, Harpal. "Crop Combination Regions in the Malwa Tract of Punjab." *Deccan Geographer*, 3, No. 1 (1965): 21-30.

Hussin, S.S. "Geographical Background of the Decline in Cotton Production in Budaun District." *Indian Geographical Journal*, 41, No. 3-4 (1966): 44-51.

Singh, Govind. "Relative Change in Area Under Jowar and Ragi in Mysore State." *National Geographical Journal of India*, 12, No. 1 (1966): 51-57.

Dayal, Edison. "Crop Combination Regions: A Case Study of the Punjab Plains." *Tijdschrift voor Economische en Sociale Geographie*, 58, No. 1 (1967): 39-47.

Roy, B.K. "Crop-Associations and Changing Pattern of Crops in the Ganga-Ghaghana Doab East." *National Geographical Journal of India*, 13, No. 4 (1967): 194-207.

Ganguli, B.N., and H. Nighet. "Crop Concentration and Crop Diversification in a Micro Region in Eastern Uttar Pradesh." *Geographical Review of India*, 30, No. 1 (1968): 14-23.

Ayyar, N.P. "Crop Regions of Madhya Pradesh: A Study in Methodology." *Geographical Review of India*, 31, No. 1 (1969): 1-19.

Kataria, M.S. "Spatial Changes in Sugar Cane Cultivation in Karnal District." *National Geographical Journal of India*, 15, No. 3-4 (1969): 224-234.

Kaur, K.S. "Changes in Net Sown Area in Amritsar Tahsil, 1951–1964: A Spatial-Temporal Analysis." *National Geographical Journal of India*, 15, No. 1 (1969): 24-37.

Mandel, B. "Crop-Combination Regions of North Bihar." *National Geographical Journal of India*, 15, No. 2 (1969): 125-137.

Bhardwaj, B.K. "Changes in Net Area Sown in Gurgaon Development Block, 1955–1964." *Geographical Viewpoint*, 1, No. 2 (1970): 1-14.

Chakravarti, A.K. "Foodgrain Sufficiency Patterns in India." *Geographical Review*, 60, No. 2 (1970): 208-228.

Husain, Majid. "Patterns of Crop Concentration in Uttar Pradesh." *Geographical Review of India*, 32, No. 3 (1970): 169-185.

Mohammad, Noor, and K.Z. Amani. "Crop Combination in the Trans-Ghagharra Plain." *Geographical Review of India*, 32, No. 1 (1970): 47-59.

Chakravarti, A.K. "Changes in the Patterns of Foodgrain Production and Sufficiency Level in India, 1921 to 1951." *Journal of Tropical Geography*, 32 (1971): 11-30.

Gupta, S.L. "The Cropping Pattern of Haryana." *Geographical Review of India*, 33, No. 1 (1971): 23-55.

Sharma, Krishna D. "Changes in the Spatial Patterns of Net Area Sown in Uttar Pradesh (1951–1966)." *Geographical Viewpoint*, 2, No. 2 (1971): 11-14.

Sharma, S.C. "Cropping Pattern and Crop Combination Regions in Etawale, Middle Ganga-Yamuna Doab." *Deccan Geographer*, 9, No. 1 (1971): 71-92.

Husain, Majid. "Crop Combination Regions of Uttar Pradesh: A Study in Methodology." *Geographical Review of India*, 34, No. 2 (1972): 134-156.

Nand, Nitya. "Crop Combination in Rajasthan." *Geographical Review of India*, 34, No. 1 (1972): 46-60.

Sharma, Parshu R. "Crop Cultivation Intensity, Their Ranking and Cross-Association Regions in Chhattisgarh Region: A Geographical Analysis." *National Geographical Journal of India*, 18, No. 2 (1972): 91-101.

Bischoff, Erhard. *Anbau und Weiterverarbeitung von Zuckerrohr in der Wirtschaftslandschaft der Indischen Union.* Tubingen, Germany: Tubinger Geographische Studien, Heft 52, 1973.

Khan, Asjad H. "Crop Combination Regions in Rohilkhand." *The Geographer, Aligarh*, 20, No. 2 (1973): 151-162.
Shinde, Suresh D. "Horticultural Resources and Potentials." *Deccan Geographer*, 11, No. 1-2 (1973): 75-84.
Akhtar, Rais. "The Cropping Pattern in the Kumoan Region." *North Eastern Geographer*, 6, No. 1-2 (1974): 39-41.
Bagchi, K., and M.M. Jana. "The Crop Combination and Spatial Pattern of Land Utilisation in Lower Silabati Basin." *Geographical Review of India*, 36, No. 4 (1974): 323-332.
Gupta, H.S. "Foodgrains Production Efficiency in Madhya Pradesh." *National Geographical Journal of India*, 20, No. 2 (1974): 131-139.
Siddiqi, M. Farooq. "Crop Combinations and Specialisation in India." *The Geographer, Aligarh*, 21, No. 1 (1974): 76-88.
Chauhon, V.S. "Crop Diversification in the Jamuna-Hindan Tract of Western Uttar Pradesh." *Geographical Observer*, 11 (1975): 48-54.
Husain, Majid. "Ganga Yamuna Doab: A Study in the Patterns of Crop Concentration." *Geographical Review of India*, 37, No. 3 (1975): 249-258.
Kang, Darshan S. "Deprivation of Hydro-Based Crop Regions in Punjab (India)." Ph.D. Dissertation, University of Nebraska, Lincoln, Nebr., 1975.
Mann, H.S. and R.P. Singh. "Crop Production in the Arid Zone with Special Reference to Western Rajasthan." *Annals of Arid Zone*, 14, No. 4 (1975): 347-357.
Mohammad, Noor. "Crop Association Regions in Ghaghara-Rapti Doab." *North East Geographer*, 7, No. 1-2 (1975): 3-9.
Sharma, B.M.P. "The Cropping Pattern in the Upper Tons Basin." *Deccan Geographer*, 13, No. 1-2 (1975): 185-198.
Chakravarti, A.K. "The Impact of the High-Yielding Varieties Program on Foodgrain Production in India." *Canadian Geographer*, 20 (1976): 199-223.
Chauhon, V.S., and Surendra Singh. "Concentration and Variation of Crops in Uttar Pradesh: A Study in Methodology." *Geographical Observer*, 12 (1976): 41-53.
Negi, B.S. "Crop-Combination Regions in India." *Geographical Observer*, 14 (1978): 38-52.
Govind Pathek, Har. "A Study in Methodology and Crop-Association Regions and Their Role in Agricultural Regionalization: A Case Study in Uttar Pradesh Himalaya." *National Geographical Journal of India*, 23, No. 1-2 (1977): 73-85.
Gupta, Jai P. "Distributional Patterns of Multiple Cropping in Upper Ganga Khadar." *Geographical Observer*, 13 (1977): 1-6.
Pal, B.K. "Adoption Period of High Yielding Varieties of Rice Among the Farmers of Bangladesh." *Oriental Geographer*, 21, No. 1-2 (1977): 44-54.
Singh, Tapeswar. "Bihar: A Study of Crop Combination Regions." *Indian Journal of Regional Science*, 9, No. 2 (1977): 145-156.
Thakur, Rameshwar. "Spatial and Temporal Variations in Crop Levels, Bihar." *Indian Geographical Studies*, 8 (1977): 64-72.

Barman, R., and M.M. Das. "Crop Combination Regions of Assam." *North Eastern Geographer*, 10, No. 1-2 (1978): 15-22.

Dhankar, Ram Pal. "Diversification of Crops in Punjab: A Case Study in Methodology." *The Geographer, Aligarh*, 25, No. 1 (1978): 57-63.

Duggal, S.L. "Crop Maturity Index for Classification of Dry Agricultural Lands." *Deccan Geographer*, 16, No. 1 (1978): 365-372.

Duggal, S.L. "Crop Maturity Index (CMI): A Simple Method for Functional Classification of Dry Agricultural Lands." *National Geographical Journal of India*, 24, No. 1-2 (1978): 29-35.

Farooqi, M.Y., and Jamshed Nasir. "The Production and Distribution of Food-Grain Protein in Uttar Pradesh." *The Geographer, Aligarh*, 25, No. 1 (1978): 8-14.

Gupta, Jai P. "Crop Combination Regions in Upper Ganga Khadar, Uttar Pradesh." *The Geographer, Aligarh*, 25, No. 2 (1978): 14-27.

Singh, Jhujar. "A New Technique of Delimiting Agricultural Production Typology in Foodcrop Dominated Economy: Haryana Case Study." *Geographia Polonica*, 40 (1979): 209-216.

Singh, Jhujar. "Spatio-Temporal Analysis of Crop Diversification in Punjab: 1951–1976." *Indian Geographical Journal*, 54, No. 1 (1979): 65-71.

Singh, Jhujar. "A Spatio-Temporal Analysis of Cropping Patterns and Crop Association in Punjab during 1951–1976." *National Geographical Journal of India*, 25, No. 3-4 (1979): 254-262.

Waseem-Ahmad, S., and Jamshed Nasir. "Changes in Production and Distribution of Foodgrain Proteins in Uttar Pradesh (1964–1974)." *National Geographical Journal of India*, 25, No. 1 (1979): 69-73.

Kayastha, S.L., and M.B. Singh. "Eastern Uttar Pradesh: Development Patterns and Potentials of the Sugar Industry." *National Geographer*, 15 (1980): 45-52.

Skukla, S.K. "Sugar Industry in the Madhya Bharat Plateau: A Case Study." *Deccan Geographer*, 18, No. 1 (1980): 746-759.

Bishnoi, B.P., and Ram Singh. "Crops and Cropping Patterns." Pp. 363-378 in *Perspectives in Agricultural Geography*, vol. 3, edited by Noor Mohammad. New Delhi, India: Concept Publishing Company, 1981.

Bishnoi, O.P. "Rotation of Crops." Pp. 379-398 in *Perspectives in Agricultural Geography*, vol. 3, edited by Noor Mohammad. New Delhi, India: Concept Publishing Company, 1981.

Chakrabarty, Satyesh. "Crop-Combination Regions of Eastern India." *Geographical Review of India*, 43, No. 2 (1981): 138-154.

Chatterjee, Kalpana, and Radindranath Bhattocharyya. "Agro-Industry of Murshidabad District: Sericulture and the Silk Industry." Pp. 459-484 in *Perspectives in Agriculturual Geography*, vol. 5, edited by Noor Mohammad. New Delhi, India: Concept Publishing Company, 1981.

Chauhon, V.S., and V.S. Khatana. "Temporal Changes in the Cropping-Pattern of the Yamuna-Hindan Tract." *Geographical Observer*, 17 (1981): 26-39.

Das, K.K. "Spatial Perspective on the Availability of Food Grains in North Bihar Plain: West of the Kosi." *Transactions, Institute of Indian Geographers*, 3, No. 2 (1981): 183-192.

Kulkarni, K.M. "Patterns of Urban Growth in India: Some Crop Regional Profiles." *Transactions, Institute of Indian Geographers*, 3, No. 1 (1981): 13-20.

Pal, S.K. "Crops, Land Use and Remote Sensing Inventories." Pp. 341-352 in *Perspectives in Agricultural Geography*, vol. 1, edited by Noor Mohammad. New Delhi, India: Concept Publishing Company, 1981.

Prakash, Jai. "Distributional Patterns of Multiple Cropping in Saharanpur (Uttar Pradesh)." Pp. 469-484 in *Perspectives in Agricultural Geography*, vol. 3, edited by Noor Mohammad. New Delhi, India: Concept Publishing Company, 1981.

Sandhu, D.S. *Geography of Sugarcane Cultivation: A Case Study of Haryana*. Kurukshetra, India: Vishal Publishers, 1981.

Siddiqi, M. Farooq. "Crop Combinations and Specialization in India." Pp. 145-158 in *Perspectives in Agricultural Geography*, vol. 4, edited by Noor Mohammad. New Delhi, India: Concept Publishing Company, 1981.

Singh, Surinder. "Dynamics of Cropping Pattern in Northern India." Pp. 485-506 in *Perspectives in Agricultural Geography*, vol. 3, edited by Noor Mohammad. New Delhi, India: Concept Publishing Company, 1981.

Subramaniam, A.R., and K. Uma Devi. "Rainfall Variability and Crop Yield in Orissa State." *Deccan Geographer*, 19, No. 1 (1981): 19-27.

Tanaka, Koji, and T. Watabe. "Traditional Systems of Small Farmers in the Central and Southern Deccan Plateau Area" (in Japanese). *Southeast Asian Studies* (Kyoto), 19, No. 2 (1981): 205-221.

Husain, Majid, and R. Sahu. "Areal Concentration of Rice in Assam." *National Geographical Journal of India*, 28, No. 1-2 (1982): 19-27.

Champa, Mitra. "Rice Production in Eastern India: Conflicting Theory and Practice." *Geographical Review of India*, 45, No. 1 (1983): 85-99.

Dayal, Edison. "Regional Response to High Yield Varieties of Rice in India." *Singapore Journal of Tropical Geography*, 4, No. 2 (1983): 87-98.

Dikshit, Jutta. "Identification of Drought-Prone Areas and Prediction of Crop Expectations in Maharashtra." *Geographical Review of India*, 45, No. 4 (1983): 48-81.

Mishra, Pratibha, and R. Bhattacharyya. "Cropping Pattern in Alwar District, Rajasthan." *Geographical Review of India*, 45, No. 1 (1983): 72-84.

Nadkarni, M.V., and R.S. Deshpande. "Growth and Instability in Crop Yields: A Case Study of Agriculture in Karnataka, South India." *Regional Studies*, 17, No. 1 (1983): 29-40.

Reddy, K.S., and A.K. Reddy. "Crop Concentration in Karimnager District, Andhra Pradesh: Emerging Problems." *Deccan Geographer*, 20-21, No. 3 and 1 (October 1982–June 1983): 295-300.

Soundara-Valli, G. "Efficiency of Rice Cultivation in the Chittar Basin, Tirunelvelli District, Tamil Nadu." *Indian Geographical Journal*, 52, No. 2 (1983): 183-190.

Uttaraprabha, M.U., and G. Soundara-Valli. "Evaluation of Crop Combinations in Tamil Nadu, India: A Case Study." *Rural Systems*, 1, No. 4 (1983): 289-294.

Ahmad, Qazi M., and M.F. Khan. "Changes in the Cropping Patterns of the Punjab Plain (1966–1967 and 1976–1977)." *The Geographer, Aligarh*, 31, No. 1 (1984): 14-31.

Das, M.M. "Crop Combination Regions of Assam: A Quantitative Analysis." *National Geographical Journal of India*, 30, No. 3 (1984): 165-172.

Dube, R.S., and R.P. Misra. "Evaluation of Foodgrain Availability in India: An Alternative Approach." *Asian Geographer*, 3, No. 1 (1984): 51-67.

Gajja, D.L., and T.S. Chauhon. "Factors Affecting Aggregate Crop Production: A Stepwise Approach." *National Geographer*, 19, No. 1 (1984): 69-72.

Maida, I. "Recent Trends in Rice Cultivation and Plantation Economy of the Western Ghats." Pp. 47-61 in *Geographical Field Research in South Asia in 1982*, edited by Kenzo Fujiwara. Hiroshima, Japan: University of Hiroshima Research and Sources Unit for Regional Geography, 1984.

Prasad, M. "Spatial Variations of Intensity and Diversity of Cropping in Bihar." *Rural Systems*, 2, No. 1 (1984): 55-62.

Ram Krishna, G.V.S., and M.D. Yadav. "Variation in Crop Production Efficiency in Indian Arid Districts." *National Geographical Journal of India*, 30, No. 3 (1984): 173-180.

Verma, R.S.P., and Indradeo Prasad. "An Analysis of Crop Combination Regions in the Lower Ghaghora-Gandak Doab." *National Geographer*, 19, No. 2 (1984): 143-152.

Mohammad, Ali, and K.Z. Amani. "Regional Imbalances in Yield and Growth Levels of Rice Production in India." *The Geographer, Aligarh*, 32, No. 2 (1985): 56-68.

Ramachandran, Janani. "The Changing Cropping Patterns in the Tribal Talukas of Dahana and Tahseri, Thane District, Maharashtra, 1901–1977/78." *National Geographical Journal of India*, 31, No. 3 (1985): 185-193.

Ramanaiah, Y.V., and N.B.K. Reddy. "The Changing Pattern of Yield Levels of Some Important Crops in Andhra Pradesh (1964–1979)." *Deccan Geographer*, 23, No. 1 (1985): 1-29.

Sangwon, S.S. "Dynamics of Cropping Pattern in Haryana: A Supply Response Analysis." *The Developing Economies* (Tokyo), 23, No. 2 (1985): 173-186.

Sharma, Tara C., and O. Coutinho. "Disparities in Foodgrains Production in Uttar Pradesh: A Spatial Temporal Analysis." *Transactions, Institute of Indian Geographers*, 7, No. 1 (1985): 67-72.

Siddiqi, M. Farooq. "Spatio-Temporal Changes in Crop Association in Bundelkhand (Uttar Pradesh)." *The Geographer, Aligarh*, 32, No. 1 (1985): 65-76.

Van Spengen, W. "Dry Grain Farming Families in Dharmapuri District: A Case Study from the Northern Thally Block." *Indian Geographical Journal*, 60, No. 2 (1985): 117-132.

Vidyanath, V. "Crop Productivity in Relation to Cropland in Andhra Pradesh: A Spatial Analysis." *Transactions, Institute of Indian Geographers*, 7, No. 1 (1985): 49-55.

Vidyanath, V. "Cropped and Irrigated Lands in Andhra Pradesh, India: A Study on Their Association." *Indian Geographical Journal*, 60, No. 1 (1985): 21-24.

Mohi-ud-din-Mir, G.H. "Saffron Cultivation in the Kashmir Valley—A Case Study of Wasura Village: Saffron Cultivation and Its Role in Socio-Economic Development." *Annals of the National Association of Geographers, India*, 6, No. 2 (1986): 61-68.

Ramanaiah, Y.V., and N.B.K. Reddy. "Crop Regions in Andhra Pradesh." *National Geographer*, 21, No. 2 (1986): 121-130.

Ramanathan, Y.V. "A Note on the Crop Diversification in Andhra Pradesh." *Transactions, Institute of Indian Geographers*, 8, No. 2 (1986): 67-70.

Tripathi, R.S. "Cropping Pattern in Bundelkhand, Uttar Pradesh." *National Geographer*, 21, No. 2 (1986): 161-172.

Bhattacharyya, R. "Temporal and Spatial Changes in the Cultivated Areas of West Bengal." *Geographical Review of India*, 49, No. 4 (1987): 56-62.

Chakravarti, A.K. "Availability of Cattle Fodder in India." *Geographical Review*, 77, No. 2 (1987): 209-217.

Jana, M.M. "Cropping Pattern in West Bengal." *Geographical Review of India*, 49, No. 3 (1987): 13-22.

Kuniyal, J.C. "Crop Concentration and Diversification in Nainital District, Uttar Pradesh Himalaya." *National Geographical Journal of India*, 33, No. 2 (1987): 160-167.

Prasad, Indradeo. "Crop-Combination Regions in the Lower Ghaghra-Gandak Doab, Bihar." *Geographical Perspective*, 1, No. 2 (1987): 39-48.

Prasad, S.N. "Production of Major Kharif Crops in India: Problems and Prospects." *Geographical Perspective*, 1, No. 2 (1987): 33-38.

Ram, Daya, and S.P.S. Karwarsa. "Crop Maturity Patterns in Rainfed Areas of Bowal Tahsik, Haryana." *Geographical Review of India*, 49, No. 2 (1987): 19-26.

Daulay, H.S., and A. Henry. "Crop Diversification and Alternative Land Use System for Rural Land Use Planning in Indian Arid Zone." *The Geographer, Aligarh*, 35, No. 2 (1988): 13-22.

Kuniyal, J.C. "Crop Combination Regions in District Nainital, Uttar Pradesh Himalaya." *Deccan Geographer*, 26, No. 1 (1988): 271-282.

Put, M., and W. Van Spengen. "A Comparative Analysis of Cereal Production and Consumption in Semi-Arid India." *Journal of Rural Studies*, 4, No. 4 (1988): 377-388.

Reddy, M.V., and N.B.K. Reddy. "Changing Pattern of Crop Combinations in Chittoar District." *National Geographer*, 23, No. 2 (1988): 89-96.

Sharma, S.K., and Ajit Kumar Jain. "Diffusion of Innovations in the Cotton Growing Tract of Madhya Pradesh: A Case of Pesticides." *The Geographer, Aligarh*, 35, No. 2 (1988): 34-44.

Singh, Hat Ram. "Assessment and Exploitation of Horticulture in Kumaun Himalaya: Some Observations." *Geographical Viewpoint*, 17-18, No. 1-2 (1988–1989): 91-96.

Singh, Poornima S., and Anjani K. Singh. "Indian Sugar Industry: A Case Study of Bihar." *Geographical Perspective*, 2, No. 2 (1988): 34-39.

Tripathi, R.S., and J.P. Viswakarma. "Land-Use Cropping Pattern and Development Levels in Banda District." *Deccan Geographer*, 26, No. 2-3 (1988): 417-428.

Ali, Gul S., and S.M.S. Hars. "Agricultural Facilities and Food Crops Production in Surguja District." *Deccan Geographer*, 27, No. 1 (1989): 475-492.

Coutinho, O., and Tara C. Sharma. "Disparity in Crop Yields in Uttar Pradesh: A Diagnostic Inquiry." *Annals of the National Association of Geographers, India*, 9, No. 1 (1989): 28-40.

Das, Harmohan, and M.M. Das. "Impact of Population Structure on Intensity of Cropping in the Plain of Pagldia-Puthimari Basin, Assam." *Annals of the National Association of Geographers, India*, 9, No. 2 (1989): 49-62.

Hashmi, S.N.I. "Trends in the Production of Pulses in Western Uttar Pradesh." *Philippine Geographical Journal*, 33, No. 4 (1989): 191-200.

Sharma, Tara C. *Green Revolution Crops: A Geographical Analysis*. Jaipur, India: Rawat Publications, 1989.

Singh, N.P., and M.C. Joshi. "Cropping Systems in Nagaland: An Alternative Strategy." *Rural Systems*, 7, No. 2 (1989): 135-140.

Sohal, K.S., and R.S. Khaira. "Temporal Changes in Crop Combination Regions in Sangrur District, Punjab." *National Geographer*, 24, No. 2 (1989): 117-128.

Swaminathan, E. "Measuring the Change in Crop Pattern: A Study for Madurai District, Tamil Nadu." *Indian Geographical Journal*, 64, No. 1 (1989): 61-64.

Teli, B.L. "Crop Combinations in India." *Indian Geographical Journal*, 64, No. 1 (1989): 96-98.

Vaid, Varsha, and V.S. Datye. "Association Between Water Deficit and the Yields of Principal Cereal Crops in Maharashtra." *Geographical Review of India*, 51, No. 2 (1989): 68-85.

Gupta, B.R.D., and K.K. Singh. "Estimating Cotton Yield at Some Stations in India Based on Weather Parameters." *National Geographical Journal of India*, 36, No. 3 (1990): 177-182.

Santra, Syamapada, Pannelal Das, and Arabeida Das. "Crop Combination Regions of Haora District: A Spatio-Temporal Analysis." *Transactions, Institute of Indian Geographers*, 12, No. 1 (1990): 57-72.

Singh, Abha L., and S. Fazel. "Impact of Price Fluctuations on Sugarcane Area in Saharanpur District, Uttar Pradesh." *National Geographer*, 25, No. 1 (1990): 75-80.

Singh, Abha L., and Sadhana Tyagi. "A Component of the Growth of Crop Output in Aligarh District (India)." *Rural Systems*, 8, No. 2 (1990): 87-98.

Sohal, K.S. "Regional Disparities in Crop Productivity in Punjab, India." *National Geographer*, 25, No. 2 (1990): 107-114.

Sohal, K.S., and A.K. Saini. "Crop Productivity Regions in Punjab, 1985–1986." *The Geographer, Aligarh*, 37, No. 2 (1990): 5-13.
Vaidya, Balakrishna C. "Cropping Pattern in Yashoda Basin." *Deccan Geographer*, 28, No. 2-3 (1990): 699-711.
Bora, A.K., D. Kotoky, and M.C. Bhuyan. "Effective Control of Shifting Cultivation Practices in North-Eastern India." *North Eastern Geographer*, 23, No. 1-2 (1991): 23-31.
Hashmi, S.N.J. "Production of Pulses in Western Uttar Pradesh." *Indian Geographical Journal*, 66, No. 1 (1991): 20-26.
Khanna, Subha. "Change in Cropping Pattern and Its Impact on Society and Environments in Gujarat." *The Geographer, Aligarh*, 38, No. 1 (1991): 45-51.
Singh, Arun K. "Development of Commercial Plantation Agriculture in Nilgiri-Wynad, Tamil Nadu." *Geographical Review of India*, 53, No. 1 (1991): 92-100.
Vaidya, Balakrishna C. "Temporal Change in the Cropping Pattern of the Cotton Region of the Yasoda Basin, Maharashtra." *National Geographer*, 26, No. 1 (1991): 31-40.
Jansen, H.G.P. "Inter-Regional Variation in the Speed of Adoption of Modern Cereal Cultivers in India." *Journal of Agricultural Economics*, 43, No. 1 (1992): 88-95.
Landy, F. "Le developpement de la canne a sucre dans le sud de l'Inde: Le developpement pour qui?" *Annales de Geographie*, 101 (1992): 147-173.
Nagarathinam, S.R., and N. Sivagnanam. "A Study on Crop Seasons and Water Availability in Coimbatore District." *Indian Geographical Journal*, 67, No. 1 (1992): 20-29.
Saini, A.K. "Regional Differentials in Productivity Level of Wheat and Rice: A Case Study of Punjab, India." *National Geographer*, 27, No. 1 (1992): 43-50.
Singh, Abha L., and S. Fazel. "Changes in Cropping Pattern Due to Variation in Prices in Upper Ganga Yamuna Doab." *The Geographer, Aligarh*, 39, No. 1 (1992): 16-32.
Gupta, Ranjana. "Soil Fertility and Productivity of the Major Crops of the Sundarban in South 24 Parganas, West Bengal." *Geographical Review of India*, 55, No. 4 (1993): 77-84.
Kumar, Vijay, and Raj Kumar. "A Note on the Cultivation of Sunflower in Haryana: Some Recent Trends." *Transactions, Institute of Indian Geographers*, 15, No. 2 (1993): 35-44.
Reddy, G.V., and M.V. Reddy. "Spatial Distribution of Groundnut Cultivation and Industry in Rayalaseema." *National Geographer*, 28, No. 2 (1993): 149-156.
Sati, V.P. "Cropping Pattern in the Hill Environment of Garhwal Himalaya: A Case Study of a Village." *National Geographer*, 28, No. 1 (1993): 31-38.
Sohal, K.S. "Crop Combination Regions of Punjab: Pre- and Post-Green Revolution Period." *National Geographer*, 28, No. 2 (1993): 115-126.
Ashrof, Jaweed. "Maize in India: Introduction or Indigenous?" *Annals, National Association of Geographers, India*, 14, No. 2 (1994): 6-26.

Das, Sarbeswar, and M.M. Das. "Intensity of Cropping in the South Bank Region of Kamrup District." *Geographical Review of India*, 56, No. 1 (1994): 34-42.

Kumar, Raj. "Spatial and Temporal Changes in Wheat Cultivation in Haryana: 1970–1980 to 1989–1990." *Geographical Review of India*, 56, No. 2 (1994): 57-68.

Kumar, Suresh, and M.H. Qureshi. "Spatial Spread of Maize Cultivation in India." *Indian Geographical Journal*, 69, No. 1 (1994): 19-27.

Singh, Abha L., and S. Fazel. "Factors Deciding Farmers' Choice of Crops." *Asian Profile*, 22, No. 5 (1994): 413-426.

Soulance, Dominique. "La place de la sericulture dans la vie des paysans du sud du Karnataka (Inde)." *Les Cahiers d'Outre Mer*, 46 (1994): 199-224.

Yadav, Ram Niwas. "Intensity of Cropping in Narnaul Tehsil (Haryana)." *Geographical Review of India*, 56, No. 3 (1994): 75-79.

Bhagabati, A.K. "Cropping Intensity Pattern in Nalbari District, Assam: An Analysis of Its Spatial Variation." *Geographical Review of India*, 57, No. 1 (1995): 78-86.

Bhagabati, A.K., and S.K. Talukar. "Pattern of Cropping Intensity in the Command Area of Kaldiya Irrigation Project, Barpeta District, Assam." *North Eastern Geographer*, 26, No. 1-2 (1995): 5-10.

Das, Debashis. "Relationship Between Size of Farm and Productivity of Crops in Birbhum District: A Case Study." *Geographical Perspective*, 9, No. 1-2 (1995): 32-41.

Iqbal, Zafor, and Shafaat Ali Khan. "Regional Imbalances in Food Crop Production in Jammu and Kashmir." *Geographical Review of India*, 57, No. 4 (1995): 274-285.

Rao, K.B., and G. Rao. "Vegetable Cultivation in Rajasthan: A Study on Predictions of Crop Water Requirements." *Indian Geographical Journal*, 70, No. 2 (1995): 104-106.

Singh, Abha L., and Shahab Fazaj. "Factors Accounting for Changes in Area Under Selected Crops in Muzeffarnagar District." *The Geographer, Aligarh*, 42, No. 1 (1995): 9-17.

Yadav, S.C., et al. "Productivity of Crops in Red and Lateritic Soils." *Geographical Review of India*, 57, No. 4 (1995): 314-322.

Hugar, S.I. "Impact of South-West Monsoon Rainfall Trend on Cropping Pattern: A Case Study of Semi-Malnad Region in Dharwad District." *Transactions, Institute of Indian Geographers*, 18, No. 1 (1996): 59-64.

Madhuramma, C., and M. Sambasiva Rao. "Water Availability and Crop Management of the Nellore District, Andhra Pradesh, India." *Transactions, Institute of Indian Geographers*, 18, No. 2 (1996): 15-22.

Ramachandran, Kaushalya, and M.N. Reddy. "Impact of Variability of Rainfall on Foodcrop Production Systems in Andhra Pradesh." *Indian Geographical Journal*, 71, No. 2 (1996): 113-118.

Saced, M.I. "An Assessment of Cropping Intensity in Village Sia Khas: A Case Study." *Geographical Review of India*, 58, No. 2 (1996): 153-161.

Shivaramu, H.S., et al. "Crop Planning for Arid Ecosystem." *Geographical Review of India*, 58, No. 4 (1996): 340-349.

Chaudhary, J.L., et al. "Crop Productivity as Influenced by Climatic Swings: A Case Study of Chhattisgarh Region." *Geographical Review of India*, 59, No. 4 (1997): 330-339.

Desothali, Vrishali. "Crop Planning for Ahmednagar and Solapur Districts, Maharashtra." *Indian Geographical Journal*, 72, No. 1 (1997): 20-26.

Hooda, Randlur S. "Transformation of Wheat- (Iriticum sativun) Cultivation in the Drought-Prone Areas of Jul Lift Land Command in South Western Haryana." *Deccan Geographer*, 35, No. 2 (1997): 139-146.

Sharma, Tara C. "Trends in Cropping Intensity in Karnataka: A Regional Profile." *Transactions, Institute of Indian Geographers*, 19, No. 1 (1997): 17-30.

Yadav, Ram Niwas. "Oilseeds Cultivation in Narnaul Tehsil, Haryana, 1986–1990." *Geographical Review of India*, 59, No. 1 (1997): 47-53.

Barah, Bimal. "Size of the Tea Estate and Productivity of Tea: A Case Study." *Deccan Geographer*, 36, No. 1 (1998): 11-18.

Bhatt, Sangeeta, and Harsha P. Bhatt. "Crop Diversification and Crop Combination in Uttar Kashi." *Geographical Review of India*, 61, No. 4 (1999): 398-402.

ANIMAL HUSBANDRY

Singh, V.R., and Y.V. Ramanaiah. "Spatial Distribution of Livestock and Its Combinations in Rayala Suma, Andhra Pradesh." *National Geographical Journal of India*, 5, No. 1 (1959): 118-128.

Kayastha, S.L. "Animal Husbandry in the Himalayan Beas." *National Geographical Journal of India*, 7, No. 2 (1961): 116-123.

Kayastha, S.L. "Animal Husbandry and Its Development in Himachal Pradesh." *National Geographical Journal of India*, 12, No. 2 (1966): 87-95.

Nitz, Hans-Jurgen. "Regionale Formen der Viehwirtschaft zwischen der Thar und dem oberen Ganges." *Geographische Rundschau*, 20, No. 11 (1968): 414-426.

Debysingh, Molly. "Poultry and Cultural Distributions in India." Ph.D. Dissertation, Syracuse University, Syracuse, N.Y., 1970.

Tewari, A.K. "Rough Pastures and Pastoral Activities of Jaisalmer." *Indian Geographical Journal*, 45, No. 1-2 (1970): 46-51.

Prasad, Har. "The Dehra Dun: A Geonomic Study in Migratory Pastoralism." *Geographical Viewpoint*, 4, No. 1-2 (1973): 27-40.

Sharma, J.P. "Pastoral Economy and Transhumance in the Himalayan Ravi-Chenab Interfluve." *Geographical Review of India*, 35, No. 3 (1973): 232-247.

Dayal, Edison. "Factors Affecting Distribution of Cattle in India." *Geographical Bulletin*, 11 (1975): 7-19.

Hoffpauir, Robert. "The Indian Milk Buffalo: A Paradox of High Performance and Low Reputation." *Asian Profile*, 5, No. 2 (1977): 111-134.

Singh, V.R., and Y.V. Ramanaiah. "Spatial Distribution of Livestock and Its Combinations in Rayalaseema (Uttar Pradesh)." *National Geographical Journal of India*, 25, No. 2 (1979): 118-128.

Debysingh, Molly. "The Cultural Geography of Poultry Keeping in India." Pp. 105-129 in *An Exploration of India: Geographical Perspectives in Society and Culture*, edited by David E. Sopher. London, England: Longman, 1980.

Kataria, Man S. *Geography of Indian Livestock and White Revolution*. Kurukshetra, India: Vishal Publications, 1982.

Chakravarti, A.K. "Some Characteristics of Spatial Distribution of Cattle Population in India." *Rural Systems*, 2, No. 2 (1984): 67-76.

Hema Malini, B. "Climatic Control on Livestock Distribution in Andhra Pradesh." *Deccan Geographer*, 22, No. 1-2 (1984): 473-486.

Nakasato, Isuguo. "Livestock Farming in Kurubathur and Yadavarahalli." Pp. 62-75 in *Geographical Field Research in South India in 1982*, edited by Kenjo Fujiwara. Hiroshima, Japan: University of Hiroshima Research and Sources Unit for Regional Geography, 1984.

Nakasato, Isuguo. "Livestock Economy in the Tropical Heavy Rain Region." Pp. 149-165 in *Geographical Field Research in South India in 1982*, edited by Kenzo Fujiwara. Hiroshima, Japan: University of Hiroshima Research and Sources Unit for Regional Geography, 1984.

Chakravarti, A.K. "The Question of Surplus Cattle in India: A Spatial View." *Geografiska Annaler B*, 67B, No. 2 (1985): 121-130.

Chakravarti, A.K. "Cattle Development Problems and Programs in India: A Regional Analysis." *GeoJournal*, 10, No. 1 (1985): 21-45.

Das, Manamohan. "Problems of Draught Animals in Assam." *Geographical Review of India*, 47, No. 4 (1985): 19-25.

Nakasato, T. "Bovine Keeping in the Villages of Western Ghats" (in Japanese). *Annual Report, Research Center for Regional Geography* (Hiroshima), 1 (1989): 25-100.

Rajandran, S. "Livestock Composition in Thanjavur District." *Geographical Review of India*, 53, No. 2 (1991): 63-68.

Rajandran, S. "Utilization Patterns of Draught Animal Power in Cavery Delta of Thanjavar District." *Geographical Review of India*, 55, No. 1 (1993): 63-72.

Haloi, Kanak. "Livestock Sector in Assam: An Assessment of Its Present Status." *North Eastern Geographer*, 25, No. 1-2 (1994): 76-87.

Ramanaiah, Y.V. "Livestock: An Alternative Enterprise for Sustainable Agricultural Development in Andhra Pradesh." *Transactions, Institute of Indian Geographers*, 18, No. 2 (1996): 45-54.

DAIRYING

Ellefson, Richard. "The Milk Supply of Major Indian Cities." Ph.D. Dissertation, University of California, Berkeley, Calif., 1968.

Irwin, P.G. "The Dairying Industry of Kaira, Western India." *South Asia*, 1 (1971): 123-131.

Noble, William A. "The Toda and Their Dairies in the Nilgiris, South India." *East Lakes Geographer*, 24 (1989): 97-114.

Grieu, E. "A propos d'une cooperative laitiere: Pratiques paysannes et visees technocratiques, le compromis a l'indienne." *Cahiers Geographique de Rouen*, 33 (1990): 7-16.

Datye, V.S., and A.A. Devikar. "Economies of Dairy Units in Ahmednagar City, Maharashtra." *Geographical Review of India*, 55, No. 4 (1993): 12-23.

FARM LAND

Muzaffor Ali, S. "Evolution of the Field Pattern in the Indo-Gangetic Divide." *Pakistan Geographical Review*, 1 (1942): 27-35.

Ganguli, B.N. "Land Use Survey and Agricultural Planning in Uttar Pradesh." *Geographical Review of India*, 15, No. 2 (1953): 5-10.

Karan, P.P. "Land Utilization and Agriculture in an Indian Village." *Land Economics*, 33, No. 1 (1957): 55-64.

Sen, Amal K. "Land Utilization and Agricultural Planning in Matpalsa Union of Birbhum, West Bengal." *Geographical Review of India*, 19, No. 4 (1957): 27-41.

Goswami, S.B. "Misirgonda: A Prototype of Agricultural Land Use in the Central Ranchi Plateau." *Geographical Review of India*, 22, No. 2 (1960): 7-15.

Singh, Basant. "Distribution of Cultivable Waste and Possibilities of Its Reclamation in Chakia (Banaras)." *National Geographical Journal of India*, 8, No. 1 (1962): 59-72.

Banerjee, Bireswar. "Changing Cropland of West Bengal." *Geographical Review of India*, 24, No. 1 (1964): 11-22.

Shinde, Suresh D., and K.S. More. "Coastal Maharashtra: A Study in Ranking of Landuse and Cropping Pattern." *National Geographer*, 7 (1964): 37-48.

Jha, B. "Land Capability Classification of the Kosi Region." *Geographical Outlook*, 4 (1965): 75-80.

Sinha, R.N.P. "Present Land-Use Pattern in Canal Irrigated Area of Patna District, Bihar." *Geographical Outlook*, 4 (1965): 63-74.

Nitz, Hans-Jurgen. "Formen Bauerlicher Landnutzung und iher raumlicher Ordnung im Vorderen Himalaya von Kumoan (Nordwestindien)." *Heidelberger Geographische Arbeiten*, 15 (1966): 311-330.

Randhawa, N.S., and E.O. Heady. "L'utilization regionale du sol dans la planification du developpement agricole en Inde." *Cahiers Institut Scientifique Economie Applique*, 169 (1966): 19-36.

Roy, B.K. "Nature and Pattern of Cultivable Waste Lands in the Eastern Part of the Ganga-Ghaghara Doab." *Deccan Geographer*, 4, No. 2 (1966): 50-62.

Singh, B.B. "Land Use, Cropping Pattern and Their Ranking." *National Geographical Journal of India*, 13, No. 1 (1967): 1-13.

Amani, K.Z. "Harnarayanpur: A Study in the Changes in Agricultural Land Use." *Geographical Outlook*, 5 (1968): 71-79.

Vats, P.C. "Land Utilization and Farming Efficiency in Village Khatta Prahaladpur." *Deccan Geographer*, 6, No. 2 (1968): 87-100.

Bhardwaj, B.K. "Changes in Current Fallow in Gurgaon Development Block." *Geographical Observer*, 6 (1969): 104-116.

Rana, Ratna S. "A Game Theoretical Approach to Agricultural Land-Use Patterns: An Indian Example." Ph.D. Dissertation, University of Pittsburgh, Pittsburgh, Pa., 1969.

Singh, S.N. "Distribution and Changing Pattern of Cultivable Wastes in Kerakat Tahsil District, Jaunpur, Uttar Pradesh." *National Geographical Journal of India*, 15, No. 3-4 (1969): 207-223.

Bhatia, Shyam S. "Dynamic Approach to the Analysis of Change: A Case Study of Cropland Use in Uttar Pradesh, India." *Geographia Polonica*, 19 (1970): 227-248.

Jalal, D.S. "Land Utilization in Boman Gaon (Kota Dun)." *National Geographical Journal of India*, 16, No. 2 (1970): 127-149.

Singh, Kashi N., and Babba Singh. "Landuse Cropping Pattern and Their Ranking in Shahganj Tahsil: A Geographical Analysis." *National Geographical Journal of India*, 16, No. 3-4 (1970): 221-235.

Chauhon, V.S., and Surendra Singh. "Changing Agricultural Land Use and Caloric Production in V. Banwaripur (Meerul): Use of Some Working Formulae" (in Hindi). *Geographical Observer*, 7 (1971): 26-38.

Mukherjee, A.B. "Geographic Patterns of Changes in Agricultural Wastelands in Uttar Pradesh, 1951–1967." *Geographical Review of India*, 33, No. 2 (1971): 120-132.

Singh, Jasbir. "Agricultural Colonization of Cultivable Wasteland in India." *Deccan Geographer*, 9, No. 2 (1971): 135-149.

Andress, Joel M. "The Morphology of Agricultural Terraces in the Kumaon Himalayas." *Yearbook, Association of Pacific Coast Geographers*, 34 (1972): 7-20.

Digar, S. "Land Use Capability Classification of the Catchment Area Above the Dam of the Kangasabati Project for Increased Production." *Bulletin of the Indian National Science Academy*, 44 (1972): 47-58.

Lal, K.K. "Field Patterns After Consolidation: A Land Reform Measure." *National Geographical Journal of India*, 18, No. 3-4 (1972): 213-222.

Ray, Aparna. "Landuse and Major Agricultural Characteristics of the Damodar-Saraswati Doab." *Geographical Review of India*, 34, No. 1 (1972): 28-37.

Sharma, Tara C. "Patterns of Cropland Use in Uttar Pradesh." *Deccan Geographer*, 10, No. 1 (1972): 1-17.

Singh, B.B., and Govind Singh. "Spatial Analysis of Land-Use and Cropping Pattern of Nanauta Block, District Saharanpur." *Geographical Viewpoint*, 3, No. 1-2 (1972): 39-52.

Mohammad, Noor. "Misraulia: A Study in Agricultural Land Use." *Geographical Review of India*, 35, No. 3 (1973): 277-288.

Murton, Brian J. "Folk Classification of Cultivated Land and Ecology in Southern India." *Proceedings, Association of American Geographers*, 5 (1973): 199-202.

Roy, B.K. "Determination of Land Use Changes, Arable Potentials and Land Use Development in West Bengal: An Analytical Investigation in Agricultural Typology." *National Geographical Journal of India*, 19, No. 1 (1973): 15-29.

Shinde, Suresh D., and K.S. More. "Coastal Maharashtra: A Study in Ranking of Land Use and Cropping Pattern." *National Geographer*, 8 (1973): 37-48.

Siddiqi, Mohammad. "Agricultural Land Use in the Black Soil Regions of Bundelkhand." *Geographical Review of India*, 35, No. 4 (1973): 341-354.

Akhtar, Rais. "Agricultural Land Use and Nutrition in the Greater Himalaya: A Case Study." *The Geographer, Aligarh*, 21, No. 2 (1974): 37-51.

Chauhon, V.S., and Surendra Singh. "The Study of Regional Patterns and Land Use Planning in Meerut District (Uttar Pradesh)" (in Hindi). *Geographical Observer*, 10 (1974): 22-32.

Negi, B.S. "Agricultural Land Use of Garhwal District." *Geographical Observer*, 10 (1974): 66-72.

Raagaard, S. "The Tilling in a Village in Southern India." *Geografisk Tidsskrift*, 73 (1974): 66-82.

Singh, B.B. "Rural Land Resource Use of Barant Block: A Geographical Analysis." *Geographical Observer*, 11 (1975): 32-47.

Singh, V.R. "Agricultural Land Use Planning in Mirzapur: A Regional Synthesis." *National Geographical Journal of India*, 21, No. 1 (1975): 6-19.

Gupta, Jai P. "Distributional Patterns of Agricultural Waste Lands in Upper Ganga Khadar." *Geographical Viewpoint*, 7, No. 1-2 (1976): 7-10.

Siddiqi, Niaz A., and Mumtaz Ahmad. "Spatio-Temporal Changes in the Cropland Use Efficiency in the Ganga-Yamuna Doab." *The Geographer, Aligarh*, 23, No. 2 (1976): 1-10.

Singh, J.S., and Ram Singh. "Agricultural Land Efficiency in Haryana (Spatio-Temporal Analysis)." *Geographical Viewpoint*, 7, No. 1-2 (1976): 32-40.

Shafi, M. "Assessment of Von Thunen's Land Use Analysis in India." *The Geographer, Aligarh*, 24, No. 1 (1977): 1-13.

Majud, A. "Changing Patterns of Cropland Use in the Union Territory of Delhi." Master's Thesis, Aligarh Muslim University, Aligarh, India, 1978.

Mohammad, Noor. *Agricultural Land Use in India*. New Delhi, India: Concept Publishing Company, 1978.

Tawde, M.D. "Geomorphic Studies and Agricultural Land Use Planning." *National Geographer*, 13 (1978): 75-80.

Sen, Jyotirmoy, and S. Sen. "Changing Agricultural Landscape of the Bhagirathi-Churni Interfluve, West Bengal." *Geographical Review of India*, 41, No. 2 (1979): 150-159.

Bhattacharyya, R. "Changing Land Use and Cropping Pattern in Bihar." Pp. 437-468 in *Perspectives in Agricultural Geography*, vol. 3, edited by Noor Mohammad. New Delhi, India: Concept Publishing Company, 1981.

Burman, Savitri G. "Land Ownership and Agricultural Practices of the Tribes of Arunachal Prade." Pp. 319-338 in *Perspectives in Agricultural Geography*, vol. 2, edited by Noor Mohammad. New Delhi, India: Concept Publishing Company, 1981.

Chaturvedi, Arun. "The Concept of Land Capability and Suitability." Pp. 61-74 in *Perspectives in Agricultural Geography*, vol. 3, edited by Noor Mohammad. New Delhi, India: Concept Publishing Company, 1981.

Das, K.K., and K.N. Das. "Agricultural Land Use and the Pressure of Population in North Bihar Plain West of the Kosi River." Pp. 55-80 in *Perspectives in Agricultural Geography*, vol. 5, edited by Noor Mohammad. New Delhi, India: Concept Publishing Company, 1981.

Das, Pannalal, and R. Bhattacharyya. "Land Capability Classes vis-a-vis Land Use Planning Regions: A Case Study of Kaliaghai River Basin." Pp. 123-143 in *Perspectives in Agricultural Geography*, vol. 3, edited by Noor Mohammad. New Delhi, India: Concept Publishing Company, 1981.

Datye, V.S., and Jaymala Diddee. "Microlevel Analysis of Land Holding and Agriculture Practice: Case Studies." *Transactions, Institute of Indian Geographers*, 3, No. 1 (1981): 89-99.

Dutt, G.K. "Structure and Functions of Agricultural Holdings in Selected Regions of India." *Geographical Review of India*, 43, No. 1 (1981): 25-40.

Ghosh, Dhira. "Impact of Irrigation on Land Use and Cropping Intensity." Pp. 503-532 in *Perspectives in Agricultural Geography*, vol. 2, edited by Noor Mohammad. New Delhi, India: Concept Publishing Company, 1981.

Gupta, R.K. "Ecological Approach to Land Capability Classification and Regional Planning." Pp. 89-108 in *Perspectives in Agricultural Geography*, vol. 3, edited by Noor Mohammad. New Delhi, India: Concept Publishing Company, 1981.

Kumar, Anil, and G.P. Sharma. "Land Use and Nutrition Level in Paspura Anchal, Bihar." Pp. 161-178 in *Perspectives in Agricultural Geography*, vol. 5, edited by Noor Mohammad. New Delhi, India: Concept Publishing Company, 1981.

Lahiri, Sulekha. "Land Use in Eastern Himalaya Region: A Case Study." Pp. 305-332 in *Perspectives in Agricultural Geography*, vol. 3, edited by Noor Mohammad. New Delhi, India: Concept Publishing Company, 1981.

Mathur, R.N., and Rama Prasad. "Impact of Consolidation on Agricultural Land Use Pattern: A Case Study." Pp. 355-362 in *Perspectives in Agricultural Geography*, vol. 2, edited by Noor Mohammad. New Delhi, India: Concept Publishing Company, 1981.

Mohammad, Noor. "Patterns of Agricultural Land Use in Ghaghara-Rapti Doab." Pp. 245-270 in *Perspectives in Agricultural Geography*, vol. 3, edited by Noor Mohammad. New Delhi, India: Concept Publishing Company, 1981.

Pal, Indra, and Lakshmi Shukla. "Changing Agricultural Land Use in the Hilly Tracts of Rajasthan." Pp. 47-60 in *Perspectives in Agricultural Geography*, vol. 3, edited by Noor Mohammad. New Delhi: Concept Publishing Company, 1981.

Priya, Indu, and Anil Kumar. "Agricultural Land Classification in Ranchi Plateau: A Case Study of Bhandra Block." In *Perspectives in Agricul-*

tural Geography, vol. 3, edited by Noor Mohammad. New Delhi, India: Concept Publishing Company, 1981.

Qureshi, M.H., and S. Aparna. "Quest for Land Reforms: An Experience in Kerala." Pp. 273-288 in *Perspectives in Agricultural Geography*, vol. 2, edited by Noor Mohammad. New Delhi, India: Concept Publishing Company, 1981.

Rana, J.S. "A Study of the Land Use in the Dhauli Valley: Garhwal Himalaya." *Geographical Observer*, 17 (1981): 69-71.

Sen, Jyotirmoy, and Sukla Sen Dasgupta. "Manikdihi: A Study in Land Holding, Tenancy System and Agricultural Development (1925–1975)." Pp. 363-398 in *Perspectives in Agricultural Geography*, vol. 2, edited by Noor Mohammad. New Delhi, India: Concept Publishing Company, 1981.

Sethi, Madhurama. "Problems of Wastelands and Their Reclamation." Pp. 145-157 in *Perspectives in Agricultural Geography*, vol. 3, edited by Noor Mohammad. New Delhi, India: Concept Publishing Company, 1981.

Shafi, M. "Assessment of Von Thunen's Land Use Analysis in India." Pp. 31-44 in *Perspectives in Agricultural Geography*, vol. 4, edited by Noor Mohammad. New Delhi, India: Concept Publishing Company, 1981.

Sharma, S.C. "Land Capability Classification and Land Use Planning: A Case Study." Pp. 109-122 in *Perspectives in Agricultural Geography*, vol. 3, edited by Noor Mohammad. New Delhi, India: Concept Publishing Company, 1981.

Sharma, S.C. "Land Capability Classification and Land Use Planning: A Study in Methodology." Pp. 75-88 in *Perspectives in Agricultural Geography*, vol. 3, edited by Noor Mohammad. New Delhi, India: Concept Publishing Company, 1981.

Sharma, S.C. "Land Classification." Pp. 29-46 in *Perspectives in Agricultural Geography*, vol. 3, edited by Noor Mohammad. New Delhi, India: Concept Publishing Company, 1981.

Shastri, Prabha. "Changes in Land Use and Cropping Pattern in the Cotton Belt of Vidarbha." Pp. 507-522 in *Perspectives in Agricultural Geography*, vol. 3, edited by Noor Mohammad. New Delhi, India: Concept Publishing Company, 1981.

Siddiqi, Niaz A. "Land Classification for Agricultural Planning." Pp. 13-20 in *Perspectives in Agricultural Geography*, vol. 3, edited by Noor Mohammad. New Delhi, India: Concept Publishing Company, 1981.

Das, M.M. "Agricultural Land Use and Cropping Pattern in Assam." *Geographical Review of India*, 44, No. 4 (1982): 32-39.

Diddee, Jayamala, G. Kewalramani, and V.S. Datye. "Regional Changes of Agricultural Land to Non-Agricultural Uses in Bombay–Poona Corridor." *Geographical Review of India*, 44, No. 2 (1982): 48-62.

Datye, V.S., Jayamala Diddee, and D.G. Gaikwad. "Process of Land Fragmentation and Effects of Land Consolidation on Cropping Patterns and Productivity: A Case Study of Village Nigudi, Pune District." *Annals of the National Association of Geographers, India*, 3, No. 1 (1983): 9-18.

Ramanaiah, Y.V., and N.B.K. Reddy. "Carrying Capacity of Land in Andhra Pradesh." *Indian Geographical Journal*, 58, No. 2 (1983): 107-118.

Reddy, P.V. "Patterns of Crop Land Use in Chandrgiri Taluk, Chittoor District, Uttar Pradesh." *National Geographical Journal of India*, 29, No. 1-2 (1983): 70-80.

Singh, Onkar, and H.C. Singh. "Impact of Consolidation on Field Pattern and Agricultural Mechanization in Dhanapur Development Block, Varanasi District." *National Geographical Journal of India*, 29, No. 1-2 (1983): 41-49.

Datye, V.S., and S.C. Gupta. "Association Between Agricultural Land Use and Physio-Socio-Economic Phenomena: A Multivariate Approach." *Transactions, Institute of Indian Geographers*, 6, No. 2 (1984): 61-72.

Maida, I. "Agricultural Land Utilization and Productivities in Narawi." Pp. 138-148 in *Geographical Field Research in South India in 1982*, edited by Kenzo Fujiwara. Hiroshima, Japan: University of Hiroshima Research and Sources Unit for Regional Geography, 1984.

Ray, Sutapa. "Impact of Catenary Association and Topo Sequence on Agricultural Landuse of Narayampur Village, Purulia." *Geographical Review of India*, 46, No. 2 (1984): 86-92.

Sadakata, Nobaru. "Natural Environment and Agricultural Land Use of Naravi." Pp. 117-125 in *Geographical Field Research in South India in 1982*, edited by Kenzo Fujiwara. Hiroshima, Japan: University of Hiroshima Research and Sources Unit for Regional Geography, 1984.

Datta, L. "Physiographic Framework and Intensity of Agricultural Land Use in Nowgong District." *North Eastern Geographer*, 17, No. 1-2 (1985): 51-65.

Ram, Daya. "Functional Classification of Dry Agricultural Lands of Tehsil Bawanikhare, Bhiwani District (Haryana)." *National Geographical Journal of India*, 31, No. 3 (1985): 194-203.

Reddy, N.B.K., and Y.V. Ramanaiah. "Changes in Agricultural Land Use Efficiency in Andhra Pradesh." *Land Use Policy*, 2, No. 3 (1985): 210-216.

Yadav, Hridai R. "Physical Factors in Land Concentration: A Case Study of Eastern Uttar Pradesh." *Transactions, Institute of Indian Geographers*, 7, No. 1 (1985): 41-48.

Das, M.M., and L. Datta. "Regional Variation in Land Use and Agriculture in North-East India." *North Eastern Geographer*, 18, No. 1-2 (1986): 28-48.

Bhadauria, S.B.S. "Land Use Regions, Crop Combination Zones and Ranking of Crops." *Geographical Viewpoint*, 15-16, No. 1-2 (1986–1987): 79-82.

Mishra, R.J. "Impact of Consolidation on Farming: A Case of Dobhi Block, Jaunpur District, Uttar Pradesh." *National Geographical Journal of India*, 32, No. 2 (1986): 118-126.

Sen, Chandra. "Land Use Patterns of Hills of Uttar Pradesh: An Ecological Abuse." *National Geographical Journal of India*, 32, No. 2 (1986): 111-117.

Singh, Jasbir, and Vijay K. Sharma. *Determinants of Agricultural Productivity: A Sample Study of Operational Holdings for Land Use Planning*. Kurukshetra, India: Vishal, 1986.

Bhattacharyya, R. "Temporal and Spatial Changes in the Cultivated Areas of West Bengal." *Geographical Review of India*, 49, No. 4 (1987): 56-62.
Shariff, Ismail. "Agricultural Development and Land Tenure in India." *Land Use Policy*, 4, No. 3 (1987): 321-330.
Sharma, J.K. "Agricultural Land Utilization, Planning and Development of Dharwad Taluk (Karnataka)." *National Geographical Journal of India*, 33, No. 2 (1987): 145-153.
Yadov, Hanuman S. "Determinants of Culturable Waste in Madhya Pradesh." *Indian Geographical Journal*, 62, No. 2 (1987): 91-96.
Ramanaiah, Y.V., and Naveen K. Charles. "Spatio-Temporal Distribution of Cultivable Wastelands in Anantapur District, Andhra Pradesh." *North Eastern Geographer*, 20, No. 1-2 (1988): 17-25.
Tripathi, R.S., and J.P. Viswakarma. "Landuse Cropping Pattern and Development Levels in Banda District." *Deccan Geographer*, 26, No. 2-3 (1988): 417-428.
Subbiala, S., D. Dhanopal, and G. Parimala. "Incidence of the Long Fallow Land in Tamil Nadu." *Geographical Review of India*, 51, No. 2 (1989): 35-50.
Goswami, D.B. "Challenging Agro-Landuse and Growth of Population in India." *Geographical Review of India*, 52, No. 4 (1990): 16-26.
Narayanamma, N. "Land Capability Classification: A Study of Southern Andhra Pradesh." *Rural Systems*, 8, No. 1 (1990): 49-60.
Prasad, N., and N.H. Khadse. "Correlation Between Slope and General and Agricultural Landuse in the Pus Dam Catchment of Akola and Yavatmal District, Maharashtra." *Geographical Perspective*, 4, No. 1-2 (1990): 1-12.
Kaur, Dhian. *Changing Patterns of Agricultural Land Use: A Spatial Analysis of Bist Doab*. Jaipur, India: Rawat Publications, 1991.
Pont, B.R., R.C. Joshi, and D.S. Jalal. "Agricultural Landuse and Nutrition in Kotadun, Kuaun Himalaya." *Geographical Review of India*, 53, No. 4 (1991): 8-18.
Gengaje, Rajan K. "Administration of Farmland Transfer in Urban Fringes: Lessons from Maharashtra, India." *Land Use Policy*, 9, No. 4 (1992): 272-286.
Mohammad, Noor, and S.K. Bandooni. "Monitoring of Landuse and Agriculture in a Hilly Region: A Case Study of Pauri Garhwal District." *The Geographer, Aligarh*, 39, No. 2 (1992): 7-20.
Reddy, N.R.S., and S. Srinivasula. "Agricultural Land Use Efficiency in Cuddapah District." *National Geographer*, 27, No. 2 (1992): 109-120.
Thakur, Baleshwar, and Phoal S. Bhushan. "Land–Man Relationship and Distribution of Agricultural Land in East Sikkim." *Transactions, Institute of Indian Geographers*, 14, No. 1 (1992): 13-32.
Sanra, Syamapada. "Factors Determining Present Agricultural Land Use of Haora." *Geographical Review of India*, 55, No. 3 (1993): 53-61.
Dhillon, Rime. "Changing Patterns of Agricultural Land Use in Punjab (India), 1960–1992." Master's Thesis, University of Montana, Missoula, Mont., 1994.

Malakar, B., and S. Basu. "Decline in Net Cropped Area in West Bengal: Configuration and Causes." *Geographical Review of India*, 58, No. 4 (1996): 330-339.

Nagabhusanam, N., K. Harinath Reddy, and H. Ugandra Raju. "Agricultural Land Use Efficiency of Nellore District, Andhra Pradesh." *Transactions, Institute of Indian Geographers*, 18, No. 2 (1996): 71-74.

Tiwari, Ram Kumar. "Measurement of Agricultural Landuse in Jharkhand." *Geographical Review of India*, 62, No. 2 (2000): 135-141.

AGRICULTURAL POPULATIONS AND SOCIAL ISSUES

Malik, Rashid A. "Agricultural Occupance of West Punjab." Master's Thesis, University of Chicago, Chicago, Illinois, 1958.

Kayastha, S.L. "Girths of Kangra Valley: A Study in Agricultural Society." *National Geographical Journal of India*, 5, No. 1 (1959): 25-33.

Mukerji, A.B. "Female Participation in Agricultural Labour in Uttar Pradesh: Spatial Variations." *National Geographer*, 6 (1963): 13-18.

Chakravarti, A.K. "The Areal Relationships of Foodgrain Production and Agricultural Population in India, 1921–1951." Ph.D. Dissertation, University of Wisconsin, Madison, Wis., 1967.

Branger, Dirk. "Agrarstruktur und Kasten wesen im sudlichen Indien (Andrah Pradesh): Sozialgeographische Strukturdiagnose zu einem Entwicklungen." *Erdkunde*, 24, No. 3 (1970): 194-206.

Sharma, Rajendra P. "Pattern of Agriculture Among the Tribals of Madhya Pradesh." *Geographical Review of India*, 33, No. 3 (1971): 175-182.

Singh, Jasbir. "Agricultural Colonization of Cultivable Wasteland in India (1950–1951 to 1965–1966)." *Deccan Geographer*, 9, No. 2 (1971): 135-149.

Chapman, G.P. "Perception and Regulation: A Case Study of Farmers in Bihar." *Transactions, Institute of British Geographers*, 62 (1974): 71-94.

Farmer, Bertram H. *Agricultural Colonization in India Since Independence*. London, England: Oxford University Press, 1974.

Birch, Brian P. "Pattern and Process in Frontier Farm Settlement." *National Geographical Journal of India*, 21, No. 3-4 (1975): 166-171.

Ishida, Hiroshi. "Geographical Investigations into Villages and Agriculture in North India by the University of Hiroshima" (in Japanese). Pp. 1-11 in *Dynamic Regional Geographical Studies of the Punjab, India*, edited by Hiroshi Ishida. Hiroshima, Japan: University of Hiroshima, Centre for Research in World Regional Geography, 1975.

Ohji, Toshiaki. "The Evolution of a Hamlet of Undeveloped Agriculture in Punjab, India: A Case Study of Salamatpur, Ropar District" (in Japanese). Pp. 68-86 in *Dynamic Regional Geographical Studies of the Punjab, India*, edited by Hiroshi Ishida. Hiroshima, Japan: University of Hiroshima, Centre for Research in World Regional Geography, 1975.

Mukerji, A.B., and S. Mehta. "Female Participation in Agricultural Labour in India: Patterns and Associations." *Tijdschrift voor Economische en Sociale Geographie*, 66, No. 2 (1975): 103-108.

Sopher, David E. "Indian Pastoral Castes and Livestock Ecologies: A Geographic Analysis." Pp. 183-208 in *Pastoralists and Nomads in South Asia*, edited by Lawrence S. Leshnik and Gunter Dietz Sibtheirmer. Wiesbaden, Germany: Otto Horrassowitz, Schriftenreibe des Sudasien-Institut der Unitersitat Heidelberg, 1975.

Mitra, Sevati. "Development Problems of Tribal Agriculture." *Geographical Review of India*, 39, No. 1 (1977): 107-116.

Joshi, Y.G. "Evaluation of the Impact of the Growth of Urbanization on Farm Productivity, Degree of Commercialization and Farm Population Density in Madhya Pradesh." *Geographical Review of India*, 40, No. 2 (1978): 175-181.

Mohammad, Noor. "Impact of Economic Factors on Diffusion of Agricultural Innovations in Central Trans-Ghaghara Plain." *Geographical Review of India*, 40, No. 3 (1978): 266-280.

Mohammad, Noor. "Impact of Social Factors on Diffusion of Agricultural Innovations in Central Trans-Ghaghara Plain." *National Geographical Journal of India*, 24, No. 1-2 (1978): 36-46.

Das, M.M. "Population Pressure and Intensity of Cropping in Assam." *Geographical Review of India*, 41, No. 2 (1979): 105-114.

Ali, A.M.S. "Farmers' Perception Toward Economic Benefits and Changing Cropping Pattern in the Savar Area." *Environment and Region*, 1 (1981): 32-48.

Bagchi, Jahnabi. "Population Pressure and Agricultural Change." Pp. 45-54 in *Perspectives in Agricultural Geography*, vol. 5, edited by Noor Mohammad. New Delhi, India: Concept Publishing Company, 1981.

Mohammad, Noor. "Socio-Demographic Determinants of Diffusion of Agricultural Innovations." Pp. 289-318 in *Perspectives in Agricultural Geography*, vol. 2, edited by Noor Mohammad. New Delhi, India: Concept Publishing Company, 1981.

Sen, Jyotirmay, and R. Bhattacharyya. "Population Pressure and Agricultural Change in the Bhagirathi: Jalangi Interfluve, 1850–1975." Pp. 81-104 in *Perspectives in Agricultural Geography*, vol. 5, edited by Noor Mohammad. New Delhi, India: Concept Publishing Company, 1981.

Tewari, A.K. "Perception of Drought Hazard Among the Farmers of Indian Desert." Pp. 223-240 in *Perspectives in Agricultural Geography*, vol. 5, edited by Noor Mohammad. New Delhi, India: Concept Publishing Company, 1981.

Joshi, Y.G. "Spatial Impact of Tribal Development Blocks on the Agricultural Development." *Geographical Review of India*, 44, No. 4 (1982): 52-62.

Singh, Jhujar. "Spatial Organization of Agriculture in a North Indian Village: A Behavioral Approach." *Transactions, Institute of Indian Geographers*, 4, No. 2 (1982): 143-152.

Joshi, Y.G. "Spatial Impact of Tribal Development Blocks as Centres of Innovation of Agricultural Development: A Study of Jhabua District, Madhya Pradesh." *Geographical Review of India*, 45, No. 2 (1983): 35-44.

Kelly, Kenneth. "Farmer Impoverishment in Birbhum District (West Bengal) During the Second Half of the 19th Century." *Geographical Review of India*, 45, No. 3 (1983): 4-15.

Misrah, V.C., and S.K. Sharma. "Population Growth and Agricultural Change in Madhya Pradesh: Spatial Analysis of Their Correlates." *National Geographer*, 18, No. 2 (1983): 141-152.

Munir, Abdul, and Farasat Ali Siddiqi. "Population Pressure and Agricultural Productivity in Ganda District." *The Geographer, Aligarh*, 30, No. 1 (1983): 94-99.

Singh, B.B., and K. Ram. "Agricultural Population Dynamics and Rural Development Typology in Roorkee Tahsil (Saharanpur)." *National Geographer*, 18, No. 1 (1983): 115-125.

Singh, M.B. "Population Pressure and Mixed Farming in India." *Rural Systems*, 2, No. 3 (1984): 147-156.

Singh, Abha L. "Agricultural Populations in the Urban Centres of Varanasi City Region: A Geographical Analysis." *National Geographical Journal of India*, 32, No. 1 (1986): 40-48.

Abuzar, M. "Agricultural Responses to Population Differentials in the Punjab-Haryana Plain." *The Geographer, Aligarh*, 34, No. 2 (1987): 59-66.

Mitra, Champa, and Goutam Sarkar. "Some Social Aspects of Agricultural Development in India." *Geographical Review of India*, 49, No. 4 (1987): 69-76.

Khan, Yusaf, and S.B. Sawant. "Population in Relation to Food Production in Western Maharashtra." *Geographical Review of India*, 50, No. 2 (1988): 44-52.

Tiwari, P.D. "A Geographical Analysis of the Calories Available from Different Crops in Madhya Pradesh." *Geographical Perspective*, 2, No. 2 (1988): 1-8.

Jha, V.C. "Role of Women in Agriculture: A Case Study of Godda District." *Geographical Perspective*, 3, No. 1 (1989): 9-12.

Bhagabati, A.K. "Social Structure and Agricultural Performance in Dolabari Irrigated Area of Sonitpur District, Assam." *Geographical Review of India*, 52, No. 4 (1990): 71-79.

Kaur, Dhiam. "Distance from Village Settlement and Agricultural Location in Bist Doab, Punjab." *National Geographical Journal of India*, 36, No. 3 (1990): 170-176.

Khatri, L.C. *Agricultural Typology: A Tribal Region of Southern Rajasthan*. Udaipur, India: Himanshu Publications, 1990.

Nasir, Jamshed, and Tasneem Fatima. "Distribution of Vitamins Obtained from Food Crops of Uttar Pradesh." *The Geographer, Aligarh*, 37, No. 2 (1990): 58-62.

Roy, B.K. "Water in India with Reference to Agriculture and Population: Some Issues and Patterns." *GeoJournal*, 20, No. 3 (1990): 271-284.

Singh, L.R., and R.N. Dubey. "Impact of Population Growth on Agricultural Land Use in Rohilkhand, Uttar Pradesh." *National Geographer*, 25, No. 1 (1990): 45-54.

Bose, Ashis. "Demographic Zones and Sub-Zones of India: Policy Implications." *Geographical Review of India*, 54, No. 1 (1992): 1-18.

Ojha, Brahm S. *Technological and Socio-Economic Changes in Agriculture in India: A Case Study: A Study at Spatio-Temporal Scale*. New Delhi, India: Oxford University Press, 1992.

Gatade, D.G., and A.P. Kumbhar. "A Geographical Analysis of Rural Population and Food Production in Upper Krishna Basin, Maharashtra." *Geographical Review of India*, 55, No. 2 (1993): 69-77.

Khanna, Sulbha. "Diffusion of Agricultural Information: A Case Study in Gujarat." *The Geographer, Aligarh*, 40, No. 1 (1993): 35-42.

Raghuram, Parvati. "Invisible Female Agricultural Labour in India." Pp. 109-119 in *Different Places, Different Voices: Africa, Asia and Latin America*, edited by Janet H. Momsen and Vivian Kinnaird. London, England: Routledge, 1993.

Sankaran, Vasumathi. "Cultural Practices as Determinants of Agricultural Efficiency." *Indian Geographical Journal*, 68, No. 2 (1993): 58-67.

Singh, Abha L., and Shahab Fazaj. "Farmers' Decision Making Process in Traditional Versus Commercial Farming: A Case Study of Upper Ganga-Yamuna Doab." *The Geographer, Aligarh*, 40, No. 2 (1993): 9-20.

Bano, Hanifa, et al. "Horticultural Land Use Analysis of Pulwoma District, Jammu and Kashmir, Using Satellite Remote Sensing Techniques." *Annals, National Association of Geographers, India*, 15, No. 2 (1995): 110-120.

Talib, Abu, Abdul Munir, and M. Farooq Siddiqi. "Agricultural Development and Poverty Alleviation in Trans-Ghaghara Plain of Uttar Pradesh, India." *Indian Geographical Journal*, 70, No. 2 (1995): 92-97.

Vaidya, Balakrishna C. "Population Growth and Agricultural Land Use in Semiarid Region: A Case Study of Ahmadnagar District." *National Geographer*, 30, No. 1 (1995): 37-44.

MacDonald, Kenneth. "Rationality, Representation and the Rich Mediating Characteristics of a Karakoram Mountain Farming System." *Human Ecology*, 26, No. 2 (1998): 287-322.

Sharma, S.K. "Social Structure and Adoption of Agricultural Innovation in Madhya Pradesh." *Geographical Review of India*, 61, No. 1 (1999): 156-164.

ENVIRONMENTAL AND CONSERVATION ISSUES

Sita, K. "Dry Farming in Peninsular India." *Indian Geographical Journal*, 35, No. 1-2 (1960): 19-36.

Subramaniam, A.R., and K.B. Rao. "Agro-Climatic Classifications of Rajasthan." *Indian Geographical Journal*, 62, No. 1 (1967): 46-52.

Siddiqi, Niaz A. "A Regional Study of the Soil Type and the Problems of Land Utilization in the Central Trans-Ghaghara Region." *Geographical Outlook*, 6 (1969): 48-54.

Farooqi, M.Y. "Climo-Edaphic Factors of Sugarcane Cultivation in Garakhpu and Deoria Districts." *Geographical Observer*, 6 (1970): 18-26.

Husain, Majid. "Variability of Rainfall in Relation to Agriculture in the Upper Ganga-Yamuna Doab." *National Geographical Journal of India*, 16, No. 1 (1970): 71-78.

Singh, S.P. "Dholri: A Case Study in Soils, Land Capability and Farm Planning." *National Geographical Journal of India*, 16, No. 1 (1970): 1-30.

Harris, Barbara. "The Green Revolution in Ludhiana District, India." *Geography*, 56, No. 3 (1971): 243-246.

Misra, G.K. "A Land Climate Complex Approach to Evolve a Suitable Cropping Pattern." *Geographical Review of India*, 33, No. 4 (1971): 253-262.

Chakravarti, A.K. "Green Revolution in India." *Annals of the Association of American Geographers*, 63, No. 3 (1973): 319-330.

Banerjee, Bireswar. "Green Revolutions in India: A Geographical Analysis." *Science and Culture*, 40 (1974): 2-3.

Singh, Jasbir. *The Green Revolution in India: How Green It Is!* Kurukshetra, India: Vishal Publications, 1974.

Banerjee, Bireswar. "Green Revolution in India: A Geographical Analysis." Pp. 88-104 in *Dynamic Regional Geographical Studies of the Punjab, India*, edited by Hiroshi Ishida. Hiroshima, Japan: Centre for Research in World Regional Geography, University of Hiroshima, 1975.

Patel, Govindbhai V. "Agriclimatology of the Semi-Arid and Arid Areas of Northwest India: Rajasthan, Gujarat, and Punjab States." Master's Thesis, Mankato State University, Mankato, Minn., 1975.

Farmer, Bertram H., ed. *Green Revolution? Technology and Change in Rice Growing Areas of Tamil Nadu and Sri Lanka.* London, England: Macmillan, 1977.

Ishida, Hiroshi, ed. *Green Revolutions in the Punjab: Techniques, Background and Effects* (in Japanese). Hiroshima, Japan: University of Hiroshima Department of Geography, Special Publication No. 4, 1977.

Jayachandran, S. "Dry Farming: A Breakthrough in Indian Agriculture." *Indian Geographical Journal*, 52, No. 1 (1977): 10-14.

Ohji, Toshiaki. "A Comparative Study of the Agricultural Techniques in the Punjab in Relation to the Green Revolution" (in Japanese). Pp. 1-28 in *Green Revolutions in the Punjab: Techniques, Background and Effects*, edited by Hiroshi Ishida. Hiroshima, Japan: University of Hiroshima Department of Geography, Special Publication No. 4, 1977.

Rozlucki, W. "The Green Revolution and the Development of Traditional Agriculture: A Case Study of India." *Geographia Polonica*, 35 (1977): 111-126.

Sharma, R.L., and Jasbir Singh. "The Dry Farming Regions of Haryana: Its Extent and Problems." *Geographical Review of India*, 39, No. 2 (1977): 164-172.

Biswas, Arabinda. "Physiological Determinism in Birbhum Agriculture." *Geographical Review of India*, 40, No. 1 (1978): 1-14.

Ahmad, Aijazuddin, and Aslam Mahmood. "Determination of Critical Drought Limits to Crop Production in Dry Regions: An Exploratory Analysis with

Reference to Western Rajasthan." *The Geographer, Aligarh*, 26, No. 2 (1979): 11-24.

Rozlucki, W. "Modernization of Traditional Agriculture: An Essay on the Green Revolution in India" (in Polish). *Prace Geograficzne*, Instytutu Geografii i Przestrzenuego Zagospodarowania PAN, 133 (1979).

Subramaniam, A.R., and K.U. Devi. "Water Balance and Cropping Pattern of Orissa State." *Geographical Review of India*, 41, No. 4 (1979): 362-369.

Akhtar, Rais. *Environment, Agriculture and Nutrition in Kumoan Region*. New Delhi, India: Morwah, 1980.

Ahmad, Enayat. "Drainage and Agriculture." Pp. 183-192 in *Perspectives in Agricultural Geography*, vol. 2, edited by Noor Mohammad. New Delhi, India: Concept Publishing Company, 1981.

Basu, S. "Physical Factors Affecting Agriculture: A Geographical Analysis." Pp. 15-24 in *Perspectives in Agricultural Geography*, vol. 2, edited by Noor Mohammad. New Delhi, India: Concept Publishing Company, 1981.

Das, M.M. "Climatic Pattern and Agro-Climatic Regions of Assam: A Geographical Analysis." Pp. 123-137 in *Perspectives in Agricultural Geography*, vol. 2, edited by Noor Mohammad. New Delhi, India: Concept Publishing Company, 1981.

Jana, M.M. "Physical Factors Affecting Agriculture in a Backward Region." Pp. 41-58 in *Perspectives in Agricultural Geography*, vol. 2, edited by Noor Mohammad. New Delhi, India: Concept Publishing Company, 1981.

Kar, Amal. "Land, Man and Desert Degradation: Geomorphic Problems of Agriculture in Western Rajasthan." Pp. 77-88 in *Perspectives in Agricultural Geography*, vol. 2, edited by Noor Mohammad. New Delhi, India: Concept Publishing Company, 1981.

Mehrotra, Naresh. "The Constraints in the Transfer of Rainfed Technology and the Priority in Agricultural Development." Pp. 427-444 in *Perspectives in Agricultural Geography*, vol. 2, edited by Noor Mohammad. New Delhi, India: Concept Publishing Company, 1981.

Mehrotra, Naresh. "The Role of Prospective Agricultural Research in Increasing the Productivity of Rainfed Regions." Pp. 463-484 in *Perspectives in Agricultural Geography*, vol. 4, edited by Noor Mohammad. New Delhi, India: Concept Publishing Company, 1981.

Mooley, D.A. "The Indian Summer Monsoon, Its Economic Aspects, Its Vagaries and Remedial Measures." Pp. 89-112 in *Perspectives in Agricultural Geography*, vol. 2, edited by Noor Mohammad. New Delhi, India: Concept Publishing Company, 1981.

Ram Krishna, G.V.S., S.P. Malhotra, and M.D. Yadav. "Agricultural Efficiency in Indian Arid Districts." *National Geographical Journal of India*, 27, No. 3-4 (1981): 115-119.

Reddy, N.B.K., and D.N. Reddy. "Agriculture in Arid and Semi-Arid Regions of India." *Transactions, Institute of Indian Geographers*, 3, No. 2 (1981): 115-136.

Singh, Harjit. "Environmental Constraints on Agriculture in a Cold Desert: A Case Study of Ladakh." Pp. 25-40 in *Perspectives in Agricultural Geography*, vol. 2, edited by Noor Mohammad. New Delhi, India: Concept Publishing Company, 1981.

Subramaniam, A.R., and K. Uma Devi. "Rainfall and Crop Yields in Orissa State." *Indian Geographical Journal*, 57, No. 2 (1982): 71-76.

Subramaniam, A.R., A.V.R. Keshava Rao, and K.B. Rao. "Climate and Horticulture in India." *Transactions, Institute of Indian Geographers*, 4, No. 2 (1982): 191-216.

Dikshit, Jutta. "Identification of Drought Prone Areas and Prediction of Crop Expectations in Maharashtra." *Geographical Review of India*, 45, No. 4 (1983): 48-81.

Mishra, S. "Agricultural Drought over Gangetic West Bengal in Relation to Aman Rice." *Geographical Review of India*, 45, No. 2 (1983): 78-82.

Ram Mohan, H.S., and V.P. Subrahmanyan. "Application of Water Balance in Agricultural Planning and Operations in South India." *National Geographical Journal of India*, 29, No. 1-2 (1983): 30-34.

Rao, M.S., and K. Kalavathy. "Water Balance and Cropping Pattern in Madurai District, Tamil Nadu." *Indian Geographical Journal*, 58, No. 2 (1983): 172-182.

Bayliss-Smith, Timothy, and S. Wanmali, eds. *Understanding Green Revolution: Agrarian Change and Development Planning in South Asia.* Cambridge, England: Cambridge University Press, 1984.

Chambers, R. "Beyond the Green Revolution: A Selective Essay." Pp. 362-379 in *Understanding Green Revolution: Agrarian Change and Development Planning in South Asia: Essays in Honour of B.H. Farmer*, edited by Timothy Bayliss-Smith and S. Wanmali. Cambridge, England: Cambridge University Press, 1984.

Farooqi, M.Y., Jamshed Nasir, and Tasneen Fatima. "Distribution of Mineral Nutrients Obtained from the Foodcrops of Uttar Pradesh." *The Geographer, Aligarh*, 31, No. 2 (1984): 86-92.

Subramaniam, A.R., and A.S. Rao. "Agroclimatic Classifications of Maharashtra." *Transactions, Institute of Indian Geographers*, 6, No. 2 (1984): 19-26.

Subramaniam, M.J., and V. Vidyanath. "Moisture Adequacy and Cropping Pattern in Andhra Pradesh, India." *Indian Geographical Journal*, 59, No. 2 (1984): 268-272.

Patnaik, Manjurani, and Jayant K. Routray. "Intensity of Drought and Spatial Variation of Agricultural Development: A Case Study of Kalahandi District in Orissa." *Transactions, Institute of Indian Geographers*, 5, No. 2 (1985): 165-170.

Banerjee, Smriti. "Identification of Drought and Its Impact on Agriculture in Uttar Pradesh." *Geographical Review of India*, 50, No. 3 (1986): 50-56.

Hemanalini, B. "Water Budgeting and Crop Yield: A Case Study of Groundnut Yields in Anantpur District, Andhra Pradesh." *National Geographer*, 21, No. 2 (1986): 179-186.

Subramaniam, A.R., and P.S. Rao. "Rainfall and Agriculture in Prakasam District, Andhra Pradesh." *Deccan Geographer*, 24, No. 1-2 (1986): 11-21.

Daulay, H.S., and A. Henry. "Desert Ecological Environments and Their Management Systems for Optimizing Crop Plant Production." *The Geographer, Aligarh*, 34, No. 2 (1987): 26-38.

Subramaniam, A.R., and K.B. Rao. "Agro-Climatic Classifications of Rajasthan." *Indian Geographical Journal*, 62, No. 1 (1987): 46-52.

Banerjee, Smriti. "Identification of Drought and Its Impact on Agriculture in Uttar Pradesh." *Geographical Review of India*, 50, No. 3 (1988): 50-56.

Mookerjee, Saswati. "Bio-Ecological Factors in Moriculture and Silkworm Rearing in West Bengal." *Annals of the National Association of Geographers, India*, 8, No. 2 (1988): 37-40.

Singh, Abha L., and Smriti Banerjee. "Pattern of Energy Imput and Its Impact on Agricultural Production in Agra Divis." *National Geographical Journal of India*, 34, No. 2 (1988): 126-133.

Sivagnamon, N., and K. Kumaraswamy. "Ground Water Quality Mapping for Agricultural Planning in the Vaippar Basin, Tamil Nadu." *Deccan Geographer*, 26, No. 2-3 (1988): 407-416.

Basu, Shila, and B.N. Chanda. "Agro-Climatic Zones of West Bengal." *Geographical Review of India*, 51, No. 2 (1989): 19-29.

Chowdhury, C.R., and D.K. Mandal. "Agro-Climatic Zones of West Bengal." *Geographical Review of India*, 51, No. 2 (1989): 13-18.

Jha, Ganga Prasad. *Agriculture in a Flood Prone Region: A Geographical Perspective*. New Delhi, India: Inter-India, 1989.

Joshi, Hamlata. "Import of Fertilizers in India." *Geographical Review of India*, 51, No. 2 (1989): 51-61.

Naidu, L.G.K., T. Lal, and S. Pandy. "Agro-Economic Studies of Black Soils vs. Alluvial Soils." *National Geographical Journal of India*, 35, No. 3-4 (1989): 231-236.

Qureshi, Salahuddin. *Regional Perspective on Dry Farming*. Jaipur, India: Rawat, 1989.

Singh, V.R., and D.P. Rai. "Use and Requirement of Water Resources in Agriculture: A Case Study of Azamparh District (Uttar Pradesh)." *Rural Systems*, 7, No. 4 (1989): 241-256.

Vaid, Varsha, and V.S. Datye. "Association Between Water Deficit and the Yields of Principal Cereal Crops in Maharashtra." *Geographical Review of India*, 51, No. 2 (1989): 68-85.

Dhillan, S.S. "Dimensions of Green Revolution in Haryana." *Geographical Review of India*, 52, No. 1 (1990): 91-98.

Raza, Mehdi, and M.L.S. Khan. "Ecological Impacts of Fertilizers and Pesticides in the Ganga Valley." *The Geographer, Aligarh*, 37, No. 1 (1990): 57-71.

Reddy, P.R.K., and M.S. Rao. "Soil Moisture Elements and Crop Management in Cuddapah District, Andhra Pradesh." *National Geographer*, 25, No. 1 (1990): 1-16.

Schgal, J.L., et al. *Agro-Ecological Regions of India*. Nagpur, India: National Bureau of Soil Survey and Land Use Planning, Indian Council of Agricultural Research, 1990.

Siva Prasad, C.R., et al. "Soils of Vellare Taluk, North Arcot District, Tamil Nadu and Their Suitability for Growing Rice." *Indian Geographical Journal*, 65, No. 2 (1990): 130-144.

Basu, Shila. "Meteorological Approach for Estimation of Water Availability Period in Cooch Behar District vis-a-vis Judicious Crop Planning." *Geographical Review of India*, 53, No. 1 (1991): 42-49.

Chatterjee, Nandini. "Agro-Climatic Perspectives on Irrigation in West Bengal, India: A District Level Analysis." *Singapore Journal of Tropical Geography*, 12, No. 1 (1991): 12-30.

Dhillan, S.S. "Dimensions of Green Revolution in Haryana." *Geographical Review of India*, 53, No. 2 (1991): 80-86.

Pandey, Vyas, and B.D. Gupta. "An Agro-Climatic Study of Some Districts of Gujarat." *Geographical Review of India,* 53, No. 4 (1991): 1-7.

Sharma, B.L. *Dryland Farming: Perspectives and Prospects*. Delhi, India: Daya Publications, 1991.

Jayakumar, R., and M. Siraz. "Groundwater Suitability for Agriculture in Attar Valley, Tamil Nadu." *Indian Geographical Journal*, 67, No. 2 (1992): 75-79.

Khanna, Sulbha. "Impact of Changing Agricultural Environments on Farmers." *Indian Geographical Journal*, 67, No. 1 (1992): 83-86.

Marius-Gnonan, K. "Le revolution verte et la maitrise de l'espace agricole: Le cas de la region de Pondichery (Inde meridionale)." *Les Cahiers d'Outre Mer*, 45 (1992): 189-212.

Nagarathinam, S.R., and N. Jivagnanam. "A Study on Crop Seasons and Water Availability in Coimbatore District." *Indian Geographical Journal*, 67, No. 1 (1992): 20-29.

Qureshi, Azim V., M. Irfan Sabir, and Akram Ali. "Groundwater Resource Potential of Dhanipur Block, Aligarh District, India, and Its Role in Agriculture." *Philippine Geographical Journal*, 36, No. 1-2 (1992): 27-39.

Ram, Daya. "Functional Classification of Rainfed Sown Areas of Sirsa Block." *Geographical Review of India*, 54, No. 2 (1992): 53-63.

Ram, Daya. "Relationships Between Pre-Sowing Rainfall and Rabi Crops Failure in Dryland Agriculture of Hisar Tehsil, Haryana." *Geographical Review of India*, 54, No. 1 (1992): 68-71.

Ramachandran, Kaushalya. "Impact of Drought on Foodgrain Production in India." *Indian Geographical Journal*, 67, No. 2 (1992): 29-36.

Tiwari, P.S. "Green Revolution in India." *Indian Geographical Journal*, 67, No. 1 (1992): 1-8.

Chaudhury, J.L., and A. Sastri. "Climatic Shifts and Water Balance: A Case Study for Two Extreme Farming Situations at Jagdalpur, Bastor." *Geographical Review of India*, 55, No. 1 (1993): 54-62.

Madhuramma, C., and M.S. Rao. "Water Balance and Cropping Pattern of Nellore District, Andhra Pradesh." *National Geographer*, 28, No. 1 (1993): 1-20.

Raagaard, Svend. "Ecology and Agricultural Development in Four Villages of Central Karnataka." *Geografisk Tidsskrift*, 93 (1993): 69-79.

Sati, V.P., and M.S.S. Rawat. "Problems of Agricultural Ecology and Their Management in the Pinder Basin." *National Geographer*, 28, No. 2 (1993): 143-148.

Shafi, M. "Ecosystem and Management of Agriculture." *The Geographer, Aligarh*, 40, No. 1 (1993): 1-8.

Singh, R.P., and C.S. Devi. "Impact of Green Revolution in North East India." *Geographical Review of India*, 55, No. 1 (1993): 47-52.

Singh, Ram, and Lakshmi Shukla. "Crop Failure Index Under Rainfed Conditions in Haryana: A Geographical Analysis." *Geographical Review of India*, 55, No. 1 (1993): 24-32.

Tamilarasan, V. "Impact of Rainfall on Irrigation and Cropping Area in Tamilnadu." *Deccan Geographer*, 31 (1993): 37-48.

Saseendran, S.A., and A.K. Murali. "Agrometeorological Studies for Optimal Utilisation of Water Resources in Kozhikode District of Kerala State." *Transactions, Institute of Indian Geographers*, 16, No. 2 (1994): 107-114.

Shafi, M. "Ecology and Land Use/Cropping System with Special Reference to India." *The Geographer, Aligarh*, 41, No. 2 (1994): 31-34.

Siddi Raju, S., S. Srinivasulu, and K. Kesava Raja. "Climatic Water Balance and Agriculture in the Pulang River Basin." *National Geographer*, 29, No. 1 (1994): 19-54.

Mandal, C., and K.T. Kolte. "Agro-Ecological Zones: A Case Study of Vidarbha." *Geographical Review of India*, 57, No. 4 (1995): 336-344.

Ram, Daya, and S.S. Yadav. "Effect of Rabi Rains and Conservation of Soil Moisture on Crop Failure." *Geographical Review of India*, 57, No. 2 (1995): 120-125.

Ram Mohan, H.S., and Haseena Raghavan. "An Agro-Climatic Appraisal of Droughts over Kerala State." *Geographical Review of India*, 57, No. 1 (1995): 39-47.

Uhlig, Harold. "Persistence and Change in High Mountain Agricultural Systems." *Mountain Research and Development*, 15, No. 3 (1995): 199-212.

Chaudhary, J.L., Ajay K. Srivastava, and D. Naidu. "Climatic Variations and Their Impact on Agricultural Productivity in Baster District." *Geographical Review of India*, 58, No. 1 (1996): 82-89.

Das, A.L., Dipak Sarkar, and Aruna Goswami. "Study of Soils in Little Andaman for Oil Palm Cultivation." *Geographical Review of India*, 58, No. 4 (1996): 367-376.

Kumar, Pradeep. "Impact of Water Table on Soil and Crops: Case Study of a Village of Haryana." *Transactions, Institute of Indian Geographers*, 18, No. 2 (1996): 75-78.

Madhava Chandran, K., M.D. Nandeshwar, and T. Valson. "Assessment of Farmers' Attitude Towards Water Users' Associations Under Cada: A Case Study in Kerala." *Transactions, Institute of Indian Geographers*, 18, No. 1 (1996): 45-52.

Maji, Pranad K., and Pannelal Das. "Impact of Soil on Crop Productivity in Nadia District." *Geographical Review of India*, 58, No. 2 (1996): 129-138.

Chaudhary, J.L. "Moisture Index and Its Impact on Rice Crop Yield: A Case Study of Bastar District, Chhattisgarh Region." *Geographical Review of India*, 59, No. 1 (1997): 34-40.

Mandal, C., and D.K. Mandal. "Use of Length of Growing Period Concept in Agro-Ecological Zones Delineation and Crop Planning Under Indian Perspective." *Geographical Review of India*, 59, No. 4 (1997): 320-329.

Sharma, Tara C. "How Green Is the Green Revolution in Karnataka? A Regional Inquiry." *Transactions, Institute of Indian Geographers*, 19, No. 2 (1997): 41-54.

Singh, Yadvindar. "Eco-Degradation of Productive Life Support Environment as Correlated to Agricultural Development in Puadh Region, Punjab." *Transactions, Institute of Indian Geographers*, 19, No. 1 (1997): 57-70.

Bohle, H.G. "Grenzen der Grunen Revolution in Indien." *Geographische Rundschau*, 51, No. 3 (1999): 111-117.

IRRIGATION

Thompson, J.P. "The Murrumbidgee Water Conservation and Irrigation Schemes." *Scottish Geographical Magazine*, 24, No. 8 (1908): 393-414.

McFarlane, John. "The Burinjuck Dam and the Murrumbidgee Irrigation Area." *Scottish Geographical Magazine*, 32, No. 1 (1916): 24-30.

Williamson, A.V. "Irrigation in the Indo-Gangetic Plain." *Geographical Journal*, 65, No. 2 (1925): 141-152.

Williamson, A.V. "Indigenous Irrigation Works in Peninsular India." *Geographical Review*, 21, No. 4 (1931): 613-626.

Dutt, P.K. "The Ganges Canal Hydro-Electric Scheme as an Aid to Agriculture in Western Uttar Pradesh." *Indian Geographical Journal*, 23, No. 4 (1948): 12-23.

Unnisa, Aziz. "Irrigation Landscape in Hyderbad, East Taluka." *Geographical Review of India*, 24, No. 2 (1962): 39-59.

Anderson, A.G. "Agriculture and Irrigation in Mysore State." Pp. 225-236. in *Proceedings, Fourth New Zealand Geography Conference*, edited by R.G. Lister and R.P. Hargreaves. Dunedin, New Zealand: New Zealand Geographical Society, 1964.

Chaturvedi, B.N., and K.S. Reddy. "Irrigation in Telengana: Sources and Problems." *Deccan Geographer*, 2 (1964): 154-170.

Hore, P.N. "Rainfall, Rice Yields and Irrigation Needs in West Bengal." *Geography*, 49, No. 2 (1964): 114-121.

Shankar, Kripa. "Irrigation in Telangana: Sources and Problems." *Deccan Geographer*, 2, No. 2 (1964): 154-170.

Anderson, A.G. "Agriculture and Irrigation in Mysore State." Pp. 225-235 in *Proceedings of the Fourth New Zealand Geography Conference*, edited by R.G. Lister and R.P. Hargreaves. Dunedin, New Zealand: New Zealand Geographical Society, 1965.

Browne, Percis. "Implications of Irrigation and Water Power for Indian Agricultural Development." Master's Thesis, Northeastern Illinois University, Chicago, Ill., 1965.

Durand-Dastes, F. "Problemes de l'irrigation en regions tropicales: Le cas de l'inde." *L'Information Geographique*, 30, No. 1 (1966): 1-13.

Adiceam, E. "The Geography of Irrigation in Tamiland." *Indian Geographical Journal*, 42, No. 1-2 (1967): 7-12.

Dupuis, Jacques. "L'irrigation dans le pays tamoul." *Les Cahiers d'Outre Mer,* 20 (1967): 94-98.

Chaturvedi, B.N. "The Origin and Development of Tank Irrigation in Peninsular India." *Deccan Geographer*, 6, No. 2 (1968): 57-86.

Nitz, Hans Jurgen. "Zur Geographie der kunstilichen Bewasserung im mittleren Nordindien." *Geographische Zeitschrift*, 56 (1968): 307-326.

Bagchi, K., and R. Bhattacharyya. "Problems of Water Supply for Agricultural Operations in Lower West Bengal." *Geographical Outlook*, 6 (1969): 26-29.

Chaudhari, B.N. "Mid-Slope Reservoir Irrigation in the Don Lands of Chatanagpur and Its Impact on Landuse Patterns." *Geographical Outlook*, 6 (1969): 82-89.

Husain, Majid. "Geographic Basis of Tube Well Irrigation in the Upper Ganga-Yamuna Doab." *Geographical Observer*, 6 (1969): 95-103.

Gupta, S.L. "The Geography of Irrigation in Haryana." *Geographical Review of India*, 32, No. 1 (1970): 29-40.

Ahmad, Enayat. "Geography of Irrigation in India." *Geographical Outlook*, 8 (1971–1972): 9-28.

Goeldner, Roland. "Techniques d'irrigation et paysages agraires dans la campagne d'Hyderbad." *Les Cahiers d'Outre Mer*, 25 (1972): 5-26.

Palanivela, C. "Some Aspects of Well Irrigation in Dry Farming Regions of Tamil Nadu." *Indian Geographical Journal*, 49, No. 2 (1974): 69-77.

Singh, Jasbir. "Regional Imbalances and Temporal Developments in Irrigation Facilities in Haryana." *Geographical Review of India*, 36, No. 2 (1974): 102-113.

Dayal, Edison. "Impact of Irrigation Expansion on Multiple Cropping in India." *Tijdschrift voor Economische en Sociale Geographie*, 68, No. 2 (1977): 100-109.

Singh, R.P. "Development of Irrigation Potential and Its Utilization." *North Eastern Geographer*, 9, No. 1-2 (1977): 32-37.

Ghosh, Dhira. "Evaluation of Draft Irrigation Programme at Sonamukhi, C.A.D.P. Area: An Approach to Benefit-Cost Analysis." *Geographical Review of India*, 40, No. 1 (1978): 62-73.

Dhillan, S.S., and D.S. Sandhu. "Irrigation Development in Punjab: Its Potential and Limitations, 1951–1952 to 1975–1976." *Geographical Review of India*, 41, No. 2 (1979): 115-122.

Pawar, C.T., and Suresh D. Shinde. "Well Irrigation in Upland District of South Maharashtra Region: A Spatio-Temporal Perspective." *Geographical Review of India*, 41, No. 4 (1979): 314-320.

Bharara, L.P. "Social Context of a Minor Irrigation System in the Arid Zone of Rajasthan." Pp. 339-354 in *Perspectives in Agricultural Geography*, vol. 2, edited by Noor Mohammad. New Delhi, India: Concept Publishing Company, 1981.

Ghosh, Dhira. "Impact of Irrigation on Land Use and Cropping Intensity." Pp. 503-532 in *Perspectives in Agricultural Geography*, vol. 2, edited by Noor Mohammad. New Delhi, India: Concept Publishing Company, 1981.

Sharma, S.K., and C.K. Jain. "Use of Water Resources for Irrigation and Agricultural Development: A Case Study of Mahhya Pradesh." Pp. 463-484 in *Perspectives in Agricultural Geography*, vol. 2, edited by Noor Mohammad. New Delhi, India: Concept Publishing Company, 1981.

Singh, S.K. "Electricity Operated Sources of Irrigation in Eastern Uttar Pradesh: Problems and Prospects." Pp. 485-502 in *Perspectives in Agricultural Geography*, vol. 2, edited by Noor Mohammad. New Delhi, India: Concept Publishing Company, 1981.

Mathur, R.N., I.P. Pandey, and V.K. Rai. "Ground Water Resources in the Development of Irrigation in Saidpur Tahsil, Ghazipur District, Uttar Pradesh." *Geographical Review of India*, 45, No. 1 (1983): 33-39.

Sharma, Tara C., and O. Coutinho. "Growth of Irrigation and Its Impact on Crop Landuse and Crop Yield in Karnataka, 1960–61 to 1976–77." *Annals of the National Association of Geographers, India*, 2, No. 1 (1983): 13-22.

Basu, Swapna. "Impact of D.Y.C. Irrigation in Changing the Landuse and Cropping Pattern of the Lower Damodar Valley." *Geographical Review of India*, 46, No. 4 (1984): 70-78.

Dikshit, J.A., and K.R. Dikshit. "An Assessment of the Need for Irrigation: A Case Study of Maharashtra." *Indian Geographical Journal*, 59, No. 2 (1984): 214-231.

Misra, B.N. "Impact of Irrigation on Farming in Mirzapur District." *Geographical Review of India*, 46, No. 4 (1984): 24-33.

More, K.S., and F.R. Mustafa. "Irrigation Requirements and Developments in Maharashtra." *Transactions, Institute of Indian Geographers*, 6, No. 1 (1984): 73-78.

Wallach, Bret. "Irrigation Developments in the Krishna Basin Since 1947." *Geographical Review*, 74, No. 2 (1984): 127-144.

Bilas, Ram. "Irrigation Water Requirement for Cultivable Land in Varanasi District: An Analysis of Suitability and Effect on Land Use." *National Geographical Journal of India*, 31, No. 3 (1985): 217-227.

Kumar, Vijay, and M. Pal. "Development of Irrigation in Haryana: Limitations and Prospects—A Review." *Transactions, Institute of Indian Geographers*, 7, No. 2 (1985): 177-185.

Mukerji, S., and S. Raju. "The Role of Pumpset Irrigation in Agricultural Practice in India." *Annals of the National Association of Geographers, India*, 5, No. 1 (1985): 45-50.

Singh, Usha. "Impact of Water Resources on Agricultural Productivity of Lucknow District, Uttar Pradesh." *National Geographical Journal of India*, 31, No. 3 (1985): 228-236.

Wallach, Bret. "British Irrigation Works in India's Krishna Basin." *Journal of Historical Geography*, 11, No. 2 (1985): 155-173.

Chandrakant, T.P., and Suresh D. Shinde. "Irrigation in Maharashtra: Spatio-Temporal Perspective." *National Geographical Journal of India*, 32, No. 2 (1986): 105-110.

Chatterjee, Nandini. "Impact of Irrigation on Agriculture in Southern West Bengal." *Geographical Review of India*, 48, No. 1 (1986): 73-81.

Singh, Abha L., and Farooq Azam. "Correlation of Irrigation and Crop Output in Western Uttar Pradesh." *National Geographical Journal of India*, 32, No. 2 (1986): 93-104.

Chatterjee, N. "Necessity of Irrigation in West Bengal: An Agro-Climatic Approach." *Geographical Review of India*, 49, No. 3 (1987): 60-66.

Sharma, B.L. *Problems and Perspectives of Watering the Crops: A Geographical Analysis of Rajasthan*. New Delhi, India: Concept Publishing Company, 1987.

Yadav, D. "Impact of Irrigation on the Crops of Haryana." *GeoJournal*, 14, No. 1 (1987): 115-118.

Gitabali, Devi H. "Impact of Irrigation on Agricultural Productivity in Manipur." *North Eastern Geographer*, 20, No. 1-2 (1988): 48-51.

Singh, Abha L., S.F. Azam, and S.N.J. Hashmi. "Inter-District Variations in the Development of Irrigation in Western Uttar Pradesh." *National Geographer*, 23, No. 1 (1988): 45-54.

Pawar, C.T. *Impact of Irrigation: A Regional Perspective*. Bombay, India: Himalaya Publication House, 1989.

Raza, Mehdi. "Irrigation Potential of Groundwater Resources of Uttar Pradesh." *The Geographer, Aligarh*, 36, No. 1 (1989): 24-36.

Siddiqi, Shamsul H. "The Pattern of Irrigation in the North Bihar Plain." *Geographical Review of India*, 51, No. 4 (1989): 55-64.

Chatterjee, Nandini. "Impact of Irrigation on Cropping Intensity in West Bengal." *Transactions, Institute of Indian Geographers*, 12, No. 2 (1990): 109-120.

Dhillon, S.S. "Irrigation in Haryana: Potentials and Limitations." *Asian Profile*, 18, No. 3 (1990): 247-264.

Gadgil, Alka, and Vrishali Desothali. "Determination of Net-Irrigation Requirement with Reference to Hybrid Jowar." *Annals of the National Association of Geographers, India*, 10, No. 1 (1990): 23-28.

Gurjar, Ram E., ed. *Geographical Perspectives on Irrigation*. Jaipur, India: Rawat Publications, 1990.

Hooda, Randlur S. "Sprinkler Irrigation and Agricultural Transformation in Bhiwani District, Haryana." *Geographical Review of India*, 52, No. 3 (1990): 81-88.

Jadhav, M.G., and B.A. Ajagekar. "Impact of Irrigation on Crop Productivity and Farm Technology." *Deccan Geographer*, 28, No. 2-3 (1990): 657-670.

Leclerc, E. "Strategies et logiques de l'irrigation en Inde du Sud: Une maitrise conservee de l'espace." *Les Cahiers Geographiques de Rouen*, 33 (1990): 17-29.

Mishra, Pratibha. "Water Management in Arid Agriculture: A Case Study of District Barmer, Rajasthan." *Annals of the National Association of Geographers, India*, 10, No. 1 (1990): 29-46.

Prasad, Houshila. "Role of Irrigation in Agricultural Development of Gyanpur Tahsil, Varanasi District." *National Geographical Journal of India*, 36, No. 3 (1990): 190-198.

Reddy, Udaya B. "Implication of Major Irrigation Projects on Rural Economy and Society: Need for Effective Policy Measures." *Rural Systems*, 8, No. 1 (1990): 33-40.

Singh, Abha L., and S.F. Azam. "Economies of Irrigation from Canals and Tubewells in Aligarh District." *The Geographer, Aligarh*, 37, No. 1 (1990): 23-34.

Singh, G.B. "Agro-Irrigational Planning Regions: Bases for Agricultural Development (A Case of Indian Punjab)." *National Geographer*, 25, No. 2 (1990): 87-94.

Jayakumar, R., and M. Siraz. "Groundwater Suitability for Agriculture in Attur Valley, Tamil Nadu." *Indian Geographical Journal*, 67, No. 2 (1992): 75-79.

Reddy, G.V., and M.V. Reddy. "Irrigation Intensity and Orientation in Rayalaseema, Andhra Pradesh." *National Geographer*, 27, No. 1 (1992): 15-26.

Singh, Abha L. "Impact of Different Sources of Irrigation on Cropping Pattern, Yields and Farm Practices." *Geographical Review of India*, 54, No. 1 (1992): 19-30.

Singh, Jhujar. "Tubewell Irrigation and Spatial Organisation of Agriculture." *Geographical Review of India*, 56, No. 4 (1994): 43-46.

Singh, M.S., and Rajendon S. Chandel. "Sustainability of Irrigation Based Agriculture in India: A Case Study of Uttar Pradesh." *Annals, National Association of Geographers, India*, 14, No. 2 (1994): 45-61.

Soundaram, S.V. "Need for Irrigation and Its Development in Tamil Nadu." *Geographical Review of India*, 56, No. 3 (1994): 60-64.

Tiwari, R.C., M. Ghosh, and R.K. Srivastava. "Strategies for Irrigation and Agricultural Planning: A Case Study of Yamunapor Region of Allahabad District." *National Geographer*, 30, No. 2 (1995): 149-162.

Das Vishwakarma, Dharam. "Inter-Community Variation in Irrigation Practices: A Case Study of Betul-Chhindwara Plateau, Madhya Pradesh." *Geographical Review of India*, 59, No. 2 (1997): 173-179.

Hooda, Randlur S. "Lift Irrigation and Commercial Crops in the Drought Prone Areas of South West Haryana." *Geographical Review of India*, 59, No. 2 (1997): 151-158.

Omkar, A.C., and N. Ramkrishan Rao. "Impact of Irrigation on Land Use Pattern: A Study of Sriramsagar Command Area." *Deccan Geographer*, 35, No. 2 (1997): 147-156.

Pragathi, P.S., and Y.V. Ramanaiah. "Irrigation Orientation and Combinations in Andhra Pradesh, India." *Transactions, Institute of Indian Geographers*, 19, No. 2 (1997): 19-26.

AGRICULTURAL MARKETING AND DISTRIBUTION

Duggal, L.S., and H.S. Malik. "A Classification of Agricultural Markets in Haryana." *Geographical Viewpoint*, 3, No. 1-2 (1972): 35-38.

Tamarskar, B.G. "Livestock Marketing System in Chhattisgarh Plain." *North Eastern Geographer*, 11, No. 1-2 (1979): 48-57.

Shrivastava, V.K. "Agricultural Marketing in Uttar Pradesh." Pp. 407-428 in *Perspectives in Agricultural Geography*, vol. 5, edited by Noor Mohammam. New Delhi, India: Concept Publishing Company, 1981.

Dube, R.S. "A Rational Approach to the Marketing of Foodgrains in Madhya Pradesh." *National Geographer*, 21, No. 2 (1986): 131-142.

Jain, C.K., and S.K. Sharma. "Agricultural Marketing in Madhya Pradesh." *Geographical Review of India*, 48, No. 3 (1986): 59-69.

Khan, Najma. "Regional Imbalances in Per Capita Agricultural Incomes in India." *Rural Systems*, 4, No. 3 (1986): 137-146.

Khan, Nizamuddin. "Direct Marketing in Agriculture in Faizabad District." *Indian Geographical Journal*, 35, No. 2 (1988): 68-79.

Maidamwar, G.T. "System of Livestock Market: Places and Network." *Deccan Geographer*, 28, No. 2-3 (1990): 671-682.

Laksmi, Abha, and S. Fazel. "Marketing of Agricultural Produce by the Farmer in Upper Ganga-Yamuna Doab Region, India." *The Geographer, Aligarh*, 41, No. 1 (1994): 1-12.

Ahmad, Atteque, and Fazher Rahman. "Irrigation and Variation in Wheat Productivity: Are They Co-Terminous? A Study of Regional Dimensions in Wheat Production Potential in Uttar Pradesh (India)." *Asian Profile*, 24, No. 4 (1996): 321-332.

TRANSPORTATION AND MECHANIZATION

Singh, Surendra. "The Impact of Electricity on Agricultural Development in the Banwaripur Group of Villages (Varanasi): A Case Study." *National Geographical Journal of India*, 17, No. 2-3 (1971): 99-112.

Kaushal, M.P. "The Mechanization of Agriculture and Its Effects on the Jajmani System in the Punjab." *Pacific Viewpoint*, 13, No. 1 (1972): 49-61.

Wilbanks, Thomas J. "Accessibility and Technological Change in Northern India." *Annals of the Association of American Geographers*, 62, No. 3 (1972): 427-436.

Mohammad, Noor. "Technological Change and Diffusion of Agricultural Innovations." *The Geographer, Aligarh*, 23, No. 1 (1976): 4-14.

Rehman, Hifzur. "Mechanization of Farming and Its Impact on Food Crop Productivity in Uttar Pradesh." *The Geographer, Aligarh*, 23, No. 2 (1976): 43-56.

Ienga, Y. "A Study of Technological Innovation of Indian Agriculture with Special Reference to Gaggar Bhana." Pp. 29-31 in *Green Revolutions in the Punjab: Techniques, Background and Effects*, edited by Hiroshi Ishida. Hiroshima, Japan: University of Hiroshima Centre for Research in World Regional Geography, 1977.

Shinde, Suresh D., and M.G. Jadhav. "Use of Energy in Agriculture in Sangli District: A Geographical Analysis." *Geographical Review of India*, 40, No. 2 (1978): 101-107.

Maidamwar, G.T. "Transportation of Trade Livestock: A Regional Case Study of Yavatmal District." *Deccan Geographer*, 19, No. 2-3 (1981): 95-102.

Mohammad, Noor. "Technological Change and Spatial Diffusion of Agricultural Innovations in Trans-Ghaghara Plain." Pp. 305-342 in *Perspectives in Agricultural Geography*, vol. 5, edited by Noor Mohammad. New Delhi, India: Concept Publishing Company, 1981.

Singh, Onkar, and S.P. Maurya. "Progress of Agricultural Mechanisation in Chakia Development Block (Varanasi District, Uttar Pradesh)." *National Geographical Journal of India*, 27, No. 3-4 (1981): 141-155.

Vyas, Anjana, and C. Thangavel. "Diffusion of Agricultural Technology and Its Implication for Future Adoption: A Case Study." *Transactions, Institute of Indian Geographers*, 5, No. 1 (1983): 23-28.

Shafi, M. "Farm Power and Productivity in Indian Agriculture." *Annals of the National Association of Geographers, India*, 5, No. 2 (1985): 10-15.

Amman, N.Q. "Impact of New Technology on Agricultural Production in Aligarh District, Rasulpur Village: A Case Study." *National Geographer*, 21, No. 2 (1986): 173-178.

Singh, Abha L., and S. Banerjee. "Pattern of Energy Input and Its Impact on Agricultural Production in Agra Divis." *National Geographical Journal of India*, 34, No. 2 (1988): 126-133.

Khanna, Sulbha. "Resource Utilization or Resource Exploitation: An Issue in Farming Technology of Gujarat." *Deccan Geographer*, 29, No. 2 (1991): 29-36.

Munir, Abdul, Shamsul H. Siddiqui, and Abu Sufiyan. "Agricultural Modernization in Uttar Pradesh." *The Geographer, Aligarh*, 41, No. 1 (1994): 13-19.

Najmul Islam Hashami, S. "Impact of New Agricultural Technology on the Agricultural Development of Haryana." *The Geographer, Aligarh*, 41, No. 2 (1994): 60-69.

Singh, B.N., and Alkeshwari Singh. "Adoptation of Innovations and Agricultural Mechanisation in District Pratapgarh, Uttar Pradesh." *Transactions, Institute of Indian Geographers*, 17, No. 2 (1995): 87-96.

AGRICULTURAL LABOR STUDIES

Schwartzburg, Joseph E. "Agricultural Labor in India: A Regional Analysis with Particular Reference to Population Growth." *Economic Development and Cultural Change*, 16 (1963): 337-352.

Sahir, A.H. "Idealized Work Calendar of a Grain Farmer." *The Geographer, Aligarh*, 22, No. 2 (1975): 1-7.

Prasad, S.N. "Problems of Increasing Employment Possibilites and Earnings of Agricultural Population in Ranchi District." Pp. 485-492 in *Perspectives in Agricultural Geography*, vol. 5, edited by Noor Mohammad. New Delhi, India: Concept Publishing Company, 1981.

Wilken, G.C. "The Fallacy of Labour/Land Comparisons of Agricultural Productivity." Pp. 455-462 in *Perspectives in Agricultural Geography*, vol. 4, edited by Noor Mohammad. New Delhi, India: Concept Publishing Company, 1981.

Khan, Najma. "Concentration of Agricultural Laborers in India." *Rural Systems*, 4, No. 4 (1986): 199-210.

Sinha, B.R., and R.N.P. Sinha. "The Relationship Between Agricultural Manpower and Land Resources in South Bihar Plain." *Rural Systems*, 4, No. 2 (1986): 85-90.

Shanthi, K. "Agricultural Wages in Tamil Nadu: A Study of Inter-District Variations and Sex-Discriminations." *Indian Geographical Journal*, 62, No. 1 (1987): 35-45.

Prasad, Rajendra. "Characteristics of Agricultural Labour Force in Hoshiarpur Tehsil (Punjab)." *Geographical Viewpoint*, 17-18, No. 1-2 (1988–1989): 51-58.

Shastri, Prabha S. "Variations in Growth of Agricultural Workers in Vidarbha." *Deccan Geographer*, 28, No. 1 (1990): 589-603.

Vaid, Varsha, and V.S. Datye. "Agricultural Labour Productivity in Maharashtra." *Geographical Review of India*, 53, No. 3 (1991): 6-20.

Joshi, Kailosh C. "Employment in Agriculture in Uttar Pradesh Hills." *The Geographer, Aligarh*, 40, No. 1 (1993): 67-74.

Saikia, B. "Problem of Surplus Labour in the Tea Plantations of Assam with Special Reference to Sonitpur District." *North Eastern Geographer*, 24, No. 1-2 (1992–1993): 64-70.

Singh, Abha L., S Fazel, and H. Sajjad. "Role of Landless Labourers in the Agricultural Economy of Aligarh Districts." *National Geographer*, 30, No. 1 (1995): 25-36.

Mauritius, Réunion, and the Seychelles Islands

GENERAL TEXTS, ARTICLES, AND THESES

Dupon, Jean-Francois. "Aspects de l'agriculture aux Seychelles: L'exemple de l'ile Silhouette." *Les Cahiers d'Outre Mer*, 24 (1971): 217-244.

Guebourg, Jean-Louis. "Une commune agricole de l'ile de la Reunion en mutation." *Les Cahiers d'Outre Mer*, 42 (1989): 189-210.

CROPS AND CROPPING PATTERNS

Robertson, C.J. "The Sugar Industry of Mauritius." *Economic Geography*, 6, No. 4 (1930): 338-351.

Caubet, Paul. "La canne a sucre a l'ile Maurice." *Annales de Geographie*, 42 (1933): 516-528.

Brookfield, H.C. "Problems of Monoculture and Diversification in a Sugar Island: Mauritius." *Economic Geography*, 35, No. 1 (1959): 25-40.

Defos du Rau, J. "Deux experiences agricoles a la Reunion: Geranium et the." *Travaux et Documents de Geographie Tropicale*, CEGET-CNRS, 20 (1975): 277-298.

Dupon, Jean-Francois. "Les petits planteurs de l'ile Maurice." *Travaux et Documents de Geographie Tropicale*, CEGET-CNRS, 20 (1975): 299-322.

Schweinfurth-Marby, H., and U. Schweinfurth. "Der Teeanbau auf Mauritius." *Geographische Zeitschrift*, 63, No. 1 (1975): 31-55.

Lefevre, D. "Quelques aspects de l'evoution recente de l'agriculture reunionnaise." Pp. 505-521 in *Geographes d'aujourd'hui*, edited by A. Dagorne and J.P. Paulet. Nice, France: Universite de Nice, 1984.

de Jong, Ali, Annelet Harts-Broekhuis, and Robert Von Bommel. "Petits perimetres irrigues, une alternative ou un complement aux rizieres sur mersibles." *Les Cahiers d'Outre Mer*, 42 (1989): 271-292.

Huetz de Lemps, Alain. "L'evolution recente de la culture de la canne a sucre a Maurice." *Les Cahiers d'Outre Mer*, 42 (1989): 155-188.

Nepal and Bhutan

ATLASES AND GRAPHIC PRESENTATIONS

Huke, Robert, Eleanor H. Huke, and T. Woodhead. *Rice-Wheat Atlas of Nepal*. Manila, Philippines: International Rice Research Institute, 1994.

GENERAL TEXTS, ARTICLES, AND THESES

Joshi, Tulsi R. "Factorial Agricultural Patterns of the Far Western Region of Nepal." *Himalayan Review*, 8 (1976): 10-19.

Douglas, Michael. "Agropolitan Development: An Alternative for Regional Development in Asia." *Himalayan Review*, 13 (1981): 37-71.

Zurik, David N. "The Urban Tiller: An Investigation into the Urban Agricultural System of Kathmadu, Nepal." Master's Thesis, Michigan State University, East Lansing, Mich., 1981.

Wagner, H.G. "Innovative Wandlungen der Agrarstrukturam Golf von Neapel, 1965–1989." *Erdkunde*, 44, No. 3 (1990): 180-192.

Adhikari, J. "Ethnicity, Off-Farm Income and Resource Use in the Semi-Subsistence Farming System of Kaski District, Nepal." Ph.D. Dissertation, Australian National University, Canberra, Australia, 1992.

CROPS AND CROPPING PATTERNS

Rana, Ratna S., and Tulsi R. Joshi. "Nepal's Foodgrain Surplus and Deficit Regions." *National Geographical Journal of India*, 14, No. 2-3 (1968): 165-175.

Rana, Ratna S. "Regional Variations in the Growth Structure of Crop Production in Nepal." *Himalayan Review*, 4 (1971): 20-27.
Amatya, Soorya L. "The Growth and General Distribution of Cash Crops in Nepal." *Himalayan Review*, 5 (1972): 1-22.
Amatya, Soorya L. "A Study of Agricultural Crop Combination in Nepal." *National Geographical Journal of India*, 19, No. 1 (1973): 66-76.
Amatya, Soorya L. "A Study of Agricultural Combination in Nepal." *Himalayan Review*, 6 (1973–1974): 1-18.
Amatya, Soorya L. *Cash Crop Farming in Nepal*. Kirtipur, Nepal: Geography Instruction Committee, Institute of Humanites and Social Sciences, Tribhuvan University, 1975.
Subedi, Bhim P. "Large Cardoman Farming in Some Villages of Ilam District." *Himalayan Review*, 14 (1982): 25-37.

ANIMAL HUSBANDRY

Bjonnes, J.M. "Animal Husbandry and Grazing: A Conservation and Management Problem in Sagarmatha National Park, Nepal." *Norsk Geografisk Tidsskrift*, 34, No. 2 (1980): 54-76.
Shrestha, R.L.J. "The Relationship Between the Forest and the Farming System in Chautara, Nepal, with Special Reference to Livestock Production." Ph.D. Dissertation, Australian National University, Canberra, Australia, 1982.
Murthy, H.P.P., and K. Ranganath. "Role of Livestock in the Rural Development of Bhutan." *National Geographer*, 20, No. 1 (1985): 75-80.
Brown, Barbara. *Sherpa of Khumba: People, Livestock and Landscape*. New York, New York: Oxford University Press, 1993.

FARM LAND

Shrestha, S.H. "A Review of Land Use Pattern in Nepal." *Himalayan Review*, 7 (1975): 33-42.
Blamont, Denis. "L'ocupation de l'espace par un systeme agro-pastoral dans les montagnes du Centre-Est du Nepal." *Les Cahiers d'Outre Mer*, 36 (1983): 201-229.

ENVIRONMENTAL AND CONSERVATION ISSUES

Shrestha, Ram K. "Impacts of Altitude and Slope on Land Use in Khandabari." *Himalayan Review*, 8 (1976): 20-32.
Banskota, Mahesh. "Interaction of Food Production, Household Energy and Environmental Change in Hill Environments: A Simulation Approach." *Himalayan Review*, 11 (1979): 45-64.
Fetzer, Karl D. "The Agricultural Suitability of Natural Units in the Kathmandu Valley, Nepal." *Applied Geography and Development*, 16 (1980): 80-91.
Webster, Peter. "To Plough or Not to Plough? A Newer Dilemma, Tabu and Technology in Kathmandu Valley, Nepal." *Pacific Viewpoint*, 22, No. 2 (1981): 99-131.

Nayava, J. "Climates of Nepal and Their Implications for Agricultural Development." Ph.D. Dissertation, Australian National University, Canberra, Australia, 1983.

Krishna, P.P., K.C. Krishna, and Kedar Basnet. "A Study of Hillslope Terrace Cultivation: Begnas Watershed, Kashi, Nepal." *National Geographical Journal of India*, 35, No. 3-4 (1989): 256-267.

Gillmor, Desmond A., and M.C. Nurse. "Farmer Initiatives in Increasing Tree Cover in Central Nepal." *Mountain Research and Development*, 11, No. 4 (1991): 329-337.

Young, C.J. "Agricultural Changes in Bhutan: Some Environmental Questions." *Geographical Journal*, 157, No. 2 (1991): 172-178.

Wichmann, Peter. *Jungquartare randtropische Verwitterung: Ein bodengeographischer Beitrag zur landschaftsentwicklung von Sudwest-Nepal.* Kiel, Germany: Kieler Geographische Schriften, Heft 88, 1993.

Chhetri, N. "A Climate Sensitive Measure of Agricultural Intensity: An Application to Nepal." Master's Thesis, Pennsylvania State University, University Park, Pa., 1999.

Pakistan

ATLASES AND GRAPHIC PRESENTATIONS

Woodhead, T., Robert Huke, and Eleanor H. Huke. *Rice-Wheat Atlas of Pakistan.* Manila, Philippines: International Rice Research Institute, 1993.

GENERAL TEXTS, ARTICLES, AND THESES

Elahi, M.K. "Efficiency of Agriculture in West Pakistan." *Pakistan Geographical Review*, 20, No. 2 (1965): 77-116.

Gotsch, Carl H. "Regional Agricultural Growth: The Case of West Pakistan." *Asian Survey*, 8, No. 3 (1968): 188-205.

Dettmann, K. "Pakistan's Ackerbau." *Geographische Rundschau*, 24, No. 8 (1972): 321-331.

Siddiqi, Akhtar H. "Some Structural Characteristics of Agriculture in Pakistan." *Indiana State University Department of Geography Papers*, 6 (1974): 19-39.

Ali, R., and F. Auriac. "Agriculture et organisation rurale: Une experience de developpement rural integre dans la region de Peshawar." *Bulletin de la Societe Languedocienne de Geographie*, 10, No. 4 (1976): 253-286.

Siddiqi, Akhtar H. "Problems and Performance of Agriculture in Pakistan." *Asian Profile*, 5, No. 4 (1977): 355-370.

Siddiqi, Akhtar H. "Agricultural Changes in Punjab in the Nineteenth Century: 1850–1900." *GeoJournal*, 12, No. 1 (1986): 43-56.

Nuesser, Marcus, and Jurgen Clemens. "Impacts on Mixed Mountain Agriculture in the Rupal Valley, Nanga Parbut, Northern Pakistan." *Mountain Research and Development*, 16, No. 2 (1996): 117-133.

CROPS AND CROPPING PATTERNS

Mahmud, F.B. "Cotton-Growing in West Pakistan." *Pakistan Geographical Review*, 4, No. 2 (1949): 36-45.

Whitehall, S.B. "Consideration of the Causes of Fluctuations in the Acreage of American Type Cotton in the Punjab (Pakistan)." *Pakistan Geographical Review*, 7, No. 2 (1952): 70-89.

Rahman, Amina. "Distribution of Rice in West Pakistan." *Pakistan Geographical Review*, 10, No. 2 (1955): 49-54.

Rather, A.H. "Tobacco in Choch." *Pakistan Geographical Review*, 10, No. 1 (1955): 37-46.

Kazi, S. Ahmad. "Distribution of Wheat in Pakistan." *Pakistan Geographical Review*, 12, No. 2 (1957): 70-74.

Arefin, Sirajul. "Sericulture in Rajshahi." *Pakistan Geographical Review*, 15, No. 1 (1960): 25-31.

Gulick, Luther H. "A Cotton-Wheat Farm in Hyderbad District, West Pakistan." *Pakistan Geographical Review*, 16, No. 1 (1961): 25-35.

Kureshy, K.U., and M.K. Elahi. "Farm Size–Crop Relationships in West Pakistan." *Pakistan Geographical Review*, 24, No. 1 (1969): 58-73.

Rahman, Mushtaqur. "Jute Cultivation in the Indus Delta, West Pakistan." *Oriental Geographer*, 14, No. 1 (1970): 35-43.

Kureshy, K.U., and M.K. Elahi. "Cropping Pattern and Crop Association in West Pakistan." *Pakistan Geographical Review*, 26, No. 2 (1971): 16-37.

Elahi, M.K. "Wheat Cultivation in West Pakistan: Its Water Budget." *Pakistan Geographical Review*, 27, No. 2 (1972): 1-31.

Siddiqi, Akhtar H. "Cropping Patterns in West Pakistan." *Journal of Geography*, 71, No. 2 (1972): 96-108.

Alizai, Mohammas Y. "Pattern of Sugarbeet Concentration in Peshawar Valley." *Pakistan Geographical Review*, 30, No. 1-2 (1975): 3-12.

Elahi, M.K. "Evolution of Cropping Pattern in the Sub-Urban Areas in Pakistan: A Case Study of Lahore." *Pakistan Geographical Review*, 31, No. 1-2 (1976): 7-18.

Witcombe, J.R. "The Distribution of Cropping Systems in Northern Pakistan." *Agro-Ecosystems*, 3, No. 4 (1977): 285-290.

Allan, Nigel J.R. "Changing Crop Complexes in the North Pakistan Foothills." *Geography*, 69, No. 3 (1984): 261-263.

Butz, David. "A Note on Crop Distribution and Micro-Environmental Conditions in Holshal and Ghoshushal Villages, Pakistan." *Mountain Research and Development*, 14, No. 1 (1994): 89-97.

ANIMAL HUSBANDRY

Siddiqi, Akhtar H. "Spatial Distribution of Livestock in Pakistan." *Geoforum*, 8, No. 1 (1977): 33-38.

Herbers, H., and G. Stober. "Bergbauerliche Vietshaltung in Yasin (Northern Areas, Pakistan): Organisatarische und rechtliche Aspekte der Hochweidenutzung." *Petermann's Geographische Mitteilungen*, 139, No. 2 (1995): 87-104.

Clemens, Jurgen, and Marcus Nuesser. "Animal Husbandry and Socio-Economic Development: Persistence and Transformation—A Nanga Parbat Case Study." *Pakistan Journal of Geography*, 5-6, No. 1-2 (1995–1996): 45-60.

FARM LAND

Elahi, M.K. "Agricultural Land Use in West Pakistan." *Pakistan Geographical Review*, 18, No. 2 (1963): 25-67.

Zaidi, I.H. "The Spatial Pattern of Farm Sizes in the Punjab Region of West Pakistan." *Pakistan Geographical Review*, 22, No. 2 (1967): 61-76.

Malik, Rashid A. "Changes in the Land Use Pattern of the Upper Indus Basin During the British Period." *Pakistan Geographical Review*, 29, No. 1 (1974): 1-43.

Khan, Fazle K., and Muhammad R. Dhanani. "The Role of 'Reported' Cultivated Area in the Green Revolution in Pakistan." *Pakistan Journal of Geography*, 4, No. 1-2 (1994): 61-68.

AGRICULTURAL POPULATIONS AND SOCIAL ISSUES

Fautz, Bruno. *Sozialstruktur und Bodennutzung in der Kulterlandschaft des Swat (Nordwesthimalaya)*. Giessen, Germany: Giessener Geographische Schriften, Heft 3, 1963.

ENVIRONMENTAL AND CONSERVATION ISSUES

Ahmad, K.S., and M. Lal Khan. "Variability of Rainfall and Its Bearing on Agriculture in the Arid and Semi-Arid Zones of West Pakistan." *Pakistan Geographical Review*, 16, No. 1 (1961): 35-50.

Rafiq, Mohammad. "Crop Ecological Zones of the Indus Plains." *Pakistan Geographical Review*, 26, No. 2 (1971): 38-49.

Mullick, M.A.H. "Die grune Revolution in Pakistan: Segen oder unheil?" *Geographische Rundschau*, 24, No. 8 (1972): 332-337.

Elahi, M.K., and A.A. Abbasi. "Crop Water Requirement in Jhelum District." *Pakistan Geographical Review*, 36, No. 1-2 (1981): 15-26.

Allan, Nigel J.R. "Deforestation et agropatoralisme dans le Pakistan de Nord." *Revue de Geographie Alpine*, 74, No. 4 (1986): 405-420.

Rahman, Mushtaqur. *Agriculture and Environment: The Physical Geography of Temperate Agriculture in Pakistan*. Budapest, Hungary: Akademiai Kiado, 1988.

Scholz, Fred. "Indus-Left-Bank Outfall Drain: Ein Versuch Pakistans zur bleibonden Sicherung der landwirtschaft im Industrieland." *Erdkunde*, 41, No. 3 (1988): 246-249.

IRRIGATION

Gulick, Luther H. "Irrigation Systems of the Former Sino Province, West Pakistan." *Geographical Review*, 53, No. 1 (1963): 79-99.

Whittington, G.W. "The Irrigated Lands of the Indus Plain in West Pakistan." *Tijdschrift voor Economische en Sociale Geographie*, 55, No. 1 (1964): 13-19.

Ahmad, Qazi M. "Some Geographical Aspects of the Indus Waters Treaty and Development of Irrigation in West Pakistan." *Pakistan Geographical Review*, 20, No. 1 (1965): 1-30.

Rahman, Mushtaqur. "Probleme der Be- und Entwasserung, versalzung und Vernassung im Sind (West Pakistan)." *Geographische Rundschau*, 19, No. 7 (1967): 261-265.

Rahman, Mushtaqur. "Formen der Bewasserung im Sind, West-Pakistan." *Die Erde*, 99 (1968): 163-175.

Alizai, Mohammas Y. "The Role of Irrigation in Agriculture in West Pakistan." Master's Thesis, Kansas State University, Manhattan, Kans., 1969.

Malik, Rashid A. "Antecedents of Irrigation Development and Settlement Pattern in the Punjab." *Pakistan Geographical Review*, 28 (1973): 22-41.

Dettman, K. "Agriculture in the Canal Irrigation Region of the Pakistan Punjab: Its Historical Background and Current Problems." *Applied Geography and Development*, 18 (1981): 111-126.

Zafar, Fareeha. "Irrigation and Cropping Patterns in the Punjab Canal Colony Districts, 1890–1891 to 1940–1941." *Pakistan Geographical Review*, 36, No. 1-2 (1981): 41-60.

Andreae, Bernd. "Intensitatsstufen des Bewasserungsfeldbaus im Peshawar-Bechen, Pakistan: Beson der heiten der Zuckerproduktion im konkurrieren den Anbaugebiet von Zuckerrohrund Zuckerrube." *Erdkunde*, 38, No. 1 (1985): 29-35.

Scholz, Fred. "Irrigation in Pakistan: An Analysis of the Significance of the Most Recently Available Statistics." *Applied Geography and Development*, 26 (1985): 98-115.

Kreutzmann, Hermann. "Oases of the Karakorum: Evolution of Irrigation and Social Organization in Hunza, North Pakistan." Pp. 243-254 in *Human Impact on Mountains*, edited by Nigel J.R. Allen, Gregory W. Knapp, and Christoph Stadel. Lanham, Md.: Rowman & Littlefield, 1988.

Kreutzmann, Hermann. "Oasenbewasserung im Karakorum: Antochtone Techniken und exogene Uberpragung in der Hochgebirgslandwirtschaft Nordpakistans." *Erdkunde*, 44, No. 1 (1990): 10-22.

Gill, Z.A., and R.K. Sampath. "Inequality in Irrigation Distribution in Pakistan." *Pakistan Development Review*, 31, No. 1 (1992): 75-100.

Rahman, Mushtaqur. "Irrigation and Farm Water Management in Pakistan." *GeoJournal*, 31, No. 4 (1993): 363-372.

Khan, Fazle K., and M. Nawaz. "Karez Irrigation in Pakistan." *GeoJournal*, 37, No. 1 (1995): 91-100.

TRANSPORTATION AND MECHANIZATION

Rahman, Mushtaqur. "Tracterization of Agriculture in Pakistan." *GeoJournal*, 15, No. 4 (1987): 387-392.

Sri Lanka/Ceylon

GENERAL TEXTS, ARTICLES, AND THESES

Farmer, Bertram H. "Agriculture in Ceylon." *Geographical Review*, 40, No. 1 (1950): 42-66.

Foster, Gerard J. "Oriental Agriculture, the Case of Ceylon." Ph.D. Dissertation, University of California, Los Angeles, Calif., 1961.

Baker, Simon. "Shifting Agriculture in Ceylon." *Journal of Geography*, 67, No. 1 (1968): 564-568.

Manogaron, Chelvadurai. "Traditional Versus Modern Agriculture: A Study of Peasant Farming in Ceylon." *Geografiska Annaler B*, 56B, No. 1 (1974): 68-77.

Farmer, Bertram H. "Some Thoughts on the Place of Field Work in Agrarian Studies." *Sri Lanka Journal of Agrarian Studies*, 1 (1980): 1-11.

CROPS AND CROPPING PATTERNS

McCune, Shannon. "Sequence of Plantation Agriculture in Ceylon." *Economic Geography*, 25, No. 3 (1949): 226-235.

Subramanyam, N. "The Tea Industry in Ceylon." *Indian Geographical Journal*, 39, No. 4 (1954): 89-105.

Chandraratne, M.F. "Cotton in Ceylon." *Ceylon Geographer*, 10, No. 3-4 (1956): 82-91.

Singh, Harpal. "Plantation Agriculture in Ceylon." Master's Thesis, University of Tennessee, Knoxville, Tenn., 1960.

Swan, S.B. "Crop Failure and Variable Yields in Ceylon." *Pacific Viewpoint*, 8, No. 2 (1967): 159-174.

Baker, Simon. "Tea Production in Ceylon." *Journal of Geography*, 67, No. 1 (1968): 49-55.

Hung, F. "The Tea Plantation System in Ceylon: Persistence and Change." *Professional Geographer*, 20 (1968): 321-325.

Manogaron, Chelvadurai. "Changes in the Production of Tobacco and Its Impact on Peasant Farming in the Jaffna Peninsula, Ceylon." Master's Thesis, Clark University, Worcester, Mass., 1968.

Attanayake, Dona C. "Changing Patterns of Rice Production in Ceylon: 1945–1964." Master's Thesis, University of Wisconsin, Milwaukee, Wis., 1969.

Hausherr, Klaus. "Traditionnelle Brandrodunsfeldbau und moderne Erschlieszungsprojekts in der 'Trockenzone' im Sudosten Ceylons: Versuche der Wiederturstellung alten Kulturlandes um Buttla, Monoragah District."

Geographische Zeitschrift. Beiheft, Erdkundliches Wissen, Heft 27 (1971): 167-204.

Marby, Heidrun. "Die Teelandschaft der Insel Ceylon: Versuch einer raumlichen und zeitlichen Differenzierung." *Geographische Zeitschrift*. Beiheft, Erdkundliches Wissen, Heft 27 (1971): 23-101.

Marby, Heidrun. *Tea in Ceylon: An Attempt at a Regional and Temporal Differentiation of the Tea Growing Areas in Ceylon*. Wiesbaden, Germany: F. Steiner, 1972.

Dangalle, Nimal. "A Decade of Success in Food Production: Patterns and Trends in the Level of Sufficiency in Rice Production in Sri Lanka, 1971–1981." *Malaysian Journal of Tropical Geography*, 6 (1981): 1-16.

Schweinfurth-Marby, H. "Die tea: Smallholders auf der Insel Ceylon." *Geographische Zeitschrift*, Beiheft 54 (1981): 117-141.

Wanasinghe, Ananda. *Small Rice Farms and Agricultural Credit in the Wet Zone of Sri Lanka*. Armidale, New South Wales: University of New England, Department of Geography, Research Series in Applied Geography, 1982.

Hesselberg, J. "The Failure of Crop Diversification in the Uda Walawe Settlement Scheme, Southern Sri Lanka." *Norsk Geografisk Tidsskrift*, 40 (1986): 117-128.

Daleus, E., et al. "Management and Environmental Constraints to Rice Yield within a Village Irrigation System: A Case Study from Sri Lanka." *GeoJournal*, 17, No. 3 (1988): 401-412.

DAIRYING

Nelson, Lloyd G. "The Dairy Industry in Sri Lanka: Problems and Prospects." Ph.D. Dissertation, University of Victoria, Victoria, British Columbia, 1976.

AGRICULTURAL POPULATIONS AND SOCIAL ISSUES

Farmer, Bertram H. "Peasant and Plantation in Ceylon." *Pacific Viewpoint*, 4, No. 1 (1963): 9-16.

Manogaron, Chelvadurai. "Traditional Versus Modern Agriculture: A Study of Peasant Farming in Ceylon." *Geografiska Annaler B*, 56B, No. 1 (1974): 68-77.

Farmer, Bertram H. *Pioneer Peasant Colonization in Ceylon: A Study in Asian Agrarian Problems*. London, England: Oxford University Press, 1957; reprinted by Greenwood Press, 1976.

Storm-Furru, Inger, and Berit H. Vandsemt. "Agricultural Intensification in the Dry Zone of Sri Lanka: Impacts on Farming Practices, Welfare and Women's Work." *Norsk Geografisk Tidsskrift*, 43, No. 2 (1989): 55-65.

Brown, David J. "Science vs. Shifting Cultivation in Sri Lanka: A Social Study of Agricultural Science." Master's Thesis, Pennsylvania State University, University Park, Pa., 1997.

ENVIRONMENTAL AND CONSERVATION ISSUES

Domros, Manfred. *The Agroclimate of Ceylon: A Contribution Towards the Ecology of Tropical Crops*. Wiesbaden, Germany: F. Steiner, 1974.

Domros, Manfred. "An Agroclimatological Land Classification of Sri Lanka (Ceylon) for Tea, Rubber, and Coconut Palms." *Applied Sciences and Development*, 7 (1976): 39-65.

Peters, Wolfgang. *Agrarklimatologische Theorie und okonomische Praxis des Teeanbaus in Sri Lanka*. Mainz, Germany: Mainzer Geographische Studien, Heft 16, 1978.

Domros, Manfred. "Dry Years and Their Relationship to Crop Production in Sri Lanka." *GeoJournal*, 5, No. 2 (1981): 133-138.

Suppiah, R. "Four Types of Relationships Between Rainfall and Paddy Production in Sri Lanka." *GeoJournal*, 10, No. 1 (1985): 109-110.

Wickramasinghe, Anoja. "Ecological Impacts of the Dry Zone Farming Systems in Sri Lanka." *Malaysian Journal of Tropical Geography*, 15 (1987): 49-57.

Wickramasinghe, Anoja. "Soils of Forest and Agricultural Systems: A Case Study in Sri Lanka." *Malaysian Journal of Tropical Geography*, 23, No. 1 (1992): 45-50.

Sandell, K. "Farmers' Eco-Strategies with Regard to Water, Nutrients, and Sustainability: A Case Study of Low-Resource Agriculture in the Dry Zone of Sri Lanka." *Geografiska Annaler B*, 75B, No. 3 (1993): 163-176.

Domros, Manfred, J.M. Gunadasa, and E.R. Ranatunge. "Intra-Seasonal Rainfall Variation and the Cultivation of Rice Under Rainfed Farming in Sri Lanka." *Peterman's Geographische Mitteilungen*, 142, No. 2 (1998): 105-114.

IRRIGATION

Cook, E.K. "A Note on Irrigation in Ceylon." *Geography*, 35, No. 2 (1950): 75-85.

Perera, Mangalage P. "Geographic Aspects of Irrigation and Settlement in the Dry Zone of Ceylon." Master's Thesis, Ohio State University, Columbus, Ohio, 1962.

Foster, Gerald J. "The Concept of Regional Development in the Indigenous Irrigation Systems of Ceylon." *Yearbook, Association of Pacific Coast Geographers*, 31 (1969): 91-100.

TRANSPORTATION AND MECHANIZATION

Ulluwishewa, R. "Factors Affecting the Mechanization of the Tillage Operation of Paddy Fields in Sri Lanka: A Geographical Perspective." *GeoJournal*, 15, No. 4 (1987): 393-398.

Southeast Asia as a Whole

GENERAL TEXTS, ARTICLES, AND THESES

Elahi, M.K. "Agriculture in South East Asia." *Pakistan Geographical Review*, 14, No. 2 (1959): 76-92.

Rizvi, Najma A. "Agricultural System of Southeast Asia." Master's Thesis, University of Florida, Gainesville, Fla., 1961.

Reed, Robert R. "Swidden in Southeast Asia." *Lipunan*, 1, No. 1 (1965): 24-52.

Spencer, Joseph E. *Shifting Cultivation in Southeastern Asia*. Berkeley: Berkeley, Calif.: University of California, California Publications in Geography, Vol. 19, 1965.

Fisher, Charles A. "Social and Economic Consequences of Western Commercial Agriculture in South-East Asia." Pp. 47-154 in *Land Use and Resources: Studies in Applied Geography*, edited by C. Embleton and J.T. Coppock. London, England: Institute of British Geographers, Special Publication No. 1, 1968.

Bruneau, Michel, ed. *Types de cultures commerciales paysannes en Asie du Sud-Est et dans le monde insulindien*. Talence, France: Travaux et documents de geographie tropicale, No. 20, 1975.

Spencer, Joseph E. *Shifting Cultivation in Southeast Asia*. Berkeley, Calif.: University of California Press, 1976.

Courtenay, P.P. "Commercial Agriculture." Pp. 108-132 in *South-East Asia: A Systematic Geography*, edited by Ronald D. Hill. Kuala Lumpur, Malaysia: Oxford University Press, 1979.

Hill, Ronald D. "Tribal and Peasant Agriculture and Fisheries." Pp. 78-107 in *South-East Asia: A Systematic Geography*, edited by Ronald D. Hill. Kuala Lumpur, Malaysia: Oxford University Press, 1979.

Gibbons, David J., Rodolphe De Koninck, and Ibrahim Hasan. *Agricultural Modernization, Poverty and Inequality: The Distributional Impact of the Green Revolution in Regions of Malaysia and Indonesia*. Farnborough, England: Saxon House, 1980.

Hill, Ronald D. "Integration in Agricultural Geography with Some Southeast Asian Examples." *Philippine Geographical Journal*, 25, No. 3 (1981): 98-107.

Barton, Thomas F. "Agricultural Framework." Pp. 65-74 in *Southeast Asia: Realm of Contrast*, edited by Ashok K. Dutt. Boulder, Colorado: Westview, 1985.

Fryer, Donald W. "Agriculture and Fisheries." Pp. 168-192 in *South East Asian Development: Geographical Perspectives*, edited by Denis Dwyer. New York, N.Y.: John Wiley, 1991.

Hill, Ronald D. "Southeast Asian Agriculture: Status and Change, and Ecological Overview." *Bulletin de la Societe Geographique de Liege*, 33 (1997): 45-66.

Hill, Ronald D. "Status and Change in Forty Years of Southeast Asian Agriculture." *Singapore Journal of Tropical Geography*, 19, No. 1 (1998): 1-25.

Hafner, James A. "Perspectives on Agriculture and Rural Development." Pp. 133-159 in *Southeast Asia: Diversity and Development*, edited by Thomas R. Leinbach and Richard Ulack. Upper Saddle River, N.J.: Prentice-Hall, 2000.

CROPS AND CROPPING PATTERNS

Spencer, Joseph E. "The Migration of Rice from Mainland Southeast Asia to Indonesia." Pp. 83-89 in *Plants and the Migration of Pacific Peoples*, edited by Jacques Barrau. Honolulu, Hawaii: Bishop Museum Press, 1963.

Dobby, E.H.G. "The Changing Significance of Padi-Growing in Southeast Asia." *Journal of Tropical Geography*, 6 (1966): 81-88.

Johnson, B.L.C. "Recent Development in Rice Breeding in Some Implications for Tropical Asia." *Geography*, 57, No. 4 (1972): 307-320.

Keong, Voon Phin. *Western Rubber Planting in South East Asia*. Kuala Lumpur, Malaysia: University of Malaya Press, 1976.

Hohnholz, Jurgen H. "Manioc Cultivation in Southeast Asia: An Agricultural Geographical Survey." *Applied Geography and Development*, 16 (1980): 117-135.

Hill, Ronald D. "Crops and People in Southeast Asia." Pp. 399-436 in *Perspectives in Agricultural Geography*, vol. 3, edited by Noor Mohammad. New Delhi, India: Concept Publishing Company, 1981.

Scholz, U. "Decrease and Revival of Shifting Cultivation in the Tropics of Southeast Asia: The Examples of Sumatra and Thailand." *Applied Geography and Development*, 17 (1982): 32-45.

Uhling, Harold. "Reishausysteme und- okotope in Sudoistasien: Geowissenschaftliche Methoden in der Reisbau forschung und die Okosysteme des Uberschwemmungsreishaus." *Erdkunde*, 37, No. 4 (1983): 269-281.

Huke, Robert E., and Eleanor H. Huke. *The Human Geography of Rice in Southeast Asia*. Manila, Philippines: International Rice Research Institute, 1990.

Pantural, S., and A. Eddy. "Some Impacts on Rice Yield from Changes in the Variance of Precipitation." *Climatological Bulletin*, 24, No. 1 (1990): 16-27.

Scholz, U. "Grune Revolution im Riesbau Sudostasiens: Eine Bilanz der letzten 35 Jahre." *Geographische Rundschau*, 50, No. 9 (1998): 531-536.

DAIRYING

Wheatley, Paul. "A Note on the Extension of Milking Practices into Southeast Asia During the First Millennium AD." *Anthropos*, 60 (1965): 577-590.

AGRICULTURAL POPULATIONS AND SOCIAL ISSUES

Pelzer, Karl J. *Pioneer Settlement in the Asiatic Tropics: Studies in Land Utilization and Agricultural Colonization in South-Eastern Asia*. New York, N.Y.: American Geographical Society, Special Publication No. 29, 1945.

Spencer, Joseph E. "Agriculture and Population in Relation to Economic Planning." *Annals of the American Academy of Political and Social Sciences*, 321 (1959): 62-70.

Fisher, Charles A. "Les consequences sociales et economiques de l'agriculture commerciale europeenne en Asie du Sud-Est." *Les Cahiers d'Outre Mer*, 18 (1965): 295-305.

Rambo, A.T. "Human Ecology Research on Tropical Agroecosystems in Southeast Asia." *Singapore Journal of Tropical Geography*, 3, No. 1 (1982): 86-99.

De Koninck, Rodolphe. "Pourquoi les paysans? Interrogations sur la territorialite de l'agriculture familiale." *Cahiers de Geographie de Quebec*, 29, No. 72-73 (1984): 261-274.

Hill, Ronald D. "Notes on Chinese Agricultural Colonization in Southeast Asia." *Erdkunde*, 42, No. 2 (1988): 123-135.

De Koninck, Rodolphe, and Steven Dery. "Agricultural Expansion as a Tool of Population Redistribution in Southeast Asia." *Journal of Southeast Asian Studies*, 28, No. 1 (1997): 1-26.

ENVIRONMENTAL AND CONSERVATION ISSUES

Petzer, Karl J. "Swidden Agriculture in Southeast Asia: Historical and Ecological Perspectives." Pp. 2761-286 in *Farmers in the Forest*, edited by Peter Kunststadter, et al. Honolulu, Hawaii: University of Hawaii, East-West Center, 1978.

Yoshimo, Masatoshi. "Climatic Change and Agriculture: Problems for the Asian Tropics." Pp. 179-189 in *Southeast Asia's Environmental Future: The Search for Sustainability*, edited by Harold C. Brookfield and Yvonne Byron. Tokyo, Japan: U.N. University Press, 1993.

Dery, Steven. "Expansion agricole et deforestation: Le modile sud-est asiatique." Master's Thesis, Universite Laval, Ste. Foy, Quebec, 1995.

Dery, Steven. "Expansion agricole et deforestation: Le modile sud-est asiatique." *Cahiers de Geographie de Quebec*, 40, No. 109 (1996): 29-48.

Cambodia

GENERAL TEXTS, ARTICLES, AND THESES

Blanadet, Raymond. "Types de cultures commerciales dans la province Battambang (Cambodge)." *Travaux et Documents de Geographie Tropicale*, CEGET-CNRS, 20 (1975): 47-65.

FARM LAND

Toporova, S.I. "Land Fund of Cambodia and Its Agricultural Utilization" (in Russian). *Vestnik Leningradskogo Universiteta-Geologiya-Geographiya*, 12, No. 2 (1975): 116-122.

Indonesia

GENERAL TEXTS, ARTICLES, AND THESES

Pelzer, Karl J. "The Agrarian Conflict in East Sumatra." *Pacific Affairs*, 30, No. 2 (1957): 151-159.

Pelzer, Karl J. "The Agricultural Foundation." Pp. 118-154 in *Indonesia*, edited by Ruth McVey. New Haven, Conn.: HRAF Press, 1963.

Cordonnier, J.C. "Agriculture et commerce au Sumatra Barat." *Travaux et Documents de Geographie Tropicale*, CEGET-CNRS, 20 (1975): 127-218.

Uhlig, Harold. "Die Agrarlandschaft im Tropenkarst-Beispiele ihrer geo-okoligischen Differenzierung aus Java und Sulawesi." *Geografski Glasnik*, 38 (1976): 313-335.

Scholz, U. *Minangkabau: Die Agrarstruktur in West-Sumatra und Moglichkeiten ihrer Entwicklung*. Giessen, Germany: Giessner Geographische Schriften, Heft 41, 1977.

Kaida, Yoshihiro. "Recent Trends in Agricultural Development in South Sumatra" (in Japanese). *Southeast Asian Studies*, Kyoto, 17, No. 3 (1979): 133-145.

Metzner, Joachim. "Disequilibrium of Agricultural Regions in the Eastern Lesser Sunda Islands." *Applied Geography and Development*, 15 (1980): 62-81.

Metzner, Joachim. "Innovations in Agriculture Incorporating Traditional Production Methods: The Case of Amorasi (Timor)." *Applied Geography and Development*, 17 (1981): 91-102.

Dove, Michael R. "The Ideology of Agricultural Development in Indonesia." Pp. 221-247 in *Central Government and Local Development in Indonesia*, edited by Colin MacAndrews. Singapore: Oxford University Press, 1986.

Scholz, U. *Agrargeographie von Sumatra: Eine Analyse der raumlichen Differenzierung der landwirtschaftlichen Produktion*. Giessen, Germany: Giessner Geographische Schriften, Heft 63, 1988.

Zimmermann, Gerd. "Zur Genese der agraren Nutzungssysteme in ostlichen Indonesian." *Geographische Zeitschrift*, 76, No. 1 (1988): 36-47.

Bundschu, I. "Umsiedlung und landwirtschaftliche Produktionssteigerung auf Nusa Penida (Indonesian)." *Zeitschrift fur Wirtschaftsgeographie*, 34, No. 1 (1990): 46-56.

Booth, A. "Regional Aspects of Indonesian Agricultural Growth." Pp. 36-60 in *Indonesia: Resources, Regions, and Environment*, edited by J. Hardjans. New York, N.Y.: Oxford University Press, 1992.

CROPS AND CROPPING PATTERNS

Terra, G.J.A. "Mixed-Garden Horticulture in Java." *Journal of Tropical Geography*, 3 (1954): 33-43.

Fryer, Donald W. "The Recovery of the Sugar Industry in Indonesia." *Economic Geography*, 33, No. 2 (1957): 171-181.

Bennett, D.C. "The Basic Food Crops of Java and Madura." *Economic Geography*, 37, No. 1 (1961): 75-87.

Khan, M. Halim. "The Rise and Decline of Cash Crops in Java." *Indonesian Journal of Geography*, 1, No. 2-3 (1961): 47-55.

Bintarto, R. "Man, Rice and Problems in Jogjakarata." *Indonesian Journal of Geography*, 3, No. 4-6 (1963): 34-39.

Reksohadiprodjo, Iso, and B. Sudersono. "Double Cropping in Wet-Rice Cultivation in Relation to Soil and Climate (in a Few Regions of Central Java)." *Indonesian Journal of Geography*, 3, No. 4-6 (1963): 10-14.

Lee, Yongo-Leng. "Padi Production and Further Settlement in Borneo." *World Crops*, 16, No. 3 (1964): 40-45.

Reksohadiprodjo, Iso. "Ups and Downs of the Sugar-Cane Industry in Indonesia." *Indonesian Journal of Geography*, 4 (1964): 61-63.

Withington, W.A. "Changes and Trends in Patterns of North Sumatra's Estate Agriculture." *Tijdschrift voor Economische en Sociale Geographie*, 55 (1964): 8-13.

Djenen. "Crop Combinations and Specializations in the Farm Agriculture of Java." Master's Thesis, Oregon State University, Corvallis, Oreg., 1970.

Seavoy, Ronald E. "The Transition to Continuous Rice Cultivation in Kalimanton." *Annals of the Association of American Geographers*, 63, No. 2 (1973): 218-225.

Uhlig, Harold. "Reisbauokosysteme mit kinstlicher Bewasserung und mit pluvialer Wasserzufuhr: Java und analoge Typen im ubrigen Sudostasien." *Erdkunde*, 38, No. 1 (1985): 16-28.

Chen, Xiurong. "Regional Imbalance in Exploitation of Natural Resources for Cro Farming in Indonesia: A Study in Agricultural Geography in Java, Sumatra, and Kalimantan." Ph.D. Dissertation, University of Hong Kong, Hong Kong, China, 1991.

Jackel, W. "Internationale Rohstaffabkommen und nationale Entwicklungspolitik im Widerspruch: Das Beispiel Koffee (Indonesien)." *Zeitschrift fur Wirtschaftsgeographie*, 35, No. 3-4 (1991): 158-169.

Cleary, M.C. "Plantation Agriculture and the Formation of Native Land Rights in British North Borneo, c1880–1930." *Geographical Journal*, 158, No. 2 (1992): 170-181.

Suryanata, Krisna. "Fruit Trees Under Contract: Tenure and Land Use Change in the Uplands of Java, Indonesia." *World Development*, 22, No. 10 (1994): 1567-1578.

Chen, Xiurang. *Regional Imbalance in Exploitation of Natural Resources for Crop Farming in Indonesia: A Study in Agricultural Geography in Java, Sumatra, and Kalimantan.* Jakarta, Indonesia: University of Indonesia Press, 1998.

Danoedor, Projo, Sigit Heru Murti, and Bir'ah Hidagat. "Integration of Spectral-Based and Spatio-Economical Models in Satellite Image Analysis for Estimating Annual Crops Production." *Indonesian Journal of Geography*, 30, No. 76 (1998): 51-68.

ANIMAL HUSBANDRY

Zimmermann, Gerd. "Die Viehhaltung in Indonesien zur Entwicklung der Tierbestande und der Versargungssituation der Bevolkerung mit tierischem Eiweiss." *Geographische Rundschau*, 27, No. 4 (1975): 162-168.

Metzner, Joachim. "Die Viehaltung in der Agrarlandschaft der Sumba und das Problem der saisonalen Hungersnot." *Geographische Zeitschrift*, 64, No. 1 (1976): 46-71.

Salle-Allouis, L. "L'elevage bovin a Bali." *Bulletin de l'Association de Geographes Francais*, 69, No. 4 (1992): 333-338.

FARM LAND

Doll, W. "Die Flurverfassung in Zentral-Java." *Geographische Zeitschrift*, 65 (1977): 283-301.

Takaya, Yoshikazu. "Agricultural Landscapes in the Komering River Basin, South Sumatra" (in Japanese). *Southeast Asian Studies*, Kyoto, 17, No. 3 (1979): 52-74.

Scholz, U. "Land Reserves in Southern Sumatra/Indonesia and Their Potentialities for Agricultural Utilization." *GeoJournal*, 4, No. 1 (1980): 19-30.

Dove, Michael R. "Extant Population/Land Balances and the Development of Oilpalm Estates in West Kalimantan." *Indonesian Journal of Geography*, 12, No. 43 (1982): 23-28.

Fox, Jeff. "Using Integrated Land Surveys and Agroecosystem Analysis to Study Upland Areas in East Java." *Indonesian Journal of Geography*, 14, No. 47 (1984): 1-8.

Preston, David A. "Too Busy to Farm: Underutilisation of Farmland in Central Java." *Journal of Development Studies*, 26, No. 1 (1989): 43-57.

AGRICULTURAL POPULATIONS AND SOCIAL ISSUES

Reksohadiprodjo, Iso, and Soedarsono Hadisaputro. "Trends in Population Change and Food (Paddy) Production." *Indonesian Journal of Geography*, 1, No. 1 (1960): 15-20.

Reksohadiprodjo, Iso, and Arifin Mukaddas. "Population Density and Swamp Paddy Cultivation with Irrigation Facilities in the Asian Monsoon Tropics with Special Reference to Java (Indonesia)." *Indonesian Journal of Geography*, 1, No. 2-3 (1961): 5-10.

Bowers, Ida I. "Factors Influencing Village Receptivity to Agricultural Innovations: A Case Study in Kakupaten Karawong, West Java." Ph.D. Dissertation, University of Hawaii, Manoa, Hawaii, 1973.

Oudemans, Robert. "Simalungun Agriculture: Some Ethno-Geographic Aspects of Dualism in North Sumatra Development." Ph.D. Dissertation, University of Maryland, College Park, Md., 1973.

Pelzer, Karl J. *Planter and Peasant: Colonial Policy and the Agrarian Struggle in East Sumatra*. The Hague, Netherlands: Martinus Nijhoff, 1978.

Seavoy, Ronald E. "Population Pressure and Land Use Change: From Tree Crops to Sawah in Northeast Kalimantan, Indonesia." *Singapore Journal of Tropical Geography*, 1, No. 2 (1980): 61-67.

Metzner, Joachim. *Agriculture and Population Pressure in Sikka, Isle of Flores: A Contribution to the Study of the Stability of Agricultural Systems in the Wet and Dry Topics*. Canberra, Australia: Australian National University, 1982.

Sevin, Olivier. "L'economie agricole des Dayak du centre de Kalimantan (Borneo, Indonesia)." *Bulletin de l'Association Geographes Francais*, 60, No. 2 (1983): 99-109.

Sevin, Olivier, et al. *Thematic Geography to Develop Transmigration Settlements: Lowland Rice and Water Management in the Southern Part of Kalimantan*. Jakarta, Indonesia: Indonesia-ORSTOM Transmigration Project PTA-44, 1985.

Van Schaik, A. *Colonial Control and Peasant Resources in Java: Agricultural Involutions Reconsidered*. Amsterdam, Netherlands: Karag/Institut voor Sociale Geographie, Universiteit von Amsterdam, 1986.

Palte, Jan G.L. *Upland Farming in Java: Indonesia—A Socio-Economic Study of Upland Agriculture and Subsistence Under Population Pressure*. Utrecht, Netherlands: Royal Dutch Geographical Society, 1989.

Suminar, Panji. "Intra-Household Distribution of Workload, Income, and Decision Making: Case Study of Irrigated and Rain-Fed Rice Production in West Java, Indonesia." Master's Thesis, University of Waterloo, Waterloo, Ontario, 1994.

Talfer, David J. and Geoffrey Wall. "Linkages Between Tourism and Food Production." *Annals of Tourism Research*, 23, No. 3 (1996): 635-653.

ENVIRONMENTAL AND CONSERVATION ISSUES

Resosudarmo, Sudjirao. "Climatic Water Balance and Agricultural Production in the Northern Plains of West Java." Ph.D. Dissertation, University of Hawaii, Manoa, Hawaii, 1977.

Metzner, Joachim. "Disequilibrium of Agricultural Regions in the Eastern Lesser Sunda Islands: Consequences for Geoecological Regional Planning." *Applied Geography and Development*, 15 (1980): 62-81.

Uhlig, Harold. "Agricultural Landscapes in Tropical Karst Regions: Examples of Geo-Ecological Differentiation from Java and Sulawesi." *Applied Geography and Development*, 15 (1980): 7-27.

Scholz, U., et al. *The Natural Regions of Sumatra and Their Agricultural Pattern: A Regional Approach*. Bogar, Indonesia: Ministry of Agriculture, Republic of Indonesia, 1983. 2 volumes.

Dove, Michael R. "The Agroecological Myth of the Javanese and the Political Economy of Indonesia." *Indonesia*, 39 (1985): 1-36.

Hemphill, Anna. "Indonesia's Green Revolution: Its Effect on the Extent and Nature of Poverty." *Bloomsbury Geographer*, 16 (1988): 17-21.

Scholz, U. "Okonomie und Okologie im Einklang: Kleinbauerliche Produktionssystem auf Sumatra." *Geographische Rundschau*, 41, No. 7-8 (1989): 424-430.

Fox, James J. "Ecological Policies for Sustaining High Production in Rice: Observations on Rice Intensification in Indonesia." Pp. 211-224 in Harold C. Brookfield and Yvonne Byron, eds. *South-East Asia's Environmental Future: The Search for Sustainability*. Tokyo, Japan: U.N. University Press, 1993.

Nibbering, J.W. "Rats and Droughts: Stability in a Changing Upland Farming System in Java." *Asian Geographer*, 12, No. 1-2 (1993): 19-32.

Kirono, Dewi G.C., and Nurul Khakhine. "Rainfall and El Nino Southern Oscillation: Links and Its Impact on Crop Production (A Case Study of Yogyakarta Special Province of Indonesia)." *Indonesian Journal of Geography*, 30, No. 76 (1998): 21-34.

Kirono, Dewi G.C. and Nigel J. Topper. "ENSO Rainfall Variability and Impacts on Crop Production in Indonesia." *Physical Geography*, 20, No. 6 (1999): 508-519.

IRRIGATION

Uhlig, Harold. "Reisbau okosystem mit kunstlicher Bewasserung und mit pluvialer wasserzufuhr. Java und analoge: Typen im ubrigen Sudostasien." *Erdkunde*, 38, No. 1 (1984): 16-28.

Maurer, J.L. "Irrigation, riziculture et gestion de l'eau en Indonesie: La modernization des traditions agrohydroliques a Java et Bali." *Revue de Geographie de Lyon*, 65, No. 1 (1990): 27-37.

AGRICULTURAL MARKETING AND DISTRIBUTION

von Ommeren, Caroline M., and Jan G.L. Palte. "Marketing Patterns of Agricultural Commodities in an Upland Area of Central Java." *Indonesian Journal of Geography*, 16, No. 52 (1986): 33-48.

Jackel, W. "Coffee Marketing in an Indonesian Highland Area: A Local Institution Under Changing External Conditions." *Tijdschrift voor Economische en Sociale Geographie*, 86, No. 2 (1995): 150-161.

TRANSPORTATION AND MECHANIZATION

Boerman, Boesriati. "The Role of Transportation in the Agricultural Development of Transmigration Colonies in North Lampung District, Sumatra, Indonesia." Master's Thesis, California State University, Fresno, Calif., 1984.

Laos

GENERAL TEXTS, ARTICLES, AND THESES

Taillard, C. "L'emergence des cultures commerciales au Laos." *Travaux et Documents de Geographie Tropicale*, CEGET-CNRS, 20 (1975): 11-44.

CROPS AND CROPPING PATTERNS

Roder, Wolf. "Slash-and-Burn Rice Systems in Transition: Challenges for Agricultural Development in the Hills of Northern Laos." *Mountain Research and Development*, 17, No. 9 (1997): 1-10.

Malaysia, Brunei, and Singapore

GENERAL TEXTS, ARTICLES, AND THESES

Dobby, E.H.G. *Agricultural Questions of Malaya*. Cambridge, England: Cambridge University Press, 1949.

Lee, Yongleng. "Agriculture in Sarawak." *Journal of Tropical Geography*, 21 (1965): 21-29.

Jackson, James C. "Le developpement de l'agriculture commerciale dans la presqu'ile malaise avant 1908." *Les Cahiers d'Outre Mer*, 24 (1971): 32-44.

Ang Kok Ping. "A Preliminary Study of the Agricultural Development in Singapore." *Geographical Papers, The South Seas Society, Singapore* (1972): 69-80.

Ooi, Jin-bee. "Agricultural Change in Singapore." *Asian Profile*, 1, No. 2 (1973): 365-376.

Hewitt, Bernard R. "The Geography of the State of Brunei, with Special Reference to Agriculture and Agricultural Systems." Ph.D. Dissertation, University of Hull, Hull, England, 1975.

Uhlig, Harold. "Le remplacement de la culture itinerante sur bullis: Etude geographique des transformations economiques et sociales en Asie tropicale—l'exemple de Sabah et Sarawak, Malaysie." *Travaux et Documents de Geographie Tropicale*, CEGET-NCRS, 20 (1975): 371-384.

Keong, Voon Phin. "Spatial Patterns of Agricultural Expansion in Semenyih and Ulu Semenyih Mukims, Selangar." *Journal of Tropical Geography*, 43 (1976): 69-83.

Hill, Ronald D. *Agriculture in the Malaysian Region*. Budapest, Hungary: Akademiai Kiado, 1982.

Hill, Ronald D. "The Malaysian Region and the World Typology of Agriculture." *Geografia Polonica*, 46 (1983): 21-48.

Templeman, Gert-Jan, and F.J.J. Suykerbuyk. "Agriculture in Singapore: Problems of Space and Productivity." *Singapore Journal of Tropical Geography*, 4, No. 1 (1983): 62-72.

Courtenay, P.P. "The Diversification of Malaysian Agriculture, 1950–1958: Objectives and Achievements." *Journal of Southeast Asian Studies*, 15, No. 1 (1984): 166-181.

Fee, Wan Leong. "Policies for Agricultural Development in Peninsular Malaysia." *Land Use Policy*, 2, No. 1 (1985): 30-40.

Jaschke, Dieter. "Agrarwirtschaft und agrarstruktureller Wandel in Malaysia." *Erdkunde*, 39, No. 1 (1985): 31-41.

De Koninck, Rodolphe. "Les politiques du developpement agricole en Malaysia au l'impatience de l'etat tutelaire." *Archipel*, 31 (1986): 131-153.

Short, David E. "Planning Agriculture's Last Stand in Singapore." *Geography*, 73, No. 4 (1988): 357-359.

Neville, Warwick. "Agribusiness in Singapore: A Capital-Intensive Service." *Journal of Rural Studies*, 8, No. 3 (1992): 241-255.

Brookfield, Harold. "Agriculture and Industry: Towards Vertical Integration." Pp. 49-63 in *Transformation with Industrialization in Peninsular Malaysia*, edited by Harold C. Brookfield. Kuala Lumpur, Malaysia: Oxford University Press, 1994.

Overton, John. "Agriculture and Industry in Colonial Malaya." Pp. 35-48 in *Transformation with Industrialization in Peninsular Malaysia*, edited by Harold C. Brookfield. Kuala Lumpur, Malaysia: Oxford University Press, 1994.

Windle, Jill, and R.A. Cromb. "Remoteness and Rural Development: Economic Impacts of Rural Roads on Upland Farmers in Sarawak, Malaysia." *Asia Pacific Viewpoint*, 38, No. 1 (1997): 37-54.

Sutton, Keith. "Agribusiness on a Grand Scale: Felda's Sahabat Complex in East Malaysia." *Singapore Journal of Tropical Geography*, 22, No. 1 (2001): 90-105.

CROPS AND CROPPING PATTERNS

Deasy, George E. "The Oil Palm in Malaya." *Journal of Geography*, 41, No. 1 (1942): 21-31.

Dobby, E.H.G. "The North Kedah Plain: A Study in the Environment of Pioneering for Rice Cultivation." *Economic Geography*, 27, No. 4 (1951): 287-315.

Dobby, E.H.G. "Malaya: Rubber and Rice." *Current History*, 25 (1953): 295-300.

Spencer, Joseph E. "Seasonality in the Tropics: The Supply of Fruit to Singapore." *Geographical Review*, 49 (1959): 475-484.

Lee, Yongleng. "Some Aspects of Shifting Cultivation in British Borneo." *Malayan Forester*, 24, No. 2 (1961): 102-109.

Ooi, Jin-bee. "The Rubber Industry of the Federation of Malaya." *Journal of Tropical Geography*, 15 (1961): 46-65.

Ho, Robert. "Mixed Farming and Multiple-Cropping in Malaya." *Journal of Tropical Geography*, 16 (1962): 1-17.

Gray, B.S. "The Potential of the Oil Palm in Malaysia." *Journal of Tropical Geography*, 17 (1963): 127-132.

Ho, Robert. "Mixed Farming and Multiple Cropping in Malaya." Pp. 88-104 in *Symposium on Land Use and Mineral Deposits in Hong Kong, Southern China, and South-East Asia*, edited by D.G. David. Hong Kong, China: Hong Kong University Press, 1964.

Neville, Warwick. "The Plantation in Malaya: A Case Study of a Pineapple Plantation in South Johore." *Tijdschrift voor Economische en Sociale Geographie*, 55 (1964): 57-69.

Jackson, James C. "Nutmeg Mania in Singapore in the 1840s." *Geographica*, 1 (1964–1965): 1-10.

Kandiah, P. "Some of the Major Problems of Achieving Self Sufficiency in the Rice Production in Malaya." *Geographica*, 1 (1964–1965): 55-59.

Wee, Y.C., and T.H. Tay. *Pineapple Cultivation in West Malaysia*. Pekan Neron, Johore, Malaysia: Malayan Pineapple Industry Board, 1965.

Hill, Ronald D. "Dry Rice Cultivation in Peninsular Malaya." *Oriental Geographer*, 10, No. 1 (1966): 10-15.

Rutherford, John. "Double Cropping of Wet Padi in Penang, Malaya." *Geographical Review*, 56, No. 2 (1966): 239-255.

Ho, Robert. "Rubber Production by Peasants of the Terachi Valley, Malaya." *Transactions, Institute of British Geographers*, 41 (1967): 187-201.

Jackson, James C. "Oil Palm: Malaya's Post-Independence Boom Crop." *Geography*, 52, No. 3 (1967): 319-321.

Jackson, James C. "Tapioca: A Plantation Crop Which Preceded Rubber in Malaya." *Malaysia in History*, 10, No. 2 (1967): 13-24.

Keong, Voon Phin. "The Rubber Small-Holding Industry in Selangar, 1895–1920." *Journal of Tropical Geography*, 24 (1967): 43-49.

Hill, Ronald D. "Pepper Growing in Johore." *Journal of Tropical Geography*, 28 (1969): 32-39.

Ho, Robert. "Rice Production in Malaya: A Review of Problems and Prospects." *Journal of Tropical Geography*, 29 (1969): 21-32.

Mahmud, Zaharah B. "The Pioneering of Wet Rice Growing Traditions in West Malaysia: A Restudy with Special Reference to the State of Kedah." *Geographica*, 5 (1969): 1-7.

Senftleben, W. "Landlichen Entwicklungsplanung im West Malaysia." *Zeitschrift fur Wirtschaftsgeographie*, 13, No. 7 (1969): 199-203.

Chung, Tseshung. "Problems of Increasing Rice Output in Malaya." Master's Thesis, Illinois State University, Bloomington, Ill., 1970.

Hill, Ronald D. "Peasant Rice Cultivation Systems with Some Malaysian Examples." *Geographia Polonica*, 19 (1970): 91-99.

Wee, Y.C. "The Development of Pineapple Cultivation in West Malaysia." *Journal of Tropical Geography*, 30 (1970): 68-75.

Keong, Voon Phin. "The Rubber Plantation Industry in Selangor 1899–1909: A Case Study of European Economic Penetration." *Geographica*, 7 (1971): 81-98.

Keong, Voon Phin. "Size Aspects of Rubber Small-Holdings in West Malaysia: A Case Study of Bentang, Pahang." *Journal of Tropical Geography*, 34 (1972): 65-72.

Colgan, Marcus. "Plantation Agriculture in Malaya." Master's Thesis, University of Guelph, Guelph, Ontario, 1973.

John, D., and James C. Jackson. "The Tobacco Industry of North Borneo: A Distinctive Formal Plantation Agriculture." *Journal of Southeast Asian History*, 4 (1973): 88-106.

Idris, Mustapha C. "The Sugar Industry in Peninsular Malaysia: Preliminary Comments." *Geographica*, 9 (1974): 38-45.

Persi, P. "La produzione di gomma naturale nella Penisola Malesi: Aspetti geografici e problemi." *Bolletin della Societa Geografia Italia*, Serie 10, 5 (1976): 191-206.

Hill, Ronald D. *Rice in Malaya: A Study in Historical Geography*. New York, N.Y.: Oxford University Press, 1978.

Courtenay, P.P. "Some Trends in the Peninsular Malaysian Plantation Sector, 1963–1973." Pp. 131-166 in *Issues in Malaysian Development*, edited by James C. Jackson. London, England: Heinemann, 1979.

Hin, Tan Wan. "The Development of the Tobacco Growing Industry in Peninsular Malaysia." *Geographica*, 13 (1979): 1-10.

Ng Hee Liang. "Aspects of Rice Cultivation in Kuala Pilah." *Geographica*, 14 (1980): 35-40.

Courtenay, P.P. "The Plantation in Malaysian Economic Development." *Journal of Southeast Asian Studies*, 12, No. 2 (1981): 329-348.

Courtenay, P.P. *Nearly at the Crossroads? A Review of Some Issues Raised in Recent Work in the Malaysian Padi Sector*. Canberra, Australia: Australian National University, Development Studies Centre, Occasional Paper No. 33, 1983.

Chan, Foongmae. "Home Gardens in Malaysia and Honolulu." Master's Thesis, University of Hawaii, Manoa, Hawaii, 1984.

Courtenay, P.P. "The New Malaysian Sugar Plantation Industry." *Geography*, 69, No. 4 (1985): 317-326.

Courtenay, P.P. "Malaysia's National Agricultural Policy: Experimental Systems in the Padi Sector." *Land Use Policy*, 4, No. 3 (1987): 294-304.

Fold, Niels. "Industrial Organization and Sectoral Linkages: A Study of the Malaysian Palm Oil Industry." *Geografisk Tidsskrift*, 94 (1994): 69-81.

Courtenay, P.P. *The Rice Sector of Peninsular Malaysia: A Rural Paradox*. Sydney, Australia: Allen & Unwin and Asian Studies Association of Australia, 1995.

Lim, Jennifer N.W., and Ian Douglas. "The Impact of Cash Cropping on Shifting Cultivation in Sabah, Malaysia." *Asia Pacific Viewpoint*, 39, No. 3 (1998): 315-326.

ANIMAL HUSBANDRY

Gosling, Lee A.P. "Patterns and Problems of Livestock Production in Malaya." Ph.D. Dissertation, University of Michigan, Ann Arbor, Mich., 1958.

DAIRYING

Wikkramatileke, Rudolph, and Darpal Singh. "Tradition and Change in an Indian Dairying Community in Singapore." *Annals of the Association of American Geographers*, 60, No. 4 (1970): 717-742.

FARM LAND

Bahvin, Tunku S. "Land Development in Post-Independence Malaysia: Some Lessons." *Geographica*, 4 (1968): 67-83.

Hock, Khoo Soo. "Agricultural Landuse." Pp. 74-149 in *Brunei in Transition: Aspects of Its Human Geography in the Sixties*, edited by Khoo Soo Hock, et al. Kuala Lumpur, Malaysia: University of Malaya, Department of Geography, 1976.

Courtenay, P.P. "Farm Size, Out-Migration and Abandoned Padi Land in Mukim Melekek, Melaka (Peninsular Malaysia)." *Malaysian Journal of Tropical Geography*, 17 (1986): 18-28.

Cromb, R.A. "Problems of State-Sponsored Land Schemes for Small Farmers: The Case of Sarawak, Malaysia." *Pacific Viewpoint*, 33, No. 1 (1992): 58-78.

Ali, J. "Agricultural Development and Land Tenure Changes: A Case Study of Kundasong Highlands in Sabah, East Malaysia." Pp. 216-226 in *Proceedings, Seventeenth New Zealand Geography Conference*, edited by William Whittaker. Christchurch, New Zealand: New Zealand Geographical Society, 1993.

Mertz, Ole, and H. Christensen. "Land Use and Crop Diversity in Two Iban Communities, Sarawak." *Geografisk Tidsskrift*, 97 (1997): 98-110.

AGRICULTURAL POPULATIONS AND SOCIAL ISSUES

Lim, Joojock. "Tradition and Peasant Agriculture in Malaya." *Journal of Tropical Geography*, 3 (1954): 44-47.

Jackson, James C. "Chinese Agricultural Pioneering in Singapore and Johore, 1800–1917." *Journal of the Royal Asiatic Society, Malayan Branch, Singapore*, 38, No. 1 (1963): 77-105.

Jackson, James C. "Chinese and European Agricultural Enterprise in Malaya, 1786–1921: A Geographical Study of Expansion and Change." Ph.D. Dissertation, University of Malaya, Kuala Lumpur, Malaysia, 1965.

Ho, Robert. *Farmers of Central Malaya*. Canberra, Australia: Australian National University Research School of Pacific Studies, Department of Geography, 1967.

Simmerman, Frederick L. "The Effect of Shifting Cultivation on the Agricultural Economy of the People of Sarawak, Malaysia." Master's Thesis, Miami University, Miami, Ohio, 1967.

Jackson, James C. *Planters and Speculators: Chinese and European Agricultural Enterprise in Malaya 1786–1921*. Kuala Lumpur, Malaysia: University of Malaya Press, 1968.

Kuhne, Dietrich. "Einige Entwicklungs- und Wesenzuge chinesischer Agrarkolonisation in Malaysia." *Geographische Rundschau*, 20, No. 11 (1968): 432-435.

De Koninck, Rodolphe. "Chinese Farmers of Singapore: A Study in Social Geography." Ph.D. Dissertation, University of Singapore, 1970.

De Koninck, Rodolphe. "Measuring the Cultural Evolution and the Modernization of Chinese Farmers of Singapore: An Experiment in the Use of Gamma Measure and the Guttman Scale." *Cahiers de Geographie de Quebec*, 16, No. 38 (1972): 243-262.

Mahmud, Zaharah B. "The Dilemma of Modernization of the Dual-Cropping Peasantry of Peninsular Malaysia." *Malaysian Journal of Tropical Geography*, 1 (1980): 64-71.

Hill, Ronald D., and Voo Min Shin. "Occupational and Spatial Mobility in an Overseas Chinese Agricultural Community: The Hakkas of Kudat, Sabah, Malaysia." *Sojourn*, 5, No. 2 (1990): 194-218.

Barrett, S.R.F. "Rural Poverty in Peninsular Malaysia and the Malaysian National Agricultural Policy." *Australian Geographer*, 24, No. 2 (1993): 60-71.

ENVIRONMENTAL AND CONSERVATION ISSUES

Joseph, K.T. "Malaya's Soil Potential." *Malayan Agriculturalist*, 1 (1960–1961): 41-44.

Weng, Chan Nyai. "The Reliability of Rainfall in Relation to Wet Rice Cultivation in Northwest Peninsular Malaysia." *Malaysian Journal of Tropical Geography*, 4 (1981): 14-23.

Chin, S.C. "Agriculture and Resource Utilisation in a Lowland Rainforest Kenyah Community." *Sarawak Museum Journal*, 35 (1985): 68-76.

Jaiyeoba, I.A. "Changes in Organic Matter and Nutrients of Forest Soils Due to Clearing and Cropping." *Malaysian Journal of Tropical Geography*, 27, No. 2 (1996): 15-22.

IRRIGATION

Short, David E., and James C. Jackson. "The Origins of an Irrigation Policy in Malaya: A Review of Development Prior to the Establishment of the Drainage and Irrigation Department." *Journal of the Malayan Branch, Royal Asiatic Society*, 44, No. 1 (1971): 78-103.

Short, David E. "Irrigation and Mixed Farming in Malaya." Pp. 253-270 in *Spatial Aspects of Development*, edited by Brian S. Hoyle. London, England: John Wiley, 1974.

De Koninck, Rodolphe. "Maitrise de l'espace ou maitrise des paysans: Les contradiction du pirimetre d'irrigation mudu en Malaysie." *Economie rurale*, 202-203 (1991): 48-52.

Overton, John. "Irrigation and Imperialism: Rice and the State in Colonial Malaya." Pp. 299-203 in *Proceedings, Seventeenth New Zealand Geography Conference*, edited by William Whittaker. Christchurch, New Zealand: New Zealand Geographical Society, 1993.

Johnson, C.L. "Politics and Power: Government Intervention in the Muda Irrigation Scheme, Malaysia." Ph.D. Dissertation, Middlesex University, Middlesex, England, 1999.

Johnson, C.L. "Government Intervention in the Muda Irrigation Scheme, Malaysia: Actors, Expectations, and Outcomes." *Geographical Journal*, 166, No. 3 (2000): 192-214.

TRANSPORTATION AND MECHANIZATION

Mahmud, Zaharah H. "The Manner and Mode of Technology Transfer in the Rural Rubber Smallholding Sector of Malaysia." *Geographica*, 15 (1986): 43-52.

AGRICULTURAL LABOR STUDIES

Templeman, Gert-Jan. "Off-Farm Employment: A Barrier to Agricultural Development—A Case Study from Trengganu, Peninsular Malaysia." *Malaysian Journal of Tropical Geography*, 6 (1982): 50-60.

Overton, John. "Integration or Self-Sufficiency? Peninsular Malaysia and the Rice Trade in Southeast Asia." *Singapore Journal of Tropical Geography*, 20, No. 2 (1999): 169-180.

Myanmar/Burma

CROPS AND CROPPING PATTERNS

Huke, Robert E. "Rice in Burma: A Geographical Analysis of Three Agricultural Villages." Ph.D. Dissertation, Syracuse University, Syracuse, N.Y., 1953.

Lubeigt, G. "L'introduction d'une nouvelle culture dans un etat socialiste: Le cas du jute en Birmanie." *Travaux et Documents de Geographie Tropicale*, CEGET-CNRS, 20 (1975): 235-270.

Tanaka, Koji, and Tadayo Watabe. "Crop Distribution in the Central Basin of Burma" (in Japanese). *Southeast Asian Studies*, Kyoto, 14, No. 2 (1976): 286-301.

The Philippines

GENERAL TEXTS, ARTICLES, AND THESES

Cutshall, Alden D. "The Geographic Factors Influencing Agriculture in the Philippine Islands." Master's Thesis, University of Illinois, Urbana, Ill., 1935.

Hainsworth, Reginald G., and Raymond T. Mayer. *Agricultural Geography of the Philippine Islands*. Washington, D.C.: Office of Foreign Agricultural Relations, United States Department of Agriculture, 1945.

Cutshall, Alden D. "Reconstruction of Philippine Agriculture." *Economic Geography*, 23, No. 4 (1947): 308-310.
Wernstedt, Frederick L. "Agricultural Regionalism on Negros Island, Philippines." Ph.D. Dissertation, University of California, Los Angeles, Calif., 1953.
Cutshall, Alden D. "Progress in Philippine Agriculture." *Journal of Geography*, 54, No. 7 (1955): 343-349.
Cutshall, Alden D. "Philippine Agriculture: Summary and Trends." *Professional Geographer*, 10 (1958): 2-4.
Cutshall, Alden D. "Regionalism in Philippine Agriculture." *Journal of Geography*, 61, No. 7 (1962): 289-295.
Oracion, T.S. "Kaingin Agriculture Among the Bukidnons of Southeastern Negros, Central Philippines." *Journal of Tropical Geography*, 17 (1963): 213-224.
Clawson, David L. "Small Scale Polyculture: An Alternative Development Model." *Philippine Geographical Journal*, 29, No. 3-4 (1985): 92-103.
Blanadet, Raymond. "La reforme agraire aux Philippines et resultats provisoires." *Les Cahiers d'Outre Mer*, 40 (1987): 27-46.
Walpole, Peter W. "A Typology of Philippine Agriculture." Master's Thesis, University of Georgia, Athens, Ga., 1987.
Santos, Paterno R. "Countryside Agribusiness." *Philippine Geographical Journal*, 32, No. 2-3 (1988): 52-55.
Lewis, Martin W. "Agricultural Regions in the Philippine Cordillera." *Geographical Review*, 82, No. 1 (1992): 29-42.

CROPS AND CROPPING PATTERNS

Borja, Luis J. "The Philippine Coconut Industry." *Economic Geography*, 3, No. 3 (1927): 382-390.
Kolb, Albert. "Die Reislandschaft auf den Philippinen." *Petermann's Geographische Mitteilungen*, 86 (1940): 113-124.
Kolb, Albert. "Die Tabaklandschaft auf den Philippinen." *Petermann's Geographische Mitteilungen*, 86 (1940): 353-366.
Morgan, Ted, and Alden Cutshall. "Tobacco Culture in the Philippines." *Bulletin, Illinois Geographical Society*, 3 (1948): 8-13.
Pelzer, Karl J. "Philippine Abaca Industry." *Far Eastern Survey*, 17, No. 6 (1948): 71-74.
Hawthorne, Arthur S. "Geographical Study of Rice Culture on Samar." Master's Thesis, University of Pittsburgh, Pittsburgh, Pa., 1949.
Spencer, Joseph E. "The Philippine Rice Problem." *Far Eastern Survey*, 4 (1949): 125-128.
Cutshall, Alden D. "Philippine Rubber Plantations." *Economic Botany*, 7, No. 1 (1953): 86-88.
Cutshall, Alden D. "Trinidad Valley: Middle Latitude Horticulture in the Philippines." *Philippine Geographical Journal*, 1, No. 1 (1953): 40-44.
Spencer, Joseph E. "The Abaca Plant and Its Fiber, Manila Hemp." *Economic Botany*, 7 (1953): 195-213.

Cutshall, Alden D. "Tobacco Production in the Philippines." *Transactions of the Illinois Academy of Science*, 52, No. 1-2 (1959): 34-44.

Panditharatna, B.E. "Major Cash Crops of the Philippines." *Oriental Geographer*, 3, No. 1 (1959): 1-16.

Cutshall, Alden D. "The Philippine Sugar Industry: Status and Problems." *Journal of Geography*, 60, No. 1 (1961): 5-8.

Schul, Norman W. "A Philippine Sugar Cane Production in the Haciendas." Ph.D. Dissertation, Syracuse University, Syracuse, N.Y., 1962.

Vander Meer, Canute. "Corn on the Island of Cebu, The Philippines." Ph.D. Dissertation, University of Michigan, Ann Arbor, Mich., 1962.

Vander Meer, Canute. "Corn Cultivation on Cebu: An Example of an Advanced Stage of Migratory Farming." *Journal of Tropical Geography*, 17 (1963): 172-177.

Philips, Paul G. "A Tagalog Farm and an Adal Farm: Two Units of Shifting Cultivation in the Humid Tropics." Pp. 1-6 in *Focus on Geographic Activity: A Collection of Original Studies*, edited by Richard S. Thoman and Donald G. Patton. New York, N.Y.: McGraw-Hill, 1964.

McLennan, Marshall S. "A Study of Selected Crop Distributions in the Philippines." Master's Thesis, San Francisco State University, San Francisco, Calif., 1967.

Schul, Norman W. "Hacienda Magnitude and Philippine Sugar Cane Production." *Asian Studies*, 5, No. 2 (1967): 258-273.

Schul, Norman W. "A Philippine Sugar Cane Plantation: Land Tenure and Sugar Cane Production." *Economic Geography*, 43, No. 2 (1967): 157-169.

Ulack, Richard. "Physical and Socio-Economic Factors Affecting Banana Cultivation in the Philippines." Master's Thesis, Pennsylvania State University, University Park, Pa., 1969.

Bondad, N.D. "Effect of Ethephon on Flowering, Fruiting and Slip Production of 'Smooth Cayenne' Pineapple." *Philippine Geographical Journal*, 7, No. 1 (1973): 1-10.

Spencer, Joseph E. "The Rise of Maize as a Major Crop in the Philippines." *Journal of Historical Geography*, 1, No. 1 (1975): 1-16.

Keong, Voon Phin. *American Rubber Planting Enterprise in the Philippines, 1900–1930.* London, England: London School of Oriental and African Studies, Department of Geography, Occasional Papers, 1977.

Garrett, Larry N., and C. Joanne Garrett. "Babacu: Agricultural Product with Industrial Development Potential." *Philippine Geographical Journal*, 23, No. 1 (1979): 11-16.

Bondad, N.D. "Mango and Its Relatives in the Philippines." *Philippine Geographical Journal*, 26, No. 2 (1982): 88-100.

Bondad, N.D. "Origin and Distribution of Mango." *Philippine Geographical Journal*, 26, No. 1 (1982): 44-52.

Wong, S.T., and E. Geronimo-Kueh. "Variations of Factors Affecting Rice Productivity in Central Luzon, Philippines." *GeoJournal*, 6, No. 5 (1982): 469-476.

Huetz de Lemps, Alain. "Negros, 'ile a sucre' des Philippines." *Les Cahiers d'Outre Mer*, 36 (1983): 231-266.

Bondad, N.D., et al. "Philippine Mangoes and Their Relationship to Southeast Asian Cultivars." *Philippine Geographical Journal*, 28, No. 1-2 (1984): 59-71.

Wang, Shue-Tuck. "A Model for Evaluating the Extent of Rice Farm Fragmentation in the Ilocos Region in the Philippines." *Journal of Environmental Management*, 24, No. 4 (1987): 347-358.

Bondad, N.D. "A Mango-Based Cropping System." *Philippine Geographical Journal*, 34, No. 3 (1990): 121-126.

ANIMAL HUSBANDRY

Villegas, Valente. "Live Weights, Carcass Values and Breeding Habits of Batangos and Bali Cattle." *Philippine Geographical Journal*, 16, No. 3 (1972): 69-73.

Villegas, Valente. "History of Importation of Indian Buffalos at Government Stock Farms." *Philippine Geographical Journal*, 18, No. 1-2 (1974): 35-42.

FARM LAND

Pendleton, Robert L. "Land Utilization: Agriculture of Mindanao, Philippines." *Geographical Review*, 32, No. 2 (1942): 180-210.

Janes, David F. "Agricultural Land Use in the Digos-Padada Valley, Davao Province, the Philippines." Master's Thesis, Pennsylvania State University, University Park, Pa., 1977.

AGRICULTURAL POPULATIONS AND SOCIAL ISSUES

Wong, S.T. "Perception and Attitude of Filipino Farmers Toward Padi Farm Fragmentation in the Ilocos Region, Philippines." *Geografiska Annaler B*, 69B, No. 1 (1987): 73-80.

Preston, David A. *Changes in Farming and Household Livelihood Strategies in Two Areas of the Cordillera of Luzon, Philippines*. Leeds, England: University of Leeds, School of Geography, Working Paper No. 527, 1989.

ENVIRONMENTAL AND CONSERVATION ISSUES

Kellman, M.C. "Some Environmental Components of Shifting Cultivation in Upland Mindanao." *Journal of Tropical Geography*, 28 (1969): 40-56.

Tolentino, Aurora S. "Fertilizer Usage." *Philippine Geographical Journal*, 29, No. 3-4 (1985): 124-126.

Bunoan, J.C. "Fertilizer Trials Under INSFFER in the Philippines." *Philippine Geographical Journal*, 30, No. 1-2 (1986): 1-20.

Recel, M.R. "Nutrient Management for Field and Plantation Crop Production in Philippine Farming Systems." *Philippine Geographical Journal*, 32, No. 4 (1988): 135-145.

Urich, Peter B. "Tropical Karst Management and Agricultural Development: Example from Bohol, Philippines." *Geografiska Annaler B*, 71B, No. 2 (1989): 95-108.

Falvo, Daniel J. "Polyvarietal Planting and Rice Blast Control in Philippine Uplands." *Middle States Geographer*, 33 (2000): 54-62.

Falvo, Daniel J. "Rice Blast Control and Polyvarietal Planting in the Philippines: A Study in Genotype by Environmental Biogeography." Master's Thesis, Rutgers University, New Brunswick, N.J., 2001.

IRRIGATION

Dwyer, Denis J. "Irrigation and Land Problems in the Central Plain in Luzon." *Geographica*, 1 (1964–1965): 68-80.

Urich, Peter B. "Deforestation and Declining Irrigation in Southeast Asia: A Philippine Case." *International Journal of Water Resource Development*, 12, No. 1 (1996): 49-63.

AQUACULTURE

McIntyre, W.E. "Philippine Fish Culture." *Scientific Monthly*, 78 (1954): 86-93.

Thailand

ATLASES AND GRAPHIC PRESENTATIONS

An Atlas of Thailand's Agricultural Resource. Bangkok, Thailand: Department of Commercial Intelligence, 1959.

GENERAL TEXTS, ARTICLES, AND THESES

Lin, Chan. "Agricultural Geography of Siam." Master's Thesis, Pennsylvania State University, University Park, Pa., 1950.

Loetsch, F. "Der Einflus des Brandrodunsbaus auf das Gefuge des Tropenwaldes und die Wasserfuhrung der Strome, untersucht am Beispiel Nord Thailands." *Erdkunde*, 12, No. 3 (1958): 182-204.

Chapman, E.C. *An Appraisal of Recent Agricultural Changes in the Northern Valleys of Thailand*. Bangkok, Thailand: Department of Land Development, 1967.

Behrman, Jere R. "Significance of Intracountry Variation for Asian Agricultural Prospects: Central and Northeastern Thailand." *Asian Survey*, 8, No. 3 (1968): 157-173.

O'Reilly, F.D. "The Process of Innovation in the Subsistence Agriculture of North East Thailand with Particular Reference to the Lana Pao Irrigation Area, Changwet Kalasin." Ph.D. Dissertation, University of London, London, England, 1974.

O'Reilly, F.D., and P.I. McDonald, eds. *Thailand's Agriculture*. Budapest, Hungary: Akademiai Kiado, 1983.

Takaya, Yoshikazu. *Agricultural Development of a Tropical Delta: A Study of the Chao Phrya Delta*. Honolulu, Hawaii: University of Hawaii Press, 1987.

CROPS AND CROPPING PATTERNS

Barton, Thomas F. "Growing Rice in Thailand." *Journal of Geography*, 59, No. 4 (1960): 153-164.

Ng, Ronald C.Y. "Rice Cultivation and Rural Settlement Density in North East Thailand." *Tijdschrift voor Economische en Sociale Geographie*, 59 (1968): 200-210.

Ng, Ronald C.Y. "Social and Economic Determinants in Rice Cultivation in Northeast Thailand." Pp. 173-186 in *Studies in Contemporary Thailand*, edited by Robert Ho and E.C. Chapman. Canberra, Australia: Australian National University, Department of Human Geography, 1973.

Bruneau, Michel. "Types de cultures commerciales paysannes en Thailande septentrionale." *Travaux et Documents de Geographie Tropicale*, CEGET-CNRS, 20 (1975): 67-126.

Dervaux, F. "Une culture commerciale en Thailande: Le coton dans la province de Loci." *Travaux et Documents de Geographie Tropicale*, CEGET-CNRS, 20 (1975): 219-234.

Johnston, David B. "Opening a Frontier: The Expansion of Rice Cultivation in Central Thailand in the 1890s." *Contributions to Asian Studies*, 9 (1976): 27-44.

Chapman, E.C. "Shifting Cultivation in Economic Development of Annual Cropping on Swidden Lowland Terrain in Northern Thailand." Pp. 222-235 in *Farmers in the Forest: Economic Development and Marginal Agriculture in Northern Thailand*, edited by Peter Kunstadter, E.C. Chapman, and Sanga Sabhasri. Honolulu, Hawaii: University of Hawaii, East–West Center, 1978.

Ishii, Yoneo. *Thailand, A Rice-Growing Society*. Honolulu, Hawaii: University of Hawaii Press, 1978.

Pongprayoon, Chatchai. "Determinants of Crop Choice: A Case Study of Farmers' Behavior in Kampaeng Saen Irrigation Project, Thailand." Ph.D. Dissertation, Southern Illinois University, Carbondale, Ill., 1979.

Brunel, Sylvie. "Le manioc: Un atoutpour le Nord-Est Thailandais." *L'Information Geographique*, 51, No. 1 (1987): 5-12.

Traub, Peter. "Regional Differences in Cane Sugar Production in Thailand: A Geographical Contribution." *Indian Geographical Journal*, 60 (1987): 95-109.

Laine, J.P. "L'evolution recente des cultures fruitieres en Thailande." *Bulletin de la Association de Geographes Francais*, 67, No. 3 (1990): 207-212.

Crooker, Richard A., and Robert N. Martin. "Accessibility and Illicit Drug Crop Production: Lessons from Northern Thailand." *Journal of Rural Studies*, 8, No. 4 (1992): 423-430.

Rigg, Jonathan. "Errors in the Making: Rice, Knowledge, Technological Change and Applied Research in Northeast Thailand." *Malaysian Journal of Tropical Geography*, 26, No. 1 (1995): 19-34.

Crooker, Richard A., and Robert N. Martin. "Marketing Distance-Decay and Highland Vegetable Production in Northern Thailand." *Geographical Bulletin*, 38, No. 2 (1996): 80-91.

Flaherty, M.S, P. Vandergeest, and P. Miller. "Rice Paddy or Shrimp Pond: Tough Decisions in Rural Thailand." *World Development*, 27, No. 12 (1999): 2045-2060.

ANIMAL HUSBANDRY

Falvey, L. "Some Economic Aspects of the Livestock Industry in the Highlands of North Thailand." *Journal of Tropical Geography*, 49 (1979): 11-18.

FARM LAND

Sternstein, Larry. "Aspects of Agricultural Land Tenure in Thailand." *Journal of Tropical Geography*, 24 (1967): 22-29.

Demaine, Harvey, and Chris J. Dixon. "Land Tenure Patterns and Agricultural Development in Northeast Thailand: A Case Study of the Lam Pao Irrigation Area in Changwat Kalasin, Part 2." *Journal of the Siam Society*, 60 (1972): 45-60.

Tanabe, S. "Land Reclamation in the Chao Phrya Delta." Pp. 40-83 in *Thailand: A Rice Growing Society*, edited by Yoneo Ishii. Honolulu, Hawaii: University of Hawaii Press, 1978.

Hirsch, Philip. "State, Capital and Land in Recently Cleared Areas of Western Thailand." *Pacific Viewpoint*, 33, No. 1 (1992): 36-57.

Dery, Steven. "Evolution des territoires agricoles et forestiers en Thailande: Une interpretation cartographique." *Les Cahiers d'Outre Mer*, 52 (1999): 35-58.

AGRICULTURAL POPULATIONS AND SOCIAL ISSUES

Surareks, Vanpen. "Agrargeographische und Sozial-oekonomische Struktur in Nord-thailand (unter besonerer Beruecksichtigung der Pachtverhaeltnisse)." Ph.D. Dissertation, University of Giessen, Giessen, Germany, 1975.

Demaine, Harvey. *The Roles of Farmers' Associations in Agricultural Development: A Case Study from Northeastern Thailand.* London, England: University of London, School of Oriental and African Studies, Department of Geography, 1976.

Parnwell, Michael J.G. "Migration and the Development of Agriculture: A Case Study of Northeast Thailand." Pp. 93-140 in *Rural Development in North-East Thailand: Case Studies of Migration, Irrigation, and Rural Credit*, edited by Michael J.G. Parnwell. Hull, England: University of Hull Centre for Southeast Asian Studies, Occasional Paper No. 12, 1986.

Forsyth, Timothy J. "Tourism and Agricultural Development in Thailand." *Annals of Tourism Research*, 22, No. 4 (1995): 877-900.

ENVIRONMENTAL AND CONSERVATION ISSUES

Pendleton, Robert L. "Some Interrelationships Between Agriculture and Forestry, Particularly in Thailand." *Journal of the Siam Society*, Natural History Supplement 12, No. 1 (1939): 33-52.

Kaida, Yoshihiro. "A Subdivision of the Chao Phrya Delta in Thailand Based on Hydrographical Conditions: Water Conditions in Deltaic Lowland Rice Fields" (in Japanese). *Southeast Asian Studies*, Kyoto, 11, No. 3 (1973): 403-413.

Norman, M.J.T. "The Potential and Limitations of Thailand's Environment for Agricultural Production." Pp. 111-136 in *Studies in Contemporary Thailand*, edited by Robert Ho and E.C. Chapman. Canberra, Australia: Australian National Univeristy, Department of Human Geography, 1973.

Rigg, Jonathan D. "The Problems of Agricultural Intensification in a Rainfed Environment: A Study of Farmers' Practices and Government Policies in Two Villages in Northeastern Thailand." Ph.D. Dissertation, University of London, London, England, 1985.

Hafner, James A., and Y. Apichatvullop. "Farming the Forest: Managing People and Trees in Reserved Forests in Thailand." *Geoforum*, 21, No. 3 (1990): 331-346.

Hafner, James A., and Y. Apichatvullop. "Migrant Farmers and the Shrinking Forests of Northeast Thailand." Pp. 187-219 in *Keepers of the Forest: Land Management Alternatives in Southeast Asia*, edited by Mark Poffenberger. West Hartford, Conn.: Kumarian, 1990.

Rigg, Jonathan. "Forests and Farmers, Land and Livelihood, Changing Resource Realities in Thailand." *Global Ecology and Biogeography Letters*, 3, No. 4-6 (1993): 277-289.

Schmidt-Vogt, Dietrich. "Swidden Farming and Secondary Vegetation: Two Case Studies from Northern Thailand." Pp. 47-66 in *Counting the Costs: Economic Growth and Change in Thailand*, edited by Jonathan Rigg. Singapore: Institute of Southeast Asian Studies, 1995.

Schmidt-Vogt, Dietrich. "Defining Degradation: The Impacts of Swidden on Forests in Northern Thailand." *Mountain Research and Development*, 18, No. 2 (1998): 135-149.

Schmidt-Vogt, Dietrich. *Swidden Farming and Fallow Vegetation in Northern Thailand*. Stuttgart, Germany: Franz Steiner, 1999.

Walsh, S.J. "Deforestation and Agricultural Extensification in Northeast Thailand: A Remote Sensing and GIS Study of Landscape Structure and Scale." *Papers and Proceedings of the Applied Geography Conference*, 22 (1999): 223-232.

Yoshino, M. "Environmental Change Affecting the Rice Production in Thailand and in Monsoon Asia." *Geographical Review of Japan*, 73B, No. 2 (2000): 178-190.

IRRIGATION

Parnwell, Michael J.G., ed. *Rural Development in North-East Thailand: Case Studies of Migration, Irrigation and Rural Credit.* Hull, England: University of Hull, Centre for Southeast Asian Studies, Occasional Paper No. 12, 1986.

Surarerks, Vanpen. *Historical Development and Mangement of Irrigation Systems in Northern Thailand.* Chiang Mai, Thailand: Chiang Mai University, 1986.

AQUACULTURE

Uthoff, D. "Die Entwicklung der marinen Aquakultur in Thailand: Okonomische Effekte und okologische Risilan." Pp. 209-223 in *Beitrage zur Geographie de Meere und Kuster,* edited by F. Wieneke. Munich, Germany: Vortrage der 9, Jahrestazung, 1993.

Flaherty, M.S., and C. Kornjanakesorn. "Marine Shrimp Aquaculture and Natural Resource Degradation in Thailand." *Environmental Management,* 19, No. 1 (1995): 27-37.

Chaijaroenwatana, Bussabong. "The Economic and Environmental Repercussions of Shrimp Farming in the Pak Phanang Coastal Area, Thailand." *Malaysian Journal of Tropical Geography,* 29, No. 2 (1998): 83-90.

Flaherty, M.S., and P. Vandergeest. "Low Salt Shrimp Aquaculture in Thailand: Goodbye Coastline Hello Khonkaen." *Environmental Management,* 22, No. 6 (1998): 817-830.

Miller, P., M.S. Flaherty, and B. Szuster. "Inland Shrimp Farming in Thailand." *Aquaculture Asia,* 4, No. 1 (1999): 27-32.

Vandergeest, Peter, M.S. Flaherty, and Paul Miller. "A Political Ecology of Shrimp Aquaculture in Thailand." *Rural Sociology,* 64, No. 4 (1999): 573-596.

AGRICULTURAL MARKETING AND DISTRIBUTION

Hafner, James A. "The Spatial Dynamics of Rice Milling and Commodity Flow in Central Thailand." *Journal of Tropical Geography,* 37 (1973): 30-38.

TRANSPORTATION AND MECHANIZATION

Louie, Kuion. "Transportation Development and Agricultural Marketing in Changwat Loei in Northeastern Thailand." Ph.D. Dissertation, University of Michigan, Ann Arbor, Mich., 1973.

Vietnam

GENERAL TEXTS, ARTICLES, AND THESES

Lafont, Pierre-Bernard. "L'agriculture sur brulis chez des Proto-Indochinois des hauts plateaux de Centre Viet Nam." *Les Cahiers d'Outre Mer,* 20, (1967): 37-48.

Shohan, Robert W. "Dalat: The Spatial Structure of an Elite Agriculture." Master's Thesis, University of Texas, Austin, Tex., 1969.

Nguyen Duc Nhun. "Les contradictions de l'organisation scientifique de l'espace et du travail agricoles au Nord-Vietnam 1954–1981." *L'Espace Geographique*, 11, No. 2 (1982): 82-94.

Hill, Ronald D., and Cheung Man Biu. "Vietnamese Agriculture: Rhetoric and Reality." *Contemporary Southeast Asia*, 7, No. 4 (1986): 292-305.

Lam Thanh Liem. "Les reformes agraires de l'Ancien Regime de Saigon et leurs repercussions zur la politique de collectivisation des terres au Viet-Nam de 1975 a 1985." *Acta Geographica*, 67-68 (1986): 32-50.

Lam Thanh Liem. "Principales causes des difficultes de l'economique agricole vietnamienne." *Acta Geographica*, 75 (1988): 29-38.

Chabert, O., and L. Fages. "Politiques agraires et strategies paysonnes en milieu fragile: Le piemont du Tam Dao." *Les Cahier d'Outre Mer*, 48 (1995): 157-180.

Lagree, S. "Evolution de l'agriculture vietnamienne dans un district du delta du Fleuve Range." *Les Cahiers d'Outre Mer*, 48 (1995): 139-156.

Pillot, Didier. "La fin des cooperatives: La decollectivisation agricole au North Vietnam." *Les Cahiers d'Outre Mer*, 48 (1995): 107-130.

CROPS AND CROPPING PATTERNS

Lafont, Pierre-Bernard. "The Slash-and-Burn (Ray) Agriculture System of the Mountain Populations of Central Vietnam." *Proceedings of the 9th Pacific Science Congress*, 7 (1959): 56-59.

Teulieres, Roger. "La culture du cafeier dans le Sud-Viet-Nam." *Les Cahiers d'Outre Mer*, 14 (1961): 293-316.

Teulieres, Roger. "Le the au Sud Viet-Nam." *Les Cahiers d'Outre Mer*, 14 (1961): 182-209.

Teulieres, Roger. "Le tabac au Viet-Nam du Sud." *Les Cahiers d'Outre Mer*, 19 (1966): 174-186.

Teulieres, Roger. "L'evolution recente de la riziculture au Viet-Nam du Sud." *Les Cahiers d'Outre Mer*, 19 (1966): 382-398.

Laine, J.P. "Note sure la situation des plantations de the et de cafe au Vietnam meridional." *Travaux et Documents de Geographie Tropicale*, CEGET-CNRS, 20 (1975): 357-368.

Vo Tung Xuan. "Rice Cultivation in the Mekong Delta: Present Situation and Potentials for Increased Production" (in Japanese). *Southeast Asian Studies*, Kyoto, 13, No. 1 (1975): 88-111.

Hill, Ronald D. "Industrial Crops in Vietnam." *Rural Systems*, 1, No. 3 (1983): 237-242.

Vibrig, H. "Zur agrarischen und forstlichen Bewertung des Para-Kautschukbaumes (Hevea brasiliensis) und zu seiner Bewirtschaftung in der SR Vietnam." *Petermann's Geographische Mitteilungen*, 129, No. 3 (1985): 181-196.

Balaize, C. "Images de l'hevea culture vietnamienne." *Bulletin de l'Association de Geographes Francais*, 67, No. 3 (1990): 199-206.

Bach, Trung H., and Ngoc Han Tran. "Riziculture et evolution recente des pratiques culturales au Nord Vietnam." *Les Cahiers d'Outre Mer*, 48 (1995): 229-250.

Tan, Stan B.H. "Coffee Frontiers in the Central Highlands of Vietnam: Networks of Connectivity." *Asia Pacific Viewpoint*, 41, No. 1 (2000): 51-67.

ANIMAL HUSBANDRY

Schaeffer, Delphine. "L'elevage au Nord Vietnam: Du collectivisme a la decollectivisation." *Les Cahiers d'Outre Mer*, 48 (1995): 181-196.

FARM LAND

Mal'khanova, I.A. "The Development of New Agricultural Lands in North Vietnam in 1961–1965." *Soviet Geography*, 11, No. 10 (1970): 828-831.

Burr, J. Millard. "Land to the Tiller: Revolution and Land Reform in Viet Nam." Master's Thesis, University of Oregon, Eugene, Oreg., 1973.

Hung Duong Minh. "Development of Virgin Lands and Enlargement of Agricultural Lands of Northern Vietnam" (in Russian). *Vestnik Leningradskogo Universiteta, Seriya Geologiya-Geografiya*, 3 (1989): 114-118.

Wenzel, A. "Weed Vegetation and Land Use of Upland Maize Fields in North-West Vietnam." *GeoJournal*, 50, No. 4 (2000): 349-357.

AGRICULTURAL POPULATIONS AND SOCIAL ISSUES

Papy, Louis. "L'agriculture sur brulis chez les Mois des Hauts-Plateaux du Vietnam central, d'apres G. Condominas." *Les Cahiers d'Outre Mer*, 11 (1958): 79-83.

Lafont, Pierre-Bernard. "L'agriculture sur brulis chez les Proto-Indochinois des hauts plateaux du Centre Viet-Nam." *Les Cahiers d'Outre Mer*, 20 (1967): 37-48.

Dery, Steven. "La colonisation agricole au Viet-Nam." *Cahiers de Geographie du Quebec*, 45, No. 124 (2001): 101-118.

ENVIRONMENTAL AND CONSERVATION ISSUES

Hirsch, Philip, and Nguyen Viet Thinh. "Implications of Economic Reform in Vietnam: Agrarian and Environmental Change in Hien Luong." *Australian Geographer*, 27, No. 2 (1996): 165-184.

IRRIGATION

Chassigneux, E. "La plaine et les irrigations du Thanh-Hoa." *Annales de Geographie*, 36 (1927): 232-253.

Robequain, C. "L'hydrolique agricole au Tonkin." *Annales de Geographie*, 42 (1933): 424-428.

AGRICULTURAL MARKETING AND DISTRIBUTION

Naud, Albert. "L'exportation des grands produits agricoles indochinois." *Annales de Geographie*, 39 (1930): 50-60.

AGRICULTURAL LABOR STUDIES

Nguyen Duc Nhun. "Les contradictions de l'organisation scientifique de l'espace et du travail agricoles au Nord-Vietnam 1954–1981." *L'Espace Geographique*, 11, No. 2 (1982): 81-94.

East Asia

GENERAL TEXTS, ARTICLES, AND THESES

Van Valkenburg, Samuel. "Agricultural Regions of Asia." *Economic Geography*, Part I, 7 (1931): 218-237; Part II, 8 (1932): 109-133; Part III, 9 (1933): 1-18; Part IV, 9 (1933): 109-135; Part V, 10 (1934): 14-34; Part VI, 11 (1935): 227-246; Part VIII, 11 (1935): 325-337; Part IX, 12 (1936): 27-44; Part X, 12 (1936): 231-249.

Cressy, George. "Agricultural Regions in Asia." *Economic Geography*, 10, No. 2 (1934): 109-142.

Hall, Robert B. "Agricultural Regions of Asia." *Economic Geography*: Part VII, 10 (1934): 324-347; 11 (1935): 33-52; and 11 (1935): 130-147.

Hayami, Yujiro, Vernon W. Ruttan, and Herman M. Southworth, eds. *Agricultural Growth in Japan, Taiwan, Korea and the Philippines*. Honolulu, Hawaii: University of Hawaii, East–West Center, 1979.

CROPS AND CROPPING PATTERNS

Terjung, W.H., et al. "Potential Paddy Rice Yields for Rainfed and Irrigated Agriculture in China and Korea." *Annals of the Association of American Geographers*, 75, No. 1 (1985): 83-101.

AGRICULTURAL POPULATIONS AND SOCIAL ISSUES

Calvert, Keith E. "Diseases Associated with Intensive Subsistence Farming in Japan and Korea." *Proceedings, Oklahoma Academy of Science*, 37 (1957): 132-134.

China

ATLASES AND GRAPHIC PRESENTATIONS

Trewartha, Glenn T. "Ratio Maps of China's Farms and Crops." *Geographical Review*, 28 (1938): 102-111.

Lo, Hoichun. "Land Use Mapping and Crop Identification from Landsat Data for the Nun River Drainage Region of Northeast China." Master's Thesis, University of Georgia, Athens, Ga., 1978.

Welch, R.A., C.P. Lo, and Clifton W. Pannell. "Mapping China's New Agricultural Land." *Photogrammetric Engineering and Remote Sensing*, 45 (1979): 1211-1228.

Klasse, Dorothy F. "Satellite Imagery to Monitor Oasis Agriculture within Turpon Depression, People's Republic of China." Master's Thesis, University of Hawaii, Manoa, Hawaii, 1981.

Huke, Robert, Eleanor H. Huke, T. Woodhead, and J. Huang. *Rice-Wheat Atlas of China*. Manila, Philippines: International Rice Research Institute, 1993.

Huke, Robert E., et al. *Rice-Wheat Atlas of China*. Hongzhou, China: China National Rice Research Institute, 1994.

Qian, Huaishui. "Regionalization for Large Area Crop Estimates by Remote Sensing: A Case Study of Chinese Wheat." *Chinese Geographical Science,* 8, No. 3 (1998): 204-211.

GENERAL TEXTS, ARTICLES, AND THESES

Davis, Ira C. "Agriculture in China." *Journal of Geography*, 18 (1919): 129-135.

Fleuer, Alberta, and Edwin J. Foscue. "Agricultural Production in China." *Economic Geography*, 3, No. 3 (1927): 297-308.

Bergsmark, Daniel R. "China: Natural Environment, Population and Agriculture." Pp. 454-474 in *Economic Geography of Asia*. New York, N.Y.: Prentice-Hall, 1940.

Hoenigfeld, Arline G. "A Comparison Between the Agricultural Geography of the Szechwan Basin in the Yunnan Plateau." Master's Thesis, University of Maryland, College Park, Md., 1949.

Chen, Chengsiang. *Agricultural Regions of China*. Taipai, Taiwan: National Taiwan University, Institute of Agricultural Geography, Research Report No. 45, 1954.

Zaychikov, V.T., ed. *Geography of Agriculture in Communist China*. Peking, China: U.S. Joint Publications Research Service, 1959.

Ballas, Donald J. "Some Notes on Agriculture in Han China." *Professional Geographer*, 17, No. 1 (1965): 13-14.

Chang, Sendou. "The Role of the Agricultural Geographer in Communist China." *Professional Geographer*, 18 (1966): 125-128.

Chung, Yuetping. "Applied Geography and Agricultural Development in China." Master's Thesis, Kent State University, Kent, Ohio, 1967.

Chung, Yuetping. "Geography and Agricultural Development in China." *Professional Geographer*, 20, No. 3 (1968): 163-166.

Giese, Ernst. "Agrare Betriebsformen im Vorland des Tien-Shan." *Erdkunde,* 22, No. 1 (1968): 51-62.

Chen, Chensiang. *The Agricultural Regions of China*. Silver Spring, Md.: American Institute of Crop Ecology, 1970.

Ishikawa, Shigeru. "Changes in the Structure of Agricultural Production in Mainland China." Pp. 346-377 in *Agrarian Policies and Problems in Communist and Non-Communist Countries*, edited by W.A.D. Jackson. Seattle, Wash.: University of Washington Press, 1971.

Tang, Anthony M. "Input–Output Relations in the Agriculture of Communist China, 1950–1965." Pp. 280-301 in *Agrarian Policies and Policies in Communist and Non-Communist Countries*, edited by W.A.D. Jackson. Seattle, Wash.: University of Washington, Press, 1971.

Dettwiler, Ronald H. "The People's Liberation Army's Role in Chinese Agriculture." Master's Thesis, San Diego State University, San Diego, Calif., 1973.

Lofquist, Bruce H. "Chinese Communist Party Policy and Units of Area Organization for Agricultural Production, 1921–1966." Master's Thesis, University of Guelph, Guelph, Ontario, 1974.

Tschuli, A.B. "Productivity and Orientation of Farming in Peoples' Communes" (in Japanese). *Science Reports: Tohoku University, Seventh Series, Geography*, 25, No. 1 (1975): 43-49.

Reed, Lawrence C., III. "The Expansion and Development of Agriculture in Sinkiang, 1949–1965: A Look at Changing Agricultural Systems." Master's Thesis, University of Wisconsin, Madison, Wis., 1976.

She, Zhixiang. "New Features in the Development of Agricultural Regions and New Problems Concerning the Distribution of Agricultural Production in Jiangsu Provice" (in Chinese). *Acta Geographica Sinica*, 34, No. 2 (1979): 104-117.

Trodliet, P. "Note sur l'evolution des systemes de culture en chine." *L'Information Geographique*, 44, No. 3 (1980): 111-117.

Chou, L. "Some Problems on the Agricultural Regionalization" (in Chinese). *Scientia Geographica Sinica*, 1, No. 1 (1981): 10-21.

Pannell, Clifton W., and Christopher L. Salter, eds. *China Geographer: No. 11, Agriculture*. Boulder, Colo.: Westview Press, 1981.

Wiens, T.B. "Agriculture in the Four Modernizations." Pp. 57-72 in *China Geographer: No. 11, Agriculture*, edited by Clifton W. Pannell and Christopher L. Salter. Boulder, Colo.: Westview, 1981.

Wu, Chuanchun. "Delineation of China's Agricultural Regions." Pp. 27-39 in *China Geographer: No. 11, Agriculture*, edited by Clifton W. Pannell and Christopher L. Salter. Boulder, Colo.: Westview, 1981.

Wu, Chuanchun. "The Transformation of Agricultural Landscape in China." Pp. 35-44 in *The Environment: Chinese and American Views*, edited by Laurence Ma and Allen G. Noble. New York, N.Y.: Methuen, 1981.

Ma, Xiangyang. "Preliminary Studies on Division of Agricultural Production Type" (in Chinese). *Scientia Geographica Sinica*, 2, No. 3 (1982): 247-253.

Huang, Bingwei. "Professor Chu Ko-chen and Geographical Research on Agriculture, with Special Reference to North China" (in Chinese). *Acta Geographica Sinica*, 39, No. 1 (1984): 1-10.

Wadekin, Karl-Eugen. "Modernization of Chinese Agriculture." *Land Use Policy*, 1, No. 3 (1984): 199-216.

Pannell, Clifton W. "Recent Chinese Agriculture." *Geographical Review*, 75, No. 2 (1985): 170-185.

Powell, S.G. *Xu hui Agricultural District: Rural Development in an Intermediate Area*. Leeds, England: University of Leeds, Department of Geography Working Paper No. 432, 1985.

Wein, Norbert. "Agriculture and Forestry in the Loess Plateau Region of China: The Example of the Chung-hua District." *GeoJournal*, 12, No. 1 (1986): 57-64.

Zhung, Jianfu, et al. "Optimal Structure Planning in Agriculture in Hunyuan County" (in Chinese). *Scientia Geographica Sinica*, 7, No. 1 (1987): 54-64.

Wein, Norbert. "Die landwirtschaft der sud-chinesischen Provinz Guang dong: Strukturen und Entwicklungsten denze." *Zeitschrift fur Wirtschafts-geographie*, 32, No. 1 (1988): 53-71.

Xu, Weidong. "Some Aspects of Agricultural Regionalization in Hainan Island, South China" (in Japanese). *Jimbun Chiri/Human Geography*, 42, No. 3 (1990): 195-219.

Zhou, Lisan, and Zhixiang She. "Agricultural Geography and Land Utilization" (in Chinese). *Acta Geografica Sinica*, 45, No. 2 (1990): 146-153.

Smil, Vaclav. "Mainland China's Agricultural Prospects: Some Fundamental Limitations." *Issues and Studies*, 17, No. 6 (1991): 45-67.

Veeck, Gregory. "Variations in Agricultural Production and Efficiency in Newly Emerging Extended Metropolitan Regions of Jiangsu Province, PRC." *Asian Geographer*, 10, No. 2 (1991): 129-146.

Xu, Guohua, and L.J. Peel. *The Agriculture of China*. New York, N.Y.: Oxford University Press, 1991.

Zhou, Baiping. "Vertical Zonation and Vertical Agriculture in the Kunlun Mountains." *Chinese Journal of Arid Land Research*, 4 (1991): 75-86.

Guo, Huancheng. "A Preliminary Study of Agricultural Types in China" (in Chinese). *Acta Geographica Sinica*, 47, No. 6 (1992): 507-515.

Liu, Zhiqun and Rangying Wang. "Model of Agricultural Ecologic-Economic System in the Intermontane Basins of Xyefeng Mountain Region." *Chinese Geographical Science*, 3, No. 1 (1993): 61-73.

Malamud, Frances P. "The Case for Independent Origins of Agriculture in China." Master's Thesis, University of California, Berkeley, Calif., 1993.

Zhao, Sangqiao. "The Spread of Agriculture in China." *Chinese Geographical Science*, 3, No. 3 (1993): 194-202.

Cai, Y., and Barry E. Smit. "Sustainability in Chinese Agriculture: Challenge and Hope." *Agriculture, Ecosystems and Environment*, 49 (1994): 279-288.

Cai, Y., and Barry E. Smit. "Sustainable Agriculture and the Chinese Perspective." *Acta Geographica Sinica*, Vol. 50 (1995): 97-106.

Kung, S. *Agricultural Geography of Hunan and Hubei in the Qing Dynasty* (in Chinese). Wuhan, China: Wuchang, 1996.

Liu, Xiangnan, Fang Huang, and Zhaiao Zhou. "Study on GIS for Yield Estimation by Remote Sensing in Jilun Maize Belt." *Chinese Geographical Science*, 6, No. 4 (1996): 351-358.

Chen, Jiaqi, and Gang Zong. "Analysis of Historical Formation of Agricultural Areas in Sichuan Basin and Geographical Factors." *Chinese Geographical Science*, 7, No. 1 (1997): 30-38.

Liu, H., and Z. Cai. "L'Agriculture et l'amenagement rural en chine." *Bulletin de la Societe Geographique de Liege*, 33 (1997): 37-44.

Wang, Fei. "Regional Disparities in China: The Agricultural Aspect." Master's Thesis, University of Manitoba, Winnipeg, Manitoba, 1997.

Guo, Qian. "From 'Gardens' to 'Plantations': Market Economy and Commercial Agricultural Development in Shaanbei, China." *Asian Geographer*, 17, No. 1-2 (1998): 127-140.

Liu, Zhaoli, et al. "Study on Model for Remote Sensing Estimation of Maize Yield." *Chinese Geographical Science*, 8, No. 2 (1998): 161-167.

Wang, Jiaji, et al. "Modification of Agricultural Landscape in Three River Area in Zizong." *Chinese Geographical Science*, 9, No. 2 (1999): 120-125.

Xu, Jianchu, et al. "Effects of Swidden Cultivation, State Policies and Customary Institutions on Land Cover in a Hani Village, Yunnan, China." *Mountain Research and Development*, 19, No. 2 (1999): 123-132.

CROPS AND CROPPING PATTERNS

Deasy, George F. "The Soya Bean in Manchuria." *Economic Geography*, 15, No. 3 (1939): 303-310.

Yen, Chiangkwoh. "The Tung Region of China." *Economic Geography*, 19, No. 4 (1943): 418-427.

Hu, Huanyong. "A New Cotton Belt in China." *Economic Geography*, 23, No. 1 (1947): 60-66.

Liu, Yuchih. "Sericulture in China." Master's Thesis, University of California at Berkeley, 1948.

Trewartha, Glenn T., and ShouJen Wang. "Notes on Rice Growing in China." *Annals of the Association of American Geographers*, 38 (1948): 278-281.

Wiens, Herold J. "Cultivation Development and Expansion in China's Colonial Realm in Central Asia." *Journal of Asian Studies*, 26, No. 1 (1966): 67-88.

Chen, Chengsiang. "The Sugar Industry of China." *Geographical Journal*, 137, No. 1 (1971): 29-40.

Myers, Roman H., and Norma Schroder. "A Comparison of Foodgrain and Cotton Yields by County in China: 1929–1930 and 1970." *China Geographer*, 5 (1976): 29-50.

Leeming, Frank. "Progress Towards Triple-Cropping in China." *Asian Survey*, 19, No. 5 (1979): 450-467.

Lu, Jonathan J. "Rice (Oryza Sativa) in Southern Shaanxi: The People's Republic of China." *Asian Profile*, 11, No. 2 (1983): 141-152.

Terjung, W.H., et al. "Actual and Potential Yield for Rainfed and Irrigated Wheat in China." *Agricultural and Forest Meteorology*, 31 (1984): 1-23.

Jowett, A.J. "China's Foodgrains: Production and Performance, 1949–1981." *GeoJournal*, 10, No. 4 (1985): 373-392.

Wang, S.Y. "An Approach on Planting Structure for Main Fields of the Middle Region of Jilin Province" (in Chinese). *Scientia Geographica Sinica*, 6, No. 1 (1986): 70-75.

Hodder, R. "Some Dynamics of Peri-Urban Vegetable Farming in China." Master's Thesis, University of Hong Kong, Hong Kong, China, 1987.

Jiang, Xiaoshi. "Grain Production and Demand in China." Master's Thesis, University of Southern Mississippi, Hattiesburg, Miss., 1987.

Wang, Benlin, et al. "An Approach to the Potential and Scale of Commodity Grain Production in the Sanjiang Plain Area" (in Chinese). *Scientia Geographica Sinica*, 7, No. 1 (1987): 44-53.

Lu, Jonathan J. "Rice in the Shanxi-Shaanxi Loess Plateau of China: A Preliminary Report." *Asian Profile*, 16, No. 2 (1988): 97-112.

Sun, Yuting, and Yamei Bai. "The Relationships Between Soybean Yield and Climatic Conditions in the Northeast Region of China" (in Chinese). *Scientia Geographica Sinica*, 8, No. 1 (1988): 18-26.

Liu, Li, and Ma Yange. "The Development, Distribution and Environment of Sunflower Production in China." *Asian Profile*, 17, No. 3 (1989): 209-220.

Zhou, Zhaode, and Jianfei Zhang. "Multi-objective Programme Model of Tropical Crops in Hainan Island" (in Chinese). *Acta Geographica Sinica*, 44, No. 4 (1989): 407-419.

Lu, Jonathan J. "Changes in China's Rice Area: 1966 to 1985." *Asian Profile*, 18, No. 5 (1990): 391-414.

Zhu, Zhuo. "A Study on Rational Location of the Cotton Production in China." *Chinese Geographical Science*, 1, No. 2 (1991): 129-140.

Lyons, Thomas P. "Grain in Fujian: Intra-Provincial Patterns of Production and Trade, 1952–1988." *The China Quarterly*, 129 (1992): 184-215.

Huang, Zhanbin, and Ximei Zhang. "Grain Production and Its Spatial Distribution in the Loess Plateau Region." *Chinese Environment and Development*, 4, No. 1 (1993): 53-74.

Wong, Ren-lin, et al. "Studies on the Potential on Grain Production in the Northeast Region of China." *Chinese Geographical Science*, 3, No. 1 (1993): 74-84.

Huang, Zhenguo. "On the Origin of Rice Agriculture in Southern China and Its Propagation in East Asia." *Chinese Geographical Science*, 4, No. 4 (1994): 289-294.

Shirasaka, Shigeru. "Changing Slash-and-Burn Cultivation in Xishuangbanna, Southwestern China." *Geographical Review of Japan*, 68B, No. 2 (1995): 107-118.

Zhong, Yu. "A Preliminary Study on the Variation Features of Grain Yield of China in Recent 40 Years." *Journal of Chinese Geography*, 6, No. 2 (1996): 42-49.

Cao, Yongsheng, et al. "Distribution of the Main Crop Germplasm Resources in China." *Chinese Geographical Science*, 7, No. 4 (1997): 310-316.

Chen, Xikang, Haijian Wang, and Ju'e Guo. "Forecast of China's Grain Production Development and Its Self-Support Rate Analysis." *Chinese Geographical Science*, 7, No. 2 (1997): 97-105.

Rozelle, Scott, Gregory Veeck, and Jikun Huang. "The Impact of Environmental Degradation on Grain Production in China, 1975–1990." *Economic Geography*, 73, No. 1 (1997): 44-66.

Eng, Irene. "Agglomeration and the Local State: The Tobacco Economy of Yunnan." *Transactions, Institute of British Geographers*, 24, No. 3 (1999): 315-330. (New Series)

Liu, Honghui, Xiaohuan Yang, and Naibin Wang. "Remote Sensing Based Estimation System for Winter Wheat Yield in North China Plain." *Chinese Geographical Science*, 9, No. 1 (1999): 40-48.

Xiao, P., and Q. Wong. "Changes in Grain Production in China Since 1949 and Major Reasons" (in Japanese). *Geographical Review of Japan*, 72A, No. 9 (1999): 589-599.

ANIMAL HUSBANDRY

Shaw, Earl B. "The Swine Industry of China." *Economic Geography*, 14, No. 4 (1938): 381-397.

Xu, Xuegang. "The Potential Productivity of Agriculture and Animal Husbandry in the Yellow River Delta" (in Chinese). *Acta Geographica Sinica*, 46, No. 2 (1991): 151-159.

DAIRYING

Li, Wei. "An Analysis of Location of the Dairy Industry in China" (in Chinese). *Acta Geographica Sinica*, 41, No. 3 (1986): 224-232.

FARM LAND

Baker, O.E. "Land and Food in China." *Far Eastern Review*, 24 (1928): 121-125.

Hu, Charles Y. "The Agricultural and Forestry Land Use of Szechuan Basin." Ph.D. Dissertation, University of Chicago, Chicago, Ill., 1942.

Chan, Hungkwon. "A Study of Three Major Agricultural Land Uses in Northern Yuen Long." Master's Thesis, University of Hong Kong, Hong Kong, China, 1971.

Salter, Christopher L. "The Litany of Tachi and the Foolish Old Man: Agricultural Landscape Modification in Mainland China." *Professional Geographer*, 24, No. 2 (1972): 113-117.

Gentelle, Pierre. "L'amenagement agricole du sol en Chine: La province du Jiang su." *L'Espace Geographiques*, 4, No. 1 (1975): 5-16.

Rodgers, Allan. "The Red Star Sino-Korean Friendship Commune: A Case Study in the Historical Geography of Land Utilization in the North China Plain." *China Geographer*, 2 (1975): 1-26.

Leeming, Frank. "New Farmland Terracing in Contemporary China." *China Geographer*, 10 (1978): 29-39.

Shen, Yuancun, Xiangrang Xie, and Liulin Yang. "Land Types and Agricultural Regionalization in the Chang Ping County, Beijing" (in Chinese). *Scientia Geographica Sinica*, 2, No. 3 (1982): 221-231.

Liang, Pu. "Agricultural Regionalization and Issues of Agricultural Land Use Structure: The Case of Guang Dong." *Asian Geographer*, 2, No. 1 (1983): 1-10.

Lo, C.P., and Clifton W. Pannell. "Seasonal Agricultural Land Use Patterns in China's Pearl River Delta from Multi-Data Landsat Images." *GeoJournal*, 10, No. 2 (1985): 183-196.

Welch, R., Yirang Hsu, and Clifton W. Pannell. "Assessment of Potential Agricultural Land in Western China with a Geographic Information System (GIS)." *China Geographer*, 12 (1985): 39-54.

Xu, Zhikang. "Rational Use of Cultivated Land in the Beijing-Tianjin-Tangshan Region." *Chinese Geography and Environment*, 1, No. 2 (1988): 21-34.

Gruschkke, A. "Neulanderschliessung in der Trockengebieten der Volksrepublik China: Konsolidierung und Neuorientierung seit Beginn der 80er Jahre." *Geographische Rundschau*, 43, No. 11 (1991): 672-680.

Hill, Ronald D. "Upland Development Policy in the People's Republic of China." *Land Use Policy*, 11, No. 1 (1994): 8-16.

Poehlmann, Kevin. "A Discussion of Factors Affecting Agricultural Land Use and Essential Oil Production in China in the 1990s." Master's Thesis, State University of New York, Buffalo, N.Y., 1997.

Wei, H. "Household Land Tenure Reform in China: Its Impact on Farming Land Use and Agro-Environment." *Land Use Policy*, 14, No. 3 (1997): 175-186.

Li, Jianhe, and Canute Vander Meer. "Assignments of Agricultural Land Use Rights to Households in Tongan Country, Fujian, China." *Asian Geographer*, 17, No. 1-2 (1998): 101-114.

Zhu, Gongwu, and Ruihua Liu. "Cultivated Land Resources and Its Development in the Tropic and Subtropic Mountain Areas of South China: A Case Study in Deqing County." *Chinese Geographical Science*, 8, No. 4 (1998): 362-369.

Han, Sunshing, and Xinghe Chun. "Diminishing Farmland and Urban Development in China, 1993–1996." *GeoJournal*, 49, No. 3 (1999): 247-267.

Smil, Vaclav. "China's Agricultural Land." *China Quarterly*, 158 (1999): 414-429.

Skinner, M.W., R.G. Kuhn, and A.E. Joseph. "Agricultural Land Protection in China: A Case Study of Local Governance in Zhejiang Province." *Land Use Policy*, 18, No. 4 (2001): 329-340.

AGRICULTURAL POPULATIONS AND SOCIAL ISSUES

Vandermeer, Paul. "Pacifying Space in a Chinese Village." *China Geographer*, 2 (1975): 27-37.

De Koninck, Rodolphe. "Les fondesments territoriaux de la persistance de l'agriculture familiale en Chine." *L'Espace Geographique*, 14, No. 3 (1985): 161-174.

Xu, Weidong, Shigeru Shirasaka, and Takeo Ichikawa. "Farming System and Settlements in Xishuangbanna, Yunnan Province, China." *Geographical Review of Japan*, 62B, No. 2 (1989): 104-115.

Philipponneau, Michel. "Agriculture et habitat rural en Chine." *Acta Geographica*, 83 (1991): 16-23.

Enderton, Catherine, and Chingen Wen. "Longlin Village, Hainan: Cash Crops and New Houses." Pp. 269-278 in *Chinese Landscapes: The Village as Place*, edited by Ronald G. Knapp. Honolulu, Hawaii: University of Hawaii Press, 1992.

Hill, Ronald D. "People, Land and an Equilibrium Trap: Guizhou Province, China." *Pacific Viewpoint*, 34, No. 1 (1993): 1-24.

Chen, Zhonghong, Hongqing Yang, and Guoyu Ni. "The Cold and Hot Damages to the Citrus in the Three Gorges Area of the Changjiang River." *Chinese Geographical Science*, 4, No. 1 (1994): 66-80.

Deng, Xianrue, and Hong Luo. "The Vertical Distribution of Agrotopoclimatic Resources in the Warm Sectors of the Three Gorges Area of the Changjiang River." *Chinese Geographical Science*, 4, No. 3 (1994): 259-267.

Veeck, Gregory. "Challenges to Family Farming in China." *Geographical Review*, 90, No. 1 (2000): 57-82.

ENVIRONMENTAL AND CONSERVATION ISSUES

Nuttonson, Michael Y. *Ecological Crop Geography of China and Its Agroclimatic Analogues in North America*. Washington, D.C.: American Institute of Crop Ecology, 1947.

Spencer, Joseph E. "The Development and Spread of Agricultural Terracing in China." Pp. 105-110 in *Symposium on Land Use and Mineral Deposits in Hong Kong, Southern China, and Southeast Asia*, edited by D.G. Davis. Hong Kong, China: Hong Kong University Press, 1964.

Gentelle, Pierre. "L'amenagement agricole du sol en Chine: La province du Jiangsu." *L'Espace Geographique*, 4, No. 1 (1975): 5-16.

Whitney, Joseph. "Striking a Blow Against Desertification: Cooperative Initiative in Chungwei County, Ningsia-Hui Autonomous Region, China." *Economic Geography*, 53, No. 4 (1977): 381-384.

Chaoi, Sungchiao. "Transforming Wilderness into Farmland: An Evaluation of Natural Conditions for Agricultural Developments in Hei Long Jiang Province." Pp. 41-55 in *China Geographer: No. 11, Agriculture*, edited by Clifton W. Pannell and Christopher L. Salter. Boulder, Colo.: Westview, 1981.

Huang, Pingwei. "Environmental Factors and the Potential Agricultural Productivity of China: An Analysis of Sunlight, Temperature and Soil Moisture." Pp. 45-71 in *The Environment: Chinese and American Views*, edited by Laurence J.C. Ma and Allen G. Noble. New York, N.Y.: Methuen, 1981.

Smil, Vaclav. "China's Agro-Ecosystems." *Agro-Ecosystems*, 7 (1981): 27-46.

Li, Zhiwu, and Boful Wu. "Physical Geography and Agriculture" (in Chinese). *Acta Geographica Sinica*, 37, No. 4 (1982): 341-348.

Zhu, Zhenda, Shu Liu, and Youling Yang. "The Possibilities and Realities on Re-management of Desertified Lands in the Inter-Distributing Areas for Agricultural and Grazing Practices in Northern China" (in Chinese). *Scientia Geographica Sinica*, 4, No. 3 (1984): 197-206.

Liu, Donglai. "A Preliminary Discussion on the Natural Superiority of Hainan Dao and Its Orientation for Developing Comprehensive Agriculture" (in Chinese). *Acta Geographica Sinica*, 40, No. 4 (1985): 356-366.

Wang, Wenqing. "An Analysis of Physical Conditions and Agricultural Structure for Agricultural Decisions in the South Slope in Changbi Mountains" (in Chinese). *Scientia Geographica Sinica*, 5, No. 3 (1985): 273-282.

Fang, Guangdi. "The Precipitation and Agriculture in Sanjiang Region" (in Chinese). *Acta Geographica Sinica*, 41, No. 4 (1986): 321-331.

Kan, Guisheng. "Some Methods of Analysis and Calculation in Agricultural Drought" (in Chinese). *Scientia Geographica Sinica*, 6, No. 4 (1986): 368-375.

Qiu, Baojian. "A New Discussion on the Regionalization of Agroclimate in China" (in Chinese). *Acta Geographica Sinica*, 41, No. 3 (1986): 202-209.

Wang, Duo. "Simulated Analysis for 110 Years of Seasonal Weather Impacts on Yangtze Delta Rice Harvest Yields." *Agricultural and Forest Meteorology*, 39 (1987): 193-203.

Qiu, Baojian. "New Thoughts on Agroclimatic Regionalization in China." *Chinese Geography and Environment*, 1, No. 1 (1988): 74-91.

Wohlke, W., Hengyue Gu, and Nanshan Ai. "Agriculture, Soil Erosion and Fluvial Processes in the Basin of the Jialing (Sichuan Province, China)." *GeoJournal*, 17, No. 1 (1988): 103-116.

Han, Xiangling, Weiqing Qu, and Yangzhuang Kong. "The Prospect of Dryland and Irrigated Farming in Huong-Huai-Hai Plain Based on Analysis of the Precipitation-Soil-Crop System." *Chinese Geography and Environment*, 2, No. 3 (1989): 47-63.

Takahashi, Hidenori, and Kiyotaka Nakagawa. "Micrometeorological Characterisitics of a Rubber Plantation on Hainan Island, China." *Geographical Review of Japan*, 62B, No. 2 (1989): 179-191.

Luo, Manjia. "Numerical Classification of Regional Structure of Agro-Ecosystem in Guongdong." *Chinese Geographical Science*, 1, No. 2 (1991): 154-165.

Tao, Zhan. "Agricultural Environmental Problems and Countermeasures in China." *Chinese Geographical Science*, 2, No. 2 (1992): 159-162.

Gale, William J., Robert W. McColl, and Xie Fang. "Sandy Fields Traditional Farming for Water Conservation in China." *Journal of Soil and Water Conservation*, 48 (1993): 474-477.

Gamble, Douglas. "Coevolutionary Development Theory as an Analytical Framework Within Ecodevelopment: A Case Study of Agricultural and Water Resources in Zhu Jiang Delta, Guangdong Province, China." Master's Thesis, University of Georgia, Athens, Ga., 1993.

Hamilton, H., and S.H. Luk. "Nitrogen Transfers in a Rapidly Eroding Agro-ecosystem: Loess Plateau, China." *Journal of Environmental Quality*, 22, No. 1 (1993): 133-140.

Li, Xiaofang. "Complex Afforestation Ecology Economic System: An Example of Sustainable Agriculture" (in Chinese). *Acta Geographica Sinica*, 48, No. 1 (1993): 47-54.

Lu, Qiyao. "Fluctuations of Agroclimatic Zones and Its Impact on Agricultural Production in China." *Chinese Geographical Science*, 3, No. 1 (1993): 11-21.

Zhao, Zhaobing. "The Relation Between Basic Geomorphic Character and Agricultural Production in Fujian" (in Chinese). *Acta Geographica Sinica*, 48, No. 2 (1993): 143-151.

Zhu, Siguang. "Historical Changes in the Agricultural Environment: The Greater Sha An-xi–Inner Mongolia–Shanxi Triangle." *Chinese Environment and Development*, 4, No. 1 (1993): 38-52.

Liu, Changming, and Moozheng You. "The South-to-North Water Transfer Project and Sustainable Agricultural Development on the North China Plain." *Chinese Environment and Development*, 5, No. 2 (1994): 96-95.

Li, Xiaofang. "Complex Ecologic-Economic System of Afforestation: An Attempt of Sustainable Agriculture." *Chinese Geographical Science*, 4, No. 2 (1994): 168-173.

Ohta, Shunji, Zenbein Uchijima and Yasuyuki Oshima. "Effect of 2 x CO_2 Climate Warming on Water Temperature and Agricultural Potential in China." *Journal of Biogeography*, 22, No. 4-5 (1995): 649-656.

Ma, S. "A Study of the Impact of Climatic Change on Grain Yield in Jilin." *Chinese Environment and Development*, 6, No. 3 (1995): 73-85.

Tan, Kok-Chiang, and Wei Xu. "Chinese Agriculture: An Agroecosystem Health Perspective." *Chinese Environment and Development*, 6, No. 3 (1995): 175-205.

Veeck, Gregory, Zhon Li, and Ling Gao. "Terrace Construction and Productivity on Loessal Soils in Zhaongyong County, Shanxi Province, People's Republic of China." *Annals of the Association of American Geographers*, 85, No. 3 (1995): 450-467.

Wang, Futay, and Zongci Zhao. "Impact of Climate Change on Natural Vegetation in China and Its Implications for Agriculture." *Journal of Biogeography*, 22, No. 4-5 (1995): 657-664.

Xu, Wei, and Kok-Chiang Tan. "Chinese Agriculture: An Agroecosystem Health Perspective." *Chinese Environment and Development*, 6, No. 3 (1995): 3-28.

Cai Y., and Barry E. Smit. "Sensitivity and Adaptation of Chinese Agriculture Under Global Climate Change" (in Chinese). *Acta Geographica Sinica*, 51 (1996): 202-212.

Smit, Barry E., and Y. Cai. "Climate Change and Agriculture in China." *Global Environmental Change*, 6, No. 3 (1996): 205-214.

Cai, Y. "Vulnerability and Adaptation of Chinese Agriculture to Global Climate Change." *Chinese Geographical Science*, 7, No. 4 (1997): 289-201.

Muldavin, Joshua. "Environmental Degradation in Heilongjiang: Policy Reform and Agrarian Dynamics in China's New Hybrid Economy." *Annals of the Association of American Geographers*, 87, No. 4 (1997): 579-613.

Gong, Zitang, Huang Biao, and Ouyong Tao. "Pedogeochemistry of China and Its Significance in Agirucultural Production." *Chinese Geographical Science*, 8, No. 4 (1998): 299-308.

Yun, Xie, and Jidong Liu. "Impact of Climate on Grain Yield Per Unit Area in China During the Years of 1949–1992." *Chinese Geographical Science*, 8, No. 4 (1998): 309-316.

Chen, X., and Y. Zong. "Major Impacts of Sea-Level Rise on Agriculture in the Yangtze Delta Area Around Shanghai." *Applied Geography*, 19, No. 1 (1999): 69-84.

Muller, J. "Die Arbeitsintensitat in der Landnutzung Chinas und ihre agrarokologischen Ursachen." *Petermann's Geographische Mitteilungen*, 143 (1999): 163-187.

Shan, Penfei, Jinlin Wen, and Changjun Cai. "Desertification Hazard and Farming-Grazing Sustainable Development in Lingwu-Yanchi Region of Ningxia." *Chinese Geographical Science*, 9, No. 2 (1999): 177-183.

Weng, Qihao. "Human-Environmental Interactions in Agricultural Land Use in a South China's Wetland Region: A Study on the Zhujiang Delta in the Holocene." *GeoJournal*, 51, No. 3 (2000): 191-202.

IRRIGATION

Vander Meer, Canute, and J. Li. "Water Distribution in Farmer Managed Rice Irrigation Systems, Fujian, China." *Geographical Review of Japan*, 71B, No. 2 (1998): 106-120.

AQUACULTURE

Ruddle, Kenneth, and Fuzhong Gang. *Integrated Agriculture—Aquaculture in South China: The Dike-Pond System of the Zhujong Delta.* New York, N.Y.: Cambridge University Press, 1988.

Wang, Amy C.Y. "Historical Geography of Pond Aquaculture Management in the Zhujiang Delta." *Asian Geographer*, 16, No. 1-2 (1997): 72-88.

TRANSPORTATION AND MECHANIZATION

Kuo, Leslie T.C. "Technical Transformation of Agriculture in Communist China." Pp. 250-274 in *Agrarian Policies and Problems in Communist and Non-Communist Countries*, edited by W.A.D. Jackson. Seattle, Wash.: University of Washington Press, 1971.

Veeck, Gregory. "Agricultural Mechanization in Jiangsu Province, 1979–1988." *Chinese Geography and Environment*, 3, No. 3 (1990): 24-41.

AGRICULTURAL LABOR STUDIES

Cheng, Lu. "China's Agricultural Labour Force Transfer." *GeoJournal*, 21, No. 1-2 (1990): 71-72.

Liu, Lee. "Labor Location and Agricultural Land Use in Jilin, China." *Professional Geographer*, 52, No. 1 (2000): 74-82.

Hong Kong

CROPS AND CROPPING

Wong, Chenta. "Vegetable Farming in Hong Kong: A Study in Agricultural Geography." Ph.D. Dissertation, University of Hong Kong, Hong Kong, China, 1971.

FARM LAND

Wang, C.T. "Land Use in Agriculture." Pp. 185-206 in *A Geography of Hong Kong*, 2nd edition, edited by T.N. Chiu and C.L. So. Hong Kong, China: Oxford University Press, 1986.

Jim, C. "Land Use Problems Associated with Farmland Abandonment." *Geography*, 82, No. 3 (1997): 277-280.

AQUACULTURE

Anderson, E.N., Jr. "Traditional Aquaculture in Hong Kong." *Journal of Tropical Geography*, 30 (1970): 11-16.

Japan

ATLASES AND GRAPHIC PRESENTATIONS

Rawden, Fiske F. "A Key to the Identification of Japanese Agriculture with Air Photos." Master's Thesis, Clark University, Worcester, Mass., 1960.

GENERAL TEXTS, ARTICLES, AND THESES

Davis, D.H. "Agricultural Occupation of Hokkaido." *Economic Geography*, 10, No. 4 (1934): 348-367.

Yamaguti, S. "Distribution of the Shifting Cultivation" (in Japanese). *Geographical Review of Japan*, 14 (1938): 1-23.

Carstairs, Margaret. "Intensification of Agriculture in Sub-Tropical Japan." Master's Thesis, University of Washington, Seattle, Wash., 1939.

Trewartha, Glenn T. "Agriculture and Fishing." Pp. 193-257 in *Japan: A Physical, Cultural and Regional Geography*, edited by Glenn T. Trewartha. Madison, Wis.: University of Wisconsin Press, 1945.

Thompson, John H. "Urban Agriculture in Southern Japan." *Economic Geography*, 33, No. 3 (1957): 224-237.

Scheidl, Leopold. "A European View of Japanese Agriculture." *Oriental Geographer*, 3, No. 2 (1959): 1-12.

Isida, Ryuziro. "Agriculture and Land Use." Pp. 63-77 in *Geography of Japan*, edited by the Association of Japanese Geographers. Tokyo, Japan: Kokusai Bunka Shinkokai, 1961.

Kawakami, Mitsugi. "Development of Wakane Agriculture in Sonriku District" (in Japanese). *Science Report of Tohoku University, Seventh Series, Geography*, 18, No. 2 (1969): 189-202.

Sasaki, Hiroshi, et al. "Aspects and Viewpoint of Agriculture in Miyoshi-Mura, Saitama-Ken, in a North-Western Suburb of Tokyo" (in Japanese). *Geographical Review of Japan*, 42, No. 10 (1969): 632-649.

Jones, Harding. "The Role of Culture in Perpetuating Inefficient Agriculture in Japan." *Proceedings, Pennsylvania Academy of Science*, 40 (1970): 32-37.

Jones, Harding. "The Japanese Agricultural Revolution: Real or Illusory." *Pennsylvania Geographer*, 9, No. 3 (1971): 15-20.

Sum, Kongsut. "Changing Structural Patterns in Japanese Agriculture with Special Reference to Tohoku." *Science Reports of Tohoku University, Seventh Series, Geography*, 21, No. 2 (1972): 235-246.

Sakamoto, Hideo. "Some Changes in the Agriculture of the Coastal Dune Region in the Northern Part of the Shonai Plain, Yamagata Prefecture" (in Japanese). *Geographical Review of Japan*, 46, No. 12 (1973): 778-794.

Sakurai, Akihisa. "Agricultural Regionalization in Central Kanto District by Using Factor Analysis and Numerical Taxonomy" (in Japanese). *Geographical Review of Japan*, 46, No. 12 (1973): 826-849.

DiMeo, Guy. "L'agriculture japonaise." *L'Information Geographique*, 39, No. 2 (1975): 62-79.

McCune, Shannon. *Geographical Aspects of Agricultural Change in the Ryukyu Islands*. Gainesville, Fla.: University of Florida Social Sciences Monograph, No. 54, 1975.

Yamamoto, S., Yoichi Asano, and I. Saito. "Geographical Approaches to Agriculture and Agricultural Regions in Japan." Pp. 113-142 in *Geography in Japan*, edited by Shinzo Kinchi. Tokyo, Japan: University of Tokyo Press, 1976.

Okahashi, Hidenori. "The Change of Agriculture in Mountain Villages Around Industrialized Areas: A Case Study of Mikawa Mountainous Area (Aichi Prefecture)" (in Japanese). *Jimbun Chiri/Human Geography*, Kyoto, 30, No. 2 (1978): 1-20.

Sum, Kongsut. "Japanese and Chinese Agriculture: Comparison and Appraisal." *Science Reports, Tohoku University, Seventh Series, Geography*, 28, No. 2 (1978): 379-385.

Iinuma, Jiro. "La logique spatiale de l'agriculture japonaise." *L'Espace Geographique*, 9, No. 2 (1980): 143-147.

Ishii, Motosuke. "Regional Trends in the Changing Agrarian Structure of Postwar Japan." Pp. 199-222 in *Geography of Japan*, edited by the Association of Japanese Geographers. Tokyo, Japan: Teikiku-Shoin Company, 1980.

Ishii, Motosuke. "Japans landwirtschaft auf dem Wag in den Nebererwerb." *Erdkunde*, 34, No. 2 (1980): 96-108.

Kitamura, Shuji. "Changes of Agriculture in the Lower Reaches Area of the Kiso After the High Economic Growth" (in Japanese). *Jimbun Chiri / Human Geography*, Kyoto, 32, No. 2 (1980): 26-40.

Ishida, Hiroshi, et al., eds. *Changing Agriculture and Rural Development—The World and Japan: Papers and Proceedings of the I.G.U. Nagana Symposium 1980*. Tokyo, Japan: International Geographical Union, Commission on Agricultural Productivity and World Food Supply, 1981.

Arizono, S. "Some Regional Characteristics of the Farming System in the Northern Part of Tohoku Province in the 18th and 19th Centuries" (in Japanese). *Jimbun Chiri / Human Geography*, Kyoto, 34, No. 2 (1982): 67-77.

Kada, R. "Trends and Characteristics of Part-Time Farming in Post-War Japan." *GeoJournal*, 6, No. 4 (1982): 367-372.

Kikuchi, T. "Development of Spatial Pattern of Daily Farming in the Pioneer Settlement at the Foot of Mt. Nasu, Tochigi Prefecture After World War II" (in Japanese). *Geographical Review of Japan*, 55, No. 6 (1982): 359-379.

Kitamuru, Shuji. "The Regional Structure of Japanese Agriculture in Terms of Part-Time Farming" (in Japanese). *Geographical Review of Japan*, 55, No. 11 (1982): 739-756.

Kornhauser, David. "The Agrarian Landscape." Pp. 34-57 in *Japan: Geographical Background to Urban-Industrial Development*, edited by David Kornhauser. London, England: Longman, 1982.

Sakamoto, Hideo. "The Regional Distribution of Comparatively Large Farms in Japan" (in Japanese). *Geographical Review of Japan*, 55, No. 1 (1982): 37-50.

Smith, E.M., and J.R. Smith. "Some Perceptions of Agriculture in the Seto Island Sea Region of Japan vis-a-vis the Corn Belt of the United States of America." *Hiroshima University, Department of Geography, Special Publication*, 12 (1982): 30-34.

Tezuka, A. "Regional Characteristics of Viable Farming Types in the Shimo-Otsu District, Dejima-Mura, Ibaraki Prefecture" (in Japanese). *Geographical Review of Japan*, 55, No. 12 (1982): 814-833.

Kitamuru, Shuji. "The Regional Structure of Japanese Agriculture in Terms of Part-Time Farming 1960–1980" (in Japanese). *Jimbun Chiri / Human Geography*, Kyoto, 35, No. 6 (1983): 55-67.

Ishii, Motosuke. "Verandrungen von Struktur und Producktivitat japanische Agrarraume." *Geographische Rundschau*, 36, No. 3 (1984): 138-145.

Singh, Rana P.B., and Shogo Yuihama. "Recent Agricultural Changes in Niike-Buraka, Japan." *National Geographical Journal of India*, 30, No. 3 (1984): 139-152.

Singh, Rana P.B., and Shogo Yuihama. "Changing Agrarian Indicators and Agricultural Regions of Japan." *National Geographical Journal of India*, 31, No. 3 (1985): 173-184.

Tkahashi, N., et al. "Regional Structure of Financial and Farming Activites in Yasato-Machi, Ibaraki Prefecture" (in Japanese). *Tsukuba Studies in Human Geography*, 9 (1985): 145-180.

Yamamoto, S., H. Akimoto, and Y. Muragama. "Regional Structure of Agriculture in the Kanto District" (in Japanese). *Tsukuba Studies in Human Geography*, 12 (1988): 139-163.

Moore, Richard H. *Japanese Agriculture: Patterns of Rural Development*. Boulder, Colo.: Westview, 1990.

Saito, I., and Mineaki Kanno. "Development of Private Sports Facilities as a Side Business of Urban Farmers." *Geographical Review of Japan*, 63B, No. 1 (1990): 48-59.

Uchiyama, Yukihisa. "Types of Agricultural Income in the Southwestern Part of the Tokyo Metropolitan Region: The Case of Kanagawa Prefecture." *Geographical Review of Japan*, 63B, No. 1 (1990): 60-72.

Latz, Gil. "The Persistence of Agriculture in Urban Japan: An Analysis of the Tokyo Metropolitan Area." Pp. 217-238 in *The Extended Metropolis: Settlement Transition in Asia*, edited by Norton Ginsburg, Bruce Koppel, and Terrence G. McGee. Honolulu, Hawaii: University of Hawaii Press, 1991.

Wiltshire, Richard. "Rural Development and Processes of Agricultural Change in Japan." Pp. 49-67 in *Agricultural Change, Environment and Economy: Essays in Honour of W.B. Morgan*, edited by Keith Hoggart. London, England: Mansell, 1993.

Stradford, H. Todd. "Changes in the Agriculture of Kochi Prefecture, Japan, Since 1986 as a Result of Declining Rice Subsidies and the Announcement of a New Agricultural Policy." Ph.D. Dissertation, University of Oklahoma, Norman, Okla., 1994.

Kamikihara, Shizue. "Contract Farming in Aso-Machi, Japan." *Asian Geographer*, 14, No. 2 (1995): 104-130.

Millward, Hugh. "Regional Variations in the Development and Restructuring of Farming in Hokkaido, Japan." *Scottish Geographical Magazine*, 111, No. 3 (1995): 150-158.

CROPS AND CROPPING PATTERNS

Trewartha, Glenn T. "The Suwa Basin: A Specialized Sericulture District in the Japanese Alps." *Geographical Review*, 20, No. 2 (1930): 224-244.

Merrie, A., and M.M. Merrie. "Sericulture et filature de la joie au Japan." *Annales de Geographie*, 46 (1937): 43-60.

Russell, Joseph A. "The Teas of Uji." *Economic Geography*, 16, No. 2 (1940): 211-224.

Kakiuchi, George H. "The Rise and Development of the Fruit Industry on the Okayama Plain." Ph.D. Dissertation, University of Michigan, Ann Arbor, Mich., 1958.

Manchester, Curtis A. "Igusa: A Critical Cash Crop in the Rural Economy of Okayama." *Economic Geography*, 34, No. 1 (1958): 47-63.

Oagasawara, Yoshikatsu. "The Role of Rice Paddy Development in Japan." *Bulletin of the Geographical Survey Institute, Tokyo*, 5, Part 4 (1958).

Kakiuchi, George H. "The Stone Wall Strawberry Industry on Kuno Mountain, Japan." *Economic Geography*, 36, No. 2 (1960): 171-184.

Kakiuchi, George H., and Setsutaro Murakami. "Satsuma Oranges in Ocho-Mura: A Study of Specialized Cash Cropping in Southwestern Japan." *Geographical Review*, 51, No. 4 (1961): 500-518.

Kakiuchi, George H. "Recent Developments in the Cultivation of Wet-Rice in Northeastern Japan." *Journal of Geography*, 63, No. 4 (1964): 155-160.

Okamoto, Jiro. "The Changing Limits of the Rice Fields in Hokkaido" (in Japanese). *Science Reports of Tohoku University, Seventh Series, Geography*, 14 (1965): 139-145.

Sakurai, Hidemitsu. "The Development Process of Peach Cultivation in Fukushima Basin, Tohoku District" (in Japanese). *Jimbun Chiri/Human Geography*, Kyoto, 17, No. 2 (1965): 38-56.

Fukui, Hideo. "Some Aspects of Recent Changes in Japanese Rice Farming" (in Japanese). *Science Reports of Tohoku University, Seventh Series, Geography*, 15 (1966): 25-33.

Okamoto, Jiro. "Cool Summer Damage on the Rice Crop and the Changes in Rice Field Acreage" (in Japanese). *Science Reports of Tohoku University, Seventh Series, Geography*, 15 (1966): 211-218.

Hough, Richard F. "Impact of the Decline in Raw Silk on the Suwa Basin of Japan." *Economic Geography*, 44, No. 2 (1968): 95-116.

Hough, Richard F. "Impact of the Decline on Raw Silk on Two Major Cacoon-Producting Regions in Japan." *Annals of the Association of American Geographers*, 58, No. 2 (1968): 250-272.

Doi, K. "The Re-examination of Weaver's Method of Combinations Analysis" (in Japanese). *Jimbun Chiri/Human Geography,* Kyoto, 22 (1970): 485-502.

McNab, J.W. "Cultivation and Field Form on Three Japanese Islands." *Geographical Review*, 60, No. 2 (1970): 176-191.

Sum, Kongsut. "Rice Cultivation in Japan and Its Related Problems" (in Japanese). *Science Reports of Tohoku University, Seventh Series, Geography*, 21, No. 1 (1971): 19-28.

Motoki, Yasushi. "Priority Production of Rice in Tohoku District: Especially on the Rapid Expansion of Rice Fields After World War II" (in Japanese). *Science Reports of Tohoku University, Seventh Series, Geography*, 22, No. 1 (1972): 23-28.

Berque, Augustin. "Une mutation des campagnes japonais: La riziculture a Hokkaido." *Annales de Geographie*, 82, No. 3 (1973): 291-315.

Matsui, Sadao. "Recent Change of the Green-House Viticulture Region in Okayama Prefecture, Japan" (in Japanese). *Geographical Review of Japan*, 47, No. 1 (1974): 1-20.

Miida, Keisuke. "Zenmai (Royal Fern) Production and Its Village Sustaining Function in the Tohoku District of Japan" (in Japanese). *Geographical Review of Japan*, 47, No. 6 (1974): 370-386.

Ukita, T. "Remote Cropping on the Yaheyama Islands" (in German). *Geographical Review of Japan*, 47, No. 8 (1974): 511-524.

Clark, Charles L. "The Okinawa Sugar Industry: Some Effects of Government Activities on Industrial Location and Economic Development." Master's Thesis, California State University at Long Beach, Calif., 1975.

Kakiuchi, George H., and Hiroyoshi Kobayashi. "Regionalization of Some Major Crops of Japan in Recent Years" (in Japanese). *Science Reports of Tohoku University, Seventh Series, Geography*, 25, No. 1 (1975): 51-66.

Mitsunobu, Okamura. "Cotton Cultivation in the Yamato Basin in the Late Edo Era" (in Japanese). *Geographical Review of Japan*, 48, No. 5 (1975): 365-377.

Oishi, Taizon. "Progressive High Productivity and Surplus Production of Rice in Japan: On the Background of Rice Production Control" (in Japanese). *Geographical Reports of Tokyo Metropolitan University*, 10 (1975): 43-68.

Arizono, Shoichiro. "Crop Combination Regions for 1960 and 1970 in Kinki Province" (in Japanese). *Geographical Review of Japan*, 50, No. 7 (1977): 420-429.

Fuji, Masami. "The Changes of House Horticulture in Miho District, Shimuzu City, Shizuoka Prefecture" (in Japanese). *Jimbun Chiri/Human Geography*, Kyoto, 29, No. 6 (1977): 90-111.

Ichikawa, Takeo, and N. Hasegawa. "Symposium: Fruit Growing Regions in Eastern Japan" (in Japanese). *Geographical Review of Japan*, 50, No. 3 (1977): 151-157.

Yamakawa, Mitsuo. "Producing Regions and Marketing System of Fresh Vegetables in Recent Japan." *Geographical Reports of Tokyo Metropolitan University*, 12 (1977): 143-157.

Isobe, Toshisada, and Yoshihiro Kitabayashi. "Symposium: Changes in Rice Production Under the Policy of Acreage Allotment" (in Japanese). *Geographical Review of Japan*, 52, No. 2 (1979): 87-91.

Kawakami, Makoto. "Contract Cultivation of Rice in Ohgata-Machi, Niigata Prefecture" (in Japanese). *Geographical Review of Japan*, 52, No. 12 (1979): 661-674.

Kikukawa, Konsuki. "The Reorganization of Rice-Producting Cooperatives in the Saga Plain" (in Japanese). *Geographical Review of Japan*, 52, No. 12 (1979): 675-688.

Matsui, Sadao. "The Reorganization of Vegetable-Producing Areas in Kochi Prefecture, Shikoku Island" (in Japanese). *Geographical Review of Japan*, 52, No. 2 (1979): 66-82.

Yamakawa, Mitsuo. "The Producing Conditions of Industrial Tomatoes in Atsumi, Aichi Prefecture" (in Japanese). *Geographical Review of Japan*, 52, No. 11 (1979): 607-622.

Aymans, Gerhard. "Geographische Aspekte des Algenbaus in Japan." *Erdkunde*, 34, No. 2 (1980): 109-119.

Ebato, A., and N. Matoumura. "Changes in Fruit Production in Japan." Pp. 223-245 in *Geography of Japan*, edited by the Geographical Association of Japan. Tokyo, Japan: Teikoku-Shoin for the Association of Japanese Geographers, 1980.

Matsui, Sadao. "Reorganization of Rice-Producing Cooperatives in the Nishi-Mikawa Plain" (in Japanese). *Geographical Review of Japan*, 53, No. 2 (1980): 75-92.

Takahashi, Masaaki. "The Changes and Characteristics of the Rice-Producing Cooperatives in the Suburban Areas: A Case Study in Fujiidera and Izumi-otsu" (in Japanese). *Geographical Review of Japan*, 52, No. 2 (1980): 93-107.

Yamakawa, Mitsuo. "Regional Differences of Vegetable Prices: On the View Point of Regional System of the National Economy" (in Japanese). *Geographical Reports of Tokyo Metropolitan University*, 14-15 (1980): 185-194.

Sakamoto, Hideo. "The Location of Onion Producing Areas in the Kitami District, Hokkaido" (in Japanese). *Jimbun Chiri/Human Geography*, Kyoto, 33, No. 5 (1981): 21-40.

Sugihara, Y. "A Study on the Limit of the Paddy Rice Cultivation by Means of the Climatic Productivity Index." *Geographical Reports of Tokyo Metropolitan University*, 16 (1981): 79-94.

Uchijima, Zenbein. "Yield Variability of Crops in Japan." *GeoJournal*, 5, No. 2 (1981): 151-164.

Tabayashi, A., and T. Kikuchi. "Silk-Raising Farmers in the Kasumigaura Region" (in Japanese). *Tsukuba University, Institute of Geoscience, Annual Report*, 8 (1982): 19-24.

Pitte, J.R. "Vignobles et vins du Japon." *Annales de Geographie*, 92, No. 2 (1983): 172-199.

Tabayashi, A. "Development of Rice Cultivation in the Beech Forest Belt, Japan" (in Japanese). *Tsukuba Studies in Human Geography*, 7 (1983): 233-256.

Tano, H. "The Land Conditions and Lotus Root Growing in the Alluvial Lowland Along Lake Kasumigaura" (in Japanese). *Geographical Review of Japan*, 56, No. 1 (1983): 27-34.

Zoble, Rhonda H. "Cultivation and Production of Nori, Wakame and Aonori in Selected Areas of Japan." Master's Thesis, San Francisco State University, San Francisco, Calif., 1983.

McDonald, Mary G. "Rice in Hokkaido: Establishment and Persistence at Japan's Northern Extreme." Master's Thesis, University of California, Berkeley, Calif., 1985.

McLin, Bruce C. "An Energy Analysis of Japanese Vinyl House Horticulture." Master's Thesis, San Diego State University, San Diego, Calif., 1985.

Shibusawa, Saito F., and K. Ikeda. "Development of Vegetable Production and Changes of Land Use in Miwa Peninsula, Kanagawa Prefecture" (in Japanese). *Tsukuba Studies in Human Geography*, 9 (1985): 95-124.

Yamamoto, S., and Y. Uhiyama. "Changing Patterns of Fruit and Grape Growing Regions in Japan from 1960–1980" (in Japanese). *Tsukuba Studies in Human Geography*, 9 (1985): 21-48.

Motoki, Y. "The Recent Developments of Glutinous Rice Production in Soga Prefecture, Main Paddy Rice Producing Region in Southwestern Japan" (in Japanese). *Tsukuba Studies in Human Geography*, 11 (1987): 227-245.

Sakamoto, Hideo. "The Circumstances and the Location of Carrot Production in the Furano District, Hokkaido" (in Japanese). *Jimbun Chiri/Human Geography*, Kyoto, 40, No. 1 (1988): 1-18.

Sasaki, Hiroshi. "Cyclamen Cultivation in Hinoharamura, Tokyo" (in Japanese). *Tsukuba Studies in Human Geography*, 14 (1990): 25-40.

Sakamoto, Hideo. "Vegetable Production by Women and Elderly Men in the Seba District, Shiojiri City" (in Japanese). *Geographical Review of Japan*, 65A, No. 8 (1992): 603-618.

Jeong, K. "The Production of Ginseng and Peony in Yatsukacho, Shimue Prefecture" (in Japanese). *Tohoku Chiri*, 45, No. 2 (1993): 98-110.

Moore, Richard H. "Resistence to Japanese Rice Policy: A Case Study of the Hachirogata Model Farm Project." *Political Geography*, 12, No. 3 (1993): 278-296.

Saito, I., and S. Hayashi. "Development of Kaki Cultivation and Innovation of Simple Controlled Atmosphere Storage in the Middle Chikugo Valley, Kyushu Island" (in Japanese). *Tsukuba Studies in Human Geography*, 17 (1993): 87-105.

Tabayashi, A. "Development of Market Gardening in Kyowa-machi, Ibaraki Prefecture" (in Japanese). *Tsukuba Studies in Human Geography*, 17 (1993): 1-31.

Zong, Guiman. "The Bases of Development and Characteristics of Market Gardening in Tokyo Metropolis: A Case Study of Kanahori Area of Tomisato, Chiba Prefecture." *Geographical Reports-Tokyo Metropolitan University*, 29 (1994): 39-55.

Morimoto, T. "Distribution and Cause of Cultivation Abandonment in the Kanto District, Japan, 1975–1990" (in Japanese). *Tsukuba Studies in Human Geography*, 19 (1995): 173-186.

Sakamoto, Hideo. "Vegetable Growers and Labor Management in Kitano-cho, Fukuoka Prefecture, Japan" (in Japanese). *Jimbun Chiri/Human Geography*, 47, No. 5 (1995): 439-457.

Araki, H. "Growth of Vegetable Production in Asahikawa, Hokkaido: Strategies of Shippers Union" (in Japanese). *Jimbun Chiri/Human Geography*, 48, No. 5 (1996): 427-448.

Kawakubo, A. "Changes in Japanese Mandarin-Producing Areas with Liberalization of the Orange Juice Trade" (in Japanese). *Jimbun Chiri/Human Geography*, 48, No. 1 (1996): 28-47.

Zhong, G.M. "The Development Mechanism and Its Sustainable Bases in a Vegetable Growing Area in Okabe Towa, Saitama Prefecture" (in Japanese). *Geographical Review of Japan*, 69A, No. 4 (1996): 223-241.

Goto, T. "Reorganization of Contract Farming Areas Under Trade Liberalization and the Overproduction of Agricultural Products: A Case Study of Processing Tomato in Japan" (in Japanese). *Jimbun Chiri/Human Geography*, 50, No. 2 (1998): 150-171.

Nihei, T. "Maintaining of Horticultural Management and Technical Innovation: A Case Study of Asahi-shi, Chiba Prefecture." *Geographical Review of Japan*, 71A, No. 9 (1998): 661-678.

Brucklacher, Ann D. "Apples and Regional Change: Life and Economy in Tsugaru, Japan." Ph.D. Dissertation, Louisiana State University, Baton Rouge, La., 1999.

Nihei, T. "Energy Efficiency of Crop Production in Japan, 1970–1990." *Geographical Review of Japan*, 73B, No. 1 (2000): 27-45.

Nihei, T. "Changes in the Energy Efficiency of Regional Crop Production in Japan." *Geographical Review of Japan*, 74B, No. 1 (2001): 47-61.

ANIMAL HUSBANDRY

Nagashima, Hiromichi. "The Development and the Patterns of Enterprise of Poultry Raising in Postwar Japan: In the Case of the Kanto District" (in Japanese). *Geographical Review of Japan*, 42, No. 1 (1969): 60-75.

Takahashi, S. "An Overview of New Domestication of Wild Boars for Production of Flesh in Japan" (in Japanese). *Geographical Review of Japan*, 57, No. 11 (1984): 781-790.

Yamamoto, M. "Development of Diversified Farming with Livestock and Transformation of Land Utilization in the Northern Kitakami Mountains, Japan" (in Japanese). *Geographical Review of Japan*, 59, No. 12 (1986): 706-728.

Yamamoto, S., and A. Bito. "Areal Patterns of Livestock Industry of Japan from 1960 to 1980" (in Japanese). *Tsukuba Studies in Human Geography*, 10 (1986): 17-41.

Nagasaka, M. "The Development and Formation of the Broiler Production Area in Iwate Prefecture, Japan" (in Japanese). *Geographical Review of Japan*, 61, No. 3 (1988): 239-257.

Nagasaka, M. "Formation of Production Areas in the Japanese Broiler Chicken Industry." *Geographical Review of Japan*, 64B, No. 1 (1991): 50-68.

Nagasaka, M. "The Regional Development of the Broiler Industry in Japan." *Science Reports, University of Tsukuba, Institute of Geoscience, Section A*, 12 (1991): 91-116.

Tabayashi, A. "Regional Structure of Race Horse Breeding in the Hidaka Region, Hokkaido" (in Japanese). *Tsukuba Studies in Human Geography*, 22 (1998): 79-98.

DAIRYING

Kakiuchi, George H., and Kazuo Ninomiya. "The Dairying Industry of Japan." *Yearbook, Association of Pacific Coast Geographers*, 32 (1970): 23-40.

Kasai, Yamato. "Japanese Dairying: Changes in Milch-Cow-Tendings and in Milk Transportation" (in Japanese). *Tohoku University Science Reports, Geography*, 19, No. 2 (1970): 141-159.

Oshiro, Kenji K. "Dairy Policies and the Development of Dairying in Tohoku, Japan." Ph.D. Dissertation, University of Washington, Seattle, Wash., 1972.

Oshiro, Kenji K. "Dairy Policies and the Development of Dairying in Tohoku, Japan." *Proceedings, Association of American Geographers*, 5 (1973): 202-206.

Koganezawa, T. "Changing Pattern of Tokyo's Milkshed" (in Japanese). *Geographical Reports of Tokyo Metropolitan University*, 14-15 (1980): 175-184.

Saito, I. "Development of a Dairy Region in the Northwestern Quarter of Saitama Prefecture" (in Japanese). *Tsukuba Studies in Human Geography*, 12 (1988): 25-43.

Kikuchi, T. "The Bases of Development of Viable Dairy Farming in Outer Suburban Areas of Tokyo Metropolis: A Case of Diluvial Uplands." *Science Reports, University of Tsukuba, Institute of Geoscience, Section A*, 10 (1989): 23-57.

Kikuchi, T. "The Bases of Development of Viable Dairy Farming in the Kanto District, Japan." *Geographical Reports, Tokyo Metropolitan University*, 29 (1994): 11-37.

FARM LAND

Pitts, Forest R. "Comparative Land Fertility and Potential in the Inland Sea and Peripheral Areas of Japan." Ph.D. Dissertation, University of Michigan, Ann Arbor, Mich., 1955.

Anstey, Robert L. "A Comparative Study of the Agricultural Land Use of the Aso Highland and the Kumamoto Lowland of Central Kyushu." Ph.D. Dissertation, University of Maryland, College Park, Md., 1957.

Motooka, Taheshi. "Changes in Land Use in Japanese Agriculture after the War: Especially in the Saga Plain and the Zenkoji Basin." *Oriental Geographer*, 2, No. 2 (1958): 121-130.

Nakajima, Minehiro. "The Management of the Dug-Down Paddy Fields and the Cultivation of Dry Fields on the Kashima Coastal Sand Dune, Ibaraki Prefecture" (in Japanese). *Geographical Review of Japan*, 39, No. 2 (1966): 84-102.

Sahashita, T. "High and Cool Reclaimed Land in Japan" (in Japanese). *Geographical Review of Japan*, 41, No. 5 (1968): 322-343.

Tamai, Kenzo. "Change of Land Utilization in Niijima, Izu Islands" (in Japanese). *Geographical Review of Japan*, 46, No. 10 (1973): 675-683.

Arizono, Shoichiro. "Retreat of the Cultivated Marginal Land" (in Japanese). *Jimbun Chiri/Human Geography*, Kyoto, 26, No. 2 (1974): 36-64.

Arizono, Shoichiro. "Arable Land Utilization Ratios in the Past 100 Years in Japan" (in Japanese). *Jimbun Chiri/Human Geography*, 27, No. 3 (1975): 107-122.

Imai, Toshinobu. "Changing Agricultural Land Use During the Exploitation Period 1868–1944 in Hokkaido" (in Japanese). *Science Reports, Tohoku University, Seventh Series, Geography*, 25, No. 2 (1975): 293-299.

Arai, Yasuhisa. "The Physical Change of Paddy Fields and the Local Government Policies in the Downstream Area of the Tenryu River" (in Japanese). *Geographical Review of Japan*, 49, No. 10 (1976): 669-684.

Harada, Toshiharu. "Transfer of Landownership and Change in Land Use in the Suburban Farming Settlement: Case Study of Ichikawa Municipal Area Adjacent to Tokyo" (in Japanese). *Geographical Review of Japan*, 49, No. 9 (1976): 616-632.

Ichiminami, F. "Distribution and Its Change of Productivity of Agricultural Land in Japan 1960–1975" (in Japanese). *Geographical Review of Japan*, 53, No. 12 (1980): 739-757.

Suzuki, Yasuo, and Yasukisa Arai. "Characteristics of Farmland Transfer and the Formation of Large Scale Farmers in the Lower Iwaki River Area" (in Japanese). *Jimbun Chiri, Human Geography*, Kyoto, 32, No. 6 (1980): 543-557.

Ichiminami, F. "Distribution and Recent Change of Productivity of Agricultural Land in Japan" (in Japanese). *Tsukuba University, Institute of Geoscience, Annual Report*, 7 (1981): 10-14.

Tanaka, Koji. "Changes in Landownership and Land Use of Former Farms in Nakano-ku and Musashino-shi, Tokyo" (in Japanese). *Geographical Review of Japan*, 55, No. 7 (1982): 453-471.

Kadomura, H. "Some Aspects of Large-Scale Land Transformation Due to Urbanization and Agricultural Development in Recent Japan." *Advances in Space Research*, 2, No. 8 (1983): 169-178.

Yamamoto, S., F. Ichiminami, and T. Vestima. "Spatial Structure of Agriculture in Kanto Region Based on Agricultural Land Productivity" (in Japanese). *Geographical Review of Japan*, 56, No. 9 (1983): 607-623.

Sasaki, Hiroshi. "Farmland Mobility and the Appearance of Large Size Farms in Niigata Plain, Japan" (in Japanese). *Tsukuba Studies in Human Geography*, 9 (1985): 125-143.

Ishii, H. and S. Yamamoto. "The Significance of Rikuden in the Agricultural Use of Tableland in Kanto Region" (in Japanese). *Tsukuba Studies in Human Geography*, 11 (1987): 1-19.

Maruyama, H. "Vertical Zonation of Agricultural Land Use on the Lower Southern Slopes of Mt. Asama, Japan" (in Japanese). *Geographical Review of Japan*, 60, No. 10 (1987): 643-666.

Hoshino, T. "Commuter Cultivation Through Getting Substitute Agricultural Land in Urbanizing Area: A Case Study in Tokyo" (in Japanese). *Tohoku Chiri*, 40, No. 4 (1988): 276-286.

Latz, Gil. *Agricultural Development in Japan: The Land Improvement District in Concept and Practice*. Chicago, Ill.: University of Chicago Department of Geography, Research Paper No. 225, 1989.

Yamamoto, S., et al. "The Land Use as a Manifestation of Urbanization and Agricultural Modernization in Hasaki-Machi, Ibaraki Prefecture" (in Japanese). *Tsukuba Studies in Human Geography*, 13 (1989): 147-189.

Maruyama, H. "Enlargement of Marginal Land and Vertical Production Strategy of Farm Households on the Lower Northern Slopes of Asama Volcano" (in Japanese). *Geographical Review of Japan*, 63, No. 2 (1990): 74-99.

Maruyama, H. "Agricultural Land Use Patterns on the Lower Southern Slopes of Asama Volcano" (in Japanese). *Science Reports, University of Tsukuba, Institute of Geoscience, Section A*, 11 (1990): 115-173.

Hoshino, T., and Y. Satoh. "Agricultural Land Acquisition in Suburban Areas of the Tokyo Metropolis" (in Japanese). *Jimbun Chiri/Human Geography*, Kyoto, 43, No. 6 (1991): 511-525.

Morimoto, T. "The Increase of Fallow and Abandoned Cultivated Land with the Development of Intensive Agriculture in Hasaki-machi, Ibaraki Prefecture" (in Japanese). *Geographical Review of Japan*, 64, No. 9 (1991): 613-636.

Maruyama, H. "Agricultural Land Use Patterns on the Lower Volcanic Slopes in Central Japan." *Geographical Review of Japan*, 65B, No. 2 (1992): 104-128.

Kikuchi, T. "Characteristics of Arable Land Use Change in Japan: A Case Study of Field Land Use in the Kanto District" (in Japanese). *Tohoku Chiri*, 45, No. 1 (1993): 13-29.

Tabayashi, A., and T. Nakagawa. "Temporal and Spatial Analysis of Agricultural Land Use on Alluvial Fans in Central Japan" (in Japanese). *Science Reports, University of Tsukuba, Institute of Geoscience, Section A*, 14 (1993): 31-48.

Onodera, A., Shizue Kamikihara, and T. Morimoto. "The Establishment of Turf Farming in Tsukuba City, Japan." *Geographical Bulletin*, 36, No. 1 (1994): 50-57.

McDonald, Mary G. "Agricultural Landholding in Japan: Fifty Years After Land Reform." *Geoforum*, 28, No. 1 (1997): 55-78.

AGRICULTURAL POPULATIONS AND SOCIAL ISSUES

Ando, Masuo. "Regional Development and Agrarian Problems: Measures for the Promotion of Rural and Agricultural Communities" (in Japanese). *Geographical Review of Japan*, 42, No. 3 (1969): 208-212.

Jones, Harding. "The Role of Culture in Perpetuating Inefficient Agriculture in Japan." *Proceedings, Pennsylvania Academy of Science*, 44 (1970): 32-37.

Ueno, Fukuo. "Areal Specialization in Japanese Peasant Farming and Agricultural Settlement." Pp. 399-435 in *Agrarian Policies and Problems in Communist and Non-Communist Countries*, edited by W.A.D. Jackson. Seattle, Wash.: University of Washington Press, 1971.

Oshiro, Kenji K. "The Continuity of Farm Households in Japan: A Study of Adaptation" (in Japanese). *Science Reports, Tohoku University, Seventh Series, Geography*, 28, No. 2 (1978): 339-351.

Ukita, T. "Rural Settlements on Ishigaki Island, Okinawa Prefecture, and Long-Distance Commuting Cultivation." *GeoJournal*, 4, No. 3 (1980): 267-272.

Oshiro, Kenji K. "Adjustment in Farm Household Systems of Japan." *Rural Systems*, 2, No. 3 (1984): 121-134.

Rutherford, John. *Rice Dominant Land Settlement in Japan: A Study of Systems within Systems*. Sydney, Australia: University of Sydney, Department of Geography, 1984.

Smith, Kevin R. "A Cultural Perspective on Japan and Its Rice Controversy of the 1980s." Master's Thesis, University of Arkansas, Fayetteville, Ark., 1988.

Tabayashi, A. "Trends and Themes of Geographical Studies on Agriculture and Rural Settlements of the Kurobe Alluvial Fan" (in Japanese). *Tsukuba Studies in Human Geography*, 14 (1990): 81-105.

Yagasaki, Noritaka. "Ethnic Agricultural Cooperatives as Adaptive Strategies in Japanese Overseas Communities: Diffusion, Development and Adaptation on Contextual Perspective." *Geographical Review of Japan*, 68B, No. 2 (1995): 119-136.

Waldiehuk, T., and H. Whitney. "Inhabitants' Attitudes Toward Agricultural Activities and Urban Development in an Urbanizing Konjuka Area in the Rural–Urban Fringe of Tokyo." *Geographical Review of Japan*, 70B, No. 1 (1997): 32-40.

ENVIRONMENTAL AND CONSERVATION ISSUES

Nuttonson, Michael Y. *Ecological Crop Geography and Field Practices in Japan, Japan's Natural Vegetation and Agro-Climatic Analysis in North America*. Washington, D.C.: American Institute of Crop Ecology, 1951.

Hoffman, Robert F. "Agricultural Uses of Hot-Spring Water in the Izu Peninsula, Japan." Master's Thesis, Southern Illinois University, Carbondale, Ill., 1967.

Carr, Michael W. "The Lake Biwa Watershed: Problems of Agricultural and Industrial Pollution." Master's Thesis, University of Washington, Seattle, Wash., 1981.

Yoshimo, Masatoshi. "Climatic Change and Rice Yield and Production in Japan During the Last 100 Years." *Geographical Review of Japan*, 66A, No. 1 (1993): 70-88.

IRRIGATION

Horiuchi, Yoshitaka. "A Study of the Irrigation Structure of the Villages in the Nara Basin" (in Japanese). *Geographical Review of Japan*, 39, No. 3 (1966): 159-167.

Tabayashi, A. "Spatial Structure of the Irrigation System in the Hokurika District" (in Japanese). *Geographical Review of Japan*, 54, No. 6 (1981): 295-316.

Tabayashi, A. "Regional Differences in Japanese Irrigation Systems" (in Japanese). *Tsukuba Studies in Human Geography*, 11 (1987): 115-137.

AQUACULTURE

Doumenge, F. "L'aquaculture sur la cote Pacifique de Tohoku." *Bulletin de la Societe Languedocienne de Geographie*, 9, No. 2-3 (1975): 193-240.

Kawakami, Mitsugi. "Regional Characters of Laver Aquaculture in Japan" (in Japanese). *Science Reports, Tohoku University, Seventh Series, Geography*, 25, No. 2 (1975): 267-270.

Nose, Takeshi. "Recent Advances in Aquaculture in Japan." *GeoJournal*, 10, No. 3 (1985): 261-276.

AGRICULTURAL MARKETING AND DISTRIBUTION

Eyre, John D. "Sources of Tokyo's Fresh Food Supply." *Geographical Review*, 49, No. 4 (1959): 455-474.

Eyre, John D. "Japanese Inter-Prefectural Rice Movements." *Economic Geography*, 38, No. 1 (1962): 78-86.

Masuyama, Yoko. "International Floral Trade of Japan" (in Japanese). *Science Reports, Tohoku University, Seventh Series, Geography*, 25, No. 2 (1975): 259-265.

Sum, Kongsut. "Agricultural Co-operatives in Japan: Functions, Role and Problems" (in Japanese). *Science Reports, Tohoku University, Seventh Series, Geography*, 28, No. 2 (1978): 371-378.

Arai, Yasuhisa. "Development of Local Vegetable-Fruit Markets and Formation of Vegetable–Fruit Growing Areas in the Northern Part of Saitama Prefecture During the Early Showa Period" (in Japanese). *Geographical Review of Japan*, 55, No. 7 (1982): 472-489.

TRANSPORTATION AND MECHANIZATION

Hall, Robert B. "Hand-Tractors in Japanese Paddy Fields." *Economic Geography*, 34, No. 4 (1958): 312-320.

Motoki, Yasushi. "Large Scale Mechanization in Tohoku Rice Mono-Culture Region: A Case Study of Western Piedmont Area of Kitakami Basin, Iwate Prefecture" (in Japanese). *Science Reports, Tohoku University Seventh Series, Geography*, 25, No. 2 (1975): 301-309.

Oshiro, Kenji K. "Mechanization of Paddy Cultivation in Japan and Its Effects on Farm Households" (in Japanese). *Science Reports, Tohoku University, Seventh Series, Geography*, 32, No. 1 (1982): 1-13.

Oshiro, Kenji K. "Mechanization of Rice Production in Japan." *Economic Geography*, 61, No. 4 (1985): 323-331.

AGRICULTURAL LABOR STUDIES

Yamamoto, S., and A. Tabayashi. "Transformations of the Rural Region of the Kurobe Alluvial Fan" (in Japanese). *Jimbun Chiri/Human Geography*, Kyoto, 27, No. 6 (1975): 33-59.

Okahashi, Hidenori. "Process of Agricultural Change and Rural Employment in Komitakara-mura, Northern Gifu Prefecture, Central Japan" (in Japanese). *Geographical Review of Japan*, 53, No. 8 (1980): 511-530.

McDonald, Mary G. "Farmers as Workers in Japan's Regional Economic Restructuring, 1965–1985." *Economic Geography*, 72, No. 1 (1996): 49-72.

Korea, North and South

GENERAL TEXTS, ARTICLES, AND THESES

Chen, Chengsiang. *Agricultural Geography of Korea.* Hong Kong, China: Geographical Research Centre, Chinese University of Hong Kong, Research Report No. 31, 1970.

Pak, Kihyuk. "A Comparative Study of the Agrarian Systems of South and North Korea." Pp. 436-458 in *Agrarian Policies and Problems in Communist and Non-Communist Countries*, edited by W.A.D. Jackson. Seattle, Wash.: University of Washington, Press, 1971.

Chon, Soohyun. "The Impact of a Growth Center on an Agricultural Hinterland: A Case Study of Taegu, Korea." Master's Thesis, University of Michigan, Ann Arbor, Mich., 1979.

Chon, Soohyun. "The Impact of a Growth Center on an Agricultural Hinterland: A Case Study of Taegu, Korea." *Geographical Bulletin*, 19 (1980): 27-36.

Shana, L., and Y. Pehg. "Some Aspects of Agricultural Development on Steep Slopes in South Korea." *GeoJournal*, 4, No. 3 (1980): 273-278.

Dege, Eckart. *Entwicklungsdisparitaten der Agrarregion Sudkoreas*. Kiel, Germany: Kieler Geographische Schriften, Heft 55, 1982.

Howell, Thomas J. "Agricultural Development in the Republic of Korea, 1972 to 1980." Master's Thesis, University of North Dakota, Grand Forks, N.Dak., 1984.

CROPS AND CROPPING PATTERNS

Lee Heewon. "The Spatial Variations and Changes of the Major Fruit Crops in Korea" (in Korean). *Journal of Geography Education*, 24 (1991): 86-109.

Jeong, K. "Characteristics of Ginseng Cultivation in Kimpo Peninsula, Korea" (in Japanese). *Geographical Review of Japan*, 66, No. 1 (1993): 1-25.

Mongolia

GENERAL TEXTS, ARTICLES, AND THESES

Wiens, Herold J. "Geographical Limitations to Food Production in the Mongolian People's Republic." *Annals of the Association of American Geographers*, 41, No. 3 (1951): 348-369.

CROPS AND CROPPING PATTERNS

Albano, H.K. "An Analysis of the Crop Producing Potential of the Mongolian People's Republic." Master's Thesis, University of Washington, Seattle, Wash., 1956.

ANIMAL HUSBANDRY

Njamdavaa, G. "Geographical Problems of the Animal Breeding in the Region of the Altai Mountains in Mongolia" (in Polish). *Zeszyty Instytutu Geografii I Prezestrzennego Zagosprodorowania*, 52 (1998): 107-122.

FARM LAND

Muller, Franz-Volker, and Bot-Ocir Bold. "On the Necessity of New Regulations for Pastoral Land Use in Mongolia." *Applied Geography and Development*, 48 (1996): 29-51.

Muller, Franz-Volker, and Bot-Ocir Bold. "Zur Relevanz neuer Regelungen fur die Weidel und nutzung in der Monglei." *Die Erde*, 127, No. 1 (1996): 63-82.

Rasmussen, M.S., et al. "Supporting Mongolian Pastoralists by Using GIS to Identify Grazing Limits and Opportunities from Livestock Census and Remote Sensing Data." *GeoJournal*, 47, No. 4 (1999): 563-571.

Taiwan

GENERAL TEXTS, ARTICLES, AND THESES

Wang, Shengtsu. "The Agricultural Regions of Taiwan." *The Science Reports of the National Tsing Hun University*, Series C, 1, No. 3 (1948): 175-195.

Chang, Chinghu. *Agricultural Geography of Taiwan*. Taipei, Taiwan: Chinese Geographical Research Institute, 1953–1958.

Chiu, Yuanwang. "The Geographical Basin of Agriculture in Yunlin Hsien" (in Chinese). *Science Reports, National Taiwan University, Department of Geography and Meteorology*, 2 (1965): 158-175.

Aurada, F. "Taiwan: Agrarwirtschaft und Agrarreform." *Mitteilungen der Osterreichischen Geographischen Gesellschaft*, 108, No. 1 (1966): 192-195.

Rogers, B.N. "Evolving Agricultural Patterns in Taiwan." Master's Thesis, University of Alabama, Tuscaloosa, Ala., 1969.

Hsieh, S.C. "Taiwan's Model of Agricultural Programs." Pp. 381-395 in *Agrarian Policies and Problems in Communist and Non-Communist Countries*, edited by W.A.D. Jackson. Seattle, Wash.: University of Washington Press, 1971.

Selyn, Robert M. "Urban Agriculture in Taiwan: Anachronism or Nutritional and Ecological Safety Valve?" *China Geographer*, 8 (1977): 15-28.

Chen, Chengsiang. *The Agricultural Regions of Taiwan*. Fumin, Taiwan: Fumin Geographical Institute of Economic Development, Research Report No. 70, 1979.

Liu, Chien. *Bauerliche landwirtschaft in Taiwan: Organisationsformen, Regionale Differenzierung und Entwicklungstendenzen*. Saarbrucken, Germany: Brietenback, 1982.

Tsung, Hui Yang. *Die Landwirtschaftliche Bodennutzung Taiwans: Strukturtypen des Anbaus und ihre Entwicklung*. Bonn, Germany: Bonner Geographische Abhandlungen, Heft 71, 1985.

CROPS AND CROPPING PATTERNS

Chen, Chengsiang. "The Banana Industry in Taiwan." *Journal of Tropical Geography*, 2 (1954): 48-55.

Chang, Taochang. "A Geographic Appraisal of the Taiwan Tea Industry." Master's Thesis, Oregon State University, Corvallis, Oreg., 1962.

Chen, Kuochang. "A Geographical Study of the Pineapple Cultivation in Taiwan" (in Japanese). *Geographical Review of Japan*, 39, No. 1 (1966): 31-49.

Ho, Chungru. "The Expansion and Changing Location of Banana Production in Taiwan." Master's Thesis, University of Wisconsin, Madison, Wis., 1973.

Williams, Jack F. "The Conflict Between Peasant and Public Interest in a Developing Country: A Case Study of the Taiwan Sugar Industry, 1950–1970." Ph.D. Dissertation, University of Hawaii, Manoa, Hawaii, 1973.

ANIMAL HUSBANDRY

Hung, Wen Wang. "Geographical Environments in the Development of Animal Husbandry Industry in Taiwan" (in Chinese). *National Taiwan University, Department of Geography, Science Reports*, 7 (1971): 134-186.

FARM LAND

Vandermeer, Paul. "Farm Plot Dispersal, Luliao Village, Taiwan." Ph.D. Dissertation, University of Michigan, Ann Arbor, Mich., 1973.

Lay, Ruhaw. "Land Consolidation in Taiwan: An Agricultural Land Use Study." Master's Thesis, Michigan State University, East Lansing, Mich., 1974.

Koch, Wilfrid. "Wandlungen in der Landnutzung des Beckens von Taipei und seiner ostlichen Umrahmung." *National Taiwan University, Department of Geography and Atmospheric Sciences, Science Reports*, 8 (1976): 63-72.

Williams, Jack F. "Agricultural Use of Slopelands in Taiwan." Pp. 89-111 in *China Geographer, Vol. 11: Agriculture*, edited by Clifton W. Pannell and Christopher L. Salter. Boulder, Colo.: Westview Press, 1981.

ENVIRONMENTAL AND CONSERVATION ISSUES

Chang Jen-hu. "A Climatological Analysis of Yield Difference Between the First and Second Rice Crop in Taiwan." *Agricultural and Forest Meteorology*, 36 (1985): 113-120.

IRRIGATION

Hsieh, Chiaomin. "Taoyuon and Chiaman: Two Irrigation Systems in Taiwan." Pp. 27-30 in *Focus on Geographic Activity: A Collection of Original Studies*, edited by Richard S. Thoman and Donald J. Patton. New York, N.Y.: McGraw-Hill, 1964.

Part 9

AUSTRALIA, NEW ZEALAND, AND THE PACIFIC BASIN

Australia

ATLASES AND GRAPHIC PRESENTATIONS

Massey, John S. "Computer-Mapping Victoria's Agricultural and Pastoral Land Use, 1906–1975." *Australian Geographical Studies*, 17, No. 1 (1979): 18-32.

GENERAL TEXTS, ARTICLES, AND THESES

Ruhl, Alfred. *Das standartsproblem in der landwirtschafts-geographie (das neuland Ost Australien)*. Berlin, Germany: E.D. Mittler and Son, 1929.

Taylor, Griffith. "Agricultural Regions of Australia." Pp. 1657-1662 in *Proceedings, Pan-Pacific Science Congress, Tokyo 1926*, 2. Tokyo, Japan: The Congress, 1928.

Taylor, Griffith. "Agricultural Regions of Australia." *Economic Geography*, Part 1, 6 (1930): 109-134; Part 2, 6 (1930): 213-242.

Christian, C.S. "Developments in Northern Australian Agriculture." *Proceedings, Royal Geographical Society of Australia, South Australian Branch*, 58 (1957): 37-42.

Scott, Peter. "The Agricultural Regions of Tasmania: A Statistical Definition." *Economic Geography*, 33, No. 2 (1957): 109-121.

Olsen, C.O. "Agricultural Potentialities in Northern Australia." *Journal of Tropical Geography*, 11 (1958): 70-83.

Thomson, K.W. "Urban Settlement and the Wheat Frontier in the Flinders Ranges, South Australia." *Proceedings, Royal Geographical Society of Australia, South Australian Branch*, 60 (1959): 25-37.

Golledge, Reginald G. "Agricultural Experiments in Northwestern Australia." *Geography*, 47, No. 2 (1962): 187-190.

Scott, Peter. "Sample Farms in Northern Tasmania." *Journal of Geography*, 62, No. 2 (1963): 56-65.

Davidson, B.R. *The Northern Myth: A Study of the Physical and Economic Limits to Agricultural and Pastoral Development in Tropical Australia*. Carlton, Australia: Melbourne University Press, 1966.

Laut, Peter. "Land, Dairy Production and Rural Population Prospects in Coastal New South Wales, with Special Reference to the Macleay Valley." Ph.D. Dissertation, Australian National University, 1969.

Williams, Michael. "Town-Farming in the Mallee Lands of South Australia and Victoria." *Australian Geographical Studies*, 8 (1970): 173-191.

Dahlke, Jurgen. *Der Weizengurtel in Sudwestaustralien: Anbau und Siedlung an der Trockengrenze*. Wiesbaden, Germany: F. Steiner, 1973.

Shilton, Peter. *Urban Influences on the Agricultural Areas of the Central Coast with Particular Reference to Part Time Farming*. Armidale, New South Wales: University of New England, Department of Geography, Research Series in Applied Geography, No. 35, 1973.

Smith, M.J. *A Riverina Farm*. Milton, Queensland: Jacaranda Press, 1973.

Courtenay, P.P. "Agricultural Development on the Wet Coast of North Queensland, 1880–1970." Pp. 19-27 in *Lectures on North Queensland History*, edited by B.J. Dalton. Townsville, Australia: James Cook University History Department, 1974.

Dayal, Edison. "The Present Pertinence of Von Thunen Theory in an Advanced Economy: A Case from Southeast Australia." Pp. 43-51 in *Proceedings, Eighth New Zealand Geography Conference*, edited by William Brockie, Richard B. LeHeron, and Evelyn Stokes. Christchurch, New Zealand: New Zealand Geographical Society, 1974.

McDonald, G.T. "The Fellowship of the Ring." Pp. 33-42 in *Proceedings, Eighth New Zealand Geography Conference*, edited by William Brockie, Richard B. LeHeron, and Evelyn Stokes. Christchurch, New Zealand: New Zealand Geographical Society, 1974.

Pigram, J.J. "River Basin Agricultural Systems." Pp. 92-108 in *Perspectives on Systems in Geography: Proceedings of the New England Symposium for Geography Teachers*, edited by J.F. Hobbs and J.J. Pigram. Armidale, Australia: University of New England, 1975.

Kolb, Albert, and Dieter Jaschke. "Die agrarische Tragfahigkeit Nordaustraliens." *Geographische Rundschau*, 29, No. 11 (1977): 366-375.

Courtenay, P.P. "Agricultural Geography." Pp. 95-102 in *Geographical Studies of the Townsville Area*, edited by D. Hopley. Townsville, Australia: James Cook University of North Queensland, Occasional Paper No. 2, 1978.

Courtenay, P.P. "Agriculture in North Queensland." *Australian Geographical Studies*, 16 (1978): 29-42.

Smailes, P.J. "The Effects of Changes in Agriculture upon the Service Sector in South Australian Country Towns, 1945–1974." *Norsk Geografisk Tidsskrift*, 33, No. 3 (1979): 125-142.

Jaschke, Dieter. "Die autralische landwirtschaft zwischen klimatischer und markwirtschaftlicher Herausforderung–Das Beispiel Neusudwales." *Erdkunde*, 31, No. 4 (1980): 269-280.

Vlasova, T. "Economic-Geographical Peculiarities of an Agricultural Use of the Australian Territory" (in Russian). *Izvestiya Akademii Nauk SSSR, Seriya Geografiya*, 3 (1980): 67-77.

Scott, Peter. *Australian Agriculture*. Budapest, Hungary: Akademiai Kiado, 1981.

Lloyd, P.L. "Queensland Agriculture: A Changing Scene." *Queensland Geographical Journal*, Third Series, 7 (1983): 1-12.

Scott, Peter. "The Typology of Australian Agriculture." *Geografia Polonica*, 46 (1983): 7-20.

Fautz, Bruno. *Agrarlandschaften in Queensland*. Wiesbaden, Germany: F. Steiner, 1984.

Williams, D.B., ed. *Agriculture in the Australian Economy*, third edition. London, England: Oxford University Press, 1991.

Epps, Roger. "The Sustainability of Australian Agricultural Production Systems: A Realistic Objective or Simply a Desirable Aim?" *Australian Geographer*, 26, No. 2 (1995): 173-179.

Norton, William. "State Boundaries and Agricultural Change in the South Eastern Australian Wheat Belt: Counterfactual Analysis, 1891–1911." *Australian Geographical Studies*, 33, No. 2 (1995): 228-241.

Liepins, Ruth. "Reading Agricultural Power: Media as Sites and Processes in the Construction of Meaning." *New Zealand Geographer*, 52, No. 2 (1996): 3-10.

Smailes, P.J. "Entrenched Farm Indebtedness and the Process of Agrarian Change: A Case Study and Its Implications." Pp. 301-322 in *Globalization and Agri-Food Restructuring*, edited by D. Burch, et al. London, England: Avebury, 1996.

Argent, N. "Global Finance/Local Crisis—Financial Deregulation and the Geographical Restructuring of Australian Farming and Farm Credit: The Case of Kangaroo Island." Ph.D. Dissertation, University of Adelaide, Adelaide, South Australia, 1997.

Pritchard, W. "The Emerging Contours of the Third Food Regime: Evidence from Australian Dairy and Wheat Sectors." *Economic Geography*, 74, No. 1 (1998): 64-74.

CROPS AND CROPPING PATTERNS

Starkey, Otis P. "The Effects of Governmental Aids and Regulation on the Australian Cane Sugar Industry." Master's Thesis, Columbia University, New York, N.Y., 1930.

Hungerford, T.A.G. "The Australian Sugar Industry." *Tijdschrift voor Economische en Sociale Geographie*, 45 (1954): 138-142.

Simonett, D.S. "Sugar Production in North Queensland." *Economic Geography*, 30, No. 3 (1954): 223-235.

Kelly, T.K., and A.D. Tweedie. "The Expanding Queensland Sugar Industry." *Journal of Tropical Geography*, 5 (1955): 44-56.

Meinig, Donald W. "Colonization of Wheat Lands: Some Australian Comparisons." *Australian Geographer*, 7 (1959): 145-156.

Cochrane, G.R. "The Renmark Fruit Growing District." *Proceedings, Royal Geographical Society of Australia, South Australian Branch*, 61 (1960): 21-36.

Huetz de Lemps, Alain. "Une grande entreprise suciere: La 'colonial sugar refining' autralienne." *Les Cahiers d'Outre Mer*, 13 (1960): 243-253.

Meinig, Donald W. *On the Margins of the Good Earth: The South Australian Wheat Frontier, 1869–1884*. Washington, D.C.: Association of American Geographers, 1962.

Duncan, Craig. "The Trevor Farm: A Case Study of Cane Sugar Production on the Queensland Coast." Pp. 37-44 in *Focus on Geographic Activity: A Collection of Original Studies,* edited by Richard S. Thoman and Donald J. Patton. New York, N.Y.: McGraw-Hill, 1964.

Van Evern, John R. "The Distribution of Commercial Production of the Queensland (Macadamia) Nut." Master's Thesis, Catholic University, Washington, D.C., 1964.

Driscoll, E.M. "Recent Developments in the Queensland Sugar Industry." *Geography*, 51, No. 2 (1966): 147-149.

Bridges, E.M. "Cotton in Australia." *Geography*, 51, No. 1 (1967): 73-75.

Murray, I.D. "Tobacco and Irrigation in Tropical Australia: The Mareeba-Dimbalah Scheme." *Journal of Tropical Geography*, 25 (1967): 34-47.

Smith, Derek L. "Vegetable Production in South Australia." Pp. 423-440 in *Liverpool Essays in Geography*, edited by R.W. Steel and R. Lawton. London, England: Longman, 1967.

Greer, Ronald. "De Weinbau in Sudaustralien: Von deutschen Auswanderen begrundet." *Zeitschrift fur Wirtschaftsgeographie*, 12, No. 8 (1968): 249-252.

Higman, B.W. "The Regional Impact of the Sugar Industry of New South Wales, 1870–1912." *Australian Geographical Studies*, 6, No. 1 (1968): 43-58.

Higman, B.W. "Sugar Plantations and Yeoman Farming in New South Wales." *Annals of the Association of American Geographers*, 58, No. 4 (1968): 697-719.

Huetz de Lemps, Alain. "Les vignobles en les vins d'Australie." *Les Cahiers d'Outre Mer*, 22 (1969): 225-251.

Crabb, P. "Aspects of the South Australian Citrus Industry." *Scottish Geographical Magazine*, 86, No. 1 (1970): 41-52.

Smith, Derek L. "Viticulture in the Barrosa Region: Prospects and Costs." *Australian Geographical Studies*, 8 (1970): 101-120.

Irwin, Peter G. *Cotton Systems of the Namoi Valley*. Milton, Queensland, Australia: Jacaranda Press, 1972.

Willis, Geoffrey A. *Harvesting and Transport of Sugar Cane in Australia*. Townsville, Australia: James Cook University of North Queensland, Department of Geography, 1972.

Huetz de Lemps, Alain. "L'originalite de la culture de la canne a sucre australienne." *Bulletin de l'Association de Geographes Francais*, 50 (1973): 485-494.

Uhlig, Harold. "Die Reisbaugebiete Australiens." *Geographische Zeitschrift*, Beiheft, 33 (1973): 411-424.

Buchanon, Maris E. *Responses to Agricultural Change: A Study of Tobacco Growing in Southern Queensland and Northern New South Wales*. Armidale, New South Wales: University of New England, Department of Geography, 1975.

Dahlke, Jurgen. "Evolution of the Wheat Belt in Western Australia." *Australian Geographer*, 13, No. 1 (1975): 3-14.

Thompson, J.F. "Viticulture in Tasmania: Climate, Micro-Climate and Site-Selection." *Tasmanian Journal of Agriculture*, 47, No. 3 (1976): 141-148.

Boundy, K.A., A. Ellington, and B.R. Walker. "Crop Production in High Rainfall Areas of Northern Victoria." Pp. 250-252 in *Proceedings, Tenth New Zealand Geography Conference*, edited by Warren Moran, Peter Hosking, and Graeme Aitkin. Auckland, New Zealand: New Zealand Geographical Society, 1979.

Miller, C.L. "Structural Adjustment and Regional Relocation in the Tasmanian Hop Industry." Pp. 180-184 in *Proceedings, Tenth New Zealand Geography Conference*, edited by Warren Moran, Peter Hosking, and Graeme Aitken. Auckland, New Zealand: New Zealand Geographical Society, 1979.

Reiner, Ernst. "Rebbau und Wein in Australien." *Zeitschrift fur Wirtschaftsgeographie*, 23, No. 5 (1979): 140-143.

Lloyd, P.L. "Sorghum-Queensland's Booming Grain Crop." *Queensland Geographical Journal*, Third Series, 6 (1981): 23-35.

Mearns, Linda O. "Simulation of the Effect of Climatic Variability on Wheat Yield in Australia." Master's Thesis, University of California, Los Angeles, Calif., 1982.

Wood, L.J. "Trends in the Tasmanian Orcharding Industry, 1960–1980." *Geography*, 67, No. 1 (1982): 68-71.

Gardner, W.K., R.G. Velthuis, and R.L. Amar. "Field Crop Production in Southwest Victoria: Area Description, Current Land Use and Potential for Crop Production." *Journal of the Australian Institute of Agricultural Science*, 49, No. 2 (1983): 60-70.

Rother, K. "Der Sonderkulturbau in Sudwest-Australien und seine sudeuropaische Tragerschaft." *Erdkunde*, 38, No. 1 (1985): 45-54.

Holland, Peter. "Standing Crop and Yield Estimates for a Hill Country Catchment in Central Otago." Pp. 98-105 in *Proceedings of the Fourteenth New Zealand Geography Conference*, edited by Richard B. LeHeron, Michael Roche, and Michael Shepard. Palmerston North, New Zealand: New Zealand Geographical Society, 1987.

Katahira, H., and K. Iwasaki. "Progress in Wheat Growing and the Variation of Rainfall in Semi-Arid Australia: A Case Study of the Murray Mallee Region, South Australia" (in Japanese). *Jimbun Chiri/Human Geography*, Kyoto, 40, No. 4 (1988): 297-318.

Wood, Les. "Pyrethrum and Essential Oils: New Cropping Ventures in Tasmania." *Geography*, 79, No. 4 (1994): 357-360.

Frost, David B. "The Evolving Sugar Industry of Queensland, Australia." *Proceedings, New England–St. Lawrence Valley Division, Association of American Geographers*, 25 (1995): 85-94.

Grosvenor, Shirley, Richard B. LeHeron, and Michael Roche. "Sustainability, Corporate Growers, Regionalisation and Pacific-Asia Links in the Tasmanian and Hawkes Bay Apple Industries." *Australian Geographer*, 26, No. 2 (1995): 163-172.

Robinson, Guy M. "Deregulation and Restructuring of the Australian Cane Sugar Industry." *Australian Geographical Studies*, 33, No. 2 (1995): 212-227.

Drummond, Ian. "Conditions of Unsustainability in the Australian Sugar Industry." *Geoforum*, 27, No. 3 (1996): 345-354.

Grosvenor, Shirley, and Les Wood. "Bye Europe, Buy Asia: The Reorientation of Tasmanian Apple Production." *Geography*, 81, No. 2 (1996): 170-173.

Griggs, Peter D. "The Origins and Early Development of the Small Cane Farming System in Queensland, 1870–1915." *Journal of Historical Geography*, 23, No. 1 (1997): 46-61.

Griggs, Peter D. "Sugar Plantations in Queenland, 1864–1912: Origins, Characteristics, Distribution, and Decline." *Agricultural History*, 74, No. 3 (2000): 609-647.

ANIMAL HUSBANDRY

Pye, E. "The Sheep and Wool Industry of Australia." *Journal of Geography*, 26, No. 9 (1927): 327-341.

Buchanon, R.O. *The Pastoral Industries of New Zealand*. London, England: Institute of British Geographers, Publication No. 2, 1935.

Taylor, Griffith. "The Distribution of Pasture in Australia." *Geographical Review*, 27, No. 2 (1937): 291-294.

Trumble, H.C. "The Development of Pastures in South Australia." *Scottish Geographical Magazine*, 54, No. 1 (1937): 10-15.

Davis, Charles M. "Merino Sheep on the Australian Riverina." *Geographical Review*, 44, No. 4 (1954): 475-494.

Heathcote, R.L. "Conservation or Opportune Use: The Pastorals' Problem in Semi-Arid Australia." *Advancement of Science*, 21, No. 89 (1965): 47-60.

McDonald, G.T. "Recent Pasture Development of the Northern Tablelands of New South Wales." *Australian Geographer*, 10 (1968): 382-391.

Richardson, G.A. "Trends in the Australian Beef Industry." *Geography*, 53, No. 1 (1968): 85-87.

Johnson, J.K. "Accessibility, Transport, and Land Development: Complementarity and Location Rent Adjustment in the Northern Territories Beef Cattle Industry." Ph.D. Dissertation, Australian National University, 1970.

Allen, A.C.B. "The Impact of Drought on Beef Cattle Numbers in the Channel Country of Southwest Queensland, Australia." *Yearbook, Association of Pacific Coast Geographers*, 33 (1971): 107-126.

McKnight, Tom L. "A Comparative View of Feral Hoofed Livestock in Australia and the United States." Pp. 31-49 in *Geographical Essays in Honour of Gilbert J. Butland*, edited by Ian Douglas, et al. Armidale Australia: University of New England, 1975.

McKnight, Tom L. *Friendly Vermin: A Survey of Feral Livestock in Australia*. Berkeley, Calif.: University of California Publications in Geography, No. 21, 1976.

Vickery, James, Malcolm McTaggert, and Susan Baker. "The Arid Zone of South Australia: Prospects for Its Pastoral Industry." *Proceedings, Royal Geographical Society of Australasia, South Australia Branch*, 76 (1976): 39-50.

White, B.J. *A Simulation Based-Evaluation of Queensland's Northern Sheep Industry*. Townsville, Australia: James Cook University of North Queensland, Department of Geography Monograph Series, No. 10, 1978.

Griggs, Peter D. "Feral Pigs in Australia: A Case Study." *Queensland Geographical Journal*, Third Series, 6 (1981): 37-48.

Baldwin, James A. "Pre-Cookian Pigs in Australia?" *Journal of Cultural Geography*, 4, No. 1 (1983): 17-27.

Bowling, Jill R. "Technology, Welfare and Intensive Animal Farming: Case Study of the Poultry and Pig Industry." Ph.D. Dissertation, Australian National University, 1985.

Heathcote, R.L. "Pastoral Australia." Pp. 259-300 in *Australia: A Geography, Volume 2, Space and Society*, edited by D.N. Jeans. Sydney, Australia: Sydney University Press, 1987.

Mollah, W.S. "Townsville Stylo: A Pasture Innovation for Northern Territory Beef Cattle Production." *Australian Geographical Studies*, 25, No. 1 (1987): 26-40.

Deane, William, and Gordon Waitt. "The Australian Wool Crisis: Farm Adjustment Pressures and Strategies." *New Zealand Geographer*, 50, No. 2 (1994): 22-27.

Robinson, Guy M. "Stud Farming in the Hunter Valley: Expansion and Reconstructing in the Equine Industry." *Australian Geographer*, 27, No. 1 (1996): 133-148.

Taylor, Michael. "Agriculture in Recession: The Regional Financial Performance of Australian Broadacre Livestock Farming." *New Zealand Geographer*, 52, No. 2 (1996): 46-55.

DAIRYING

Rutherford, John, and L. Dillon. "Dairy Farming in the Beriquin and Denimein Irrigation Districts." *Review of Market and Agricultural Economics*, 22 (1954): 87-164.

de Laubenfels, D.J. "The Atherton Tableland: A Dairyland in the Tropics." *Journal of Geography*, 67, No. 2 (1968): 104-107.

Wilvert, Calvin. "The Southward Migration of Dairying in Australia." Ph.D. Dissertation, University of California, Berkeley, Calif., 1971.

Wilvert, Calvin. "The Changing Evolution of Australia's Leading Subtropical Dairying Center: The Richmond River District, 1885–1935." *Proceedings, Association of American Geographers*, 5 (1973): 302-305.

Lewthwaite, Gordon R. "Australian Milk and Its Market: A Dairy System in Transition." *Queensland Geographical Journal*, 3 (1975): 1-22.

Lewthwaite, Gordon R. "Retreat, Reform and Regional Concentration: The Shifting Pattern of Australian Dairy Farming." *Queensland Geographical Journal*, 4 (1978): 1-20.

Codrington, S.B. "The Milk Zone in New South Wales." *Geography Bulletin*, 11, No. 4 (1979): 122-132.

Lewthwaite, Gordon R. "Product Diversification and Territorial Contraction: The Changing Status of Australia's Dairy States." *Queensland Geographical Journal*, 5 (1979): 21-42.

Dragovich, Deidre. "Role of Government Protection in the Growth and Decline of Dairy Farming in New South Wales." *Geography Bulletin*, 12, No. 3 (1980): 102-110.

Grosvenor, Shirley, and Les Wood. "Deregulation, Restructuring and Productivity in the Tasmanian Dairy Industry." *New Zealand Geographer*, 52, No. 2 (1996): 56-64.

FARM LAND

Rutherford, John. "Some Aspects of Land Utilization on Dairy Farms on the Lower North Coast." *Review of Market and Agricultural Economics*, 19 (1951): 179-247.

Meinig, Donald W. "Goyer's Line of Rainfall: The Role of a Geographical Concept in South Australian Land Policy and Agricultural Settlement." *Agricultural History*, 35, No. 2 (1961): 202-214.

Heathcote, R.L. "Historical Changes in the Appraisal of Pastoral Land Resources." *Review of Marketing and Agricultural Economics*, 31 (1963): 3-23.

Wadham, J.M., and R.K. Wilson. *Land Utilization in Australia*. Melbourne, Australia: Melbourne University Press, 1964.

Camm, J.C.R. "Land Settlement and the Development of Farming Under the Agricultural Lands Purchase Act of 1894 and Closer Settlement Acts, 1906–1917." *Queensland Heritage*, 1 (1968): 25-31.

McKnight, Tom L. "Biotic Influences on Australian Pastoral Land Use." *Yearbook, Association of Pacific Coast Geographers*, 32 (1970): 7-22.

Twidale, C.R. "'Lands' or Relict Strip Fields in South Australia." *Agricultural History Review*, 20, No. 1 (1972): 46-60.

Borman, Lesley, and Ann Marshall. "Land-Use in the 'Ninety-Mile Desert' in 1940." *Proceedings, Royal Geographical Society of Australasia, South Australian Branch*, 74 (1974): 53-64.

Hill, Ronald D., and Derek L. Smith. "Farm Fragmentations on Western Eyre Peninsula, South Australia." *Australian Geographical Studies*, 15, No. 2 (1977): 158-173.

Dayal, Edison. "Fragmented Farms in the Wheat Belt of New South Wales." Pp. 261-263 in *Proceedings, Tenth New Zealand Geography Conference*, edited by Warren Moran, Peter Hosking, and Graeme Aitken. Auckland, New Zealand: New Zealand Geographical Society, 1979.

Nix, H.A. "Land Resources: Area Suitable for Agriculture." Pp. 25-38 in *Land and Water Resources of Australia*, edited by E.G. Hallsworth and J.T. Woodcock. Parkville, Australia: Australian Academy of Technical Science, 1979.

Laut, Peter. "Land Evaluation for Bovine Tuberculosis Eradication in Northern Australia." *Australian Geographical Studies*, 24, No. 2 (1986): 259-271.

Epps, W.R., and Rob Crittenden. "Appraisal of Land Degradation in Australia: Rectification of Agricultural Land Management." *Land Use Policy*, 9, No. 3 (1992): 199-208.

Miller, Leigh. "Agribusiness, Contract Farmers, and Land-Use Sustainability in North-West Tasmania." *Australian Geographer*, 26, No. 2 (1995): 104-111.

Smith, C.S., G.T. McDonald, and R.N. Thwaites. "TIM: Assessing the Sustainability of Agricultural Land Management." *Journal of Environmental Management*, 60, No. 4 (2000): 267-288.

AGRICULTURAL POPULATIONS AND SOCIAL ISSUES

Scott, Peter. "Trade-Center Population Change, Centralization, and Trade-Area Farming Type." *Rural Sociology*, 33, No. 4 (1968): 424-436.

Chisholm, David. "Farmer Age, Attitude, Economic Indicators and Continued Tenure in Agriculture: A Study of Three Central Riverina Shires." Master's Thesis, University of Oklahoma, Norman, Okla., 1977.

Close, W.G. "Population Change and Agricultural Adjustment in Rural Victoria, 1961–1971." Master's Thesis, Monash University, Australia, 1979.

Mollah, W.S. "Early Agricultural Settlement at Katherine, Northern Territory." *Malaysian Journal of Tropical Geography*, 5 (1982): 40-53.

Scott, Peter. "Farm and Off-Farm Family Income in Australia." *Geographia Polonica*, 57 (1989): 137-148.

Smailes, P.J., and R. Clermont. *Agriculture, Business and Population Mobility in the Wakefield Plains District Council.* Adelaide, Australia: University of Adelaide, Department of Geography, 1994.

Liepins, Ruth. "Women in Agriculture: Advocates for a Gendered Sustainable Agriculture." *Australian Geographer*, 26, No. 2 (1995): 118-126.

Phillips, Emily, and Ian Gray. "Farming Practice As Temporally and Spatially Situated Intersections of Biography, Culture, and Social Structure." *Australian Geographer*, 26, No. 2 (1995): 127-132.

Argent, N. "The Globalized Agriculture–Finance Relation: South Australia Farm Families and Rural Communities in a New Regulatory Environment." Pp. 283-299 in *Globalization and Agri-Food Restructuring: Perspectives from the Australasian Region*, edited by D. Burch, R. Rickson, and G. Lawrence. Aldershot, England: Avebury, 1996.

Liepins, Ruth. "Women in Agriculture: A Geography of Australian Activism." Ph.D. Dissertation, University of Melbourne, Melbourne, Australia, 1996.

Teather, Elizabeth K. "Farm Women in Canada, New Zealand, and Australia Redefine Their Rurality." *Journal of Rural Studies*, 12, No. 1 (1996): 1-14.

Argent, N. "Global Finance/Local Crisis: Financial Deregulation and the Geographical Restructuring of Australian Farming and Farm Credit—The Case of Kangeroo Island." Ph.D. Dissertation, University of Adelaide, Adelaide, Australia, 1997.

Head, Lesley. "The Northern Myth Revisited? Aborigines, Environment and Agriculture in the Ord River Irrigation Scheme, States One and Two." *Australian Geographer*, 30, No. 2 (1999): 141-158.

ENVIRONMENTAL AND CONSERVATION ISSUES

Taylor, Griffith. "Agricultural Climatology of Australia." *Quarterly Journal of the Royal Meteorological Society*, 46 (1920): 331-355.

Trumble, H.C. "Climatic Factors in Relation to the Agricultural Regions of South Australia." *Transactions, Royal Society of South Australia*, 70 (1939): 313-347.

Christian, C.S. "The Eco-Complex in Its Importance for Agricultural Assessment." Pp. 587-603 in *Biogeography and Ecology in Australia*, edited by

A. Keast, R.L. Crocker, and C.S. Christian. Den Haag, Netherlands: Monographiae Biologicae No. 8, 1959.

Moore, R.M. "Ecological Observations on Plant Communities Grazed by Sheep in Australia." Pp. 500-513 in *Biogeography and Ecology in Australia*, edited by A. Keast, R.L. Crocker, and C.S. Christian. The Hague, Netherlands: Junk, 1959.

Robinson, K.W., and T.M. Burley. "Flood-Plain Farming on the Maitland Flats, Humber Valley, New South Wales." *Economic Geography*, 38, No. 3 (1962): 234-250.

Heathcote, R.L. "Conservation or Opportune Use? The Pastoralists' Problem in Semi-Arid Australia." *Advancement of Science*, 21 (1964): 47-60.

Lee, K.W. *Farm Level Drainage on the Lower Macleay*. Armisdale, Australia: University of New England, Research Series in Applied Geography, No. 23, 1968.

Twidale, C.R., and Derek L. Smith. "A Perfect Desert Transformed: The Agricultural Development of Northwestern Eyre Peninsula, South Australia." *Australian Geographer*, 11, No. 5 (1971): 437-454.

Conacher, Arthur J., and I.D. Murray. "Implications and Causes of Salinity Problems in the Western Australian Wheat Belt: The York Mawson Area." *Australian Geographical Studies*, 11 (1973): 40-61.

Fitzpatrick, Eugene. "The Utility of Water Balance Accounting Models for Improved Regional Agroclimatic Assessments: An Integrative Summary of Research on the Climatic Controls of Pasture and Crop Productivity in Australia." Ph.D. Dissertation, Rutgers University, New Brunswick, N.J., 1975.

Potter, J.C. "Drought in the Settled Districts of South Australia, 1976: A Symposium—Drought and Its Impact on Agriculture." *Proceedings, Royal Geographical Society of Australia, South Australian Branch*, 78 (1977): 25-28.

McDonald, G.T. "Rural Flood Mitigation Project Assessments." Pp. 193-204 in *Perspectives in Agricultural Geography*, vol. 2, edited by Noor Mohammad. New Delhi, India: Concept Publishing Company, 1981.

Conacher, Arthur J., and Jeanette Conacher. *Organic Farming in Australia*. Perth, Australia: University of Western Australia, Department of Geography Working Papers, 1982, Geowest No. 18, 1982.

Dragovich, Deidre. "Geographical Variation in Monthly Rainfall and Milk Production in East New South Wales." *Agricultural Meteorology*, 26 (1982): 255-264.

Conacher, Jeanette, and Arthur J. Conacher. "Some Aspects of the Agricultural Use and Abuse of Australia's Soil Resources." Pp. 123-135 in *Proceedings, Twelth New Zealand Geography Conference*, edited by Ian Owens, et al. Christchurch, New Zealand: New Zealand Geographical Society, 1983.

Rother, K. "Der sudwest- und sud-australische Agrarraum: Seine Eigenart und Stellung in den Mediterranen Subtropen der Neuen Welt." *Geographische Rundschau*, 37, No. 1 (1985): 13-20.

Hobbs, Joseph, et al. "The Effects of Climatic Variations on Agriculture in the Australian Wheatbelt." Pp. 667-747 in *The Impact of Climatic Variation on*

Agriculture, Volume 2, Assessments in Semi-Arid Regions, edited by M.L. Parry, T.R. Carter, and N.T. Kanjin. Dordrecht, Netherlands: Kluwer, 1988.

Halliday, James. "Climate and Soil in Australia." *Journal of Wine Research*, 4, No. 1 (1993): 19-34.

Pigram, J.J. "Climate Change and Irrigation: An Australian Response." *Canadian Water Resources Journal*, 20, No. 4 (1995): 227-236.

Arthington, A.H. "The Effects of Agricultural Land Use and Cotton Production on Tributaries of the Darling River, Australia." *GeoJournal*, 40, No. 1-2 (1996): 115-125.

Derrick, James. "Comparison of Agroecosytems: Organic and Conventional Broadacre Farming in South East Australia." Ph.D. Dissertation, Australian National University, Canberra, Australia, 1997.

Hill, Rosemary, and Peter D. Griggs. "Rainforests, Agriculture and Aboriginal Fire-Regimes in Wet Tropical Queensland, Australia." *Australian Geographical Studies*, 38, No. 2 (2000): 138-157.

IRRIGATION

Wills, J.T. "Rainfall in Australia: Value of Rain and Irrigation." *Scottish Geographical Magazine*, 3 (1887): 161-173.

Rutherford, John. "Integration of Irrigation and Dryland Farming in the Southern Murray Basin." *Review of Marketing and Agricultural Economics*, 28 (1960): 97-150.

Crabb, P. "Community Irrigation Projects in the Waikerie District of South Australia." *Geography*, 52, No. 4 (1967): 412-415.

Rutherford, John. "Government Irrigation and Its Physical Environment." Pp. 137-194 in *Studies in Australian Geography*, edited by G.H. Dury and M.K. Logan. London, England: Heinemann, 1968.

Davidson, B.R. *Australia Wet or Dry? The Physical and Economic Limits to the Expansion of Irrigation*. Carlton, Australia: Melbourne University Press, 1969.

Maye, B. "Efficiency and Equity in Water Resource Management: Irrigation Development in the Guydir Valley, New South Wales." Pp. 161-166 in *Proceedings, Eleventh New Zealand Geography Conference*, edited by R.F. Watters. Wellington, New Zealand: New Zealand Geographical Society, 1981.

Whitaker, C. "The Ord River Irrigation Area: A Geographer's View of Fanciful Reveries and Stark Realities in the Tropical Northwest of Australia." *Proceedings, Royal Geographical Society of Australasia, South Australia Branch,* 80 (1981): 45-69.

Pigram, J.J. "Irrigation Agriculture and the Rural System." *Rural Systems*, 1, No. 2 (1983): 121-134.

Day, D.G. "Water Allocation to Agriculture in the Hunter Valley." *Land Use Policy*, 2, No. 3 (1985): 187-204.

Pigram, J.J., and Helen K. Mulligan. "Private Sector Involvement in Irrigation Agriculture: An Australian Perspective." *Land Use Policy*, 8, No. 2 (1991): 133-142.

AGRICULTURAL MARKETING AND DISTRIBUTION

Hoyle, Brian S. "Ports and Hinterlands in an Agricultural Economy: The Case of the Australian Sugar Industry." *Geography*, 69, No. 4 (1984): 303-316.

Parsons, H. "Fresh Fruit and Vegetable Supply and Melbourne Market By-pass." Ph.D. Dissertation, Monash University, Australia, 1997.

Pritchard, W. "Transformation in Agri-Food Trade and Exports: Australia as the Supermarket to Asia? Governments, Territory and Political Economy in the Australian Agri-Food System." *Rural Sociology*, 64, No. 2 (1999): 284-301.

TRANSPORTATION AND MECHANIZATION

Hill, Ronald D., and Derek L. Smith. "Is Von Thunen Alive and Well? A Transport Cost Surface for South Australian Wheat." *Australian Geographical Studies*, 32, No. 2 (1994): 183-190.

Fiji and the Pacific Islands

GENERAL TEXTS, ARTICLES, AND THESES

Bowman, Robert C. "Army Farms and Agricultural Development in the Southwest Pacific." *Geographical Review*, 36 (1946): 420-446.

Pelzer, Karl J. "Agriculture in the Truk Islands." *Foreign Agriculture*, 2, No. 6 (1947): 75-81.

Johnston, W.B. "Pacific Island Agriculture: The Contemporary Position in Western Samoa and Some Wider Implications." *Economic Geography*, 29, No. 1 (1953): 26-38.

Barrau, Jacques. "Le milieu et l'agriculture traditionnelle en Melanesie." *Annales de Geographie*, 65 (1956): 362-382.

Barrett, Ward. "Agriculture of Western Samoa." Ph.D. Dissertation, University of California, Berkeley, Calif., 1959.

Fox, J.W., and K.B. Cumberland. *Western Samoa: Land, Life and Agriculture in Tropical Polynesia*. Christchurch, New Zealand: Whitcomb and Tombs, 1962.

Lee, David. "Problems in Tropical Agriculture: A Case Study from Guam." *Yearbook, Association of Pacific Coast Geographers*, 33 (1971): 47-64.

Brookfield, H.C. "Intensification and Disintensification in Pacific Agriculture: A Theoretical Approach." *Pacific Viewpoint*, 13 (1972): 30-48.

Gregor, Howard F. "Agricultural Intensity in the Pacific Southwest." *Proceedings, Association of American Geographers*, 8 (1976): 54-58.

Lightfoot, James E. "The Changing Agricultural Landscape of Talofofo, Guam." Master's Thesis, University of California at Riverside, 1979.

Ward, R. Gerard. "Dilemmas in South Pacific Agriculture." *South Pacific Journal of Natural Science*, 3 (1982): 9-30.

Ward, R. Gerard. "Les dilemmas de l'agriculture dans le Pacifique Sud." *L'Espace Geographique*, 11, No. 4 (1982): 269-280.

Finau, M. "Agriculture in Tonga: An Examination of Factors Influencing Participation and Production." Master's Thesis, Brigham Young University, Provo, Utah, 1987.

Murray, Warwick E. "The Second Wave of Globalization and Agrarian Change in the Pacific Islands." *Journal of Rural Studies*, 17, No. 2 (2001): 135-148.

CROPS AND CROPPING PATTERNS

Robertson, C.J. "The Fiji Sugar Industry." *Economic Geography*, 7, No. 4 (1931): 400-411.

Watters, R.F. "Cultivation in Old Samoa." *Economic Geography*, 34, No. 4 (1958): 338-351.

Watters, R.F. "Shifting Cultivation in the Southwest Pacific." Pp. 88-90 in *Proceedings of the Second New Zealand Geography Conference*. Christchurch, New Zealand: New Zealand Geography Conferernce, 1958.

Ward, R. Gerard. "The Banana Industry in Western Samoa." *Economic Geography*, 35, No. 2 (1959): 123-137.

Watters, R.F. "Some Forms of Shifting Cultivation in the Southwest Pacific." *Journal of Tropical Geography*, 14 (1960): 35-50.

Ward, R. Gerard. "Cash Cropping and The Fijian Village." *Geographical Journal*, 130, No. 4 (1964): 484-506.

Saussol, Alain. "Le cafe en Nouvelle Caledonie: Grandeur et vicissitude d'une colonisation." *Les Cahiers d'Outre Mer*, 20 (1967): 275-305.

Constantini, Otto. "Die Zuckerindustrie auf den Fidschi-Inseln." *Zeitschrift fur Wirtschaftsgeographie*, 16, No. 5 (1972): 156-159.

Thoman, Randolph R. "The Tongan Agricultural System, with Special Emphasis on Plant Assemblages." Ph.D. Dissertation, University of California, Los Angeles, Calif., 1975.

Huetz de Lemps, Alain. "La canne a sucre aux Iles Fidji." *Les Cahiers d'Outre Mer*, 31 (1978): 5-47.

Blume, Helmut. "Die Zuckerwirtschaft von Fiji und Hawaii: Konvergenzen und Divergenzen zweier tropischer pazifischen Inselraume." *Geographische Zeitschrift*, 68, No. 4 (1980): 284-304.

Cerveau, M.P. "L'horticulture commerciale dans les Iles du-Vent, archipel de l a Societe, Polynesie Francaise: Une activite d'avenir?" *Les Cahiers d'Outre Mer*, 51 (1998): 141-156.

FARM LAND

Bell, Timothy A. "The Opunohu Valley, Moorea: A Study of Land Use and Tropical Agriculture in French Polynesia." Master's Thesis, University of Oregon, Eugene, Oreg., 1964.

Bryant, Nevin A. "Change in the Agricultural Land Use in West Upali, Western Samoa." Master's Thesis, University of Hawaii, Manoa, Hawaii, 1967.

Cochrane, G.R. "Land Use Problems in the Dry Zone of Vitilevu." Pp. 111-118 in *Proceedings, Fifth New Zealand Geography Conference*, edited by A.G. Anderson. Auckland, New Zealand: New Zealand Geographical Society, 1968.

Thoman, Randolph R. "Tongan Agricultural Land Use: A Study of Plant Resources." Pp. 153-160 in *Proceedings, Eighth New Zealand Geography Conference*, edited by William Brockie, Richard B. LeHeron, and Evelyn Stokes. Christchurch, New Zealand: New Zealand Geographical Society, 1974.

Kuhlken, Robert T. "Agricultural Terracing in the Fiji Islands." Ph.D. Dissertation, Louisiana State University, Baton Rouge, La., 1994.

AGRICULTURAL POPULATIONS AND SOCIAL ISSUES

Farrell, Bryan H. "Some Human Problems in the Agriculture of Western Samoa." Pp. 99-103 in *Proceedings of the Second New Zealand Geography Conference*. Christchurch, New Zealand: New Zealand Geographical Society, 1958.

Lewthwaite, Gordon R. "Man and Land in Early Tahiti: Polynesian Agriculture Through European Eyes." *Pacific Viewpoint*, 5, No. 1 (1964): 11-34.

Johnston, Kenneth M. *Village Agriculture in Aitutaki, Cook Islands*. Wellington, New Zealand: Victoria University of Wellington, Department of Geography, Pacific Viewpoint Monograph No. 1, 1967.

Anderson, A.G. "Indian Small Farming in Fiji." Ph.D. Dissertation, University of Auckland, Auckland, New Zealand, 1969.

Maude, Alaric. "Shifting Cultivation and Population Growth in Tonga." *Journal of Tropical Geography*, 31 (1970): 57-64.

Anderson, A.G. *Indo-Fijian Small Farming: Profiles of a Peasantry*. Auckland, New Zealand: Auckland University Press, 1974.

Karolle, Bruce. "Agriculture, Population and Development in Guam: Some Options for the Future." Ph.D. Dissertation, Michigan State University, East Lansing, Mich., 1978.

Hau'ofa, Epeli, and R. Gerard Ward. "The Social Context." Pp. 49-72 in *South Pacific Agriculture Choices and Constraints: South Pacific Agricultural Survey 1979*. Canberra, Australia: Australian National University Press, 1980.

Ward, R. Gerard, and Epeli Hau'ofa. "The Demographic and Dietary Contexts." Pp. 27-48 in *South Pacific Agriculture Choices and Constraints: South Pacific Agricultural Survey 1979*. Canberra, Australia: Australian National University Press, 1980.

Breton, Stephanie N. "The Symbolism of Production and Exchange in the Agricultural Rituals of Primitive Societies: The Case Study of the Trobriand Islands." Master's Thesis, University of California, Berkeley, Calif., 1981.

Ward, R. Gerard. "The Pacific Islands in the Twenty-First Century: Land, People, Agriculture." *Journal of Pacific Society*, 14 (1982): 6-20.

Paulson, Deborah. "Forest Depletion, Village Agriculture and Social Change in Rural Western Samoa: A Case Study." Ph.D. Dissertation, University of Hawaii, Manoa, Hawaii, 1992.

Frizelle, Jackie. "Women in Agriculture in the Soloman Islands." Pp. 248-250 in *Proceedings, Seventeenth New Zealand Geography Conference*, edited by William Whittaker. Christchurch, New Zealand: New Zealand Geographical Society, 1993.

Maiava, S. "The Incorporation of Cattle into Village Life and Systems in Western Samoa." Pp. 251-256 in *Proceedings, Seventeenth New Zealand Geography Conference*, edited by William Whittaker. Christchurch, New Zealand: New Zealand Geographical Society, 1993.

DeBunce, Lincoln A. "Cultural Changes in Horticultural Practices on the High Island of Kosrae, Micronesia." Ph.D. Dissertation, University of Oregon, Eugene, Oreg., 1996.

ENVIRONMENTAL AND CONSERVATION ISSUES

Ward, R. Gerard. "The Environmental Context." Pp. 3-26 in *South Pacific Agricultural Choices and Constraints: South Pacific Agricultural Survey 1979*. Canberra, Australia: Australian National University Press, 1980.

Nullet, D., and T.W. Giambelluca. "Risk Analysis of Seasonal Agricultural Drought on Low Pacific Islands." *Agricultural and Forest Meteorology*, 42 (1988): 229-239.

AQUACULTURE

Le Guen, Tanguy, and Gilbert David. "L'aquaculture de la crevette Penaeides en Nouvelle-Caledonie (1970–1995)." *Les Cahiers d'Outre Mer*, 51 (1998): 305-322.

AGRICULTURAL MARKETING AND DISTRIBUTION

Cunningham, Glenn. "Food for Tahiti." *Economic Geography*, 37, No. 4 (1961): 347-352.

Kirkham, William S. "The Forest and the Market: Agricultural Intensification and Market Participation in Samoa." Master's Thesis, Ohio University, Athens, Ohio, 1999.

Storey, Donovan, and Warwick E. Murray. "Dilemmas of Development in Oceania: The Political Economy of the Tongan Agro-Export Sector." *Geographical Journal*, 167, No. 4 (2001): 291-304.

New Zealand

GENERAL TEXTS, ARTICLES, AND THESES

Cumberland, K.B. "The Agricultural Regions of New Zealand." *Geographical Journal*, 113, No. 1-3 (1948): 43-63.

Critchfield, Howard J. "The Agricultural Geography of Southland, New Zealand." Ph.D. Dissertation, University of Washington, Seattle, Wash., 1952.

Hargreaves, R.P. "Mission Farming Before 1830." Pp. 71-78 in *Proceedings, Third New Zealand Geography Conference*. Palmerston North, New Zealand: New Zealand Geographical Society, 1961.

Sears, P.D. "The Potential of New Zealand Grassland Farming." Pp. 65-70 in *Proceedings, Third New Zealand Geography Conference*. Palmerston North, New Zealand: New Zealand Geographical Society, 1961.

Hargreaves, R.P. "Waimate: Pioneer New Zealand Farm." *Agricultural History*, 36, No. 1 (1962): 38-45.

Sears, P.D. "Regional Differences in Grassland Farming Practice." Pp. 191-202 in *Land and Livelihood: Essays in Honour of George Jobberns*, edited by M. McCaskill. Christchurch, New Zealand: New Zealand Geographical Society, 1962.

Duncan, A.A. "Agricultural Patterns." Pp. 120-139 in *Central Otago*, edited by R.G. Lister and R.P. Hargreaves. Christchurch, New Zealand: New Zealand Geographical Society, 1965.

Hargreaves, R.P. "Farm Fences in Pioneer New Zealand." *New Zealand Geographer*, 21 (1965): 144-155.

McArthur, A.T.G., and K.T. Sanderson. "Value of Farm Development in Northland." Pp. 49-58 in *Proceedings, Fifth New Zealand Geography Conference*, edited by A.G. Anderson. Auckland, New Zealand: New Zealand Geographical Society, 1967.

Barrows, J.A. "Hokianga County: A Model of a Depressed Area." Pp. 59-66 in *Proceedings, Fifth New Zealand Geography Conference*, edited by A.G. Anderson. Auckland, New Zealand: New Zealand Geographical Society, 1968.

Stover, Stephen L. "The Government as Farmer in New Zealand." *Economic Geography*, 45, No. 4 (1969): 324-338.

Fairgay, J.D.M. "The Promotion of Innovation Adoption by Farm Advisory Officers." Pp. 188-191 in *Proceedings, Tenth New Zealand Geography Conference*, edited by Warren Moran, Peter Hosking, and Graeme Aitken. Auckland, New Zealand: New Zealand Geographical Society, 1979.

Moran, Warren. "Spatial Pattern of Agriculture on the Urban Periphery: The Auckland Case." *Tijdschrift voor Economische en Sociale Geographie*, 70 (1979): 164-176.

Story, R. "Agricultural Growth: A Constant in a Decade of Change?" Pp. 58-59 in *Proceedings, Eleventh New Zealand Geography Conference*, edited by R.F. Watters. Wellington, New Zealand: New Zealand Geographical Society, 1981.

Paulin, R.T. "Potential for Farming and Forestry on the East Cape." Pp. 154-156 in *Proceedings, Eleventh New Zealand Geography Conference*, edited by R.F. Watters. Wellington, New Zealand: New Zealand Geographical Society, 1982.

Taylor, N.W. "The Costs and Benefits of Investing in Agriculture Compared with Other Sectors." Pp. 53-57 in *Proceedings, Eleventh New Zealand Geography Conference*, edited by R.F. Watters. Wellington, New Zealand: New Zealand Geographical Society, 1982.

Willis, R. "Farming in New Zealand and the E.E.C." *New Zealand Geographer*, 40, No. 1 (1984): 3-12.

Cloke, Paul. "State Deregulation and New Zealand's Agricultural Sector." *Sociologia Ruralis*, 29, No. 1 (1989): 34-48.

LeHeron, Richard B. "New Zealand Agriculture and Changes in the Agriculture–Finance Relation During the 1980s." *Environment and Planning A*, 23, No. 11 (1991): 1653-1670.

LeHeron, Richard B. "Perspectives on Pluriactivity." Pp. 24-32 in *Family Farming in Australia and New Zealand*, edited by M. Alston. Wagga Wagga, New South Wales: Charles Stuart University, Centre for Rural Social Research, 1991.

LeHeron, Richard B., Michael Roche, and T. Johnston. "The Reregulation of New Zealand Agriculture and Its Implications for a Restatement of Policy and Planning Agendas." *Progress in Rural Policy and Planning*, 2 (1992): 266-281.

Cloke, Paul, and Richard B. LeHeron. "Agricultural Deregulation: The Case of New Zealand." Pp. 104-126 in *Regulating Agriculture*, edited by P. Lowe, T. Marsden, and S. Whatmore. London, England: David Fulton, 1994.

LeHeron, Richard B. "A 'Fresh' Place in Food's Space." *Area*, 27, No. 1 (1995): 23-32.

Roche, Michael. "New Zealand: The Colonial Farm in the Pacific Century." *Asia Pacific Viewpoint*, 36, No. 2 (1995): 115-128.

Cloke, Paul. "Looking Through European Eyes? A Re-Evaluation of Agricultural Deregulation in New Zealand." *Sociologia Ruralis*, 36 (1996): 307-330.

Journeaux, Philip. "Trends in New Zealand Agriculture." *New Zealand Journal of Geography*, 102 (1996): 1-9.

Smith, W., and Lindsay Saunders. "Agricultural Sustainability: The Cost of the 1984 Agricultural Policy Reforms." *New Zealand Geographer*, 52, No. 1 (1996): 21-28.

Bradshaw, Benjamin, Christopher Cocklin, and Barry E. Smit. "Subsidy Removal and Farm-Level Stewardship in Northland." *New Zealand Geographer*, 54, No. 2 (1998): 12-20.

Johnsen, Sarah. "Agricultural Restructuring and Response: Inter-Relationships Between Farm Adjustment Strategies in Waihemo, 1984–1997." *New Zealand Geographer*, 55, No. 1 (1999): 25-34.

LeHeron, Richard B., and Michael Roche. "Rapid Reregulation Agricultural Restructuring and the Reimaging of Agriculture in New Zealand." *Rural Sociology*, 64, No. 2 (1999): 203-218.

Argent, N. "Whither the Lender of Last Resort? The Rise and Fall of Public Farm Credit in Australia and New Zealand." *Journal of Rural Studies*, 16, No. 1 (2000): 61-78.

CROPS AND CROPPING PATTERNS

Smith, Albert W. "The Development of the Kauri-Gum Industry and Its Role in the Economy of Northland, New Zealand." Ph.D. Dissertation, University of Washington, Seattle, Wash., 1952.

Maunder, W.J. "The Effect of Climatic Variations on Oat Production in Southland County." Pp. 105-112 in *Proceedings, Fourth New Zealand Geography Conference*, edited by R.G. Lister and R.P. Hargreaves. Dunedin, New Zealand: New Zealand Geographical Society, 1964.

Slater, Frances A. "A Study of Arable Farming in the Lower Waitaki and Waiareka Areas of North Otago." Pp. 67-76 in *Proceedings, Fourth New Zealand Geography Conference*, edited by R.G. Lister and R.P. Hargreaves. Dunedin, New Zealand: New Zealand Geographical Society, 1964.

Kemp, W.S., and D.W. Wilson. "Fruit Growing." Pp. 140-155 in *Central Otago*, edited by R.G. Lister and R.P. Hargreaves. Christchurch, New Zealand: New Zealand Geographical Society, 1965.

Slater, Frances A. "A Study of Arable Farming in the Lower Waitaki and Waiareka Areas of North Otago." Pp. 67-75 in *Proceedings, Fourth New Zealand Geography Conference*, edited by R.G. Lister and R.P. Hargreaves. Dunedin, New Zealand: New Zealand Geographical Society, 1965.

Heller, Charles F. "The Role of Wheat in 19th Century Middle-Latitude Settlement: Examples from Canterbury and Michigan." *Australian Geographical Studies*, 4, No. 2 (1966): 96-118.

James, Kevin. "Early Gardening on Mona Island, Cook Strait, New Zealand." *New Zealand Geographer*, 43, No. 1 (1987): 18-32.

Perry, M., et al. "Growing Discipline Through Total Quality Management in a New Zealand Horticultural Region." *Journal of Rural Studies*, 13, No. 3 (1997): 289-304.

Hayward, D.J., et al. "Networking, Technology and Governance: Lessons from New Zealand Horticulture." *Environment and Planning A*, 30, No. 11 (1998): 2025-2040.

McKenna, M.K., Michael Roche, and Richard B. LeHeron. "Sustaining the Fruits of Labour: A Comparative Localities Analysis of the Integrated Fruit Promotion Programme in New Zealand's Apple Industry." *Journal of Rural Studies*, 14, No. 4 (1998): 393-409.

McKenna, M.K. "Growing Apples and the 'Growing Pains' of Restructuring for New Zealand's Pip Fruit Industry." *New Zealand Geographer*, 54, No. 2 (1998): 37-45.

McKenna, M.K., et al. "Core Issues in New Zealand's Apple Industry: Global–Local Challenges." *Geography*, 84, No. 3 (1999): 275-280.

Saloman, Jean-Noel. "Les vignobles et les vins neo-zelandais." *Les Cahiers d'Outre Mer*, 52 (1999): 115-142.

Moran, William. "Culture et nature dans la geographie de l'industrie vinicole neo-zelandaise." *Annales de Geographie*, 109, No. 4-5 (2000): 525-551.

ANIMAL HUSBANDRY

Harris, Otto. "Sheep in New Zealand." *Bulletin of the Geographical Society of Philadelphia*, 26 (1928): 79-92.

Buchanon, Keith. "Sheep Rearing in New Zealand." *Economic Geography*, 7, No. 4 (1931): 365-379.

Cumberland, K.B. "High-Country 'Run': The Geography of Extensive Pastoralism in New Zealand." *Economic Geography*, 20, No. 3 (1944): 204-220.

Critchfield, Howard J. "The Growth of Pastoralism in Southland, New Zealand." *Economic Geography*, 30, No. 3 (1954): 283-300.

Sellenberg, E. "Die Entwicklung der Viehwirtschaft in Neuseeland." *Erdkunde*, 14, No. 2 (1960): 115-133.

Curry, Leslie. "Regional Variation in the Seasonal Programming of Livestock Farms in New Zealand." *Economic Geography*, 39, No. 2 (1963): 95-118.

Fielding, Gordon J. "Sheep Farming in New Zealand: A Review." *New Zealand Geographer*, 19 (1963): 160-169.

Hill, Ronald D. "Pastoralism in the Wairarapa, 1844–1853." Pp. 25-49 in *Land and Society in New Zealand*, edited by R.F. Watters. Wellington, New Zealand: Reed, 1965.

Hargreaves, R.P. "Sheep Farming, 1800–1920." *New Zealand's Nature Heritage*, 3 (1974): 1022-1029.

Maunder, W.J. "The Use of Econoclimatic Models in National Supply Forecasting: New Zealand Wool Production." Pp. 22-27 in *Proceedings, Tenth New Zealand Geography Conference*, edited by Warren Moran, Peter Hosking, and Graeme Aitken. Auckland, New Zealand: New Zealand Geographical Society, 1979.

Stover, Stephen L. "Grazing and Exotic Forestry: The Development of an Interculture Experiment in New Zealand." *Agricultural History*, 53, No. 2 (1979): 438-450.

Johnson, R.W.M. "New Zealand Pastoral Agriculture in a More Market Environment." Pp. 210-215 in *Proceedings, Fourteenth New Zealand Geography Conference*, edited by Richard B. LeHeron, Michael Roche, and Michael Shepard. Palmerston North, New Zealand: New Zealand Geographical Society, 1987.

LeHeron, Richard B. "Changes in New Zealand Pastoral Farming." Pp. 1-4 in *Geography Resource Centre, Common Topic Activities, Changes in New Zealand Farming*. Christchurch, New Zealand: Christchurch Teacher's College, 1987.

LeHeron, Richard B. "A Political Economy Perspective on the Expansion of New Zealand Livestock Farming, 1960–1984: 1. Agricultural Policy." *Journal of Rural Studies*, 5, No. 1 (1989): 17-32.

LeHeron, Richard B. "A Political Economy Perspective on the Expansion of New Zealand Livestock Farming, 1960–1984: 2. Aggregate Farmer Responses: Evidence and Policy Implications." *Journal of Rural Studies*, 5, No. 1 (1989): 33-44.

Hopkins, Alan. "Recent Trends in New Zealand's Livestock Industry." *Geography*, 76, No. 3 (1991): 269-273.

DAIRYING

Belshaw, Horace. "The Dairying Industry of New Zealand." *Economic Geography*, 3, No. 3 (1928): 281-296.

Harris, Otto. "Dairying in New Zealand." *Bulletin of the Geographical Society of Philadelphia*, 26, No. 4 (1928): 293-304.

Hargreaves, R.P. "New Zealand's Dairy Industry Before 1882." Pp. 121-128 in *Proceedings, Fourth New Zealand Geography Conference*, edited by R.G. Lister and R.P. Hargreaves. Dunedin, New Zealand: New Zealand Geographical Society, 1964.

Lewthwaite, Gordon R. "Wisconsin and the Waikato: A Comparison of Dairy Farming in the United States and New Zealand." *Annals of the Association of American Geographers*, 54, No. 1 (1964): 59-87.

Bewley, J.K. "A Survey of Changes in Production, Manufacturing and Marketing in the New Zealand Dairy Industry, 1947–1948 to 1967–1968." *New Zealand Geographer*, 26 (1970): 36-49.

Lewthwaite, Gordon R. "Commonwealth and Common Market: The Dilemma of the New Zealand Dairy Industry." *Geographical Review*, 61, No. 1 (1971): 72-101.

Rowlands, Christine. "Changes in the Structure and Organization of the New Zealand Dairy Industry Since 1950." Pp. 158-163 in *Proceedings, Sixth New Zealand Geography Conference*, edited by R.J. Johnston and J.M Soons. Auckland, New Zealand: New Zealand Geographical Society, 1970.

Watson, M., and Garth Cant. "Variations in the Productivity of Waikato Dairy Farms." Pp. 165-174 in *Proceedings, Seventh New Zealand Geography Conference*, edited by Evelyn Stokes. Hamilton, New Zealand: New Zealand Geographical Society, 1972.

Lewthwaite, Gordon R. "New Zealand Milk on the Map." *Annals of the Association of American Geographers*, 70, No. 4 (1980): 475-491.

Moran, Warren, and Sarah J. Nason. "Spatio-Temporal Localization of New Zealand Dairying." *Australian Geographical Studies*, 19, No. 1 (1981): 47-66.

Anderson, A.G., and Warren Moran. *Land Use Change in Northland, Part One: Dairy Farming in the 1970s*. Auckland, New Zealand: University of Auckland, Department of Geography, 1983.

Sundstrom, Marvin T. "Dairy Industry Developments in Canada and New Zealand Since 1900." Pp. 207-211 in *Proceedings, Twelfth New Zealand Geography Conference*, edited by Ian Owens, et al. Christchurch, New Zealand: New Zealand Geographical Society, 1983.

Sundstrom, Marvin T. "Technological Transfer: The Case of Canadian Involvement in the New Zealand Dairy Industry, 1880–1920." *Journal of Rural Studies*, 2, No. 2 (1986): 103-116.

Kikuchi, T. "Sustainability of Dairy Farming Location in the Auckland Region, New Zealand" (in Japanese). *Geographical Reports, Tokyo Metropolitan University*, 31 (1996): 183-194.

Blunden, Greg, Warren Moran, and Adrian Bradly. "Archaic Relations of Production in Modern Agricultural Systems: The Example of Sharemilking in New Zealand." *Environment and Planning A*, 29, No. 10 (1997): 1759-1776.

FARM LAND

Fielding, Gordon J. "State Assisted Private Land Development in Northland." Pp. 113-120 in *Proceedings, Fourth New Zealand Geography Conference*, edited by R.G. Lister and R.P. Hargreaves. Dunedin, New Zealand: New Zealand Geographical Society, 1964.

Stover, Stephen L. "Developing New Farm Land in New Zealand." *Journal of Geography*, 68, No. 2 (1969): 106-113.

Eyles, G.D. "A Land Resource Base for Planning New Zealand Agriculture." Pp. 223-226 in *Proceedings, Tenth New Zealand Geography Conference*, edited by Warren Moran, Peter Hosking, and Graeme Aitkin. Auckland, New Zealand: New Zealand Geographical Society, 1979.

Ward, A.D., E.C.R. Warr, and D.B. Williams. "Rural Sub-Divisions: Some Critical Observations on the Basis of Smallholders and Farmlets in the Monawatu and Hawkes Bay." Pp. 166-170 in *Proceedings, Tenth New Zealand Geography Conference*, edited by Warren Moran, Peter Hosking, and Graeme Aitkin. Auckland, New Zealand: New Zealand Geographical Society, 1979.

Anderson, A.G., and Warren Moran. *Landuse Change in Northland: Part One—Dairy Farming in the 1970s: A Review*. Auckland, New Zealand: University of Auckland, Department of Geography, Occasional Paper No. 12, 1983.

Kearns, Robin A., and Warren Moran. *Land Use Change in Northland, Part Two: Irrigation Proposals and Land Use Intensification at Maungatepere*. Auckland, New Zealand: University of Auckland, Department of Geography, 1984.

LeHeron, Richard B. "State, Economy and Crisis in New Zealand in the 1980s: Implications for Land-Based Production of a New Mode of Regulation." *Applied Geography*, 8, No. 4 (1988): 273-290.

Kikuchi, T. "Sustainable Changes of Agricultural Land Use in New Zealand." *Geographical Reports, Tokyo Metropolitan University*, 30 (1995): 147-156.

Smith, William, and Lindsay Saunders. "Agricultural Policy Reforms and Sustainable Land Management: A New Zealand Case Study." *Australian Geographer*, 26, No. 2 (1995): 112-117.

Moran, Warren. "Farm Size Change in New Zealand." *New Zealand Geographer*, 53, No. 1 (1997): 3-13.

AGRICULTURAL POPULATIONS AND SOCIAL ISSUES

Hargreaves, R.P. "The Maori Agriculture of the Auckland Province in the Mid-Nineteenth Century." *Journal of the Polynesian Society*, 68 (1959): 61-79.

Hargreaves, R.P. "Maori Agriculture After the Wars." *Journal of the Polynesian Society*, 69 (1960): 354-367.

Hargreaves, R.P. "Changing Maori Agriculture in Pre-Waitangi, New Zealand." *Journal of the Polynesia Society*, 72 (1963): 101-117.

Cant, Garth. "A Case Study in Outmigration: Movement of Adult Farm Workers from the Selwyn–Ashburton Electoral District During the Period 1963–

1966." Pp. 41-48 in *Proceedings of the Fifth New Zealand Geography Conference*, edited by A.G. Anderson. Auckland, New Zealand: New Zealand Geographical Society, 1967.

Fautz, Bruno. *Die Entwicklung neuseelandischer Kulturlandschaften unter sucht in vier ausgewahlten Farmregionen*. Saarbrucken, Germany: Arbeiten aus dem Geographischen Institut der Universitat des Saarlandes, Sonderband 2, 1970.

Lewthwaite, Gordon R. "The Maori Agricultural Revolution: Tropical Crops and Temperate Land." *Yearbook, Association of Pacific Coast Geographers*, 32 (1970): 99-106.

Ishida, Hiroshi. "A Geography of Maori Agriculture: A Longitudinal Perspective." Pp. 204-209 in *Proceedings, Fourteenth New Zealand Geography Conference*, edited by Richard B. LeHeron, Michael Roche, and Michael Shepard. Palmerston North, New Zealand: New Zealand Geographical Society, 1987.

Begg, M. "Rural Women: A Case Study of Dairy Farm Women." Pp. 31-36 in *Proceedings, Sixteenth New Zealand Geography Conference*, edited by David Hawke. Auckland, New Zealand: New Zealand Geographical Society, 1991.

Keating, Norah, and Heather McCrostie-Little. "Generations New Zealand: Retirement and Succession in Farm Families." Pp. 289-293 in *Proceedings, Seventeenth New Zealand Geography Conference*, edited by William Whittaker. Christchurch, New Zealand: New Zealand Geographical Society, 1993.

Wilson, Olivia. "They Changed the Rules: Farm Family Responses to Agricultural Deregulation in Southland." *New Zealand Geographer*, 50, No. 1 (1994): 3-13.

Fairweather, John R. "Changes in the Age of Farmers in New Zealand, 1971–1991." *New Zealand Geographer*, 51, No. 1 (1995): 24-27.

Moran, Warren, Greg Blunden, and Adrian Bradly. "Empowering Family Farms Through Cooperatives and Producer Marketing Boards." *Economic Geography*, 72, No. 2 (1996): 161-177.

Workman, M., and Adrian Bradly. "Family Farmers, Real Regulation and the Experience of Food Regimes." *Journal of Rural Studies*, 12, No. 3 (1996): 245-258.

Joseph, A.E., Jacqueline M. Lidgard, and Richard Bedford. "Dealing with Ambiguity: On the Interdependence of Change in Agriculture and Rural Communities." *New Zealand Geographer*, 57, No. 1 (2001): 16-26.

ENVIRONMENTAL AND CONSERVATION ISSUES

Maunder, W.J. "Climatic Variation and Agricultural Production in New Zealand." *New Zealand Geographer*, 22, No. 1 (1966): 55-69.

Maunder, W.J. "Agroclimatological Relationships: A Review and a New Zealand Contribution." *Canadian Geographer*, 12 (1968): 73-84.

Maunder, W.J. "The Use and Value of Weather-Based Forecasts and Its Implications." Pp. 163-182 in *Perspectives in Agricultural Geography*, vol. 2, edited by Noor Mohammad. New Delhi, India: Concept Publishing Company, 1981.

Maunder, W.J. "Weather and Operational Decision-Making: The Challenge." Pp. 241-250 in *Perspectives in Agricultural Geography*, vol. 5, edited by Noor Mohammad. New Delhi, India: Concept Publishing Company, 1981.

Wilson, Geoff A. "The Pace of Indigenous Forest Clearance on Farms in the Catline District, South Island, New Zealand, 1861–1991." *New Zealand Geographer*, 49, No. 1 (1993): 15-25.

LeHeron, Richard B., et al. "The Impacts of Hail Storms on the Apple Industry of the Heretaunga Plains." *New Zealand Journal of Geography*, 97 (1994): 14-21.

Wilson, Geoff A. "Towards Sustainable Management of Natural Ecosystems on Farms? A New Zealand Perspective." *Journal of Environmental Planning and Management*, 37, No. 2 (1994): 171-187.

Fitzharris, B.B., and W. Endlicher. "Climatic Conditions for Wine Grape Growing." *New Zealand Geographer*, 52, No. 1 (1996): 1-11.

IRRIGATION

Stuart, Robert C. "Irrigation in Cantebury." Pp. 64-71 in *Proceedings of the Second New Zealand Geography Conference*. Christchurch, New Zealand: New Zealand Geographical Society, 1958.

Gillies, A. "Social and Demographic Consequences of Irrigation Development in North Otago." Pp. 46-49 in *Proceedings, Nineth New Zealand Geography Conference*, edited by T.J. Hearn and R.P. Hargreaves. Dunedin, New Zealand: New Zealand Geographical Society, 1977.

Hearn, T.J. "Riparianism, Prior Appropriation, and the Development of Irrigation in Central Otago, 1865–1920." Pp. 89-103 in *Man, Environments, and Planning: Essays in Honour of R.G. Lister*, edited by L.D.B. Heenan and G.W. Kearsley. Dunedin, New Zealand: University of Otago, Department of Geography, 1981.

Kearns, Robin A. "Sources and Implications of Policy for Irrigation: The Maungatapere Example." Pp. 140-148 in *Proceedings, Twelth New Zealand Geography Conference*, edited by Ian Owens, et al. Christchurch, New Zealand: New Zealand Geographical Society, 1983.

Collins, D.C.A., Robin A. Kearns, and Richard B. LeHeron. "Water Pressure: Irrigation, Governance and Land Use Intensification in Maungatopere, New Zealand." *Journal of Rural Studies*, 17, No. 1 (2001): 29-39.

AQUACULTURE

Hickman, R.W. "Marine Farming and Coastal Water Use Conflicts." Pp. 159-160 in *Proceedings, Eleventh New Zealand Geography Conference*, edited by R.F. Watters. Wellington, New Zealand: New Zealand Geographical Society, 1982.

AGRICULTURAL MARKETING AND DISTRIBUTION

Sinclair, J.G., and C.C. Kissling. "A Network Analysis Approach to Fruit Distribution Planning." Pp. 131-136 in *Proceedings, Sixth New Zealand Geography Conference*, edited by R.J. Johnston and J.M. Soons. Christchurch, New Zealand: New Zealand Geographical Society, 1970.

LeHeron, Richard B. "Reorganization of the New Zealand Export Meat Freezing Industry: Political Dilemmas and Spatial Impacts." Pp. 108-127 in *The State and the Spatial Management of Industrial Change*, edited by D. Rich and G.J.R. Linge. London, England: Routledge, 1990.

LeHeron, Richard B., et al. "Pluriactivity in New Zealand's Agro-Commodity Chains." *Proceedings, Rural Economy and Society Section, New Zealand Sociology Conference, Agribusiness and Economics Research Unit*. Lincoln University, Discussion Paper No. 129 (1991): 41-55.

LeHeron, Richard B., Michael Roche, and T. Johnston. "Pluriactivity: An Exploration of Issues with Reference to New Zealand's Livestock and Fruit Agro-Commodity Systems." *Geoforum*, 25, No. 2 (1994): 155-172.

McKenna, M.K., Richard B. LeHeron, and Michael Roche. "Living Local, Growing Global: Renegotiating the Export Production Regime in New Zealand's Pip Fruit Sector." *Geoforum*, 32, No. 2 (2001): 157-166.

AGRICULTURAL LABOR STUDIES

Beattie, A.J., and Richard B. LeHeron. "Farm Labour Isolation and Accessibility." Pp. 185-187 in *Proceedings, Tenth New Zealand Geography Conference*, edited by Warren Moran, Peter Hosking, and Graeme Aitken. Auckland, New Zealand: New Zealand Geographical Society, 1979.

Moran, Warren, Karl Benediktsson, and Sarah Manning. "Labour Processes and Hill Country Farms." Pp. 48-55 in *Proceedings, Fifteenth New Zealand Geography Conference*, edited by Richard Welch. Dunedin, New Zealand: New Zealand Geographical Society, 1989.

Papua New Guinea

GENERAL TEXTS, ARTICLES, AND THESES

Brass, L.J. "Stone Age Agriculture in New Guinea." *Geographical Review*, 31, No. 4 (1941): 555-569.

Powell, J.M., et al. *Agricultural Traditions of the Mount Hague Area*. Port Moresby, New Guinea: University of Papua New Guinea, Department of Geography, Occasional Paper No. 12, 1975.

Jackson, R.T. "The Trend Towards Large-Scale Agriculture in Papua New Guinea." *Singapore Journal of Tropical Geography*, 3, No. 1 (1981): 44-52.

Allen, B.J., ed. *Agricultural and Nutritional Studies on the Nembi Plateau, Southern Highlands*. Port Moresby, New Guinea: University of Papau New Guinea, Occasional Paper No. 4, 1984.

Thompson, H. "Subsistence Agriculture in Papua New Guinea." *Journal of Rural Studies*, 2, No. 3 (1986): 233-244.

Eden, M.J. "Swidden Cultivation in Forest and Savanna in Lowland Southwest Papua New Guinea." *Human Ecology*, 21, No. 2 (1993): 145-166.

CROPS AND CROPPING PATTERNS

Schweinfurth, Ulrich. "Der Teeanbau in Neuguinea: Landschaftswandel und Raumplanung im ostlichen Teil des Zentragebirges der Tropeninsel." *Erdkunde*, 24, No. 3 (1970): 220-228.

Garrett, Larry N. "The Impact of Igomoea Batatos on New Guinea Highland Agriculture and an Investigation of Its Origins." *Philippine Geographical Journal*, 18, No. 1-2 (1974): 7-20.

Chartres, C.J. "Land Resources Assessment for Sugar-Cane Cultivation in Papua New Guinea." *Applied Geography*, 1, No. 4 (1981): 259-272.

Bayliss-Smith, Timothy. "Pre-Ipomopan Agriculture in the Papua New Guinea Highlands Above 2000 Meters: Some Experimental Data on Taro Cultivation." Pp. 285-320 in *Prehistoric Intensive Agriculture in the Tropics*, edited by I.S. Farrington. Cambridge, England: British Archaeological Society, International Research Series, No. 232, 1985.

Grieve, Robert B. "The Oil Palm Industry of Papua New Guinea." *Australian Geographer*, 17, No. 1 (1986): 72-76.

Gassiat-Sanguinet, Anne. "Quel avenir por l'arachide en Haute-Guinee? Les consequences des interventions sur la production et la commercialisation d'une culture de rente." *Les Cahiers d'Outre Mer*, 46 (1993): 273-296.

ANIMAL HUSBANDRY

Johnston, J.K. "The Beef Cattle Industry in Papua New Guinea." *Journal of Tropical Geography*, 42 (1976): 35-46.

Morren, G.E.B. "From Hunting to Herding: Pigs and the Control of Energy in Montane New Guinea." Pp. 274-315 in *Subsistence and Survival: Rural Ecology in the Pacific*, edited by Timothy Bayliss-Smith. London, England: Academic Press, 1977.

Connell, J. "A King Development? Spatial and Structural Changes in the Beef Cattle Industry of Papua New Guinea." *GeoJournal*, 7, No. 6 (1983): 587-598.

FARM LAND

Brookfield, Harold C., and Paula Brown. *Struggle for Land: Agriculture and Group Territories Among the Chimbu of the New Guinea Highlands*. Melbourne, Australia: Oxford University Press, 1963.

Rutherford, G.K. "The Soils and Land Use of the Area Occupied by the Enga of the New Guinea Highlands." *Canadian Geographer*, 8, No. 2 (1964): 142-146.

Easton, P. "The Plantation Redistribution Scheme in Papua New Guinea." Pp. 52-62 in *Land Reform, Land Settlement and Cooperatives*. Rome, Italy: United Nations, FAO, 1983.

AGRICULTURAL POPULATIONS AND SOCIAL ISSUES

Spate, O.H.K. "Changing Native Agriculture in New Guinea." *Geographical Review*, 43, No. 2 (1953): 151-172.

Claval, Pierre. "Des agriculteurs montagnards en Nouvell-Guinie." *Les Cahiers d'Outre Mer*, 19 (1966): 408-412.

Waddell, Eric W. *The Mound Builders: Agricultural Practices, Environment and Society in the Central Highlands of New Guinea*. Seattle, Wash.: University of Washington Press, 1972.

Bayliss-Smith, Timothy. "Subsistence Agriculture and Nutrition in the Bimin Valley, Okasapmin Subdistrict, Papua New Guinea." *Singapore Journal of Tropical Geography*, 6, No. 2 (1985): 101-115.

ENVIRONMENTAL AND CONSERVATION ISSUES

Gagne, W.C. "Staple Crops in Subsistence Agriculture: Their Major Insect Pests, with Emphasis on Biogeographical and Ecological Aspects." Pp. 229-259 in *Biogeography and Ecology of New Guinea*, vol. 1, edited by J.L. Gressitt. The Hague, Netherlands: W. Junk, Monographiae Biologicae Series, Volume 42, 1982.

Klaer, Wendelin, and Ernst Loffler. "The Influence of Traditional Agriculture on the Natural Processes of Erosion in the Tropical Forests and Grasslands of Papua New Guinea." *Applied Geography and Development*, 21 (1983): 66-74.

Sillitoe, Paul. "Fallow and Fertility Under Subsistence Cultivation in the Papua New Guinea Highlands: I. Fallow Succession." *Singapore Journal of Tropical Geography*, 16, No. 1 (1995): 82-100.

Sillitoe, Paul. "Fallow and Fertility Under Subsistence Cultivation in the Papua New Guinea Highlands: II. Soil Fertility." *Singapore Journal of Tropical Geography*, 16, No. 1 (1995): 101-115.

Author/Editor Index

– B –

– E –

– H –

– K –

– L –

– M –

– O –

– P –

– Q –

– R –

– T –

– W –

About the Author

Thomas A. Rumney is currently a professor of geography at Plattsburgh State University in northern New York state. A native of northeastern New York, he got his degrees in geography from the State University of New York–Oneonta (1969), the University of Illinois (1971), and the University of Maryland (1980). He served in the U.S. Navy as a Vietnamese translator from 1971 to 1974. His recent publications include *A Scholar's Guide to Geographical Writing on the American and Canadian Past* (with M. Conzen and G. Wynn, University of Chicago Press), a series of seven bibliographic volumes on the geography of Canada, and articles on the geography of agriculture.

www.ingramcontent.com/pod-product-compliance
Ingram Content Group UK Ltd.
Pitfield, Milton Keynes, MK11 3LW, UK
UKHW020800290926
15470UKWH00008B/1838/J